OPERATION OF WASTEWATER TREATMENT PLANTS

Sixth Edition
Volume I

A Field Study Training Program

prepared by

Office of Water Programs
College of Engineering and Computer Science
California State University, Sacramento

in cooperation with the

California Water Environment Association

★★

Kenneth D. Kerri, Project Director
Bill B. Dendy, Co-Director
John Brady, Consultant and Co-Director
William Crooks, Consultant

★★

for the

U.S. Environmental Protection Agency
Office of Water Program Operations
Municipal Permits and Operations Division
First Edition, Technical Training Grant No. 5TT1-WP-16-03 (1970)
Second Edition, Grant No. T900690010 (1980)

2004

NOTICE

This manual is revised and updated before each printing based on comments from persons using the manual.

FIRST EDITION

First – Seventh Printings (1971 – 1979)	46,000

SECOND EDITION, Volume I

First Printing, 1980	7,000
Second Printing, 1982	8,000

THIRD EDITION, Volume I

First Printing, 1985	8,000
Second Printing, 1988	8,000
Third Printing, 1989	12,000
Fourth Printing, 1990	20,000

FOURTH EDITION, Volume I

First Printing, 1992	25,000
Second Printing, 1994	30,000
Third Printing, 1998	30,000
Fourth Printing, 2001	3,000

FIFTH EDITION, Volume I

First Printing, 2002	15,000

SIXTH EDITION, Volume I

First Printing, 2004	25,000

OPERATOR TRAINING MATERIALS

OPERATOR TRAINING MANUALS AND VIDEOS IN THIS SERIES are available from the Office of Water Programs, California State University, Sacramento, 6000 J Street, Sacramento, CA 95819-6025, phone: (916) 278-6142, e-mail: wateroffice@csus.edu, FAX: (916) 278-5959, or website: www.owp.csus.edu.

1. *OPERATION OF WASTEWATER TREATMENT PLANTS*, 2 Volumes,

2. *OPERATION AND MAINTENANCE OF WASTEWATER COLLECTION SYSTEMS*, 2 Volumes,* (also available in Spanish)

3. *COLLECTION SYSTEMS: METHODS FOR EVALUATING AND IMPROVING PERFORMANCE,*

4. *SMALL WASTEWATER SYSTEM OPERATION AND MAINTENANCE*, 2 Volumes,

5. *INDUSTRIAL WASTE TREATMENT*, 2 Volumes,

6. *ADVANCED WASTE TREATMENT,*

7. *TREATMENT OF METAL WASTESTREAMS,*

8. *PRETREATMENT FACILITY INSPECTION,***

9. *WATER TREATMENT PLANT OPERATION*, 2 Volumes,

10. *SMALL WATER SYSTEM OPERATION AND MAINTENANCE,****

11. *WATER DISTRIBUTION SYSTEM OPERATION AND MAINTENANCE*, and

12. *UTILITY MANAGEMENT.*

* *OPERATION AND MAINTENANCE TRAINING VIDEOS.* This series of six 30-minute videos demonstrates the equipment and procedures collection system crews use to safely and effectively operate and maintain their collection systems. These videos complement and reinforce the information presented in Volumes I and II of *OPERATION AND MAINTENANCE OF WASTEWATER COLLECTION SYSTEMS.*

** *PRETREATMENT FACILITY INSPECTION TRAINING VIDEOS.* This series of five 30-minute videos demonstrates the procedures to effectively inspect an industry, measure flows, and collect samples. These videos complement and reinforce the information presented in *PRETREATMENT FACILITY INSPECTION.*

*** *SMALL WATER SYSTEM TRAINING VIDEOS.* This series of nine training videos was prepared for the operators, managers, owners, and board members of small drinking water systems. Topics covered include operators' roles and responsibilities; safety, operation, and maintenance of surface and groundwater treatment, distribution, and storage facilities; monitoring; administration; financial management; and also emergency preparedness and response.

The Office of Water Programs at California State University, Sacramento, has been designated by the U.S. Environmental Protection Agency as a *SMALL PUBLIC WATER SYSTEMS TECHNOLOGY ASSISTANCE CENTER.* This recognition will provide funding for the development of training videos for the operators and managers of small public water systems. Additional training materials will be produced to assist the operators and managers of small systems.

PREFACE TO THE FIRST EDITION

The purposes of this home study program are:

a. to develop qualified treatment plant operators;
b. to expand the abilities of existing Operators, permitting better service to both their employers and the public; and
c. to prepare operators for *CERTIFICATION EXAMINATIONS.*[1]

To provide you with information needed to operate wastewater treatment plants as efficiently as possible, experienced plant Operators prepared the material on treatment plant processes. Each chapter begins with an introduction and then discusses start up, daily operation, interpretation of lab results and possible approaches to solving operational problems. This order of topics was determined during the testing program on the basis of operators' comments indicating the information they needed most urgently. Additional chapters discuss maintenance, safety, sampling, laboratory procedures, hydraulics, records, analysis and presentation of data, and report writing.

Plant influents (raw wastewater) and the efficiencies of treatment processes vary from plant to plant and from location to location. The material contained in this program is presented to provide you with an understanding of the basic operational aspects of your plant and with information to help you analyze and solve operational problems. This information will help you operate your plant as efficiently as possible.

Wastewater treatment is a rapidly advancing field. To keep pace with scientific advances, the material in this program must be periodically revised and updated. This means that you, the operator, must recognize the need to be aware of new advances and the need for continuous training beyond this program.

Originally the concepts for this manual evolved from Larry Trumbull, 1967 Chairman of the Operator Training Committee of the California Water Pollution Control Association. Bill Dendy and Kenneth Kerri, Project Directors, investigated possible means of financial support to develop and test the manual and prepared a successful application to the Federal Water Pollution Control Administration (5TT1-WP-16-03). The chapters were written, tested by pilot groups of operators and potential operators, reviewed by consultants and the Federal Water Quality Administration, and rewritten in accordance with the suggestions from these sources.

The project directors are indebted to the many operators and other persons who contributed to the manual. Every effort was made to acknowledge material from the many excellent references in the wastewater treatment field. Special thanks are due John Brady and William Crooks who both contributed immensely to the manual. F.J. Ludzack, Chemist, National Training Center, Environmental Protection Agency, Water Quality Office, offered many technical improvements. A note of thanks is also due our typists, Linda Smith, Gloria Uri, Daryl Rasmussen, Vicki Sadlem, Peggy Courtney, and Pris Jernigan. Illustrations were drawn by Martin Garrity.

Following the first year of use by over 6,500 operators and persons interested in operation, minor editing changes were necessary to correct typing errors and omissions and also to rewrite and expand questions and sections that could be clarified. Improvements suggested by operators using the manual were summarized and forwarded to a special Technical Advisory Task Force composed of operators familiar with the manual. This Task Force was formed as a subcommittee of the Water Pollution Control Federation's Personnel Advancement Committee and was chaired by Sam Warrington. We gratefully thank John Brady, Carlos Doyle, Otto Havens, Wilbur Holst, William Johnson, F.J. Ludzack and David Vandersommen for their efforts to improve our original version.

Kenneth D. Kerri
Bill Dendy

1973

---

[1] *Certification Examination. An examination administered by a state agency that operators take to indicate a level of professional competence. In most plants the Chief Operator of the plant must be "certified" (successfully pass a certification examination). In the United States, certification of operators of water treatment plants and wastewater treatment plants is mandatory.*

PREFACE TO THE SECOND EDITION

During the 1970s many people decided that something must be done to control water pollution. The United States Congress passed the "Federal Water Pollution Control Act Amendments of 1972" (PL 92-500) and subsequent amendments. The objective of this Act is to restore and maintain the quality of the Nation's waters. In order to achieve this objective, the Act contains provisions for a financial grant program to assist municipalities with the planning, construction, start up and training of personnel in publicly-owned wastewater treatment plants. Grant funds have been used to build many new plants to date and many more plants will be built in the future. These plants are becoming more complex and are requiring operators with higher levels of knowledge and skills in order to ensure that the plants produce a high quality effluent.

This manual, *OPERATION OF WASTEWATER TREATMENT PLANTS*, was used by over 40,000 persons interested in the operation of treatment plants during the 1970s. Every year when more manuals were printed, the manual was updated on the basis of comments and suggestions provided by persons using the manual. After six years of use by operators, the authors, the California Water Pollution Control Association, and the U.S. Environmental Protection Agency (EPA) decided that the contents of the manual should be reexamined, updated and revised. To accomplish this task, EPA provided the Foundation of California State University, Sacramento, with a grant to conduct the necessary studies, writing and field tests.

Recently the U.S. Environmental Protection Agency and the Association of Boards of Certification (ABC) have undertaken studies to document "need to know" tasks performed by wastewater treatment plant operators, skills required, alternative methods of training, training material needs and availability, and the development of instructional materials for certification examinations. Every effort has been made to incorporate the results of these studies in this Second Edition of *OPERATION OF WASTEWATER TREATMENT PLANTS.*

The project directors are indebted to the many operators and other persons who contributed to the Second Edition. Material from the many excellent references in the wastewater treatment field has been acknowledged wherever possible. Joe Bahnick, Ken Hay, Adelaide Lilly, Frank Lapensee and Bob Rose, U.S. Environmental Protection Agency, served ably as resource persons, consultants and advisers. Special thanks are due our project consultants, Mike Mulbarger, Carl Nagel and Al Petrasek who provided technical advice. Our education reviewers were George Gardner and Larry Hannah. Christine Umeda and Marlene Itagaki administered the national field testing program. A note of thanks was well earned by our typists Charlene Arora, Elaine Saika and Gladys Kornweibel. Illustrations were drawn by Martin Garrity.

Kenneth D. Kerri
John Brady

1980

USES OF THIS MANUAL

Originally this manual was developed to serve as a home-study course for operators in remote areas or persons unable to attend formal classes either due to shift work, personal reasons or the unavailability of suitable classes. This home-study training program used the concepts of self-paced instruction where you are your own instructor and work at your own speed. In order to certify that a person had successfully completed this program, an objective test was included at the end of each chapter and the training course became a correspondence or self-study type of program.

Once operators started using this manual for home study, they realized that it could serve effectively as a textbook in the classroom. Many colleges and universities have used the manual as a text in formal classes often taught by operators. In areas where colleges were not available or were unable to offer classes in the operation of wastewater treatment plants, operators and utility agencies joined together to offer their own courses using the manual.

Utility agencies have enrolled from three to over 300 of their operators in this training program. A manual is purchased for each operator. A senior operator or a group of operators are designated as instructors. These operators help answer questions when the persons in the training program have questions or need assistance. The instructors grade the objective tests, record scores, and notify California State University, Sacramento, of the scores when a person successfully completes this program. This approach avoids the long wait while papers are being graded and returned by CSUS.

This manual was prepared to help operators run their treatment plants. Please feel free to use it in the manner which best fits your training needs and the needs of other operators. We will be happy to work with you to assist you in developing your training program. Please feel free to contact:

Project Director
Office of Water Programs
California State University, Sacramento
6000 J Street
Sacramento, California 95819-6025
Phone (916) 278-6142

INSTRUCTIONS TO PARTICIPANTS
IN HOME-STUDY COURSE

Procedures for reading the lessons and answering the questions are contained in this section.

To progress steadily through this program, you should establish a regular study schedule. For example, many operators in the past have set aside two hours during two evenings a week for study.

The study material is contained in two volumes divided into 20 chapters. Some chapters are longer and more difficult than others. For this reason, many of the chapters are divided into two or more lessons. The time

required to complete a lesson will depend on your background and experience. Some people might require an hour to complete a lesson and some might require three hours; but that is perfectly all right. *THE IMPORTANT THING IS THAT YOU UNDERSTAND THE MATERIAL IN THE LESSON!*

Each lesson is arranged for you to read a short section, write the answers to the questions at the end of the section, check your answers against suggested answers; and then *YOU* decide if you understand the material sufficiently to continue or whether you should read the section again. You will find that this procedure is slower than reading a normal textbook, but you will remember much more when you have finished the lesson.

Some discussion and review questions are provided following each lesson in the later chapters. These questions review the important points you have covered in the lesson. Write the answers to the discussion and review questions in your notebook.

After you have completed the last chapter in each volume, you will find a final examination. This exam is provided for you to review how well you remember the material. You may wish to review the entire manual before you take the final exam. Some of the questions are essay-type questions, which are used by some states for higher-level certification examinations. After you have completed the final examination, grade your own paper and determine the areas in which you might need additional review before your next examination.

You are your own teacher in this program. You could merely look up the suggested answers at the end of each chapter or copy them from someone else, but you would not understand the material. Consequently, you would not be able to apply the material to the operation of your plant nor recall it during an examination for certification or a civil service position.

YOU WILL GET OUT OF THIS PROGRAM WHAT YOU PUT INTO IT.

SUMMARY OF PROCEDURE

A. OPERATOR (YOU)

1. Read what you are expected to learn in each chapter (the chapter objectives).

2. Read sections in the lesson.

3. Write your answers to questions at the end of each section in your notebook. You should write the answers to the questions just like you would if these were questions on a test.

4. Check your answers with the suggested answers.

5. Decide whether to reread the section or to continue with the next section.

6. Write your answers to the discussion and review questions at the end of each lesson in your notebook.

B. ORDER OF WORKING LESSONS

To complete this program you will have to work all of the chapters. You may proceed in numerical order, or you may wish to work some chapters sooner. The Arithmetic Appendix, "How to Solve Wastewater Treatment Plant Arithmetic Problems," may be worked before Chapter 4 because Chapter 4 requires the use of simple arithmetic. If you have trouble with the problems in Chapter 4 or some of the following chapters, you may find it helpful to refer to the Arithmetic Appendix or you may decide to work the Arithmetic Appendix first.

Chapter 16, "Laboratory Procedures and Chemistry," in Volume II, may be studied with Chapter 5 because the operation of sedimentation and flotation treatment processes requires some laboratory tests. Again, you may wish to refer to the lab chapter while working on Chapter 5 and the other chapters, or you may wish to work the lab chapter first.

SAFETY IS A VERY IMPORTANT TOPIC. Everyone working in a treatment plant must always be safety conscious. You must take extreme care with your personal hygiene to prevent the spread of disease to yourself and your family. Operators in treatment plants daily encounter situations and equipment that can cause a serious disabling injury or illness if the operator is not aware of the potential danger and does not exercise adequate precautions. For these reasons, you may decide to work Chapter 14, "Plant Safety," in Volume II, early in your studies. In each chapter, *SAFE PROCEDURES ARE ALWAYS STRESSED.*

C. ENROLLMENT FOR CREDIT AND CERTIFICATE

Students wishing to earn credits and a certificate for completing this course may enroll by contacting the Office of Water Programs, California State University, Sacramento, 6000 J Street, Sacramento, CA 95819-6025, (916) 278-6142. If you have already enrolled, the enrollment packet you were sent contains detailed instructions for completing and returning the objective tests. Please read these important instructions carefully before marking your answer sheets.

Following successful completion of each volume in this program, a Certificate of Completion will be sent to you. If you wish, the Certificate can be sent to your supervisor, the mayor of your town, or any other official you think appropriate. Some operators have been presented their Certificate at a City Council meeting, got their picture in the newspaper, and received a pay raise.

OPERATION OF WASTEWATER TREATMENT PLANTS, VOLUME I
COURSE OUTLINE

OPERATION OF WASTEWATER TREATMENT PLANTS, VOLUME II
COURSE OUTLINE

TECHNICAL CONSULTANTS, FIRST EDITION

William Garber   Frank Phillips
George Gardner   Warren Prentice
Joe Nagano   Ralph Stowell
Carl Nagel   Larry Trumbull

TECHNICAL CONSULTANTS, SECOND EDITION

George Gardner   Carl Nagel
Larry Hannah   Al Petrasek
Mike Mulbarger   Russ Armstrong
                      (Third Edition)

Other similar operator training programs that may be of interest to you are our courses and training manuals on advanced waste treatment, utility management, industrial waste treatment, and small wastewater system operation and maintenance.

ADVANCED WASTE TREATMENT

COURSE OUTLINE

UTILITY MANAGEMENT

COURSE OUTLINE

INDUSTRIAL WASTE TREATMENT, VOLUME I

COURSE OUTLINE

INDUSTRIAL WASTE TREATMENT, VOLUME II

COURSE OUTLINE

SMALL WASTEWATER SYSTEM OPERATION AND MAINTENANCE, VOLUME I

COURSE OUTLINE

SMALL WASTEWATER SYSTEM OPERATION AND MAINTENANCE, VOLUME II

COURSE OUTLINE

CHAPTER 1

THE TREATMENT PLANT OPERATOR

by

Larry Trumbull

and

William Crooks

TABLE OF CONTENTS

Chapter 1.   THE TREATMENT PLANT OPERATOR

OBJECTIVES

Chapter 1. THE TREATMENT PLANT OPERATOR

At the beginning of each chapter in this manual, you will find a list of *OBJECTIVES*. The purpose of this list is to stress those topics in the chapter that are most important. Contained in the list will be items you need to know and skills you must develop to operate and to maintain your plant as efficiently and safely as possible.

Following completion of Chapter 1, you should be able to:

1. List topics covered in Volume I (see Volume I Course Outline for titles of Chapters 1 through 10),

2. Explain the type of work done by treatment plant operators,

3. Describe where to look for jobs in this profession, and

4. Outline how to learn how to do the jobs performed by treatment plant operators.

CHAPTER 1.   THE TREATMENT PLANT OPERATOR

This portion of Chapter 1 was prepared especially for the
new or the potential *WASTEWATER*[1] treatment plant operator.
If you are an experienced operator, you may find some new
viewpoints.

1.0   WHAT IS A TREATMENT PLANT OPERATOR?

Before modern society entered the scene, water was puri-
fied in a natural cycle as shown below:

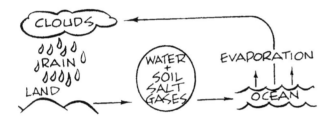

Simplified natural purification cycle

But modern society and the intensive use of the water resource and the resulting water pollution could not wait for
sun, wind, and time to accomplish the purification of soiled
water; consequently, treatment plants were built. Thus,
nature was given an assist by a team consisting of designers,
builders, and treatment plant operators. Designers and builders occupy the scene only for an interval, but operators go on
forever. They are the final and essential link in maintaining
and protecting the aquatic environment upon which all life depends.

1.00   What Does a Treatment Plant Operator Do?

Simply described, the operator keeps a wastewater (sewage) treatment plant working. Physically, the operator turns
valves, pushes switches, collects samples, lubricates equipment, reads gages, and records data. An operator may also
maintain equipment and plant areas by painting, weeding, gar-
dening, repairing machinery, and replacing parts.

Mentally, an operator inspects records, observes conditions,
makes calculations to determine that the plant is working effectively, and predicts necessary maintenance and facility
needs to ensure continued effective operation of the plant.
The operator also has an obligation to explain to supervisors,
councils, civic bodies, and the general public what the plant
does, and most importantly, why its continued and expanded
financial support is vital to the welfare of the community.

1.01   Who Does the Treatment Plant Operator Work For?

An operator's paycheck usually comes from a city, sanitation
district, or other public agency. The operator may, however, be

employed by one of the many large industries that operate
their own treatment plants. Operators also may work for private
contractors that are retained to operate and maintain municipal or industrial wastewater treatment plants. As an operator
you are responsible to your employer for maintaining an economical and efficient operating facility. An even greater obligation rests with the operator because the great numbers of
people who rely upon downstream water supplies are totally
dependent upon the operator's competence and trustworthi-
ness for their welfare. In the final analysis, the operator is really
working for these vitally affected downstream water users.

1.02   Where Does the Treatment Plant Operator Work?

Obviously the operator works in a wastewater treatment
plant. But the different types and locations of treatment plants
offer a wide range of working conditions. From the mountains
to the sea, wherever people congregate into communities,
wastewater treatment plants will be found. From a unit process
operator at a complex municipal facility to a one-person manager of a small town plant, you can select your own special
place in treatment plant operation.

1.03   What Pay Can a Treatment Plant Operator Expect?

In dollars? Prestige? Job satisfaction? Community service?
In opportunities for advancement? By whatever scale you use,
returns are what you make them. If you choose a large municipality, the pay is good and advancement prospects are tops.
Choose a small town and pay may not be as good, but job satisfaction, freedom from time-clock hours, community service,
and prestige may well add up to outstanding personal achieve-
ment. Total reward depends on you.

The operator's duties

1.04   What Does It Take To Be a Treatment Plant Operator?

Desire. First you must choose to enter this profession. You
can do it with a grammar school, a high school, or a college

---

[1] *Wastewater.   A community's used water and water carried solids that flow to a treatment plant. Storm water, surface water, and groundwater infiltration also may be included in the wastewater that enters a wastewater treatment plant. The term "sewage" usually refers to household wastes,
but this word is being replaced by the term "wastewater."*

education. The amount of education needed depends largely on the type of treatment facility and also local certification and entry-level job requirements. While some jobs will always exist for manual labor, the real and expanding need is for *TRAINED OPERATORS*. New techniques, advanced equipment, and increasing instrumentation require a new breed of operator, one who is willing to learn today, and gain tomorrow, for surely your plant will move toward newer and more effective operating procedures and treatment processes. Indeed, the truly service-minded operator assists in adding to and improving the plant performance on a continuing basis.

Tomorrow's forgotten operator stopped learning yesterday

You can be an operator tomorrow by beginning your learning today; or you can be a better operator, ready for advancement, by accelerating your learning today.

This training course, then, is your start toward a better tomorrow, both for you and for the public who will receive better water from your efforts.

QUESTIONS

Place an "X" by the correct answer or answers. After you have answered all the questions, check your answers with those given at the end of the chapter on page 9. Reread any sections you did not understand and then proceed to the next section. You are your own teacher in this training program, and *YOU* should decide when you understand the material and are ready to continue with new material.

EXAMPLE

This is a training course on
_____ A. Accounting.
_____ B. Engineering.
__X__ C. Wastewater treatment plant operation.
_____ D. Salesmanship.

1.0A Wastewater is the same thing as
_____ A. Rain.
_____ B. Soil.
_____ C. Sewage.
_____ D. Condensation.

1.0B What does an operator do?
_____ A. Collect samples.
_____ B. Lubricate equipment.
_____ C. Record data.

1.0C Who employs treatment plant operators?
_____ A. Cities.
_____ B. Sanitation districts.
_____ C. Industries.

Check your answers on page 9.

1.1 YOUR PERSONAL TRAINING COURSE

Beginning on this page, you are embarking on a training course that has been carefully prepared to allow you to improve your knowledge of and ability to operate a wastewater treatment plant. You will be able to proceed at your own pace; you will have an opportunity to learn a little or a lot about each topic. The course has been prepared this way to fit the various needs of operators, depending on what kind of plant you have or how much you need to learn about it. To study for certification examinations, you will have to cover all the material. You will never know everything about your plant or about the wastewater that flows through it, but you can begin to answer some very important questions about how and when certain things happen in the plant. You can also learn to manipulate your plant so that it operates at maximum efficiency.

1.2 WHAT DO YOU ALREADY KNOW?

If you already have some experience operating a wastewater treatment plant, you may use the first three chapters for a review. If you are relatively new to the wastewater treatment field, these chapters will provide you with the background information necessary to understand the later chapters. The remainder of this introductory chapter describes your role as a *PROTECTOR OF WATER QUALITY*, your *QUALIFICATIONS* to do your job, a little about staffing needs in the wastewater treatment field, and some information on other *TRAINING OPPORTUNITIES*.

1.3 THE WATER QUALITY PROTECTOR: *YOU*

Historically Americans have shown a great lack of interest in the protection of their water resources. We have been content to think that "the solution to pollution is dilution." For years we were able to dump our wastes with little or no treatment back into the nearest *RECEIVING WATERS*.[2] As long as there was enough dilution water to absorb the waste material, nature took care of our disposal problems for us. As more and more towns and industry sprang up, waste loads increased until the natural purification processes could no longer do the job. Many waterways were converted into open sewers. Unfortunately, for many areas this did not signal the beginning of a cleanup campaign. It merely increased the frequency of the cry: "We don't have the money for a treatment plant," or the ever-popular, "If we make industries treat their wastes they will move to another state." Thus, the pollution of our waters increased.

Water quality protector

---

[2] *Receiving Water. A stream, river, lake, ocean, or other surface or groundwaters into which treated or untreated wastewater is discharged.*

Within the last 20 years, we have seen many changes in this depressing picture. We now realize that we must give nature a hand by treating wastes before they are discharged. Adequate treatment of wastes will not only protect our health and that of our downstream neighbors; it can also increase property values, allow game fishing and various recreational uses to be enjoyed, and attract water-using industries to the area. Today we are seeing massive efforts being undertaken to control water pollution and improve water quality throughout the nation. This includes the efforts not only of your own community, county, and state, but also the federal government.

Great sums of public and private funds are now being invested in large, complex municipal and industrial wastewater treatment facilities to overcome this pollution; and you, the treatment plant operator, will play a key role in the battle. Without efficient operation of your plant, much of the research, planning, and building that has been done and will be done to accomplish the goals of water quality control in your area will be wasted. You are the difference between a facility and a performing unit. You are, in fact, a *WATER QUALITY PROTECTOR* on the front line of the water pollution battle.

The receiving water quality standards and waste discharge requirements that your plant has been built to meet have been formulated to *PROTECT* the water users downstream from your plant. These uses may include domestic water supply, industrial water supply, agricultural water supply, stock and wildlife watering, propagation of fish and other aquatic and marine life, shellfish culture, swimming and other water contact sports, boating, aesthetic enjoyment, hydroelectric power, navigation, and others.

Therefore, you have an obligation to the users of the water downstream, as well as to the people of your district or municipality. You are the *KEY WATER QUALITY PROTECTOR* and must realize that you are in a responsible position. The main benefit of a successful wastewater treatment program is the protection of public health.

QUESTIONS

Write your answers in a notebook and then compare your answers with those on page 9.

1.3A   How did many receiving waters become polluted?

1.3B   Why must municipal and industrial wastewaters receive adequate treatment?

1.4   YOUR QUALIFICATIONS

The skill and ability required for your job depend to a large degree on the size and type of treatment plant where you are employed. You may work at a large modern treatment plant serving several hundred thousand persons and employing a hundred or more operators. In this case you are probably a specialist in one or more phases of the treatment process.

On the other hand, you may operate a small plant serving only a thousand people or fewer. You may be the only operator at the plant or, at best, have only one or two additional employees. If this is the case, you must be a "jack of all trades" because of the diversity of your tasks.

1.40   Your Job

To describe the operator's duties, let us start at the beginning. Let us say that the need for a new or improved wastewater treatment plant has long been recognized by the community. The community has voted to issue the necessary bonds to finance the project, and the consulting engineers have submitted plans and specifications. It is in the best interests of the community and the consulting engineer that you be in on the ground floor planning. If it is a new plant, you should be present or at least available during the construction period in

Visitors touring a treatment plant

order to become completely familiar with the entire plant, including the equipment and machinery and their operation. This will provide you with the opportunity to relate your plant drawings to actual facilities.

You and the engineer should discuss how the treatment plant should best be run and the means of operation the designer had in mind when the plant was designed. If it is an old plant being remodeled, you are in a position to offer excellent advice to the consulting engineer. Your experience provides valuable technical knowledge concerning the characteristics of wastewater, its sources, and the limitations of the present facilities. Together with the consultant, you are a member of an expert team able to advise the district or city.

Once the plant is operating, you become an administrator. In a small plant your duties may not include supervision of personnel, but you are still in charge of records. You are responsible for operating the plant as efficiently as possible, keeping in mind that the primary objective is to protect the receiving water quality by continuous and efficient plant performance. Without adequate, reliable records of every phase of operation, the effectiveness of your operation has not been documented (recorded).

You may also be the budget administrator. Most certainly you are in the best position to give advice on budget requirements, management problems, and future planning. You should be aware of the necessity for additional expenditures, including funds for plant enlargement, equipment replacement, and laboratory requirements. You should recognize and define such needs in sufficient time to inform the proper officials to enable them to accomplish early planning and budgeting.

You are in the field of public relations and must be able to explain the purpose and operation of your plant to visitors, civic organizations, school classes, representatives of news media, and even to city council or directors of your district. Public interest in water quality is increasing, and you should be prepared to conduct plant tours that will contribute to public acceptance and support. A well-guided tour for officials of regulatory agencies or other operators may provide these people with sufficient understanding of your plant to allow them to suggest helpful solutions to operational problems.

Special care and safety must be practiced when visitors are taken through your treatment plant. An accident could spoil all of your public relations efforts.

The appearance of your plant indicates to the visitor the type of operation you maintain. If the plant is dirty and rundown with flies and other insects swarming about, you will be unable to convince your visitors that the plant is doing a good job. *YOUR RECORDS SHOWING A HIGH-QUALITY EFFLUENT[3] WILL MEAN NOTHING TO THESE VISITING CITIZENS UNLESS YOUR PLANT APPEARS CLEAN AND WELL-MAINTAINED AND THE EFFLUENT LOOKS GOOD.*

Another aspect of your public relations duties is your dealings with the downstream water user. Unfortunately, the operator is often considered by the downstream user as a polluter rather than a water quality protector. Through a good public information program, backed by facts supported by reliable data, you can correct the impression held by the downstream user and establish "good neighbor" relations. This is indeed a challenge. Again, you must understand that you hold a very responsible position and be aware that the sole purpose of the operation of your plant is to protect the downstream user, be that user a private property owner, another city or district, an industry, or a fisherman.

You are required to understand certain laboratory procedures in order to conduct various tests on samples of wastewater and receiving waters. On the basis of the data obtained from these tests, you may have to adjust the operation of the treatment plant to meet receiving stream standards or discharge requirements.

As an operator you must have a knowledge of the complicated mechanical principles involved in many treatment mechanisms. In order to measure and control the wastewater flowing through the plant, you must have some understanding of hydraulics. Practical knowledge of electric motors, circuitry, and controls is also essential.

Safety is a very important operator responsibility. Unfortunately too many operators take safety for granted. This is one reason why the wastewater treatment industry has one of the worst safety records of any industry. *YOU* have the responsibility to be sure that your treatment plant is a safe place to work and visit. Everyone must follow safe procedures and understand why safe procedures must be followed at all times. All operators must be aware of the safety hazards in and around treatment plants. You should plan or be a part of an

---

[3] *Effluent. Wastewater or other liquid—raw (untreated), partially or completely treated—flowing FROM a reservoir, basin, treatment process, or treatment plant.*

active safety program. Chief operators frequently have the responsibility of training new operators and must encourage all operators to work safely.

Clearly then, today's wastewater treatment plant operator must possess a broad range of qualifications.

QUESTIONS

Write your answers in a notebook and then compare your answers with those on page 9.

1.4A   Why is it important that the operator be present during the construction of a new plant?

1.4B   How does the operator become involved in public relations?

1.5   STAFFING NEEDS AND FUTURE JOB OPPORTUNITIES

The wastewater treatment field, like so many others, is changing rapidly. New plants are being constructed and old plants are being modified and enlarged to handle the wastewater from our growing population and to treat the new chemicals being produced by our space-age technology. Operators, maintenance personnel, foremen, managers, instrumentation experts, and laboratory technicians are sorely needed.

A look at past records and future predictions indicates that wastewater treatment is a rapidly growing field. According to the U. S. Bureau of Labor Statistics (BLS),[4] water and wastewater treatment plant operators held about 98,000 jobs in 1996. Approximately half of the jobs were in the wastewater treatment industry, and the majority of operators worked for local governments. The BLS estimates that the number of jobs in the water and wastewater treatment industries is expected to increase by an average of 51 percent in the period 1996-2006. Factors contributing to the increase include population growth, retirement of many current operators, regulatory requirements, more sophisticated treatment, and operator certification. The need for *trained* operators is increasing rapidly and is expected to continue to grow in the future.

1.6   TRAINING YOURSELF TO MEET THE NEEDS

This training course is not the only one available to help you improve your abilities. The states have offered various types of both long- and short-term operator training through their health departments and water pollution control associations have provided training classes conducted by members of the associations, largely on a volunteer basis. The Water Environment Federation (WEF) has developed two visual aid training courses to complement its Manual of Practice No. 11. State and local colleges have provided valuable training under their own sponsorship or in partnership with others. Many state, local, and private agencies have conducted both long- and short-term training as well as interesting and informative seminars. The California Water Environment Association has prepared several excellent study guides for operators. Excellent textbooks have been written by many state agencies. Those of the New York State Health Department and the Texas Water Utilities Association deserve special attention. The Canadian government has developed very good training manuals for operators.

Listed below are several very good references in the field of wastewater treatment plant operation that are frequently referred to throughout this course. The name in quotes represents the term usually used by operators when they mention the reference. Prices listed are those available when this manual was published and will probably increase in the future.

1. *"MOP 11." OPERATION OF MUNICIPAL WASTEWATER TREATMENT PLANTS* (MOP 11). Obtain from Water Environment Federation (WEF), Publications Order Department, 601 Wythe Street, Alexandria, VA 22314-1994. Order No. M05110. Price to members, $108.75; nonmembers, $157.75; price includes cost of shipping and handling.

2. *"NEW YORK MANUAL." MANUAL OF INSTRUCTION FOR WASTEWATER TREATMENT PLANT OPERATORS* (two-volume set) distributed in New York by the New York State Department of Health, Office of Public Health Education, Water Pollution Control Board. Distributed outside of New York State by Health Education Services, PO Box 7126, Albany, NY 12224. Price $20.00 for the two-volume set, plus $5.00 shipping and handling. Make checks payable to Health Education Services.

3. *"TEXAS MANUAL." MANUAL OF WASTEWATER TREATMENT.* Obtain from Texas Water Utilities Association, 1106 Clayton Lane, Suite 101 East, Austin, TX 78723-1093. Price to members, $25.00; nonmembers, $35.00; plus $3.50 shipping and handling.

These publications cover the entire field of treatment plant operation. At the end of many of the chapters yet to come, lists of other references will be provided.

---

[4] *Refer to the Bureau of Labor Statistics' website at www.bls.gov for additional information about the types of jobs available in the water and wastewater industries, working conditions, earnings potential, and the job outlook.*

SUGGESTED ANSWERS

Chapter 1. THE TREATMENT PLANT OPERATOR

You are not expected to have the exact answer suggested for questions requiring written answers, but you should have the correct idea. The numbering of the questions refers to the section in the manual where you can find the information to answer the questions. Answers to questions numbered 1.0 can be found in Section 1.0, "What is a Treatment Plant Operator?"

Answers to questions on page 5.

1.0A   C

1.0B   A, B, C

1.0C   A, B, C

Answers to questions on page 6.

1.3A   Receiving waters became polluted by a lack of public concern for the impact of waste discharges and by discharging wastewater into a receiving water beyond its natural purification capacity.

1.3B   Municipal and industrial wastewaters must receive adequate treatment to protect receiving water users.

Answers to questions on page 8.

1.4A   The operator should be present during the construction of a new plant in order to become familiar with the plant before the operator begins operating it.

1.4B   The operator becomes involved in public relations by explaining the purpose and operation of the plant to visitors, civic organizations, school classes, news reporters, and city or district representatives.

CHAPTER 2

WHY TREAT WASTES?

by

William Crooks

TABLE OF CONTENTS

Chapter 2.   WHY TREAT WASTES?

OBJECTIVES

Chapter 2.   WHY TREAT WASTES?

Following completion of Chapter 2, you should be able to:

1. Give reasons for preventing pollution,

2. Identify various types of waste discharges,

3. Recognize the effects of waste discharges on receiving waters,

4. Describe the different types of solids in wastewater,

5. Explain what happens in a natural cycle, and

6. Identify a NPDES permit.

PROJECT PRONUNCIATION KEY

by Warren L. Prentice

The Project Pronunciation Key is designed to aid you in the pronunciation of new words. While this key is based primarily on familiar sounds, it does not attempt to follow any particular pronunciation guide. This key is designed solely to aid operators in this program.

You may find it helpful to refer to other available sources for pronunciation help. Each current standard dictionary contains a guide to its own pronunciation key. Each key will be different from each other and from this key. Examples of the difference between the key used in this program and the *WEBSTER'S NEW WORLD COLLEGE DICTIONARY*[1] "Key" are shown below.

In using this key, you should accent (say louder) the syllable that appears in capital letters. The following chart is presented to give examples of how to pronounce words using the Project Key.

| | SYLLABLE | | | | |
WORD	1st	2nd	3rd	4th	5th
acid	AS	id			
coliform	COAL	i	form		
biological	BUY	o	LODGE	ik	cull

The first word, *ACID*, has its first syllable accented. The second word, *COLIFORM*, has its first syllable accented. The third word, *BIOLOGICAL*, has its first and third syllables accented.

We hope you will find the key useful in unlocking the pronunciation of any new word.

Term	Project Key	Webster Key
acid	AS-id	aś id
coliform	COAL-i-form	kō′ lə fôrm
biological	BUY-o-LODGE-ik-cull	bī ə läj′ i kəl

[1] The *WEBSTER'S NEW WORLD COLLEGE DICTIONARY*, Fourth Edition, 1999, was chosen rather than an unabridged dictionary because of its availability to the operator. Other editions may be slightly different.

WORDS

Chapter 2.   WHY TREAT WASTES?

AEROBIC BACTERIA (AIR-O-bick back-TEAR-e-ah) AEROBIC BACTERIA

Bacteria which will live and reproduce only in an environment containing oxygen which is available for their respiration (breathing), namely atmospheric oxygen or oxygen dissolved in water. Oxygen combined chemically, such as in water molecules (H_2O), cannot be used for respiration by aerobic bacteria.

ALGAE (AL-gee) ALGAE

Microscopic plants which contain chlorophyll and live floating or suspended in water. They also may be attached to structures, rocks, or other submerged surfaces. Algae produce oxygen during sunlight hours and use oxygen during the night hours. Their biological activities appreciably affect the pH, alkalinity, and dissolved oxygen of the water.

ANAEROBIC BACTERIA (AN-air-O-bick back-TEAR-e-ah) ANAEROBIC BACTERIA

Bacteria that live and reproduce in an environment containing no "free" or dissolved oxygen. Anaerobic bacteria obtain their oxygen supply by breaking down chemical compounds which contain oxygen, such as sulfate (SO_4^{2-}).

BIOCHEMICAL OXYGEN DEMAND (BOD) BIOCHEMICAL OXYGEN DEMAND (BOD)

The rate at which organisms use the oxygen in water or wastewater while stabilizing decomposable organic matter under aerobic conditions. In decomposition, organic matter serves as food for the bacteria and energy results from its oxidation. BOD measurements are used as a measure of the organic strength of wastes in water.

BIOCHEMICAL OXYGEN DEMAND (BOD) TEST BIOCHEMICAL OXYGEN DEMAND (BOD) TEST

A procedure that measures the rate of oxygen use under controlled conditions of time and temperature. Standard test conditions include dark incubation at 20°C for a specified time (usually five days).

COLIFORM (COAL-i-form) COLIFORM

One type of bacteria. The presence of coliform-group bacteria is an indication of possible pathogenic bacterial contamination. The human intestinal tract is one of the main habitats of coliform bacteria. They may also be found in the intestinal tracts of warm-blooded animals, and in plants, soil, air, and the aquatic environment. Fecal coliforms are those coliforms found in the feces of various warm-blooded animals; whereas the term "coliform" also includes other environmental sources.

DISINFECTION (dis-in-FECT-shun) DISINFECTION

The process designed to kill or inactivate most microorganisms in wastewater, including essentially all pathogenic (disease-causing) bacteria. There are several ways to disinfect, with chlorination being the most frequently used in water and wastewater treatment plants. Compare with STERILIZATION.

EFFLUENT EFFLUENT

Wastewater or other liquid—raw (untreated), partially or completely treated—flowing *FROM* a reservoir, basin, treatment process, or treatment plant.

EVAPOTRANSPIRATION (ee-VAP-o-TRANS-purr-A-shun) EVAPOTRANSPIRATION

(1)   The process by which water vapor passes into the atmosphere from living plants. Also called TRANSPIRATION.

(2)   The total water removed from an area by transpiration (plants) and by evaporation from soil, snow and water surfaces.

IMHOFF CONE IMHOFF CONE

A clear, cone-shaped container marked with graduations. The cone is used to measure the volume of settleable solids in a specific volume (usually one liter) of wastewater.

INORGANIC WASTE INORGANIC WASTE

Waste material such as sand, salt, iron, calcium, and other mineral materials which are only slightly affected by the action of organisms. Inorganic wastes are chemical substances of mineral origin; whereas organic wastes are chemical substances usually of animal or plant origin.

MILLIGRAMS PER LITER, mg/L (MILL-i-GRAMS per LEET-er) MILLIGRAMS PER LITER, mg/L

A measure of the concentration by weight of a substance per unit volume. For practical purposes, one mg/L of a substance in water is equal to one part per million parts (ppm). Thus a liter of water with a specific gravity of 1.0 weighs one million milligrams. If it contains 10 milligrams of dissolved oxygen, the concentration is 10 milligrams per million milligrams, or 10 milligrams per liter (10 mg/L), or 10 parts of oxygen per million parts of water, or 10 parts per million (10 ppm).

NUTRIENT CYCLE NUTRIENT CYCLE

The transformation or change of a nutrient from one form to another until the nutrient has returned to the original form, thus completing the cycle. The cycle may take place under either aerobic or anaerobic conditions.

NUTRIENTS NUTRIENTS

Substances which are required to support living plants and organisms. Major nutrients are carbon, hydrogen, oxygen, sulfur, nitrogen and phosphorus. Nitrogen and phosphorus are difficult to remove from wastewater by conventional treatment processes because they are water soluble and tend to recycle. Also see NUTRIENT CYCLE.

ORGANIC WASTE ORGANIC WASTE

Waste material which comes mainly from animal or plant sources. Organic wastes generally can be consumed by bacteria and other small organisms. Inorganic wastes are chemical substances of mineral origin.

PATHOGENIC (PATH-o-JEN-ick) ORGANISMS PATHOGENIC ORGANISMS

Bacteria, viruses, cysts, or protozoa which can cause disease (giardiasis, cryptosporidiosis, typhoid, cholera, dysentery) in a host (such as a person). There are many types of organisms which do *NOT* cause disease and which are *NOT* called pathogenic. Many beneficial bacteria are found in wastewater treatment processes actively cleaning up organic wastes.

pH (pronounce as separate letters) pH

pH is an expression of the intensity of the basic or acidic condition of a liquid. Mathematically, pH is the logarithm (base 10) of the reciprocal of the hydrogen ion activity.

$$pH = Log \frac{1}{[H^+]}$$

The pH may range from 0 to 14, where 0 is most acidic, 14 most basic, and 7 neutral. Natural waters usually have a pH between 6.5 and 8.5.

POLLUTION POLLUTION

The impairment (reduction) of water quality by agricultural, domestic or industrial wastes (including thermal and radioactive wastes) to a degree that the natural water quality is changed to hinder any beneficial use of the water or render it offensive to the senses of sight, taste, or smell or when sufficient amounts of wastes create or pose a potential threat to human health or the environment.

PRIMARY TREATMENT PRIMARY TREATMENT

A wastewater treatment process that takes place in a rectangular or circular tank and allows those substances in wastewater that readily settle or float to be separated from the water being treated.

RECEIVING WATER RECEIVING WATER

A stream, river, lake, ocean, or other surface or groundwaters into which treated or untreated wastewater is discharged.

SECONDARY TREATMENT SECONDARY TREATMENT

A wastewater treatment process used to convert dissolved or suspended materials into a form more readily separated from the water being treated. Usually the process follows primary treatment by sedimentation. The process commonly is a type of biological treatment process followed by secondary clarifiers that allow the solids to settle out from the water being treated.

SEPTIC (SEP-tick) SEPTIC

A condition produced by anaerobic bacteria. If severe, the sludge produces hydrogen sulfide, turns black, gives off foul odors, contains little or no dissolved oxygen, and the wastewater has a high oxygen demand.

STABILIZE STABILIZE

To convert to a form that resists change. Organic material is stabilized by bacteria which convert the material to gases and other relatively inert substances. Stabilized organic material generally will not give off obnoxious odors.

STERILIZATION (STARE-uh-luh-ZAY-shun) STERILIZATION

The removal or destruction of all microorganisms, including pathogenic and other bacteria, vegetative forms and spores. Compare with DISINFECTION.

TRANSPIRATION (TRAN-spur-RAY-shun) TRANSPIRATION

The process by which water vapor is released to the atmosphere by living plants. This process is similar to people sweating. Also see EVAPOTRANSPIRATION.

CHAPTER 2.   WHY TREAT WASTES?

2.0 PREVENTION OF POLLUTION

The operator's main job is to protect the many users of receiving waters. As an operator you must do the best you can to remove any substances that will unreasonably affect these users.

Many people think *ANY* discharge of waste to a body of water is pollution. However, with our present system of using water to carry away the waste products of homes and industries, it would be impossible and perhaps unwise to prohibit the discharge of all wastewater to oceans, streams, and groundwater basins. Today's technology is capable of treating wastes in such a manner that existing or potential receiving water uses are not unreasonably affected. Definitions of pollution include any interference with the beneficial reuse of water or failure to meet water quality requirements. Any questions or comments regarding this definition must be settled by the appropriate enforcement agency.

2.1 WHAT IS PURE WATER?

Water is a combination of two parts hydrogen and one part oxygen, or H_2O. This is true, however, only for "pure" water such as might be manufactured in a laboratory. Water as we know it is not "pure" hydrogen and oxygen. Even the distilled water we purchase in the store has measurable quantities of various substances in addition to hydrogen and oxygen. Rainwater, even before it reaches the earth, contains many substances. These substances, since they are not found in "pure" water, may be considered "impurities." When rain falls through the atmosphere, it gains nitrogen and other gases. As soon as the rain flows over land, it begins to dissolve from the earth and rocks such substances as calcium, carbon, magnesium, sodium, chloride, sulfate, iron, nitrogen, phosphorus, and many other materials. Organic matter (matter derived from plants and animals) is also dissolved by water from contact with decaying leaves, twigs, grass, or small insects and animals. Thus, a fresh flowing mountain stream may pick up many natural "impurities," some possibly in harmful amounts, before it ever reaches civilization or is affected by the waste discharges of society. Many of these substances, however, are needed in small amounts to support life and are useful to humans. Concentrations of impurities must be controlled or regulated to prevent harmful levels in receiving waters.

Water + Impurities

QUESTIONS

Write your answers in a notebook and then compare your answers with those on page 26.

2.1A   How does water pick up dissolved substances?

2.1B   What are some of the dissolved substances in water?

2.2 TYPES OF WASTE DISCHARGES

The waste discharge that first comes to mind in any discussion of stream pollution is the discharge of domestic wastewater. Wastewater contains a large amount of *ORGANIC WASTE.*[2] Industry also contributes substantial amounts of organic waste. Some of these organic industrial wastes come from vegetable and fruit packing; dairy processing; meat packing; tanning; and processing of poultry, oil, paper and fiber (wood), and many other industries. All organic materials have one thing in common—they all contain carbon.

Another classification of wastes is *INORGANIC WASTES.*[3] Domestic wastewater contains inorganic material as well as

---

[2] *Organic Waste.   Waste material which comes mainly from animal or plant sources. Organic wastes generally can be consumed by bacteria and other small organisms. Inorganic wastes are chemical substances of mineral origin.*

[3] *Inorganic Waste.   Waste material such as sand, salt, iron, calcium, and other mineral materials which are only slightly affected by the action of organisms. Inorganic wastes are chemical substances of mineral origin; whereas organic wastes are chemical substances usually of animal or plant origin.*

organic, and many industries discharge inorganic wastes which add to the mineral content of receiving waters. For instance, a discharge of salt brine (sodium chloride) from water softening will increase the amount of sodium and chloride in the receiving waters. Some industrial wastes may introduce inorganic substances such as chromium or copper, which are very toxic to aquatic life. Other industries (such as gravel washing plants) discharge appreciable amounts of soil, sand, or grit, which also may be classified as inorganic wastes.

There are two other major types of wastes that do not fit either the organic or inorganic classification. These are heated (thermal) wastes and radioactive wastes. Waters with temperatures exceeding the requirements of the enforcing agency may come from cooling processes used by industry and from thermal power stations generating electricity. Radioactive wastes are usually controlled at their source, but could come from hospitals, research laboratories, and nuclear power plants.

QUESTIONS

Write your answers in a notebook and then compare your answers with those on page 26.

2.2A Several of the following contain significant quantities of organic material. Which are they?

    a. Domestic wastewater
    b. Cooling water from thermal power stations
    c. Paper mill wastes
    d. Metal plating wastes
    e. Tanning wastes

2.2B List four types of pollution.

2.3 EFFECTS OF WASTE DISCHARGES

Certain substances not removed by wastewater treatment processes can cause problems in receiving waters. This section reviews some of these substances and discusses why they should be treated.

2.30 Sludge and Scum

If certain wastes (including domestic wastewater) do not receive adequate treatment, large amounts of solids may accumulate on the banks of the receiving waters, or they may settle to the bottom to form sludge deposits or float to the surface and form rafts of scum. Sludge deposits and scum are not only unsightly but, if they contain organic material, they may also cause oxygen depletion and be a source of odors. *PRIMARY*

TREATMENT[4] units in the wastewater treatment plant are designed and operated to remove the sludge and scum before they reach the receiving waters.

2.31 Oxygen Depletion

Most living creatures need oxygen to survive, including fish and other aquatic life. Although most streams and other surface waters contain less than 0.001% *DISSOLVED OXYGEN* (10 milligrams of oxygen per liter of water, or 10 mg/L[5]), most fish can thrive if there are at least 5 mg/L and other conditions are favorable. When oxidizable wastes are discharged to a stream, bacteria begin to feed on the waste and decompose or break down the complex substances in the waste into simple chemical compounds. These bacteria also use dissolved oxygen (similar to human respiration or breathing) from the water and are called *AEROBIC BACTERIA.*[6] As more organic waste is added, the bacteria reproduce rapidly; and as their population increases, so does their use of oxygen. Where waste flows are high, the population of bacteria may grow large enough to use the entire supply of oxygen from the stream faster than it can be replenished by natural diffusion from the atmosphere. When this happens, fish and most other living things in the stream that require dissolved oxygen die.

Oxygen depletion

Therefore, one of the principal objectives of wastewater treatment is to prevent as much of this "oxygen-demanding" organic material as possible from entering the receiving water. The treatment plant actually removes the organic material the same way a stream does, but it accomplishes the task much more efficiently by removing the wastes from the wastewater. *SECONDARY TREATMENT*[7] units are designed and operated to use natural organisms such as bacteria *IN THE PLANT* to *STABILIZE*[8] and to remove organic material.

Another effect of oxygen depletion, in addition to the killing of fish and other aquatic life, is the problem of odors. When all the dissolved oxygen has been removed, *ANAEROBIC BACTERIA*[9] begin to use the oxygen that is combined chemically

---

[4] *Primary Treatment.   A wastewater treatment process that takes place in a rectangular or circular tank and allows those substances in wastewater that readily settle or float to be separated from the water being treated.*

[5] *Milligrams per Liter, mg/L (MILL-i-GRAMS per LEET-er).   A measure of the concentration by weight of a substance per unit volume. For practical purposes, one mg/L of a substance in water is equal to one part per million parts (ppm). Thus a liter of water with a specific gravity of 1.0 weighs one million milligrams. If it contains 10 milligrams of dissolved oxygen, the concentration is 10 milligrams per million milligrams, or 10 milligrams per liter (10 mg/L), or 10 parts of oxygen per million parts of water, or 10 parts per million (10 ppm).*

[6] *Aerobic Bacteria (AIR-O-bick back-TEAR-e-ah).   Bacteria which will live and reproduce only in an environment containing oxygen which is available for their respiration (breathing), namely atmospheric oxygen or oxygen dissolved in water. Oxygen combined chemically, such as in water molecules (H_2O), cannot be used for respiration by aerobic bacteria.*

[7] *Secondary Treatment.   A wastewater treatment process used to convert dissolved or suspended materials into a form more readily separated from the water being treated. Usually the process follows primary treatment by sedimentation. The process commonly is a type of biological treatment process followed by secondary clarifiers that allow the solids to settle out from the water being treated.*

[8] *Stabilize.   To convert to a form that resists change. Organic material is stabilized by bacteria which convert the material to gases and other relatively inert substances. Stabilized organic material generally will not give off obnoxious odors.*

[9] *Anaerobic Bacteria (AN-air-O-bick back-TEAR-e-ah).   Bacteria that live and reproduce in an environment containing no "free" or dissolved oxygen. Anaerobic bacteria obtain their oxygen supply by breaking down chemical compounds which contain oxygen, such as sulfate (SO_4^{2-}).*

with other elements in the form of chemical compounds, such as sulfate (sulfur and oxygen), which are also dissolved in the water. When anaerobic bacteria remove the oxygen from sulfur compounds, hydrogen sulfide (H_2S), which has a "rotten egg" odor, is produced. This gas is not only very odorous, but it also erodes (corrodes) concrete and can discolor and remove paint from homes and structures. Hydrogen sulfide also may form explosive mixtures with air and is a toxic gas capable of paralyzing your respiratory system. Other products of anaerobic decomposition (putrefaction: PEW-truh-FACK-shun) also can be objectionable.

QUESTIONS

Write your answers in a notebook and then compare your answers with those on page 26.

2.3A   What causes oxygen depletion when organic wastes are discharged to the water?

2.3B   What kind of bacteria cause hydrogen sulfide gas to be released?

2.32   Human Health

Up to now we have discussed the physical or chemical effects that a waste discharge may have on the uses of water. More important, however, may be the effect on human health through the spread of disease-causing bacteria and viruses. Initial efforts to control human wastes evolved from the need to prevent the spread of diseases. Although untreated wastewater contains many billions of bacteria per gallon, most of these are not harmful to humans, and some are even helpful in wastewater treatment processes. However, humans who have a disease that is caused by bacteria or viruses may discharge some of these harmful organisms in their body wastes. Many serious outbreaks of communicable diseases have been traced to direct contamination of drinking water or food supplies by the body wastes from a human disease carrier.

Some known examples of diseases that may be spread through wastewater discharges are

Diseases

Fortunately the bacteria that grow in the intestinal tract of diseased humans are not likely to find the environment in the wastewater treatment plant or receiving waters favorable for their growth and reproduction. Although many *PATHOGENIC ORGANISMS*[10] are removed by natural die-off during the normal treatment processes, sufficient numbers can remain to cause a threat to any downstream use involving human contact or consumption. If such uses exist downstream, the treatment plant must also include a *DISINFECTION*[11] process.

The disinfection process most often used is the addition of chlorine. In most cases proper chlorination of a *WELL-TREATED WASTE* will result in essentially a complete kill of the pathogenic organisms. Operators must realize, however, that the breakdown or malfunction of equipment could result in the discharge at any time of an effluent that contains pathogenic bacteria. To date no one working in the wastewater collection or treatment field is known to have become infected by the AIDS virus due to conditions encountered while working on the job. Good personal hygiene is an operator's best defense against infections and diseases.

2.33   Other Effects

Some wastes adversely affect the clarity and color of the receiving waters, making them unsightly and unpopular for recreation.

Many industrial wastes are highly acid or alkaline (basic), and either condition can interfere with aquatic life, domestic use, and other uses. An accepted measurement of a waste's acidic or basic condition is its *pH*.[12] Before wastes are discharged, they should have a pH similar to that of the receiving water.

Waste discharges may contain toxic substances, such as heavy metals (lead, mercury, cadmium, and chromium) or cyanide, which may affect the use of the receiving water for domestic purposes or for aquatic life. Plant effluents chlorinated for disinfection purposes may have to be dechlorinated to protect receiving waters from the toxic effects of residual chlorine.

---

[10] *Pathogenic (PATH-o-JEN-ick) Organisms.   Bacteria, viruses, cysts, or protozoa which can cause disease (giardiasis, cryptosporidiosis, typhoid, cholera, dysentery) in a host (such as a person). There are many types of organisms which do NOT cause disease and which are NOT called pathogenic. Many beneficial bacteria are found in wastewater treatment processes actively cleaning up organic wastes.*

[11] *Disinfection (dis-in-FECT-shun).   The process designed to kill or inactivate most microorganisms in wastewater, including essentially all pathogenic (disease-causing) bacteria. There are several ways to disinfect, with chlorination being the most frequently used in water and wastewater treatment plants.*

[12] *pH.   Technically, this is the logarithm of the reciprocal of the hydrogen ion concentration, which is explained in Chapter 16, "Laboratory Procedures and Chemistry."   pH expresses the intensity of the acidic or basic condition of a liquid. The pH may range from 0 to 14, where 0 is most acidic, 14 is most basic, and 7 is neutral. Most natural waters have a pH between 6.5 and 8.5.*

Taste- and odor-producing substances may reach levels in the receiving water that are readily detectable in drinking water or in the flesh of fish.

Treated wastewaters contain NUTRIENTS[13] capable of encouraging excess ALGAE[14] and plant growth in receiving waters. These growths hamper domestic, industrial, and recreational uses. Conventional wastewater treatment plants do not remove a major portion of the nitrogen and phosphorus nutrients.

QUESTIONS

Write your answers in a notebook and then compare your answers with those on page 26.

2.3C Where do the disease-causing bacteria in wastewater come from?

2.3D What is the term that means "disease-causing"?

2.3E What is the most frequently used means of disinfecting treated wastewater?

2.4 SOLIDS IN WASTEWATER (Figure 2.1)

One of the primary functions of a treatment plant is the removal of solids from wastewater.

2.40 Types of Solids

In Section 2.2 you read about the different TYPES of pollution: organic, inorganic, thermal, and radioactive. For a normal municipal wastewater that contains domestic wastewater as well as some industrial and commercial wastes, the concerns of the treatment plant designer and operator usually are to remove the organic and inorganic SUSPENDED SOLIDS, to remove the DISSOLVED ORGANIC SOLIDS (the treatment plant does little to remove DISSOLVED INORGANIC SOLIDS), and to kill or inactivate the PATHOGENIC ORGANISMS by disinfection. THERMAL and RADIOACTIVE wastes require special treatment processes.

Since the main purpose of the treatment plant is removal of solids from the wastewater, a detailed discussion of the types of solids is in order. Figure 2.1 will help you understand the different terms.

2.41 Total Solids

For discussion purposes assume that you obtain a one-liter sample of raw wastewater entering the treatment plant. Heat this sample enough to evaporate all the water and weigh all the solid material left (residue); it weighs 1,000 milligrams. Thus, the TOTAL SOLIDS concentration in the sample is 1,000 milligrams per liter (mg/L). This weight includes both DISSOLVED and SUSPENDED solids.

2.42 Dissolved Solids

How much is dissolved and how much is suspended? To determine this you could take an identical sample and filter it through a very fine mesh filter such as a membrane filter or fiberglass. The suspended solids will be caught on the filter, and the dissolved solids will pass through with the water. You can now evaporate the water and weigh the residue to determine the weight of DISSOLVED SOLIDS. In Figure 2.1 the amount is shown as 800 mg/L. The remaining 200 mg/L is suspended solids.

2.43 Suspended Solids

Suspended solids are composed of two parts: settleable and nonsettleable. The difference between settleable and nonsettleable solids depends on the size, shape, and weight per unit volume of the solid particles; large-sized particles tend to settle more rapidly than smaller particles. The amount of settleable solids in the raw wastewater should be estimated in order to design settling basins (primary units), sludge pumps, and sludge handling facilities. Also, measuring the amount of settleable solids entering and leaving the settling basin allows you to calculate the efficiency of the basin for removing the settleable solids. A device called an IMHOFF CONE[15] is used to measure settleable solids in milliliters per liter, mL/L. (The example in Figure 2.1 shows a settleable solids concentration of 130 mg/L. The settled solids in the Imhoff cone had to be dried and weighed by proper procedures to determine their weight.)

You may calculate the weight of nonsettleable solids by subtracting the weight of dissolved and settleable solids from the weight of total solids. In Figure 2.1 the nonsettleable solids concentration is shown as 70 mg/L.

2.44 Organic and Inorganic Solids

For total solids or for any separate type of solids, such as dissolved, settleable, or nonsettleable, the relative amounts of organic and inorganic matter can be determined. Procedures for the determination of amounts of organic and inorganic solids are provided in Chapter 16, "Laboratory Procedures and Chemistry." This information is important for estimating solids handling capacities and for designing treatment processes for removing the organic portion in waste. The organic portion can be very harmful to receiving waters.

2.45 Floatable Solids

Treatment units are designed to remove the solids in raw wastewater and treated effluent. There is no standard method for the measurement and evaluation of floatable solids. Floatable solids are undesirable in the plant effluent because the sight of floatables in receiving waters indicates the presence of inadequately treated wastewater.

---

[13] Nutrients. Substances which are required to support living plants and organisms. Major nutrients are carbon, hydrogen, oxygen, sulfur, nitrogen and phosphorus. Nitrogen and phosphorus are difficult to remove from wastewater by conventional treatment processes because they are water soluble and tend to recycle.

[14] Algae (AL-gee). Microscopic plants which contain chlorophyll and live floating or suspended in water. They also may be attached to structures, rocks, or other submerged surfaces. Algae produce oxygen during sunlight hours and use oxygen during the night hours. Their biological activities appreciably affect the pH, alkalinity, and dissolved oxygen of the water.

[15] Imhoff Cone. A clear, cone-shaped container marked with graduations. The cone is used to measure the volume of settleable solids in a specific volume (usually one liter) of wastewater.

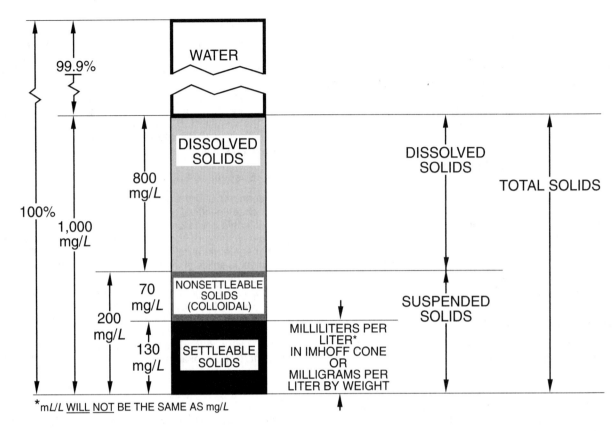

*mL/L WILL NOT BE THE SAME AS mg/L

Fig. 2.1   Typical composition of solids in raw wastewater
(floatable solids not shown)

QUESTIONS

Write your answers in a notebook and then compare your answers with those on page 26.

2.4A   Total solids consist of _____ and _____ solids, both of which contain organic and inorganic matter.

2.4B   Why is it necessary to measure settleable solids?

2.4C   An Imhoff cone is used to measure _____ solids.

2.5   NATURAL CYCLES

When the treated wastewater from a plant is discharged into *RECEIVING WATERS*[16] such as streams, rivers, or lakes, natural cycles in the aquatic (water) environment may become upset. Whether any problems are caused in the receiving waters depends on the following factors:

1. Type or degree of treatment,

2. Size of flow from the treatment plant,

3. Characteristics of wastewater from treatment plant,

4. Amount of flow in the receiving stream or volume of receiving lake that can be used for dilution,

5. Quality of the receiving waters,

6. Amount of mixing between *EFFLUENT*[17] and receiving waters, and

7. Uses of receiving waters.

Natural cycles of interest in wastewater treatment include the natural purification cycles such as the cycle of water from evaporation or *TRANSPIRATION*[18] to condensation to precipitation to runoff and back to evaporation, the life cycles of aquatic organisms, and the cycles of nutrients. These cycles are occurring continuously in wastewater treatment plants and in receiving waters at different rates depending on environmental conditions. Treatment plant operators control and accelerate these cycles to work for their benefit in treatment plants and in receiving waters rather than have these cycles cause plant operational problems and disrupt downstream water uses.

NUTRIENT CYCLES[19] (Figure 2.2) are a special type of natural cycle because of the sensitivity of some receiving

---

[16] *Receiving Water.   A stream, river, lake, ocean, or other surface or groundwaters into which treated or untreated wastewater is discharged.*

[17] *Effluent.   Wastewater or other liquid—raw (untreated), partially or completely treated—flowing FROM a reservoir, basin, treatment process or treatment plant.*

[18] *Transpiration (TRAN-spur-RAY-shun).   The process by which water vapor is released to the atmosphere by living plants. This process is similar to people sweating.*

[19] *Nutrient Cycle.   The transformation or change of a nutrient from one form to another until the nutrient has returned to the original form, thus completing the cycle. The cycle may take place under either aerobic or anaerobic conditions.*

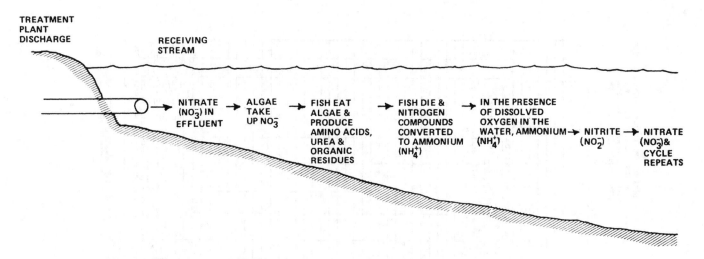

TREATMENT PLANT DISCHARGE

RECEIVING STREAM

NITRATE (NO_3^-) IN EFFLUENT → ALGAE TAKE UP NO_3^- → FISH EAT ALGAE & PRODUCE AMINO ACIDS, UREA & ORGANIC RESIDUES → FISH DIE & NITROGEN COMPOUNDS CONVERTED TO AMMONIUM (NH_4^+) → IN THE PRESENCE OF DISSOLVED OXYGEN IN THE WATER, AMMONIUM (NH_4^+) → NITRITE (NO_2^-) → NITRATE (NO_3^-) & CYCLE REPEATS

Fig. 2.2   Simplified illustration of nitrogen cycle

waters to nutrients. Important nutrients include carbon, hydrogen, oxygen, sulfur, nitrogen, and phosphorus. All of the nutrients have their own cycles, yet each cycle is influenced by the other cycles. These nutrient cycles are very complex and involve chemical changes in living organisms.

To illustrate the concept of nutrient cycles, a simplified version of the nitrogen cycle will be used as an example (Figure 2.2). A wastewater treatment plant discharges nitrogen in the form of nitrate (NO_3^-) in the plant effluent to the receiving waters. Algae take up the nitrate and produce more algae. The algae are eaten by fish, which convert the nitrogen to amino acids, urea, and organic residues. If the fish die and sink to the bottom, these nitrogen compounds can be converted to ammonium (NH_4^+). In the presence of dissolved oxygen and special bacteria, the ammonium is converted to nitrite (NO_2^-) then to nitrate (NO_3^-), and finally the algae can take up the nitrate and start the cycle all over again.

If too much nitrogen is discharged to receiving waters, too many algae could be produced. Water with excessive algae can be unsightly. Bacteria decomposing dead algae from occasional die-offs can deplete the dissolved oxygen and cause a fish kill. Thus, the nitrogen cycle has been disrupted, as well as the other nutrient cycles. If no dissolved oxygen is present in the water, the nitrogen compounds are converted to ammonium (NH_4^+), the carbon compounds to methane (CH_4), and the sulfur compounds to hydrogen sulfide (H_2S). Ammonia (NH_3) and hydrogen sulfide are odorous gases. Under these conditions the receiving waters are *SEPTIC*[20]; they stink and look terrible. Throughout this manual you will be provided information on how to control these nutrient cycles in your treatment plant in order to treat wastes and to control odors, as well as to protect receiving waters.

2.6 NPDES PERMITS (Figure 2.3)

NPDES stands for **N**ational **P**ollutant **D**ischarge **E**limination **S**ystem. NPDES permits are required by the Federal Water Pollution Control Act Amendments of 1972 with the intent of making the nation's waters suitable for swimming and for fish and wildlife. The permits regulate discharges into navigable waters from all point sources of pollution, including industries, municipal wastewater treatment plants, sanitary landfills, large agricultural feedlots, and return irrigation flows. An industry discharging into municipal collection and treatment systems need not obtain a permit but must meet certain specified pretreatment standards. These permits may outline a schedule of compliance for a wastewater treatment facility such as dates for the completion of plant design, engineering, construction, and/or treatment process changes. Instructions for completing NPDES reporting forms and the necessary forms are available from the regulatory agency issuing the permit.

Your main concern as an operator is the effluent (discharge) limitations specified in the NPDES permit for your plant. The permit may specify monthly average and maximum levels of suspended solids, *BIOCHEMICAL OXYGEN DEMAND (BOD)*,[21] and the most probable number (MPN) of *COLIFORM*[22] group bacteria. Larger plants must report effluent temperatures because of the impact of temperature changes on natural cycles. Also, average and maximum flows may be identified as well as an acceptable range of pH values. Almost all effluents are expected to contain virtually no substances that would be toxic to organisms in the receiving waters. NPDES permits have effluent limit restrictions on toxic substances. The NPDES permit will specify the frequency of collecting samples and the methods of reporting the results. Details on how to comply with NPDES permits will be provided throughout this manual.

---

[20] *Septic (SEP-tick).   A condition produced by anaerobic bacteria. If severe, the sludge produces hydrogen sulfide, turns black, gives off foul odors, contains little or no dissolved oxygen, and the wastewater has a high oxygen demand.*

[21] *Biochemical Oxygen Demand (BOD).   The rate at which organisms use the oxygen in water or wastewater while stabilizing decomposable organic matter under aerobic conditions. In decomposition, organic matter serves as food for the bacteria and energy results from its oxidation. BOD measurements are used as a measure of the organic strength of wastes in water.*

[22] *Coliform (COAL-i-form).   One type of bacteria. The presence of coliform-group bacteria is an indication of possible pathogenic bacterial contamination. The human intestinal tract is one of the main habitats of coliform bacteria. They may also be found in the intestinal tracts of warm-blooded animals, and in plants, soil, air, and the aquatic environment. Fecal coliforms are those coliforms found in the feces of various warm-blooded animals; whereas the term "coliform" also includes other environmental sources.*

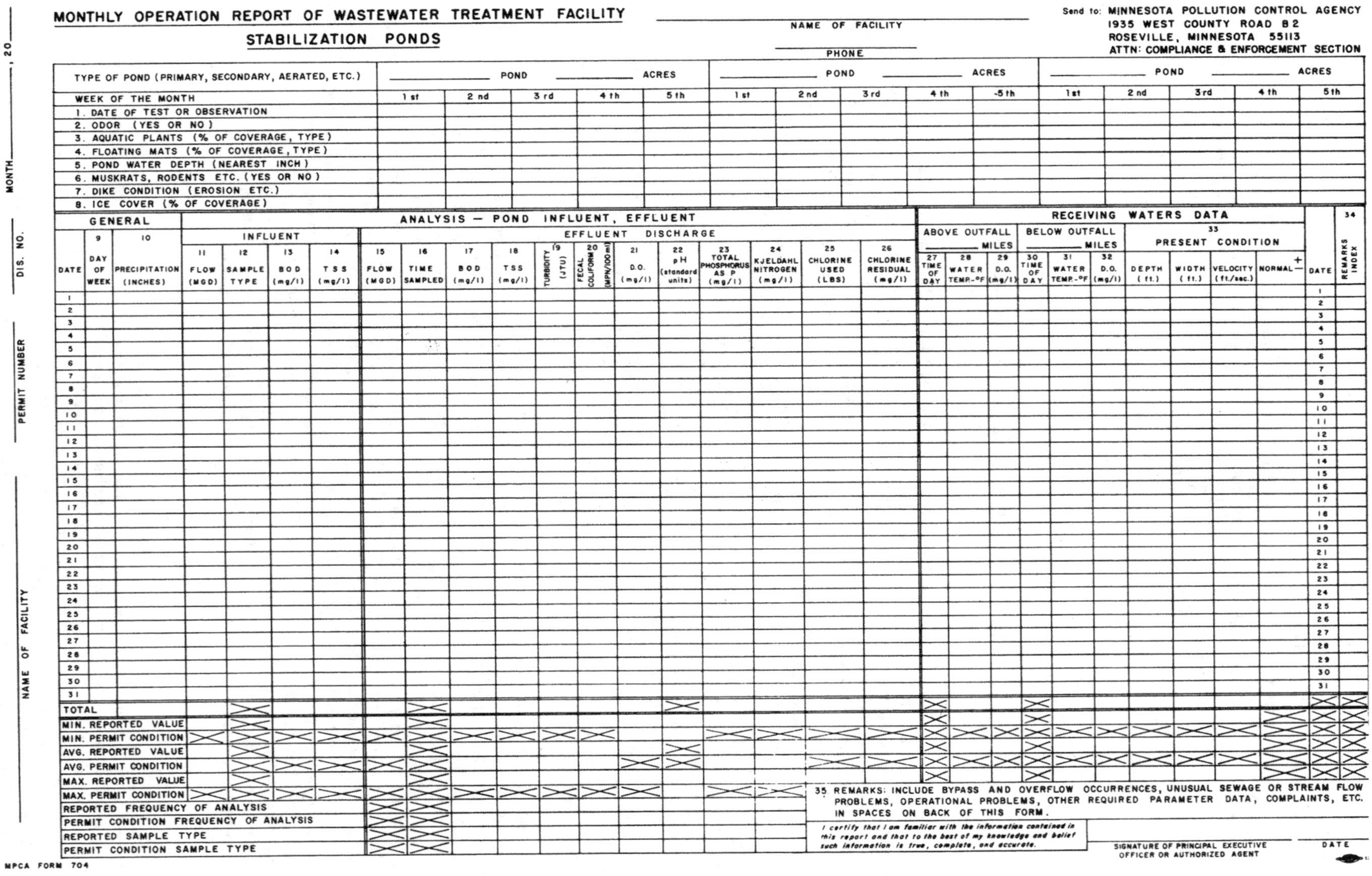

MONTHLY OPERATION REPORT OF WASTEWATER TREATMENT FACILITY

STABILIZATION PONDS

NAME OF FACILITY

PHONE

Send to: MINNESOTA POLLUTION CONTROL AGENCY
1935 WEST COUNTY ROAD B2
ROSEVILLE, MINNESOTA 55113
ATTN: COMPLIANCE & ENFORCEMENT SECTION

MONTH ____, 20__ DIS. NO. ____ PERMIT NUMBER ____ NAME OF FACILITY ____

TYPE OF POND (PRIMARY, SECONDARY, AERATED, ETC.)	____ POND ____ ACRES					____ POND ____ ACRES					____ POND ____ ACRES				
WEEK OF THE MONTH	1st	2nd	3rd	4th	5th	1st	2nd	3rd	4th	5th	1st	2nd	3rd	4th	5th
1. DATE OF TEST OR OBSERVATION															
2. ODOR (YES OR NO)															
3. AQUATIC PLANTS (% OF COVERAGE, TYPE)															
4. FLOATING MATS (% OF COVERAGE, TYPE)															
5. POND WATER DEPTH (NEAREST INCH)															
6. MUSKRATS, RODENTS ETC. (YES OR NO)															
7. DIKE CONDITION (EROSION ETC.)															
8. ICE COVER (% OF COVERAGE)															

GENERAL — ANALYSIS – POND INFLUENT, EFFLUENT — RECEIVING WATERS DATA

DATE	9 DAY OF WEEK	10 PRECIPITATION (INCHES)	INFLUENT 11 FLOW (MGD)	12 SAMPLE TYPE	13 BOD (mg/l)	14 TSS (mg/l)	EFFLUENT DISCHARGE 15 FLOW (MGD)	16 TIME SAMPLED	17 BOD (mg/l)	18 TSS (mg/l)	19 TURBIDITY (JTU)	20 FECAL COLIFORM (MPN/100 ml)	21 D.O. (mg/l)	22 pH (standard units)	23 TOTAL PHOSPHORUS AS P (mg/l)	24 KJELDAHL NITROGEN (mg/l)	25 CHLORINE USED (LBS)	26 CHLORINE RESIDUAL (mg/l)	ABOVE OUTFALL ____ MILES 27 TIME OF DAY	28 WATER TEMP.-°F	29 D.O. (mg/l)	BELOW OUTFALL ____ MILES 30 TIME OF DAY	31 WATER TEMP.-°F	32 D.O. (mg/l)	33 PRESENT CONDITION DEPTH (ft.)	WIDTH (ft.)	VELOCITY (ft./sec.)	NORMAL +/–	DATE	34 REMARKS INDEX
1																													1	
2																													2	
3																													3	
4																													4	
5																													5	
6																													6	
7																													7	
8																													8	
9																													9	
10																													10	
11																													11	
12																													12	
13																													13	
14																													14	
15																													15	
16																													16	
17																													17	
18																													18	
19																													19	
20																													20	
21																													21	
22																													22	
23																													23	
24																													24	
25																													25	
26																													26	
27																													27	
28																													28	
29																													29	
30																													30	
31																													31	
TOTAL																														

MIN. REPORTED VALUE
MIN. PERMIT CONDITION
AVG. REPORTED VALUE
AVG. PERMIT CONDITION
MAX. REPORTED VALUE
MAX. PERMIT CONDITION
REPORTED FREQUENCY OF ANALYSIS
PERMIT CONDITION FREQUENCY OF ANALYSIS
REPORTED SAMPLE TYPE
PERMIT CONDITION SAMPLE TYPE

35 REMARKS: INCLUDE BYPASS AND OVERFLOW OCCURRENCES, UNUSUAL SEWAGE OR STREAM FLOW PROBLEMS, OPERATIONAL PROBLEMS, OTHER REQUIRED PARAMETER DATA, COMPLAINTS, ETC. IN SPACES ON BACK OF THIS FORM.

I certify that I am familiar with the information contained in this report and that to the best of my knowledge and belief such information is true, complete, and accurate.

SIGNATURE OF PRINCIPAL EXECUTIVE OFFICER OR AUTHORIZED AGENT

DATE

MPCA FORM 704

Fig. 2.3 Typical NPDES permit reporting form

QUESTIONS

Write your answers in a notebook and then compare your answers with those on page 26.

2.5A   Why should an operator have an understanding of natural cycles?

2.5B   What can happen when nutrient cycles are disrupted and there is no dissolved oxygen in the receiving water?

2.6A   What does NPDES stand for?

2.7   ARITHMETIC ASSIGNMENT

A good way to learn how to solve arithmetic problems is to work on them a little bit at a time. In this operator training manual we are going to make a short arithmetic assignment at the end of every chapter. If you will work this assignment, you can easily learn how to solve wastewater treatment arithmetic problems.

Turn to the Appendix, "How to Solve Wastewater Treatment Plant Arithmetic Problems," at the back of this manual and read the following sections:

1. *OBJECTIVES,*

2. *A.0, HOW TO STUDY THIS APPENDIX,* and

3. *A.1, BASIC ARITHMETIC.*

Solve all of the problems in Sections A.10, Addition; A.11, Subtraction; A.12, Multiplication; A.13, Division; A.14, Rules for Solving Equations; A.15, Actual Problems; A.16, Percentage; and A.17, Sample Problems Involving Percent. Try to use an electronic pocket calculator.

2.8   ADDITIONAL READING

For a detailed discussion of the physical and chemical composition of wastewater, you may wish to refer to the following sources:

1. *MOP 11,* Chapter 17, "Characterization and Sampling of Wastewater."*

2. *NEW YORK MANUAL,* Chapter 1, "Wastewater."*

3. *TEXAS MANUAL,* Chapter 1, "Wastewater, Its Composition, Chemistry and Biology."*

* Actual chapter number and title may depend on edition.

2.9   REVIEW

In this chapter you have read why it is necessary to treat wastewater, something about the types of waste discharges and their effects, and a brief description of the different kinds of solids in wastewater. This is intended to be only a general discussion of these subjects; you will find more detail in later chapters.

You are now ready to go on to Chapter 3, which describes the basic elements of wastewater collection and treatment systems. Chapter 3 actually begins the discussion of *HOW* to treat wastewater. Chapter 2 has told you *WHY* you need to do so.

Operators of wastewater treatment plants operate their plants with the objectives of providing the best possible treatment of wastes to protect the receiving waters, downstream users, and neighbors. They accomplish these objectives by:

1. Removing wastes from the wastewater to protect the receiving waters,

2. Meeting NPDES Permit requirements,

3. Minimizing odors to avoid nuisance complaints,

4. Minimizing costs,

5. Minimizing energy consumption, and

6. Maintaining an effective preventive maintenance program.

The remaining chapters in this manual were prepared for you by operators with the intent of providing you with the knowledge and skills necessary to be a wastewater treatment plant operator.

SUGGESTED ANSWERS

Chapter 2.   WHY TREAT WASTES?

Answers to questions on page 18.

2.1A   Water picks up dissolved substances as it falls as rain, flows over land, and is used for domestic, industrial, agricultural, and recreational purposes.

2.1B   Some of the dissolved substances in water include hydrogen, oxygen, calcium, carbon, magnesium, sodium, chloride, sulfate, iron, nitrogen, phosphorus, and organic material.

Answers to questions on page 19.

2.2A   The following contain significant quantities of organic material: domestic wastewater, paper mill wastes, and tanning wastes.

2.2B   Four types of pollution are: organic, inorganic, thermal, and radioactive pollution.

Answers to questions on page 20.

2.3A   Organic wastes in water provide food for the bacteria. These bacteria require oxygen to survive and consequently deplete the oxygen in the water in a way similar to the way oxygen is removed from air when people breathe.

2.1B   Hydrogen sulfide gas is produced by anaerobic bacteria.

Answers to questions on page 21.

2.3C   Disease-causing bacteria in wastewater come from the body wastes of humans who have a disease.

2.3D   Pathogenic means disease-causing.

2.3E   Chlorination is the most frequently used means of disinfecting treated wastewater.

Answers to questions on page 22.

2.4A   Total solids consist of *DISSOLVED* and *SUSPENDED* solids, both of which contain organic and inorganic matter.

2.4B   Settleable solids must be measured to determine the efficiency of settling basins. This amount must also be known to calculate loads on settling basins, sludge pumps, and sludge handling facilities for design and operational purposes. (You should have recognized the need to know the efficiency of settling basins.)

2.4C   An Imhoff cone is used to measure *SETTLEABLE* solids.

Answers to questions on page 25.

2.5A   Operators need an understanding of natural cycles in order to control wastewater treatment processes and odors and also to protect receiving waters.

2.5B   When nutrient cycles become disrupted and there is no dissolved oxygen in the receiving water, these waters become septic, stink, and look terrible.

2.6A   NPDES stands for **N**ational **P**ollutant **D**ischarge **E**limination **S**ystem.

ATIONAL
OLLUTANT
ISCHARGE
LIMINATION
YSTEM

CHAPTER 3

WASTEWATER TREATMENT FACILITIES

by

John Brady

and

William Crooks

TABLE OF CONTENTS

Chapter 3.   WASTEWATER TREATMENT FACILITIES

OBJECTIVES

Chapter 3.   WASTEWATER TREATMENT FACILITIES

Following completion of Chapter 3, you should be able to:

1. Describe wastewater collection systems,

2. Describe various types of wastewater treatment plants,

3. Draw schematic plan layouts of typical plants,

4. List the major wastewater treatment processes and the purpose of each process,

5. Identify various methods of effluent disposal, and

6. Identify various methods of solids disposal.

WORDS

Chapter 3.   WASTEWATER TREATMENT FACILITIES

BIOCHEMICAL OXYGEN DEMAND (BOD)   BIOCHEMICAL OXYGEN DEMAND (BOD)

The rate at which organisms use the oxygen in water or wastewater while stabilizing decomposable organic matter under aerobic conditions. In decomposition, organic matter serves as food for the bacteria and energy results from its oxidation. BOD measurements are used as a measure of the organic strength of wastes in water.

BIOCHEMICAL OXYGEN DEMAND (BOD) TEST   BIOCHEMICAL OXYGEN DEMAND (BOD) TEST

A procedure that measures the rate of oxygen use under controlled conditions of time and temperature. Standard test conditions include dark incubation at 20°C for a specified time (usually five days).

BIOSOLIDS   BIOSOLIDS

A primarily organic solid product, produced by wastewater treatment processes, that can be beneficially recycled. The word biosolids is replacing the word sludge.

COMBINED SEWER   COMBINED SEWER

A sewer designed to carry both sanitary wastewaters and storm or surface water runoff.

COMMINUTION (com-mi-NEW-shun)   COMMINUTION

Shredding. A mechanical treatment process which cuts large pieces of wastes into smaller pieces so they won't plug pipes or damage equipment. COMMINUTION and SHREDDING usually mean the same thing.

DETENTION TIME   DETENTION TIME

The time required to fill a tank at a given flow or the theoretical time required for a given flow of wastewater to pass through a tank.

DEWATER   DEWATER

(1) To remove or separate a portion of the water present in a sludge or slurry. To dry sludge so it can be handled and disposed of.

(2) To remove or drain the water from a tank or a trench.

EFFLUENT   EFFLUENT

Wastewater or other liquid—raw (untreated), partially or completely treated—flowing *FROM* a reservoir, basin, treatment process, or treatment plant.

FACULTATIVE (FACK-ul-TAY-tive) POND   FACULTATIVE POND

The most common type of pond in current use. The upper portion (supernatant) is aerobic, while the bottom layer is anaerobic. Algae supply most of the oxygen to the supernatant.

GRIT   GRIT

The heavy material present in wastewater, such as sand, coffee grounds, eggshells, gravel and cinders.

HEADWORKS   HEADWORKS

The facilities where wastewater enters a wastewater treatment plant. The headworks may consist of bar screens, comminutors, a wet well and pumps.

INFILTRATION (IN-fill-TRAY-shun)   INFILTRATION

The seepage of groundwater into a sewer system, including service connections. Seepage frequently occurs through defective or cracked pipes, pipe joints, connections or manhole walls.

INFLOW

Water discharged into a sewer system and service connections from sources other than regular connections. This includes flow from yard drains, foundation drains and around manhole covers. Inflow differs from infiltration in that it is a direct discharge into the sewer rather than a leak in the sewer itself.

INFLUENT

Wastewater or other liquid—raw (untreated) or partially treated—flowing *INTO* a reservoir, basin, treatment process, or treatment plant.

MEDIA

The material in a trickling filter on which slime accumulates and organisms grow. As settled wastewater trickles over the media, organisms in the slime remove certain types of wastes thereby partially treating the wastewater. Also the material in a rotating biological contactor or in a gravity or pressure filter.

PHOTOSYNTHESIS (foe-toe-SIN-thuh-sis)

A process in which organisms, with the aid of chlorophyll (green plant enzyme), convert carbon dioxide and inorganic substances into oxygen and additional plant material, using sunlight for energy. All green plants grow by this process.

PRIMARY TREATMENT

A wastewater treatment process that takes place in a rectangular or circular tank and allows those substances in wastewater that readily settle or float to be separated from the water being treated.

SANITARY SEWER

A pipe or conduit (sewer) intended to carry wastewater or waterborne wastes from homes, businesses, and industries to the POTW (**P**ublicly **O**wned **T**reatment **W**orks). Storm water runoff or unpolluted water should be collected and transported in a separate system of pipes or conduits (storm sewers) to natural watercourses.

SECONDARY TREATMENT

A wastewater treatment process used to convert dissolved or suspended materials into a form more readily separated from the water being treated. Usually the process follows primary treatment by sedimentation. The process commonly is a type of biological treatment process followed by secondary clarifiers that allow the solids to settle out from the water being treated.

SHREDDING

Comminution. A mechanical treatment process which cuts large pieces of wastes into smaller pieces so they won't plug pipes or damage equipment. SHREDDING and COMMINUTION usually mean the same thing.

SLUDGE (sluj)

(1) The settleable solids separated from liquids during processing.

(2) The deposits of foreign materials on the bottoms of streams or other bodies of water.

STORM SEWER

A separate pipe, conduit or open channel (sewer) that carries runoff from storms, surface drainage, and street wash, but does not include domestic and industrial wastes. Storm sewers are often the recipients of hazardous or toxic substances due to the illegal dumping of hazardous wastes or spills created by accidents involving vehicles and trains transporting these substances. Also see SANITARY SEWER.

SUPERNATANT (sue-per-NAY-tent)

Liquid removed from settled sludge. Supernatant commonly refers to the liquid between the sludge on the bottom and the scum on the surface of an anaerobic digester. This liquid is usually returned to the influent wet well or to the primary clarifier.

WEIR (weer)

(1) A wall or plate placed in an open channel and used to measure the flow of water. The depth of the flow over the weir can be used to calculate the flow rate, or a chart or conversion table may be used to convert depth to flow.

(2) A wall or obstruction used to control flow (from settling tanks and clarifiers) to ensure a uniform flow rate and avoid short-circuiting.

WET OXIDATION WET OXIDATION

A method of treating or conditioning sludge before the water is removed. Compressed air is blown into the liquid sludge. The air and sludge mixture is fed into a pressure vessel where the organic material is stabilized. The stabilized organic material and inert (inorganic) solids are then separated from the pressure vessel effluent by dewatering in lagoons or by mechanical means.

WET WELL WET WELL

A compartment or tank in which wastewater is collected. The suction pipe of a pump may be connected to the wet well or a submersible pump may be located in the wet well.

CHAPTER 3.   WASTEWATER TREATMENT FACILITIES

3.0 COLLECTION, TREATMENT, DISPOSAL

Facilities for handling wastewater are usually considered to have three major components or parts: collection, treatment, and disposal or reuse. For a municipality, these components make up the wastewater facilities; for an individual industry that handles its own wastewater, the same three components are also necessary. This training course is directed primarily to plant operators for municipalities, so the discussion in this and later chapters will be in terms of municipal wastewater treatment facilities. Treatment of industrial wastes is discussed in the *INDUSTRIAL WASTE TREATMENT* manuals in this series of manuals.

3.1 COLLECTION OF WASTEWATER

Collection and transportation of wastewater to the treatment plant is accomplished through a complex network of pipes and pumps of many sizes.

Major water-using industries that contribute waste to the collection system may affect the efficiency of a wastewater treatment plant, especially if there are periods during the day or during the year when these industrial waste flows are a major load on the plant. For instance, canneries are highly seasonal in their operations; therefore, it is possible to predict the time of year to expect large flows from them. A knowledge of the location, amount, and types of wastes from commercial and industrial dischargers to the collection system may enable an operator or an industrial pretreatment facility inspector to locate the source of a problem in the plant *INFLUENT*,[1] such as oil from a refinery or a gas station.

The length of time required for wastes to reach your plant can also affect treatment plant efficiency. Hydrogen sulfide gas (rotten egg odor) may be released by anaerobic bacteria feeding on the wastes if the flow time is quite long and the weather is hot; this can cause odor problems, damage concrete in your plant, and make the wastes more difficult to treat. (Solids won't settle easily, for instance.) Wastes from isolated subdivisions located far away from the main collection network often have this "aging" problem.

3.10 Sanitary, Storm, and Combined Sewers

For most sewerage systems, the sewer coming into the treatment plant carries wastes from households and commer-

cial establishments in the city or district, and possibly some industrial wastes. This type of sewer is called a *SANITARY SEWER*.[2] All storm runoff from streets, land, and roofs of buildings is collected separately in a *STORM SEWER*,[3] which normally discharges to a watercourse without treatment. In some areas only one network of sewers has been laid out beneath the city to pick up both sanitary wastes and storm water in a *COMBINED SEWER*.[4] Treatment plants that are designed to handle the sanitary portion of the wastes sometimes may become overloaded during storms due to inadequate capacity, allowing partially treated wastes to be discharged into receiving waters, as a result of attempting to treat extreme hydraulic (flow) overloads. These problems are very challenging for operators. Separation of combined sewers into sanitary and storm sewers is very costly and difficult to accomplish.

Even in areas where the sanitary and storm sewers are separate, *INFILTRATION*[5] of groundwater or storm water into sanitary sewers through breaks, open joints, or corrosion-damaged pipes can cause high flow problems at the treatment

Manholes allow access for inspection and maintenance of the collection system

---

[1] *Influent.   Wastewater or other liquid—raw (untreated) or partially treated—flowing INTO a reservoir, basin, treatment process, or treatment plant.*
[2] *Sanitary Sewer.   A pipe or conduit (sewer) intended to carry wastewater or waterborne wastes from homes, businesses, and industries to the POTW (**P**ublicly **O**wned **T**reatment **W**orks). Storm water runoff or unpolluted water should be collected and transported in a separate system of pipes or conduits (storm sewers) to natural watercourses.*
[3] *Storm Sewer.   A separate pipe, conduit or open channel (sewer) that carries runoff from storms, surface drainage, and street wash, but does not include domestic and industrial wastes. Storm sewers are often the recipients of hazardous or toxic substances due to the illegal dumping of hazardous wastes or spills created by accidents involving vehicles and trains transporting these substances. Also see SANITARY SEWER.*
[4] *Combined Sewer.   A sewer designed to carry both sanitary wastewaters and storm or surface water runoff.*
[5] *Infiltration (IN-fill-TRAY-shun).   The seepage of groundwater into a sewer system, including service connections. Seepage frequently occurs through defective or cracked pipes, pipe joints, connections or manhole walls.*

plant. Replacement or sealing of leaky sections of sewer pipe is called for in these cases. The treatment plant operator is generally the first to know about infiltration problems because of the unusually high flows observed at the plant during periods of storm water runoff.

Sanitary sewers are normally placed at a slope sufficient to produce a water velocity (speed) of approximately two feet per second when flowing full. This velocity will usually prevent the deposition of solids that may clog the pipe or cause odors. Manholes are placed every 300 to 500 feet to allow for inspection and cleaning of the sewer.

When low areas of land must be sewered or where pipe depth under the ground surface becomes excessive, pump stations (Figure 3.1) are normally installed. These pump stations lift the wastewater to a higher point from which it may again flow by gravity, or the wastewater may be pumped under pressure directly to the treatment plant. A large pump station located just ahead of the treatment plant can create problems by periodically sending large volumes of flow to the plant one minute, and virtually nothing the next minute. These fluctuating flows can be reduced by using variable speed pumps or short pumping cycles.

For additional information on collection systems and wastewaters entering your plant see *OPERATION AND MAINTENANCE OF WASTEWATER COLLECTION SYSTEMS*, Volumes I and II, and *PRETREATMENT FACILITY INSPECTION* in this series of manuals.

QUESTIONS

Write your answers in a notebook and then compare your answers with those on page 56.

3.1A   Why should the operator be familiar with the wastewater collection and transportation network?

3.1B   What problem may occur when it takes a long time for wastewater to flow through the collection sewers to the treatment plant?

3.1C   List three types of sewers.

3.1D   Why are combined sewers a problem?

3.2 TREATMENT PLANTS

Upon reaching a wastewater treatment plant, the wastewater flows through a series of treatment processes (Figure 3.2) that remove the wastes from the water and reduce its threat to the public health before it is discharged from the plant. The number of treatment processes and the degree of treatment usually depend on the uses of the receiving waters. Treated wastewaters discharged into a small stream used for a domestic water supply and swimming will require considerably more treatment than wastewater discharged into water used solely for navigation.

To provide you with a general picture of treatment plants, the remainder of this chapter will follow the paths wastewater might travel as it passes through a plant. You will be introduced to the names of the treatment processes, the kinds of wastes the processes treat or remove, and the location of the processes in the flow path. Not all treatment plants are alike; however, there are certain typical flow patterns that are similar from one plant to another. Figures 3.3, 3.4, and 3.5 show some possible flow patterns through treatment plants.

When wastewater enters a treatment plant, it usually flows through a series of pretreatment or preliminary treatment processes—screening, shredding, and grit removal. These processes remove the coarse material from the wastewater. Flow measuring devices are usually installed after preliminary treatment processes to record the flow rates and volumes of wastewater treated by the plant. Pre-aeration is used to "freshen" the wastewater and to help remove oils and greases.

Next the wastewater will generally receive primary treatment. During primary treatment, some of the solid matter carried by the wastewater will settle out or float to the water surface where it can be separated from the wastewater being treated.

Secondary treatment processes usually follow primary treatment and commonly consist of biological processes. This means that organisms living in the controlled environment of the process are used to partially stabilize (oxidize) organic matter not removed by previous treatment processes and to convert it into a form that is easier to remove from the wastewater.

Waste material removed by the treatment processes goes to solids handling facilities and then to ultimate disposal.

Waste treatment ponds may be used to treat wastes remaining in wastewater after pretreatment, primary treatment, or secondary treatment. Ponds are frequently constructed in rural areas where there is sufficient available land.

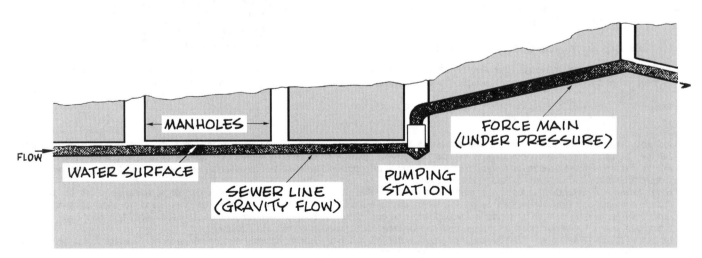

Fig. 3.1 Collection sewer profile

<u>TREATMENT PROCESS</u>   <u>FUNCTION</u>

<u>PRELIMINARY TREATMENT</u>

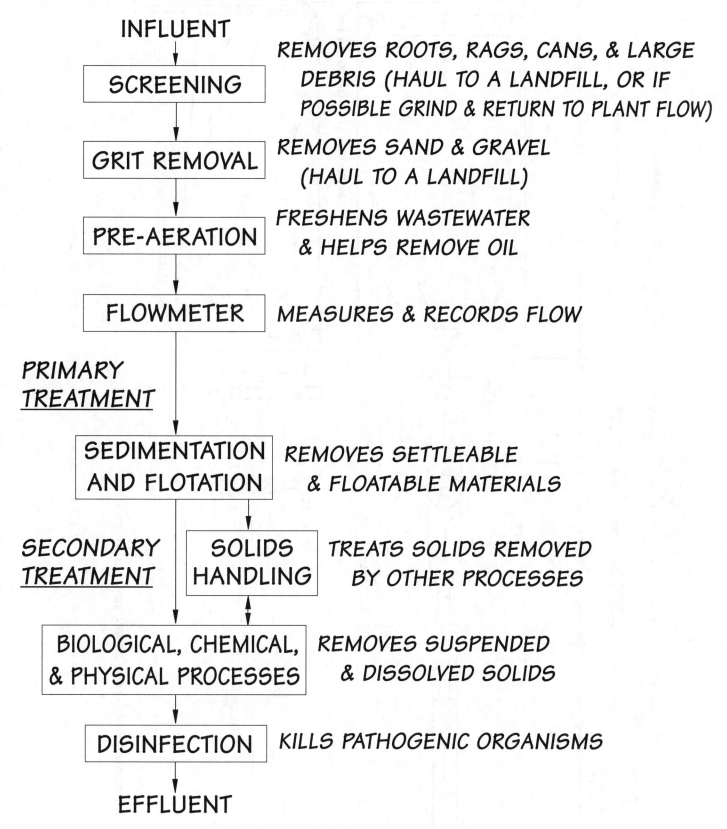

INFLUENT

SCREENING — REMOVES ROOTS, RAGS, CANS, & LARGE DEBRIS (HAUL TO A LANDFILL, OR IF POSSIBLE GRIND & RETURN TO PLANT FLOW)

GRIT REMOVAL — REMOVES SAND & GRAVEL (HAUL TO A LANDFILL)

PRE-AERATION — FRESHENS WASTEWATER & HELPS REMOVE OIL

FLOWMETER — MEASURES & RECORDS FLOW

<u>PRIMARY TREATMENT</u>

SEDIMENTATION AND FLOTATION — REMOVES SETTLEABLE & FLOATABLE MATERIALS

<u>SECONDARY TREATMENT</u>   SOLIDS HANDLING — TREATS SOLIDS REMOVED BY OTHER PROCESSES

BIOLOGICAL, CHEMICAL, & PHYSICAL PROCESSES — REMOVES SUSPENDED & DISSOLVED SOLIDS

DISINFECTION — KILLS PATHOGENIC ORGANISMS

EFFLUENT

Fig. 3.2   Flow diagram of wastewater treatment plant processes

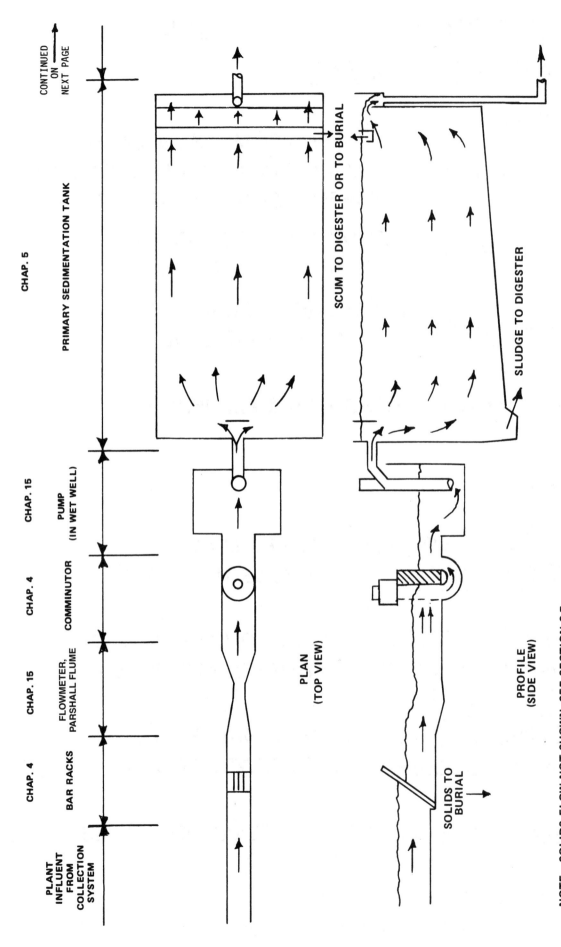

Fig. 3.3   Possible flow pattern through a trickling filter plant

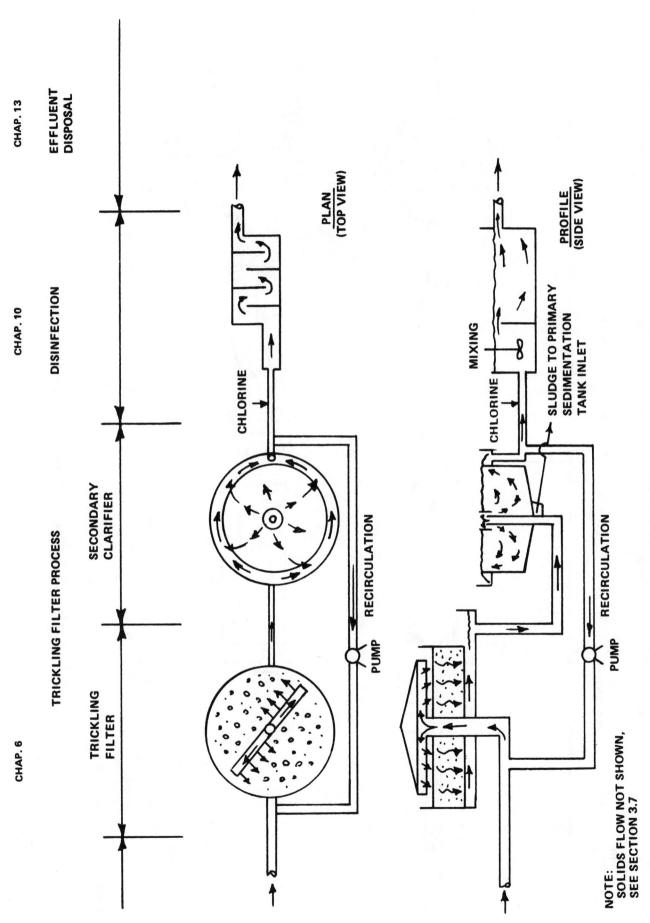

CHAP. 6   TRICKLING FILTER PROCESS

CHAP. 10   DISINFECTION

CHAP. 13   EFFLUENT DISPOSAL

TRICKLING FILTER

SECONDARY CLARIFIER

CHLORINE

RECIRCULATION

PUMP

PLAN (TOP VIEW)

CHLORINE

MIXING

SLUDGE TO PRIMARY SEDIMENTATION TANK INLET

RECIRCULATION

PUMP

PROFILE (SIDE VIEW)

NOTE:
SOLIDS FLOW NOT SHOWN, SEE SECTION 3.7

Fig. 3.3   Possible flow pattern through a trickling filter plant (continued)

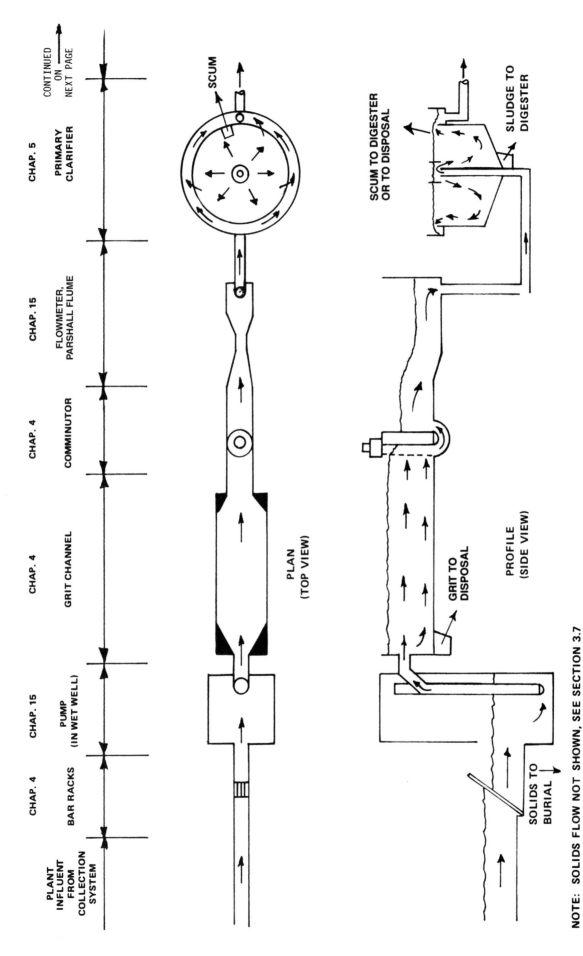

CHAP. 4   CHAP. 15   CHAP. 4   CHAP. 4   CHAP. 15   CHAP. 5

PLANT INFLUENT FROM COLLECTION SYSTEM   BAR RACKS   PUMP (IN WET WELL)   GRIT CHANNEL   COMMINUTOR   FLOWMETER, PARSHALL FLUME   PRIMARY CLARIFIER

CONTINUED ON NEXT PAGE

SCUM

PLAN (TOP VIEW)

SCUM TO DIGESTER OR TO DISPOSAL

SLUDGE TO DIGESTER

GRIT TO DISPOSAL

SOLIDS TO BURIAL

PROFILE (SIDE VIEW)

NOTE: SOLIDS FLOW NOT SHOWN, SEE SECTION 3.7

Fig. 3.4   Possible flow pattern through an activated sludge plant

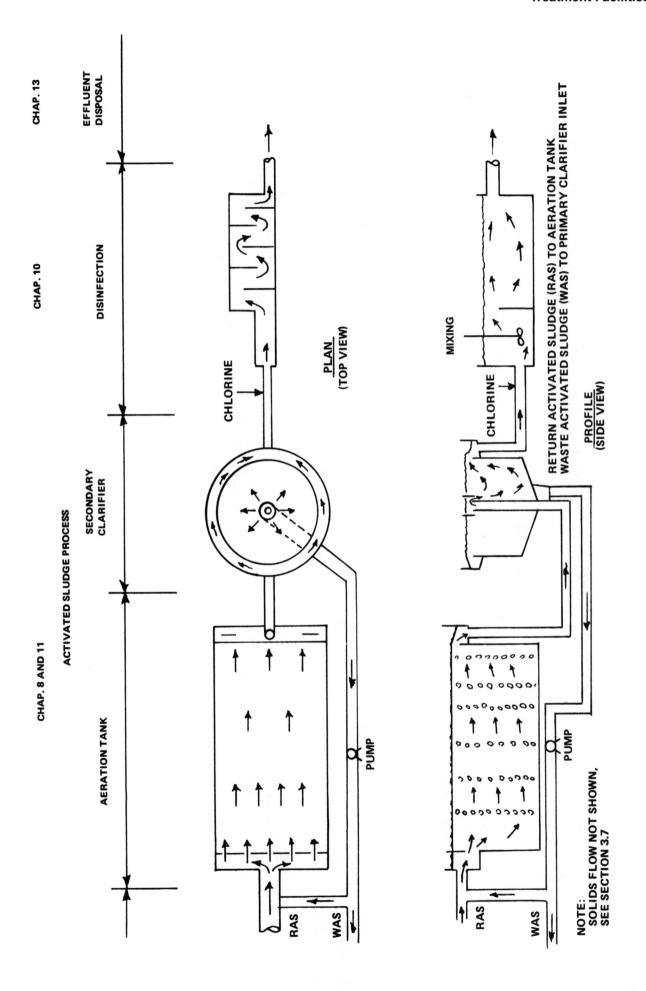

CHAP. 8 AND 11

ACTIVATED SLUDGE PROCESS

AERATION TANK

SECONDARY CLARIFIER

CHAP. 10

DISINFECTION

CHAP. 13

EFFLUENT DISPOSAL

CHLORINE

PLAN (TOP VIEW)

PUMP

RAS

WAS

MIXING

CHLORINE

PROFILE (SIDE VIEW)

RETURN ACTIVATED SLUDGE (RAS) TO AERATION TANK
WASTE ACTIVATED SLUDGE (WAS) TO PRIMARY CLARIFIER INLET

PUMP

RAS

WAS

NOTE:
SOLIDS FLOW NOT SHOWN,
SEE SECTION 3.7

Fig. 3.4   *Possible flow pattern through an activated sludge plant (continued)*

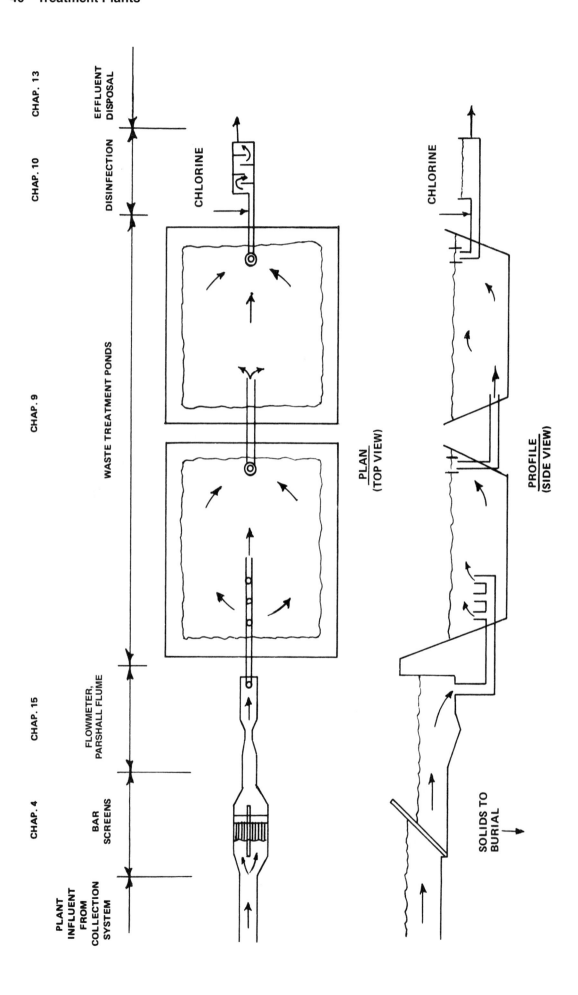

Fig. 3.5   Possible flow pattern through a pond treatment plant

Advanced methods of waste treatment are being developed for general cleanup of wastewater or removal of substances not removed by conventional treatment processes. They may follow the treatment processes previously described, or they may be used instead of them. See *ADVANCED WASTE TREATMENT* in this series of manuals for more details.

Before treated wastewater is discharged to the receiving waters, it should be disinfected to prevent the spread of disease. Chlorine is usually added for disinfection purposes. After the chlorine contact basin, sulfur dioxide (SO_2) may be added to the *EFFLUENT*[6] to neutralize the chlorine and thus detoxify the effluent.

In the following sections these treatment processes will be briefly discussed to provide an overall concept of a treatment plant. Details will be presented in later chapters to provide complete information on each of these processes.

3.3 PRELIMINARY TREATMENT (Chapter 4)

3.30 Purpose

Preliminary treatment (or pretreatment) processes commonly consist of screening, *SHREDDING*,[7] and grit removal to separate coarse material from the wastewater being treated.

Screened and ground

3.31 Screening

Wastewater flowing into the treatment plant will occasionally contain pieces of wood, roots, rags, and other debris. To protect equipment and reduce any interference with in-plant flow, debris and trash are usually removed by a bar screen (Figure 3.6). Most screens in treatment plants consist of parallel bars placed at an angle in a channel in such a manner that the wastewater flows through the bars. Trash collects on the bars and is periodically raked off by hand or by mechanical means. In most plants these screenings are disposed of by burying or burning. In some cases they are automatically ground and returned to the wastewater flow for removal by a later process.

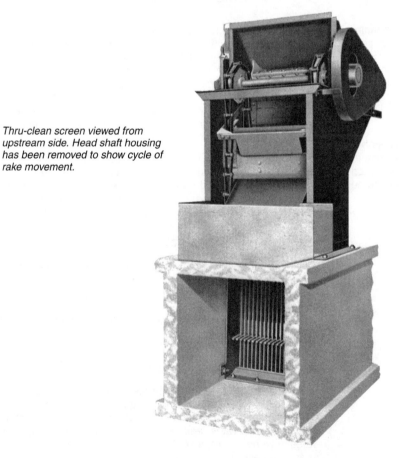

Thru-clean screen viewed from upstream side. Head shaft housing has been removed to show cycle of rake movement.

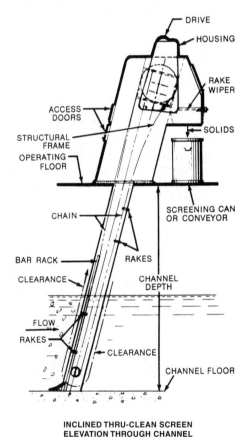

INCLINED THRU-CLEAN SCREEN
ELEVATION THROUGH CHANNEL

Fig. 3.6   Mechanically cleaned bar screen
(Permission of Smith & Loveless, Inc.)

---

[6] *Effluent.   Wastewater or other liquid—raw (untreated), partially or completely treated—flowing FROM a reservoir, basin, treatment process, or treatment plant.*

[7] *Shredding.   Comminution. A mechanical treatment process which cuts large pieces of wastes into smaller pieces so they won't plug pipes or damage equipment. Shredding and comminution usually mean the same thing.*

3.32   Shredding

Devices are also available that cut up or shred material while it remains in the wastewater stream. The most common of these are the barminutor (Figure 3.7) and the comminutor (Figure 3.8). One of these devices usually follows a bar screen.

Fig. 3.7   Barminutor   Fig. 3.8   Comminutor
(Courtesy Chicago Pump)

3.33   Grit Chambers or Grit Channels

Most sewer pipes are laid at a slope steep enough to maintain a wastewater flow of two feet per second (fps) (0.6 m/sec) to prevent *GRIT*[8] from settling in the pipes. If the velocity is reduced slightly below that, say to 1.5 fps (0.45 m/sec), some of the larger, heavier particles will settle out. If the velocity is reduced to about 1 fps (0.3 m/sec), heavy inorganic material such as sand, eggshells, and cinders will settle, but the lighter organic material will remain in suspension.

Grit should be removed early in the treatment process because it is abrasive and will rapidly wear out pumps and other equipment. Since it is mostly inorganic, it cannot be broken down by any biological treatment process and thus should be removed as soon as possible.

DON'T REMOVE GRIT WITH YOUR BARE HANDS!

Fig. 3.9   Grit channel
(WEF *MOP NO. 11,*
OPERATION OF WASTEWATER TREATMENT PLANTS)

Grit is usually removed in a long, narrow trough called a "grit channel" (Figure 3.9). A grit channel is designed to provide a flow-through velocity of 1 fps (0.3 m/sec). The settled grit may be removed either manually or mechanically. Since there is normally some organic solid material deposited along with the grit, it is usually buried to avoid nuisance conditions. Some plants are equipped with "grit washers" that remove some of the organic material from the grit so that organic solids can remain in the main waste flow to be treated.

Many treatment plants have aerated grit chambers in which compressed air is added through diffusers to provide better separation of grit and other solids. Aeration in this manner also "freshens" a "stale" or septic wastewater, helping to prevent odors and to assist the biological treatment process.

QUESTIONS

Write your answers in a notebook and then compare your answers with those on page 56.

3.3A   Why is grit removed early in the treatment process?

3.3B   What is usually done with grit that has been removed from the wastewater?

3.4   FLOW MEASURING DEVICES

Although flow measuring devices are not for *TREATING* wastes, it is necessary to know the quantity of wastewater flow so adjustments can be made on pumping rates, chlorination rates, aeration rates, and other processes in the plant. Flow rates must be known, also, for calculation of loadings on treatment processes and treatment efficiency. Most operators prefer to have a measuring device at the *HEADWORKS*[9] of their treatment plant.

---

[8] *Grit.   The heavy material present in wastewater, such as sand, coffee grounds, eggshells, gravel and cinders.*
[9] *Headworks.   The facilities where wastewater enters a wastewater treatment plant. The headworks may consist of bar screens, comminutors, a wet well and pumps.*

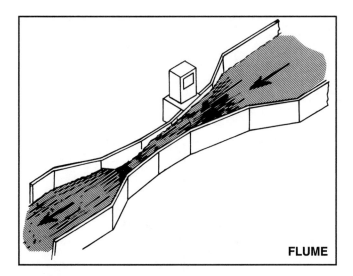

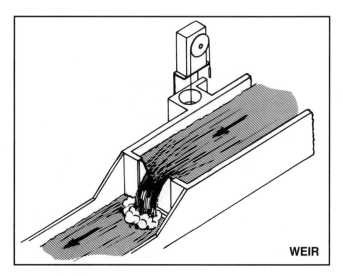

Fig. 3.10 Flow measuring devices

The most common measuring device is a Parshall flume (Figure 3.10). Basically it is a narrow place in an open channel which allows the quantity of flow to be determined by measuring the depth of flow. This method is widely used for measuring wastewater because its smooth constriction does not offer any protruding sharp edges or areas where wastewater particles may catch or collect behind the metering device.

Another measuring device used in open channels is a WEIR[10] (Figure 3.10). The weir, which is placed across the channel, is a wall over which the wastewater may fall. Weirs are usually made of a thin material and may have either a rectangular or V-notch opening. The flow over the weir is determined by the depth of wastewater going through the opening. A disadvantage of a weir is the relatively dead water space that occurs just upstream of the weir. If the weir is used at the head end of the plant, organic solids may settle out in this area. When this occurs, odors and unsightliness can result. Also, as the solids accumulate, the flow reading may become incorrect.

A good measuring device for flows of treated or untreated wastewater in pipelines is a Venturi meter. This meter has a special section of contracting pipe, and it measures flow by measuring the difference in pressure in the pipeline before the constriction and at the constriction. There are no sharp obstructions to catch rags and debris. Magnetic flowmeters also are being used successfully to measure wastewater flow.

QUESTIONS

Write your answers in a notebook and then compare your answers with those on page 56.

3.4A   Why is a Parshall flume widely used for measuring wastewater flow?

3.4B   Why are weirs not frequently used to measure the influent to a plant?

3.5 PRIMARY TREATMENT (Chapter 5)

We have previously discussed the reduction in velocity of the incoming waste to approximately one foot per second in order to settle out heavy inorganic material or grit. The next step in the treatment process is normally called *SEDIMENTATION* or *PRIMARY TREATMENT*. In this process the waste is directed into and through a large tank or basin. Flow velocity in these tanks is reduced to about 0.03 foot per second (0.009 m/sec), allowing the settleable solids to fall to the bottom of the tank and lighter material to float to the water surface, thus making the wastewater much clearer. For this reason these sedimentation tanks are called "clarifiers." The first clarifier that the wastewater flows into is called a *PRIMARY* clarifier. We will discuss later the need for another clarifier after the biological treatment process. This second clarifier is called a "secondary clarifier."

Clarifiers normally are either rectangular (Figure 3.11) or circular (Figure 3.12). Primary clarifiers are usually designed to provide 1.5 to 2 hours of *DETENTION TIME*.[11] Secondary clarifiers usually provide slightly more time.

Generally the longer the detention time, the greater the solids removal. In a tank with two hours' detention time, approximately 60 percent of the suspended solids in the raw wastewater will either settle to the bottom or float to the surface and be removed. Removal of these solids will usually reduce the *BIOCHEMICAL OXYGEN DEMAND (BOD)*[12] of the waste approximately 30 percent. The exact removal depends on the amount of BOD contained in the settled material.

---

[10] *Weir (weer).   (1) A wall or plate placed in an open channel and used to measure the flow of water. The depth of the flow over the weir can be used to calculate the flow rate, or a chart or conversion table may be used to convert depth to flow. (2) A wall or obstruction used to control flow (from settling tanks and clarifiers) to ensure a uniform flow rate and avoid short-circuiting.*

[11] *Detention Time.   The time required to fill a tank at a given flow or the theoretical time required for a given flow of wastewater to pass through a tank.*

[12] *Biochemical Oxygen Demand (BOD).   The rate at which organisms use the oxygen in water or wastewater while stabilizing decomposable organic matter under aerobic conditions. In decomposition, organic matter serves as food for the bacteria and energy results from its oxidation. BOD measurements are used as a measure of the organic strength of wastes in water.*

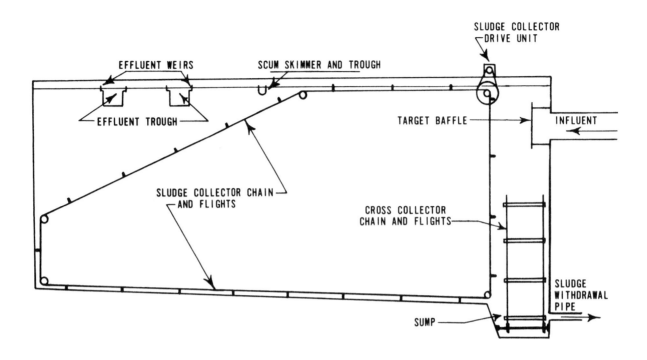

Fig. 3.11 Rectangular clarifier

(Courtesy Jeffrey Mfg. Co.)

Fig. 3.12   Circular clarifier

All primary clarifiers, no matter what their shape, must have a means for collecting the settled solids (called *SLUDGE*[13]) and the floating solids (called scum). In rectangular tanks sludge and scum collectors are usually wooden beams (flights) attached to endless chains. The collector flights travel on the surface, in the direction of the flow, conveying grease and floatable solids down to the scum trough to be skimmed off to the solids (sludge) handling facilities. The flights then drop below the surface and return to the influent end along the bottom, moving the settled raw sludge to the sludge hopper. The sludge is periodically pumped from the hopper to the sludge handling facilities (such as digesters).

In circular tanks, scrapers (plows) attached to a rotating arm rotate slowly around the bottom of the tank. The plows push the settled sludge toward the center and into the sludge hopper. Scum is collected by a rotating blade at the surface. As in the case of the rectangular tank, both scum and sludge are usually pumped to the solids or sludge handling facilities.

The clear surface water of the primary tank flows from the tank by passing over a weir. The weir must be long enough to allow the treated water to leave at a low velocity; if it leaves at a high velocity, particles settling to the bottom or those already on the bottom may be picked up and carried from the tank.

QUESTIONS

Write your answers in a notebook and then compare your answers with those on page 56.

3.5A   What is the purpose of flights or plows in a clarifier?

3.5B   What happens to the sludge and scum collected in a primary clarifier?

3.6   SECONDARY TREATMENT

3.60   Purpose

In many treatment plants the wastewater flows from the primary clarifier into another unit where it receives *SECONDARY* or *BIOLOGICAL TREATMENT*. This means that the wastewater is exposed to living organisms (such as bacteria) that eat the dissolved and nonsettleable organic material remaining in the waste. The two processes used almost universally for biological treatment are the *TRICKLING FILTER* and *ACTIVATED SLUDGE*. These are both *AEROBIC* biological treatment processes, which means the organisms require dissolved oxygen in order to live, eat, and reproduce.

3.61   Trickling Filters (Chapter 6)

The trickling filter (Figure 3.13) is one of the oldest and most dependable of the biological treatment processes. Most of these plants are designed to remove 70 to 85% of the BOD-causing wastes and suspended solids present in the influent.

The trickling filter is a bed of 1½- to 5-inch rock, slag blocks, or specially manufactured *MEDIA*[14] over which settled wastewater from the primary clarifier is distributed (Figure 3.14). The settled wastewater is usually applied by an overhead rotating distributor and trickles over and around the media as it

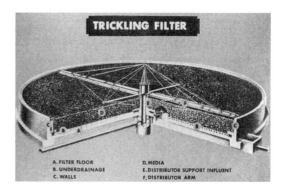

Fig. 3.13   Trickling filter
(Courtesy Water Environment Federation)

flows downward to the effluent collection channel. Since the media and the spaces (voids) in between them are large (usually 2½- to 4-inch [6- to 10-cm] diameter) and since the applied wastewater no longer has any large particles (they settled out in the clarifier), the trickling filter does not remove solids by a filtering action. A more correct term would be to call the filter a "biological contact bed" or "biological reactor" since this is the function it performs. The filter bed offers a place for aerobic bacteria and other organisms to attach themselves and multiply as they feed on the passing wastewater. This process of feeding on or decomposing waste is exactly the same as the process occurring in the stream when waste is discharged to it. In the trickling filter, however, the organisms use the oxygen that enters the waste from the surrounding air, rather than using up the stream's supply of dissolved oxygen. Thus the voids between the media must be large so sufficient oxygen can be supplied by circulating air.

Fig. 3.14   Trickling filter
(Courtesy Water Environment Federation)

The wastewater being distributed on the filter usually has passed through a primary clarifier, but it still contains approximately 70 percent of its original organic matter, which represents food for organisms. For this reason a tremendous population of organisms develops on the media. This population

---

[13] *Sludge (sluj).   (1) The settleable solids separated from liquids during processing. (2) The deposits of foreign materials on the bottoms of streams or other bodies of water.*

[14] *Media.   The material in a trickling filter on which slime accumulates and organisms grow. As settled wastewater trickles over the media, organisms in the slime remove certain types of wastes thereby partially treating the wastewater. Also the material in a rotating biological contactor or in a gravity or pressure filter.*

continues to grow as more waste is applied. Eventually the layer of organisms on the media gets so thick that some of it breaks off (sloughs off) and is carried into the filter effluent channel. This material is normally called *HUMUS*. Since humus is principally organic matter, its presence in a stream would be undesirable. This humus is usually removed by settling in a *SECONDARY CLARIFIER*. Humus sludge from the secondary clarifier is usually returned to the primary clarifier to be resettled and pumped to the sludge handling facilities along with the "raw" sludge, which settles out as previously described.

3.62 Rotating Biological Contactors (Chapter 7)

Rotating biological contactors (RBCs) are similar to trickling filters and are located after primary clarifiers. Biological contactors have a rotating "shaft" surrounded by plastic discs called the "media." A biological slime grows on the media when conditions are suitable. This process is very similar to a trickling filter where the biological slime grows on rock or other media and settled wastewater (primary clarifier effluent) is applied over the media. With rotating biological contactors, the biological slime grows on the surface of the plastic-disc media. The slime is rotated into the settled wastewater and then into the atmosphere to provide oxygen for the organisms. The wastewater being treated usually flows parallel to the rotating shaft, but may flow perpendicular to the shaft as it flows from stage to stage or tank to tank. Effluent from the rotating biological contactors flows through secondary clarifiers for removal of suspended solids and dead slime growths.

3.63 Activated Sludge (Chapters 8 and 11)

Another biological treatment unit that is used in secondary treatment, following the primary clarifier, is the aeration tank. When aeration tanks are used with the sedimentation process, the treatment plant is called an *ACTIVATED SLUDGE* plant. The activated sludge process is widely used by large cities and communities where land is expensive and where large volumes must be highly treated, economically, without creating a nuisance to neighbors. The activated sludge plant is probably the most popular biological treatment process being built today for larger installations or small package plants. These plants are capable of BOD and suspended solids reductions of 90 to 99 percent. The activated sludge process is a biological process, and it serves the same function as a trickling filter or rotating biological contactor. Effluent from a primary clarifier is piped to a large aeration tank (Figure 3.15). Oxygen is supplied to the tank by either introducing compressed air or pure oxygen into the bottom of the tank and letting it bubble through the wastewater and up to the top, or by churning the surface mechanically to introduce atmospheric oxygen.

Aerobic bacteria and other organisms thrive as they travel through the aeration tank. With sufficient food and oxygen they multiply rapidly, as in a trickling filter. By the time the waste reaches the end of the tank (usually 4 to 8 hours), most of the organic matter in the waste has been used by the bacteria for producing new cells. The effluent from the tank, usually called *MIXED LIQUOR*, consists of suspended material containing a large population of organisms and a liquid with very little soluble BOD. The activated sludge forms a lacy network that captures pollutants.

The organisms are removed in the same manner as they were in the trickling filter plant. The mixed liquor is piped to a secondary clarifier; the organisms settle to the bottom of the tank while the clear effluent flows over the top of the effluent weirs. This effluent is usually clearer than a trickling filter effluent because the suspended material in the mixed liquor settled to the bottom of the clarifier more readily than the material in a trickling filter effluent. The settled organisms are known as *ACTIVATED SLUDGE*. They are extremely valuable to the treatment process. *IF THEY ARE REMOVED QUICKLY FROM THE SECONDARY CLARIFIER, THEY WILL BE IN GOOD CONDITION AND HUNGRY FOR MORE FOOD* (organic wastes). They are therefore pumped back (recirculated) to the influent end of the aeration tank where they are mixed with the incoming wastewater. Here they begin all over again to feed on the organic material in the waste, decomposing it and creating new organisms.

Left uncontrolled, the number of organisms would eventually be too great, and therefore some must periodically be removed. This is accomplished by pumping a small amount of the activated sludge to the primary clarifier or directly to the sludge handling facilities. If the organisms are pumped to the clarifier, they settle along with the raw sludge and then are removed to the sludge handling facilities.

There are many variations of the conventional activated sludge process, but they all involve the same basic principle. These variations will be discussed in Chapters 8 and 11, "Activated Sludge."

3.64 Secondary Clarifiers (Chapter 5)

As previously mentioned, trickling filters, rotating biological contactors, and activated sludge tanks produce effluents that contain large populations of microorganisms and associated materials (humus). These microorganisms must be removed from the flow before it can be discharged to the receiving waters. This task is usually accomplished by a secondary clarifier. In this tank the trickling filter or biological contactor humus or activated sludge separates from the liquid and settles to the bottom of the tank. This sludge is removed to the primary clarifier to be resettled with the primary sludge or returned to the beginning of the secondary process to continue treating the wastewater. In most activated sludge plants the waste activated sludge is pumped to waste sludge handling facilities instead of to the primary clarifier. The clear effluent flows over a weir at the top of the secondary clarifier.

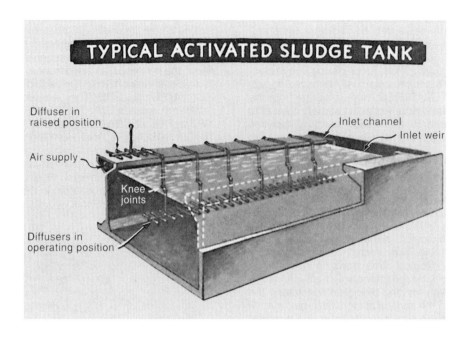

(Courtesy Water Environment Federation)

Fig. 3.15   Aeration tank

QUESTIONS

Write your answers in a notebook and then compare your answers with those on page 56.

3.6A   Would it be a good idea to use trickling filter media of various sizes so it could pack together better?

3.6B   Why is a secondary clarifier needed after a trickling filter, rotating biological contactor, or aeration tank?

3.6C   Activated sludge can be pumped from the secondary clarifier to_____or_____.

3.7 SOLIDS HANDLING AND DISPOSAL (Chapter 12)

3.70 Purpose

Solids removed from wastewater treatment processes are commonly broken down by a biological treatment process called *SLUDGE DIGESTION*. After digestion and removal of water (*DEWATERING*[15]), the remaining material (*BIOSOLIDS*[16]) may be used as fertilizer or soil conditioner. Some solids, such as scum from a clarifier, may be disposed of by burning or burial. Possible solids handling systems are shown on Figures 3.16 and 3.17.

3.71 Digestion and Dewatering

Settled sludge from the primary clarifier and occasionally settled sludge from the secondary clarifier are periodically pumped to a digestion tank. The tank is usually completely sealed to keep air from getting inside (Figures 3.18 (below) and 3.19). This type of digester is called an *ANAEROBIC DIGESTER* because of the anaerobic bacteria that abound in the tank. Anaerobic bacteria thrive in an environment without dissolved oxygen by using the oxygen that is chemically combined with their food supply.

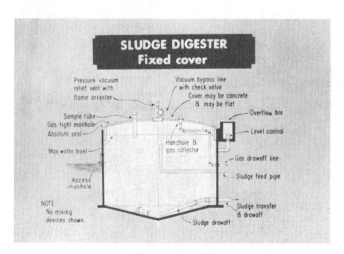

Fig. 3.18 Sludge digester
(Courtesy Water Environment Federation)

Two major types of bacteria are present in the digester. The first group starts eating the organic portion of the sludge to form organic acids and carbon dioxide gas. These bacteria are called *ACID FORMERS*. The second group breaks down the organic acids to simpler compounds and forms methane and carbon dioxide gas. These bacteria are called *GAS FORMERS*. The methane gas is usually used to heat the digester or to run engines in the plant. The production of gas indicates that organic material is being eaten by the bacteria. A sludge is usually considered properly digested when 50 percent of the organic matter has been destroyed and converted to gas. This normally takes approximately 30 days if the temperature is kept at about 95°F (35°C).

Most digestion tanks are mixed continuously to bring the food to the organisms, to provide a uniform temperature, and to avoid the formation of thick scum blankets. When a digester is not being mixed, the solids usually settle to the bottom, leaving a liquid known as *SUPERNATANT* above the sludge. In many plants, however, there is no separation of solids and liquids after two days of sitting without mixing due to the type of sludge. The supernatant is displaced from the tank each time a fresh charge of raw sludge is pumped from the primary clarifier. The displaced supernatant usually is returned from the digester back to the plant headworks and mixed with incoming raw wastes. Supernatant return should be slow to prevent overloading or shock loading of the plant.

A scum blanket will usually develop above the supernatant level. Scum blankets consist of grease, soap, rubber goods, hair, petroleum products, plastics, and filter tips from cigarettes. These scum blankets may contain most of the added food or sludge. Organisms that digest the sludge are usually below the supernatant and little digestion will occur if the organisms and food don't get together. Control of scum blankets consists of mixing the digester contents and burning or burying skimmings instead of pumping them to the digester.

Above the scum blanket or normal water level is the gas collection area. Digester gas is usually around 70 percent methane and about 30 percent carbon dioxide. *WHEN MIXED WITH AIR, IN THE PROPER PROPORTION, DIGESTER GAS IS EXTREMELY EXPLOSIVE. DON'T ALLOW DIGESTER GAS AND AIR TO MIX.*

In most newer plants, sludge digestion takes place in two tanks. The first or primary digester is usually heated and mixed. Rapid digestion takes place along with most of the gas production. In the secondary tank, the digested sludge and supernatant are allowed to separate, thus producing a clearer supernatant and better-digested sludge.

Digester sludge from the bottom of the tank is periodically removed for dewatering. This is accomplished in sand drying beds (Figure 3.20), lagoons, centrifuges, vacuum filters (Figure 3.21), and filter presses. The sludge is then burned, buried, or used as fertilizer on certain crops (not on crops that are eaten without cooking). Sludge that has been adequately digested drains readily and is not offensive.

Some of today's activated sludge treatment plants are equipped with aerobic digesters. An aerobic digester is usually an open tank with compressed air being blown through the sludge. Destruction of organic matter is accomplished by aerobic bacteria, which require dissolved oxygen to survive. The

---

[15] *Dewater.   (1) To remove or separate a portion of the water present in a sludge or slurry. To dry sludge so it can be handled and disposed of. (2) To remove or drain the water from a tank or a trench.*

[16] *Biosolids.   A primarily organic solid product, produced by wastewater treatment processes, that can be beneficially recycled. The word biosolids is replacing the word sludge.*

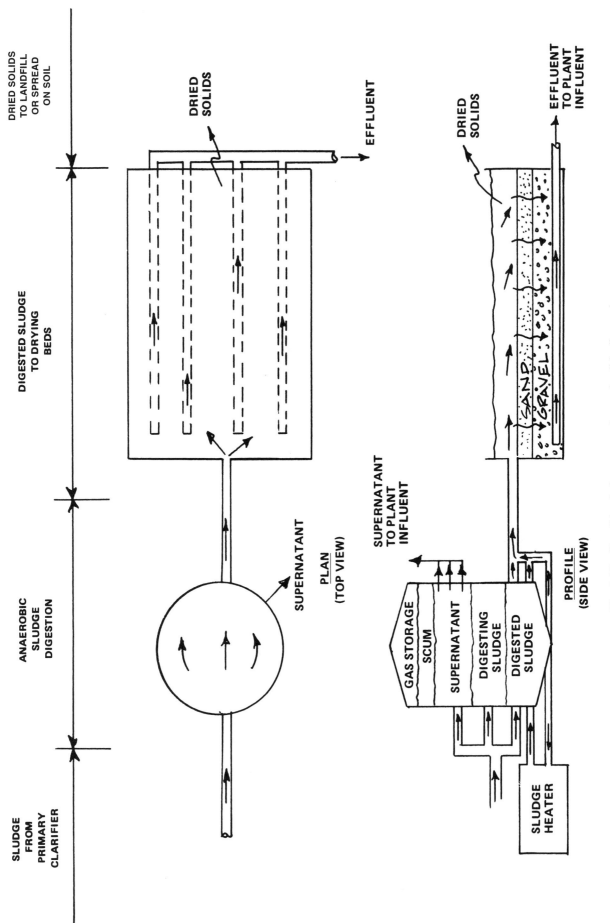

Fig. 3.16   Possible sludge processing and solids disposal flow pattern (sludge drying beds)

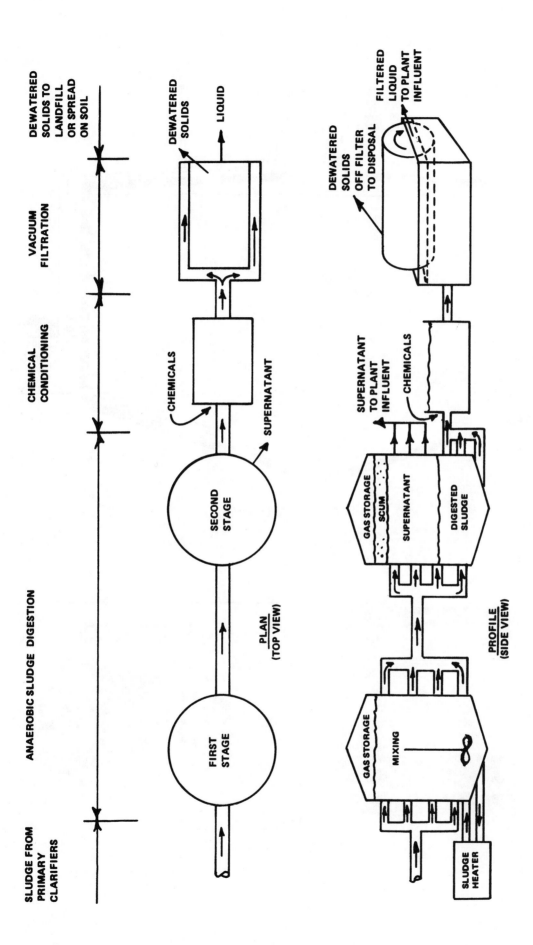

Fig. 3.17   Possible sludge processing and solids disposal
flow pattern (vacuum filtration)

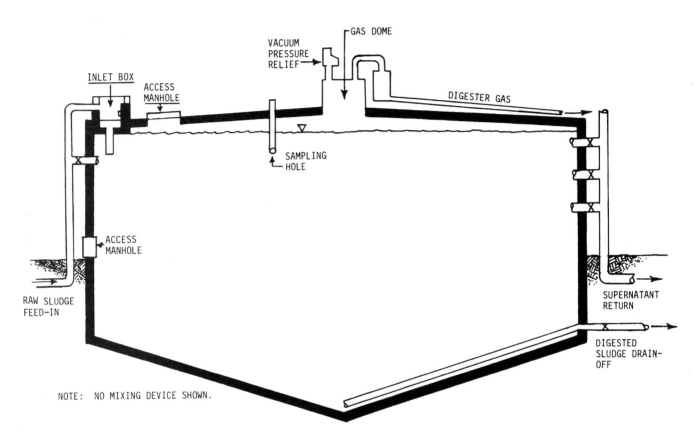

GAS DOME

VACUUM
PRESSURE
RELIEF

INLET BOX

ACCESS
MANHOLE

DIGESTER GAS

SAMPLING
HOLE

ACCESS
MANHOLE

RAW SLUDGE
FEED-IN

SUPERNATANT
RETURN

DIGESTED
SLUDGE DRAIN-
OFF

NOTE: NO MIXING DEVICE SHOWN.

Fig. 3.19   Section of sludge digester

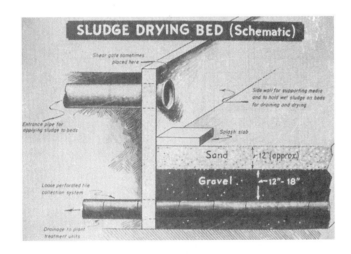

Fig. 3.20   Sludge drying bed
(Courtesy Water Environment Federation)

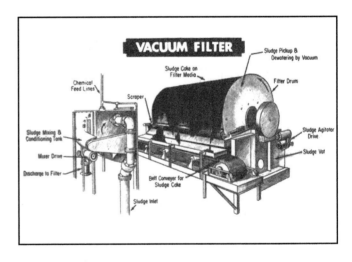

Fig. 3.21   Vacuum filter
(Courtesy Water Environment Federation)

aeration equipment is turned off to allow time for the solids to separate from the water before the supernatant is removed.

One advantage of aerobic digestion is that no explosive gas is produced. On the other hand, this is also a disadvantage since the anaerobic digester gas is used as a fuel for boilers and engines around the plant. Aerobic sludge from an aerobic digester doesn't thicken as readily as sludge from an anaerobic digester. Aerobic sludge does not filter as well as an equivalent concentration of anaerobic sludge. Supernatant from aerobic digesters is not as strong or difficult to treat as supernatant from anaerobic digesters.

3.72 Incineration

Sludge may be disposed of by burning if the process does not create an air pollution problem. Sludge that has not been dewatered previously can be conditioned by *WET OXIDATION*,[17] dewatered, and then burned. Incineration or burial of skimmings from the clarifiers will prevent treatment plant operational problems.

QUESTIONS

Write your answers in a notebook and then compare your answers with those on page 56.

3.7A What two basic types of bacteria are present in an anaerobic digester?

3.7B Why are digesters mixed?

3.7C List some of the ways to dewater and dispose of digested sludge.

3.8 WASTE TREATMENT PONDS (Chapter 9)

The waste treatment pond (Figure 3.22), or stabilization pond, is a special method of biological treatment deserving attention. Ponds do not resemble the concrete and steel structures or the mechanical devices previously discussed, but these simple depressions in the ground are capable of producing an effluent comparable to some of the most modern plants with respect to BOD and bacteria reduction.

The types of treatment processes and the locations of ponds are determined by the design engineer on the basis of economics and the degree of treatment required to meet the water quality standards of the receiving waters. In some treatment plants, wastewater being treated may flow through a coarse screen and flowmeter before it flows through a series of ponds. In other types of plants, ponds are located after primary treatment, or after trickling filters.

When wastewater is discharged to a pond, the settleable solids fall to the bottom just as they do in a primary clarifier. The solids begin to decompose and soon use up all the dissolved oxygen in the nearby water. A population of anaerobic bacteria then continues the decomposition, much the same as in an anaerobic digester. As the organic matter is destroyed, methane and carbon dioxide are released. When the carbon dioxide rises to the surface, some of it is used by algae, which convert it to oxygen by the process of *PHOTOSYNTHESIS*.[18] This is the same process used by living plants. Aerobic bacteria, algae, and other microorganisms feed on the dissolved solids in the upper layer of the pond similar to the way they do in a trickling filter or aeration tank. Algae produce oxygen for the other organisms to use.

Some shallow ponds (3 to 6 feet deep) have dissolved oxygen throughout their entire depth. These ponds are called aerobic ponds. They usually have a mechanical apparatus (surface aerator) adding oxygen as well as their oxygen supply from algae. Another type of aerated pond has oxygen delivered by a diffused air system similar to the system used in activated sludge plants.

Deep (8 to 12 feet), heavily loaded ponds may be without oxygen throughout their depth. These ponds are called anaerobic ponds. At times these ponds can be quite odorous and therefore they are used only in sparsely populated areas.

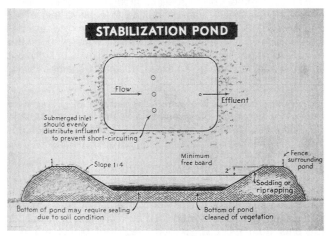

(Courtesy Water Environment Federation)

(Courtesy Water and Sewage Works Magazine)

Fig. 3.22 Pond

---

[17] *Wet Oxidation. A method of treating or conditioning sludge before the water is removed. Compressed air is blown into the liquid sludge. The air and sludge mixture is fed into a pressure vessel where the organic material is stabilized. The stabilized organic material and inert (inorganic) solids are then separated from the pressure vessel effluent by dewatering in lagoons or by mechanical means.*

[18] *Photosynthesis (foe-toe-SIN-thuh-sis). A process in which organisms, with the aid of chlorophyll (green plant enzyme), convert carbon dioxide and inorganic substances into oxygen and additional plant material, using sunlight for energy. All green plants grow by this process.*

Ponds that contain an aerobic top layer and an anaerobic bottom layer are called *FACULTATIVE PONDS*.[19] These are the ponds normally seen in most areas. If they are properly designed and operated, they are virtually odor-free and produce a well-oxidized (low BOD) effluent.

Occasionally ponds are used after a primary treatment unit. In this case, they are usually called "oxidation ponds." When they are used to treat raw wastewater, they are called "raw wastewater lagoons" or "waste stabilization ponds."

The effluent from ponds is usually moderately low in bacteria. This is especially true when the effluent runs from one pond to another or more (series flow). A long detention time, usually a month or more, is required in order for harmful bacteria and undesirable solids to be removed from the pond effluent. If the receiving waters are used for water supply or body contact sports, chlorination of the effluent may still be required.

QUESTION

Write your answer in a notebook and then compare your answer with the one on page 56.

3.8A   How are facultative ponds similar to the following:

　　　1. A clarifier?

　　　2. A digester?

　　　3. An aeration tank?

3.9   ADVANCED METHODS OF TREATING WASTEWATER *(ADVANCED WASTE TREATMENT)*

The treatment processes described so far in this chapter are considered *CONVENTIONAL* treatment processes. As our population grows and industry expands, more effective treatment processes will be required. Advanced methods of waste treatment may follow conventional processes, or they may be used instead of these processes. Sometimes advanced methods of waste treatment are called tertiary (TER-she-AIR-ee) treatment because they frequently follow secondary treatment. Advanced methods of waste treatment include coagulation-sedimentation (used in water treatment plants), adsorption, and electrodialysis. Other treatment processes that may be used include reverse osmosis, chemical oxidation, and the use of polymers.

Advanced methods of treatment are used to reduce the nutrient content (nitrate and phosphate) of wastewater to prevent blooms of algae in lakes, reservoirs, or streams. Carbon filters are used to reduce the last traces of organic materials. In some parts of the arid West, advanced methods are used to enable the use of the plant effluent for recreational reservoirs.

QUESTION

Write your answer in a notebook and then compare your answer with the one on page 56.

3.9A   If wastewater from a secondary treatment plant were coagulated with alum or lime and settled in a clarifier, would this be considered a method of advanced waste treatment?

3.10   DISINFECTION (Chapter 10)

Although the settling process and biological processes remove a great number of organisms from the wastewater flow, there remain many thousands of bacteria in every milliliter of wastewater leaving the secondary clarifier. If there are human wastes in the water, it is possible that some of the bacteria are *PATHOGENIC*, or harmful to humans. Therefore, if the treated wastewater is discharged to a receiving water that is used for a drinking water supply or swimming or wading, the water pollution control agency or health department will usually require disinfection of the effluent prior to discharge.

Disinfection is usually defined as a process designed to kill or inactivate *PATHOGENIC* organisms. The killing of *ALL* organisms is called sterilization.

Sterilization is not accomplished in treatment plants as the final effluent always contains some living organisms after disinfection due to the inefficiency of the killing process.

Disinfection can be accomplished by almost any process that will create a harsh environment for the organisms. Strong light, heat, oxidizing chemicals, acids, alkalies, poisons, and many other substances will disinfect. Most disinfection in wastewater treatment plants is accomplished by applying chlorine, which is a strong oxidizing chemical.

Chlorine gas is used in most treatment plants although some of the smaller plants use a liquid chlorine solution as their source. The dangers in using chlorine gas, however, have prompted some of the larger plants to switch to hypochlorite solution (bleach) even though it is more expensive.

Chlorine gas, which is withdrawn from pressurized cylinders containing liquid chlorine, is mixed with water or treated wastewater to make up a strong chlorine solution. Liquid hypochlorite solution can be used directly. The strong chlorine solution is then mixed with the effluent from the secondary clarifier. Proper and adequate mixing are very important. The effluent is then directed to a chlorine contact basin or tank. The basin can be any size or shape, but better results are obtained if the basin is long and narrow. This shape prevents rapid movement or short-circuiting through the basin. Square or rectangular basins can be baffled to achieve this effect (Figure 3.23). Basins are usually designed to provide approximately 20 to 30 minutes theoretical contact time, although the trend is to longer times. If the plant's outfall line is of sufficient length, it may function as an excellent contact chamber since short-circuiting will not occur.

In some areas the effluent must be dechlorinated or detoxified before discharge to the receiving waters to protect aquatic life. Sulfur dioxide (SO_2) can be added after the chlorine contact basin to neutralize the remaining residual chlorine.

3.11   EFFLUENT DISPOSAL (Chapter 13)

Ultimately the effluent from a wastewater treatment plant must be disposed of in the environment. This can be into water, onto land, or the water can be reclaimed and reused. Effluents from most wastewater treatment plants are discharged into receiving waters such as streams, rivers, and lakes. With water becoming scarcer due to increased demands and with required higher degrees of treatment, plant effluent is becoming a valuable resource. Both industry and agriculture are discovering that treated effluent may be the most economical source of additional water.

---

[19] *Facultative (FACK-ul-TAY-tive) Pond.   The most common type of pond in current use. The upper portion (supernatant) is aerobic, while the bottom layer is anaerobic. Algae supply most of the oxygen to the supernatant.*

CHLORINE CONTACT BASIN

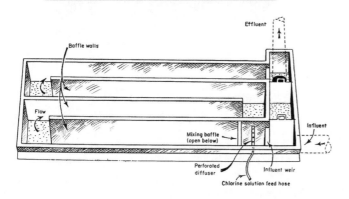

(Courtesy Water Environment Federation)

Fig. 3.23 Chlorine contact basin

Land disposal is another method of ultimate disposal and can be a means of recharging groundwater basins or storing water for future use. Evaporation ponds are used to dispose of effluents to the atmosphere. Regardless of the method of ultimate effluent disposal, operators must carefully operate wastewater treatment plants so that plant effluent will not cause any adverse impacts on the method of ultimate disposal or on the environment.

3.12 SOLIDS DISPOSAL (Chapter 3, *ADVANCED WASTE TREATMENT*)

Final solids disposal is one of the major problems facing many operators today. Solids removed from wastewater by pretreatment processes such as bar racks, screens, and grit removal systems may be disposed of by dewatering and then direct burial in an approved sanitary landfill or incineration with the remaining ash disposed of in a landfill. Grease and scum from primary and secondary treatment processes are usually pumped to anaerobic digesters or disposed of in incinerators or in sanitary landfills.

Both aerobic and anaerobic sludge digestion processes produce stabilized or digested solids that ultimately must be disposed of in the environment. Disposal methods include composited with another material such as leaves, farm land application as a soil conditioner or fertilizer, burial in a sanitary landfill, or incineration with ash disposal in a landfill.

QUESTIONS

Write your answers in a notebook and then compare your answers with those on page 56.

3.10A Does disinfection usually kill all organisms in the plant effluent?

3.10B Which would provide better chlorine contact, a 10,000-gallon cubical tank or a length of 10-inch pipe flowing full and containing the same volume as the cubical tank?

3.13 ARITHMETIC ASSIGNMENT

Turn to the Arithmetic Appendix at the back of this manual and read the following sections:

1. A.2, *AREAS,*

2. A.3, *VOLUMES,*

3. A.10, *BASIC CONVERSION FACTORS,*

4. A.11, *BASIC FORMULAS, AND*

5. A.12, *HOW TO USE THE BASIC FORMULAS.*

Check all of the arithmetic in Sections A.2, *AREAS* (A.20, A.21, A.22, A.23, A.24, A.25, and A.26), and A.3, *VOLUMES* (A.30, A.31, A.32, A.33, and A.34), on an electronic pocket calculator. You should be able to get the same answers.

3.14 ADDITIONAL READING

Some books you can read to obtain further information on the treatment plant and the various processes involved are:

1. *MOP 11,*

2. *NEW YORK MANUAL,*

3. *TEXAS MANUAL,* and

4. *ADVANCED WASTE TREATMENT* in this series of manuals.

SUGGESTED ANSWERS

Chapter 3.   WASTEWATER TREATMENT FACILITIES

Answers to questions on page 34.

3.1A   The operator should know the origin of wastes reaching the plant, the time it takes, and how the wastes are transported (flow by gravity or by gravity and pumped). Such knowledge will help you to spot troubles and take corrective action.

3.1B   If the flow time to reach the plant is very long, hydrogen sulfide gas may develop and cause corrosion damage to concrete in the transportation system and in the plant. Also undesirable odors develop and solids are difficult to settle.

3.1C   Three types of sewers are sanitary, storm, and combined sewers.

3.1D   In a system using combined sewers, flows sometimes may become overloaded during storms because the plant does not have the capacity to handle the additional storm water runoff.

Answers to questions on page 42.

3.3A   Grit should be removed early in the treatment process because it is abrasive and will wear out pumps and other equipment.

3.3B   Grit removed from the wastewater is usually buried to avoid causing a nuisance.

Answers to questions on page 43.

3.4A   Parshall flumes are widely used for measuring wastewater flow because they have no obstructions.

3.4B   Weirs are not frequently used to measure influent flows because solids may collect behind the weir causing odors and inaccurate flow measurements.

Answers to questions on page 46.

3.5A   "Flights" in rectangular tanks move scum along the surface to a scum trough and push sludge along the bottom to a hopper for removal to the sludge handling facility. "Plows" scrape sludge along the bottom of circular tanks to a hopper for removal.

3.5B   Sludge and scum are usually pumped to sludge handling facilities such as digesters.

Answers to questions on page 49.

3.6A   No. If the media were packed together, air could not circulate and the organisms on the media would not get enough oxygen.

3.6B   A secondary clarifier is needed after a trickling filter, rotating biological contactor, or aeration tank to allow organisms in treated wastewater to be removed by settling.

3.6C   Activated sludge can be pumped from the secondary clarifier to either the primary clarifier or waste sludge handling facilities.

Answers to questions on page 53.

3.7A   The two types of bacteria in an anaerobic digester are: (1) a group that eats organic sludge to form organic acids and carbon dioxide gas (acid formers); and (2) a group that breaks down the organic acids into simpler compounds and forms methane and carbon dioxide gas (gas formers).

3.7B   Digesters are mixed to bring food and organisms together, provide a uniform temperature, and prevent the formation of a scum blanket.

3.7C   Digested sludge may be dewatered by using sand drying beds, lagoons, centrifuges, vacuum filters, or filter presses. Ultimately the dried sludge may be used as a soil conditioner or it may be buried.

Answer to question on page 54.

3.8A   A facultative pond acts like a clarifier by allowing solids to settle to the bottom, like a digester because solids on the bottom are decomposed by anaerobic bacteria, and like an aeration tank because of the action of aerobic bacteria in the upper layer of the pond.

Answer to question on page 54.

3.9A   Yes, if wastewater from a secondary treatment plant were coagulated with alum or lime and settled in a clarifier, this would be considered a method of advanced waste treatment.

Answers to questions on page 55.

3.10A   No, disinfection is designed to kill or inactivate pathogenic organisms; sterilization kills all organisms in wastewater.

3.10B   The pipe would provide better chlorine contact because water cannot short-circuit (take a short route) through a pipe, while it might not move evenly through a tank and thus some of the water would have a shorter contact time.

CHAPTER 4

RACKS, SCREENS, COMMINUTORS, AND GRIT REMOVAL
(PRELIMINARY TREATMENT)

by

Larry Bristow

TABLE OF CONTENTS
Chapter 4.   RACKS, SCREENS, COMMINUTORS, AND GRIT REMOVAL

OBJECTIVES

Chapter 4.   RACKS, SCREENS, COMMINUTORS, AND GRIT REMOVAL

Following completion of Chapter 4, you should be able to:

1. Explain the purposes of racks, screens, comminutors, grit channels, grit separators, and pre-aeration,

2. Properly start up, operate, shut down, and maintain the preliminary treatment process,

3. Identify potential safety hazards and conduct preliminary treatment duties using safe procedures,

4. Determine the volume of screenings and how long a disposal site will last before it is full,

5. Measure the flow velocity in a grit channel,

6. Regulate flow velocities in a grit channel,

7. Develop an operational strategy for preliminary treatment processes, and

8. Review plans and specifications for preliminary treatment processes.

NOTE: Information on maintenance of equipment can be found in Chapter 15, "Maintenance."

WORDS

Chapter 4.   RACKS, SCREENS, COMMINUTORS, AND GRIT REMOVAL

AEROBIC (AIR-O-bick) DECOMPOSITION   AEROBIC DECOMPOSITION

The decay or breaking down of organic material in the presence of "free" or dissolved oxygen.

ALKALINITY (AL-ka-LIN-it-tee)   ALKALINITY

The capacity of water or wastewater to neutralize acids. This capacity is caused by the water's content of carbonate, bicarbonate, hydroxide, and occasionally borate, silicate, and phosphate. Alkalinity is expressed in milligrams per liter of equivalent calcium carbonate. Alkalinity is not the same as pH because water does not have to be strongly basic (high pH) to have a high alkalinity. Alkalinity is a measure of how much acid must be added to a liquid to lower the pH to 4.5.

ANAEROBIC (AN-air-O-bick) DECOMPOSITION   ANAEROBIC DECOMPOSITION

The decay or breaking down of organic material in an environment containing no "free" or dissolved oxygen.

BUFFER   BUFFER

A solution or liquid whose chemical makeup neutralizes acids or bases without a great change in pH.

BUFFER CAPACITY   BUFFER CAPACITY

A measure of the capacity of a solution or liquid to neutralize acids or bases. This is a measure of the capacity of water or wastewater for offering a resistance to changes in pH.

CLARIFIER (KLAIR-uh-fire)   CLARIFIER

Settling Tank, Sedimentation Basin. A tank or basin in which wastewater is held for a period of time during which the heavier solids settle to the bottom and the lighter materials float to the water surface.

DECOMPOSITION, DECAY   DECOMPOSITION, DECAY

Processes that convert unstable materials into more stable forms by chemical or biological action. Waste treatment encourages decay in a controlled situation so that material may be disposed of in a stable form. When organic matter decays under anaerobic conditions (putrefaction), undesirable odors are produced. The aerobic processes in common use for wastewater treatment produce much less objectionable odors.

DETRITUS (dee-TRY-tus)   DETRITUS

The heavy, coarse mixture of grit and organic material carried by wastewater. Also called GRIT.

DIFFUSER   DIFFUSER

A device (porous plate, tube, bag) used to break the air stream from the blower system into fine bubbles in an aeration tank or reactor.

DIGESTER (die-JEST-er)   DIGESTER

A tank in which sludge is placed to allow decomposition by microorganisms. Digestion may occur under anaerobic (more common) or aerobic conditions.

DISSOLVED OXYGEN   DISSOLVED OXYGEN

Molecular (atmospheric) oxygen dissolved in water or wastewater, usually abbreviated DO.

EFFLUENT   EFFLUENT

Wastewater or other liquid—raw (untreated), partially or completely treated—flowing *FROM* a reservoir, basin, treatment process, or treatment plant.

EXPLOSIMETER   EXPLOSIMETER

An instrument used to detect explosive atmospheres. When the **L**ower **E**xplosive **L**imit (LEL) of an atmosphere is exceeded, an alarm signal on the instrument is activated. Also called a combustible gas detector.

GRIT   GRIT

The heavy material present in wastewater, such as sand, coffee grounds, eggshells, gravel and cinders.

GRIT REMOVAL   GRIT REMOVAL

Grit removal is accomplished by providing an enlarged channel or chamber which causes the flow velocity to be reduced and allows the heavier grit to settle to the bottom of the channel where it can be removed.

HEAD   HEAD

The vertical distance, height or energy of water above a point. A head of water may be measured in either height (feet or meters) or pressure (pounds per square inch or kilograms per square centimeter).

HEAD LOSS   HEAD LOSS

An indirect measure of loss of energy or pressure. Flowing water will lose some of its energy when it passes through a pipe, bar screen, comminutor, filter or other obstruction. The amount of energy or pressure lost is called "head loss." Head loss is measured as the difference in elevation between the upstream water surface and the downstream water surface and may be expressed in feet or meters.

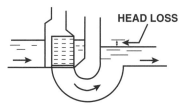

HYDROGEN SULFIDE GAS (H_2S)   **HYDROGEN SULFIDE GAS (H_2S)**

Hydrogen sulfide is a gas with a rotten egg odor. This gas is produced under anaerobic conditions. Hydrogen sulfide gas is particularly dangerous because it dulls the sense of smell so that you don't notice it after you have been around it for a while. In high concentrations, hydrogen sulfide gas is only noticeable for a very short time before it dulls the sense of smell. The gas is very poisonous to the respiratory system, explosive, flammable, colorless and heavier than air.

INFLUENT   INFLUENT

Wastewater or other liquid—raw (untreated) or partially treated—flowing *INTO* a reservoir, basin, treatment process, or treatment plant.

INORGANIC WASTE   INORGANIC WASTE

Waste material such as sand, salt, iron, calcium, and other mineral materials which are only slightly affected by the action of organisms. Inorganic wastes are chemical substances of mineral origin; whereas organic wastes are chemical substances usually of animal or plant origin.

LIMIT SWITCH   LIMIT SWITCH

A device that regulates or controls the travel distance of a chain or cable.

O & M MANUAL   O & M MANUAL

Operation and **M**aintenance Manual. A manual that describes detailed procedures for operators to follow to operate and maintain a specific wastewater treatment or pretreatment plant and the equipment of that plant.

ORGANIC WASTE   ORGANIC WASTE

Waste material which comes mainly from animal or plant sources. Organic wastes generally can be consumed by bacteria and other small organisms. Inorganic wastes are chemical substances of mineral origin.

OZONATION (O-zoe-NAY-shun)   OZONATION

The application of ozone to water, wastewater, or air, generally for the purposes of disinfection or odor control.

POTW   POTW

Publicly **O**wned **T**reatment **W**orks. A treatment works which is owned by a state, municipality, city, town, special sewer district or other publicly owned and financed entity as opposed to a privately (industrial) owned treatment facility. This definition includes any devices and systems used in the storage, treatment, recycling and reclamation of municipal sewage (wastewater) or industrial wastes of a liquid nature. It also includes sewers, pipes and other conveyances only if they carry wastewater to a POTW treatment plant. The term also means the municipality (public entity) which has jurisdiction over the indirect discharges to and the discharges from such a treatment works.

PRE-AERATION   PRE-AERATION

The addition of air at the initial stages of treatment to freshen the wastewater, remove gases, add oxygen, promote flotation of grease, and aid coagulation.

PRELIMINARY TREATMENT   PRELIMINARY TREATMENT

The removal of metal, rocks, rags, sand, eggshells, and similar materials which may hinder the operation of a treatment plant. Preliminary treatment is accomplished by using equipment such as racks, bar screens, comminutors, and grit removal systems.

PRETREATMENT FACILITY   PRETREATMENT FACILITY

Industrial wastewater treatment plant consisting of one or more treatment devices designed to remove sufficient pollutants from wastewaters to allow an industry to comply with effluent limits established by the US EPA General and Categorical Pretreatment Regulations or locally derived prohibited discharge requirements and local effluent limits. Compliance with effluent limits allows for a legal discharge to a POTW.

PUTREFACTION (PEW-truh-FACK-shun)   PUTREFACTION

Biological decomposition of organic matter with the production of foul-smelling products associated with anaerobic conditions.

PUTRESCIBLE (pew-TRES-uh-bull)   PUTRESCIBLE

Material that will decompose under anaerobic conditions and produce nuisance odors.

RACK   RACK

Evenly spaced parallel metal bars or rods located in the influent channel to remove rags, rocks, and cans from wastewater.

RAW WASTEWATER   RAW WASTEWATER

Plant influent or wastewater *BEFORE* any treatment.

SCREEN   SCREEN

A device used to retain or remove suspended or floating objects in wastewater. The screen has openings that are generally uniform in size. It retains or removes objects larger than the openings. A screen may consist of bars, rods, wires, gratings, wire mesh, or perforated plates.

SEPTIC (SEP-tick)   SEPTIC

A condition produced by anaerobic bacteria. If severe, the sludge produces hydrogen sulfide, turns black, gives off foul odors, contains little or no dissolved oxygen, and the wastewater has a high oxygen demand.

SHEAR PIN   SHEAR PIN

A straight pin that will fail when a certain load or stress is exceeded. The purpose of the pin is to protect equipment from damage due to excessive loads or stresses.

SLUDGE (sluj)   SLUDGE

(1) The settleable solids separated from liquids during processing.

(2) The deposits of foreign materials on the bottoms of streams or other bodies of water.

SLUDGE DIGESTION   SLUDGE DIGESTION

The process of changing organic matter in sludge into a gas or a liquid or a more stable solid form. These changes take place as microorganisms feed on sludge in anaerobic (more common) or aerobic digesters.

SLURRY (SLUR-e)   SLURRY

A thin, watery mud or any substance resembling it (such as a grit slurry or a lime slurry).

SPECIFIC GRAVITY   SPECIFIC GRAVITY

(1) Weight of a particle, substance or chemical solution in relation to the weight of an equal volume of water. Water has a specific gravity of 1.000 at 4°C (39°F). Wastewater particles or substances usually have a specific gravity of 0.5 to 2.5.

(2) Weight of a particular gas in relation to the weight of an equal volume of air at the same temperature and pressure (air has a specific gravity of 1.0). Chlorine has a specific gravity of 2.5 as a gas.

WEIR (weer), PROPORTIONAL   WEIR, PROPORTIONAL

A specially shaped weir in which the flow through the weir is directly proportional to the head.

CHAPTER 4. RACKS, SCREENS, COMMINUTORS, AND GRIT REMOVAL

4.0 CAUTION

Many wastewater treatment plant operators have been seriously injured due to avoidable accidents. According to regular surveys by the Water Environment Federation, the wastewater treatment and pollution control industry has a higher accident rate than most industries reporting to the National Safety Council. Working in a wastewater treatment plant is not necessarily more dangerous than working in other industries. The poor record of the past may have been caused by operators not being aware of unsafe conditions, not immediately correcting these conditions when they became obvious, and not knowing safe procedures.

There are many potential safety hazards around a wastewater treatment plant. *ACCIDENTS CAN BE REDUCED BY THINKING SAFETY.* You should protect yourself from injury by maintaining firm footing, keeping walk areas clear, immediately cleaning up spills, and shutting off, tagging, and locking out the electrical power before working on equipment.

You must take adequate precautions to prevent becoming infected with waterborne diseases such as dysentery or hepatitis. At any given moment, some people in your community are ill; disease bacteria and viruses from these people are in the wastewater reaching your plant. (*NOTE:* There is no evidence that anyone has contracted AIDS as a result of contact with wastewater.) When cleaning equipment such as pumps, bar screens, and grit channels, you often must place your hands in raw wastewater. Also, the tools used to work on equipment frequently become contaminated. Consequently, *GOOD PERSONAL HYGIENE MUST BE OBSERVED BY ALL OPERATORS AT ALL TIMES. ALWAYS WASH YOUR HANDS THOROUGHLY BEFORE EATING OR SMOKING.*

A good practice is to change work clothing before going home. Any clothing that has been worn at the wastewater treatment plant must be laundered separately from the family wash.

QUESTIONS

Write your answers in a notebook and then compare your answers with those on page 99.

TRUE OR FALSE:

4.0A The wastewater treatment and pollution control industry has a higher accident rate than most other industries reporting to the National Safety Council.

4.0B Electrical power must always be shut off before working on equipment.

4.0C Operators must wash their hands thoroughly before eating, smoking, or going home.

4.1 PRELIMINARY TREATMENT

In various ways, a little or a lot of almost everything finds its way into sewers and ends up at the wastewater treatment plant. Cans, bottles, pieces of scrap metal, sticks, rocks, bricks, plastic toys, plastic lids, caps from toothpaste tubes, towels and other rags, sand—all are found in the plant *INFLUENT.*[1]

---

[1] *Influent. Wastewater or other liquid—raw (untreated) or partially treated—flowing INTO a reservoir, basin, treatment process or treatment plant.*

These materials are troublesome in various ways. Pieces of metal, rocks, and similar items will cause pipes to plug, may damage or plug pumps, or jam sludge collector mechanisms in settling tanks (*CLARIFIERS*[2]). Sand, eggshells, and similar materials (grit) can plug pipes, cause excessive wear in pumps, and use up valuable space in the sludge *DIGESTERS*.[3]

If a buried or otherwise inaccessible pipe is plugged, or a sludge collector mechanism jams, or a critical pump is put out of commission, serious consequences can result. A plant operating at reduced efficiency increases the pollution level of the effluent discharged into receiving waters. This can cause health hazards to downstream water users, sludge deposits in a stream or lake (with resultant odors and unsightliness), and sometimes the death of fish and other aquatic life. Also, repairs of this type involve a good deal of hard (sometimes rather unpleasant) work and usually result in large, unbudgeted expenses.

With these things in mind, it is evident that an important part of a wastewater treatment plant is the equipment used to remove rocks, large debris, grit, and other materials as early as possible. These items of equipment include screens, racks, comminutors, and grit removal devices and are called "pretreatment or preliminary treatment facilities."

SCREENING is that part of the pretreatment or preliminary treatment facilities that removes the larger debris (rocks, cans, bottles, rags). This equipment may consist of parallel bars or slotted drums through which the wastewater must pass. Accumulated debris is removed from the screens either manually or mechanically. In some plants, cutters shred the rags and debris accumulated on the screens or drums, return the shredded materials to the flow, and allow them to pass through the screens to continue in the wastewater flow.

GRIT REMOVAL involves removing the heavy *INORGANIC WASTES*,[4] such as sand and gravel, from the wastewater. This is done by reducing the velocity of the wastewater enough so that the heavier particles settle to the bottom of special channels or hoppers. The grit may be further processed (separated or dewatered) in cyclone separators or grit classifiers (wash-

ers). Grit usually flows from the hoppers, to the cyclone separator, and then to the classifier. Final disposal of the grit is usually by burial which can be rather unpleasant and costly work.

PRE-AERATION[5] is used to freshen the wastewater and separate oils and grease. This process tends to increase the overall efficiency of solids and BOD removal.

See Figure 4.1 for the location of these processes in a typical plant. Your plant may differ in some respects from the arrangement shown because of differences in one or more of the following factors: quantity of wastewater; characteristics of the wastewater; seasonal variations in quantity and characteristics; discharge requirements; construction, operating, and maintenance costs of various processes; and the designer's preferences.

QUESTIONS

Write your answers in a notebook and then compare your answers with those on page 99.

4.1A   Which of the following items may be found in a treatment plant influent?

　　　a. Cans
　　　b. Toys
　　　c. Rubber Goods
　　　d. Pieces of Wood
　　　e. All of These

4.1B   Why should coarse debris (rocks, boards, metal) be removed at the plant entrance or headworks?

4.1C   What items of equipment are used to remove rocks, pieces of wood, metal, and rags from wastewater?

4.2 BAR SCREENS AND RACKS

Parallel bars may be placed at an angle or set vertically in a channel in such a manner that the wastewater will flow through the bars, but large solids and debris will be caught on the bars. These bars are commonly called "racks" when the spacing between them is 3 to 4 inches (7.6 to 10.2 cm) or more. When the spacing is about $3/8$ inch to 2 inches (0.95 to 5.1 cm), they are called "bar screens." Bar screens are used to screen the influent flow on a continuous basis and are usually mechanically cleaned. Usually racks are found in bypass channels where flows are diverted when bar screens are being serviced or repaired. Bar racks are manually cleaned due to their infrequent use.

Various other mechanical screening methods are in use involving actual coarse screens or perforated sheet metal. These units are automatically cleaned with scrapers, rotating brushes, water sprays, or air jets. The screens may be in the form of belts, discs, or drums set in a channel so that the wastewater flows through the submerged portion. The collected debris is removed as it passes the brushes or sprays.

---

[2] *Clarifier (KLAIR-uh-fire).   Settling Tank, Sedimentation Basin. A tank or basin in which wastewater is held for a period of time, during which the heavier solids settle to the bottom and the lighter material will float to the water surface.*

[3] *Digester (die-JEST-er).   A tank in which sludge is placed to allow decomposition by microorganisms. Digestion may occur under anaerobic (more common) or aerobic conditions.*

[4] *Inorganic Waste.   Waste material such as sand, salt, iron, calcium, and other mineral materials which are only slightly affected by the action of organisms. Inorganic wastes are chemical substances of mineral origin; whereas organic wastes are chemical substances usually of animal or plant origin.*

[5] *Pre-aeration.   The addition of air at the initial stages of treatment to freshen the wastewater, remove gases, add oxygen, promote flotation of grease, and aid coagulation.*

TREATMENT PROCESS   FUNCTION

PRELIMINARY TREATMENT

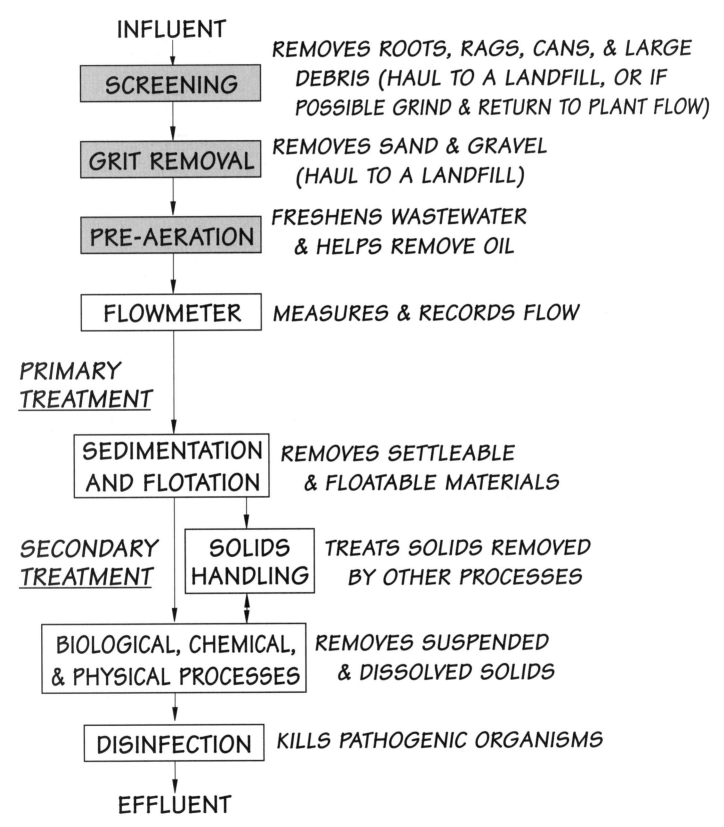

INFLUENT

SCREENING — REMOVES ROOTS, RAGS, CANS, & LARGE DEBRIS (HAUL TO A LANDFILL, OR IF POSSIBLE GRIND & RETURN TO PLANT FLOW)

GRIT REMOVAL — REMOVES SAND & GRAVEL (HAUL TO A LANDFILL)

PRE-AERATION — FRESHENS WASTEWATER & HELPS REMOVE OIL

FLOWMETER — MEASURES & RECORDS FLOW

PRIMARY TREATMENT

SEDIMENTATION AND FLOTATION — REMOVES SETTLEABLE & FLOATABLE MATERIALS

SECONDARY TREATMENT

SOLIDS HANDLING — TREATS SOLIDS REMOVED BY OTHER PROCESSES

BIOLOGICAL, CHEMICAL, & PHYSICAL PROCESSES — REMOVES SUSPENDED & DISSOLVED SOLIDS

DISINFECTION — KILLS PATHOGENIC ORGANISMS

EFFLUENT

Fig. 4.1   Flow diagram of typical plant

These devices usually are not used in preliminary treatment, but are installed at the outlets of clarifiers or chlorine contact basins and at industrial pretreatment facilities. To maintain screening equipment, look for specific information in the manufacturer's literature, the plant *O & M MANUAL*,[6] or in Chapter 15, "Maintenance."

4.20 Safety Around Bar Screens and Racks

Whenever you work around open channels or tanks, be careful not to trip or slip and fall into the wastewater or moving machinery. Falling into wastewater exposes you to disease, to the possibility of drowning, and obviously to a very unpleasant bath. Stay behind guardrails whenever possible. When working on mechanical equipment, tag operating controls and lock power off, and keep the key with *YOU*. Run mechanical equipment only when the guards are in place over moving parts. If you must clean or service operating equipment that may expose you to a hazard, use an extension tool, *NOT YOUR HAND!* The area should be posted "No Smoking" because of the possibility of explosive materials and gases from industrial discharges in the plant influent.

Before starting to rake material from a manually cleaned bar screen, examine the area for objects or structures that might interfere with the rake handle and knock you off balance. Determine if there are any guardrails, corners of buildings or diversion structures, light posts or overhead lights, or electrical wires that the end of the rake might hit. Do not stand on a slippery surface while raking material.

Back injuries, hernias, and muscle strains can occur from pulling too hard when lifting inlet or outlet gates or pulling heavy, water-logged debris from the racks. Never attempt to lift gates or rake debris that requires more strength than you can exert safely. When lifting heavy objects, always keep your back straight, bend at your knees, and lift with your leg muscles.

4.21 Manually Cleaned Bar Screens

A manually cleaned bar screen is shown in Figures 4.2 and 4.3, and the purposes of the parts are summarized in Table 4.1. Bar screens require frequent attention. As debris collects on the bars, it blocks the channel and causes wastewater to back up into the sewer. The more debris that collects on the bars, the greater the *HEAD LOSS*[7] through the bar screens. As the flow backs up, *ORGANIC WASTES*[8] tend to settle out in the channel and sewer, depleting the *DISSOLVED OXYGEN*[9] in the wastewater. Subsequently, *SEPTIC*[10] conditions develop.

The septic conditions produce *HYDROGEN SULFIDE*[11] which has a rotten egg odor; causes corrosion to concrete, metal, and paint; and also sometimes produces a toxic and explosive atmosphere in a poorly ventilated room or pit outdoors. If cleaning of the bar screens is infrequent, wastewater can back up and overflow the influent channel or flow through the bypass channel and bar rack. In either case, larger debris will be allowed to enter downstream treatment units. When the bar screens are cleaned, a sudden rush of septic wastewater can create a "shock load" on the treatment processes. The sudden high flow may carry grit into the clarifiers or carry additional solids over the clarifier weirs, reducing the efficiency of the clarifiers and secondary treatment units. Also increased odors will most likely result. Thus, failure to keep the bar screens clean will result in the lowering of the quality of the *EFFLUENT*[12] whenever a shock load of septic wastewater is released.

Cleaning of manually cleaned bar screens may be accomplished with a rake with tines (prongs) that will fit between the bars. The debris is raked to the top of the rack or into the drainage trough (whichever is provided). After draining, the debris is placed into the container provided (a bin or garbage can).

---

[6] *O & M Manual.* **O**peration and **M**aintenance Manual. A manual that describes detailed procedures for operators to follow to operate and maintain a specific wastewater treatment plant and the equipment of that plant.

[7] *Head Loss.* An indirect measure of loss of energy or pressure. Flowing water will lose some of its energy when it passes through a pipe, bar screen, comminutor, filter or other obstruction. The amount of energy or pressure lost is called "head loss." Head loss is measured as the difference in elevation between the upstream water surface and the downstream water surface and may be expressed in feet or meters. In this case, the HEAD LOSS is the height to which the water surface must build up in front of the bar screens higher than the water surface downstream from the bar screens. The buildup of water is caused by debris collecting on the bar screen (Figures 4.3 and 4.6).

[8] *Organic Waste.* Waste material which comes mainly from animal or plant sources. Organic wastes generally can be consumed by bacteria and other small organisms. Inorganic wastes are chemical substances of mineral origin.

[9] *Dissolved Oxygen.* Molecular (atmospheric) oxygen dissolved in water or wastewater, usually abbreviated DO.

[10] *Septic (SEP-tick).* A condition produced by anaerobic bacteria. If severe, the sludge produces hydrogen sulfide, turns black, gives off foul odors, contains little or no dissolved oxygen, and the wastewater has a high oxygen demand.

[11] *Hydrogen Sulfide Gas (H_2S).* Hydrogen sulfide is a gas with a rotten egg odor. This gas is produced under anaerobic conditions. Hydrogen sulfide gas is particularly dangerous because it dulls the sense of smell so that you don't notice it after you have been around it for a while. In high concentrations, hydrogen sulfide gas is only noticeable for a very short time before it dulls the sense of smell. The gas is very poisonous to the respiratory system, explosive, flammable, colorless and heavier than air.

[12] *Effluent.* Wastewater or other liquid—raw (untreated), partially or completely treated—flowing FROM a reservoir, basin, treatment process, or treatment plant.

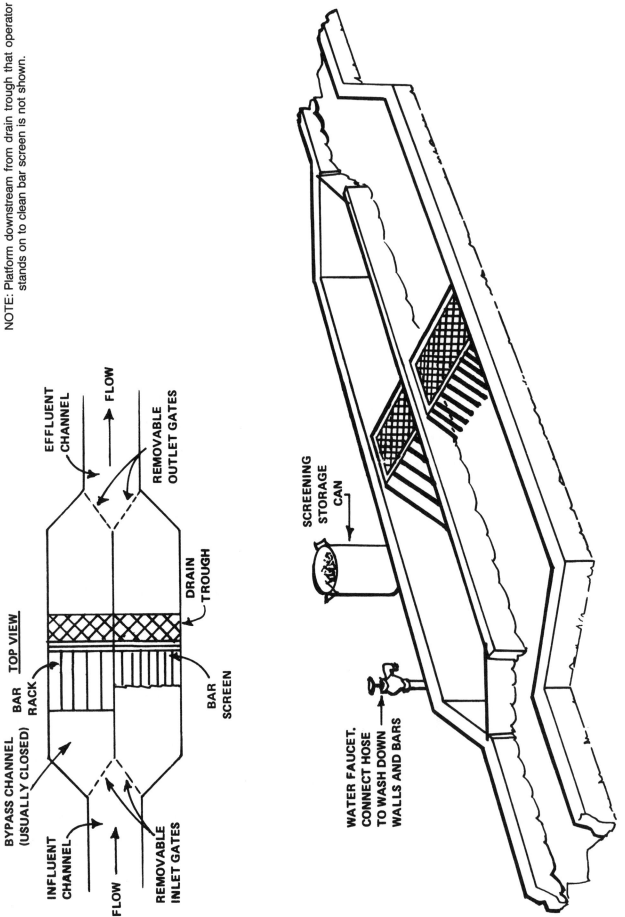

NOTE: Platform downstream from drain trough that operator stands on to clean bar screen is not shown.

EFFLUENT CHANNEL

FLOW

REMOVABLE OUTLET GATES

DRAIN TROUGH

BAR RACK   TOP VIEW

BYPASS CHANNEL (USUALLY CLOSED)

BAR SCREEN

INFLUENT CHANNEL

FLOW

REMOVABLE INLET GATES

SCREENING STORAGE CAN

WATER FAUCET. CONNECT HOSE TO WASH DOWN WALLS AND BARS

Fig. 4.2   Manually cleaned bar screen installation

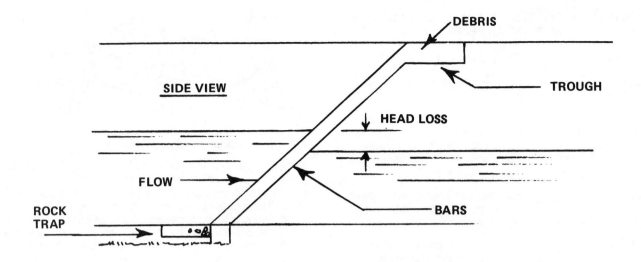

Fig. 4.3 Manually cleaned bar screen

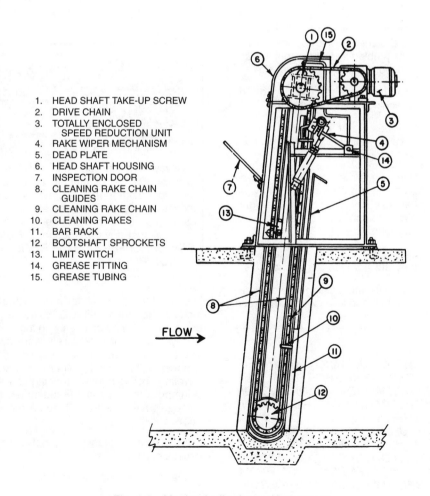

1. HEAD SHAFT TAKE-UP SCREW
2. DRIVE CHAIN
3. TOTALLY ENCLOSED
   SPEED REDUCTION UNIT
4. RAKE WIPER MECHANISM
5. DEAD PLATE
6. HEAD SHAFT HOUSING
7. INSPECTION DOOR
8. CLEANING RAKE CHAIN
   GUIDES
9. CLEANING RAKE CHAIN
10. CLEANING RAKES
11. BAR RACK
12. BOOTSHAFT SPROCKETS
13. LIMIT SWITCH
14. GREASE FITTING
15. GREASE TUBING

Fig. 4.4 Mechanically cleaned bar screen
(Permission of Smith & Loveless, Inc.)

TABLE 4.1   PURPOSE OF MANUALLY CLEANED BAR SCREENS AND PARTS

Part	Purpose
Influent or Inlet Channel	Conveys wastewater to bar racks.
Inlet Gate	Selects channel to be used. Inserted in influent channel to prevent flows from reaching screens or racks when they are being maintained or repaired.
Bar Screen	Prevents large solids from damaging pumps, plugging pipes, or reaching other treatment processes.
Bar Rack	Prevents large solids from damaging pumps, plugging pipes, or reaching other treatment processes while bar screen is out of service.
Outlet Gate	Selects channel to be used. Inserted in effluent channel to help isolate screens or racks from flows when they are being maintained or repaired.
Drainage Trough	Allows screenings or solids raked from bar screens to drain in order to reduce volume, weight, and moisture content of solids.
Screening Storage Can	Container in which to store dewatered screenings to reduce fly and odor problems until screenings are transported to the disposal site.
Rake	Tool used to remove solids or screenings from bar screens.
Hose Bib (Water Faucet)	Provides water under pressure to hose for washing down walls, floor and bar screens.
Disposal of Screenings	Screenings may be disposed of by burning, burial on site, or may be hauled off by a waste removal (garbage) company for burial in a sanitary landfill.
Gate Storage Rack	Stores inlet and outlet gates when they are not inserted in channels.
Platform	Provides safe location where operator can stand and clean bar screen with a rake.

EXTREME CAUTION SHOULD BE TAKEN WHEN RAKING THE SCREEN—FOOTING MAY BE POOR DUE TO THE WATER AND GREASE UNDERFOOT, LACK OF ENOUGH ROOM TO STAND, AND LOCATION OF THE TROUGH OR BUCKET FOR THE DEBRIS. YOU SHOULD INSPECT THIS AREA VERY CAREFULLY TO SPOT HAZARDS AND TAKE CORRECTIVE ACTION. GOOD HOUSEKEEPING, A GUARDRAIL, A HANGER OR OTHER STORAGE FOR THE RAKE, GOOD FOOTING, AND ATTENTION TO OTHER IMPORTANT ITEMS WILL GREATLY REDUCE THE POSSIBILITY OF INJURY.

4.22 Mechanically Cleaned Screens

Mechanically cleaned screens (Figure 4.4, page 69) overcome the problem of wastewater backing up and greatly reduce the time required to take care of this part of your plant. There are various types of mechanisms in use, the more common being traveling rakes that bring the debris up out of the channel and into hoppers or other debris containers. You should keep these units well lubricated and adjusted. Lubrication and adjustment procedures usually will be found in the manufacturer's literature furnished with the equipment. Further information may be found in the plant O & M manual or in the "Maintenance" chapter of this manual (Volume II). A few minutes spent in proper maintenance procedures can save hours or days of trouble and help to keep your plant operating efficiently.

Occasionally some debris that the equipment cannot remove will be present. Periodic checks should be made so that these materials can be removed manually. To determine if some material is stuck in the screen, *LOCK OUT POWER* to the unit and divert the flow through another channel or "feel" across the screen with a rake or similar device.

Always shut the unit off *FIRST. NEVER* reach into the operating range of machinery while it is running. Slow-moving equipment is especially hazardous. Because it moves slowly, it does not appear dangerous. However, most geared-down machinery is so powerful that it can crush almost any obstruction. *A HUMAN HAND, FOR INSTANCE, OFFERS LITTLE RESISTANCE TO THIS TYPE OF EQUIPMENT.*

Other items to watch for with mechanically cleaned screens include the drive units; adjustments are necessary if the cables do not wind up evenly on the drums of cable machines. A frayed cable or excessively worn chain must be replaced. The equipment should be adjusted if one end of the rake unit is riding higher than the other (causing jamming), if the unit is traveling past its normal stopping position, or if the equipment is jumping or chattering. Frequent hosing or washing down of this equipment will prevent the buildup of slimes with resultant odors and flies.

If the rake mechanism will not move, look for two possible causes of the problem:

1. Rake mechanism jammed, or

2. Equipment broken.

When the rake will not move, try resetting the circuit breaker. If nothing happens, turn or push equipment switches to OFF, lock out, and tag the electrical breakers. Divert wastewater to another channel or screen. Look for and remove obstructions that have jammed the rake mechanism.

Whenever a mechanically cleaned bar rack has stalled or the *SHEAR PIN*[13] snapped, be very careful when you attempt to uncouple the chain drive. Be sure to remove a link from the driven side (upper length). This will allow the chain to fly outward and downward, toward the floor. *NEVER* remove a link from the lower length because it will fly outward and upward toward *YOU*.

Look for broken equipment when the motor runs but the rack mechanism does not operate. The problem may be caused by a broken chain, cable, or shear pin. If the motor is running, the *LIMIT SWITCH*[14] would not be the cause of the problem. To repair or replace a broken part, turn or push equipment switches to the OFF position, lock out, and tag the electrical breakers. Divert wastewater to another channel or screen. Perform the necessary repairs and place the unit in service again. Observe the unit for proper operation.

4.23 Operational Procedures

Routine operation of screens and racks will depend on the size of the treatment plant (number of screens and racks), amount of debris in the wastewater, quantity of wastewater, and the head loss across the unit. If the allowable backup (head loss) of wastewater is not specified in the plant O & M manual, a good starting point is a limit of 3 inches (7.6 cm). Cleaning the screens or racks and changing the number of units in service are basic ways to keep the head loss below or near the desired level and at a minimum. These methods should be used daily and adjustments made to match the flow.

Mechanically cleaned screens may incorporate automatic controls[15] that operate the cleaning device whenever the head loss reaches or exceeds a preselected level. Other screen-cleaning devices may operate on a timer that starts the device, allows it to run for a specified time, and then shuts it off for a selected time period. These units usually have a "continuous run" position that allows the device to operate continuously when necessary.

To place a bar rack or screen in service, start up the unit, open the outlet gate, and then open the inlet gate (Figure 4.2).

To remove a bar rack or screen from service, close or insert the inlet gate and then the outlet gate. Turn the mechanical unit off, drain the channel, and wash off the unit. If your plant has two screens in series, one screen may be removed using a hoist, washed off, and returned to service without diverting the flow.

Storms or sewer cleaning operations by maintenance crews may cause a sudden surge of wastewater and debris and a resultant greater head loss through the screens. Under these conditions, quick action by the operator in adjusting the clean-ing frequency and the number of channels or screens in service can prevent such problems as the channels overflowing or wastewater backing up into the sewers. If enough wastewater backs up into the sewers, manhole covers may even pop open and untreated wastewater will be discharged into the streets or back up into homes. Operators must be very alert in these situations.

4.24 Disposal of Screenings

The material removed from the screens is very offensive and hazardous. This material stinks and attracts rats and flies. Burial or incineration are two common methods of disposal. *Be sure that whatever method of disposal your plant uses will not cause any adverse impact on groundwater or surface waters. Contact your regulatory agency to obtain any necessary permits or approvals for disposal procedures.*

The practice of using grinders (shredders, disintegrators) to cut up screenings and return them to the wastewater imposes an increased load on downstream treatment processes, especially surface skimming devices. This problem can be very serious for plants treating combined storm and wastewater flows when the first winter storm transports large amounts of leaves and debris to the treatment plant.

Depending on plant location and surroundings, you may find it necessary to plan ahead to locate appropriate sites for disposal of screenings. If burial is used, you should estimate how long a certain area can be used before you must find additional space for disposal. The disposal site volume divided by the daily volume of screenings produced will tell you how many days the site will last. For example, assume your plant has a flow of two million gallons per day (MGD) and that over a two-week period you remove an average of *30 GALLONS* of screenings daily. This figures out to *FOUR CUBIC FEET* (cu ft) per day. You bury the screenings each day in a pit that you estimate will hold 15 cubic yards of screenings *IN ADDITION* to the 6 inches (15 cm) of soil used to cover up the screenings.

---

[13] *Shear Pin. A straight pin that will fail when a certain load or stress is exceeded. The purpose of the pin is to protect equipment from damage due to excessive loads or stresses.*

[14] *Limit Switch. A device that regulates or controls the travel distance of a chain or cable.*

[15] *Automatic controls are discussed in Volume II, Chapter 15, "Maintenance," and Chapter 9, "Instrumentation," ADVANCED WASTE TREATMENT.*

EXAMPLE 1

(1 cu ft = about 7.5 gallons for practical purposes)[16]

Thus:

$$\text{Volume (or Filling Rate), cu ft/day} = \frac{\text{Volume, gal/day}}{7.5 \text{ gal/cu ft}}$$

$$= \frac{30 \text{ gal/day}}{7.5 \text{ gal/cu ft}}$$

$$= 4 \text{ cu ft/day}$$

You should convert gallons to cubic feet (ft^3 or cu ft) or cubic yards (yd^3, cu yd) because earthwork is figured on this basis. With this information, you are now prepared to estimate how long before the pit will be filled.

FIRST, convert the 15 cu yd (pit) capacity to cu ft:

$$\text{Pit Capacity, cu ft} = \text{Capacity, cu yd} \times 27 \frac{\text{cu ft}}{\text{cu yd}}$$

$$= 15 \text{ cu yd} \times 27 \frac{\text{cu ft}}{\text{cu yd}}$$

$$= 405 \text{ cu ft}$$

SECOND, divide the pit capacity by the daily volume of screenings to find the estimated time before the pit is full as follows:

$$\text{Time, days} = \frac{\text{Pit Capacity, cu ft}}{\text{Filling Rate, cu ft/day}}$$

$$= \frac{405 \text{ cu ft}}{4 \text{ cu ft/day}}$$

$$= \text{About 101 days}$$

Thus you have about 101 days from the time you begin to bury screenings in the pit until you will have to dig another pit provided you consider the 6 inches (15 cm) of soil used to cover up the screenings. You should keep daily records of the volume of screenings buried to be sure the daily amount stays about the same. If it increases very much, it could be due to an increase in daily wastewater flow or perhaps some unnecessary disposal of rags or other material into the sewer. You may get anywhere from 0.5 to 12 cubic feet of screenings per million gallons (3.7 to 90 liters of screenings per million liters) of wastewater flow depending on what people or industries dump into the sewers. This volume per million gallons should stay fairly constant for your plant unless something unusual is happening.

You can check on the daily flow in MGD, or during any time period, by reading your flow totalizer. The totalizer records the total flow through the plant. If you record the totalizer reading at the start and at the end of any time period, the difference is the total flow for that time period.

QUESTIONS

Write your answers in a notebook and then compare your answers with those on page 99.

4.2A Manually cleaned bar screens should be cleaned frequently to prevent which of the following:

    a. The screen from breaking
    b. Septic conditions from developing upstream
    c. A shock load on the plant when eventually cleaned
    d. Formation of hydrogen sulfide, which has a rotten egg odor and causes corrosion of concrete and paints
    e. All of these

4.2B What safety precautions should be taken when cleaning a bar screen?

4.2C What should be done *FIRST* if a problem develops in a mechanically cleaned screen?

4.2D How can screenings be disposed of?

4.2E A plant receives a flow of 4.4 million gallons (MG) on a certain day. The day's screenings are calculated to be 11 cubic feet. How many cubic feet of screenings were removed per MG of flow?

4.3 COMMINUTION (com-mi-NEW-shun)

4.30 Comminutors

Comminutors are devices that act both as a cutter and a screen. Their purpose is to shred (comminute) the solids and leave them in the wastewater. This overcomes problems of screenings disposal. Some models (Figure 4.5) are mounted in a channel and the wastewater flows through them; other models (Figure 4.6) can be mounted directly in a pipeline. The rags and other cuttable debris are shredded by cutters (teeth) until they can pass through the openings. Pieces of wood and plastic are rejected and remain on the water surface in front of the comminutor. This debris must be removed manually. Most of these units have a shallow pit in front of them to catch rocks and scrap metal. The flow to the comminutor should be shut off periodically and the debris removed from the trap. The frequency of inspecting the trap can be determined from experience. However, it is not wise to allow more than a few days between inspections.

A comminutor consists of a rotating drum with slots for the wastewater to pass through (Figures 4.7 and 4.8 and Table 4.2). Other types of comminutors have stationary slotted screens with oscillating cutters mounted on a shaft (Figures 4.9 and 4.10). Cutting teeth are mounted in rows on the drum. The teeth pass through cutter bars or "combs" with very small clearances so that a shearing action is obtained. The wastewater passes into the vertically mounted drum through the slots in the drum and flows out the bottom. A rubber seal, held in place by a bolted-down ring, prevents leakage under the drum. This seal should be checked whenever the rock and scrap metal trap is checked.

---

[16] *1 cu ft = about 6.2 Imperial gallons for practical purposes for operators in Canada and other nations using the Imperial gallon.*

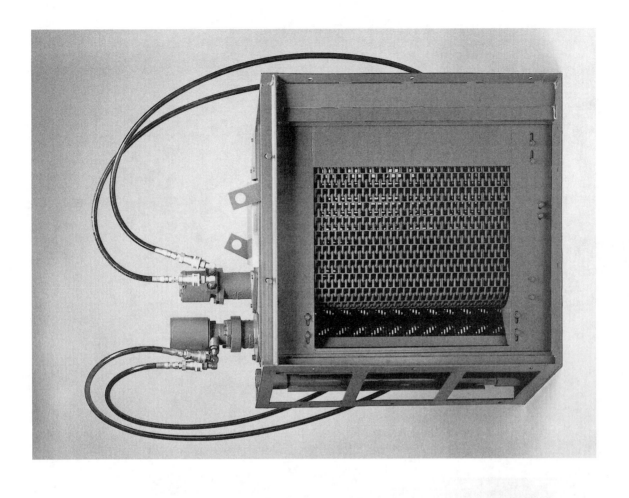

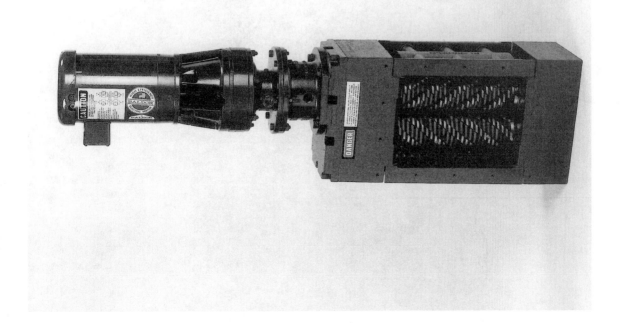

Fig. 4.5 In-channel Muffin Monster® and Channel Monster® (solids grinders)

(Courtesy of JWC Environmental)

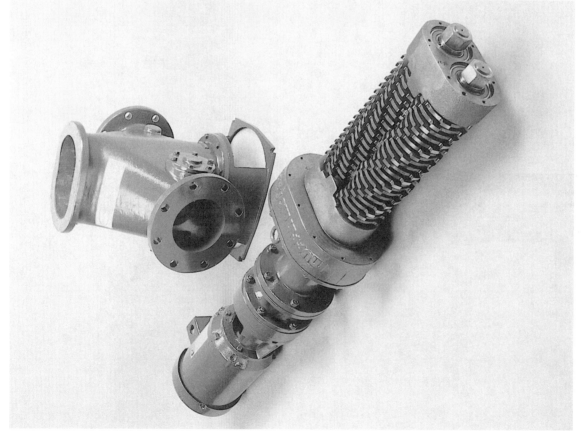

Fig. 4.6   In-line Muffin Monsters® (solids grinders)
(Courtesy of JWC Environmental)

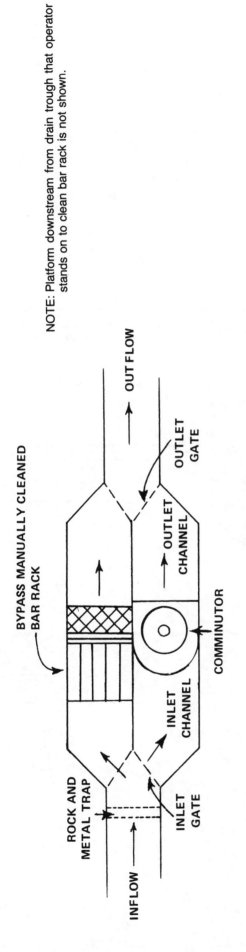

NOTE: Platform downstream from drain trough that operator stands on to clean bar rack is not shown.

OUT FLOW

BYPASS MANUALLY CLEANED
BAR RACK

OUTLET GATE

OUTLET CHANNEL

PLAN OR TOP VIEW

COMMINUTOR

INLET CHANNEL

ROCK AND METAL TRAP

INLET GATE

INFLOW

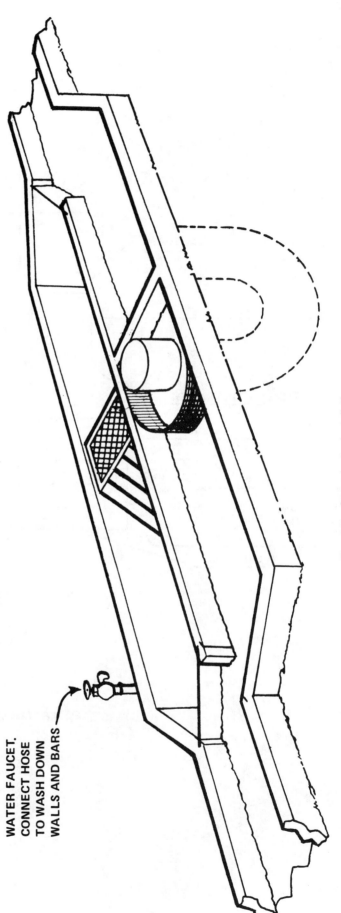

WATER FAUCET.
CONNECT HOSE
TO WASH DOWN
WALLS AND BARS

Fig. 4.7   Comminutor installation

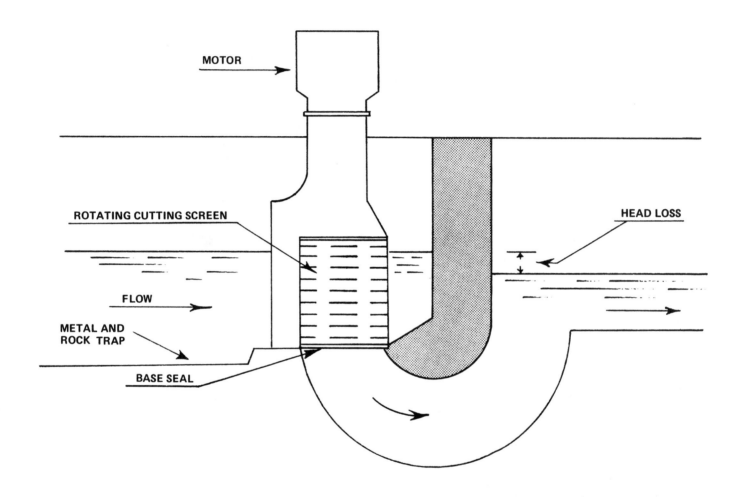

Fig. 4.8   Comminutor

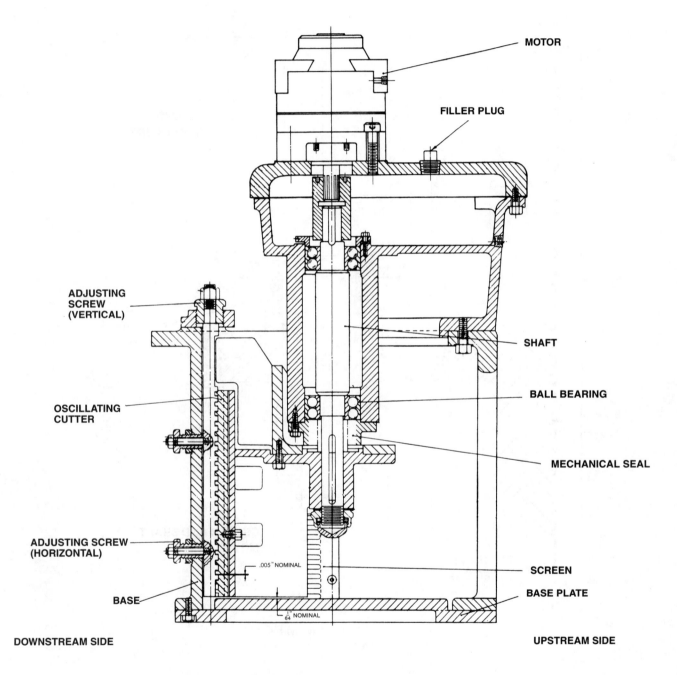

Fig. 4.9   Comminutor sectional drawing
(Courtesy Worthington Marine & Industrial Products, Inc.)

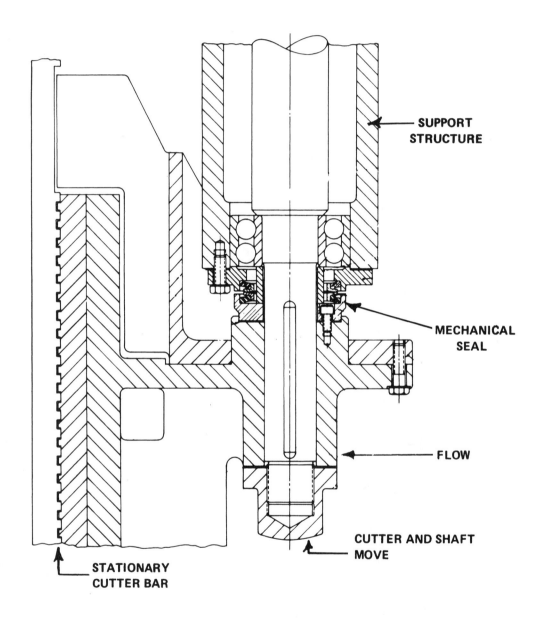

SUPPORT
STRUCTURE

MECHANICAL
SEAL

FLOW

CUTTER AND SHAFT
MOVE

STATIONARY
CUTTER BAR

Fig. 4.10   J-Ring bellows type seal
(Courtesy Worthington Marine & Industrial Products, Inc.)

TABLE 4.2 PURPOSE OF COMMINUTORS AND PARTS

Part	Purpose
Comminutor	Shreds (comminutes) solids and leaves them in wastewater.
Inlet Channel	Conveys wastewater and solids to comminutor.
Inlet Gate	Selects channel to be used, diverts wastewater and solids to manually cleaned bar screen, and prevents wastewater flowing to comminutor so equipment can be inspected, maintained, repaired, or replaced.
Outlet Gate	Prevents wastewater from flowing back to comminutor when equipment is being inspected, maintained, repaired, or replaced.
Oscillating and Stationary Cutters	Devices containing sharp blades that shred solids.
Adjusting Screw	Raises, lowers, or moves the cutting blades sideways to provide specified clearances so that a shearing action is achieved.
Mechanical Seal	Prevents wastewater from reaching and damaging bearings (Figure 4.9).
Motor and Gear Reducer	Moves cutting blades.
Bypass Manually Cleaned Bar Rack	Prevents large solids from damaging pumps, plugging pipes, or reaching other plant treatment processes while comminutor is not in service or during peak storm flows.
Metal and Rock Trap	Prevents metal and rocks from damaging cutting blades or teeth of comminutor.
Platform	Provides safe location where operator can stand and clean bar rack with a rake.

SPECIAL NOTE:

Some older comminutors outside the USA may have a *MERCURY SEAL*[17] to keep water out of the bearings.

Mercury in water can be oxidized to mercury ions which are toxic to living organisms. Consequently, the Environmental Protection Agency (EPA) has banned the use of mercury seals in comminutors and other applications.

The various manufacturers of equipment using mercury seals have designed conversion kits to provide the necessary water seals by mechanical means. Detailed drawings and instructions are furnished with each conversion kit. It is important that these kits be installed on all mercury seal-type comminutors to replace the mercury seal with a mechanical seal. Details will vary with different companies. Equipment manufacturers furnish service manuals and parts lists with each unit.

CAUTION CAUTION CAUTION

Mercury is poisonous. Routes of mercury entry to your body include inhalation, skin absorption, and eye and skin contact. Wear appropriate clothing to prevent repeated or prolonged skin contact. Wash up thoroughly and promptly when your skin is wet or contaminated with mercury. Remove gold rings and other jewelry from your hands first because they may become coated with mercury. If your ring is thus coated, it will have to be heated to burn off the mercury. If you must handle or work with mercury, be sure to work over a large tray in order to catch any spills. Breathing the fumes can be fatal or cause loss of hair and teeth. Plenty of fresh air ventilation is an absolute *MUST*. Use a respirator if you might be exposed to mercury concentrations exceeding allowable limits.

---

[17] *Mercury seals have been outlawed in the USA because of the toxic effects of mercury in the environment. All plants should have replaced all mercury seals with mechanical seals (Figures 4.9 and 4.10).*

4.31 Barminutors

There are many variations of comminuting devices. One of them has the trade name of "Barminutor" (Figures 4.11, 4.12, and 4.13 and Table 4.3). This unit consists of a bar screen made of U-shaped bars and a rotating drum with teeth and "shear bars." The rotating drum travels up and down the bar screen. Careful attention must be given to maintaining the oil level in these machines; otherwise, water may get into the bearings. Cutting edges on the cutter should be sharpened or replaced at regularly scheduled intervals. The frequency of maintenance depends on the type and abrasiveness of the wastewater being treated. Normal and abnormal operational procedures as well as start-up and shutdown procedures for barminutors are similar to those used for comminutors.

Careful attention to preventive maintenance procedures will save you a lot of trouble. Use the manufacturer's instructions, the Maintenance chapter, and the plant O & M manual to become thoroughly familiar with the equipment in your plant. When *ALL* lubrication points and oil levels are properly maintained and *ALL* adjustments and clearances are correct, the equipment may be expected to give long and trouble-free service.

If you are operating an old plant and the manufacturer's literature is no longer available, try to develop a maintenance program that appears suited for the equipment. Old records should provide you with an indication of how the equipment was maintained in the past and if the procedures worked. If nothing is available, try inspecting oil levels weekly and changing oil every 90 days with SAE 90 oil.

TABLE 4.3 PURPOSE OF BARMINUTORS AND PARTS

Part	Purpose
Barminutor	Shreds solids and leaves them in wastewater.
Bar Screen	Retains large solids in wastewater for shredding by rotating cutter.
Rotating Cutter	Moves up and down bar screen and shreds solids retained on bar screen.
Motor and Gears	Turns rotating cutter and moves rotating cutter up and down bar screen.
Counterweight (not all models have counterweights)	Reduces the energy required by the motor to move the rotating cutter up and down the bar screen.

4.32 Safety Around Comminutors and Barminutors

When working around comminutors and barminutors, use safe procedures and be aware of safety hazards. Wet walkways and dewatered channels are slippery. Take appropriate precautions so you don't slip and fall while attempting to remove floatables that did not pass through the comminutor or barminutor from the surface of the water.

Never attempt to unplug or unjam the cutter blades without *FIRST* bypassing the unit and then turning off electrical power,

locking out the breaker, and placing a tag on the breaker indicating who did it and when. The moving parts of a comminutor are especially dangerous and can quickly cut off your finger and the rotating drum of a barminutor could easily cut you. Treat this equipment with respect. Do not attempt to repair or troubleshoot electrical equipment and controls unless you are qualified and authorized to do so.

4.33 Operational Procedures

Routine operation of comminutors and barminutors (comminution units) is basically the same as the operation of screens and racks. The main factors are the number of units, the amount of debris in the wastewater, and the head loss through the unit. Look up the allowable head loss through the unit in the manufacturer's literature or the plant O & M manual. Where two or more comminution units are available, keep enough units operating to stay within the head loss limits.

Sudden high flows or heavy amounts of debris (resulting from industrial dumps, sewer line maintenance, or storm flows) may require prompt action to avoid having wastewater back up into the sewers or overflow the channels. Automatic controls on some equipment may take care of most of these situations.

In some plants a bypass channel (with a bar rack) is provided to relieve high flow conditions and problems during downtime on the comminution unit while it is being cleaned or serviced. The debris from the bar rack may be returned to the comminution unit's influent (a little at a time) after the unit is back in service.

Starting up or shutting down comminutors and barminutors is done the same way as for mechanically cleaned screens. When starting up a unit, first turn on the equipment, then open the outlet gate, and finally open the inlet gate. Always read the manufacturer's instructions first because some units require the inlet gate to be open before starting. Observe the unit to verify that it sounds normal and appears to be operating properly.

To shut down a comminution unit, reverse the procedure. After shutdown, drain the channel and hose down the channel and the equipment to reduce problems from odors and flies.

Daily, or more often if experience indicates, shut down each unit to look for cans and other debris and to check for "ropes" of rags hanging from the slotted drums or "U" bars. Presence of this type of debris indicates that the cutters may be worn or out of adjustment. If the motor stops (kicks out), it may be another indication of dull or improperly adjusted cutters. If the debris is not cut when the cutters mesh, this causes an overload on the drive motor.

On barminutors, look for accumulations of sand and gravel that could interfere with the travel of the drum. Other possible problems include overtravel and, on some models, difficulties with the cables. Overtravel occurs when the cutter drum travels too far up or too far down and reaches the bottom of the channel. Usually overtravel occurs when the limit switches on the cable drums fail to work properly and permit excessive travel and cable unwinding from the drum. If this happens cables may not wind back up on the drum evenly after unwinding. This will cause the cutter drum to travel at an angle across the screen, thus damaging both the cutters and the screen bars. Be sure that the unit stays level when it moves up and down, that the cables wind evenly on the drums, and that the cables are not frayed. A cable malfunction can result in major damage to the barminutor.

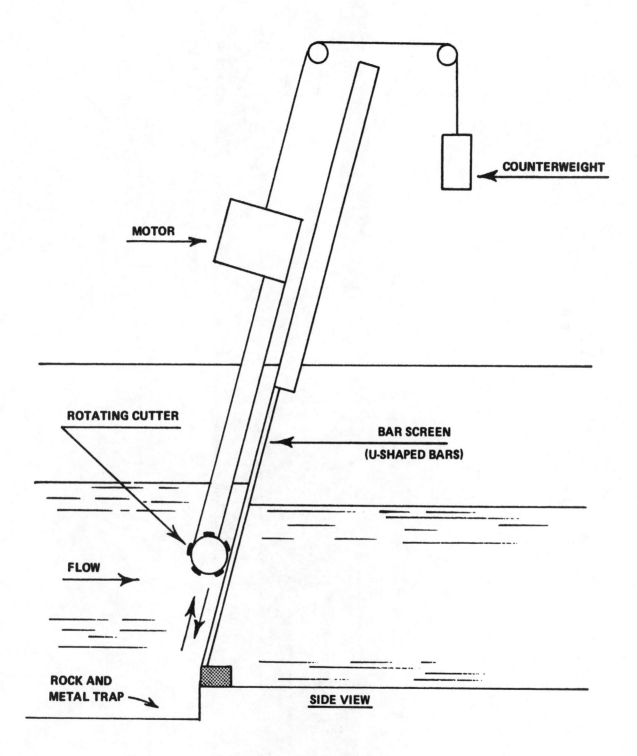

COUNTERWEIGHT

MOTOR

ROTATING CUTTER

BAR SCREEN
(U-SHAPED BARS)

FLOW

ROCK AND
METAL TRAP

SIDE VIEW

Fig. 4.11 Barminutor

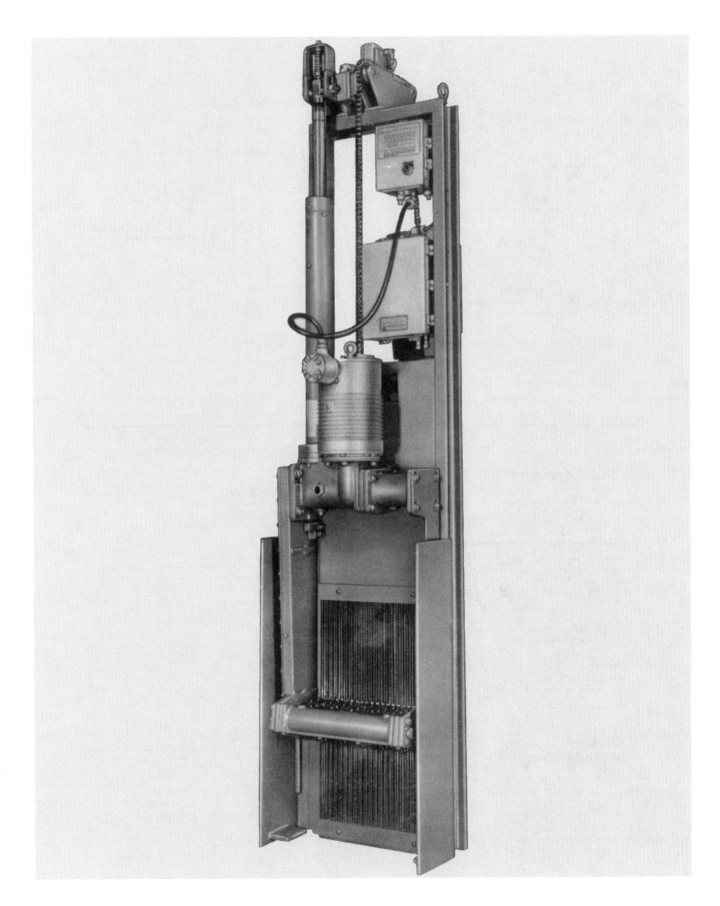

Fig. 4.12   Barminutor
(Permission of FMC Corporation, Environmental Equipment Division)

Fig. 4.13   Barminutor installation
(Permission of FMC Corporation, Environmental Equipment Division)

QUESTIONS

Write your answers in a notebook and then compare your answers with those on page 99.

4.3A   How are comminution units different from bar racks and bar screens?

4.3B   What are the advantages of comminuting machines over screens?

4.3C   What has replaced mercury seals in comminutors?

4.3D   Why is it hazardous to handle mercury?

        a. It is poisonous.
        b. Breathing fumes may be fatal.
        c. Breathing fumes may cause loss of hair and teeth.
        d. All of the above.

4.4   GRIT REMOVAL

Grit (sand, eggshells, and cinders) is the heavier mineral matter that is found in wastewater and will not decompose or "break down." This material causes excessive wear in pumps. A mixture of grit, tar, grease, and other cementing materials can form a solid mass in pipes and digesters. This mass will not move and cannot be removed by ordinary means. Consequently, grit should be removed as soon as possible after reaching the plant.

4.40   Grit Channels (Figure 4.14 and Table 4.4)

The simplest means of removing grit from the wastewater flow is to pass it through channels or tanks that allow the velocity of flow to be reduced to a range of 0.7 to 1.4 ft/sec (0.2 to 0.4 m/sec). The objective is to allow the grit to settle to the bottom while keeping the lighter organic solids moving along to the next treatment unit. Experience has shown that a flow-through velocity of one foot per second (ft/sec) is best.

Velocity is controlled by several means. With multiple-channel installations, the operator may vary the number of channels (chambers) in service at any one time to maintain a flow velocity of approximately one ft/sec in the grit channels. Another method of controlling flow velocity involves the use of proportional weirs (Figure 4.15) at the outlet for automatic regulation.

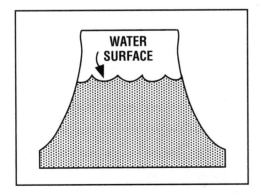

Fig. 4.15   Proportional weir

TABLE 4.4   PURPOSE OF GRIT CHANNELS AND PARTS

Part	Purpose
Grit Channel	Provides a reduction in flow velocity. The lower velocity allows grit to settle to the bottom while keeping the lighter organic solids moving along to the next treatment unit.
Grit Settling Area	Provides space for grit to settle, accumulate and be removed.
Center Wall	Separates grit channels.
Slide or Inlet Gate	Regulates number of grit channels in service in order to maintain desirable flow velocities in grit channels.
Stop or Outlet Gate	Prevents backflow. Insert when cleaning.
Weir	Controls velocity in grit channel.
Grit Storage or Grit Hopper	Accumulates and stores grit before removal and disposal.
Dewatering Drain	Drains grit channel for inspection, cleaning, and repairs.
Drain Valve	Allows draining of grit channel.

The proportional weir in Figure 4.15 will tend to *MAINTAIN* the velocity in the grit channels when the flows increase. This happens because the exit area will decrease, thus increasing the depth of water flow in the channel in direct proportion to the flow.

Flow velocities also may be regulated by the shape of the grit channels. Some grit channels have cross-sectional shapes similar to that of a proportional weir. The operator also may regulate the velocities in a grit channel by using bricks or cinder blocks to change cross-sectional shape or area. However, this can cause cleaning, maintenance, and operational problems.

A simple method for estimating the velocity is to place a stick in the channel and record the time it takes for the stick to travel a measured distance. Calculate as follows:

$$\text{Velocity, ft/sec} = \frac{\text{Distance Traveled, ft}}{\text{Time, sec}}$$

EXAMPLE 2

A stick travels 25 feet in 20 seconds; estimate its velocity.

Solution:

$$\text{Velocity, ft/sec} = \frac{\text{Distance, ft}}{\text{Time, sec}}$$

$$= \frac{25 \text{ ft}}{20 \text{ sec}}$$

$$= 1.25 \text{ ft/sec (0.38 m/sec)}$$

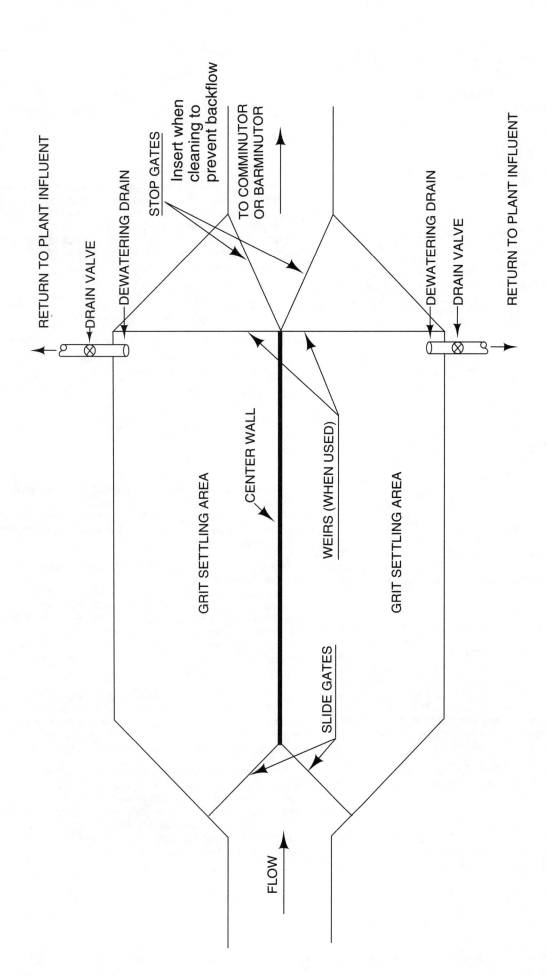

Fig. 4.14   Grit channel

The actual velocity probably will be slightly higher than your estimate, but this is a very quick way to estimate the grit channel velocity.

A more accurate method for determining the average velocity in the grit channel is based on:

1. The cross-sectional area of the flow in the grit channel, and

2. The quantity of flow (from the flowmeter).

The formula for calculating grit channel velocity is shown in the following example.

EXAMPLE 3

Assume your grit channel is two feet wide. The wastewater is flowing at a depth of one foot, and the flowmeter registers a flow of 1 MGD. The cross-sectional area of the flow is 2 sq ft (depth, ft x width, ft = 1 ft x 2 ft = 2 sq ft). The flow must be converted into cubic measure. We learned that one cubic foot equals 7.5 gallons. Thus, from calculations below, 1 MGD = 1.55 cu ft/sec (cubic feet per second, CFS):

$$1 \text{ MGD} = \frac{\left(1,000,000 \ \frac{gal}{day}\right)}{\left(7.5 \ \frac{gal}{cu \ ft} \ x \ 24 \ \frac{hr}{day} \ x \ 60 \ \frac{min}{hr} \ x \ 60 \ \frac{sec}{min}\right)} = 1.55 \text{ CFS}$$

Using this new conversion factor:

$$\text{Average Velocity, ft/sec} = \frac{\text{Flow Rate, cu ft/sec}}{\text{Area, sq ft}}$$

$$= \frac{1.55 \text{ cu ft/sec}}{2 \text{ sq ft}}$$

$$= 0.77 \text{ ft/sec} \ (0.23 \text{ m/sec})$$

To obtain this answer, we converted the flow from MGD to cu ft/sec. Then we divided the flow (1.55 cu ft/sec) by the cross-sectional area of the wastewater in the channel (2 sq ft).

EXAMPLE 4

Since we have checked the velocity, we should now determine if the length of the channel is appropriate for our flow conditions. All particles settle at different rates based on their size and weight. Most grit channels are designed to remove 0.2 mm (millimeter) size sand and all other heavier materials. Experiments have shown this size particle will settle downward at about 0.075 ft/sec (0.023 m/sec). This means that if wastewater is flowing in a channel at a depth of one foot and a particle of 0.2 mm is introduced at the surface, it will take:

$$\text{Settling Time, sec} = \frac{\text{Depth, ft}}{\text{Settling Rate, ft/sec}}$$

$$= \frac{1 \text{ ft}}{0.075 \text{ ft/sec}}$$

$$= 13.3 \text{ seconds to settle}$$

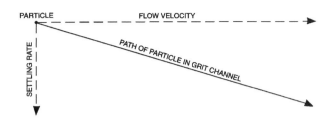

If this waste were flowing at one foot per second, it would travel for 13.3 seconds, or a distance of 13.3 feet, before the particle reached the channel bottom. If the waste were flowing at a depth of three feet in the channel, it would take 13.3 seconds/ft x 3 ft = 39.9 seconds or 39.9 feet before the particle reached the bottom. Therefore, the required length of any grit channel can be checked by using the formula:

$$\text{Length, ft} = \frac{\text{(Depth of Channel, ft)(Flow Velocity, ft/sec)}}{\text{Settling Rate, ft/sec}}$$

and for 0.2 mm sand traveling at a flow velocity of 1 ft/sec:

$$\text{Length, ft} = \frac{\text{(Depth, ft)(1.0 ft/sec)}}{0.075 \text{ ft/sec}}$$

$$= \frac{1.0 \text{ x Depth, ft}}{0.075}$$

$$= 13.3 \text{ x Depth, ft}$$

In case of dead spots where organic materials settle out and become *PUTRESCIBLE*,[18] a deflector (Figure 4.16) installed at one side may cure the trouble. (Be sure you don't create a new dead spot.) Also, certain trouble spots could be filled in with concrete.

Methods of grit removal range from use of a scoop shovel to various types of collectors and conveyors. For manually cleaned channels, the frequency of cleaning is determined by experience. If the grit builds up too much, it may interfere with the flow-through velocity, cause the wastewater to back up into the sewer, or may cause an overflow. The channel should be removed from service during the cleaning operation. This makes the job easier and no grit is washed into the downstream treatment processes.

Since there is always a small amount of organic matter in the grit channel, disposal of grit should be treated the same as disposal of screenings. Burial is the most satisfactory disposal method. Failure to quickly cover grit with six inches (15 cm) of soil results in odors and attracts flies and rats.

There are many types of mechanical grit-collector mechanisms. Common ones are chain-driven scrapers called "flights" (Figure 4.17) that are moved slowly along the bottom and up an incline out of the water to a hopper, or along the bottom to an underwater trough where a screw conveyor lifts the grit to a storage hopper or truck. Some designs use conveyor belts with buckets attached.

---

[18] *Putrescible (pew-TRES-uh-bull). Material that will decompose under anaerobic conditions and produce nuisance odors.*

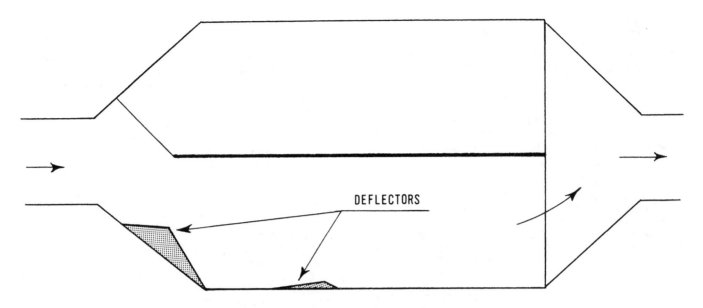

Fig. 4.16 Deflectors installed in a grit channel

CLEANING GRIT CHANNELS MANUALLY CAN BE QUITE HAZARDOUS. TAKE PRECAUTIONS AGAINST SLIPPING AND BACK STRAIN. BEWARE OF DANGEROUS GASES WHEN WORKING IN GRIT CHANNELS.

Before starting or diverting wastewater into a grit channel, be sure the channel or settling area is clear of debris, tools, and people. If the grit channel has scrapers or flights, start them and observe smooth and proper operation before diverting the wastewater into the channel. Once the wastewater is flowing through the grit channel, measure the velocity. Make any necessary adjustments in the number of channels in service and adjust the weirs or other velocity-regulating devices.

Normal operation consists of measuring velocities at low, average, and high flows. To achieve desired velocities, maintain the proper number of grit channels in service and adjust the weirs.

Grit should be removed on a daily basis. Inspect the grit for organic material which could indicate that velocities are too low. If less than expected amounts of grit are removed and the grit consists mainly of large or heavy sand, the velocities could be too high. A little organic matter in the grit usually indicates that proper velocities are being maintained.

Abnormal operating conditions can develop during rainstorms or periods of heavy snow melt. Peak canning seasons or periodic industrial dumps also can create abnormal conditions. Operating problems also occur during periods of high flows and when solids loadings become excessive. During these conditions, attempt to maintain velocities as close to 1.0 ft/sec (0.3 m/sec) as possible. Grit will have to be removed more frequently when the grit and heavy solids loadings are higher than usual.

Try to schedule grit channel shutdowns during periods of low flows. Shut down grit channels for maintenance or repairs by placing alternate or additional grit channels in service if possible. Insert slide or inlet gate and outlet stop gate. Open the drain valve and dewater the grit channel. Be careful the drain does not plug. Wash out or hose down the floor of the

Fig. 4.17 Chain-driven scrapers (flights)
(Courtesy Jeffrey Mfg. Co.)

grit channel before entering. Dewatered grit channels have slippery bottoms, so walk and work carefully.

Maintenance of grit channels consists of keeping the facilities clean and inspecting for corrosion damage and cracks in the walls and floor. Corrosion rates can be slowed by the application of protective coatings. In channels with chain and flight collectors, inspect chains and sprockets for excessive wear at least twice a year and inspect the bearings and anchor bolts every time the channel is dewatered. Motors must be lubricated and greased in accordance with the manufacturer's recommendations.

An *AERATED* grit chamber is actually a tank with a sloping bottom and a hopper or trough in the lower end (Figure 4.18 and Table 4.5). Air is injected through *DIFFUSERS*[19] located along the wall of the tank above the trough. The mixture of air and water has a lower *SPECIFIC GRAVITY*[20] than water so the grit settles out better. The rolling action of the water in the tank moves the grit along the bottom to the grit hopper. Grit is removed from the hopper by pumps or a conveyor system.

Aerated grit chambers are most frequently found at activated sludge plants where there is a readily available air supply. The pre-aeration usually helps to "freshen" the wastewater. The older wastewater becomes, the more difficult it is for aerobic organisms to treat wastes and for solids to settle. A freshening process tends to make downstream treatment processes more effective.

To start an aerated grit chamber, be sure all equipment works properly and that the grit chamber contains no debris or tools. Start air passing through the diffuser. Allow wastewater to enter the aerated chamber slowly. Adjust the air flow rates to obtain the desired circular motion in the grit chamber. Air rates also are regulated to control the size of grit and the volume of grit removed from the wastewater.

TABLE 4.5   PURPOSE OF AERATED GRIT CHAMBER AND PARTS

Part	Purpose
Aerated Grit Chamber	Removes grit.
Diffuser	Disperses air into wastewater to reduce velocities. The mixture of air and water has a lower specific gravity than water so the grit settles out better. The circular motion created by the air moves the grit to the bottom of the chamber where it can settle out.
Grit Hopper	Collects settled grit before removal.
Collector Mechanism	Removes grit from chamber for disposal. Other grit collector mechanisms include screw conveyors, buckets, or pumps.

The velocity of roll or agitation regulates the size of particles of a given specific gravity that will be removed in the grit chamber. If the velocity of roll is too great, grit will be carried out of the chamber. If the velocity of roll is too low, organic material will be removed with the grit. With proper adjustment of the quantity of air, almost 100 percent grit removal can be obtained and the grit should be well washed. Wastewater should move through the grit chamber in a circular (helical) path and should make two to three passes across the bottom of the tank at peak flows. There should be more passes across the bottom at lower flows. Wastewater should be introduced into the grit chamber in the direction of flow.

Aerated grit chambers usually have a detention time of approximately three minutes at peak flows with detention times ranging from three to five minutes.

Normal operation consists of maintaining hydraulic capacity (design flow) through each chamber and the desired circular motion. Remove grit at regular intervals or continuously from the grit hopper depending on the equipment and also on the grit loading. If scum and floating debris accumulate on the surface in dead areas, remove this material twice a day, or more often if needed.

When abnormal operating conditions exist due to high flows or heavy solids loadings, adjust air flow to the tank to maintain the desired grit removal at the operating conditions experienced. If grit and solids are not removed, they may be carried over to the primary clarifiers. If so, increased primary sludge pumping rates or more frequent or longer pumping times may be necessary.

Shutdown procedures consist of diverting wastewater flows around the aerated grit chamber, draining the facility, and stopping the air supply. Draining is not necessary during temporary shutdown, such as taking a unit out of service during low flows.

4.41   Cyclone Grit Separators

Another method of separating grit from organic matter is by the use of cyclone grit separators (Figure 4.19 and Table 4.6).

---

[19] *Diffuser.   A device (porous plate, tube, bag) used to break the air stream from the blower system into fine bubbles in an aeration tank or reactor.*
[20] *Specific Gravity.   (1) Weight of a particle, substance or chemical solution in relation to the weight of an equal volume of water. Water has a specific gravity of 1.000 at 4°C (39°F). Wastewater particles or substances usually have a specific gravity of 0.5 to 2.5. (2) Weight of a particular gas in relation to the weight of an equal volume of air at the same temperature and pressure (air has a specific gravity of 1.0). Chlorine has a specific gravity of 2.5 as a gas.*

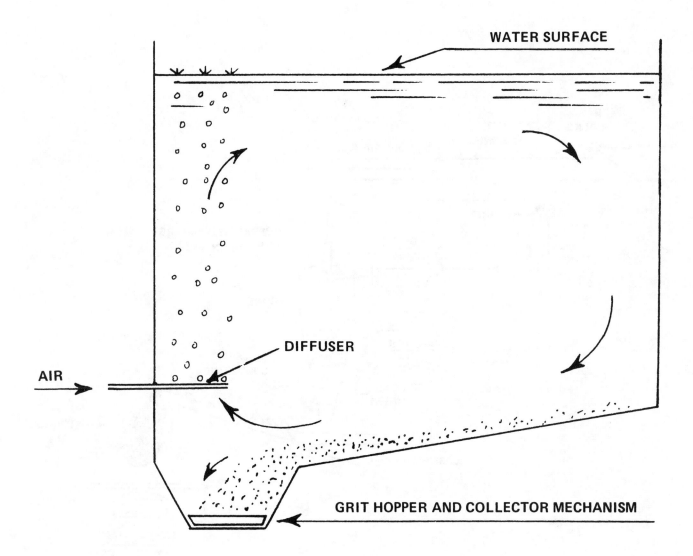

WATER SURFACE

DIFFUSER

AIR

GRIT HOPPER AND COLLECTOR MECHANISM

NOTE:
Aerated grit chambers often have agitation air systems in the grit hopper to prevent compaction of grit when grit removal is intermittent.

Fig. 4.18   Aerated grit chamber

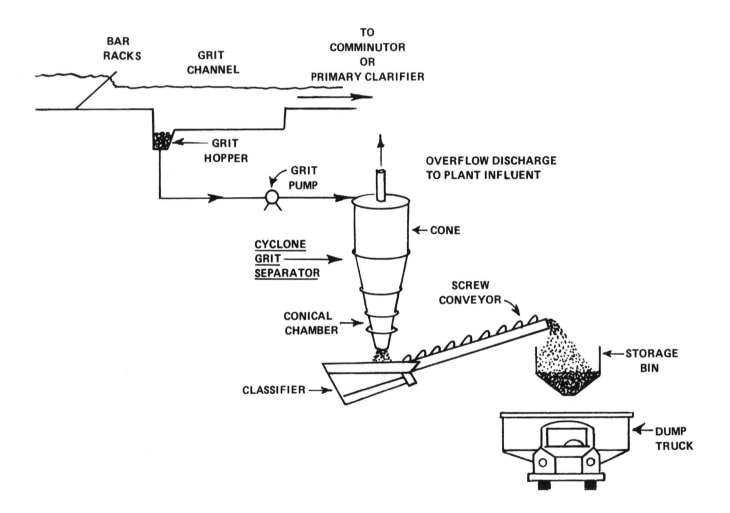

Fig. 4.19 Cyclone grit separator

Grit from mechanically cleaned grit channels or other grit removal facilities is pumped as a *SLURRY*[21] in water to the cyclone. The velocity of the slurry as it enters along the wall of the cyclone causes the slurry to spin or swirl around the inside of the cyclone. This is called the "primary vortex" (Figure 4.20). The particles of grit, being heavier than the wastewater, are forced outward to the casing of the cyclone. The grit spirals downward toward the apex or bottom of the cyclone. At the bottom, the heavier particles (grit) pass out of the cyclone through a small hole called an "orifice." The remaining lighter particles and water are carried upward (still spinning in what is called the "secondary vortex") and out the overflow discharge pipe. The primary vortex causes the heavy particles to move toward the wall of the cyclone and out the bottom. The secondary vortex moves the lighter particles toward the center of the cyclone and out the top.

The grit is usually removed from a grit classifier at the bottom of the cyclone by means of a screw conveyor to storage hoppers. Ultimately the grit may be hauled away and buried at a disposal site or incinerated.

4.410 Safety

Working around and on cyclone grit separators requires the exercise of extreme caution. Always shut off, lock out, and tag equipment before attempting to remove objects causing stoppages or to work on the equipment. Slowly moving screw conveyors and other equipment can seriously injure you by crushing or tearing your hand or leg. Sometimes the lifting of heavy cyclone parts is required from very difficult positions, which could result in sprains or strains if you are not careful. In these situations, hoists or ropes should be used whenever possible. Be very cautious on slippery surfaces because a fall may result in a serious injury.

4.411 Start-Up

Prior to initial start-up of a cyclone grit separator, inspect the area and facilities for tools and debris. Pump water to the unit to see that the piping is clear and that the pressure gage and other equipment all function properly.

Normal operation start-ups involve starting the grit conveyor and then the pumps. Once the separator is operating, make the necessary adjustments as discussed in the next section under *OPERATION*.

4.412 Operation

FEED: The amount of water being pumped with the grit is important. In general, dilute feed slurries will allow separation of smaller particles. Thicker slurries will result in a slower rotational motion and reduced capacity. Inlet pressure is critical and should be maintained as close as possible to manufacturer's recommended pressure.

UNDERFLOW DISCHARGE: The underflow discharge from the bottom (apex) should appear as a hollow cone shape. If the discharge is too heavy, it appears as a rotating solid spiral (looks like a rope) and a heavily overloaded condition will appear as a straight stream lacking spiral motion. No underflow could mean the feed slurry is too thick or debris (such as rags) has plugged the apex.

FEED INLET PRESSURE AND OVERFLOW DISCHARGE: The inlet pressure affects efficiency and must be maintained within the limits prescribed by the manufacturer. The overflow discharge is through an adjustable orifice called the "vortex

TABLE 4.6 PURPOSE OF CYCLONE GRIT SEPARATOR AND PARTS

Part	Purpose
Cyclone Grit Separator	Separates grit from wastewater and organic material.
Grit Pump	Pumps grit from hopper to cyclone grit separator.
Entrance Chamber	Introduces grit to cyclone.
Conical Chamber	Provides enclosed vessel (tank) where grit is separated from water and organic matter.
Overflow Discharge Pipe	Carries wastewater and organic matter back to plant influent.
Grit Classifier	Removes organics from grit by washing.
Screw Conveyor	Moves grit from grit hopper under cyclone to grit storage bin.
Grit Storage Bin	Stores grit until loaded into dump truck for hauling to disposal site.
Screw Conveyor Drive Motor	Turns screw conveyor.

finder." Inlet pressure and overflow discharge are closely related functions. The optimum settings may be found quickly by making the adjustments just discussed while observing the operation of the unit.

4.413 Shutdown

To remove the cyclone grit separator from service, shut down the facility by turning off the supply pump that feeds the cyclone.

Allow the cyclone to drain. Hose down and wash out the overflow discharge, grit classifier and screw conveyor. Turn off the grit classifier and screw-drive motor.

4.414 Maintenance

Maintenance consists of applying procedures outlined in your plant O & M manual and manufacturer's literature for maintenance of the cyclone, grit pump, screw conveyor, and screw conveyor drive motor. The cyclone liner and screw conveyor must be replaced when excessive wear prevents proper adjustments and the equipment no longer functions as intended.

---

[21] *Slurry (SLUR-e). A thin, watery mud or any substance resembling it (such as a grit slurry or a lime slurry).*

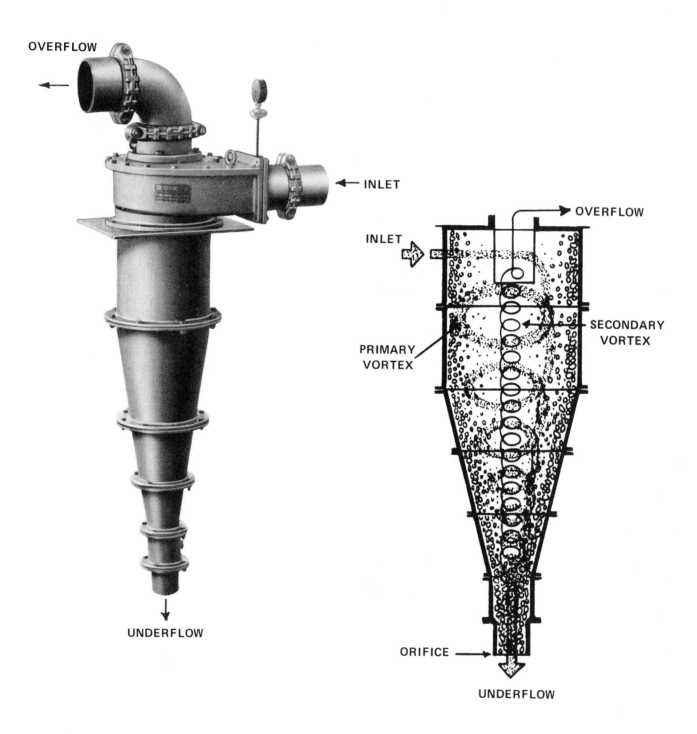

Fig. 4.20 How a grit cyclone separator works
(Permission of WEMCO Division, Envirotech Corporation)

QUESTIONS

Write your answers in a notebook and then compare your answers with those on pages 99 and 100.

4.4A Grit is composed mostly of which of the following substances?

a. Grease
b. Sand
c. Rubber Goods
d. Eggshells
e. Wood

4.4B Why bother to remove grit?

4.4C How can you control the velocity in a grit channel in order to maintain velocities within a range of approximately 0.7 to 1.4 fps?

4.4D A stick travels 20 feet in 40 seconds in a grit channel.

a. What is the velocity in the channel?
b. What corrective action should be taken, if any?

4.4E Assume you wish to calculate the velocity in the grit channel at your plant's peak flow. Examining the flow charts, you determine that peak flows are usually about 2.75 MGD. The grit channel is three feet wide, and the flow depth is 17 inches at peak flow. What is the velocity in the grit channel under these conditions?

4.4F List the safety hazards that might be encountered while manually cleaning a grit channel.

4.4G What is the purpose of a cyclone grit separator?

4.4H List the possible safety hazards an operator may encounter when working around a cyclone grit separator.

4.42 Grit Washing

A grit channel with a slower flow velocity than recommended may allow large quantities of organic matter to settle out with the grit. This heavy, coarse mixture of grit and organic material is called *DETRITUS*.[22] In some plants grit channels are called "detritus tanks." Organic matter may be separated from the grit by washing the detritus to re-suspend the organic matter.

In some cases the grit is used as fill material. Under these conditions it is necessary to wash the grit. Figure 4.21 shows a typical grit washer, and the purposes of the parts are summarized in Table 4.7. Grit settles to the bottom and is removed by a screw conveyor (or other device), while the velocity created by the impeller suspends the lighter organic materials so that they flow over the outlet weir.

When working around a grit washer, avoid slippery areas where you could slip and fall or even slip into the grit washer. Always turn off, lock out, and tag the controls and main breaker of the screw conveyor before attempting to remove stuck material and to inspect or repair the screw conveyor.

Before pumping grit or wash water to a grit washer, be sure both the screw conveyor and screw-type impeller work properly. Under normal operating conditions, pump grit to the washer

at regular intervals as necessary. Operate the screw conveyor during the grit washing operation. When the grit washing procedure is completed, turn off the equipment.

To shut down the grit washer for inspection, maintenance, or repairs, drain the facility and wash down the walls and conveyor. Treat the empty washer as a confined space and provide adequate ventilation. Be sure two people carefully observe and maintain verbal or visual contact with anyone who enters an empty grit washer for any reason in order to rescue this person if necessary. Before entering, test the area for oxygen deficiency/enrichment, combustible gases, and toxic gases (hydrogen sulfide). Monitor the atmosphere at frequent enough intervals to ensure that safe conditions are maintained while anyone is in the grit washer. Wear a safety harness and have a self-contained breathing apparatus nearby for rescue purposes. If the development of a dangerous atmosphere (oxygen deficiency/enrichment, explosive conditions, or toxic gases) is likely, additional confined space safety precautions must be used and a confined space entry permit may be required. See Volume II, Chapter 14, "Plant Safety," for additional confined space safety procedures.

Maintenance consists of inspecting the entire facility for corrosion damage and examining moving parts for wear. Oil and grease the equipment in accordance with the plant O & M manual and the manufacturer's instructions.

TABLE 4.7   PURPOSE OF GRIT WASHER AND PARTS

Part	Purpose
Grit Washer	Washes organic material out of grit and sand.
Inlet Chamber	Mixes grit and wash water and introduces mixture into grit washer.
Wash Water	Helps separate organic material from grit and sand.
Screw-Type Impeller	Circulates contents of grit washer and brings water and separated organics to surface of washer for removal.
Motor	Turns screw-type impeller.
Outlet	Removes water and organics from grit washer.
Screw Conveyor	Conveys grit from bottom of grit washer to hopper for transfer to truck for disposal.

QUESTIONS

Write your answers in a notebook and then compare your answers with those on page 100.

4.4I Why is it necessary or desirable to "wash" grit?

4.4J List the maintenance steps for a grit washer.

---

[22] *Detritus* (dee-TRY-tus).   *The heavy, coarse mixture of grit and organic material carried by wastewater. Also called grit.*

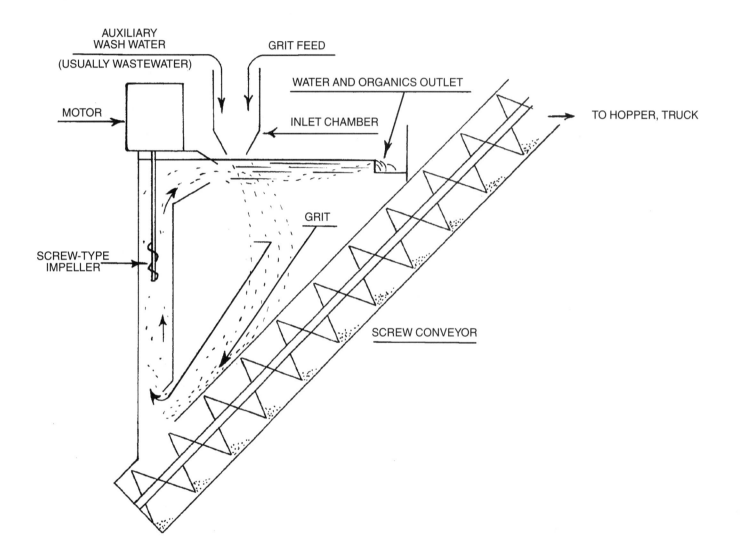

AUXILIARY
WASH WATER

(USUALLY WASTEWATER)

GRIT FEED

MOTOR

WATER AND ORGANICS OUTLET

INLET CHAMBER

TO HOPPER, TRUCK

GRIT

SCREW-TYPE
IMPELLER

SCREW CONVEYOR

Fig. 4.21 Grit washer

4.43 Quantities of Grit

Treatment plants having well-constructed separate wastewater collection systems can usually expect to average 1 to 4 cu ft of grit per million gallons (7.5 to 30 liters of grit per million liters).[23] These quantities have been rising in recent years due to household garbage grinders. They can also be expected to increase during storm periods.

Plants receiving waste from combined collection systems can expect to average 4 to 15 cu ft of grit per million gallons (30 to 110 *L* per million liters) with peaks many times higher during storm periods. Grit collected during storm periods has been reported at over 500 cu ft per million gallons (4 cu m per million liters), probably the result of flow from broken sewers or open channels.

Records of grit quantities should be kept in the same manner as for screenings.

QUESTION

Write your answer in a notebook and then compare your answer with the one on page 100.

4.4K   Your plant has an average flow of 2.0 MGD. An average of 4 cu ft of grit is removed each day. How many cu ft of grit per MG of flow are removed?

4.44 Disposal of Grit

Final disposal of grit is by burial. This may be in a sanitary landfill or other types of landfill operations. Regardless of the method, at least 6 inches (15 cm) of soil should be placed over the grit to keep out rats and flies. *Be sure that whatever method of disposal your plant uses will not cause any adverse impact on groundwater or surface waters. Contact your regulatory agency to obtain any necessary permits or approvals for disposal procedures.*

4.45 Pre-Aeration

Pre-aeration is a wastewater treatment process used to improve grit removal efficiency, to freshen wastewater, to remove gases, to add oxygen, to promote flotation of grease, and to aid coagulation. The freshening of wastewater improves the effectiveness of downstream treatment processes. The pre-aeration process is usually located before primary sedimentation (Figure 4.1). Other processes used to accomplish freshening include *OZONATION*[24] and prechlorination.

Pre-aeration consists of aerating wastewater in a channel or separate tank for 10 to 45 minutes. Aeration may be accomplished by either mechanical surface aeration units or diffused air systems.[25] Air application rates with a diffused air system normally range from 0.5 to 1.0 cu ft of air per gallon of wastewater (3.75 to 7.5 cu m air/cu m wastewater) treated.

4.5 OPERATIONAL STRATEGY

This chapter has covered basic concepts in the operation and maintenance of equipment used to remove grit, rags, coarse materials, and other debris from the wastewater before it enters the treatment plant. Now let's consider some things to do and to watch for when a piece of equipment breaks down or the plant becomes overloaded by excessive flows.

If screens or comminutors are overloaded or bypassed, you can expect the following problems:

1. Sticks and rags can foul the raw sludge pumps for the primary clarifiers;

2. Debris can plug the orifices in trickling filter distributors;

3. Debris can interfere with air diffusers in the aeration tanks of activated sludge plants;

4. Floating debris can appear in the chlorine contact basin and leave the plant in the final effluent; and

5. Solids can plug return sludge pumps and flowmeters in activated sludge plants.

If grit channels are bypassed or overloaded, grit can reach the primary clarifiers. To reduce resulting problems, increase the pumping of raw sludge to keep the clarifier hopper, piping, and pumps from plugging.

You can get into other types of trouble by increasing pumping unless your plant is equipped with sludge thickeners. Overpumping from the clarifier will affect the digesters. The grit will occupy valuable space; the extra water will lower the temperature of the digesters or require more energy for heating and may wash the *ALKALINITY*[26] *BUFFER*[27] out of the digester. Also, excessive amounts of supernatant from the digesters will add to the organic load on the plant. Therefore, if you are faced with this problem, you should look for ways to relieve

---

[23] *Uniform reporting of results is important. Everyone should use the same units.*

[24] *Ozonation (O-zoe-NAY-shun).   The application of ozone to water, wastewater, or air, generally for the purposes of disinfection or odor control.*

[25] *See Chapter 8, "Activated Sludge," for a discussion of aeration facilities.*

[26] *Alkalinity (AL-ka-LIN-it-tee).   The capacity of water or wastewater to neutralize acids. This capacity is caused by the water's content of carbonate, bicarbonate, hydroxide, and occasionally borate, silicate, and phosphate. Alkalinity is expressed in milligrams per liter of equivalent calcium carbonate. Alkalinity is not the same as pH because water does not have to be strongly basic (high pH) to have a high alkalinity. Alkalinity is a measure of how much acid must be added to a liquid to lower the pH to 4.5.*

[27] *Buffer.   A solution or liquid whose chemical makeup neutralizes acids or bases without a great change in pH.*

and/or to avoid the problem in the future. For example, the raw sludge or digester supernatant could be pumped to a standby tank or pond whenever the grit channels must be bypassed or are overloaded.

To remove floating debris from chlorine contact basins and the final effluent, try hand skimming with hardware cloth nets. Also hardware cloth screens may be installed in the clarifier effluent troughs and final effluent channel.

These problems and possible actions to take show how important it is for the operator to do everything possible to keep preliminary treatment equipment well maintained. Obviously, breakdowns of preliminary treatment equipment may be expected to cause a lot of problems throughout the plant. In spite of good operation and maintenance, failures may still occur.

You should consider what to do when you have a failure at your plant. You must plan in advance what to do to correct the situation or to adjust the other treatment processes. For example, you can have screens made and emergency storage facilities prepared to handle wastewater bypasses and overflows. Advance preparation for emergencies can save a lot of extra work and make it possible to keep your plant operating efficiently under unfavorable circumstances.

4.6 DESIGN REVIEW

When reviewing plans and specifications, operators can be very helpful to design engineers by thinking of how they plan to operate and maintain each treatment process and facility. This section lists a few items operators should study on plans for expansion of existing facilities or construction of new treatment plants.

4.60 Racks and Screens

1. Is there room for a rake when removing screenings? Will the handle of the rake hit any buildings, overhead wires, light posts, or overhead lights?

2. Is there provision for a good standing place that won't become slippery?

3. Is there some place to drain and store screenings?

4. Would a screw press be effective for dewatering screenings?

5. Where is the disposal site for the screenings?

6. How many channels and racks or screens are there, and what is the capacity of each? What will happen during peak storm flows?

7. What are the standby units?

8. What are the Operation and Maintenance (O & M) requirements of the facilities?

9. Are there provisions for a hoist for removal of screenings if facility is located below grade?

10. Is there an adequate dock facility for loading containers on a truck for disposal?

11. Is there adequate water under sufficient pressure to hose down machinery and area?

12. Are sufficient spare parts provided in accordance with manufacturer's recommendations?

4.61 Grit Removal Facilities

1. Are there guardrails around the grit chambers or grit channels?

2. Are there provisions for taking the grit channels or grit chambers out of service and for draining?

3. How many channels will be required for average and wet weather flows? What will happen during peak storm flows?

4. Are there standby units if there are high flows or if one unit is out for repairs during a storm?

5. How easily can the units be cleaned?

6. Can the grit hoppers be flushed easily?

7. Are there grit storage facilities and are they adequate?

8. Are there grit dewatering capabilities?

9. Are there provisions for skimming floatables from water surfaces?

10. Have items 9 through 12 in Section 4.60, "Racks and Screens" been checked for this section too?

4.62 Wet Wells

1. Explosion-proof wiring should be installed in areas such as the wet well, screening room, and any other enclosed space. Floating fuels such as gasoline or fuel oil can enter a wet well, evaporate, and form an explosive mixture with air. Significant amounts of fuel may enter a sanitary sewer system unintentionally by being dumped or drained into a storm sewer system that is part of a combined collection system or through underground fuel line leaks. *EXPLOSIMETERS*[28] are strongly suggested because they can sound an alarm before explosive conditions are reached.

2. Ventilation is very important to control toxic gases and corrosion, as well as explosive conditions. Forced air circulation creates a positive air pressure on wastewater in a wet well and tends to keep toxic gases in solution. This procedure is considered superior to an exhaust gas system.

3. If floating oils can be expected in influent wastewaters, skimmers can be helpful in wet wells.

QUESTIONS

Write your answers in a notebook and then compare your answers with those on page 100.

4.5A   What kinds of problems can be created for the operator when screens or comminutors are overloaded or bypassed?

4.6A   What items would you check when reviewing the plans and specifications for racks and screens?

---

[28] *Explosimeter.   An instrument used to detect explosive atmospheres. When the **L**ower **E**xplosive **L**imit (LEL) of an atmosphere is exceeded, an alarm signal on the instrument is activated. Also called a combustible gas detector.*

4.7 ARITHMETIC ASSIGNMENT

Turn to the Arithmetic Appendix at the back of this manual and read all of Section A.4, *METRIC SYSTEM*.

In Section A.13, *TYPICAL WASTEWATER TREATMENT PLANT PROBLEMS*, read and work the problems in the following sections:

1. A.130, Flows, and

2. A.131, Grit Channels.

4.8 ADDITIONAL READING

1. *MOP 11*, Chapter 18, "Preliminary Treatment." *

2. *NEW YORK MANUAL*, Chapter 3, "Preliminary Treatment." *

3. *TEXAS MANUAL*, Chapter 9, "Screens, Grinders, Grit Chambers and Grease Traps."*

* Depends on edition.

4.9 METRIC CALCULATIONS

This section contains the solutions to all problems in this chapter using metric calculations.

4.90 Conversion Factors

MGD x 3,785	= cu m/day
cu m/day x 0.000264	= MGD
gallons x 3.785	= liters
liters x 0.264	= gallons
1,000 L	= 1 cu m
ft x 0.3048	= m
m x 3.281	= ft
cu ft x 28.32	= liters
L x 0.035315	= cu ft

4.91 Problem Solutions

EXAMPLE 1—METRIC

A wastewater treatment plant has an average flow of 8,000 cubic meters per day. An average of 120 liters of grit is removed each day. How many liters of grit are removed per cubic meter of flow?

Known	Unknown
Flow, cu m/day = 8,000 cu m/day	Grit Removed, L/cu m
Grit, L/day = 120 L/day	

1. Calculate the grit removed in liters of grit per cubic meter of flow.

$$\text{Grit Removed, } L/\text{cu m} = \frac{\text{Grit Removed, } L/\text{day}}{\text{Flow, cu m/day}}$$

$$= \frac{120 \ L/\text{day}}{8,000 \ \text{cu m/day}}$$

$$= 0.015 \ L/\text{cu m}$$

$$\text{or} \quad = 1.5 \times 10^{-2} \ L/\text{cu m}$$

Screenings are buried in a pit that will hold 12 cubic meters of screenings *IN ADDITION* to the soil used to cover up the screenings. How many days will the site last?

Known	Unknown
Flow, cu m/day = 8,000 cu m/day	Time to Fill Pit, days
Screenings, L/day = 120 L/day	
Pit Capacity, cu m = 12 cu m	

1. Calculate the filling rate of the pit in cubic meters per day.

$$\text{Filling Rate, cu m/day} = \text{Screenings, } L/\text{day} \times \frac{1 \text{ cu m}}{1,000 \ L}$$

$$= 120 \ L/\text{day} \times \frac{1 \text{ cu m}}{1,000 \ L}$$

$$= 0.12 \text{ cu m/day}$$

2. Estimate the time to fill the pit.

$$\text{Time to Fill Pit, days} = \frac{\text{Pit Capacity, cu m}}{\text{Filling Rate, cu m/day}}$$

$$= \frac{12 \text{ cu m}}{0.12 \text{ cu m/day}}$$

$$= 100 \text{ days}$$

EXAMPLE 2—METRIC

Estimate the velocity in a grit channel if a stick travels 10 meters in 25 seconds.

Known	Unknown
Distance, m = 10 m	Velocity, m/sec
Time, sec = 25 sec	

Estimate the velocity in the grit channel.

$$\text{Velocity, m/sec} = \frac{\text{Distance, m}}{\text{Time, sec}}$$

$$= \frac{10 \text{ m}}{25 \text{ sec}}$$

$$= 0.4 \text{ m/sec}$$

EXAMPLE 3—METRIC

A grit channel is one meter wide and the wastewater is flowing at a depth of 0.5 meter. The flow is 10,000 cubic meters per day. Estimate the velocity in the grit channel in meters per second.

Known	Unknown
Width, m = 1 m	Velocity, m/sec
Depth, m = 0.5 m	
Flow, cu m/day = 10,000 cu m/day	

1. Convert flow from cubic meters per day to cubic meters per second.

$$\text{Flow, } \frac{\text{cu m}}{\text{sec}} = \text{Flow, } \frac{\text{cu m}}{\text{day}} \times \frac{1 \text{ day}}{24 \text{ hr}} \times \frac{1 \text{ hr}}{60 \text{ min}} \times \frac{1 \text{ min}}{60 \text{ sec}}$$

$$= 10,000 \ \frac{\text{cu m}}{\text{day}} \times \frac{1 \text{ day}}{24 \text{ hr}} \times \frac{1 \text{ hr}}{60 \text{ min}} \times \frac{1 \text{ min}}{60 \text{ sec}}$$

$$= 0.116 \text{ cu m/sec}$$

2. Determine area of grit channel.

   Area, sq m = Width, m x Depth, m

   $$= 1 \text{ m} \times 0.5 \text{ m}$$

   $$= 0.5 \text{ sq m}$$

3. Estimate velocity in meters per second.

   $$\text{Velocity, m/sec} = \frac{\text{Flow, cu m/sec}}{\text{Area, sq m}}$$

   $$= \frac{0.116 \text{ cu m/sec}}{0.5 \text{ sq m}}$$

   $$= 0.23 \text{ m/sec}$$

EXAMPLE 4—METRIC

Determine the desired length of a grit channel if the depth is 0.5 meter and the particle settling velocity is 0.023 meter per second. The flow velocity is 0.23 meter per second.

Known		Unknown
Depth, m	= 0.5 m	Grit Channel Length, m
Settling Vel, m/sec	= 0.023 m/sec	
Flow Vel, m/sec	= 0.23 m/sec	

Calculate the desired length of the grit channel.

$$\text{Length, m} = \frac{(\text{Depth of Channel, m}) (\text{Flow Vel, m/sec})}{\text{Settling Velocity, m/sec}}$$

$$= \frac{(0.5 \text{ m}) (0.23 \text{ m/sec})}{0.023 \text{ m/sec}}$$

$$= 5 \text{ m}$$

Please answer the discussion and review questions next.

DISCUSSION AND REVIEW QUESTIONS
Chapter 4.   RACKS, SCREENS, COMMINUTORS, AND GRIT REMOVAL

Write the answers to these questions in your notebook. The purpose of these questions is to indicate to you how well you understand the material in the chapter.

1. Why do you think the wastewater treatment and pollution control industry has a higher accident rate than most other industries reporting to the National Safety Council?

2. Why should coarse material (rocks, boards, metal) be removed at the plant entrance?

3. What is the main advantage of comminutors over screens?

4. How can an operator regulate the velocity in a grit channel?

5. A stick travels 30 feet in 50 seconds in a grit channel. What is the flow velocity in the grit channel? Please show your calculations in a neat fashion so someone can help you if necessary.

6. What precautions should be taken when cleaning a grit channel?

7. Calculate the grit removed from a grit channel in cubic feet per million gallons if during a 24-hour period the average flow was 3 MGD and 4.5 cu ft of grit were removed.

8. What would you do if the screens or comminutor in your plant were overloaded or bypassed?

9. Why should an operator be given the opportunity to review the plans and specifications for the expansion of existing facilities or construction of a new wastewater treatment plant?

SUGGESTED ANSWERS

Chapter 4. RACKS, SCREENS, COMMINUTORS, AND GRIT REMOVAL

Answers to questions on page 64.

4.0A True. The wastewater treatment and pollution control industry has a higher accident rate than most other industries reporting to the National Safety Council.

4.0B True. Electrical power must always be shut off before working on equipment.

4.0C True. Operators must wash their hands thoroughly before eating, smoking, or going home.

Answers to questions on page 65.

4.1A Cans, toys, rubber goods, and pieces of wood may all be found in treatment plant influent.

4.1B Coarse debris must be removed at the plant entrance to prevent damage to pumps, plugging of pipes, and filling of digesters.

4.1C Large pieces of material, such as rocks, boards, metal, and rags, are removed by racks, screens, comminutors, and grit removal devices.

Answers to questions on page 72.

4.2A (b), (c), and (d). Manually cleaned bar screens should be cleaned frequently to prevent septic conditions from developing upstream, a shock load on the plant when eventually cleaned, or formation of hydrogen sulfide, which has a rotten egg odor and causes corrosion of concrete and paints. Usually bar screens are very sturdy and will not collapse under the load from a blockage or an uncleaned screen.

4.2B When cleaning a bar screen, check to make sure that your footing will be secure by removing any slippery substances such as water and grease. Be certain there is adequate space to safely lift the screenings and that there is a receptacle for the screenings (debris).

4.2C Visually identify what appears to be the cause of the problem, then shut off the machine if you must work on the equipment. Any moving equipment is hazardous, regardless of its speed.

4.2D Screenings may be disposed of by covering them with a minimum of six inches of earth or by incineration.

4.2E A plant receives a flow of 4.4 million gallons (MG) on a certain day. The day's screenings are calculated to be 11 cubic feet. How many cubic feet of screenings were removed per MG of flow?

$$\text{Quantity Removed, cu ft/MG} = \frac{\text{Volume Removed, cu ft/day}}{\text{Average Flow, MGD}}$$

$$= \frac{11 \text{ cu ft/day}}{4.4 \text{ MGD}}$$

$$= 2.5 \text{ cu ft/MG}$$

Answers to questions on page 84.

4.3A Bar racks and screens remove debris from the wastewater while comminution units grind up debris and leave it in the wastewater.

4.3B Advantages of comminuting machines over screens include the elimination of screenings disposal, flies, and odor problems. A disadvantage is that plastic and wood may be rejected and must be removed manually.

4.3C Mechanical seals have replaced mercury seals in comminutors.

4.3D (d). Mercury must be handled with caution at all times because it is poisonous and breathing the fumes may be fatal or cause loss of hair and teeth.

Answers to questions on page 93.

4.4A (b) and (d). Grit is composed of heavy material such as sand and eggshells that will settle in the grit chamber at proper flow velocities.

4.4B Grit must be removed to prevent wear in pumps, plugged lines, and the occupation of valuable space in digesters.

4.4C The flow velocity in a grit channel can be controlled by:

1. Varying the number of channels in service in a multiple-channel installation,
2. Use of proportional weirs, and
3. Lining sides with bricks or cinder blocks if velocity is too low. This could occur in a new plant.

You have the right idea if your answer includes possible adjustments of the cross-sectional area of the flow channel.

4.4D (a) The velocity of a stick that travels 20 feet in 40 seconds in a grit channel is 0.5 ft/sec.
(b) Increase velocity to keep lighter organic solids moving.

4.4E Assume you wish to calculate the velocity in the grit channel at your plant's peak flow. Examining the flow charts, you determine that peak flows are usually about 2.75 MGD. The grit channel is three feet wide, and the flow depth is 17 inches at peak flow. What is the velocity in the grit channel under these conditions?

1. Convert the flow of 2.75 MGD to cu ft/sec.

$$\text{Flow, cu ft/sec} = \text{Flow, MGD} \times \frac{1.55 \text{ cu ft/sec}}{\text{MGD}}$$

$$= (2.75 \text{ MGD}) \frac{1.55 \text{ cu ft/sec}}{\text{MGD}}$$

$$= 4.26 \text{ cu ft/sec}$$

2. Convert depth of flow from 17 inches to feet.

$$\text{Depth, ft} = \frac{17 \text{ in}}{12 \text{ in/ft}}$$

$$= 1.4 \text{ ft}$$

3. Calculate cross-sectional area of channel.

$$\text{Area, sq ft} = \text{Depth, ft} \times \text{Width, ft}$$

$$= 1.4 \text{ ft} \times 3 \text{ ft}$$

$$= 4.2 \text{ sq ft}$$

4. Calculate velocity.

$$\text{Average Velocity, ft/sec} = \frac{\text{Flow, cu ft/sec}}{\text{Area, sq ft}}$$

$$= \frac{4.26 \text{ cu ft/sec}}{4.2 \text{ sq ft}}$$

$$= 1.01, \text{ or } 1 \text{ ft/sec}$$

4.4F   Slipping or a back injury are hazards that might be encountered when manually cleaning a grit channel. Also, beware of dangerous gases when working in a covered grit channel. There have been instances of gasoline or similar material leaking into the sewer and creating a potentially explosive hazard.

4.4G   The purpose of a cyclone grit separator is to separate grit from organic material and wastewater.

4.4H   Safety hazards an operator may encounter when working around a cyclone grit separator include:

1. Electrical hazards,

2. Slowly moving screw conveyors and other equipment,

3. Lifting heavy parts or materials, and

4. Slippery surfaces.

Answers to questions on page 93.

4.4I   Grit is "washed" to remove organic material before disposal. If the organic matter is not removed, then odors could develop. If used as fill material, the fill could settle when the organics decompose.

4.4J   Maintenance for a grit washer consists of:

1. Inspecting facility for corrosion damage;

2. Examining moving parts for wear; and

3. Oiling and greasing equipment in accordance with the plant O & M manual and the manufacturer's instructions.

Answer to question on page 95.

4.4K   Grit removals should be recorded as cubic feet of grit per million gallons of flow.

Answer: 2 cu ft of grit/million gallons.

Answers to questions on page 96.

4.5A   Problems created when screens or comminutors are overloaded or bypassed include:

1. Sticks and rags can foul the raw sludge pumps for the primary clarifiers,

2. Debris can plug the orifices in trickling filter distributors,

3. Debris can interfere with air diffusers in the aeration tanks of activated sludge plants,

4. Floating debris can appear in chlorine contact basin and the final effluent, and

5. Solids can plug return sludge pumps and flowmeters in activated sludge plants.

4.6A   When checking the plans and specifications for racks and screens, determine if there is:

1. Room for a rake when removing screenings and space so rake handle will not hit any buildings, overhead wires, light posts, or overhead lights;

2. Provision for a good standing place that won't become slippery;

3. Some place to drain and store screenings;

4. Need for a screw press to dewater screenings;

5. A disposal site for the screenings;

6. Sufficient number of channels and capacity of racks and screens and provision to handle peak storm flows;

7. A sufficient number of standby units;

8. Adherence to the Operation and Maintenance (O & M) requirements of the facilities;

9. Hoist to remove screenings if necessary;

10. Dock facility for loading containers of screenings;

11. Adequate water under sufficient pressure for hosing down machinery and area; and

12. Sufficient spare parts.

CHAPTER 5

SEDIMENTATION AND FLOTATION

by

Elmer Herr

TABLE OF CONTENTS

Chapter 5. SEDIMENTATION AND FLOTATION

OBJECTIVES

Chapter 5.   SEDIMENTATION AND FLOTATION

Following completion of Chapter 5, you should be able to:

1. Inspect new sedimentation and flotation equipment for proper installation and operation,

2. Place new facilities in service,

3. Schedule and conduct operation and maintenance duties,

4. Sample influent and effluent, interpret lab results, and make appropriate adjustments in treatment process (procedures for selecting sample location and analysis of samples are presented in Volume II, Chapter 16),

5. Recognize factors that indicate a clarifier is not performing properly, identify the source of the problem, and take corrective action,

6. Determine when, how often, and how much sludge should be pumped,

7. Conduct your duties in a safe fashion,

8. Explain the principles of the sedimentation and flotation processes,

9. Determine loadings on a clarifier,

10. Keep accurate and appropriate records on the operation of the process,

11. Develop an operating strategy for clarifiers, and

12. Review plans and specifications for clarifiers.

WORDS

Chapter 5.   SEDIMENTATION AND FLOTATION

ACTIVATED SLUDGE (ACK-ta-VATE-ed sluj) PROCESS   ACTIVATED SLUDGE PROCESS

A biological wastewater treatment process which speeds up the decomposition of wastes in the wastewater being treated. Activated sludge is added to wastewater and the mixture (mixed liquor) is aerated and agitated. After some time in the aeration tank, the activated sludge is allowed to settle out by sedimentation and is disposed of (wasted) or reused (returned to the aeration tank) as needed. The remaining wastewater then undergoes more treatment.

AEROBIC BACTERIA (AIR-O-bick back-TEAR-e-ah)   AEROBIC BACTERIA

Bacteria which will live and reproduce only in an environment containing oxygen which is available for their respiration (breathing), namely atmospheric oxygen or oxygen dissolved in water. Oxygen combined chemically, such as in water molecules (H_2O), cannot be used for respiration by aerobic bacteria.

ANAEROBIC BACTERIA (AN-air-O-bick back-TEAR-e-ah)   ANAEROBIC BACTERIA

Bacteria that live and reproduce in an environment containing no "free" or dissolved oxygen. Anaerobic bacteria obtain their oxygen supply by breaking down chemical compounds which contain oxygen, such as sulfate (SO_4^{2-}).

BULKING (BULK-ing)   BULKING

Clouds of billowing sludge that occur throughout secondary clarifiers and sludge thickeners when the sludge does not settle properly. In the activated sludge process, bulking is usually caused by filamentous bacteria or bound water.

COAGULANTS (co-AGG-you-lents)   COAGULANTS

Chemicals that cause very fine particles to clump (floc) together into larger particles. This makes it easier to separate the solids from the water by settling, skimming, draining or filtering.

COLLOIDS (CALL-loids)   COLLOIDS

Very small, finely divided solids (particles that do not dissolve) that remain dispersed in a liquid for a long time due to their small size and electrical charge. When most of the particles in water have a negative electrical charge, they tend to repel each other. This repulsion prevents the particles from clumping together, becoming heavier, and settling out.

COMPOSITE (come-PAH-zit) (PROPORTIONAL) SAMPLE   COMPOSITE (PROPORTIONAL) SAMPLE

A composite sample is a collection of individual samples obtained at regular intervals, usually every one or two hours during a 24-hour time span. Each individual sample is combined with the others in proportion to the rate of flow when the sample was collected. The resulting mixture (composite sample) forms a representative sample and is analyzed to determine the average conditions during the sampling period.

DENSITY (DEN-sit-tee)   DENSITY

A measure of how heavy a substance (solid, liquid or gas) is for its size. Density is expressed in terms of weight per unit volume, that is, grams per cubic centimeter or pounds per cubic foot. The density of water (at 4°C or 39°F) is 1.0 gram per cubic centimeter or about 62.4 pounds per cubic foot.

DETENTION TIME   DETENTION TIME

The time required to fill a tank at a given flow or the theoretical time required for a given flow of wastewater to pass through a tank.

EMULSION (e-MULL-shun)   EMULSION

A liquid mixture of two or more liquid substances not normally dissolved in one another; one liquid is held in suspension in the other.

FLIGHTS   FLIGHTS

Scraper boards, made from redwood or other rot-resistant woods or plastic, used to collect and move settled sludge or floating scum.

FLOCCULATION (FLOCK-you-LAY-shun)   FLOCCULATION

The gathering together of fine particles after coagulation to form larger particles by a process of gentle mixings.

FREEBOARD   FREEBOARD

The vertical distance from the normal water surface to the top of the confining wall.

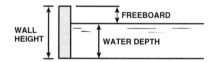

GASIFICATION (GAS-i-fi-KAY-shun)   GASIFICATION

The conversion of soluble and suspended organic materials into gas during aerobic or anaerobic decomposition. In clarifiers the resulting gas bubbles can become attached to the settled sludge and cause large clumps of sludge to rise and float on the water surface. In anaerobic sludge digesters, this gas is collected for fuel or disposed of using a waste gas burner.

HYDRAULIC LOADING   HYDRAULIC LOADING

Hydraulic loading refers to the flows (MGD or cu m/day) to a treatment plant or treatment process. Detention times, surface loadings and weir overflow rates are directly influenced by flows.

HYDROSTATIC SYSTEM   HYDROSTATIC SYSTEM

In a hydrostatic sludge removal system, the surface of the water in the clarifier is higher than the surface of the water in the sludge well or hopper. This difference in pressure head forces sludge from the bottom of the clarifier to flow through pipes to the sludge well or hopper.

LAUNDERS (LAWN-ders)   LAUNDERS

Sedimentation tank effluent troughs, consisting of overflow weir plates.

LINEAL (LIN-e-al)   LINEAL

The length in one direction of a line. For example, a board 12 feet long has 12 lineal feet in its length.

MPN (pronounce as separate letters)   MPN

MPN is the **M**ost **P**robable **N**umber of coliform-group organisms per unit volume of sample water. Expressed as a density or population of organisms per 100 mL of sample water.

MASKING AGENTS   MASKING AGENTS

Substances used to cover up or disguise unpleasant odors. Liquid masking agents are dripped into the wastewater, sprayed into the air, or evaporated (using heat) with the unpleasant fumes or odors and then discharged into the air by blowers to make an undesirable odor less noticeable.

MILLIMICRON (MILL-uh-MY-kron)   MILLIMICRON

A unit of length equal to $10^{-3}\mu$ (one thousandth of a micron), 10^{-6} millimeters, or 10^{-9} meters; correctly called a nanometer, nm.

MOLECULE (MOLL-uh-KULE)   MOLECULE

The smallest division of a compound that still retains or exhibits all the properties of the substance.

OSHA (O-shuh)   OSHA

The Williams-Steiger **O**ccupational **S**afety and **H**ealth **A**ct of 1970 (OSHA) is a federal law designed to protect the health and safety of industrial workers and treatment plant operators. It regulates the design, construction, operation and maintenance of industrial plants and wastewater treatment plants. The Act does not apply directly to municipalities, *EXCEPT* in those states that have approved plans and have asserted jurisdiction under Section 18 of the OSHA Act. *HOWEVER, CONTRACT OPERATORS AND PRIVATE FACILITIES DO HAVE TO COMPLY WITH OSHA REQUIREMENTS.* Wastewater treatment plants have come under stricter regulation in all phases of activity as a result of OSHA standards. OSHA also refers to the federal and state agencies which administer the OSHA regulations.

PACKAGE TREATMENT PLANT   PACKAGE TREATMENT PLANT

A small wastewater treatment plant often fabricated at the manufacturer's factory, hauled to the site, and installed as one facility. The package may be either a small primary or a secondary wastewater treatment plant.

REPRESENTATIVE SAMPLE   REPRESENTATIVE SAMPLE

A sample portion of material or wastestream that is as nearly identical in content and consistency as possible to that in the larger body of material or wastestream being sampled.

RETENTION TIME

RETENTION TIME

The time water, sludge or solids are retained or held in a clarifier or sedimentation tank. See DETENTION TIME.

ROTATING BIOLOGICAL CONTACTOR (RBC)

ROTATING BIOLOGICAL CONTACTOR (RBC)

A secondary biological treatment process for domestic and biodegradable industrial wastes. Biological contactors have a rotating "shaft" surrounded by plastic discs called the "media." The shaft and media are called the "drum." A biological slime grows on the media when conditions are suitable and the microorganisms that make up the slime (biomass) stabilize the waste products by using the organic material for growth and reproduction.

SEPTIC (SEP-tick)

SEPTIC

A condition produced by anaerobic bacteria. If severe, the sludge produces hydrogen sulfide, turns black, gives off foul odors, contains little or no dissolved oxygen, and the wastewater has a high oxygen demand.

SEPTICITY (sep-TIS-it-tee)

SEPTICITY

Septicity is the condition in which organic matter decomposes to form foul-smelling products associated with the absence of free oxygen. If severe, the wastewater produces hydrogen sulfide, turns black, gives off foul odors, contains little or no dissolved oxygen, and the wastewater has a high oxygen demand.

SHORT-CIRCUITING

SHORT-CIRCUITING

A condition that occurs in tanks or basins when some of the flowing water entering a tank or basin flows along a nearly direct pathway from the inlet to the outlet. This is usually undesirable since it may result in shorter contact, reaction, or settling times in comparison with the theoretical (calculated) or presumed detention times.

SLOUGHINGS (SLUFF-ings)

SLOUGHINGS

Trickling filter slimes that have been washed off the filter media. They are generally quite high in BOD and will lower effluent quality unless removed.

SLUDGE GASIFICATION

SLUDGE GASIFICATION

A process in which soluble and suspended organic matter are converted into gas by anaerobic decomposition. The resulting gas bubbles can become attached to the settled sludge and cause large clumps of sludge to rise and float on the water surface.

SLUDGE VOLUME INDEX (SVI)

SLUDGE VOLUME INDEX (SVI)

This is a calculation which indicates the tendency of activated sludge solids (aerated solids) to thicken or to become concentrated during the sedimentation/thickening process. SVI is calculated in the following manner: (1) allow a mixed liquor sample from the aeration basin to settle for 30 minutes; (2) determine the suspended solids concentration for a sample of the same mixed liquor; (3) calculate SVI by dividing the measured (or observed) wet volume (mL/L) of the settled sludge by the dry weight concentration of MLSS in grams/L.

$$\text{SVI, m}L\text{/gm} = \frac{\text{Settled Sludge Volume/Sample Volume, m}L/L}{\text{Suspended Solids Concentration, mg/}L} \times \frac{1,000 \text{ mg}}{\text{gram}}$$

SPECIFIC GRAVITY

SPECIFIC GRAVITY

(1) Weight of a particle, substance or chemical solution in relation to the weight of an equal volume of water. Water has a specific gravity of 1.000 at 4°C (39°F). Wastewater particles or substances usually have a specific gravity of 0.5 to 2.5.

(2) Weight of a particular gas in relation to the weight of an equal volume of air at the same temperature and pressure (air has a specific gravity of 1.0). Chlorine has a specific gravity of 2.5 as a gas.

SURFACE LOADING SURFACE LOADING

One of the guidelines for the design of settling tanks and clarifiers in treatment plants. Used by operators to determine if tanks and clarifiers are hydraulically (flow) over- or underloaded. Also called overflow rate.

$$\text{Surface Loading, GPD/sq ft} = \frac{\text{Flow, gallons/day}}{\text{Surface Area, sq ft}}$$

TOXIC (TOX-ick) TOXIC

A substance which is poisonous to a living organism.

TRICKLING FILTER TRICKLING FILTER

A treatment process in which the wastewater trickles over media that provide the opportunity for the formation of slimes or biomass which contain organisms that feed upon and remove wastes from the water being treated.

WEIR (weer) DIAMETER WEIR DIAMETER

Many circular clarifiers have a circular weir within the outside edge of the clarifier. All the water leaving the clarifier flows over this weir. The diameter of the weir is the length of a line from one edge of a weir to the opposite edge and passing through the center of the circle formed by the weir.

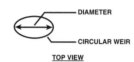

DIAMETER

CIRCULAR WEIR

TOP VIEW

DIAMETER

CROSS SECTION

CHAPTER 5.   SEDIMENTATION AND FLOTATION

(Lesson 1 of 3 Lessons)

NOTE: This chapter is divided into three lessons. The purpose is to divide the material into specific subject areas to make the information easier to study.

5.0   PURPOSE OF SEDIMENTATION AND FLOTATION

Raw or untreated wastewater contains materials that will either settle to the bottom or float to the water surface readily when the wastewater velocity is allowed to become very slow. Sewers are designed to keep the raw wastewater flowing rapidly to prevent solids from settling out in the collection system lines. Grit channels (see Chapter 4) are designed to allow the wastewater to flow at a slightly slower rate than in the sewers so that heavy, inorganic grit will settle to the bottom where it can be removed. Settling tanks decrease the wastewater velocity far below the velocity in a collection sewer.

In most municipal wastewater treatment plants, the treatment unit that immediately follows the grit channel (see Figures 5.1 and 5.2 for a typical plant layout) is the *SEDIMENTATION AND FLOTATION UNIT.* This unit may be called a settling tank, sedimentation tank, or clarifier. The most common name is *PRIMARY CLARIFIER*, since it helps to clarify or clear up the wastewater.

A typical plant (Figures 5.1 and 5.2) may have clarifiers located at two different points. The one that immediately follows the bar screen, comminutor, or grit channel (some plants don't have all of these) is called the *PRIMARY CLARIFIER*, merely because it is the first clarifier in the plant. The other, which follows other types of treatment units, is called the *SECONDARY CLARIFIER* or the *FINAL CLARIFIER.* The two types of clarifiers operate almost exactly the same way. The reason for having a secondary clarifier is that other types of treatment following the primary clarifier convert more solids to the settleable form, and they have to be removed from the treated wastewater. Because of the need to remove these additional solids, the secondary clarifier is considered part of these other types of processes.

The main difference between primary and secondary clarifiers is in the density of the sludge handled. Primary sludges are usually denser than secondary sludges. Effluent from a secondary clarifier is normally clearer than primary effluent.

Solids that settle to the bottom of a clarifier are usually scraped to one end (in rectangular clarifiers) or to the middle (in circular clarifiers) into a sump. From the sump the solids are pumped to the sludge handling or sludge disposal system. Systems vary from plant to plant and include sludge digestion, vacuum filtration, filter presses, incineration, land disposal, lagoons, and burial. Figures 5.3, 5.4, and 5.5 show detailed sketches of rectangular and circular clarifiers. Tables 5.1 and 5.2 list the purposes of the parts of the clarifiers.

Disposal of skimmed solids varies from plant to plant. Skimmed solids may be buried with material cleaned off the bar screen, incinerated, or pumped to the digester. Even though pumping skimmed solids to a digester is not considered good practice because skimmings can cause operational problems in digesters, it is a common practice.

This chapter contains information on start-up, daily operation, shutdown, and maintenance procedures; sampling and laboratory analyses; some problems to look out for; safety; and basic principles of sedimentation and flotation. You may wish to refer to Volume II, Chapter 16, which contains details of laboratory analyses, for further information.

TREATMENT PROCESS   FUNCTION

PRELIMINARY TREATMENT

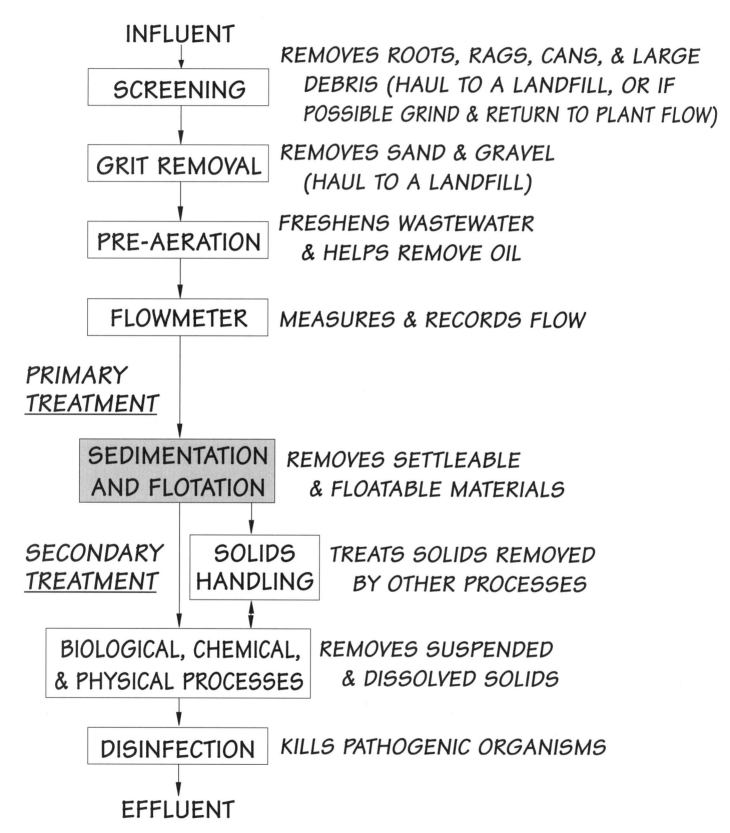

INFLUENT

SCREENING — REMOVES ROOTS, RAGS, CANS, & LARGE DEBRIS (HAUL TO A LANDFILL, OR IF POSSIBLE GRIND & RETURN TO PLANT FLOW)

GRIT REMOVAL — REMOVES SAND & GRAVEL (HAUL TO A LANDFILL)

PRE-AERATION — FRESHENS WASTEWATER & HELPS REMOVE OIL

FLOWMETER — MEASURES & RECORDS FLOW

PRIMARY TREATMENT

SEDIMENTATION AND FLOTATION — REMOVES SETTLEABLE & FLOATABLE MATERIALS

SECONDARY TREATMENT

SOLIDS HANDLING — TREATS SOLIDS REMOVED BY OTHER PROCESSES

BIOLOGICAL, CHEMICAL, & PHYSICAL PROCESSES — REMOVES SUSPENDED & DISSOLVED SOLIDS

DISINFECTION — KILLS PATHOGENIC ORGANISMS

EFFLUENT

Fig. 5.1   Flow diagram of typical plant

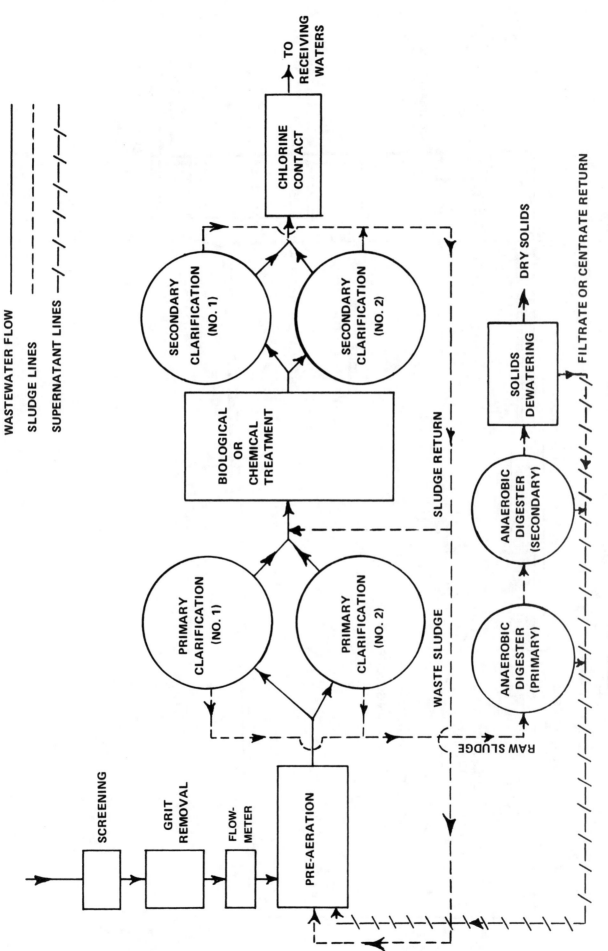

WASTEWATER FLOW ─────

SLUDGE LINES ─ ─ ─ ─

SUPERNATANT LINES ─/─/─/─/─

SCREENING

GRIT REMOVAL

FLOW-METER

PRE-AERATION

PRIMARY CLARIFICATION (NO. 1)

PRIMARY CLARIFICATION (NO. 2)

BIOLOGICAL OR CHEMICAL TREATMENT

SECONDARY CLARIFICATION (NO. 1)

SECONDARY CLARIFICATION (NO. 2)

CHLORINE CONTACT

TO RECEIVING WATERS

WASTE SLUDGE

SLUDGE RETURN

RAW SLUDGE

ANAEROBIC DIGESTER (PRIMARY)

ANAEROBIC DIGESTER (SECONDARY)

SOLIDS DEWATERING

DRY SOLIDS

FILTRATE OR CENTRATE RETURN

Fig. 5.2   Plan diagram of typical clarifiers in a wastewater treatment plant

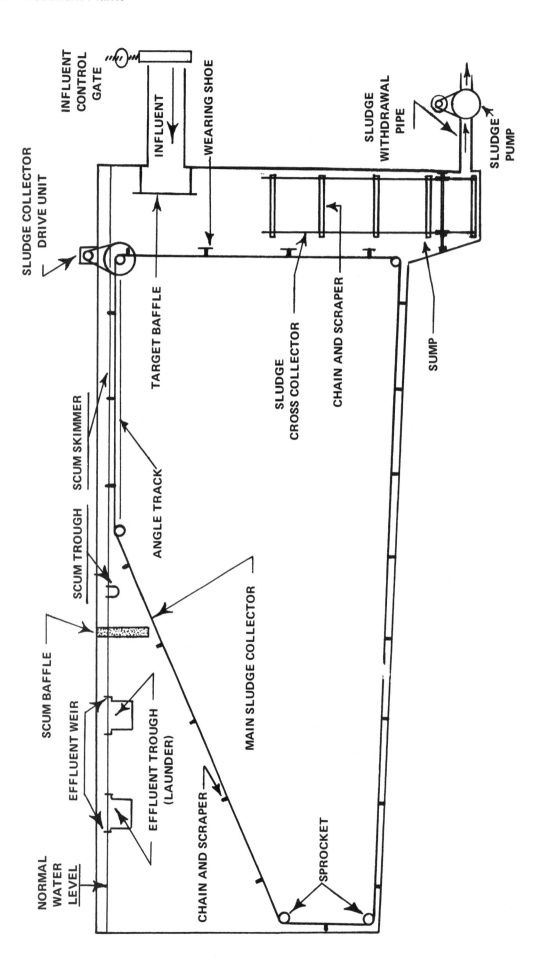

Fig. 5.3   Side-view section of a rectangular sedimentation basin

TABLE 5.1   PURPOSE OF RECTANGULAR SEDIMENTATION BASIN PARTS

Part	Purpose	Part	Purpose
Influent Control Gate	Throttles or stops the flow to the sedimentation basin or clarifier.	Cross Collector	Drags sludge to deep end of sump for removal by pumping. Also prevents bridging of sludge in sump.
Influent Channel or Pipe	Transports wastewater to the clarifier.	Sump	Receives settled sludge from the floor of the sedimentation basin. Stores sludge in sufficient quantity to avoid frequent (less than once per hour) removal by pumping, but of sufficient volume to maintain sludge thickness and to exclude water in the sedimentation basin from being pumped out during the pumping cycle.
Target Baffle or Deflector Plate	Spreads the wastewater evenly across the width of the clarifier for even distribution and prevents short-circuiting.		
Effluent Weir	Ensures equal flow over all weirs. Designed for small surface elevation (water level) adjustments in the clarifier provided the plate is designed for vertical movement (up or down).		
Effluent Trough (Launder)	Collects the settled wastewater flowing over the weirs and conveys it from the sedimentation basin.	Sludge Withdrawal Pump	Removes the sludge from the sump (pit).
Main Sludge Collector	Drags settled solids (sludge) to the sump. A continuous chain with crosspieces (flights or scrapers) attached.	Wearing Shoe	Prevents wear on the scraper crosspieces. Usually a piece of iron attached close to the outer ends of the scrapers.
Sprocket	Supports chain, adjusts tension, or forces the chain to move. A Wheel with teeth around the outside that fit in the chain link.	Scum Skimmer or Collector	Skims or collects floating material from the surface of the wastewater and moves it to the scum trough.
Angle Track	Provides a track on which the main collector crosspieces ride.	Scum Trough	Receives the floating material from the scum skimmer for removal.
Sludge Collector Drive	Provides power that causes the main and cross collector units to move.	Scum Baffle	Extends above the water surface and prevents the floating material from reaching the effluent trough.

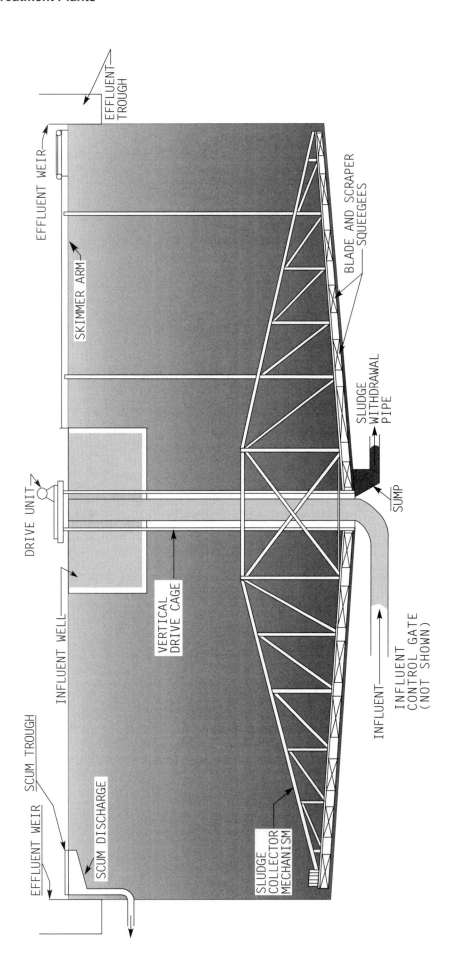

EFFLUENT
TROUGH

EFFLUENT WEIR

SKIMMER ARM

BLADE AND SCRAPER
SQUEEGEES

SLUDGE
WITHDRAWAL
PIPE

SUMP

DRIVE UNIT

INFLUENT WELL

VERTICAL
DRIVE CAGE

INFLUENT

INFLUENT
CONTROL GATE
(NOT SHOWN)

SCUM TROUGH

EFFLUENT WEIR

SCUM DISCHARGE

SLUDGE
COLLECTOR
MECHANISM

*Fig. 5.4   Side-view section of a circular clarifier with blades
and scraper squeegees*

TABLE 5.2   PURPOSE OF CIRCULAR CLARIFIER PARTS

Part	Purpose	Part	Purpose
Influent Control Gate	Throttles or stops the flow to the clarifier.	Scum Pipe	Allows the collected scum to flow from the skimmer box to a scum tank or a pump.
Influent Channel or Pipe	Transports wastewater to the clarifier.	Drive Unit	Causes the collector to rotate. A power unit that is connected to the vertical drive cage and causes the collector to rotate.
Influent Well	Receives the flow from the influent pipe, reduces flow velocities and distributes flow evenly across the upper portion of the clarifier contents. A small circular compartment in the top center of the clarifier.	Vertical Drive Cage	Transmits power from drive unit to the sludge collector mechanism.
Effluent Weir	Ensures equal flow over all weirs. Designed for small surface elevation (water level) adjustments in the clarifier, provided the plate is designed for vertical movement (up or down).	Sludge Collector Mechanism	Drags settled solids across clarifier bottom to a sludge collection pit or sump. A mechanism that rotates around the bottom of the clarifier and consists of squeegee-type scrapers.
Effluent Trough (Launder)	Collects the settled wastewater flowing over the weirs and conveys it from the clarifier.	Blades and Scraper Squeegees	Scrape sludge from bottom of clarifier to sump.
Scum Skimmer Arm	Skims or collects floating material from the surface of the wastewater and moves it to the scum trough.	Sump	Collects the sludge before withdrawal.
Scum Trough	Receives the floating material scraped from the surface by the scum skimming arm.	Sludge Withdrawal Pipe	Removes the sludge from the clarifier. Usually connected to a sludge pump.

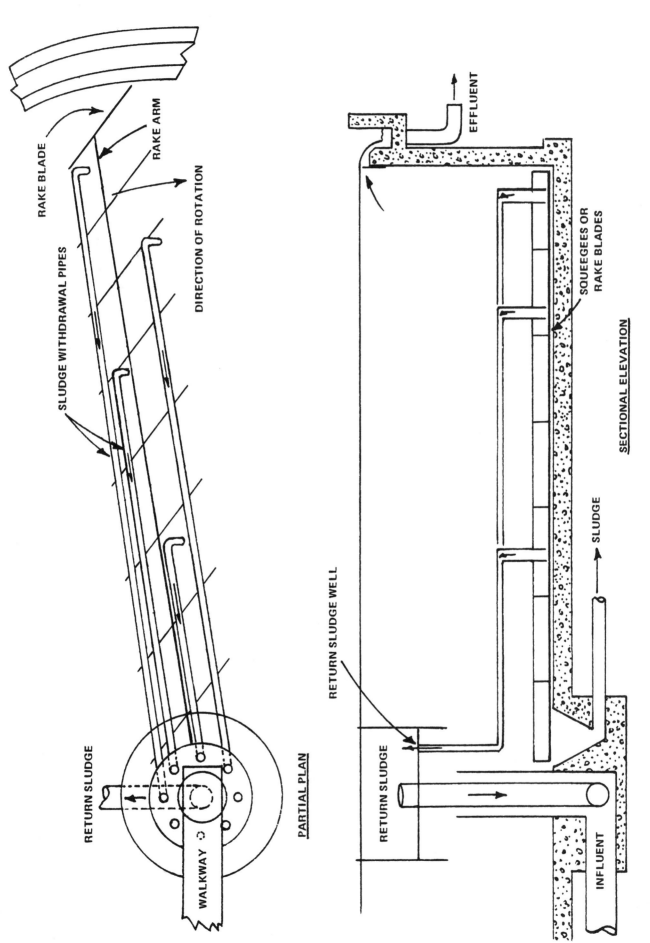

RAKE BLADE

RAKE ARM

DIRECTION OF ROTATION

SLUDGE WITHDRAWAL PIPES

RETURN SLUDGE

WALKWAY

PARTIAL PLAN

EFFLUENT

SQUEEGEES OR
RAKE BLADES

SECTIONAL ELEVATION

RETURN SLUDGE WELL

RETURN SLUDGE

SLUDGE

INFLUENT

Fig. 5.5   Side-view section of a circular clarifier with riser suction pipes

5.1   OPERATION AND MAINTENANCE

5.10   Start-Up Procedures

Before starting a new unit or one that has been out of service for cleaning or repair, inspect the tank carefully as outlined in this section. Now is a good time to become familiar with the "internal workings" of the clarifier because they are usually under water.

A. Circular Clarifiers

Check items:

1. Control gates for proper operation.

2. Clarifier tank for sand and debris.

3. Collector drive mechanism for lubrication, oil level, drive alignment, and complete assembly.

4. Gaskets, gears, drive chain sprockets, and drive motor (usually single speed) for proper installation and rotation.

5. Squeegee blades on the collector plows for proper distance from the floor of the tank.

6. All other mechanical items below the waterline for proper installation and operation.

7. Tank sumps or hoppers and return lines for debris and obstructions.

8. Tank structure for corrosion, cracks, and other indications of structural failure.

If everything checks out properly, turn on the mechanism. Let it make several revolutions while checking that the squeegee does not travel high or low, missing the bottom or scraping in some areas. The scraping action should control the entire area from the outside wall to the sludge hopper. Also be certain that the mechanism runs smoothly without jerks or jumps. Improper movement may be caused by problems with the drive unit, squeegees that have too much drag, or an uneven clarifier floor. If the unit is water lubricated, be sure sufficient water is in the tank to cover the center bearing.

If the unit is equipped with a stall alarm, test it to see if the mechanism will stop when overloaded. With the unit running, time the period it takes for the plows to make one complete revolution around the tank and *RECORD* the time for later reference.

Check and *RECORD* the amperage that the motor draws. Let the unit operate for several hours. If no problems develop, it should be okay.

Always be safety conscious, even during start-up.

1. If you are working down in the tank, wear a hard hat for protection from falling objects.

2. Keep hands away from moving equipment.

3. When working on equipment, be sure to tag and use a lock-out device on the main circuit breaker and influent control gates to prevent equipment from starting unexpectedly and causing equipment damage and/or personal injury. If the lockout device has a key, keep the key in your pocket.

B. Rectangular Clarifiers

The tank hoppers, channels, control gates, weirs, bearings, grease lines, and drive alignment should be checked the same as for the circular clarifiers. The sludge collectors are different in rectangular clarifiers. Wooden crosspieces or *FLIGHTS*[1] are laid across the tank and each end is attached to an endless chain along both sides of the tank. The chains and crosspieces connected together as a unit are called the "collector mechanism." The collector chains are driven by a connecting shaft and sprockets that drag the crosspieces along rails imbedded in the floor of the tank and along each side just under the surface of the water.

Each wooden flight is equipped with metal wearing shoes to ride the rails. A metal wearing shoe usually is a flat metal plate attached to the end of the wooden flight to prevent excessive wear on the wood portion. Plastic wearing shoes may be used instead of metal ones.

Check to ensure that the flights are straight across the tank, and that the chain on one side is not one or two links longer or shorter than the chain on the opposite side. If this occurs, the wooden flights will run at an angle across the tank. This either will cause sludge to pile higher on the trailing side or cause the flights to hang up with resultant severe damage to the flights.

Caution should be exercised before starting the sludge collectors in an empty clarifier if they have not been operational for several weeks. The wearing shoes on the flights may have started rusting where they are sitting on the rails. A good practice is to lift each individual flight off the rail to be certain it is free and to apply a light grease or SAE 90 gear oil to the shoe and rail. If these precautions are not taken before the collector is turned on, the flight (wearing) shoes could stick to the rails and the whole collector system could be pulled down to the floor of the tank. Before the collectors are started in a new tank, each flight should be checked for a clearance of one to two inches (2.5 to 5 cm) between the wall and the end of the flight. If flights are too long, they may rub the tank wall

---

[1] *Flights.   Scraper boards, made from redwood or other rot-resistant woods or plastic, used to collect and move settled sludge or floating scum.*

and break the flight. Do not run a sludge collector very long in an empty tank if the lower shaft bearings of the collector mechanism depend on water rather than oil or grease for lubrication.

Safety precautions for the start-up of rectangular clarifiers are similar to those for circular clarifiers.

5.11   Daily Operation and Maintenance

During normal operations, you should schedule the following daily activities:

1. *INSPECTION.* Make several daily inspections with a stop, look, listen, and think routine.

2. *CLEANUP.* Using water under pressure, wash off accumulations of solid particles, grease, slime, and other material from walkways, handrails, and all other exposed parts of the structure and equipment.

3. *LUBRICATION.* Grease all moving equipment according to manufacturer's specifications and check oil levels in motors where appropriate.

4. *PREVENTIVE MAINTENANCE.* Follow the manufacturer's specifications.

5. *FLIGHTS.* Examine bolts for looseness, corrosion, and excessive wear on those parts that can be inspected above the waterline.

6. *CHAIN AND SPROCKET.* Check for wear because 0.05 inch (1.3 mm) wear on each of 240 link pins will cause about one foot (0.3 m) of extra slack.

7. *RECORDKEEPING.* Write in your pocket notebook any unusual observations and transfer these notes to the plant record sheet (typical sheet is shown in the Appendix, page 161).

8. *SAMPLING AND LABORATORY ANALYSIS.* Details are in the next section (5.2).

9. *SLUDGE AND SCUM PUMPING.* See Section 5.3.

5.12   Cleaning Weirs and Troughs

Algae can be a problem on weirs and in effluent troughs. It is unsafe for operators to get into the trough and the weirs and troughs can be too far away from walkways to use a brush on a pole to remove algae. Chlorine can be added between the baffle and the weir to control algae. The chlorine solution pipe can be completely looped around the clarifier using a 3/4-inch (19-mm) PVC pipe with 1/8-inch (3-mm) holes drilled where the V-notch is lowest. Other approaches include adding chlo-

rine at the high point in the trough opposite the effluent discharge gate or placing chlorine tablets in the trough at several spots every day. Adding chlorine in the trough controls algae in the trough. Using a fire hose or high-pressure water jet to remove algae from the weirs and troughs also can be effective.

5.13   Shutdown Procedures

Annually during periods of low flow, each clarifier should be shut down for inspection, routine maintenance, and any necessary repairs. Even though the clarifier and all equipment are working properly, an annual inspection helps to prevent serious problems and failures in the future when harmful consequences can result.

1. Divert the flow to other clarifiers and close the influent and effluent control gates of the clarifier being shut down.

2. Pump all remaining sludge to digester or to solids handling system.

3. Dewater clarifier by draining remaining wastewater to headworks or pumping remaining wastewater to other clarifiers.

4. Hose down walls, floor, and equipment inside clarifier while draining clarifier.

5. Inspect clarifier in accordance with Section 5.10, "Start-Up Procedures" (page 117).

6. Repair or replace all broken or defective equipment.

7. Repaint metal surfaces that have lost their protective coating or are showing signs of corrosion.

QUESTIONS

Write your answers in a notebook and then compare your answers with those on page 158.

5.0A   What is the main difference between the sludge from primary and secondary clarifiers?

5.0B   What is the main difference between the effluent from primary and secondary clarifiers?

5.1A   List the significant items to check before start-up of a circular clarifier.

5.1B   What safety precautions should be taken during start-up of a clarifier?

5.1C   What happens when the crosspieces in a rectangular clarifier are not straight across the tank?

5.14   Operational Strategy

The purpose of a sedimentation tank or clarifier is to remove settleable and floatable solids by sedimentation and flotation. The factor most often reported as influencing clarifier performance is the flow into the plant. Both the *SURFACE LOADING*[2] and *DETENTION TIME*[3] are directly related to flow (see Section 5.61, "Primary Clarifiers," page 134). In most plants the surface loading and detention time vary widely throughout the

---

[2] *Surface Loading.* One of the guidelines for the design of settling tanks and clarifiers in treatment plants. Used by operators to determine if tanks and clarifiers are hydraulically (flow) over- or underloaded. Also called overflow rate.

$$\text{Surface Loading, GPD/sq ft} = \frac{\text{Flow, gallons/day}}{\text{Surface Area, sq ft}}$$

[3] *Detention Time.* The time required to fill a tank at a given flow or the theoretical time required for a given flow of wastewater to pass through a tank.

day due to the hourly changes in plant inflow resulting from the activities of the people and industries in the community. In spite of these great fluctuations, most clarifiers produce fairly consistent removals of BOD and suspended solids.

Most clarifiers that do not produce an acceptable effluent (see page 129) usually fail due to operator errors, equipment failures, or excessive hydraulic loadings (shock loads). The operator's job is very simple. Be sure that accumulated settled solids are removed from the bottom of the clarifier before *SEPTICITY*[4] and *GASIFICATION*[5] take place. Also be sure that surface floatables (oil and grease) are continuously or regularly skimmed and removed from the water surface to prevent floatables from reaching downstream secondary treatment and disinfection processes.

Equipment or process failures caused by operator errors include:

1. Insufficient frequency or time for removing sludge;

2. Poor equipment maintenance and housekeeping; and

3. Insufficient knowledge of equipment and/or treatment processes as related to:

    a. Laboratory analyses,

    b. Clarifier loadings,

       (1) Flows, MGD
       (2) Detention time, hr
       (3) Surface loading, GPD/sq ft
       (4) Weir overflow, GPD/ft
       (5) Solids, lbs solids/day/sq ft (secondary clarifiers)
       (6) Solids balance, lbs/day in = lbs/day out

    c. Inability to recognize a mechanical-electrical problem, and

    d. Not restarting a drive mechanism that tripped out during a momentary power failure, such as one caused by a thunderstorm.

The best operational strategy for a clarifier is to develop and implement a good preventive maintenance program, to closely monitor operating conditions, and to respond to any lab results that indicate problems are developing (see Section 5.24, "Response to Poor Clarifier Performance" on page 129). Any other clarifier problems usually result from abnormal conditions.

In large treatment plants with four or more clarifiers, plant performance may be improved by diverting flows to more or fewer clarifiers under certain circumstances. For example, during low flow periods from midnight until 6:00 A.M., fewer clarifiers may be needed on the process flow line. To determine the number of clarifiers that should be on line, try not to allow unfavorable detention times (less than half an hour or longer than three hours) to last longer than four to six hours without placing more or fewer clarifiers on line. Try to prevent detention times in primary clarifiers from becoming too long in order to keep the wastewater that is flowing to aerobic biological treatment processes fresh. To place an additional clarifier on line, or to take one off line, use the following procedures:

1. When placing a clarifier on line, open the inlet gate to the clarifier.

2. Operate the sludge collector, skimmer, and pumps according to normal daily operating procedures.

3. When taking a clarifier off line, divert the wastewater being treated to other clarifiers by closing the clarifier inlet gate.

4. Collect the sludge on the bottom of the clarifier and pump the sludge from the hopper if it is thick enough to pump. Thickness of the sludge can be determined by the sound of the sludge pump, the use of thickness measuring sensors, or the observation of sludge through a sight glass.

5. The skimmer and sludge collector mechanism may be left on or turned off depending on the conditions of tank contents with regard to sludge left in the clarifier or scum on the water surface.

These procedures assume a clarifier being placed on line has been off line for less than two days. Also when a clarifier is taken off line, the assumption is made that the clarifier will be back on line within two days. Usually the number of secondary clarifiers on line remains constant on a daily or weekly basis; however, the same procedures apply if necessary. The number of secondary clarifiers on line may change with increases or decreases in seasonal flows or the condition of the solids in the secondary system. During extremely cold weather, be careful that clarifiers that are not covered do not freeze when taken off the line. You may be better off leaving them on the line so they won't freeze during severe freezing conditions.

Operational strategies for secondary clarifiers following chemical and biological treatment processes are discussed in the following chapters related specifically to these processes. Also see Section 5.62, "Secondary Clarifiers," page 137.

5.15 Abnormal Conditions

Abnormal conditions influencing clarifier performance consist of:

1. *TOXIC*[6] wastes from industrial spills or dumps,

2. Storm flows and hydraulic overloads, and

3. Septicity from collection system problems.

---

[4] *Septicity (sep-TIS-it-tee).* Septicity is the condition in which organic matter decomposes to form foul-smelling products associated with the absence of free oxygen. If severe, the wastewater produces hydrogen sulfide, turns black, gives off foul odors, contains little or no dissolved oxygen and the wastewater has a high oxygen demand.

[5] *Gasification (GAS-i-fi-KAY-shun).* The conversion of soluble and suspended organic materials into gas during aerobic or anaerobic decomposition. In clarifiers the resulting gas bubbles can become attached to the settled sludge and cause large clumps of sludge to rise and float on the water surface. In anaerobic sludge digesters, this gas is collected for fuel or disposed of using a waste gas burner.

[6] *Toxic (TOX-ick).* A substance which is poisonous to a living organism.

There is not much an operator can do in terms of clarifier operation to improve or maintain clarifier performance under these conditions. If any of these events occurs, corrective action should be taken. If a toxic waste dump is suspected or identified as the cause of a plant upset, immediate action should be taken to identify the source and prevent future dumps. Development and enforcement of a sewer-use ordinance and an industrial pretreatment facility inspection program is an effective long-term approach for reducing plant upsets due to toxic wastes. The installation of monitoring devices, instrumentation, control structures, and chemical feed systems will also provide the operator with helpful tools to maintain clarifier performance under abnormal conditions.

Shock loads from toxic wastes are best treated or controlled (after they are identified) by the addition of proper chemicals such as coagulants or chlorine at the plant headworks or at the preliminary treatment area. Some adjustments, such as increased recirculation rates with trickling filters to dilute the toxic wastes, may be successful in the secondary treatment processes.

In activated sludge plants, the impact of toxic wastes may be reduced if higher aeration rates will strip (drive out) the toxicants out of the mixed liquor. Usually this procedure is ineffective because the toxic waste has already passed through the plant before its impact is discovered. Reduction or a complete halting of the solids wasting rate may help if some resistant bacteria have survived. If possible, change the mode of operation to contact stabilization or step aeration, thereby exposing only a small portion of the organism population to the toxic waste. If all the bacteria have been destroyed or if the toxicant is bound in the sludge, get rid of the solids. Do not dispose of toxic solids in a digester. They may be disposed of in an approved sanitary landfill, provided you receive permission from the proper regulatory authorities.

If storm flow infiltration is a frequent problem, the sealing of sanitary sewers and/or the use of a flow equalization basin may improve the quality of clarifier effluent. There may not be much that can be done to prevent the development of septic wastewater in a collection system. Septic conditions can develop during hot weather when wastewater travel times in the collection system to the treatment plant become too long. Chemical treatment with chlorine or hydrogen peroxide added at a pump station may improve the condition of septic wastewater and protect structures from corrosion damage.

5.16   Troubleshooting

The charts on the next seven pages list indicators of possible process or component failure in a circular or rectangular clarifier, the probable causes of failure, and how to remedy and prevent the failure. Also see Section 5.24, "Response to Poor Clarifier Performance," page 129.

QUESTIONS

Write your answers in a notebook and then compare your answers with those on page 158.

5.1D   Describe a good operational strategy for a clarifier.

5.1E   What types of abnormal conditions could affect clarifier performance?

5.1F   What steps could be taken to improve clarifier effluent quality when excessive storm flow infiltration is a frequent problem?

Troubleshooting Guide—Primary Clarification

(Adapted from *PERFORMANCE EVALUATION AND TROUBLESHOOTING AT MUNICIPAL WASTEWATER TREATMENT FACILITIES*, Office of Water Program Operations, US EPA, Washington, DC.)

INDICATOR/OBSERVATION	PROBABLE CAUSE	CHECK OR MONITOR	SOLUTION
1. Floating sludge (*BULKING*[7]).	1a. Sludge decomposing in tank, lifted by gasification.	1a. Sludge mats or chunks with a thickness of 1/2" - 3" or more floating on surface of clarifier.	1a. Remove sludge more frequently or at higher rate.
	1b. Sludge collection mechanism, or flights, are OFF.	1b. Check to see if sludge collectors are running.	1b. Restart collectors, observe unit through one complete revolution, ensuring that the unit is not jerking, jumping, or tripping out due to overload.
	1c. Insufficient sludge removal.	1c. Check sludge pump operation.	1c. Monitor sludge pump operation through one complete cycle for: a. Duration of ON time b. Pump output c. Thickness of sludge pumped (percent solids or density). If pump operates as it should, and the sludge solids remain at the same thickness during the cycle, then increase pumping time at least 5 min/hr; if pump does not discharge as it should, clean pump. If pump runs normally but sludge being pumped is thick, continue on through this checklist.
	1d. Sludge blanket too deep.	1d. Sound across clarifier for sludge blanket depth and level.	1d. If sludge blanket is higher than normally maintained, increase pumping cycle time.
	1e. Sludge pump runs but discharges thin or no sludge.	1e. Sludge line blocked or plugged.	1e. 1. Check pump suction line to clarifier for closed valves. If valves are closed, open and recheck pump output. 2. If all valves are open, backflush sludge suction line to clarifier to remove blockage. 3. If sludge line cannot be cleared and restored to service, take clarifier out of service, dewater, and remove blockage from clarifier sump or sludge line.
	1f. Sludge collector is damaged or needs adjusting.	1f. Remove clarifier from service, dewater tank.	1f. 1. Check collector flights. Replace broken flights and worn shoes. 2. On circular clarifiers check and adjust squeegees on plows to proper clearance and mechanism for level, so that plows don't ride high on one side of tank floor and drag on the opposite side.

---

[7] Bulking (BULK-ing).   *Clouds of billowing sludge that occur throughout secondary clarifiers and sludge thickeners when the sludge does not settle properly. In the activated sludge process, bulking is usually caused by filamentous bacteria or bound water.*

INDICATOR/OBSERVATION	PROBABLE CAUSE	CHECK OR MONITOR	SOLUTION
1. Floating sludge (bulking). (continued)	1g. Rectangular tanks, damaged or missing inlet baffles.	1g. Sludge blanket higher (deeper) on one side of tank than other.	1g. Repair or replace inlet baffles to ensure even distribution of flow across tank to produce even solids distribution.
	1h. Return of well-nitrified waste activated sludge.	1h. Pea size or leathery, thin, foamy layer of sludge floating on clarifier surface.	1h. 1. Change activated sludge from extended aeration by reducing aeration time or sludge age. 2. If plant effluent is required to be highly nitrified, add surface water sprays to primary clarifiers to break up froth or move it nearer to scum collectors. 3. If possible, do not route waste activated sludge to primary clarifiers, but to separate treatment unit such as flotation thickener or separate clarifier.
	1i. Solids load increased to plant.	1i. Monitor plant influent for settleable solids.	1i. If increased load is related to industrial contributor, have industry implement pretreatment system.
	1j. Toxic discharge or spill.	1j. Plant influent for pH level.	1j. 1. Neutralize toxic waste if possible. 2. Pump sludge (dewatered, if possible) to tank truck and dispose of it in an approved sanitary landfill.
2. Black and odorous septic wastewater entering clarifier.	2a. Malfunction of in-plant pretreatment.	2a. Check operation of prechlorination and pre-aeration units.	2a. Increase pretreatment units to compensate for load.
	2b. Inadequate pretreatment of industrial waste.	2b. Sample plant influent flows, identify industrial contributors, particularly high temperature dischargers.	2b. Review contributors' discharge requirements to collection system, enforce or upgrade pretreatment requirements to correct problem.
	2c. Disposal of wastewater from septic tank pumpers into collection system.	2c. Monitor plant influent; septic disposal would be batch and not continuous load.	2c. 1. Review discharge from septic tank pumping firms in area and enforce sewer-use ordinance. 2. Increase prechlorination during duration of discharge. 3. May require secondary treatment adjustment of additional air or oxygen for activated sludge plant, or increase recycle flows on trickling filters or rotating biological contactors.
	2d. Solids decomposing in collection system.	2d. Monitor BOD and H_2S during low flows at various collection system locations.	2d. Decomposition due to low velocity or extended wastewater travel time may be reduced by use of chlorine or hydrogen peroxide injected at one or more locations in a collection system, usually most available location is a wastewater lift station.
	2e. Wastewater lift station.	2e. Check detention time in lift station inlet sump or wet well, and how long wastewater is in force main.	2e. Long wet well detention time—increase pump cycle or add chlorine to station effluent.

INDICATOR/OBSERVATION	PROBABLE CAUSE	CHECK OR MONITOR	SOLUTION
2. Black and odorous septic wastewater entering clarifier. (continued)	2f. Collection system siphons.	2f. Check velocity through siphons.	2f. If low velocity and multiple barrels are in operation, gate out one barrel during low flows to maintain wastewater velocity to prevent solids deposition in barrels that will be scoured out at peak flows.
	2g. Sewer maintenance cleaning activities.	2g. Require collection maintenance crews to inform wastewater treatment plant when large blockages are cleared, or cleaning activities produce exceptionally septic loads, or the cleaning is close to the plant where collection system dilution will not be available.	2g. 1. Increase pretreatment units to maximum or reasonable levels, for example, increase pre-aeration, increase prechlorination by 20 to 25 percent when septic flow arrives at headworks. 2. Adjust secondary units correspondingly to meet additional load.
	2h. Recycle of excessively strong digester supernatant.	2h. Monitor inflow and solids content of digester's supernatant discharge.	2h. 1. Return strong supernatant at low plant flows. 2. Stop mixing digester to obtain higher quality supernatant. 3. Maintain proper sludge inventory so digesters are not overloaded. 4. Pump thicker raw sludge to digester to reduce hydraulic load.
3. Black and odorous septic wastewater leaving primary clarifier.	3a. Improper sludge removal.	3a. Sound clarifiers for sludge blanket thickness. Determine if collector mechanism is operating.	3a. 1. Maintain sludge blankets at plant's optimum levels that produce a thick raw sludge, but prevent sludge bulking and loss of solids to secondary system. 2. If sludge blanket is thick, increase sludge removal rate. 3. If available, chlorinate clarifier effluent to reduce load on secondary system, but do not over-chlorinate. 4. Where multiple clarifiers are in service and only one is bulking, remove it from service until bulking is corrected in that tank.
	3b. Sludge pump.	3b. Check pump operation 1. Run time 2. Output 3. Thickness of sludge pumped.	3b. 1. Clean sludge pump if plugged. 2. Increase pump cycle if sludge was still thick at end of test cycle.
	3c. Sludge withdrawal line plugged.	3c. Check sludge pump output.	3c. 1. Check sludge line suction valves. 2. Backflush sludge line. 3. If cannot clear line, take clarifier out of service and pump down to clear obstruction.
	3d. Sludge collector worn or damaged.	3d. Remove clarifier from service, dewater tank.	3d. 1. Check collector flights. Replace broken flights and worn shoes. 2. On circular clarifiers check and adjust squeegees or plows to proper clearance and mechanism for level, so that plows don't ride high on one side of tank floor and drag on the opposite side.

INDICATOR/OBSERVATION	PROBABLE CAUSE	CHECK OR MONITOR	SOLUTION
3. Black and odorous septic wastewater leaving primary clarifier. (continued)	3e. Short-circuiting through tank.	3e. Flow into and out of clarifier.	3e. 1. Reduce solids recycle load to clarifiers by chemicals or pretreatment. 2. Adjust pH to nearly neutral. 3. Cool wastewater flows from heat exchangers or other units before returning flow to clarifiers. 4. If problem recycle is only during daily high flow periods, try to schedule bulk of "recycle" flows at daily low flow periods (early mornings) to reduce load on clarifiers.
4. Scum escaping in clarifier effluent.	4a. Frequency of removal inadequate.	4a. 1. Rectangular tanks. Frequency of scum skimmer operation. 2. Circular clarifiers. Is collector mechanism operating?	4a. 1. Operate skimmers more frequently. 2. Restart collector mechanism. Check that skimmer blade is down.
	4b. Plugged scum trough.	4b. Scum trough filled with scum.	4b. Pump scum pit, clear and hose down scum trough.
	4c. Water or air skimmers are OFF.	4c. Check that water sprays or air blowers are working.	4c. 1. Restart spray system. 2. Unplug nozzles and adjust so scum on surface is pushed to skimmers for removal from clarifier.
	4d. Hydraulic overload.	4d. Amount of flow through tank too high.	4d. If possible reduce flow through tank: 1. Put additional clarifier in service. 2. Reduce influent flow by collection system storage. CAUTION: Monitor collection system to prevent overflow.
	4e. Worn or damaged skimmer.	4e. Check rubber wiper blades and alignment of skimmer.	4e. Replace worn or torn wiper blades, adjust skimmer to proper clearance and tension.
	4f. Heavy industrial waste contribution.	4f. Plant influent.	4f. Limit industrial waste contributions. Implement/enforce new or existing sewer-use ordinances.
5. Sludge hard to pump from hopper.	5a. Pump worn.	5a. Check pump output.	5a. Repair pump to discharge at rated capacity.
	5b. Sludge line restricted.	5b. Sludge pump suction pressure or vacuum.	5b. Clean sludge line with hydrocleaner, pig, or rod to remove grease and sludge deposits.
	5c. Excessive grit, clay, or other easily compacted material.	5c. Total solids percent organic/inorganic in influent.	5c. 1. Check operation of preliminary treatment units (grit chamber). 2. Check possible industrial contributors for proper pretreatment. 3. Jet sludge hopper with air or water to resuspend material so it may be pumped out. CAUTION: During jetting turn off sludge collectors and cross collectors. 4. Normally occurs after high storm flows.

INDICATOR/OBSERVATION	PROBABLE CAUSE	CHECK OR MONITOR	SOLUTION
5. Sludge hard to pump from hopper. (continued)	5d. Rags or other large debris in sludge hopper.	5d. When hopper is probed, rag balls and sticks are found.	5d. 1. Check headworks screening system for proper operation and removal of screenings from flow. 2. Repair comminutors, barminutors, or disintegrators to proper operation. 3. Limit use of coarse rack bypass screens. 4. May have to dewater tank to clear hopper.
6. Undesirably low solids content in sludge.	6a. Overpumping of sludge.	6a. Frequency and duration of sludge pumping; solids concentration.	6a. Reduce frequency and duration of pumping cycles.
	6b. Hydraulic overload.	6b. Influent flow rate.	6b. Provide more even flow distribution in all tanks, if multiple tanks.
	6c. Short-circuiting of flow through tanks.	6c. Dye or other flow tracers.	6c. 1. Change weir setting. 2. Repair or replace baffles.
	6d. Inoperable or malfunctioning collector mechanism.	6d. Collector mechanism.	6d. Return collector to service or repair as necessary.
7. Short-circuiting of flow through tanks.	7a. Uneven weir settings.	7a. Weir settings.	7a. Change weir settings.
	7b. Damaged or missing inlet line baffles.	7b. Damaged baffles.	7b. Repair or replace baffles.
8. Surging flow.	8. Poor influent pump programming.	8. Pump cycling.	8. Modify pumping cycle.
9. Excessive sedimentation in inlet channel.	9. Velocity too low.	9. Velocity.	9. Increase velocity or agitate channel with air or water to prevent settling or deposits of solids.
10. Excessive slime growth on surfaces and weirs.	10. Accumulations of wastewater solids and resultant growth.	10. Inspect surfaces.	10. Frequent and thorough cleaning of surfaces.
11. Circular clarifiers: Frequent stall alarms, overload trips, shear pin failures.	11a. Excessive sludge accumulation.	11a. Mechanism load indicator and solids concentration sludge blanket depth.	11a. Increase frequency of sludge pumping to reduce sludge blanket level.
	11b. Rags or debris entangled in collector mechanism.	11b. Unit is off, when restarted mechanism jerks or immediately stalls or shears pin.	11b. 1. Take clarifier out of service, dewater, and then remove rags and debris from tank. Check squeegees and plow for damage and proper clearances. 2. When repaired, run mechanism through at least two complete revolutions to check level, clearances, and smoothness of operation before refilling. 3. Check headworks screening equipment for proper debris removal and condition.
Rectangular clarifiers: Frequent shear pin, flights, chain breakage.	11a. Excessive sludge accumulation.	11a. Smooth operation, solids concentration, and sludge blanket depth.	11a. Increase frequency of sludge pumping to reduce sludge blanket level.

INDICATOR/OBSERVATION	PROBABLE CAUSE	CHECK OR MONITOR	SOLUTION
11. Rectangular clarifiers: Frequent shear pin, flights, chain breakage. (continued)	11b. Rags or debris entangled in collector mechanism.	11b. Collector jerks or jumps, flights offset.	11b. 1. Take clarifier out of service, dewater, remove rags and debris. 2. Replace broken or damaged flights. 3. Realign collector chains to obtain parallel flights. 4. Check and replace worn wearing shoes. 5. Check rails for damage. 6. Lube rails and wearing shoes before restarting collectors. 7. Run collectors at least two revolutions, check for smooth travel on floor and rails, proper clearance of flight ends to walls and floor before restoring tank to service.
	11c. Chain wear.	11c. Chain on one side travels faster than chain on opposite side.	11c. Check chain wear. Replace entire chain, not just a few links.
	11d. Improper shear pin size.	11d. Shear pin fails but no apparent cause.	11d. Change shear pin size to meet manufacturer's specifications.
12. Excessive corrosion on unit.	12a. Septic wastewater. Inappropriate/inadequate protective coating.	12a. Color and odor of wastewater. Protective coating.	12a. Paint surfaces with corrosion-resistant paint. See Section 2, "Black and odorous septic wastewater entering clarifier."
	12b. Electrolysis.	12b. Pitting and tuberculation.	12b. Sacrificial anodes or cathodic protection.
13. Noisy chain drive.	13a. Moving parts rub stationary parts.	13a. Alignment.	13a. Tighten and align casing and chain. Remove dirt or other interfering matter.
	13b. Loose chain.	13b. Drive components.	13b. Maintain taut chain at all times.
	13c. Faulty lubrication.	13c. Lubrication.	13c. Lubricate properly.
	13d. Misalignment or improper assembly.	13d. Alignment and assembly.	13d. Correct alignment and assembly of drive.
	13e. Worn parts.	13e. Drive components.	13e. Replace worn chain or bearings. Reverse worn sprockets before replacing.
	13f. Chain does not fit sprockets (should only occur on a new or rebuilt unit).	13f. Drive components.	13f. Replace with correct parts.
14. Rapid wear of chain drive.	14a. Faulty lubrication.	14a. Lubrication.	14a. Lubricate properly.
	14b. Loose or misaligned parts.	14b. Alignment.	14b. Align and tighten entire drive.
15. Chain climbs sprockets.	15a. Loose chain.	15a. Chain and sprockets.	15a. Tighten.
	15b. Worn out chain or worn sprockets.	15b. Chain and sprockets.	15b. Replace chain. Reverse or replace sprockets.
	15c. Chain does not fit sprockets.	15c. Chain and sprockets.	15c. Replace chain. Reverse or replace sprockets.

INDICATOR/OBSERVATION	PROBABLE CAUSE	CHECK OR MONITOR	SOLUTION
16. Stiff chain.	16a. Faulty lubrication.	16a. Lubrication.	16a. Lubricate properly.
	16b. Rust or corrosion.	16b. Drive chain.	16b. Clean and lubricate.
	16c. Misalignment or improper assembly.	16c. Alignment and assembly.	16c. Correct alignment and assembly of drive.
	16d. Worn out chain or worn sprockets.	16d. Drive chain/sprockets.	16d. Replace chain. Reverse or replace sprockets.
17. Broken chain or sprockets in chain drive system.	17a. Shock or overload.	17a. When tank has been out of service, make inspection of collector system, particularly rectangular tanks. [See 11b]	17a. Avoid shock and overload or isolate through couplings. [See 11b also]
	17b. Worn chain or chain that does not fit sprockets.	17b. Drive chain/sprockets.	17b. Replace chain. Reverse or replace sprockets.
	17c. Rust or corrosion.	17c. Drive chain/sprockets.	17c. Replace parts. Correct corrosive conditions.
	17d. Misalignment.	17d. Alignment.	17d. Correct alignment.
	17e. Interference.	17e. Drive chain/sprockets.	17e. Make sure no debris interferes between chain and sprocket teeth. Loosen chain if necessary for proper clearance over sprocket teeth.
	17f. Improperly installed/rated shear pin.	17f. Shear pin.	17f. Replace pin with one meeting manufacturer's specifications. Install pin properly.
18. Oil seal leak.	18a. Oil seal failure.	18a. Oil seal.	18a. Replace seal.
	18b. Misalignment.	18b. Shaft alignment.	18b. Correct alignment.
	18c. Scored shaft.	18c. Shaft surface.	18c. Repair or replace shaft.
19. Bearing or universal joint failure.	19a. Excessive wear.	19a. Wear.	19a. Replace joint or bearing.
	19b. Lack of lubrication.	19b. Lubrication.	19b. Lubricate joint and/or bearings.
	19c. Misalignment.	19c. Shaft alignment.	19c. Correct alignment.
20. Binding of sludge pump shaft.	20a. Improper adjustment of packing.		20a. Adjust packing.
	20b. Bent shaft due to excessive discharge pressure (closed valve).		20b. Replace shaft.
	20c. Binding of a pump shaft could be caused by debris in the impeller/casing or rotor/stator area of certain types of sludge pumps.		20c. Remove debris.

5.2 SAMPLING AND LABORATORY ANALYSIS

5.20 Need for Sampling and Analysis

Proper analysis of representative samples is the only conclusive method of measuring the efficiency of clarifiers. Tests may be conducted at the plant site where the sample is collected or in the laboratory. The tests performed depend on the downstream treatment processes.

Detailed procedures for performing control tests on primary treatment and sedimentation processes are outlined in Volume II, Chapter 16, "Laboratory Procedures and Chemistry." The frequency of testing and the expected ranges will vary from plant to plant. Strength of the wastewater, freshness, characteristics of the water supply, weather, and industrial wastes will all serve to affect the "common" range of the various test results.

Tests	Frequency	Location	Common Range For Primary Clarifiers
1. Dissolved Oxygen (DO)	Daily	Effluent	0 - 2 mg/L
2. Settleable Solids	Daily	Influent Effluent	5 - 15 mL/L 0.3 - 3 mL/L
3. pH	Daily	Influent Effluent	6.5 - 8.0* 6.5 - 8.0*
4. Temperature	Daily	Influent	50 - 85°F* 10 - 30°C*
5. BOD	Weekly (Minimum)	Influent Effluent	150 - 400 mg/L 50 - 150 mg/L
6. Suspended Solids	Weekly (Minimum)	Influent Effluent	150 - 400 mg/L** 50 - 150 mg/L**

Where discharge requirements permit primary treatment only, items 7 and 8 may be appropriate.

7. Chlorine Residual (if needed)	Daily	Plant Effluent	0.5 - 3.0 mg/L Depends on effluent requirement
8. Coliform Group Bacteria (if needed)	Weekly	Effluent	5,000 - 2,000,000 per 100 mL Depends on effluent requirement

* Depends on region, water supply, and discharges to the collection system.

** Also may be recorded as packed volume from centrifuge test.

5.21 Sampling

Samples of the influent to the clarifier and the effluent from it will give you information on the clarifier efficiency. As with all sampling, the purpose is to collect samples that represent the true nature of the wastewater or stream being sampled. The amount of solids, BOD, bacteria, and pH and temperature levels will probably vary throughout the day, week, and year. You must determine these variations in order to understand how well your clarifier is doing its job. Details on laboratory analysis and data recording are contained in Volume II, Chapter 16, "Laboratory Procedures and Chemistry."

Test results obtained from collected samples may determine the number of clarifier units that should be in operation in order to obtain the best degree of treatment. Plants with more than one clarifier can have too many as well as not enough clarifiers in operation.

5.22 Calculation of Clarifier Efficiency

To calculate the efficiency of any wastewater treatment process, you need to collect samples of the influent and the effluent of the process, preferably *COMPOSITE SAMPLES*[8] for a 24-hour period. Next, measure the particular water quality indicators (for example, BOD, suspended solids) you are interested in and calculate the treatment efficiency. Calculations of treatment efficiency are for process control purposes. Your main concern must be the quality of the plant effluent, regardless of percent of wastes removed.

EXAMPLE 1

The influent BOD to a primary clarifier is 200 mg/L, and the effluent BOD is 140 mg/L. What is the efficiency of the primary clarifier in removing BOD?

Known	Unknown
Influent BOD, mg/L = 200 mg/L	Efficiency, %
Effluent BOD, mg/L = 140 mg/L	

Calculate the BOD removal efficiency.

$$\text{Efficiency, \%} = \frac{(\text{In} - \text{Out})}{\text{In}} (100\%)$$

$$= \frac{(200 \text{ mg/}L - 140 \text{ mg/}L)}{200 \text{ mg/}L} (100\%)$$

$$= \left(\frac{60 \text{ mg/}L}{200 \text{ mg/}L}\right) (100\%)$$

$$= (.30)(100\%)$$

$$= 30\% \text{ BOD Removal}$$

NOTE: The same formula is used to calculate clarifier removal efficiency for all the water quality indicators (parameters) listed in Section 5.23.

---

[8] *Composite (come-PAH-zit) (Proportional) Sample.   A composite sample is a collection of individual samples obtained at regular intervals, usually every one or two hours during a 24-hour time span. Each individual sample is combined with the others in proportion to the rate of flow when the sample was collected. The resulting mixture (composite sample) forms a representative sample and is analyzed to determine the average conditions during the sampling period.*

5.23 Typical Clarifier Efficiencies

Following is a list of some typical percentages for primary clarifier efficiencies:

Water Quality Indicator	Expected Removal Efficiency
Settleable solids	95% to 99%
Suspended solids	40% to 60%
Total solids	10% to 15%
Biochemical oxygen demand	20% to 50%
Bacteria	25% to 75%

pH generally will not be affected significantly by a clarifier. You can expect wastewater to have a pH of about 6.5 to 8.0, depending on the region, water supply, and wastes discharged into the collection system.

Clarifier efficiencies are affected by many factors, including:

1. Types of solids in the wastewater, especially if there is a significant amount of industrial wastes.

2. Age (time in collection system) of wastewater when it reaches the plant. Older wastewater becomes stale or septic; solids do not settle properly because gas bubbles cling on the particles and tend to hold them in suspension.

3. Rate of wastewater flow as compared to design flow. This is called the "hydraulic loading."

4. Mechanical conditions and cleanliness of clarifier.

5. Proper sludge withdrawal. If sludge is allowed to remain in the tank, it tends to gasify and the entire sludge blanket (depth) may rise to the water surface in the clarifier.

6. Suspended solids that are returned to the primary clarifiers from other treatment processes may not settle completely. Sources of these solids include waste activated sludge, digester supernatant, and sludge dewatering facilities (centrate from centrifuges and filtrate from filters).

5.24 Response to Poor Clarifier Performance

If laboratory analysis or visual inspection indicates that a clarifier is not performing properly, the source of the problem must be identified and corrective action taken. Listed below are clarifier problems with related items to be checked to identify the source of the problem. (Also see Section 5.16, "Troubleshooting.")

Problem	Check Items (pages 129 and 130)
1. Floating chunks of sludge	1, 2, 3, 4, 5
2. Large amounts of floating scum	2.3*, 2.4*, 2.5*
3. Loss of solids over effluent weirs	1, 2, 3, 4, 5, 2.7*, 2.8*
4. Low removal efficiencies	5, 6a
5. Low pH plus odors	1, 2, 3, 4, 5, 6
6. Deep sludge blanket, but pumping thin sludge	3, 2.1*, 2.2*, 2.3*, 2.6*
7. Sludge collector mechanism jerks or jumps	6
8. Sludge collector mechanism will not operate. Drive motor thermal overloads, or overload protective switches keep tripping.	6

* Check Item 2 is divided into two parts, (a) circular clarifier, and (b) rectangular clarifier. If you have a floating scum problem (Problem 2 above), check under 2. COLLECTOR MECHANISM (page 130) either section (a) circular, or (b) rectangular, items 3, 4, and 5, depending on the type of clarifier in your plant.

CHECK ITEMS

1. SLUDGE PUMP[9]

    a. *PISTON PUMPS*
      (1) Ball-check seating
      (2) Shear pin
      (3) Packing adjustment
      (4) Drive belts
      (5) High pressure switch
      (6) Pumping time

    b. *POSITIVE DISPLACEMENT SCRU (SCREW) PUMPS*
      (1) Pump gas bound
      (2) Rotor plugged
      (3) Drive belt
      (4) Packing adjustment
      (5) Pumping time

    c. *CENTRIFUGAL PUMPS*
      (1) Pump gas bound
      (2) Packing adjustment
      (3) Impeller plugged
      (4) Pumping time

---

[9] *For more information on pumps, see Volume II, Chapter 15, "Maintenance."*

d. *AIR INJECTOR*
   (1) Air supply
   (2) Foot valves
   (3) Slide valves
   (4) Electrodes
   (5) Pumping time

2. COLLECTOR MECHANISM

   a. *CIRCULAR CLARIFIER*
      (1) Drive motor
      (2) Shear pin
      (3) Overload switch
      (4) Skimmer dump arm
          (a) operation
          (b) rubber squeegee
      (5) Scum trough
      (6) Scum box

   b. *RECTANGULAR CLARIFIER*
      (1) Drive motor
      (2) Shear pin
      (3) Clutch and drive gear
      (4) Flights
      (5) Scum trough
      (6) Skimmer operation
      (7) Cross collector
      (8) Inlet line or slot
      (9) Target baffle

3. PIPES AND SLUDGE SUMP

   Sometimes pipes or sumps may be cleaned by back-flushing.

4. QUALITY OF SUPERNATANT RETURN FROM DIGESTER

5. INFLUENT

   a. *CHANGE IN COMPOSITION OR TEMPERATURE*

   b. *CHANGE IN FLOW RATE*
      An increase in flow rate can cause hydraulic overload. This can be determined by calculating the detention time, weir overflow rate, and surface loading rate (Section 5.61). If a tank is hydraulically underloaded, effluent could be recirculated back to the primary clarifier to reduce the length of detention time.

6. JERKING, JUMPING, OR STALLED COLLECTOR MECHANISM

   a. *SLUDGE BLANKET TOO DEEP*
      Pump out sludge if mechanism is all right.

   b. *DRIVE UNIT MAY HAVE BAD SPROCKET OR DEFECTIVE CHAIN LINK*

   c. *BROKEN FLIGHT, OR A ROCK OR STICK JAMMED BETWEEN FLIGHT OR SQUEEGEE BLADE AND FLOOR OF TANK*

      If item (b) or (c) occurs, or the mechanism won't operate properly, the tank must be dewatered. Never attempt to back up or help pull a collector mechanism because severe equipment damage will result.

Your corrective action will depend on the source of the problem and the facilities available in your plant.

QUESTIONS

Write your answers in a notebook and then compare your answers with those on page 158.

5.2A   At what two points should samples be collected for measuring clarifier efficiency?

5.2B   List five basic laboratory measurements used to determine clarifier efficiency.

5.2C   About what percentage of settleable solids should you expect to be removed by your clarifier?

5.2D   What is the suspended solids removal efficiency of a primary clarifier if the influent concentration is 300 mg/L and the effluent is 120 mg/L?

5.3 SLUDGE AND SCUM PUMPING

The particles that settle to the floor of the clarifier are called "sludge." The accumulated sludge should be removed frequently. This is accomplished by mechanical cleaning devices and pumps in most tanks. (See Figures 5.3, 5.4, and 5.5.) Mechanically cleaned tanks need not be shut down for cleaning. *SEPTIC CONDITIONS* may develop rapidly in primary clarifiers if sludge is not removed at regular intervals. The proper interval is dependent on many conditions and may vary from thirty minutes to eight hours, and as much as twenty-four hours in a few instances. Experience will dictate the proper frequency of removal. Sludge septicity can be recognized when *SLUDGE GASIFICATION*[10] causes large clumps of sludge to float on the water surface. Septic sludge is generally very odorous and acid (has a low pH).

As thick a sludge as possible should be pumped from the clarifier sump with the least amount of water. The amount of sludge solids in the water affects the volume of sludge pumped and the digester operation. A good, thick primary sludge will contain from 4.0 to 8.0 percent dry solids as indicated by the Total Solids Test in the laboratory. Conditions that may affect sludge concentration are the *SPECIFIC GRAVITY*,[11] size and shape of the particles, temperature of wastewater, and turbulence in the tank.

Withdrawal (pumping) rates should be slow in order to prevent pulling too much water with the sludge. While the sludge is being pumped, take samples frequently and examine them visually for excess water. If the samples show a "thin" sludge, it is time to stop pumping. Practice learning to recognize the differences between thin or concentrated sludges. There are several methods for determining "thick" or "thin" sludge without a laboratory analysis:

1. Sound of the sludge pump. The sludge pump will usually have a different sound when the sludge is thick than when it is thin.

---

[10] *Sludge Gasification.   A process in which soluble and suspended organic matter are converted into gas by anaerobic decomposition. The resulting gas bubbles can become attached to the settled sludge and cause large clumps of sludge to rise and float on the water surface.*

[11] *Specific Gravity.   (1) Weight of a particle, substance or chemical solution in relation to the weight of an equal volume of water. Water has a specific gravity of 1.000 at 4°C (39°F). Wastewater particles or substances usually have a specific gravity of 0.5 to 2.5. (2) Weight of a particular gas in relation to the weight of an equal volume of air at the same temperature and pressure (air has a specific gravity of 1.0). Chlorine has a specific gravity of 2.5 as a gas.*

2. Pressure gage readings. Pressure will be higher on the discharge side of the pump and vacuum will be greater on the inlet (suction) side of the pump when sludge is thick.

3. Sludge density gage readings.

4. Visual observation of a small quantity (gallon or less).

5. While sludge is being pumped, watch through a sight glass in the sludge line.

When you learn to use the indicators listed above, you should compare them frequently with lab tests. The laboratory Total Solids Test is the only accurate method for determining exact density. However, this analytical procedure is too slow for controlling a routine pumping operation. Many operators use the centrifuge test (see Volume II, Chapter 16, "Laboratory Procedures and Chemistry") to obtain quick results.

Floating material (scum) may leave the clarifier in the effluent unless a method has been provided for holding it back. To collect scum, a baffle is generally provided at some location in the tank. Primary clarifiers often have a scum collection area where the scum is skimmed off by some mechanical method, usually a skimming arm or a paddle wheel. If mechanical methods are not provided, use hand tools such as a skimming dipper attached to a broom handle.

Frequently check the scum trough to be sure it is working properly. Clean the box with a brush and hot water. Scum may be disposed of by burning or burial.

See Volume II, Chapter 15, "Maintenance," Section 15.8 for details on how to unplug pipes and pumps.

QUESTIONS

Write your answers in a notebook and then compare your answers with those on page 158.

5.3A   How often should sludge be removed from a clarifier?

5.3B   How can you tell when to stop pumping sludge?

5.3C   How can floating material (scum) be kept from the clarifier effluent?

5.4   GENERAL MAINTENANCE

Following are some hints to help you keep your clarifiers operating properly:

1. Maintain a record-and-file system for future reference. This should contain sheets to write down a description and date for all repairs and regular maintenance activities such as lubrication. Other items to be kept in the file are operating instruction manuals; brochures; and names, addresses, and telephone numbers of manufacturers' representatives. Some plants keep this information on a computer.

2. *ALWAYS* lubricate equipment at the intervals recommended by the manufacturer and use the *PROPER* lubricants (follow manufacturers' recommendations). Be sure that you do not overlubricate.

3. Clean all equipment and structures regularly. Remove floating material and algae from inlet baffles and effluent weirs and launders. Keeping scum removal equipment clean and properly adjusted will help prevent odors.

4. Inspect and correct (if possible) all peculiar noises, leaks, pressure and vacuum gage irregularities, belts, electrical systems, and safety devices.

5. When a sedimentation tank must be drained for inspection or repairs, keep wooden flights moist by periodic sprinkling with a hose to prevent cracking and warping.

6. Keep the weirs level. This helps prevent short-circuiting which reduces the efficiency of the clarifier.

5.5   SAFETY

1. *GASES*

Any enclosed area, such as a wet well for a pump, may contain accumulated poisonous, asphyxiating, or explosive gases if ventilation is not proper. The most common of these gases are:

a. Hydrogen Sulfide (H_2S). Causes a "rotten egg" odor. This strong odor, noticeable at low concentrations, is a poor warning sign because it may quickly dull your sense of smell (this is called olfactory fatigue), and some people are physically unable to smell hydrogen sulfide. Hydrogen sulfide readily combines with oxygen to form sulfuric acid which will dissolve concrete. If you breathe too much hydrogen sulfide, respiratory irritative effects may occur (bronchitis, pulmonary edema). These effects occur due to the formation of alkali sulfide when the gas comes in contact with moist tissue. Remember, hydrogen sulfide in gaseous form is extremely toxic and it can be flammable and explosive when mixed with the proper amount of air (oxygen).

b. Chlorine (Cl_2). Very irritating to the eyes, skin, and mucous membranes. Causes death by suffocation (asphyxiation) and formation of acid in the lungs. Chlorine acts as an asphyxiant by causing cramps in the muscles of the larynx (choking), swelling of the mucous membranes, and suspension of respiration (syncope). Chlorine is extremely dangerous (toxic). The proper type of breathing equipment (self-contained air) should be readily available when working with chlorine.

c. Carbon Dioxide (CO_2). Odorless, tasteless. This can cause asphyxiation by displacing oxygen in a poorly ventilated area. Concentrations of 10 percent (100,000 ppm) can produce unconsciousness and death from oxygen deficiency.

d. Carbon Monoxide (CO). Colorless, odorless, nonirritating, flammable, explosive. Look out for carbon monoxide around gas engines or leaky gas systems in poorly ventilated places. Interferes with the blood's ability to carry oxygen; may cause asphyxiation.

e. Gasoline and other petroleum products. May cause fires or explosions, or displace oxygen and asphyxiate you.

f. Methane (CH_4). Explosive, odorless, and may cause asphyxiation.

For a detailed discussion of the hazards and safety precautions when dangerous gases may be present, refer to Volume II, Chapter 14, "Plant Safety." The *NEW YORK MANUAL*, Volume II, pages 13-9 to 13-11 (depending on edition), Table 13.1, "Common Dangerous Gases Encountered in Sewers and at Sewage Treatment Plants," contains information on methods of testing for gases. Contact your local *OSHA*[12] office for approved procedures and equipment.

2. *FALLS*

Avoid falls by:

a. Cleaning up oil and grease slicks on walkways *PROMPTLY.*

b. *WALKING, NOT RUNNING*, when near open tanks.

c. Avoiding clutter. Pick up and store hoses, ropes, cables, tools, buckets, and lumber.

d. Not sitting on, climbing through, or hanging over guardrails or handrails.

e. Providing gratings, deck covers, or safety chains on or around openings to pits below floor level.

3. *DROWNING*

To prevent drowning:

a. Put handrails and proper walkways by all open tanks.

b. Cover open pits with gratings and deck plates.

c. Have approved life preservers and life lines handy to throw to anyone who may fall in. Appropriate equipment should be worn when necessary.

d. Use the buddy system when working around or across a water surface. (Example: Skimming scum manually.)

4. *STRAINS AND OVEREXERTION*

Use proper tools or equipment:

a. To move stuck or reluctant valves.

b. To lift heavy objects.

5. *ELECTRIC SHOCK*

a. Do not use water for cleaning electrical panels, electric motors, or other electrical equipment.

b. Use rubber floor mats in front of electrical panels.

c. Do not work on electrical equipment unless you are a qualified electrician and authorized to do so.

on
SEDIMENTATION AND FLOTATION

Please answer the discussion and review questions next.

---

[12] *OSHA (O-shuh).   The Williams-Steiger **O**ccupational **S**afety and **H**ealth **A**ct of 1970 (OSHA) is a federal law designed to protect the health and safety of industrial workers and treatment plant operators. It regulates the design, construction, operation and maintenance of industrial plants and wastewater treatment plants. The Act does not apply directly to municipalities, EXCEPT in those states that have approved plans and have asserted jurisdiction under Section 18 of the OSHA Act. HOWEVER, CONTRACT OPERATORS AND PRIVATE FACILITIES DO HAVE TO COMPLY WITH OSHA REQUIREMENTS. Wastewater treatment plants have come under stricter regulation in all phases of activity as a result of OSHA standards. OSHA also refers to the federal and state agencies which administer the OSHA regulations.*

DISCUSSION AND REVIEW QUESTIONS

Chapter 5.   SEDIMENTATION AND FLOTATION

(Lesson 1 of 3 Lessons)

At the end of each lesson in this chapter you will find some discussion and review questions. The purpose of these questions is to indicate to you how well you understand the material in the lesson. Write the answers to these questions in your notebook before continuing.

1. What is the function of a primary clarifier?

2. What items should be inspected before starting a clarifier or when a clarifier has been taken out of service for cleaning or repair?

3. How would you dispose of sludge in a primary clarifier containing a toxic waste?

4. Calculate the efficiency of a clarifier removing BOD if the influent BOD is 260 mg/L and the effluent is 155 mg/L. Show your work.

5. What would you do if the solids- and BOD-removal efficiencies of a primary clarifier suddenly dropped and the effluent appeared to contain more solids than usual?

6. How often should sludge be pumped from a primary clarifier?

7. What safety precautions should you take to avoid accidents when working around a treatment plant? List five that you consider most important.

CHAPTER 5.   SEDIMENTATION AND FLOTATION

(Lesson 2 of 3 Lessons)

5.6   PRINCIPLES OF OPERATION

5.60   Types of Units

Sedimentation and flotation units are designed to remove physically those solids that will settle easily to the bottom or float easily to the top. Sedimentation is usually the principal basis of design in such units and will be discussed in more detail in this section. Flotation of fats, oils, hair, and other light material also is very important to protect the aesthetics (appearance) of receiving waters.

The sedimentation and flotation units commonly found are:

1. Primary clarifiers
2. Secondary clarifiers
3. Combined sedimentation-digestion units
4. Flotation units
5. Imhoff tanks

This section will describe each unit individually as it relates to another process or as a process by itself.

5.61   Primary Clarifiers

The most important function of the primary clarifier is to remove as much settleable and floatable material as possible. Removal of organic settleable solids is very important because they cause a high demand for oxygen (BOD) in receiving waters or subsequent biological treatment units in the treatment plant.

Many factors influence the design of clarifiers. Settling characteristics of suspended particles in water are probably the most important considerations. The design engineer must consider the speed at which particles will settle in order to determine the correct dimensions for the tank. Rapid movement of water (velocity) will hold most particles in suspension and carry them along until the velocity of water is slowed sufficiently for the particles to settle. The tendency of particles to float and the rate of downward travel (settling) of a particle are dependent on the weight of the particle in relation to the weight of an equal volume of water (specific gravity), the particle size and shape, and the temperature of the liquid. Organic settleable solids are seldom more than 1 to 5 percent heavier than water; therefore, their settling rates are slow.

If the horizontal velocity of water is slowed to a rate of 1.0 to 2.0 feet (0.3 to 0.6 m) of travel per minute (grit channel velocities were around 1 ft/sec or 0.3 m/sec), most particles with a specific gravity of 1.05 (5% more than water) will settle to the

bottom of the container. Specific gravity of water is 1.000 at 4.0 degrees Celsius (formerly Centigrade) or 39°F; it weighs 8.34 pounds per gallon (1 kg/L). Wastewater solids with a specific gravity of 1.05 will weigh 8.76 pounds per gallon (1.05 times 8.34 pounds equals 8.76 pounds per gallon). The relationship of the particle settling rate to liquid velocity may be explained very simply by use of a sketch (Figure 5.6).

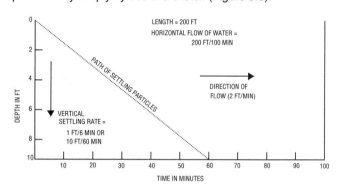

Fig. 5.6   Path of settling particle

Suppose the liquid velocity is horizontal at the rate of 2.0 feet per minute (0.6 m/min) and the tank is 200 feet (60 meters) long. It will take 100 minutes (200 ft divided by 2.0 ft/min) to travel through the tank. If the particle, during its diagonal course of travel, settles vertically toward the bottom of the tank at a rate of 1.0 foot (0.3 m) in 6 minutes, it will rest on the floor of the tank in 60 minutes if the tank is 10 feet (3 meters) deep. If the particle settles at the rate of 10 feet (3 meters) in 60 minutes, it should settle in the first 60 percent portion of the tank because the liquid surrounding it requires 100 minutes to flow through the tank.

There are many factors that will influence settling characteristics in a particular clarifier. A few of the more common ones are temperature, short circuits, detention time, weir overflow rate, surface loading rate, and solids loading. These factors are discussed in the following paragraphs.

1. *TEMPERATURE.* Water expands as temperature increases (above 4°C) or contracts as temperature decreases (down to 4°C). Below 4°C, the opposite is true. In general, as water temperature increases, the settling rate of particles increases; as temperature decreases, so does the settling rate. *MOLECULES*[13] of water react to temperature changes. They are closer together when liquid temperature is lower; thus, *DENSITY*[14] increases and water

---

[13] *Molecule (MOLL-uh-KULE).   The smallest division of a compound that still retains or exhibits all the properties of the substance.*

[14] *Density (DEN-sit-tee).   A measure of how heavy a substance (solid, liquid or gas) is for its size. Density is expressed in terms of weight per unit volume, that is, grams per cubic centimeter or pounds per cubic foot. The density of water (at 4°C or 39°F) is 1.0 gram per cubic centimeter or about 62.4 pounds per cubic foot. If one cubic centimeter of a substance (such as iron) weighs more than 1.0 gram (higher density), it will sink or settle out when put in water. If it weighs less (lower density, such as oil), it will rise to the top and float. Sludge density is normally expressed in grams per cubic centimeter.*

becomes heavier per given volume because there is more of it in the same space. As water becomes more dense, the density difference between water and solid particles becomes less; therefore the particles settle more slowly. This is illustrated in Figure 5.7.

WATER MOLECULES ARE EXPANDED. THIS ALLOWS FOR EASY SETTLING.

WATER MOLECULES ARE CLOSE. PARTICLE SETTLING DIFFICULT.

WARM WATER
100°C (LESS DENSE)
(7.989 LBS/GAL)
(0.96 kg/L)

COLD WATER
4°C (MORE DENSE)
(8.335 LBS/GAL)
(1.0 kg/L)

Fig. 5.7   Influence of temperature on settling

2. *SHORT CIRCUITS.* As wastewater enters the settling tank, it should be evenly dispersed across the entire cross section of the tank and should flow at the same velocity in all areas toward the discharge end. When the velocity is greater in some sections than in others, serious short-circuiting may occur. The high velocity area may decrease the detention time in that area, and particles may be held in suspension and pass through the discharge end of the tank because they do not have time to settle out. On the other hand, if velocity is too low, undesirable septic conditions may occur. Short-circuiting may easily begin at the inlet end of the sedimentation tank (Figure 5.8). This is usually prevented by the use of weir plates, baffles, port openings, and by proper design of the inlet channel.

Short-circuiting also may be caused by turbulence and stratification of density layers due to temperature or salinity. Temperature layers can cause short-circuiting when a warm influent flows across the top of cold water in a settling tank or when a cold influent flows under warm water in a settling tank.

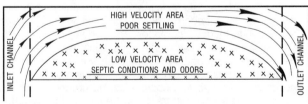

Top View Looking Down

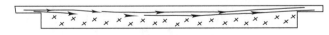

Side View—Warm Influent

Side View—Cold Influent

Fig. 5.8   Short-circuiting

3. *DETENTION TIME.* Wastewater should remain in the clarifier long enough to allow sufficient settling time for solid particles. If the tank is too small for the quantity of flow and the settling rate of the particles, too many particles will be carried out the effluent of the clarifier. The relationship of *DETENTION TIME* to *SETTLING RATE* of the particles is important. Most engineers design settling tanks for about 2.0 to 3.0 hours of detention time. This is, of course, flexible and dependent on many circumstances. Detention time can be calculated by use of two known factors:

1. Flow in gallons per day (GPD), and
2. Tank dimensions or volume.

EXAMPLE 2

The flow is 3.0 million gallons per day (MGD), or 3,000,000 gal/day. Tank dimensions are 60 feet long by 30 feet wide by 10 feet deep. What is the detention time?

FORMULAS:

$$\text{Detention Time, hr} = \frac{\text{Tank Volume, cu ft} \times 7.5 \text{ gal/cu ft} \times 24 \text{ hr/day}}{\text{Flow, gal/day}}$$

Tank Volume, cu ft = Length, ft x Width, ft x Depth, ft [15]

CALCULATIONS:

Tank Volume, cu ft = Length, ft x Width, ft x Depth, ft

$$= 60 \text{ ft} \times 30 \text{ ft} \times 10 \text{ ft}$$

$$= 18,000 \text{ cu ft}$$

$$\text{Detention Time, hr} = \frac{\text{Tank Volume, cu ft} \times 7.5 \text{ gal/cu ft} \times 24 \text{ hr/day}}{\text{Flow, gal/day}}$$

$$= \frac{18,000 \text{ cu ft} \times 7.5 \text{ gal/cu ft} \times 24 \text{ hr/day}}{3,000,000 \text{ gal/day}}$$

$$= \frac{3,240,000 \text{ gal-hr/day}}{3,000,000 \text{ gal/day}}$$

$$= 1.08 \text{ hours}$$

EVALUATION. If detention time is only 1.08 hours and if laboratory tests indicate poor removal of solids, then additional tank capacity should be placed into operation (if available) in order to obtain additional detention time. Keep in mind, however, that flows fluctuate considerably during the day and night and any calculated detention time is for a specific flow.

DISCUSSION. The formula given in this section allows you to calculate the *THEORETICAL* detention time. *ACTUAL* detention time is often less than the detention time calculated using the formula and can be measured by the use of dyes, tracers, or floats.

---

[15] *For a circular clarifier, Tank Volume, cu ft = 0.785 x (Diameter, ft)^2 x Depth, ft.*

4. *WEIR OVERFLOW RATE.* Wastewater leaves the clarifier by flowing over weirs and into effluent troughs (*LAUNDERS*[16]) or some type of weir arrangement. The number of *LINEAL*[17] feet of weir in relation to the flow is important to prevent short circuits or high velocity near the weir or launder which might pull settling solids into the effluent. The weir overflow rate is the number of gallons of wastewater that flow over one lineal foot of weir per day. Most designers recommend about 10,000 to 20,000 gallons per day per lineal foot of weir. Higher weir overflow rates have been used for materials with a high settling rate or for intermediate treatment. Secondary clarifiers and high effluent quality requirements generally need lower weir overflow rates than would be acceptable for primary clarifiers. The calculation for weir overflow rate requires two known factors:

1. Flow in GPD, and
2. Lineal feet of weir.

EXAMPLE 3

The flow is 5.0 MGD in a circular tank with a 90-foot *WEIR DIAMETER.*[18] What is the weir overflow rate?

FORMULAS:

$$\text{Weir Overflow, GPD/ft} = \frac{\text{Flow Rate, GPD}}{\text{Length of Weir, ft}}$$

Length of Circular Weir = 3.14 x Weir Diameter, ft

CALCULATIONS:

$$
\begin{aligned}
\text{Length of Circular Weir, ft} &= 3.14 \times (\text{Weir Diameter, ft}) \\
&= 3.14 \times 90 \text{ ft} \\
&= 283 \text{ Lineal Feet of Weir}
\end{aligned}
$$

$$
\begin{aligned}
\text{Weir Overflow, GPD/ft} &= \frac{\text{Flow Rate, GPD}}{\text{Length of Weir, ft}} \\
&= \frac{5{,}000{,}000 \text{ gal/day}}{283 \text{ ft}} \\
&= 17{,}668 \text{ GPD/ft}
\end{aligned}
$$

5. *SURFACE SETTLING RATE OR SURFACE LOADING RATE.* This rate is expressed in terms of gallons per day per square foot (GPD/sq ft) of tank surface area. Some designers and operators have indicated that the *SURFACE LOADING RATE* has a direct relationship to the settleable solids removal efficiency in the settling tank. The suggested loading rate varies from 300 to 1,200 GPD/sq ft, depending on the nature of the solids and the treatment requirements. Low loading rates are frequently used in small plants in cold climates. In warm regions, low rates may cause excessive detention which could lead to septicity. The calculation for surface loading rate requires two known factors:

1. Flow in GPD, and
2. Square feet of liquid surface area.

EXAMPLE 4

The flow into a clarifier is 4.0 MGD in a tank 90 feet long and 35 feet wide. What is the surface loading rate?

FORMULA:

$$\text{Surface Loading Rate, GPD/sq ft} = \frac{\text{Flow Rate, GPD}}{\text{Area, sq ft}}$$

CALCULATIONS:

$$
\begin{aligned}
\text{Surface Area, sq ft} &= \text{Length, ft} \times \text{Width, ft}[19] \\
&= 90 \text{ ft} \times 35 \text{ ft} \\
&= 3{,}150 \text{ sq ft}
\end{aligned}
$$

$$
\begin{aligned}
\text{Surface Loading Rate, GPD/sq ft} &= \frac{\text{Flow Rate, GPD}}{\text{Area, sq ft}} \\
&= \frac{4{,}000{,}000 \text{ GPD}}{3{,}150 \text{ sq ft}} \\
&= 1{,}270 \text{ GPD/sq ft}
\end{aligned}
$$

6. *SOLIDS LOADING.* The term "solids loading" is used to indicate the amount of solids that can be removed daily by a clarifier for each square foot of clarifier liquid surface area. If the solids loading increases above design values, you can expect an increase in effluent solids. This concept can be applied to secondary clarifiers and gravity or flotation sludge thickeners. Loading rates are expressed in pounds per day per square foot (lbs/day/sq ft) and depend on the nature of the solids and treatment requirements. To calculate the solids loading requires three known factors:

1. Flow in MGD,
2. Suspended solids concentration in mg/L, and
3. Liquid surface area in square feet.

---

[16] Launders (LAWN-ders).   Sedimentation tank effluent troughs, consisting of overflow weir plates. When the flow leaves a sedimentation unit, it usually flows into a trough after it leaves the tank. The top edge of the trough, over which wastewater flows as it enters the trough, is considered a weir.

[17] Lineal (LIN-e-al).   The length in one direction of a line. For example, a board 12 feet long has 12 lineal feet in its length.

[18] Weir (weer) Diameter.   Many circular clarifiers have a circular weir within the outside edge of the clarifier. All the water leaving the clarifier flows over this weir. The diameter of the weir is the length of a line from one edge of a weir to the opposite edge and passing through the center of the circle formed by the weir.

[19] For a circular clarifier, Surface Area, sq ft = 0.785 x (Diameter, ft)^2.

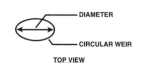

EXAMPLE 5

A circular secondary clarifier with a diameter of 100 feet treats a flow of 4.5 MGD (3.5 MGD inflow and 1.0 MGD return sludge flow) with a mixed liquor suspended solids concentration of 4,200 mg/L. What is the solids loading in lbs/day/sq ft?

FORMULAS:

Solids Applied, lbs/day = Flow, MGD x Conc, mg/L x 8.34 lbs/gal

$$\text{Solids Loading,} \atop \text{lbs/day/sq ft} = \frac{\text{Solids Applied, lbs/day}}{\text{Surface Area, sq ft}}$$

CALCULATIONS:

$$\begin{aligned}\text{Solids Applied,} \atop \text{lbs/day} &= \text{Flow, MGD x Conc, mg/L x 8.34 lbs/gal}^{20}\\ &= 5 \text{ MGD x 4,200 mg/L x 8.34 lbs/gal}\\ &= 157,626 \text{ lbs/day}\end{aligned}$$

$$\begin{aligned}\text{Surface Area,} \atop \text{sq ft} &= (0.785)(\text{Diameter, ft})^2\\ &= (0.785)(100 \text{ ft})^2\\ &= 7,850 \text{ sq ft}\end{aligned}$$

$$\begin{aligned}\text{Solids Loading,} \atop \text{lbs/day/sq ft} &= \frac{\text{Solids Applied, lbs/day}}{\text{Surface Area, sq ft}}\\ &= \frac{157,620 \text{ lbs/day}}{7,850 \text{ sq ft}}\\ &= 20.1 \text{ lbs/day/sq ft}\end{aligned}$$

TYPICAL SOLIDS LOADINGS:

Primary Clarifiers	Usually not a design consideration
Secondary Clarifiers (activated sludge)	12 to 30 lbs/day/sq ft
Dissolved-Air Flotation	5 to 40 lbs/day/sq ft
Sludge Thickening	5 to 20 lbs/day/sq ft

DETENTION TIME, WEIR OVERFLOW RATE, SURFACE LOADING RATE, and *SOLIDS LOADING* are four mathematical methods of checking the performance of existing facilities against the design values. However, laboratory analysis of samples is the only reliable method of measuring clarifier efficiency. *IF LABORATORY RESULTS INDICATE A POORLY OPERATING CLARIFIER, THE MATHEMATICAL METHODS MAY HELP YOU TO IDENTIFY THE PROBLEM.*

QUESTIONS

Write your answers in a notebook and then compare your answers with those on pages 158 and 159.

5.6A   What is "short-circuiting" in a clarifier?

5.6B   Why is "short-circuiting" undesirable?

5.6C   How can "short-circuiting" be corrected?

5.6D   A circular clarifier has a diameter of 80 feet and an average depth of 10 feet. The flow of wastewater is 4.0 MGD and the suspended solids concentration is 190 mg/L. Calculate the following:

1. Detention Time, in hours
2. Weir Overflow Rate, in GPD/ft
3. Surface Loading Rate, in GPD/sq ft

5.6E   A circular clarifier has a diameter of 80 feet and an average depth of 10 feet. The clarifier treats 4.0 MGD from the plant inflow plus 1.2 MGD of return sludge flow. The mixed liquor suspended solids concentration is 2,700 mg/L. Calculate the solids loading in lbs/day/sq ft.

5.62   Secondary Clarifiers

Secondary clarifiers usually are located after a biological process in the flow pattern of a treatment plant. (See Figure 5.2.) The most common biological processes are the *ACTIVATED SLUDGE PROCESS*,[21] the *TRICKLING FILTER*,[22] and *ROTATING BIOLOGICAL CONTACTORS*.[23]

---

[20] *The units of this formula can be proved by remembering that 1 liter equals or weighs one million milligrams.*

$$\frac{mg}{L} = \frac{mg}{1{,}000{,}000 \ mg} = \frac{mg}{M \ mg}$$

Therefore,

$$Flow, \ MGD \ x \ Conc, \ mg/L \ x \ 8.34 \ lbs/gal = \frac{M \ gal}{day} \ x \ \frac{mg \ SS}{M \ mg} \ x \ \frac{lbs}{gal} = \frac{lbs \ SS}{day}$$

[21] *Activated Sludge (ACK-ta-VATE-ed sluj) Process.   A biological wastewater treatment process which speeds up the decomposition of wastes in the wastewater being treated. Activated sludge is added to wastewater and the mixture (mixed liquor) is aerated and agitated. After some time in the aeration tank, the activated sludge is allowed to settle out by sedimentation and is disposed of (wasted) or reused (returned to the aeration tank) as needed. The remaining wastewater then undergoes more treatment.*

[22] *Trickling Filter.   A treatment process in which the wastewater trickles over media that provide the opportunity for the formation of slimes or biomass which contain organisms that feed upon and remove wastes from the water being treated.*

[23] *Rotating Biological Contactor (RBC).   A secondary biological treatment process for domestic and biodegradable industrial wastes. Biological contactors have a rotating "shaft" surrounded by plastic discs called the "media." The shaft and media are called the "drum." A biological slime grows on the media when conditions are suitable and the microorganisms that make up the slime (biomass) stabilize the waste products by using the organic material for growth and reproduction.*

In some plants a chemical process may be used instead of a biological process, but the latter is far more common for municipal treatment plants.

5.620 Trickling Filter Clarifiers

A secondary clarifier is used after a trickling filter to settle out SLOUGHINGS[24] from the filter media. Filter sloughings are a product of biological action in the filter; the material is generally quite high in BOD and will lower the effluent quality unless it is removed. A detailed description of trickling filters can be found in Chapter 6.

Secondary tanks following trickling filters may be either circular or rectangular and have sludge collector mechanisms similar to primary clarifiers. Clarifier detention times are about the same as for primary clarifiers, but the surface loading and weir overflow rates are generally lower due to the less dense characteristics of secondary sludges. The following are ranges of loading rates for secondary clarifiers used after trickling filters:

Detention Time — 2.0 to 3.0 hours
Surface Loading — 800 to 1,200 GPD/sq ft
Weir Overflow — 5,000 to 15,000 GPD/lineal ft

The amount of solids settling out in a secondary clarifier following a trickling filter will be very irregular due to a number of varying conditions in the biological treatment process. In general, you can expect to pump about 30 to 40 percent more sludge from the secondary clarifier than from the primary; thus total sludge pumping to the digester will increase by that amount. These figures indicate how the trickling filter "creates" settleable solids which were not present in the raw wastewater in settleable form.

The sludge in the secondary settling tank will usually have characteristics and appearance completely different than the sludge collected in a primary settling tank. This sludge will usually be much darker in color, but should not be gray or black. Sludge will turn black if it is allowed to stay in the secondary clarifier too long. If this happens, then the sludge pumping rate should be increased or the time of pumping lengthened or made more frequent. These sludges generally require frequent pumping. Pumps used to pump trickling filter sludges from secondary clarifiers are similar to raw sludge pumps and may be piston, progressive cavity, or centrifugal-type pumps.

The sludge particle sizes may be very irregular with generally good (rapid) settling characteristics. The sludge may appear to be a fluffy, humus-type material and usually will have little or no odor if sludge removal occurs at regular intervals. Disposal of sludge collected in the final settling tanks depends on the particular plant design and the characteristics of the sludge. Sometimes disposal is accomplished by transferring the sludge to a primary settling tank to be settled with primary sludge. Other times it is transferred directly to the digestion system.

5.621 Activated Sludge Clarifiers

Secondary clarifier tanks, which follow the activated sludge process, are designed to handle large volumes of sludge. They are more conservative in design because the sludge tends to be less dense. The following are ranges of loading rates for secondary clarifiers used after aeration tanks in the activated sludge process:

Detention Time — 2.0 to 3.0 hours
Surface Loading — 300 to 1,200 GPD/sq ft
Weir Overflow — 5,000 to 15,000 GPD/lineal ft
Solids Loading — 24 to 30 lbs/day/sq ft

Their purpose is identical, except that the particles to be settled are received from the aeration tank rather than the trickling filter. Most secondary sedimentation tanks used with the activated sludge process are equipped with mechanisms capable of quickly removing the sludge due to the importance of rapidly returning sludge to the aeration tank. The sludge volume in the secondary tank will be greater from the activated sludge process than from the trickling filter process.

Sludge removal mechanisms in secondary tanks have tended to differ from most primary clarifier mechanisms, especially those in circular clarifiers. These secondary circular clarifiers are designed for continuous sludge removal by HYDROSTATIC SYSTEMS,[25] with the activated sludge being pumped back to the aeration tanks by large-capacity pumps. These pumps usually are of the centrifugal type with variable-speed controls or are of the large air-lift type.

Figures 5.9, 5.10, and 5.11 illustrate three variations of sludge removal mechanisms for secondary clarifiers used in the activated sludge process. These mechanisms are designed to remove the settled activated sludge as rapidly as possible, thus reducing the sludge RETENTION TIME[26] in the clarifiers. Several of these mechanisms have valves or adjustable rings to control the return sludge rates from the different collection points in the clarifier mechanism.

Flows may be regulated from each pipe removing activated sludge from the clarifier in order to control the activated sludge process. The reason for the ability to regulate each pipe is that different activated sludge densities will develop different settling patterns in a particular clarifier. For example, an activated sludge with an SVI[27] of 100 will develop a bell pattern sludge blanket with most of the activated sludge solids settling near

---

[24] Sloughings (SLUFF-ings). Trickling filter slimes that have been washed off the filter media. They are generally quite high in BOD and will lower effluent quality unless removed.

[25] Hydrostatic System. In a hydrostatic sludge removal system, the surface of the water in the clarifier is higher than the surface of the water in the sludge well or hopper. This difference in pressure head forces sludge from the bottom of the clarifier to flow through pipes to the sludge well or hopper.

[26] Retention Time. The time water, sludge or solids are retained or held in a clarifier or sedimentation tank.

[27] Sludge Volume Index (SVI). This is a calculation which indicates the tendency of activated sludge solids (aerated solids) to thicken or to become concentrated during the sedimentation/thickening process. SVI is calculated in the following manner: (1) allow a mixed liquor sample from the aeration basin to settle for 30 minutes; (2) determine the suspended solids concentration for a sample of the same mixed liquor; (3) calculate SVI by dividing the measured (or observed) wet volume (mL/L) of the settled sludge by the dry weight concentration of MLSS in grams/L.

$$SVI, mL/gm = \frac{Settled\ Sludge\ Volume/Sample\ Volume,\ mL/L}{Suspended\ Solids\ Concentration,\ mg/L} \times \frac{1,000\ mg}{gram}$$

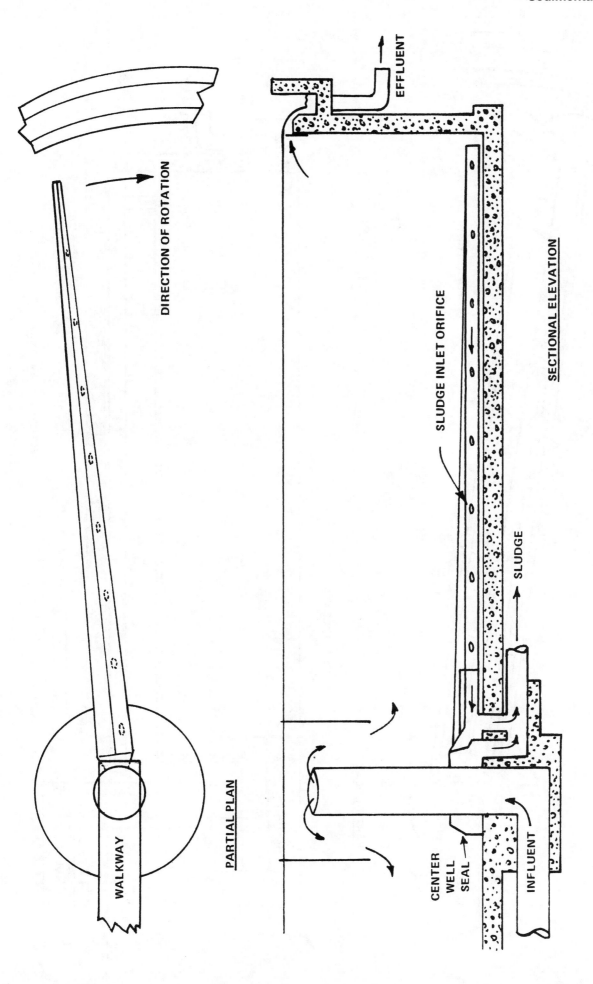

DIRECTION OF ROTATION

WALKWAY

PARTIAL PLAN

EFFLUENT

SLUDGE INLET ORIFICE

SLUDGE

SECTIONAL ELEVATION

CENTER WELL SEAL

INFLUENT

Fig. 5.9   Secondary clarifier sludge removal mechanism
(Link Belt)

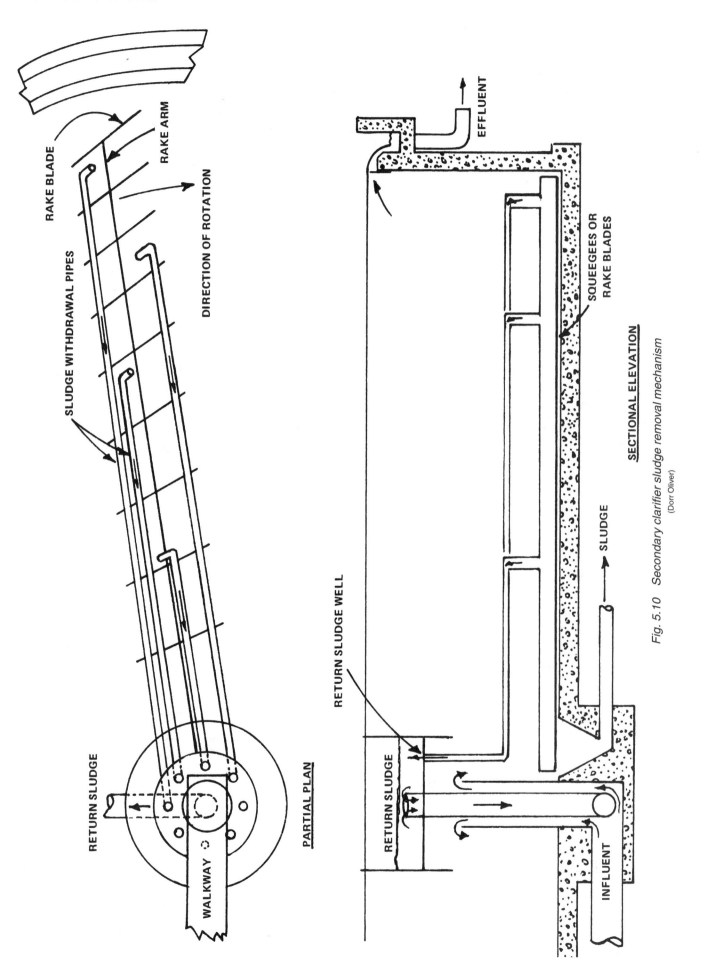

RAKE BLADE

RAKE ARM

DIRECTION OF ROTATION

SLUDGE WITHDRAWAL PIPES

RETURN SLUDGE

WALKWAY

PARTIAL PLAN

EFFLUENT

SQUEEGEES OR RAKE BLADES

RETURN SLUDGE WELL

SLUDGE

RETURN SLUDGE

INFLUENT

SECTIONAL ELEVATION

Fig. 5.10   Secondary clarifier sludge removal mechanism

(Dorr Oliver)

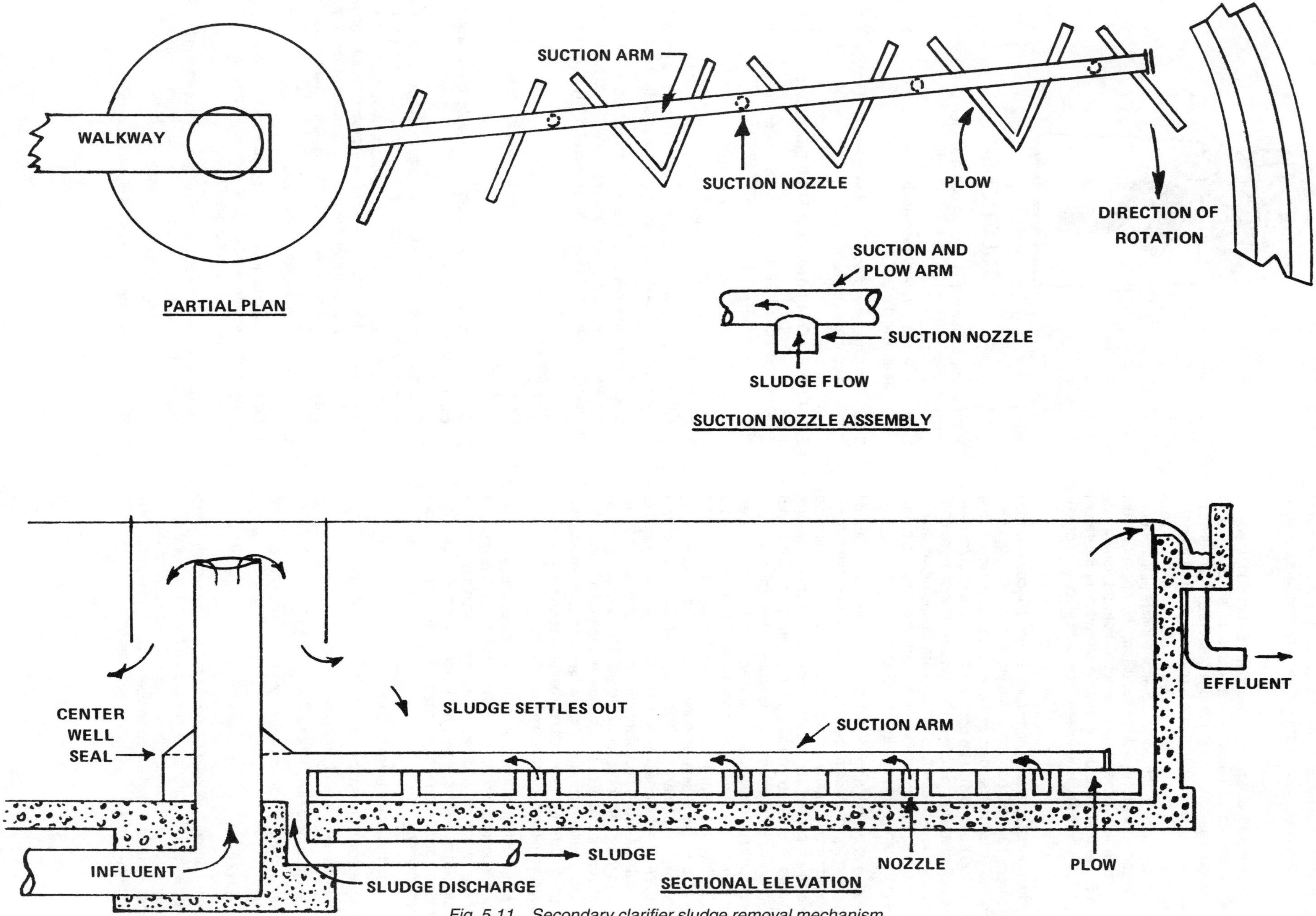

DIRECTION OF ROTATION

PLOW

SUCTION NOZZLE

SUCTION ARM

WALKWAY

PARTIAL PLAN

SUCTION AND PLOW ARM

SUCTION NOZZLE

SLUDGE FLOW

SUCTION NOZZLE ASSEMBLY

EFFLUENT

PLOW

NOZZLE

SUCTION ARM

SLUDGE SETTLES OUT

SLUDGE

SLUDGE DISCHARGE

CENTER WELL SEAL

INFLUENT

SECTIONAL ELEVATION

Fig. 5.11   Secondary clarifier sludge removal mechanism

(Walker Process)

the tank inlet, thus requiring most of the sludge to be removed from the center or inner quarter of the tank floor area (Figure 5.12).

If the activated sludge degrades to an SVI of 500, then the solids settling curve takes on the shape of a bowl (Figure 5.13). Under this condition the sludge gathers at the outside of the clarifier and requires higher return flow rates. Unfortunately the outside sludge return nozzles or pipes must handle a much larger clarifier bottom area than the inside pipes or nozzles. Regardless of the sludge condition, the operator must adjust return flow rates to remove solids from the tank areas where the activated sludge is settling by reducing the return sludge flows from areas where the activated sludge is not settling.

Under normal operating conditions, return sludge rates may range from 10 to 50 percent of the plant inflow. During times when the activated sludge process is upset, return sludge rates of 100 percent may be desirable in order to maintain sufficient activated sludge solids in the system. Under these conditions you must be careful that the return sludge rate does not become too high. During high return rates, the resulting turbulence in the tank can upset the sludge blanket.

Wasting of excess activated sludge from the system should be to some liquid-solids separation process other than the primary clarifier. In many plants, excess or waste activated sludge is processed by separate gravity or flotation sludge thickeners in order to concentrate the sludges to 3.0 to 4.5 percent solids before pumping the solids to the digester or dewatering system for disposal. Waste activated sludge pumped to the primary clarifiers in large plants usually develops into a solids buildup problem in the plant. This buildup consists of a cycle of ever increasing amounts of activated sludge being wasted to the primary clarifier which produces more raw sludge. When additional volumes of raw sludge are pumped to the digester, more supernatant carrying digester solids is returned to the headworks for treatment. These solids produce greater volumes of activated sludge and the cycle continues. This solids buildup creates a solids handling problem and deterioration of the effluent until one of these sources of solids is removed from the cycle to solve the problem.

Of all the different types of clarifiers that an operator must regulate, secondary clarifiers in the activated sludge process are the most critical and require the most attention from the operator. To help the operator regulate clarifier operation, aids have been developed which consist of instrumentation capable of monitoring:

1. Levels of sludge blanket in clarifier,

2. Concentration of suspended solids in clarifier effluent,

3. Control and pacing of return sludge flows,

4. Level of turbidity in clarifier effluent,

5. Concentration of dissolved oxygen (DO) in clarifier effluent, and

6. Level of pH.

Laboratory tests should be conducted to measure all of the above items and to provide a check on the accuracy of the instrumentation. Other tests that should be conducted on the clarifier effluent include biochemical oxygen demand (BOD) and ammonia nitrogen (NH_3-N) measurements.

QUESTIONS

Write your answers in a notebook and then compare your answers with those on page 159.

5.6F   Why are secondary clarifiers needed in secondary treatment plants?

5.6G   What usually is done with the sludge that settles out in secondary clarifiers?

5.7   REVIEW OF PLANS AND SPECIFICATIONS

Plans and specifications for a wastewater treatment plant should be reviewed by operators so they can:

1. Become familiar with a proposed plant,

2. Learn what will be constructed, and

3. Offer suggestions on how the plant can be designed for easier and more effective operation and maintenance.

When reviewing plans and specifications, carefully study those areas influencing how the plant will be operated and maintained. Also look carefully for potential safety hazards.

5.70   Operation

1. Control gates must be suitably located in order to isolate each clarifier.

2. Baffles, weirs, or skirts should be capable of controlling clarifier inlet velocities.

3. Collector mechanisms and drive units must have protective devices such as shear pins, clutches, stall alarms, and ON/OFF switches. Be sure that clarifiers with suction-type mechanisms for removing sludge have provisions for sludge removal from the center floor of the clarifier in case the suction mechanism fails and also for draining the clarifier.

4. Surface skimmer for scum removal.

5. Scum box should be located with consideration given to direction of prevailing winds and to removal of any accumulation of floatables on the water surface.

6. Hose bibs (high-pressure water faucets) conveniently located with respect to scum boxes, launders, weirs, and clarifier center columns for easy washdown.

7. Sampling equipment should be easily accessible.

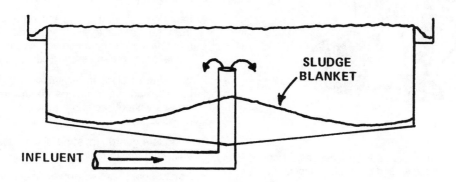

Fig. 5.12   Activated sludge settling near center of clarifier (bell-shaped pattern)

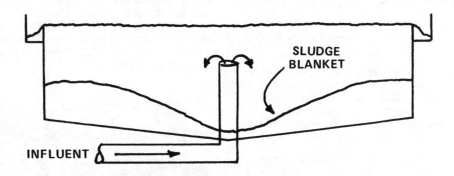

Fig. 5.13   Activated sludge settling near outer edge of clarifier (bowl-shaped pattern)

8. Clarifier flow and level controllers (if installed) must be properly located for operational purposes.

9. Grease or scum weirs must be at proper depth for scum control and removal of floating material from water surface.

5.71   Maintenance

1. Drive mechanisms, lubrication points, locations for changing oil in gear boxes (cases), and turntables must be accessible.

2. Weirs, launders, and control boxes must be accessible for cleaning, painting, and other maintenance activities.

3. Sludge pumps must be conveniently located and capable of backflushing pipelines or pumping down clarifiers.

4. Provisions should be made for connections and/or locations for portable pumps to dewater clarifiers if clarifiers are not connected to plant drainage system.

5. Influent and effluent pipelines, conduits, or channels must be installed so that each end can be isolated and dewatered by gravity drain or portable pump.

6. Sludge and scum lines to pump suctions must be kept as short as possible and free of fittings (90-degree bends and reducers).

7. Cleanouts are required on sludge and scum lines to provide access for cleaning equipment such as sewer rods and high-velocity cleaners. Cleanouts should be installed in the lines at locations that allow the lines to be worked on while the clarifier remains in service, instead of having to dewater the clarifier to clear a stoppage or clean a line.

8. Auxiliary service lines (water, air, electrical, instrumentation, sample, and chemical feed) should be studied. These lines should have isolation valves (to valve off portions of lines) at appropriate locations and should be accessible for repairs when necessary. Conduits for instrumentation, electrical wiring, and cables should be equipped with pull boxes that are *WATERTIGHT*. Sample lines should have cleanouts and valving to allow for periodic flushing of the lines. Air lines must be equipped with condensate drains at all low points, including the ends of the line.

9. Covered clarifiers should contain lightweight openings to provide easy access to scum channels, skimmers, launders, and drive mechanism units.

5.72   Safety

1. Clarifiers must be equipped with adequate access by stairs, ladders, ramps, catwalks, and bridges with railings that meet all state and OSHA requirements.

2. Catwalks and bridges must have floor plates or grates firmly secured and equipped with toeboards and nonskid surfaces.

3. Adequate lighting must be provided on the clarifier.

4. Launders, channels, and effluent pipelines that carry flow from the clarifier to another conduit, channel, or structure must have safety grates over the entrance to prevent accidental entry into the system caused by slipping or falling.

5. In a circular clarifier, turntables, adjustable inlet deflection baffles, and return sludge control valves must have safe access without requiring the operator to leave a bridge or catwalk.

6. Adequate guards must be placed over chain drives, belts, and other moving parts.

7. Safety hooks, poles, and/or floats should be stationed at strategic locations near every basin to rescue anyone who falls into a basin.

8. Do not allow any pipes or conduits to cross on top of catwalks or bridges.

9. Adequate offset of drive units, motors, and other equipment must be provided to allow unobstructed access to all areas.

5.8   FLOTATION PROCESSES

After primary sedimentation, wastewater always contains some suspended solids that neither settle nor float to the surface and therefore remain in the liquid as it passes through the clarifier. Dissolved solids will, of course, travel through the clarifiers because they are unaffected by these units. *COLLOIDS* and *EMULSIONS* are two other forms of solids that are very difficult to remove.

Colloids (CALL-loids) are very small, finely divided solids (particulates that do not dissolve) that remain dispersed in liquid for a long time due to their small size and electrical charge. Colloids are usually less than 200 *MILLIMICRONS*[28] in size, and generally will not settle readily. Organic colloids exert a high oxygen demand, so their removal is desirable.

---

[28] Millimicron (MILL-uh-MY-kron).   A unit of length equal to $10^{-3}\mu$ (one thousandth of a micron), 10^{-6} millimeters, or 10^{-9} meters; correctly called a nanometer, nm.

An emulsion is a liquid mixture of two or more liquid substances not normally dissolved in one another, but one liquid held in suspension in the other. It usually contains suspended globules of one or more of the substances. The globules usually consist of grease, oil, fat, or resinous substances. This material also exerts a high oxygen demand.

One method for removing emulsions and colloids is by a "flotation process," pumping air into the mixture to cause the suspended material to float to the surface where it can be skimmed off.

The particles can be FLOCCULATED[29] with air or chemical COAGULANTS[30] and forced or carried to the liquid surface by minute air bubbles. Figure 5.14 shows the chain of events in the flotation process.

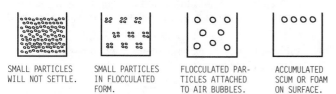

SMALL PARTICLES WILL NOT SETTLE.

SMALL PARTICLES IN FLOCCULATED FORM.

FLOCCULATED PARTICLES ATTACHED TO AIR BUBBLES. BUBBLES CARRY PARTICLES TO SURFACE.

ACCUMULATED SCUM OR FOAM ON SURFACE. MOST AIR BUBBLES ARE RELEASED.

Fig. 5.14   Flotation process

Most of the air bubbles are released at the liquid surface. Particles in the form of scum or foam are removed by skimming.

There are two common flotation processes in practice today:

1. VACUUM FLOTATION. The wastewater is aerated for a short time in a tank where it becomes saturated with dissolved air. The air supply is then cut off and large air bubbles pass to the surface and into the atmosphere. The wastewater then flows to a vacuum chamber which pulls out dissolved air in the form of tiny air bubbles. The bubbles then float the solids to the top.

2. PRESSURE FLOTATION. Air is forced into the wastewater in a pressure chamber where the air becomes dissolved in the liquid. The pressure is then released from the wastewater, and the wastewater is returned to atmospheric pressure. Because of the change in pressure, the dissolved air is released from solution in the form of tiny air bubbles. These air bubbles rise to the surface and, as they rise, carry solids to the surface.

Any flotation process is based upon release of gas bubbles in the liquid suspension (Figure 5.14) under conditions in which the bubbles and solids will associate with each other to form a combination with a lower specific gravity than the surrounding liquid. They must stay together long enough for the combination to rise to the surface and be removed by skimming.

For more detailed information on the operation of flotation processes and gravity thickeners, see Chapter 3, "Residual Solids Management," in the ADVANCED WASTE TREATMENT manual.

QUESTIONS

Write your answers in a notebook and then compare your answers with those on page 159.

5.7A   What safety items should be considered when reviewing plans and specifications for clarifiers?

5.8A   Would you place the flotation process BEFORE or AFTER primary sedimentation?

5.8B   Give a very brief description of:

1. Colloid
2. Emulsion

5.8C   Why is the "flotation process" used in some wastewater treatment plants?

5.8D   Give a brief description of the vacuum flotation process.

END OF LESSON 2 OF 3 LESSONS
on
SEDIMENTATION AND FLOTATION

Please answer the discussion and review questions next.

---

[29] Flocculation (FLOCK-you-LAY-shun).   The gathering together of fine particles after coagulation to form larger particles by a process of gentle mixing.

[30] Coagulants (co-AGG-you-lents).   Chemicals that cause very fine particles to clump (floc) together into larger particles. This makes it easier to separate the solids from the water by settling, skimming, draining or filtering.

DISCUSSION AND REVIEW QUESTIONS

Chapter 5.   SEDIMENTATION AND FLOTATION

(Lesson 2 of 3 Lessons)

Write the answers to these questions in your notebook before continuing. The question numbering continues from Lesson 1.

8. Why should floatable solids be removed from wastewater?

9. Explain how temperature influences clarifier performance.

10. Draw a clarifier and indicate what is meant by short-circuiting.

11. A circular primary clarifier has a diameter of 60 feet and an average depth of 8 feet. The flow of wastewater is 2.0 MGD. Calculate the following:

    1. Detention Time, in hours
    2. Weir Overflow Rate, in GPD/ft
    3. Surface Loading Rate, in GPD/sq ft
    4. Comment on the surface loading on the clarifier.

12. If a circular secondary clarifier similar to problem 11 receives a plant flow of 2.0 MGD plus a return sludge flow of 0.4 MGD, what is the solids loading in lbs/day/sq ft if the mixed liquor suspended solids concentration is 3,600 mg/L?

13. What safety items would you study when reviewing plans and specifications for clarifiers?

CHAPTER 5.   SEDIMENTATION AND FLOTATION

(Lesson 3 of 3 Lessons)

5.9   COMBINED SEDIMENTATION-DIGESTION UNIT

5.90   Purpose of Unit

A combined sedimentation-digestion unit consists of a small clarifier constructed over a sludge digester (Figures 5.15 and 5.16 and Table 5.3). Treatment units of this type have been designed and constructed to serve small populations such as schools, campgrounds, and subdivisions. Usually they are installed instead of Imhoff tanks or septic tank systems. Wastewater treatment efficiencies are similar to primary clarifiers with approximately 65 percent of the suspended solids and 35 percent of the biochemical oxygen demand removed from the influent.

BAR SCREEN

PRIMARY SETTLING → EFFLUENT TO
1. TREATMENT SYSTEM
2. OXIDATION PONDS
3. LAND TREATMENT

SLUDGE DIGESTION

DIGESTED SOLIDS
TO DRYING BEDS

SANITARY
LANDFILL

Fig. 5.15   Flow pattern for combined sedimentation-digestion unit

5.91   How the Unit Works

The combined sedimentation-digestion unit is considered a *PACKAGE TREATMENT PLANT.*[31] Plant influent usually passes through some type of flowmeter to record flows. A bar screen is often the first treatment unit of the package. Coarse solids are caught by the bar screen and removed manually on a daily basis or more often if necessary. Wastewater enters the clarifier near the surface in the center and the circular influent well directs the flow and solids toward the bottom of the clarifier. Settled wastewater slowly flows through the clarifier and leaves over the effluent weir around the outside of the clarifier. The effluent leaves the unit by the effluent trough (launder) and usually receives additional treatment in a secondary package plant (Chapter 8, "Activated Sludge"), ponds (Chapter 9, "Wastewater Stabilization Ponds"), or land treatment disposal system (Chapter 8, "Wastewater Reclamation and Recycling," *ADVANCED WASTE TREATMENT).*

Solids settling to the bottom of the clarifier (tray) are scraped to the center of the unit. A slot in the center of the tray allows the solids to flow into the digestion compartment. Below the slot is a sludge seal or boot which prevents gas from digestion and digested sludge from floating up into the clarifier. In the digester, sludge undergoes anaerobic decomposition (explained in Volume II, Chapter 12). Digested sludge is removed from the bottom of the digester by pumping or by gravity flow to drying beds.

---

[31] *Package Treatment Plant.   A small wastewater treatment plant often fabricated at the manufacturer's factory, hauled to the site, and installed as one facility. The package may be either a small primary or a secondary wastewater treatment plant.*

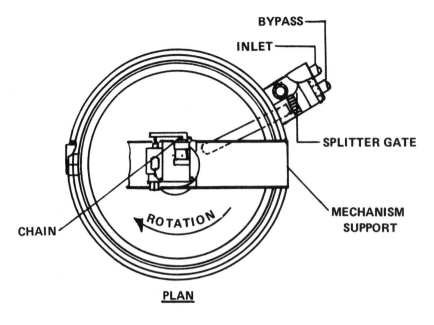

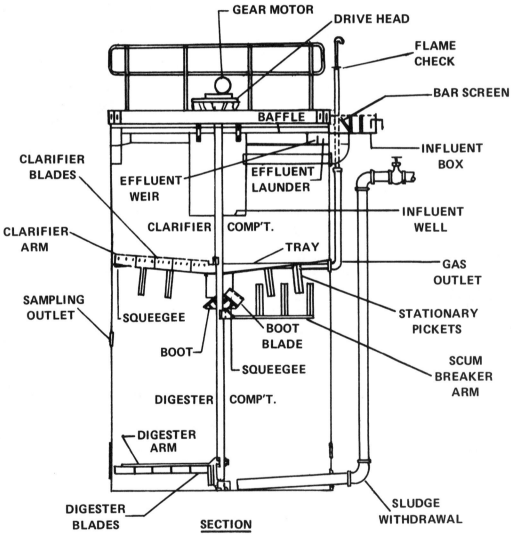

Fig. 5.16   Combined sedimentation-digestion unit
(Permission of Dorr-Oliver Incorporated)

TABLE 5.3 PURPOSE OF COMBINED SEDIMENTATION-DIGESTION UNIT PARTS

Part	Purpose
Bar Screen	Removes coarse material to prevent clogging of pipes or interference with mechanical units such as scrapers and pumps.
Flame Check	Prevents flame from traveling down gas outlet pipe to top of digester compartment where a flame could cause an explosion.
Gear Motor	Provides power to turn the scraper blades in the bottom of both the clarifier and the digester.
Clarifier Arm and Blades	Scrapes the settled solids (sludge) to a center hole where the sludge enters the digester.
Clarifier Compartment	Provides storage space for wastewater being treated and allows heavy solids to settle to the bottom and light solids to float to the surface.
Digester Arm and Blades	Scrapes the digested sludge on the digester bottom to the sludge withdrawal pipe.
Effluent Launder or Effluent Trough	Conveys the settled wastewater away from the clarifier.
Scum Breaker Arm and Stationary Pickets	Breaks up scum accumulation in the top of the digester compartment.
Digester Compartment	Provides storage space for sludge, allows digestion to occur and the separation of liquids and solids for disposal.
Sampling Outlet	Allows withdrawal of digester sludge for laboratory testing.
Sludge Withdrawal	Allows digested sludge to be removed from digester.

Scum is skimmed from the surface of the clarifier into a scum trough. From the trough, the scum flows to the scum pit (Figure 5.17). A submersible pump moves the scum to the digestion compartment.

Supernatant (the liquid in the top portion of the digester) is removed from the digester by lowering an adjustable overflow tube (Figure 5.17) and allowing the liquid to flow into the scum pit. Again the submersible pump moves the supernatant to the clarifier influent well and the supernatant flows through the clarifier.

Gas from the digestion process rises to the top of the digester as tiny bubbles. The tray (bottom of clarifier) slopes upward to the outside. This slope helps move sludge to the center slot and allows gas to accumulate along the outside edge of the digestion portion of the unit. The gas is collected in a gas dome and usually burned by a waste-gas burner. Sufficient gas is not produced to serve as a reliable source of energy for power or for heating.

5.92 Sampling and Analysis

Sampling locations and laboratory tests performed depend upon NPDES (National Pollutant Discharge Elimination System) permit requirements, available time and capability of facility to make operational changes on the basis of the interpretation of test results. Typical tests, their purposes, and expected ranges of test results are listed in Table 5.4.

5.93 Operation

5.930 Start-Up Procedures

Always thoroughly inspect the entire unit before allowing wastewater to enter. Any corrections will be more difficult after the unit is full of wastewater.

1. Remove all debris and tools from unit.

2. Follow the wastewater and solids flow paths through the unit (Section 5.91). Be sure you understand how the unit works and what happens when all the valves and switches are either open or closed.

3. Lubricate all equipment.

4. Allow unit to run for two hours.

5. Observe operation of unit. Be sure all equipment has proper alignment and clearance.

6. Keep all bolts tight.

7. Fill unit with water.

8. Allow unit to fill and effluent to flow over effluent weirs.

9. Inspect tank and pipes for leaks. Repair any leaks. If unit has not been previously tested for leaks, allow unit to sit for 24 hours without any influent. If water level drops more than one inch, find leak and repair it.

10. Make sure pipes are not plugged.

11. Operate valves to be sure they operate freely and are watertight.

12. Divert wastewater to unit. Leave dome vent open until gas production starts. Vent must be open to atmosphere to prevent gases from accumulating in confined spaces. Time for gas production to start will depend on temperature of sludge under digestion. The warmer the sludge, the shorter the time. Gas will start being produced after approximately three weeks.

13. Chlorinate final effluent.

14. Add lime to digestion unit if pH is below 7.0. Sample supernatant by lowering adjustable overflow tube in scum pit and measuring pH (see Chapter 16 for procedure). If the pH is closely controlled, these units may be started without seeding with digested sludge. Lime should be added on the *VERY FIRST DAY DURING START-UP* and daily until the pH remains above 7.0. Recommended lime doses are as follows (see page 152):

**TABLE 5.4   COMBINED SEDIMENTATION-DIGESTION
UNIT TESTS, PURPOSE, AND RESULTS[a]**

Test	Typical Results	Purpose
1. Inflow, MGD	Depends on population served. May range from 80 to 130 gallons per person per day.	Determine hydraulic loading on facility. Identify flow trends and when facility is approaching design capacity.
Clarifier		
2. Suspended Solids	(Infl) 100 - 300 mg/L (Effl) 50 - 100 mg/L	Indicates efficiency of clarifier removal.
3. Settleable Solids	(Infl) 50 - 100 mL/L (Effl) 5 - 15 mL/L	Indicates efficiency of clarifier removal.
4. Biochemical Oxygen Demand (BOD) (Optional)	(Infl) 150 - 300 mg/L (Effl) 100 - 200 mg/L	Indicates efficiency of clarifier removal.
Digestion		
5. Temperature	Depends on location, season, and whether sludge is heated.	Forecasts digestion rates which depend on temperature of digesting sludge.
6. pH	7.0 - 7.8	Determines if effective digestion is taking place. Too low a pH value indicates poor digestion.
7. Quantity of sludge withdrawn	Depends on population, detention time in digester, and temperature.	Determines effectiveness of clarifier in removing solids and effectiveness of digester in reducing solids.
Effluent (from final treatment process) 8. Chlorine residual	Depends on NPDES permit requirements.	Determines if sufficient chlorine is being applied to effluent to achieve adequate disinfection.

[a] Laboratory tests should be conducted on a regular basis from twice a week to daily depending on NPDES permit requirements. Results should be plotted (Figure 5.18) immediately to identify any trends that need correcting.

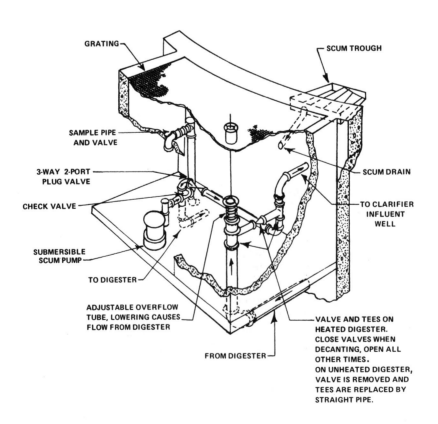

Fig. 5.17   Arrangement of scum pit piping

(Permission of Dorr-Oliver Incorporated)

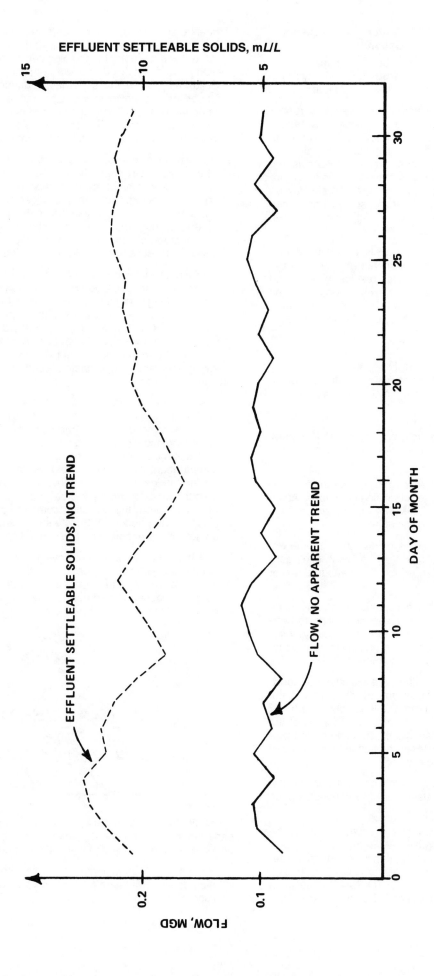

Fig. 5.18   Plot of lab results to identify any trends

Diameter of Unit, ft	Hydrated Lime, lbs	Water, Gallons	Frequency
10-18	25	30	Every 2 days
20-28	25	50	Every day
30-40	50	100	Every day

Mix the lime with water in a barrel until milky. Use rubber gloves and eye/face protection. Lime is a moderately caustic irritant to all exposed surfaces of the body. Pour mixture into scum pit and then pump into digestion unit using submersible pump. Be sure valves are properly set so mixture is not pumped into clarifier. Allow some wastewater to flow from influent of clarifier into scum pit (or washdown hose may be used to add water to scum pit). Pump out scum pit and wash down to remove any remaining lime to digestion unit. Stop adding lime when pH remains above 7.0.

Another chemical used to raise the pH is anhydrous ammonia. Be sure to handle anhydrous ammonia very carefully because it is a hazardous chemical. Use respiratory, eye, and skin protection if ammonia hazards may exist in concentrations above federal and/or local standards (Threshold Limit Value, 25 ppm; Short Term Inhalation Limit, 50 ppm for five minutes).

5.931 Normal Operation

Normal operation consists of inspecting the unit on a daily basis. Observe the flow through the facility and the operation of the equipment. Look and listen for anything unusual.

DAILY

1. Remove debris and solids from bar screen and properly dispose of them by burial. Hose down screen and channel walls.

2. Hose down baffles, weirs, scum trough, and scum pit to remove any grease, scum, or other floating debris. Accumulations of this material are unsightly and usually produce odors and flies if not removed immediately.

3. Measure pH in digester. If pH is below 7.0, add lime according to start-up instructions. If pH remains fairly constant and above a pH of 7.0, pH may be measured two or three times per week.

4. Withdraw supernatant once or twice a day. Analysis and interpretation of lab test results of suspended solids and settleable solids in effluent can indicate the frequency of supernatant withdrawal. Solids in the effluent may indicate that supernatant should be removed more frequently.

Supernatant is withdrawn from the digestion compartment by lowering the sleeve on the overflow tube in the scum box. When the top of the sleeve is below the water surface in the clarifier, the supernatant will flow out of the digestion compartment. Removal of the supernatant provides more space in the digester for sludge being scraped from the bottom (tray) of the clarifier.

Lower the sleeve so approximately one-half inch of water will flow over the top of the sleeve for 15 to 30 minutes twice a day.

5. After supernatant withdrawal to the scum pit, pump the clear supernatant to the clarifier. When most of the supernatant has been pumped out of the pit, only scum will remain. Pump the scum to the digester. *DO NOT ALLOW THE SUBMERSIBLE SCUM PUMP TO RUN DRY* because the pump motor can be damaged. Pump the scum fre-

quently enough to prevent odors from developing or flies becoming nuisances. If the supernatant is high in solids, return the supernatant to the digester. Try to prevent supernatant solids from flowing out with the clarifier effluent. Solids may be in the supernatant during initial start-up. Under these conditions, the supernatant may be pumped to drying beds until the digester solids in the supernatant can indicate that the digested solids should be removed from the digestion compartment (digester).

6. Digested sludge should be withdrawn from the digester when solids start appearing in the supernatant. This will occur several months after start-up and regularly during normal operation. Frequency of digested sludge removal will depend on the solids loading on the unit, design capacity, and effectiveness of the digestion process in the unit. The scraper mechanism on the bottom of the digester helps to prevent sludge coning during sludge withdrawal.

Withdraw the sludge slowly so the sand and other grit on the bottom of the digestion unit will be removed too. Do not remove sludge so fast that the clarified effluent will stop flowing over the effluent weirs. If the digested sludge is allowed to flow out too quickly, the sludge directly above the sludge withdrawal pipe will flow out. A cone will develop in the sludge and the supernatant will flow out rather than allowing the remaining digested sludge to flow toward the withdrawal pipe.

Always leave some digested sludge remaining in the digestion compartment. As soon as the sludge starts to run thin, stop removing sludge. If a cone has developed and a lot of digested sludge remains, try removing some supernatant to the scum pit and pumping it back to the digestion compartment. One way to avoid withdrawing too much digested sludge is to only remove a portion of the sludge from the digester during one day. For example, if 1/40th of the sludge from the digestion compartment will cover the sludge drying beds with 3 inches (7.5 cm) of sludge, then do not withdraw more than this amount on any given day.

When you are through removing digested sludge, wash out the line with plant effluent. Shut the valve by the digestion compartment and leave the remainder of the line open so gas produced from digestion will not cause a pressure buildup which could damage the pipe or valves.

Usually the digested sludge is discharged to sand drying beds (Volume II, Chapter 12). After the sludge is dried, it can be removed and disposed of in a sanitary landfill.

Do not smoke around drying beds while the digested sludge is being applied. Methane gas can mix with air to form explosive conditions.

If odors develop from the drying bed, a neutral *MASKING AGENT*[32] may be applied. Lime also has been used to control odors.

7. In colder climates the contents of the digester may be heated in order to obtain better and faster digestion. Try to maintain the temperature of the digester contents between 80 and 95°F (27 and 35°C). The closer to 90°F (32°C), the faster the rate of digestion.

8. Gas production will not start until solids digestion starts in the digester. The gas line should be equipped with a moisture trap and a flame arrester, and connected to a waste gas burner.

Be sure the gas line is clear. Scum and undigestible material can prevent gas from flowing from the digestion compartment. If necessary, remove this scum and debris.

Once sufficient methane gas is produced, the waste gas should be burned.

9. *PROBLEM.* Floating sludge on clarifier surface.

Floating sludge may result from sludge not passing from the bottom of the clarifier, around the sludge seal, and into the sludge digester.

a. Try withdrawing more supernatant so the sludge can flow into the digester.

b. Shut off the sludge scraper. Check the sludge seal by "feeling" with a pole or rod to remove any screenings or other objects which might be plugging the hole.

c. Pump less scum or supernatant into the digester. Excessive pumping may be forcing sludge out of the digestion compartment.

5.932 Abnormal Operation

Abnormal operation occurs when:

1. Inflows are higher than design flows due to storm water inflow and infiltration,

2. Solids loadings are high due to seasonal or industrial discharges, and

3. Toxic substances or high or low pH liquids are released into the collection system.

With a single combined sedimentation-digestion unit there is little an operator can do in terms of adjusting valves or directing flows. If abnormal conditions occur occasionally and upset the clarifier and/or the digestion unit, provisions should be made to construct an emergency pond to hold the abnormal flows and substances until they can be treated by the unit during low flow periods.

5.933 Shutdown Procedures

The unit should be shut down annually for inspection, maintenance, and repair. If there is only one unit for treating the wastewater, a standby emergency pond should be available to contain the wastewater during shutdown. Schedule shutdown during periods of expected low flows. Shutdown procedures are as follows:

1. Divert flow to other units or to standby pond.

2. Drain clarifier and digester to drying beds.

3. Wash down inside of unit.

4. Inspect facility. Be sure facility is adequately ventilated and test for oxygen deficiency/enrichment, combustible gases, and toxic gases. Appropriate confined space entry procedures should be implemented to accomplish entry/work in a combined sedimentation-digestion unit. See Volume II, Chapter 14, "Plant Safety," for additional confined space safety procedures. Look for:

a. Corrosion damage,
b. Unpainted surfaces,
c. Worn parts, and
d. Cracks and leaks.

5. Make necessary repairs.

6. Follow start-up procedures to place unit back on line.

5.934 Operational Strategy

With only one unit, the operator must try to treat the entire flow. If inflows do not exceed design capacity, problems should not develop.

1. Follow normal operational procedures.

2. Collect and analyze samples on a regular basis.

3. Plot results of tests and look for trends.

4. Make any necessary adjustments.

5. Maintain all equipment according to schedule.

5.94 Maintenance

1. Lubricate all equipment in accordance with manufacturers' recommendations.

2. If any tools or other objects fall into clarifier, stop rotation and remove the tool or object.

3. If scraper mechanism stops moving, determine cause and remove it before attempting to start mechanism again. *DO NOT TAMPER WITH THE OVERLOAD SWITCH ADJUSTMENTS IN AN ATTEMPT TO FORCE THE MACHINE TO OPERATE AGAINST THE OVERLOAD.*

5.95 Safety

1. Be careful when removing debris from open tanks. Secure firm footing so you won't slip or fall. Do not try to lift more than you can safely lift.

2. Be sure all moving machinery parts have covers or guards.

3. Be aware of the fact that *DIGESTER GAS* is *VERY TOXIC OR POISONOUS*. When mixed with air this gas can *BURN* or *EXPLODE*. Everyone must strictly observe the following rules:

a. Post a danger sign near the gas dome indicating "DANGER. NO SMOKING OR OPEN FLAMES."

b. Keep all lighted cigars, cigarettes, pipes, or any open fire away from the digester or digester gas at all times.

---

[32] *Masking Agents. Substances used to cover up or disguise unpleasant odors. Liquid masking agents are dripped into the wastewater, sprayed into the air, or evaporated (using heat) with the unpleasant fumes or odors and then discharged into the air by blowers to make an undesirable odor less noticeable.*

c. Do not inhale digester gas.

d. Do not enter the settling or digestion compartments unless they are empty of all sludge and forced ventilation has cleared them of any atmospheric hazards. Compliance with federal/local confined space regulations must be attained. Remember that any respirator must have a self-contained supply of air for you to breathe. See Volume II, Chapter 14, "Plant Safety," for additional safety procedures.

e. Remove all oil and grease spills and other slippery matter from surfaces.

5.96 Acknowledgment

The authors wish to thank Dorr-Oliver Incorporated for allowing the use of their material for the preparation of this section.

QUESTIONS

Write your answers in a notebook and then compare your answers with those on page 160.

5.9A   What is a combined sedimentation-digestion unit?

5.9B   What abnormal operating conditions might the operator of a combined sedimentation-digestion unit encounter?

5.9C   List the major maintenance items for a combined sedimentation-digestion unit.

5.10 IMHOFF TANKS

Imhoff tanks are rarely constructed today. Your plant may consist of *ONLY* an Imhoff tank if it serves a very small community or if it was constructed many years ago. It is quite possible that you may never have operating responsibility for one of these units. They will be discussed for general knowledge and for the few operators who will have operating responsibility for them.

The Imhoff tank combines sedimentation and sludge digestion in the same unit. There is a top compartment where sedimentation occurs and a bottom compartment for digestion of settled particles (sludge). The two compartments are separated by a floor with a slot designed to allow settling particles to pass through to the digestion compartment (Figure 5.19).

Wastewater flows slowly through the upper tank as in any other standard rectangular sedimentation unit. The settling solids pass through the slot to the bottom sludge digestion tank. Anaerobic digestion of solids is the same as in a separate digester. Gas bubbles are formed in the digestion area by bacteria. As the gas bubbles rise to the surface, they carry

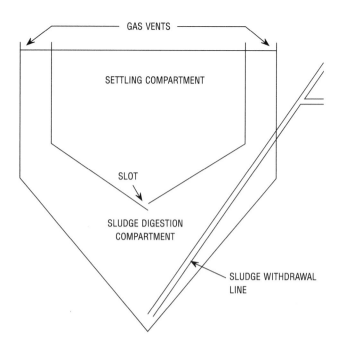

Fig. 5.19   Imhoff tank

solid particles with them. The slot is designed to prevent solids from passing back into the upper sedimentation area as a result of gasification. Solids would flow from the unit with the effluent if they were permitted to pass back into the upper sedimentation area.

The same calculations previously used for clarifiers can be used to determine loading rates for the settling area of the Imhoff tank. (Volume II, Chapter 12, "Sludge Digestion and Solids Handling," will explain the anaerobic process in the sludge digestion area of this unit.) Some typical values for design and operation of Imhoff tanks are:

Settling Area

Wastewater Detention Time—1.0 to 4.0 hours
Surface Settling Rate—600 to 1,200 GPD/sq ft
Weir Overflow Rate—10,000 to 20,000 GPD/ft
Suspended Solids Removal—45% to 65%
BOD Removal—25% to 35%

Digestion Area

Digestion Capacity—1.0 to 3.0 cu ft/person
Sludge Storage Time—3 to 12 months

Here are a few operational suggestions:

1. In general, there is no mechanical sludge scraping device for removing settled solids from the floor of the settling areas. Solids may accumulate before passing through the slot to the digestion area. It may be necessary to push the accumulation through the slot with a squeegee or similar device attached to a long pole. Dragging a chain on the floor and allowing it to pass through the slot is another method for removing the sludge accumulation.

2. Scum from the sedimentation area is usually collected with hand tools and placed in a separate container for disposal. Scum also may be transferred to the gas venting area where it will work down into the digestion compartment. Scum in the gas vents should be kept soft and broken up by soaking it periodically with water or by punching holes in it and mixing it with the liquid portion of the digestion compartment. The addition of hydrated lime may be helpful for

controlling odors from the gas vent area and also for adjusting the chemical balance of the scum for easier digestion if necessary.

3. Some Imhoff tanks have the piping and valving to reverse the direction of flow from one end toward the other end. If possible, the flow should be reversed periodically for the purpose of maintaining an even sludge depth in the digestion compartment. The sludge level in the digestion area must be lower than the slot in the floor of the settling area to prevent plugging of the slot. A line of gas bubbles directly over the slot indicates the sludge level in the digestion chamber is too high.

4. The explanation of sludge digestion in Volume II, Chapter 12 will supply information that can be applied to the digestion area in the Imhoff tank. Neither sludge mixing nor heating devices are used in an Imhoff tank. Sludge loading rates, withdrawal rates, laboratory tests, and visual appearance of sludges are very similar to what they are in an *UNHEATED* digester. If visual appearance and smell are the only methods you have of judging the sludge, it is safe to assume that the process is working satisfactorily if sludge in the digestion area is relatively odorless or has a musty smell and is black or very dark in color.

The laboratory testing program for an Imhoff tank should be complete enough to identify operational problems and to supply necessary information to regulatory agencies. The following minimum program is suggested, assuming adequate laboratory facilities, personnel, and size of the system.

SUGGESTED ANALYSIS	USUAL RANGE	TYPICAL REMOVAL %
Settling Area		
Settleable Solids	3.0 - 10.0 mL/L	75 - 90
Suspended Solids	200 - 400 mg/L	45 - 65
pH	6.7 - 7.3	
Alkalinity	1 - 300 mg/L	
BOD	200 - 500 mg/L	25 - 35
Digestion Area		
pH	6.7 - 7.3	
Alkalinity	1,000 - 3,000 mg/L	
Volatile Acids	100 - 500 mg/L	

Efficiency of operation can be determined by measuring the settleable solids, suspended solids, or BOD of the influent and effluent.

QUESTIONS

Write your answers in a notebook and then compare your answers with those on page 160.

5.10A   What are the two components of an Imhoff tank?

5.10B   How can you force settled material into the digestion compartment?

5.10C   How could you maintain a fairly level sludge blanket in the digester portion of an Imhoff tank?

5.10D   Describe the sludge from an Imhoff tank that is operating properly.

5.11 SEPTIC TANKS

Septic tanks are used mostly for treating the wastewater from individual homes or from small populations (such as camps) where sewers have not been provided. They operate very much like an Imhoff tank except there is not a separate digestion compartment. Detention time is usually long (12 to 24 hours) and most settleable solids will remain in the tank.

They must be pumped out and disposed of periodically to prevent the tank from filling. Part of the solids in the septic tank are liquified and discharged with the wastewater into a subsurface soil leaching system. Conditions are not favorable for rapid gasification and most waste stabilization occurs in the soil.

Septic tank effluent is usually disposed of in underground perforated pipes called *LEACH LINES*, and sampling of effluent may be impossible. The ability of the soil to leach the septic tank effluent is the critical factor in subsurface waste disposal systems.

One method of operating septic tank effluent leaching systems is to apply effluent to half of the system while the other half rests. Monthly switch the flow from one half to the other half. This procedure gives the leaching system a chance to recover its percolation ability.

For additional information, see *SMALL WASTEWATER SYSTEM OPERATION AND MAINTENANCE*, Volume I, Chapter 4, "Septic Tanks and Pumping Systems," in this series of operator training manuals, available from the Office of Water Programs, California State University, Sacramento.

5.12 ARITHMETIC ASSIGNMENT

Turn to the Arithmetic Appendix at the back of this manual and read the following sections:

1. A.5, *WEIGHT-VOLUME RELATIONS*, and
2. A.132, Sedimentation Tanks and Clarifiers.

Check all of the arithmetic in these two sections on an electronic pocket calculator. You should be able to get the same answers.

5.13 ADDITIONAL READING

1. *MOP 11*, Chapter 17, "Characterization and Sampling of Wastewater," and Chapter 19, "Primary Treatment."*

2. *NEW YORK MANUAL*, Chapter 4, "Primary Treatment."*

3. *TEXAS MANUAL*, Chapter 11, "Sedimentation."*

* Depends on edition.

5.14 METRIC CALCULATIONS

This section contains the solutions to all problems in this chapter using metric calculations.

5.140 Conversion Factors

MGD x 3,785	= cu m/day
cu m/day x 0.000264	= MGD
gallons x 3.785	= liters
1,000 L	= 1 cu m
ft x 0.3048	= m
m x 3.281	= ft
cu ft x 28.32	= liters
L x 0.035315	= cu ft

5.141 Problem Solutions

1. The influent BOD to a primary clarifier is 200 mg/L, and the effluent BOD is 140 mg/L. What is the efficiency of the primary clarifier in removing BOD?

Known	Unknown
Infl BOD, mg/L = 200 mg/L	Prim Clar Eff, %
Effl BOD, mg/L = 140 mg/L	

CALCULATIONS:

1. Calculate the primary clarifier efficiency, %, in removing BOD.

$$\text{Efficiency, \%} = \frac{(\text{In} - \text{Out})}{\text{In}} \times 100\%$$

$$= \frac{(200 \text{ mg/}L - 140 \text{ mg/}L)}{200 \text{ mg/}L} \times 100\%$$

$$= \frac{60 \text{ mg/}L}{200 \text{ mg/}L} \times 100\%$$

$$= (0.30)(100\%)$$

$$= 30\% \text{ BOD Removal}$$

NOTE: This problem solution is exactly like the solution in the text because the influent and effluent BODs are given in mg/*L* which is the Metric System.

2. The flow to a rectangular sedimentation tank is 12,000 cu m per day. Tank dimensions are 20 meters long by 10 meters wide by 3 meters deep. What is the detention time?

Known		Unknown
Flow, cu m/day	= 12,000 cu m/day	Detention Time, hr
Length, m	= 20 m	
Width, m	= 10 m	
Depth, m	= 3 m	

FORMULAS:

$$\text{Detention Time, hr} = \frac{\text{Tank Volume, cu m} \times 24 \text{ hr/day}}{\text{Flow, cu m/day}}$$

Tank Vol, cu m   = Length, m x Width, m x Depth, m[33]

CALCULATIONS:

1. Calculate the tank volume in cubic meters.

Tank Vol, cu m = Length, m x Width, m x Depth, m

$$= 20 \text{ m} \times 10 \text{ m} \times 3 \text{ m}$$

$$= 600 \text{ cu m}$$

2. Estimate the detention time in hours.

$$\text{Detention Time, hr} = \frac{\text{Tank Volume, cu m} \times 24 \text{ hr/day}}{\text{Flow, cu m/day}}$$

$$= \frac{600 \text{ cu m} \times 24 \text{ hr/day}}{12,000 \text{ cu m/day}}$$

$$= 1.2 \text{ hours}$$

3. The flow is 20,000 cu m per day in a circular tank with a 30-meter weir diameter. What is the weir overflow rate in cubic meters per day per meter of weir length?

Known		Unknown
Flow, cu m/day	= 20,000 cu m/day	Weir Overflow Rate, cu m/day/m
Weir Diameter, m	= 30 m	

FORMULAS:

$$\text{Weir Overflow Rate, cu m/day/m} = \frac{\text{Flow, cu m/day}}{\text{Length of Weir, m}}$$

Length of Circular Weir, m = π x Weir Diameter, m

CALCULATIONS:

1. Calculate the length of circular weir in meters.

Length of Circular Weir, m = π x Weir Diameter, m

$$= 3.14 \times 30 \text{ m}$$

$$= 94.2 \text{ m}$$

2. Estimate the weir overflow rate in cubic meters per day per meter of weir length.

$$\text{Weir Overflow, cu m/day/m} = \frac{\text{Flow, cu m/day}}{\text{Length of Weir, m}}$$

$$= \frac{20,000 \text{ cu m/day}}{94.2 \text{ m}}$$

$$= 212 \text{ cu m/day/m of Weir}$$

4. The flow into a rectangular clarifier is 15,000 cu m per day in a tank 30 meters long and 10 meters wide. What is the surface loading rate in cubic meters per day per square meter of surface area?

Known		Unknown
Flow, cu m/day	= 15,000 cu m/day	Surface Loading, cu m/day/sq m
Length, m	= 30 m	
Width, m	= 10 m	

FORMULA:

$$\text{Surface Loading, cu m/day/sq m} = \frac{\text{Flow, cu m/day}}{\text{Surface Area, sq m}}$$

CALCULATIONS:

1. Calculate the surface area in square meters.

Surface Area, sq m = Length, m x Width, m[34]

$$= 30 \text{ m} \times 10 \text{ m}$$

$$= 300 \text{ sq m}$$

2. Estimate the surface loading in cubic meters per day per square meter of surface area.

$$\text{Surface Loading, cu m/day/sq m} = \frac{\text{Flow, cu m/day}}{\text{Surface Area, sq m}}$$

$$= \frac{15,000 \text{ cu m/day}}{300 \text{ sq m}}$$

$$= 50 \text{ cu m/day/sq m}$$

---

[33] *For a circular clarifier, Tank Vol, cu m = 0.785 x (Diameter, m)^2 x Depth, m*
[34] *For a circular clarifier, Surface Area, sq m = 0.785 x (Diameter, m)^2*

5. A circular secondary clarifier with a diameter of 30 meters treats a flow of 17,000 cubic meters per day (13,000 cu m/day inflow and 4,000 cu m/day return sludge flow) with a mixed liquor suspended solids concentration of 4,200 mg/*L*. What is the solids loading in kilograms of solids per day per square meter of surface area?

Known	Unknown
Flow, cu m/day = 17,000 cu m/day	Solids Loading,
Diameter, m = 30 m	kg/day/sq m
MLSS, mg/*L* = 4,200 mg/*L*	

FORMULAS:

$$\text{Solids Applied, kg/day} = \text{Flow, } \frac{\text{cu m}}{\text{day}} \times \text{MLSS, } \frac{\text{mg}}{L} \times \frac{1{,}000\ L}{1\ \text{cu m}} \times \frac{1\ \text{kg}}{1{,}000{,}000\ \text{mg}}$$

$$\text{Solids Loading, kg/day/sq m} = \frac{\text{Solids Applied, kg/day}}{\text{Surface Area, sq m}}$$

CALCULATIONS:

1. Calculate the solids applied in kilograms per day.

$$\text{Solids Applied, kg/day} = \text{Flow, } \frac{\text{cu m}}{\text{day}} \times \text{MLSS, } \frac{\text{mg}}{L} \times \frac{1{,}000\ L}{1\ \text{cu m}} \times \frac{1\ \text{kg}}{1{,}000{,}000\ \text{mg}}$$

$$= 17{,}000\ \frac{\text{cu m}}{\text{day}} \times 4{,}200\ \frac{\text{mg}}{L} \times \frac{1{,}000\ L}{1\ \text{cu m}} \times \frac{1\ \text{kg}}{1{,}000{,}000\ \text{mg}}$$

$$= 71{,}400\ \text{kg/day}$$

2. Calculate the surface area in square meters.

$$\text{Surface Area, sq m} = \frac{(\pi)}{4}\ (\text{Diameter, m})^2$$

$$= \frac{(3.14)}{4}\ (30\ \text{m})^2$$

$$= 706.5\ \text{sq m}$$

3. Estimate the surface loading in kilograms of solids per day per square meter of surface area.

$$\text{Solids Loading, kg/day/sq m} = \frac{\text{Solids Applied, kg/day}}{\text{Surface Area, sq m}}$$

$$= \frac{71{,}400\ \text{kg/day}}{706.5\ \text{sq m}}$$

$$= 101\ \text{kg/day/sq m}$$

END OF LESSON 3 OF 3 LESSONS
on
SEDIMENTATION AND FLOTATION

Please answer the discussion and review questions next.

DISCUSSION AND REVIEW QUESTIONS

Chapter 5.   SEDIMENTATION AND FLOTATION

(Lesson 3 of 3 Lessons)

Write the answers to these questions in your notebook. The question numbering continues from Lesson 2.

14. How does a combined sedimentation-digestion unit work?

15. What would you do if floating sludge appeared on the surface of a combined sedimentation-digestion unit?

16. What is the critical factor in subsurface wastewater disposal systems?

SUGGESTED ANSWERS

Chapter 5.   SEDIMENTATION AND FLOTATION

ANSWERS TO QUESTIONS IN LESSON 1

Answers to questions on page 118.

5.0A   The main difference between the sludge from primary and secondary clarifiers is that primary sludge is usually denser than secondary sludge.

5.0B   The main difference between the effluent from primary and secondary clarifiers is that the effluent from a secondary clarifier is normally clearer than primary effluent.

5.1A   Significant check items before starting a circular clarifier include:

1. Control gates for operation,
2. Clarifier tank for sand and debris,
3. Collector drive mechanism for lubrication, oil level, drive alignment, and complete assembly,
4. Gaskets, gears, drive chain sprockets, and drive motor for proper installation and rotation,
5. Squeegee blades on the collector plows for proper distance from the floor of the tank,
6. All other mechanical items below the waterline for proper installation and operation,
7. Tank sumps or hoppers and return lines for debris and obstructions, and
8. Tank structure for corrosion, cracks, and other indications of structural failure.

5.1B   Safety precautions that should be taken during start-up of a clarifier include:

1. Wear a hard hat when down in the tank for protection from falling objects;
2. Keep hands away from moving equipment; and
3. When working on equipment, be sure to tag and use a lockout device on the main circuit breaker and influent control gates to prevent equipment from starting unexpectedly and causing equipment damage and/or personal injury.

5.1C   When the crosspieces in a rectangular clarifier are not straight across the tank, sludge will be piled higher on the trailing side and/or the crosspieces will hang up and cause severe damage to the flight.

Answers to questions on page 120.

5.1D   The best operational strategy for a clarifier is to develop and implement a good preventive maintenance program, to closely monitor operating conditions, and to respond to any lab results that indicate problems are developing.

5.1E   Abnormal conditions that could affect clarifier performance include:

1. Toxic wastes from industrial spills or dumps,
2. Storm flows and hydraulic overloads, and
3. Septicity from collection system problems.

5.1F   Steps that could be taken to improve clarifier effluent quality when infiltration is a frequent problem are sealing of the sanitary sewers and use of a flow equalization basin.

Answers to questions on page 130.

5.2A   To measure clarifier efficiency, sample the influent and effluent of the clarifier.

5.2B   The basic laboratory measurements used to determine clarifier efficiency are settleable solids, suspended solids, total solids, BOD, and coliform group bacteria.

5.2C   A clarifier should be able to remove 95 to 99 percent of settleable solids.

5.2D   What is the suspended solids removal efficiency of a primary clarifier if the influent concentration is 300 mg/L and the effluent is 120 mg/L?

$$\text{Efficiency, \%} = \frac{(\text{In} - \text{Out})}{\text{In}} \, (100\%)$$

$$= \frac{(300 \text{ mg/}L - 120 \text{ mg/}L)}{300 \text{ mg/}L} \, (100\%)$$

$$= 60\%$$

Answers to questions on page 131.

5.3A   Remove sludge from a clarifier often enough to prevent septic conditions or sludge gasification.

5.3B   Stop pumping sludge when it becomes thin. Thin sludge can be detected by the sound of the sludge pump, differences in sludge pumping pressure gage readings, and by visual observation of the sludge.

5.3C   Scum can be kept out of the clarifier effluent by a baffle placed around the inside edge of the overflow weir.

ANSWERS TO QUESTIONS IN LESSON 2

Answers to questions on page 137.

5.6A   Short-circuiting occurs in a clarifier when the flow is not uniform throughout the tank. In this situation the water flows too rapidly in one or more sections of the clarifier to allow sufficient time for settling to occur.

5.6B   Short-circuiting is undesirable because where the velocity is too high, particles will not have time to settle. Where the velocity is too low, undesirable septic conditions may develop.

5.6C   Short-circuiting may be corrected by installing weir plates, baffles, port openings, and by proper design of the inlet channel.

5.6D A circular clarifier has a diameter of 80 feet and an average depth of 10 feet. The flow of wastewater is 4.0 MGD and the suspended solids concentration is 190 mg/L. Calculate the following:

1. Detention Time, in hours
2. Weir Overflow Rate, in GPD/ft
3. Surface Loading Rate, in GPD/sq ft

$$\text{Tank Volume, cu ft} = \frac{\pi}{4} \times (\text{Diameter, ft})^2 \times \text{Depth, ft}$$

$$= (0.785)(80 \text{ ft})^2 \times 10 \text{ ft}$$

$$= 0.785 \times 6,400 \times 10$$

$$= 0.785 \times 64,000$$

$$= 50,240 \text{ cu ft}$$

$$\text{Tank Volume, gal} = 50,240 \text{ cu ft} \times 7.5 \text{ gal/cu ft}$$

$$= 376,800 \text{ gal}$$

1. $$\text{Detention Time, hr} = \frac{\text{Tank Volume, gal} \times 24 \text{ hr/day}}{\text{Flow, gal/day}}$$

$$= \frac{376,800 \text{ gal} \times 24 \text{ hr/day}}{4,000,000 \text{ gal/day}}$$

$$= .376800 \times 6$$

$$= 2.2608$$

$$= 2.3 \text{ hr}$$

2. $$\text{Weir Overflow Rate, GPD/ft} = \frac{\text{Flow Rate, GPD}}{\text{Length of Weir, ft}}$$

$$= \frac{4,000,000 \text{ GPD}}{3.14 \times 80 \text{ ft}}$$

$$= \frac{4,000,000 \text{ GPD}}{251.2 \text{ ft}}$$

$$= 15,923 \text{ GPD/ft}$$

3. Surface Loading Rate

Calculate Surface Area, sq ft

$$\text{Surface Area, sq ft} = \frac{\pi}{4} \times (\text{Diameter, ft})^2$$

$$= (0.785)(80 \text{ ft})^2$$

$$= 0.785 \times 6,400$$

$$= 5,024 \text{ sq ft}$$

$$\text{Surface Loading Rate, GPD/sq ft} = \frac{\text{Flow Rate, GPD}}{\text{Surface Area, sq ft}}$$

$$= \frac{4,000,000 \text{ GPD}}{5,024 \text{ sq ft}}$$

$$= 800 \text{ GPD/sq ft (close enough)}$$

NOTE: The suspended solids concentration of 190 mg/L was not needed to solve this problem. Try to determine the information to solve problems and forget the unimportant data.

5.6E A circular clarifier has a diameter of 80 feet and an average depth of 10 feet. The clarifier treats 4.0 MGD from the plant inflow plus 1.2 MGD of return sludge flow. The mixed liquor suspended solids concentration is 2,700 mg/L. Calculate the solids loading in lbs/day/sq ft.

Calculate Solids Applied, lbs/day.

$$\text{Solids Applied, lbs/day} = \text{Total Flow, MGD} \times \text{Conc, mg/L} \times 8.34 \text{ lbs/gal}$$

$$= (4.0 \text{ MGD} + 1.2 \text{ MGD}) \times 2,700 \text{ mg/L} \times 8.34 \text{ lbs/gal}$$

$$= 117,094 \text{ lbs/day}$$

$$\text{Solids Loading, lbs/day/sq ft} = \frac{\text{Solids Applied, lbs/day}}{\text{Surface Area, sq ft}}$$

$$= \frac{117,094 \text{ lbs/day}}{5,024 \text{ sq ft}} \quad \text{(from 5.6D)}$$

$$= 23.3 \text{ lbs/day/sq ft}$$

Answers to questions on page 142.

5.6F Secondary clarifiers are needed in secondary treatment plants to remove solids from the secondary process.

5.6G Sludge settling in the secondary clarifier may be returned to the primary clarifier to be settled with the primary sludge, pumped to the beginning of the biological process for recycling, or pumped directly to the sludge handling facilities.

Answers to questions on page 145.

5.7A Safety items that should be considered when reviewing plans and specifications for clarifiers include:

1. Access to clarifier;
2. Toeboards and nonskid surfaces on catwalks and bridges;
3. Adequate lighting;
4. Safety grates over entrances to launders, channels, and effluent pipelines;
5. In a circular clarifier, baffles and valves should be accessible without having to leave a bridge or catwalk;
6. Guards over moving parts;
7. Safety hooks, poles, or floats at strategic locations for rescue operations;
8. Catwalks and bridges free of overhead obstructions; and
9. Unobstructed access to drive units, motors, and other equipment.

5.8A The flotation process should be placed *AFTER* primary sedimentation.

5.8B Colloids—Very small, finely divided solids (particles that do not dissolve) that remain dispersed in a liquid for a long time due to their small size and electrical charge.

Emulsion—A liquid mixture of two or more liquid substances not normally dissolved in one another, but one liquid held in suspension in the other.

5.8C The flotation process is used to remove colloids and emulsions.

5.8D The vacuum flotation process consists of aerating the wastewater and creating a vacuum to pull out the air which will carry the solids to the water surface.

ANSWERS TO QUESTIONS IN LESSON 3

Answers to questions on page 154.

5.9A A combined sedimentation-digestion unit consists of a small clarifier constructed over a sludge digester. Treatment units of this type have been designed and constructed to serve small populations.

5.9B Abnormal operating conditions that might be encountered include:

1. Inflows higher than design flows,
2. High solids loadings, and
3. Toxic substances or high or low pH levels.

5.9C Major maintenance items for a combined sedimentation-digestion unit include:

1. Lubricate all equipment in accordance with manufacturers' recommendations;
2. If a tool or object falls into the clarifier, stop rotation and remove tool or object; and
3. If scraper mechanism stops moving, determine cause and remove it before attempting to start mechanism again.

Answers to questions on page 155.

5.10A The two components of an Imhoff tank are (1) settling area, and (2) sludge digestion area.

5.10B Settled material may be forced into the digestion compartment by pushing it through the connecting slot with a squeegee. Dragging a chain on the floor and allowing it to pass through the slot is another method for removing the sludge accumulation.

5.10C A fairly level sludge blanket is maintained by reversing the flow at regular intervals.

5.10D Digested sludge in an Imhoff tank is relatively odorless or has a musty smell, and it is black or very dark in color.

APPENDIX

Monthly Data Sheet

CLEANWATER, U.S.A.
WATER POLLUTION CONTROL PLANT

OPERATOR: _____

MONTHLY RECORD _____ 20____

DATE	DAY	WEATHER	FLOW MGD	TEMP.	pH	SETT. SOLIDS	SUSP. SOLIDS	BOD	GRIT–YD³	pH	BOD	SUSP SOLIDS	D O	CL₂ RES.	RAW SLUDGE GAL/DAY	SCUM GAL/DAY	% CO₂	GAS PROD. FT³/DAY	MIXING–HRS.	REMARKS
1	M	FAIR	1.200	70	7.3	8	208	190	1.0	7.2	118	110	1.0	2.1	4750	100	32	10400	4	
2	T	"	1.051	69	7.2	10	218	205	1.0	7.1	128	108	1.1	2.4	4108	110	32	10900	1	
3	W	"	1.120	69	7.3	11	222	220	1.5	7.1	154	111	0.9	3.0	4302	90	33	10800	8	
4	T	"	0.987	70	7.1	9	201	184	1.0	7.0	130	97	1.2	1.8	3810	120	33	11200	8	Sludge to #1 Bed – 16,000 Gal.
5	F	CL.	1.008	68	7.0	7	248	232	0.75	7.0	140	112	0.7	1.9	4005	115	34	10600	8	
6	S	"	1.102	68	7.2	9	210	211	0.75	7.1	138	100	0.8	2.1	4190	120	32	11000	4	
7	S	FAIR	0.974	69	7.3	8	215	199	0.75	7.1	140	101	0.9	2.0	3915	110	32	10800	4	
8																				
9																				
10																				
11																				
12																				
13																				
14																				
15																				
16																				
17																				
18																				
19																				
20																				
21																				
22																				
23																				
24																				
25																				
26																				
27																				
28																				
29																				
30																				
31																				
MAX.																				
MIN.																				
AVG.			1.016	69	7.1	9	210	195	1.0	7.1	135	106	0.9	2.0	4154	100	33	10810	5	

Headings grouping: RAW WASTEWATER (TEMP., pH, SETT. SOLIDS, SUSP. SOLIDS, BOD, GRIT–YD³); EFFLUENT (pH, BOD, SUSP SOLIDS, D O, CL₂ RES.); DIGESTION (RAW SLUDGE GAL/DAY, SCUM GAL/DAY, % CO₂, GAS PROD. FT³/DAY, MIXING–HRS.)

SUMMARY DATA

% REMOVAL	BOD	S S
INF. – EFF.	30	49

SLUDGE DATA

% SOLIDS – AVG.	4.8
LBS. DRY SOLIDS / DAY	1663
% VOL. SOLIDS – AVG.	76
LBS. VOL. SOLIDS / DAY	1264
LBS. VOL. SOL. / 1000 FT³ / DAY	25.3
GAL. SLUDGE TO BEDS	48,000
CU. YDS. CAKE REMOVED	22
FT³ GAS / LB. VOL. SOLIDS	8.5
FT³ GAS / MG. FLOW	10,640
CU. YDS. GRIT / MG. FLOW	1.0

COST DATA

MAN DAYS 44

PAYROLL	$1250 —
POWER PURCHASED	250 —
OTHER UTILITIES (GAS ETC.)	60 —
GASOLINE, OIL, GREASE	30 —
CHEMICALS & SUPPLIES	60 —
MAINTENANCE	130 —
VEHICLE COSTS	70 —
OTHER	20 —
TOTAL	$1870 —
OPERATING COST / MG.	$ 59.37
OPERATING COST / CAPITA / MG.	$ 0.19

FLOW METER:
LAST 445237
1st 413749
TOTAL 31.488 MG

ELECTRIC METER:
LAST 5029
1st 4821
MULT. 80 x 208 = 16,640 KWH.

RAW SLUDGE:
LAST 828588
1st 699814
STROKES–SCUM 3100
TOTAL 128,774 x 1.0 = 128,774 GALS.

GAS METER:
LAST 718406
1st 383296
TOTAL 335,110 FT³

CHAPTER 6

TRICKLING FILTERS

by

Larry Bristow

TABLE OF CONTENTS

Chapter 6.   TRICKLING FILTERS

OBJECTIVES

Chapter 6.   TRICKLING FILTERS

Following completion of Chapter 6, you should be able to:

1. Explain the principles of the trickling filter treatment process and the operation of the process,

2. Inspect a new trickling filter for proper installation,

3. Place a new filter into service,

4. Schedule and safely conduct operation and maintenance duties,

5. Sample influent and effluent, interpret lab results, and make appropriate adjustments in treatment process,

6. Recognize factors that indicate a trickling filter is not performing properly, identify the source of the problem, and take corrective action,

7. Develop an operating strategy for a trickling filter,

8. Conduct your duties in a safe fashion,

9. Identify the different types of trickling filters,

10. Determine hydraulic and organic loadings on a trickling filter, and

11. Keep records for a trickling filter plant.

WORDS

Chapter 6.   TRICKLING FILTERS

AEROBIC (AIR-O-bick)   AEROBIC

A condition in which atmospheric or dissolved molecular oxygen is present in the aquatic (water) environment.

AEROBIC (AIR-O-bick) PROCESS   AEROBIC PROCESS

A waste treatment process conducted under aerobic (in the presence of "free" or dissolved oxygen) conditions.

ANAEROBIC (AN-air-O-bick)   ANAEROBIC

A condition in which atmospheric or dissolved molecular oxygen is *NOT* present in the aquatic (water) environment.

BIOMASS (BUY-o-MASS)   BIOMASS

A mass or clump of organic material consisting of living organisms feeding on the wastes in wastewater, dead organisms and other debris. Also see ZOOGLEAL FILM.

COLLOIDS (CALL-loids)   COLLOIDS

Very small, finely divided solids (particles that do not dissolve) that remain dispersed in a liquid for a long time due to their small size and electrical charge. When most of the particles in water have a negative electrical charge, they tend to repel each other. This repulsion prevents the particles from clumping together, becoming heavier, and settling out.

DISTRIBUTOR   DISTRIBUTOR

The rotating mechanism that distributes the wastewater evenly over the surface of a trickling filter or other process unit. Also see FIXED SPRAY NOZZLE.

FIXED SPRAY NOZZLE   FIXED SPRAY NOZZLE

Cone-shaped spray nozzle used to distribute wastewater over the filter media, similar to a lawn sprinkling system. A deflector or steel ball is mounted within the cone to spread the flow of wastewater through the cone, thus causing a spraying action. Also see DISTRIBUTOR.

FORCE MAIN   FORCE MAIN

A pipe that carries wastewater under pressure from the discharge side of a pump to a point of gravity flow downstream.

HUMUS SLUDGE   HUMUS SLUDGE

The sloughed particles of biomass from trickling filter media that are removed from the water being treated in secondary clarifiers.

LOADING   LOADING

Quantity of material applied to a device at one time.

MASKING AGENTS   MASKING AGENTS

Substances used to cover up or disguise unpleasant odors. Liquid masking agents are dripped into the wastewater, sprayed into the air, or evaporated (using heat) with the unpleasant fumes or odors and then discharged into the air by blowers to make an undesirable odor less noticeable.

MICROORGANISMS (MY-crow-OR-gan-IS-zums)   MICROORGANISMS

Very small organisms that can be seen only through a microscope. Some microorganisms use the wastes in wastewater for food and thus remove or alter much of the undesirable matter.

NITRIFICATION (NYE-truh-fuh-KAY-shun)   NITRIFICATION

An aerobic process in which bacteria change the ammonia and organic nitrogen in wastewater into oxidized nitrogen (usually nitrate). The second-stage BOD is sometimes referred to as the "nitrogenous BOD" (first-stage BOD is called the "carbonaceous BOD").

ORIFICE (OR-uh-fiss)   ORIFICE

An opening (hole) in a plate, wall, or partition. In a trickling filter distributor, the wastewater passes through an orifice to the surface of the filter media. An orifice flange or plate placed in a pipe consists of a slot or a calibrated circular hole smaller than the pipe diameter. The difference in pressure in the pipe above and at the orifice may be used to determine the flow in the pipe.

PARALLEL OPERATION   PARALLEL OPERATION

Wastewater being treated is split and a portion flows to one treatment unit while the remainder flows to another similar treatment unit. Also see SERIES OPERATION.

PHYSICAL WASTE TREATMENT PROCESS   PHYSICAL WASTE TREATMENT PROCESS

Physical waste treatment processes include use of racks, screens, comminutors, clarifiers (sedimentation and flotation) and filtration. Chemical or biological reactions are important treatment processes, but *NOT* part of a physical treatment process.

PONDING   PONDING

A condition occurring on trickling filters when the hollow spaces (voids) become plugged to the extent that water passage through the filter is inadequate. Ponding may be the result of excessive slime growths, trash, or media breakdown.

PROTOZOA (pro-toe-ZOE-ah)   PROTOZOA

A group of motile microscopic organisms (usually single-celled and aerobic) that sometimes cluster into colonies and generally consume bacteria as an energy source.

RECIRCULATION   RECIRCULATION

The return of part of the effluent from a treatment process to the incoming flow.

SECONDARY TREATMENT   SECONDARY TREATMENT

A wastewater treatment process used to convert dissolved or suspended materials into a form more readily separated from the water being treated. Usually the process follows primary treatment by sedimentation. The process commonly is a type of biological treatment process followed by secondary clarifiers that allow the solids to settle out from the water being treated.

SERIES OPERATION   SERIES OPERATION

Wastewater being treated flows through one treatment unit and then flows through another similar treatment unit. Also see PARALLEL OPERATION.

SHOCK LOAD (TRICKLING FILTERS)   SHOCK LOAD

The arrival at a plant of waste which is toxic to organisms in sufficient quantity or strength to cause operating problems. Possible problems include odors and sloughing off of the growth or slime on the trickling filter media. Organic or hydraulic overloads also can cause a shock load.

TRICKLING FILTER   TRICKLING FILTER

A treatment process in which the wastewater trickles over media that provide the opportunity for the formation of slimes or biomass which contain organisms that feed upon and remove wastes from the water being treated.

TRICKLING FILTER MEDIA   TRICKLING FILTER MEDIA

Rocks or other durable materials that make up the body of the filter. Synthetic (manufactured) media have been used successfully.

TWO-STAGE FILTERS   TWO-STAGE FILTERS

Two filters are used. Effluent from the first filter goes to the second filter, either directly or after passing through a clarifier.

ZOOGLEAL (ZOE-glee-al) FILM   ZOOGLEAL FILM

A complex population of organisms that form a "slime growth" on the trickling filter media and break down the organic matter in wastewater. These slimes consist of living organisms feeding on the wastes in wastewater, dead organisms, silt, and other debris. "Slime growth" is a more common term.

CHAPTER 6. TRICKLING FILTERS

(Lesson 1 of 3 Lessons)

6.0 HOW A TRICKLING FILTER WORKS

6.00 Description of a Trickling Filter

In the initial chapters of this course, you have learned about physical methods of wastewater treatment. In general, these techniques (processes) consist of the screening of large particles, settling of heavy material, and floating of light material by preliminary and primary treatment units (screen, grit channel, clarifier). Although primary treatment is very efficient for removing settleable solids and scum or floatable solids, it is not capable of removing lighter suspended solids or dissolved solids which may exert a strong oxygen demand on the receiving waters.

In order to remove the very small suspended solids (colloids) and dissolved solids, waste treatment plants include *SECONDARY TREATMENT*.[1] This process produces an overall plant removal of suspended solids and BOD of 90 percent or more. The three most common secondary treatment processes are trickling filters, rotating biological contactors, and activated sludge. This chapter will deal with *TRICKLING FILTERS*.[2]

Figures 6.1 and 6.2 show where a trickling filter usually is located in a plant.

Most trickling filters are large-diameter, shallow, cylindrical structures filled with stone and having an overhead distributor. (See Figure 6.3 and Table 6.1.) Many variations of this design have been built. When natural media (stones) are used, the trickling filter is usually cylindrical with a shallow bed; when synthetic media (plastics) are used, the filter could be cylindrical or rectangular with a much deeper bed. Structures containing deep beds of synthetic media may be called filter towers or biofilter towers. Square or rectangular filters have been constructed with fixed sprinklers for wastewater distribution. Another type of filter called the "rotating biological contactor" is discussed in detail in Chapter 7. This contactor treats the

wastewater using methods similar to a trickling filter except instead of applying water over the media, the media are rotated through the wastewater being treated. The structures for trickling filters or rotating biological contactors may be covered and forced-air ventilated for odor control purposes, or covered to prevent freezing in some areas.

6.01 Principles of Treatment Process

Trickling filters, biological oxidation beds, and rotating biological contactors consist of three basic parts:

1. The media (and retaining structure),
2. The underdrain system, and
3. The distribution system.

The media provide a large surface area upon which a biological slime growth develops. This slime growth, sometimes called a *ZOOGLEAL FILM*,[3] contains the living organisms that break down the organic material. The media may be rock, slag, coal, bricks, redwood blocks, molded plastic (Figure 6.4), or any other sound, durable material. The media should be of such sizes and stacked in such a fashion as to provide empty spaces (voids) for air to ventilate the filter and keep conditions aerobic. For rock, the size will usually be from about two inches to four inches (5 to 10 cm). Although actual size is not too critical, it is important that the media be uniform in size to permit adequate ventilation. The media depth ranges from about three to eight feet (1 to 2.5 meters) for rock media trickling filters and 15 to 30 feet (5 to 10 meters) for synthetic media.

The underdrain system of a trickling filter has a sloping bottom. This leads to a center channel which collects the filter effluent. The underdrain system also supports the media and permits air flow. Common materials and methods for constructing underdrain systems include the use of spaced redwood stringers and of prefabricated blocks constructed of concrete, vitrified clay, or other suitable material.

The distribution system, in the vast majority of cases, is a rotary-type distributor which consists of two or more horizontal pipes supported a few inches above the filter media by a central column. The wastewater is fed from the column through the horizontal pipes and is distributed over the media through orifices located along one side of each of these pipes (or arms). Rotation of the arms is due either to the rotating water-sprinkler reaction from wastewater flowing out the orifices ("jet like")

---

[1] *Secondary Treatment. A wastewater treatment process used to convert dissolved or suspended materials into a form more readily separated from the water being treated. Usually the process follows primary treatment by sedimentation. The process commonly is a type of biological treatment process followed by secondary clarifiers that allow the solids to settle out from the water being treated.*

[2] *Trickling Filter. A treatment process in which the wastewater trickles over media that provide the opportunity for the formation of slimes or biomass which contain organisms that feed upon and remove wastes from the water being treated. Trickling filters are sometimes called biofilters, accelo filters or aerofilters, depending on the recirculation pattern.*

[3] *Zoogleal (ZOE-glee-al) Film. A complex population of organisms that form a "slime growth" on the trickling filter media and break down the organic matter in wastewater. These slimes consist of living organisms feeding on the wastes in wastewater, dead organisms, silt, and other debris. "Slime growth" is a more common term.*

TREATMENT PROCESS

FUNCTION

PRELIMINARY TREATMENT

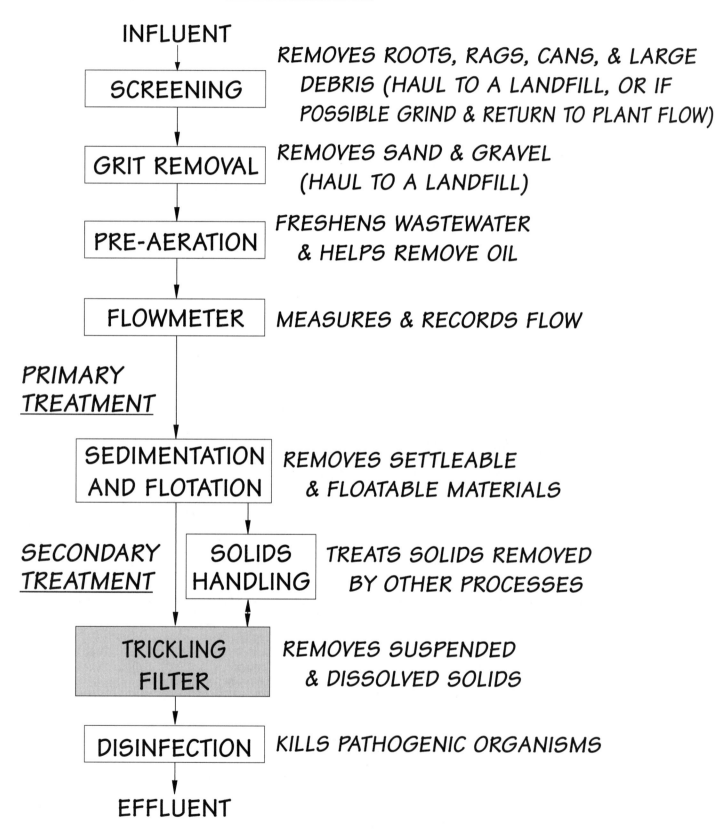

INFLUENT

SCREENING — REMOVES ROOTS, RAGS, CANS, & LARGE DEBRIS (HAUL TO A LANDFILL, OR IF POSSIBLE GRIND & RETURN TO PLANT FLOW)

GRIT REMOVAL — REMOVES SAND & GRAVEL (HAUL TO A LANDFILL)

PRE-AERATION — FRESHENS WASTEWATER & HELPS REMOVE OIL

FLOWMETER — MEASURES & RECORDS FLOW

PRIMARY TREATMENT

SEDIMENTATION AND FLOTATION — REMOVES SETTLEABLE & FLOATABLE MATERIALS

SECONDARY TREATMENT

SOLIDS HANDLING — TREATS SOLIDS REMOVED BY OTHER PROCESSES

TRICKLING FILTER — REMOVES SUSPENDED & DISSOLVED SOLIDS

DISINFECTION — KILLS PATHOGENIC ORGANISMS

EFFLUENT

Fig. 6.1   Flow diagram of treatment plant

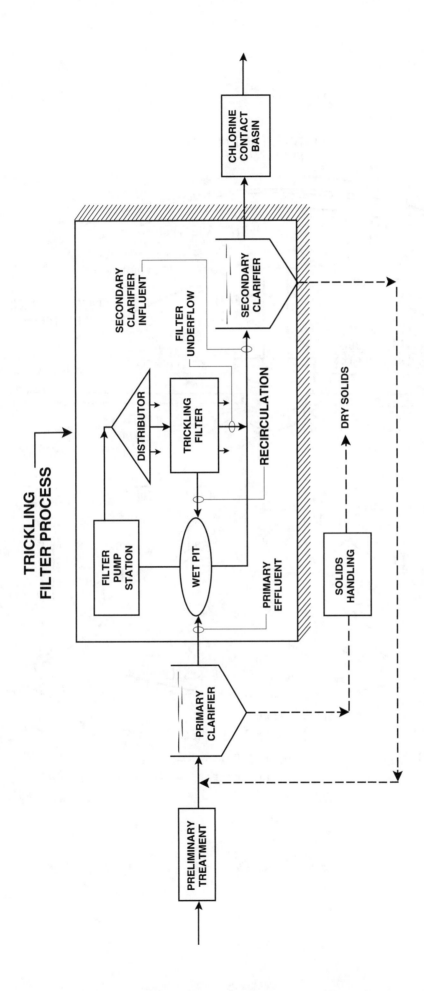

Fig. 6.2 Typical trickling filter plant

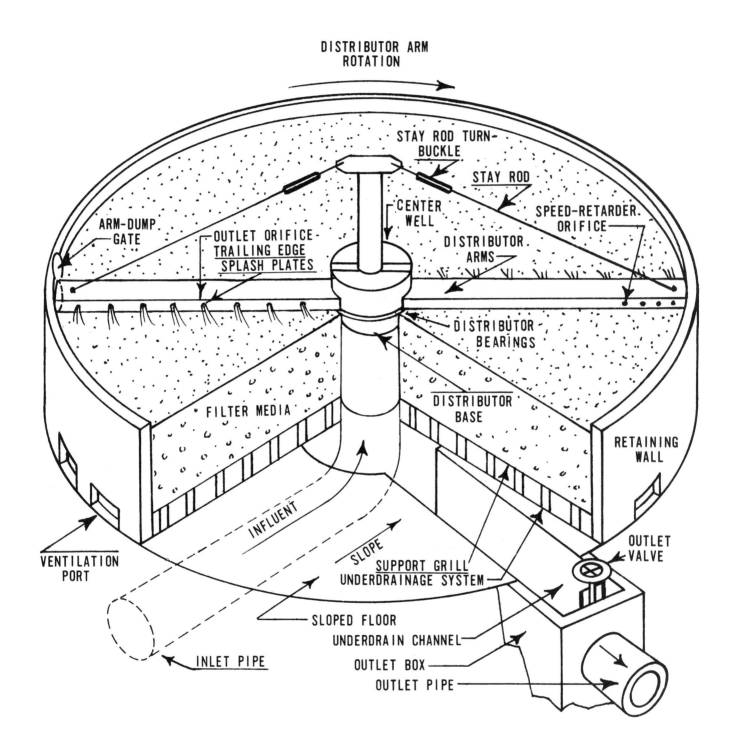

Fig. 6.3   Trickling filter

TABLE 6.1   PURPOSE OF TRICKLING FILTER PARTS

Part	Purpose	Part	Purpose
Inlet Pipe	Conveys wastewater to be treated to trickling filter.	Underdrain Channel	Drains filter effluent to outlet box.
Distributor Base	Supports rotating distributor arms.	Outlet Box	Collects filter effluent before it flows to next process.
Distributor Bearings	Allow distributor arms to rotate.	Outlet Valve	Regulates flow of filter effluent from outlet box into outlet pipe. Closed when filter is to be flooded.
Distributor Arm	Conveys wastewater to outlet orifices located along arms.		
Outlet Orifice	Controls flow to filter media. Adjustable to provide even distribution of wastewater to each square foot of filter media.	Outlet Pipe	Conveys filter effluent to next treatment process.
		Retaining Wall	Holds filter media in place.
Speed-Retarder Orifice	Regulates speed of distributor arms.	Ventilation Port	Allows air to flow through media.
		Stay Rod	Supports distributor arm.
Splash Plate	Distributes flow from orifices evenly over filter media.	Turnbuckle on Stay Rod	Permits adjusting and leveling of distributor arm in order to produce even distribution of wastewater over the media.
Arm Dump Gate	Drains distributor arm and controls filter flies along filter retaining wall. Also used for flushing distributor arm to remove accumulated debris which might block outlet orifices.		
		Center Well	Provides for higher water head to maintain equal flow to distributor arms. Usually a head of 18 to 24 inches (45 to 60 cm) is maintained on the orifices.
Filter Media	Provide a large surface area upon which the biological slime growth develops.		
Support Grill	Keeps filter media in place and out of underdrain system.	Splitter Box	Divides flow to trickling filters, for recirculation or to secondary clarifiers.
		AUXILIARY EQUIPMENT	
Underdrain System	Collects treated wastewater from under filter media and conveys it to underdrain channel. Also permits air flow through media.	Recirculation Pump	Returns or recirculates flows to trickling filters.

Fig. 6.4   *Installation of synthetic media in trickling filter*
(Courtesy of The Dow Chemical Company)

or by some mechanical means. The distributors are equipped with mechanical-type seals at the center column to prevent leakage and protect the bearings, guy- or stay rods for seasonal adjustment of the distributor arms to maintain an even distribution of wastewater over the media, and quick-opening or arm dump gates at the end of each arm to permit easy flushing.

The fixed-nozzle distribution system is not as common as the rotary type and is commonly found only at small plants. Disadvantages of fixed nozzles include difficult access for nozzle cleaning and high pumping requirements to maintain good hydraulic distribution. The nozzles are located on the surface of the filter like a lawn sprinkler system. Each fixed-nozzle consists of a circular orifice with an inverted cone-shaped deflector mounted above the center. The deflector breaks the flow into a spray. Some types have a steel ball in the inverted cone. (See Figure 6.5.) The fixed-nozzle system requires an elaborate piping system to ensure even distribution of the wastewater. The nozzles extend six to twelve inches (15 to 30 cm) above the media and are shaped so that an overlapping spray pattern exists at the start of dosing when the head in the dosing tank is the greatest. The pattern is carefully worked out to provide a relatively even distribution of the wastewater.

Flow is usually intermittent and is controlled by automatic siphons which regulate the flow from dosing tanks. (See Figure 6.5.) Dosing tanks and siphons should be constructed to facilitate cleaning and reduce problems caused by corrosion. Attempt to record the time required to fill and discharge the dosing chamber. If this time becomes shorter, this could indicate that grease and solids are accumulating in the siphon and pipes and should be removed.

6.02 Principles of Operation

The maintenance of a good growth of organisms on the filter media is crucial to successful operation.

The term "filter" is rather misleading because it indicates that solids are separated from the liquid by a straining action. This is not the case. Passage of wastewater through the filter causes the development of a gelatinous coating of bacteria, *PROTOZOA*,[4] and other organisms on the media. This growth of organisms absorbs and uses much of the suspended, colloidal, and dissolved organic matter from the wastewater as it passes over the growth in a rather thin film. Part of this material is used as food for production of new cells, while another portion is oxidized to carbon dioxide and water. Partially decomposed organic matter together with excess and dead film is continuously or periodically washed (sloughed) off and passes from the filter with the effluent.

For the oxidation (decomposition) processes to be carried out, the biological film requires a continuous supply of dissolved oxygen which may be absorbed from the air circulating through the filter voids (spaces between the rocks or other

media). Adequate ventilation of the filter must be provided; therefore the voids in the filter media must be kept open. Ventilation may be by either natural ventilation or by a forced air ventilation system. Clogged void space can create operational problems including ponding and reduction in overall filter efficiency. The void space provided by synthetic (plastic) media is about 95 percent of the total filter volume, thus providing space for biological slimes to slough and pass through the media. Rock media contains about 35 percent void space. Trickling filters with plastic media may be loaded at much higher rates than rock media without developing plugging, ponding, fly, or odor problems. Highly loaded filters may be called roughing filters and are commonly combined with other biological treatment processes (activated sludge, rotating biological contactors (RBCs), ponds) to achieve higher levels of BOD removal.

A method of increasing the efficiency of trickling filters is to add recirculation. Recirculation is a process in which filter effluent is recycled and brought into contact with the biological film more than once. Recycling of filter effluent increases the contact time with the biological film and helps to seed the lower portions of the filter with active organisms. Due to the increased flow rate per unit of area, these higher flows tend to cause more continuous and uniform sloughing of excess or aged growths. Uniform and continuous sloughing of growths is important because this provides a more aggressive surface of new growths to treat the wastewater. Sloughing of growths prevents ponding and improves ventilation through the filter. Increased hydraulic loadings also decrease the opportunity for snail and filter fly breeding. The thickness of the biological growth has been observed to be directly related to the organic strength of the wastewater (the higher the BOD, the thicker the layers of organisms). By the use of recirculation, the strength of wastewater applied to the filter can be diluted, thus helping to prevent excessive buildup of growths.

Recirculation may be constant or intermittent and at a steady or fluctuating rate. Sometimes recirculation (recycling) is practiced only during periods of low flow to keep rotary distributors in motion, to prevent drying of the filter growths, or to prevent freezing. Recirculation in proportion to flow may be used to reduce the strength of the wastewater applied to the filter, while steady recirculation tends to even out the highs and lows of organic loading. Steady recirculation, however, requires the use of more energy. Some plants operate intermittently at high recirculation rates (all recirculation pumps on) for two or three hours each week. This high rate will cause sloughing on a regular basis rather than allowing the slime growths to build up and slough under uncontrolled conditions.

Almost any organic waste that can be successfully treated by other aerobic biological processes can be treated on trickling filters. This includes, in addition to domestic wastewater, such wastewaters as might come from food processing, textile, carbonated beverage, dairy and fermentation industries, and certain pharmaceutical processes. Industrial wastewaters that cannot be treated are those that contain excessive concentrations of toxic materials such as pesticide residues, heavy metals, and highly acidic or alkaline wastes.

For maximum efficiency, the slime growths on the filter media should be kept fairly aerobic. This can be accomplished by proper design of the wastewater collection system, proper operation of primary clarifiers, or by pretreatment of the waste-

---

[4] *Protozoa (pro-toe-ZOE-ah). A group of motile microscopic organisms (usually single-celled and aerobic) that sometimes cluster into colonies and generally consume bacteria as an energy source.*

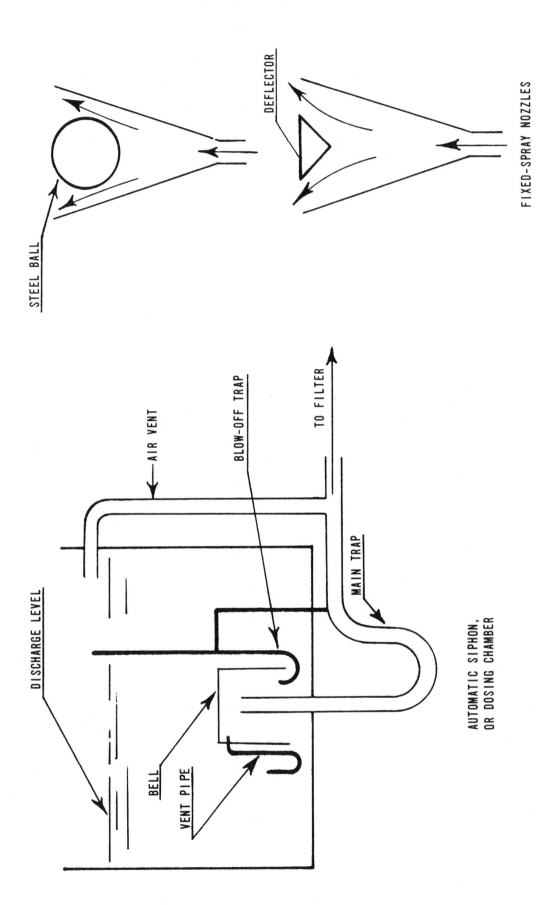

STEEL BALL

DEFLECTOR

FIXED-SPRAY NOZZLES

AIR VENT

BLOW-OFF TRAP

TO FILTER

MAIN TRAP

DISCHARGE LEVEL

BELL

VENT PIPE

AUTOMATIC SIPHON,
OR DOSING CHAMBER

Fig. 6.5 Siphon and nozzle details for fixed-spray filters

water by aeration or addition of recycled filter effluent. The air supply to the slimes may be improved by increased air or wastewater recirculation. The thin slime growth may be aerobic on the surface, but anaerobic next to the media. A trickling filter media of rock or slag can accumulate slimes only on the outside surface, but manufactured media provide considerably more surface area per unit of dead space.

The temperatures of the wastewater and of the climate also affect filter operation, with temperature of the wastewater being the more important. Of course, temperature of the wastewater will vary with the weather. Within limits, activity of the organisms increases as the temperature rises. Therefore, higher loadings and greater efficiency are possible in warmer climates if aerobic conditions can be reasonably maintained in the filter.

QUESTIONS

Write your answers in a notebook and then compare your answers with those on page 204.

6.0A   Primary treatment is effective in removing (a)_____ _____, and (b)_____, but is not nearly as effective in removing (c)_____.

6.0B   What is the purpose of "secondary treatment"?

6.0C   What causes the distributor arms or pipes to rotate on a trickling filter?

6.0D   How does the trickling filter process work?

6.0E   How does recirculation increase the efficiency of a trickling filter?

6.1   CLASSIFICATION OF FILTERS

6.10   General

Depending upon the hydraulic and organic loadings applied, filters are classified as standard-rate, high-rate, or roughing filters. Further designations such as single-stage, two-stage, and series or parallel are used to indicate the flow pattern of the plant. The hydraulic loading applied to a filter is the total volume of liquid, including recirculation, expressed as gallons per day per square foot of filter surface area (GPD/sq ft). The organic loading is expressed as the pounds of BOD applied

per day per 1,000 cubic feet of filter media (lbs BOD/day/ 1,000 cu ft). Where recirculation is used, an additional organic loading will be placed on the filter; however, this added loading is omitted in most calculations because it was included in the influent load. Procedures for calculating the hydraulic and organic loadings are given in Section 6.7, "Loading Criteria."

6.11   Standard-Rate Filters

The standard-rate filter is operated with a hydraulic loading range of 25 to 100 gal/day/sq ft (1 to 4 MGD/acre or 1,020 to 4,075 L/day/sq m or 9.4 to 37.5 MLD/ha), and an organic BOD

loading of 5 to 25 lbs/day/1,000 cu ft (200 to 1,000 lbs/day/ac-ft or 80 to 400 kg/day/1,000 cu m or 740 to 3,690 kg/day/ha-ft). The filter media is usually rock with a depth of 6 to 8 feet (1.8 to 2.4 m), with application to the filter by a rotating distributor. Many standard-rate filters are equipped to provide some recirculation during low flow periods.

The filter growth is often heavy and, in addition to the bacteria and protozoa, many types of worms, snails, and insect larvae can be found. The growth usually sloughs off at intervals, noticeably in spring and fall. The effluent from a standard-rate filter treating municipal wastewater is usually quite stable with BODs as low as 20 to 25 mg/L.

6.12   High-Rate Filters

High-rate filters were the result of trying to reduce costs associated with standard-rate filters and attempting to treat increased waste loads with the same facility. Studies indicated that essentially the same BOD reductions could be obtained at the higher design loadings.

High-rate filters usually have rock media with a depth of 3 to 5 feet (0.9 to 1.5 m) or synthetic media with a depth of 15 to 30 feet (4.6 to 9.1 m). Recommended loadings range from 100 to 1,000 gal/day/sq ft (4,075 to 40,750 L/day/sq m) for rock and 350 to 2,100 gal/day/sq ft (14,250 to 85,575 L/day/sq m) for synthetic media and 25 to 100 lbs BOD/day/1,000 cu ft (400 to 1,600 kg BOD/day/1,000 cu m) for rock and 50 to 300 lbs BOD/day/1,000 cu ft (800 to 4,800 kg BOD/day/1,000 cu m) for synthetic media. These filters are designed to receive wastewater continually, and practically all high-rate installations use recirculation. Loadings may be higher for synthetic media.

Due to the heavy flow of wastewater over the media, more uniform sloughing of the filter growths occurs from high-rate filters. This sloughed material is somewhat lighter than from a standard-rate unit and therefore more difficult to settle. Effluents with BODs as low as 20 to 50 mg/L are sometimes produced by high-rate plants treating municipal wastewater.

6.13   Roughing Filters

A roughing filter is actually a high-rate filter receiving a very high organic loading. Any filter receiving an organic loading of 100 to over 300 lbs of BOD/day/1,000 cu ft (1,600 to 4,800 kg BOD/day/1,000 cu m) of media volume is considered to be in this class. This type of filter is used primarily to reduce the organic load on subsequent oxidation processes such as a second-stage filter or activated sludge process. Many times they are used in plants that receive strong organic industrial wastes. They are also used where an intermediate (50-70 percent BOD removal) degree of treatment is satisfactory. Most roughing filters have provisions for recirculation.

Operation of the filter is basically the same as for the high-rate filters with recirculation. *OVERALL* BOD reductions are much lower than the high-rate filters, but reductions *PER UNIT VOLUME OF FILTER MEDIA* are greater.

6.14   Filter Staging

Figure 6.6 shows various filter and clarifier layouts. The decision as to the number of filters (or stages) required is one of

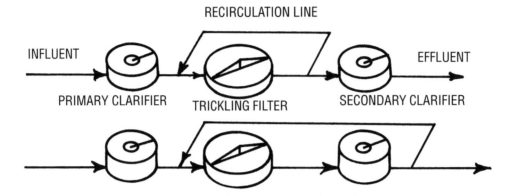

Typical Single-Stage Recirculation Patterns

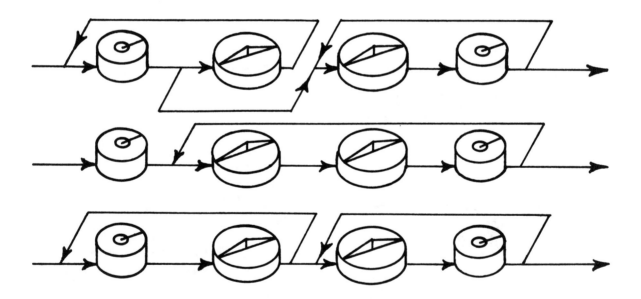

Typical Two-Stage Recirculation Patterns

Fig. 6.6   Trickling filter recirculation patterns

design rather than operation. In general, however, at smaller plants where the flow is fairly low, the strength of the raw wastewater is average, and effluent quality requirements are not too strict, a single-stage plant (one filter) is often sufficient and most economical. In slightly overloaded plants, the addition of some recirculation capability can sometimes improve the effluent quality enough to meet receiving water standards and NPDES permit requirements without the necessity of adding more stages.

In two-stage filter plants, two filters are operated in series. Sometimes a secondary clarifier is installed between the two filters. Recirculation is almost universally practiced at two-stage plants with many different arrangements being possible. The choice of a recirculation scheme is based on consideration of which arrangement produces the best effluent under the particular conditions of wastewater strength and other characteristics. (See Figure 6.6.)

QUESTIONS

Write your answers in a notebook and then compare your answers with those on page 205.

6.1A   What are the three general classifications of trickling filters?

6.1B   What are the principal differences between standard-rate and high-rate filters?

6.2   STARTING, OPERATING, AND SHUTTING DOWN A FILTER

6.20   Pre-Start[5]

A new plant is seldom started up without some unexpected, frustrating problems. Some careful inspecting ahead of time can prevent many of these situations.

Filter bearings often come packed in a special grease to prevent damage during transportation. This packing grease must be removed and replaced with the proper grease before start-up.

If at all possible, you should arrange to be present when your new equipment is serviced. You should see that the correct oil is used in all oil reservoirs. Many contractors will put motor oil in everything and consider it serviced. For future reference, record the amount and type of oil each reservoir holds.

After the oil has been installed in a distributor, check the arms for even adjustment and level. Rotate the unit by hand and observe for smooth turning. Any vibration or roughness should be corrected before putting the unit in service.

If the distributor has adjustable orifices, get the design specifications and a rule and check out the orifice settings. File the specification sheet for future reference.

In a trickling filter plant with fixed-spray nozzles, each nozzle should be checked to ensure that it is free of foreign objects.

In order to prevent damage to pumps, crawl into the underdrain system of the filter and remove any debris (rocks, pieces of wood, and other debris). Check painted surfaces for damaged areas. Touch these up before they get wet to prevent corrosion and further damage. A few nicks and scratches in a distributor arm can seriously affect the life of the original protective coatings.

Check all valves in the system for smooth operation. On sliding-gate valves, see that the gates seat properly. There are adjustable wedges and stops on this type of valve. With the valve adjusted, set the lock nut on the stem to prevent jamming or closing the gate too tightly. These small precautions will yield years of trouble-free valve operation.

In addition to the general items covered in this section, you should be certain that the correct manufacturer's manual has been furnished for each piece of equipment. Read each manual carefully and follow the given recommendations. Obtain the oils and greases recommended; or, if you buy from one oil company, have their representative furnish you a *WRITTEN LIST* of the company's products that are equivalent to those recommended by the equipment manufacturer.

6.21   Placing Filter in Service

Try to schedule the starting of trickling filters during late April through early June (depends on local conditions). This procedure will produce the most growth during the shortest period of time. Problems avoided will include wet weather flows in the spring, odors in the summer, and dormant bacteria in the winter.

When you have checked out all equipment mechanically, starting up the trickling filter portion of the plant is very simple. Start the wastewater flow to the filters, observing the rotating arms carefully for smooth operation, speed of rotation, and even distribution of the waste over the media. Time the speed of rotation, record the flow rate, and log them for future reference.

NOTE: Starting up recirculation may be tricky in some plants. The pump may run out of water before the return from the filter has begun. You may have to block the channels (launders) in the clarifier and build up extra water before starting the pump. Conversely, shutting off recirculation will result in a surge of water because the pump is no longer removing water, but water is still returning from the filter.

For fixed nozzles, observe the spray pattern. Usually some debris will show up to plug some of the nozzles, the amount depending on how thoroughly the plant was checked out prior to start-up. Be sure to keep the nozzles clear so that the wastewater is distributed over all of the filter media. Regular care is required to keep fixed nozzles working properly.

---

[5] *Contracts for treatment plant construction usually include services of the consultant and contractor to assist in start-up of new facilities. The operator should make full use of these services.*

Several days will pass before any growth starts to develop on the filter media, and up to several weeks will pass before full development occurs. Time of year, weather conditions, and strength of the waste are all factors that will affect the time needed for growth development.

Growth may be accelerated by recirculating wastewater through the trickling filter prior to treating the main wastewater flow stream. Waste activated sludge may also be added to the recirculated flow to encourage growth development.

During this period of growth development, an unstable effluent will be produced. This effluent will exert a pollutional load on the receiving waters. Heavy chlorination is usually used during this time to reduce the pollutional load and the health hazard to some extent.

In some locations, such as where fish are threatened, the use of chlorine in this manner may be restricted. If an older plant is being phased out, it may be possible to load the new facilities lightly or intermittently until a full growth is established.

Procedures for starting up pumps, clarifiers, and other equipment are discussed in other chapters.

6.22 Daily Operation

Once growth on the media has been established and the plant is in "normal operation," very little routine operational control is required. Careful daily observation is important. Items to be checked daily are:

1. Any indication of ponding,
2. Filter flies,
3. Odors,
4. Plugged orifices,
5. Roughness or vibration of the distributor arms,
6. Leakage past the distributor turntable seal,
7. Splash beyond the filter media, and
8. Cleanup of slimes not on media.

Refer to the appropriate paragraphs in the following lesson on operational problems for procedures to correct these conditions.

Operation of clarifiers is interconnected with trickling filter operation. If the recirculation pattern permits, it is a good idea to return filter effluent to the primary clarifier. This is a very effective odor-control measure because it adds oxygen to incoming wastewater that is often septic. In some plants, increasing the recirculation rate will increase the hydraulic loading on the clarifier. *BE SURE THE HYDRAULIC LOADING REMAINS WITHIN THE ENGINEERING DESIGN LIMITS.* If the hydraulic loading is too low, septic conditions

may develop in the clarifier. Excessively high loadings may wash solids out of the clarifier. Hydraulic loadings for primary clarifiers (including recirculation flows) should not be greater than 900 to 1,200 GPD/sq ft (36,700 to 48,900 *L*/day/sq m) for average conditions and should be less than 2,000 to 3,000 GPD/sq ft (81,500 to 122,250 *L*/day/sq m) for peak conditions.

Recirculation during low inflow periods of the day and night may help to keep the slime growths wet, minimize fly development, and wash off excessive slime growths. Reduce or stop recirculation during high flow periods, if necessary, to avoid clarifier problems from hydraulic overloading. Recirculation of final clarifier effluent dilutes influent wastewater and recirculation improves slime development on the media. Proper recirculation rates help to control snail populations on the media.

You should, by evaluating your own operating records, adjust the process to obtain the best possible results for the least cost. Power costs are a large item in a plant budget. In order to conserve energy, use the lowest recirculation rates that will yield good results. Be careful not to cause ponding, reduced BOD removal efficiency, or other problems that result from recirculation rates that are too low. Also, reduced hydraulic loadings mean better settling in the clarifiers. This results in less chlorine usage in plants that disinfect the final effluent, since organic matter exerts a high chlorine demand. If filter effluent, rather than secondary clarifier effluent, is recirculated, the hydraulic loading on the secondary clarifier is not affected.

QUESTIONS

Write your answers in a notebook and then compare your answers with those on page 205.

6.2A Prepare a checklist of items that should be inspected before a trickling filter is placed in service.

6.2B During start-up of a trickling filter, why should the plant effluent be heavily chlorinated?

6.2C Prepare a checklist of items needing daily inspection during "normal operation" of trickling filters.

6.2D What may happen to a clarifier effluent if the clarifier is not operated within design hydraulic loadings?

6.23 Shutdown of a Filter

Always take a few minutes to plan what you are going to do before shutting down a major plant process or piece of equipment, such as a trickling filter, regardless of the seriousness of the problem or the need for immediate action. Items that must be considered are listed below.

1. What is the incoming flow? Could a shutdown be scheduled at a better time such as during lower flows or when more operators are available to perform the work?

2. How will a shutdown affect the rest of the plant? When the process or equipment is placed back on line after a shutdown, will it cause development of a hydraulic surge which will overload other processes (clarifiers) or equipment (such as chlorinators)?

3. If the filter is to be shut down for maintenance, are the necessary tools and other items (such as funnels, buckets, and lubricants) available?

4. Is there any other task that should be performed while the unit is off the line? For example, does one of the recirculation pumps need repacking?

To shut down a trickling filter, consider the following step-by-step procedures if they apply to your treatment plant:

1. Inspect your plant to be sure there are no abnormal conditions hindering the effectiveness of other operating areas and process units.

2. If the filter to be taken out of service has filter influent and recirculation pumps that supply *ONLY* the filter being shut down, reduce the pump speed to the minimum range. Reducing the speed of a pump will tend to relieve a part of the surge created to the remaining process units when the filter is shut down. Also, due to the reduced load when the pump is started again, the life of variable-speed pumps using belt drives will be extended.

3. Stop the influent flow (feed) and recirculation pumps for the filter. Allow the distributor arms to stop moving. Secure the distributor arms. Open the end gates. Restart the pump in order to flush the arms for a few minutes. Do not try to open the gates or stop the arms when the arms are still moving.

4. Stop the influent flow (feed) and recirculation pumps for the filter and close the pump discharge valves. Tag and lock out the pump motor starters. The filter distributor will stop rotating soon because no water is flowing out the outlet orifices.

WARNING. NEVER ATTEMPT TO STOP A ROTATING DISTRIBUTOR BY STANDING IN FRONT OF IT OR GRABBING IT WITH YOUR HANDS.

5. Check the remaining plant parts for proper operation, particularly wet wells and distribution or diversion structures between the other filters and clarifiers for normal water levels and position of flow control valves.

6. Once the distributor arm has stopped rotating, remove debris and rags from the distributor arm orifice plates. Also remove from the top of the media any debris and rags which could have been dumped during flushing of the distributor arms.

If the filter is to be left out of service for several days or longer, the following steps should be taken.

1. Close the filter underdrain outlet gates to prevent flow from other units entering the underdrain channel.

2. Drain or pump down the underdrain channel to prevent odors and insects from developing in the captured (stagnant) wastewater.

3. Hose down the distributor arms, side walls, vent ducts, and underdrain channels.

4. Remove any grit or debris from the main underdrain collection channel. Inspect the underdrains and remove any debris in order to prevent stoppages.

5. Check the oil level in the distributor turntable for proper level and the possible presence of water.

6. Inspect the turntable seal.

7. Consider removal of material (biomass) from media if growths are very heavy. If not removed, excessive growths may cause ponding when the filters are restarted. After drying, the material can be removed by the use of a leaf rake. Most of the remaining material will be flushed out when the unit is put back in service.

These steps take a small amount of extra time, but they can prevent unnecessary mistakes or prevent your plant effluent from violating discharge requirements.

QUESTIONS

Write your answers in a notebook and then compare your answers with those on page 205.

6.2E   What is the first thing an operator should do before shutting down a trickling filter?

6.2F   What items should be considered when planning to shut down a trickling filter?

6.3   SAMPLING AND ANALYSIS

6.30   Important Considerations

The trickling filter is a biological treatment unit and therefore loadings and efficiencies of the unit are normally determined on the basis of influent characteristics (inflow and biochemical oxygen demand) and required quality of effluent or receiving waters (dissolved oxygen and solids). Detailed procedures for performing the trickling filter control tests are given in Volume II, Chapter 16, "Laboratory Procedures and Chemistry." The frequency of each test and expected ranges will vary from plant to plant. Strength of the wastewater (BOD), freshness, characteristics of the water supply, weather, and industrial wastes will all affect the "common" range of the various test results.

6.31   Typical Trickling Filter Plant Lab Results
(Also see NOTES on page 182.)

Test	Frequency	Location	Common Range
1. Dissolved Oxygen	Daily	Final Effl.	1.5 - 2.0 mg/L
		Filter Underflow	3.0 - 8.0 mg/L
2. Settleable Solids	Daily	Influent	5 - 15 mL/L
		Final Effl.	0 - 3 mL/L
3. pH	Daily	Influent	6.8 - 8.0
		Final Effl.	7.0 - 8.5
4. Temperature	Daily	Influent	-
5. BOD	Weekly (Minimum)	Influent	150 - 400 mg/L
		Prim. Effl.	100 - 260 mg/L
		Final Effl.	15 - 40 mg/L
6. Suspended Solids	Weekly (Minimum)	Influent	150 - 400 mg/L
		Prim. Effl.	60 - 150 mg/L
		Final Effl.	15 - 40 mg/L
7. Chlorine Residual	Daily	Final Effl.	0.5 - 2.0 mg/L
8. Coliform Bacteria	Weekly (Minimum)	Final Effl., Chlorinated	50 - 700/100 mL
9. Clarity	Daily	Final Effl.	1 - 3 ft (0.3 - 1 m)

NOTES: Results of tests listed above as "Primary Effluent" may vary at different plants due to the many variations in recirculation patterns and activities of those discharging wastes into the collection system.

Settleable solids tests of the effluent are usually required by regulatory agencies. If your plant is operating efficiently, the settleable solids will be so low as to be unreadable. In this case, record as "Trace" or less than 0.1 mL/L.

Tests of trickling filter effluent for dissolved oxygen, settleable solids, and clarity are sometimes useful in evaluating problems when they occur. Operators should know what range is "common" for their plants.

Frequency of testing may vary widely from that shown in the table. In some locations (near water supply intakes or recreational areas), a much higher frequency may be required by regulatory agencies. For example, a chlorine residual analyzer with recording chart may be required for continuous monitoring.

An easy test that should be made periodically by the operator is to check the distribution of wastewater over the filter. Pans of the same size are placed level with the rock surface at several points along the radius of a circular filter. The distributor arm should then be run long enough to almost fill the pans. The arm is then stopped and the amount in each pan should not differ from the average by more than 5 percent. If the distribution is not uniform, the orifices and/or the stay rod turnbuckles must be adjusted.

6.32 Response to Poor Trickling Filter Performance

There are several operational procedures an operator can follow to correct deficiencies in plant performance. The ability to make corrections will depend on your alertness and ingenuity, as well as the design of the collection system and treatment plant. In Section 6.31, the common ranges are listed for a number of lab test results. If your plant is not operating within or near the common ranges for your plant, then you may have problems.

SUSPENDED SOLIDS. An effluent that is high in suspended solids may be expected to affect all the other test results listed. Ordinarily this will be due to four principal factors:

1. Heavy sloughing from the filters,

2. High hydraulic loading or short-circuiting through the secondary or final clarifier,

3. Shock loading caused by toxic wastes or hydraulic or organic overloads, and

4. Gasification (denitrification) caused by septic sludge in the secondary clarifier.

Heavy sloughing may be due to seasonal weather changes, a period of heavy organic loading on the filters, or corrective action taken to overcome ponding, filter flies, or other problems (see Section 6.4). High hydraulic loading or short-circuiting in the secondary clarifier will carry the light solids from the filters over the weirs. If a plant is not receiving more

flow than it was designed to handle, you may be able to adjust recirculation rates or the flow pattern (see Figure 6.6, page 178) to reduce the clarifier loading. Refer to Chapter 5, "Sedimentation and Flotation," for solutions to problems created in clarifiers by short-circuiting and sludge withdrawal.

BOD. The effluent BOD will generally go up or down along with the suspended solids. This is not always the case, however. Anything you can do to ensure that the wastewater arrives at the plant in an aerobic condition will reduce the organic load (and odors); consequently, the effluent BOD will be lower. Aeration has been used in *FORCE MAINS*[6] with some degree of success when applied properly.

The recirculation rate and flow pattern will affect effluent quality. These can be varied experimentally, keeping in mind that too low a recirculation rate leads to filter flies and ponding, while too high a rate may cause excessive sloughing or hydraulically overload the clarifiers. Biological systems respond to a change in their environment and establish a balance with the existing conditions, but it takes time. If you change your operation, give your plant a couple of weeks to reach an equilibrium state (level out) before you decide whether or not you have helped the situation.

The shortcomings of the BOD test must be recognized. This test is difficult to use as a daily operational tool unless the influent BOD remains fairly constant. If an industry dumped some wastewater with a high BOD, you could not measure the BOD and base your operational adjustments on the test results because they will not be available for five days. You will have to adjust your operation on the basis of your experience and the probable BOD. Use the chemical oxygen demand (COD) test to estimate rapid changes in the influent load. For control purposes, the COD test procedure will produce results in about four hours.

SETTLEABLE SOLIDS. High settleable solids in the effluent mean that solids are being carried over the clarifier weir. This also means that the suspended solids will be high. Refer to the paragraph in this section on suspended solids for corrective action.

DISSOLVED OXYGEN (DO). One of the principal functions of a trickling filter plant is to stabilize the oxygen-demanding substances in the wastewater being treated. This is achieved by the addition of dissolved oxygen to the water. If the suspended solids and BOD are within range, the DO is almost certain to be in range also. Increased recirculation will increase the DO. In plants with very low inflows, excessive detention time in the clarifiers may cause a problem of low DO in the effluent. If this is the case, remember that any agitation of the effluent will cause it to pick up dissolved oxygen. If the elevation is available, a staircase type of effluent discharge will help; otherwise, it may be necessary to aerate the effluent using compressed air or paddle-type aerators (see Chapter 8, "Activated Sludge"). In plants where low DO results in rising sludge (denitrification), the sludge blanket in the secondary clarifier should be lowered to less than 0.5 foot by increasing the sludge pumping rate.

CHLORINE DEMAND. Difficulty in maintaining a chlorine residual in the effluent (assuming normal detention period) will be due primarily to excessive solids in the effluent. Refer to the paragraph in this section on suspended solids. The loss of ammonia and the inability to form chloramines may also result in increased chlorine demands and poor coliform reduction.

---

[6] *Force Main.* A pipe that carries wastewater under pressure from the discharge side of a pump to a point of gravity flow downstream.

CLARITY. Clarity of the effluent also is related primarily to the amount, size, shape, and characteristics of the suspended solids in the effluent. In some cases, industrial or food processing wastes may cause discoloration. Trickling filter effluents tend to be slightly turbid. (Excessive turbidity can interfere with disinfection.)

pH. The pH of the effluent should move from whatever value is found in the influent toward neutral (a pH of 7.0). Normally the influent pH will be somewhat acidic (a pH of less than 7.0) and will move up to 7.0 or slightly higher. Other than pH changes caused by industrial waste dumps or other unusual wastes entering the plant, the pH should remain "normal" as long as the suspended solids and BOD are within reasonable limits. An effective industrial waste monitoring and regulation (pretreatment facility inspection) program will decrease the probability of harmful industrial discharges. If the pH varies beyond the range between 7 to 9, then corrective action (chemical neutralization) may be required to ensure good treatment.

NUTRIENTS. Trickling filter bacteria need an adequate supply of nutrients (nitrogen and phosphorus) to grow and properly treat wastes. Adequate nutrients are usually available in domestic wastewater. Industrial wastewaters are often nutrient deficient (especially food processing wastes). When industrial wastewaters dominate the treatment system, the composition of the influent should be analyzed to ensure that for each incoming 100 pounds of BOD there are five pounds of ammonia nitrogen and one pound of orthophosphate. If the nutrient levels are insufficient, nutrients must be added to achieve good treatment.

COLIFORM COUNT. When the bacterial count requirement must be met, excessive solids in the effluent are a serious problem. Even with high chlorine residuals, sporadic results occur because some particles are not penetrated completely by the chlorine. If in-plant corrections do not solve the solids carryover problem, some type of water treatment plant technique, such as coagulation and settling, or sand or diatomaceous earth filters, may have to be used. Good disinfection is achieved if the previous treatment processes do their job. However, nitrification/denitrification is also a very common cause of poor coliform reduction. Some plants have either taken treatment units off line (to reduce nitrification) or added about one mg/L ammonia prior to chlorination to form chloramines to improve disinfection.

QUESTIONS

Write your answers in a notebook and then compare your answers with those on page 205.

6.3A   How would you determine if the distribution of wastewater over a trickling filter is even?

6.3B   List the laboratory tests used to measure the efficiency of a trickling filter.

6.3C   (1) Calculate the efficiency of a trickling filter plant if the suspended solids of the plant influent is 200 mg/L and the plant effluent suspended solids is 20 mg/L.

   (2) What is the efficiency of the trickling filter only if the effluent suspended solids from the primary clarifier (wastewater applied to filter) is 140 mg/L and the effluent suspended solids remains at 20 mg/L?

END OF LESSON 1 OF 3 LESSONS
on
TRICKLING FILTERS

Please answer the discussion and review questions next.

DISCUSSION AND REVIEW QUESTIONS

Chapter 6.   TRICKLING FILTERS

(Lesson 1 of 3 Lessons)

At the end of each lesson in this chapter you will find some discussion and review questions. The purpose of these questions is to indicate to you how well you understand the material in the lesson. Write the answers to these questions in your notebook before continuing.

1. Draw a sketch of a trickling filter and label the essential parts.

2. Why is recirculation important in the operation of a trickling filter?

3. Why should a trickling filter be checked carefully before a new one is started or an existing one is placed in service again?

4. Why would the efficiency of waste removal by a new trickling filter be low during the first few days?

5. What steps would you follow when shutting down a trickling filter?

6. Why do laboratory test results for trickling filter plants vary from

   a. plant to plant?
   b. month to month within a plant?

7. What would you do if laboratory results or visual inspection indicated a sudden drop in efficiency of a trickling filter?

8. Calculate the suspended solids removal efficiency of a trickling filter plant if the influent suspended solids were 360 mg/L and the effluent suspended solids were 40 mg/L.

CHAPTER 6.   TRICKLING FILTERS

(Lesson 2 of 3 Lessons)

6.4   OPERATIONAL STRATEGY

In actual operation, the trickling filter is one of the most trouble-free types of secondary treatment. This process requires less operating attention and control than other types. Where recirculation is used, difficulties due to *SHOCK LOADS*[7] are less frequent and recovery is faster. This is because the filter can act like a sponge and treat great amounts of BOD for short time periods without a severe upset. Suspended solids in the trickling filter effluent tend to make the effluent somewhat turbid; thus a poorer quality effluent due to shock loads may not be visibly evident. Recirculation is used to maintain a constant load on the filter and thus produce a better quality of effluent. However, there are some problems which include ponding, odors, insects, and, in colder climates, freezing. These problems are all controllable and, in most cases, preventable.

6.40   Daily Operating Procedures

Daily operation of a trickling filter plant is described in Section 6.22, "Daily Operation." *USE THE LOWEST RECIRCULATION RATES THAT WILL PRODUCE GOOD RESULTS* (meet NPDES permit requirements), but not cause ponding or other problems. Low recirculation rates conserve energy and minimize plant operating costs. To produce good results on a continuous basis, the operator must observe and record those items that are critical to the wastewater treatment process.

Successful operation of a trickling filter plant requires routine observation of the process units, analysis of plant inflows to obtain wastewater characteristics, and determination of the water quality of the plant effluent. An alert operator will note changes in process units by observing various physical factors such as the flow rate or height over weirs, launder levels, amount of scum on a clarifier, the appearance of the effluent, the rotation of filter distributor arms, the spray pattern, the color of the media, and odors which indicate a change in the biological treatment system. Changes in any of these factors require investigation to identify the cause and to determine necessary corrective action.

Measure and record all important process factors. Plant inflows are measured by the plant influent flowmeter. Usually these flows produce similar daily patterns on the recording chart during dry weather. At some plants, similar patterns will be produced during certain storm conditions. Laboratory analyses of samples from various process stages, including plant influent, primary effluent, and secondary clarifier effluent, will indicate the water quality changes taking place in the plant. Most samples should be analyzed to determine the temperature, pH, dissolved oxygen, BOD, COD, and settleable solids. Effluent samples should be analyzed for chlorine residual and the most probable number of coliform group organisms in addition to the usual water quality indicators. All of this information is used by the operator to adjust and control the treatment processes.

Trickling filter plants can be operated on the basis of two loading criteria:

1. Hydraulic loading, gallons per day per square foot (L/day/ sq m); and

2. Organic loading, pounds of BOD per day per 1,000 cubic feet (kg BOD/day/1,000 cu m) of media.

Section 6.7, "Loading Criteria," shows how to calculate these loadings. Operators should attempt to maintain a fairly constant hydraulic loading on each trickling filter by adjusting the recirculation rate. The recirculation rate is also adjusted to maintain a DO from 3 to 6 mg/L in the filter effluent from rock media and 4 to 8 mg/L from synthetic media. The recirculation rate should not be so low as to allow ponding to develop. Organic loadings should be calculated on a weekly basis and compared with plant effluent suspended solids and BOD. If the plant effluent BOD or suspended solids are changing, look for changes in the hydraulic and organic loadings.

Plant effluent is the main indicator of how effectively a trickling filter plant is working. If the quality of the plant effluent starts to drop as shown by increases in the effluent BOD or suspended solids, then changes must be made in the operation of the plant. Operational changes available to the operator include changes in the recirculation rates and change in the operation of the trickling filters from parallel to series operation or vice versa.

One way to detect changes in plant effluent and select corrective action is by the use of a trend chart such as the one shown in Figure 6.7. To use a trend chart, record the days of the month and day of the week (not shown in Figure 6.7) along the bottom. Determine what you consider are the most important operational factors for your plant (such as plant influent

---

[7] *Shock Load (Trickling Filters).   The arrival at a plant of waste which is toxic to organisms in sufficient quantity or strength to cause operating problems. Possible problems include odors and sloughing off of the growth or slime on the trickling filter media. Organic or hydraulic overloads also can cause a shock load.*

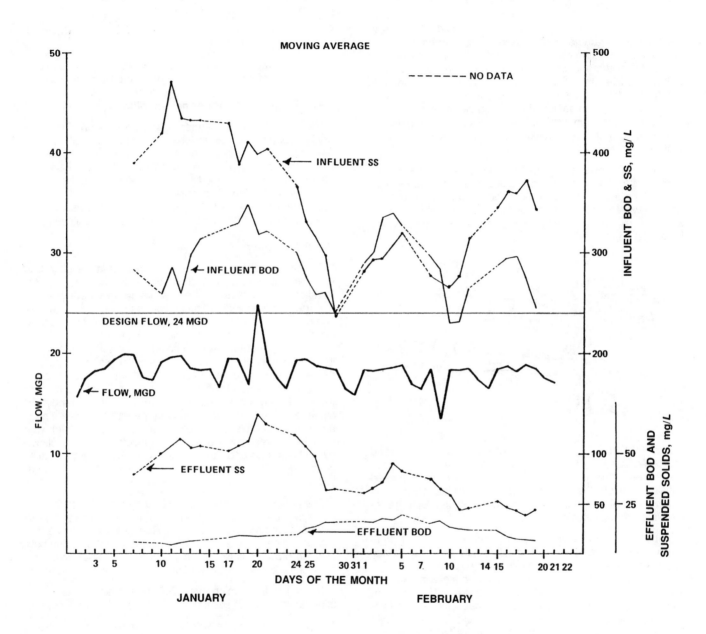

Fig. 6.7   Typical trend chart for a trickling filter plant

flow and effluent suspended solids) and draw a scale for each factor along the side. Plot the results on a daily basis and look for any trends. If the effluent suspended solids start to increase, look for the cause (increased inflow or possibly too low a hydraulic loading) and select corrective action.

Seasonal changes will have an impact on biological treatment processes and the quality of your plant effluent. Changes will occur when the weather changes from cold to warm and again when it changes from warm to cold. Changes in plant influent characteristics caused by discharges from canneries, meat packing houses, or metal plating processes may require the operator to change recirculation rates or mode (sequence) of filter operations. Operators have several methods of adjusting treatment processes, but you must be able to recognize what you can adjust and how you can control the treatment processes in your plant. The next section, "Response to Abnormal Conditions," provides details on how to correct problems.

Figure 6.7 shows the results from the operation of an actual trickling filter. The influent and effluent data plotted are seven-day moving averages in order to smooth out daily fluctuations and to reveal trends. Procedures for calculating moving averages are explained in Volume II, Chapter 18, "Analysis and Presentation of Data."

Examination of Figure 6.7 reveals the value of using trend charts and plotting the moving averages. During the week starting on Monday, January 24, the effluent BOD increased considerably. On the first of February the effluent BOD was still high. What happened? How could the effluent BOD be reduced to previous levels?

Influent BOD and suspended solids values were fairly high during January and fluctuated considerably from day to day (the actual variation was smoothed out by plotting the moving average). Some of this variation was due to a storm on January 11 (high influent suspended solids) and another storm on January 20 (flows slightly above design capacity). These fluctuations probably caused an excessive loss of the biomass (high effluent suspended solids from January 10 to 26). Also a period of cold weather during the latter half of January reduced the organism activity and thus the ability of the organisms in the biomass to remove the BOD from the effluent.

What should the operator do during these conditions? Recirculation should be increased to reduce the effluent BOD. Influent conditions should be watched more closely in order to maintain more constant hydraulic and organic loadings on the trickling filters. These actions were instituted on Monday, February 7, after reviewing data and plotting the moving averages. By Tuesday, February 15, the plot of the effluent BOD moving average was back down to typical values. Perhaps if effluent

COD values were used for operational control, the trend of increasing effluent BOD values could have been identified and corrected sooner.

QUESTIONS

Write your answers in a notebook and then compare your answers with those on page 205.

6.4A   What is the major consideration for daily operation of a trickling filter?

6.4B   Recirculation rates should be adjusted to maintain a DO of from ____ to ____ mg/L in the filter effluent.

6.41   Response to Abnormal Conditions

Every wastewater treatment plant will face unusual or abnormal conditions. How successfully these unusual situations are handled depends on the advance planning and preparations taken by the plant operator. In many cases, what is an abnormal operating condition in one plant may be handled as a routine operating procedure in another plant. This is because the operator took the time to review the potential situation and developed a plan to cope with the unusual event.

6.410   Ponding

Ponding results from a loss of open area in the filter. If the voids are filled, flow tends to collect on the surface in ponds. Ponding can be caused by excessive organic loading without a corresponding high recirculation rate. Perhaps the most common source of ponding is from the lack of good primary clarification prior to the filter. Another cause of ponding can be the use of media that are too small or not sufficiently uniform in size. In nonuniform media, the smaller pieces fit between the larger ones and thus make it easier for the slimes to plug the filter. If this condition exists, replacement of the media is the most satisfactory solution. Other causes of ponding include a poor or improper media permitting cementing or breakup, accumulation of fibers or trash in the filter voids (spaces between media), a high organic growth rate followed by a shock load and rapid, uncontrolled sloughing, or an excessive growth of insect larvae or snails which may accumulate in the voids.

The cause of ponding must be located since it may increase rapidly and take over large areas of the filter. Increasing the hydraulic loading by increasing the recirculation rate or adjusting the orifices on the distributor assembly so that it distributes flow more evenly is likely to flush off some of the heavier portions of the biological film and may slowly cure this condition.

Minor ponding, which may occur from time to time, can be eliminated by any of several methods, including the following:

1. Spray filter surface with a high-pressure water stream. Sometimes stopping a rotary distributor over the ponded area will flush the growth from the voids. One way to do this is to shut off the flow momentarily, wait for the distributor to stop, move the distributor to the problem area, and then restart the flow while keeping the distributor over the ponded area.

2. Hand turn or stir the filter surface with a rake, fork, or bar. Remove any accumulation of leaves or other debris.

3. Dose the filter with chlorine at about 5 mg/L for several hours. If done during a period of low flow, the amount of chlorine used is held to a minimum.

4. If it is possible to flood the filter, keeping the media submerged for 24 hours will cause the growth to slough somewhat. Keep the surface of the media covered, but don't let the water rise high enough to get into the distributor bearings. Under these conditions, the growths tend to become anaerobic and loosen or liquify. After the holding period carefully release the wastewater in order to avoid violating NPDES effluent discharge requirements.

5. Shut off flow to the filter for several hours. The growth will dry and can be removed by the use of a leaf rake. Most of the remaining material will be flushed out when the unit is put back in service.

Be sure to keep in mind that your primary purpose is to turn out an effluent of consistently good quality. With this in mind, the above corrective actions are listed in order, starting with procedures that will least affect the effluent. If at all possible, ponding should be corrected *BEFORE* it becomes serious. Items 4 and 5 are drastic measures. However, the job must be done so that full efficiency of the filter is restored. In cases 4 and 5, more chlorine will be needed for effluent disinfection. Where dechlorination is required to protect fish in the receiving stream, more chemicals also will be needed for this purpose until the filters are again operating normally.

6.411 Odors

Since operation of trickling filters is an *AEROBIC*[8] process, no serious odors should exist unless odor-producing compounds are present in the wastewater in high concentrations. The presence of foul odors indicates that *ANAEROBIC*[9] conditions are predominant. Anaerobic conditions are usually present under that portion of slime growth that is next to the media surface. As long as the surface of the slime growth (zoogleal film) is aerobic, odors should be minor. Corrective measures should be taken immediately if foul odors develop. The following are guidelines for maintaining trickling filters to prevent odor problems.

1. Do everything possible (such as prechlorination or preaeration) to maintain aerobic conditions in the sewer collection system and in the primary treatment units.

2. Check ventilation in the filter. Heavy biological growths or obstructions in the underdrain system will cut down ventilation. Examine ventilation facilities such as the draft tube or other inlets for stoppages. If necessary, force air into underdrains using mechanical equipment such as fans or compressors. Natural ventilation through a filter will occur if the vents are open and the *DIFFERENCE* between air temperature and filter temperature is greater than 3°F (2°C).

3. Increase the recirculation rate to provide more oxygen to the filter bed and increase sloughing.

4. Keep the wastewater splash from the distributor away from exposed structures, grass, and other surfaces. If slime growths appear on sidewalks, inside walls of the filter or distributor splash plates, remove them immediately.

5. In some cases during hot weather, odors will be noticeable from filters in good condition. If these odors are a serious problem (close neighbors), the situation can sometimes be resolved with one of the commercially available *MASKING AGENTS*.[10]

6. For covered filters, a forced-air ventilation system and odor control of the exhaust air stream is usually provided. Refer to the plant O & M manual and/or the manufacturer's literature for proper operation of this equipment. A covered filter and odor control system do not substitute for good operation and housekeeping procedures. The other points covered in this section will still apply. However, where uncovered filters have become a problem (such as in a nearby housing development), the addition of a cover and an odor control system could solve the problem. Before a filter is covered, an investigation of potential corrosion control and odor scrubbing controls must be undertaken to prevent further problems.

6.412 Filter Flies

The tiny, gnat-size filter fly (psychoda) is the primary nuisance insect connected with trickling filter operations. They are occasionally found in great numbers and can be an extremely difficult problem to plant operating personnel as well as nearby neighbors. Preferring an alternately wet and dry environment for development, the flies are found most frequently in low-rate filters and are usually not much of a problem in high-rate filters. Control usually can be accomplished by the use of one or more of the following methods.

1. Increase recirculation rate. A continuous hydraulic loading of 200 GPD/sq ft (8 cu m/day/sq m) or more will keep filter fly larvae washed out of rock filter media. Synthetic media will require higher hydraulic loadings or the use of weekly "flushings" by turning on all the filter pumps.

2. Keep orifice openings clear, including end gates of distributor arms. The gates can be opened slightly to obtain a flushing action on the walls.

3. Apply approved insecticides with caution to filter walls and to other plant structures.

---

[8] *Aerobic (AIR-O-bick).   A condition in which atmospheric or dissolved molecular oxygen is present in the aquatic (water) environment.*

[9] *Anaerobic (AN-air-O-bick).   A condition in which atmospheric or dissolved molecular oxygen is NOT present in the aquatic (water) environment.*

[10] *Masking Agents.   Substances used to cover up or disguise unpleasant odors. Liquid masking agents are dripped into the wastewater, sprayed into the air, or evaporated (using heat) with the unpleasant fumes or odors and then discharged into the air by blowers to make an undesirable odor less noticeable. "Neutral" odors may be the most desirable.*

4. Flood the filter for 24 hour at intervals frequent enough to prevent completion of the life cycle. This cycle is as short as seven days in hot weather. A poor effluent will result from this practice so it should be carefully monitored.

5. Dose with about 1 mg/L chlorine for a few hours each week. The chlorine will cause some of the slime layer to slough off. Too much chlorine will remove too much of the slime layer, reducing BOD removal and lowering the effluent quality of the plant.

6. Shrubbery, weeds, and tall grass provide a natural sanctuary for filter flies. *GOOD GROUNDS MAINTENANCE AND CLEANUP PRACTICES WILL HELP TO MINIMIZE FLY PROBLEMS.*

6.413   Sloughing

One of the most common problems with trickling filter operation is the periodic uncontrolled sloughing of biological slime growths from the filter media. Increasing the recirculation pumping rate to the filter on a weekly basis may help to induce controlled sloughing rather than to allow the slime growths to build up. During this flushing process, slow down the rotation of the distributor arms by adding more speed-retarder orifices to produce a slower rotation and thus cause a longer flush that penetrates deeper into the filter media. This will help prevent the buildup of the slime biomass.

6.414   Poor Effluent Quality

Check the organic load on the filter when the treated effluent quality is poor. Measure both the soluble and total BOD in the final effluent. The results will indicate if the poor effluent is caused by BOD associated with escaping solids (high total BOD) or whether the poor effluent results from the trickling filter BOD removal capacity being exceeded (high soluble BOD).

6.415   Cold Weather Problems

Cold weather usually does not offer much of a problem to wastewater flowing in a pipe or through a clarifier. Occasionally, however, wastewater sprayed from distributor nozzles or exposed in thin layers on the media may reach the freezing point

and cause a buildup of ice on the filter. Several measures can be taken to reduce ice problems on the filter.

1. Decrease the amount of recirculation (because influent is usually warmer than recycled flows), provided sufficient flow will remain to keep the filter working properly.

2. Operate *TWO-STAGE FILTERS*[11] in parallel rather than in series.

3. Adjust or remove orifices and splash plates to reduce the spray effect.

4. Construct wind screens, covers, or canopies to reduce heat losses.

5. Break up and remove the larger areas of ice buildup.

6. Partially open end gates to provide a stream rather than a spray along the retaining wall.

7. Add hot water or steam to the filter influent if necessary.

Although the efficiency of the filter unit is reduced during periods of icing, it is important to keep this unit running. Taking the unit out of service will not only reduce the quality of the effluent but may lead to additional maintenance problems, such as ice forming, with the possibility of structural damage. Also, moisture may condense in the oil and damage the bearings.

QUESTIONS

Write your answers in a notebook and then compare your answers with those on pages 205 and 206.

6.4C   What are some of the causes of ponding?

6.4D   How would you correct a ponding problem?

6.4E   How would you control odor problems in a trickling filter?

6.4F   Trickling filter flies can be controlled by what methods?

6.4G   Why should a trickling filter not be taken out of service during icing conditions?

6.416   Plant Inflow

Plant inflows may be considered abnormal when there are high flow rates, extreme levels of suspended solids or biochemical oxygen demand (BOD), or inflow of a septic influent. High flow rates (greater than 2.5 times the **A**verage **D**ry **W**eather **F**low (ADWF)) usually result from one of four sources:

1. Storm water inflow and groundwater infiltration,

2. Broken collection system pipe that permits excess inflow from groundwater or a creek or stream,

3. Clearance of a main line sewer stoppage and the release of the backed-up wastewater, or

4. Industrial discharges.

The plant operator will always be aware when condition one will occur; however, conditions two, three, and four may occur without warning, yet the plant operator is often the first to know. Each of the four conditions has a common characteristic, *EXCESS FLOW.* This excess flow may load the plant to its hydraulic capacity or exceed it. Conditions three and four

---

[11] *Two-Stage Filters.   Two filters are used. Effluent from the first filter goes to the second filter, either directly or after passing through a clarifier. (See Figure 6.6, page 178.)*

can impose other loads which the operator must also consider. These loads include heavy solids and BOD, septicity, and, in the case of the industrial discharge, a toxic material or load harmful to the biological treatment processes.

In a trickling filter plant, the operator usually has three alternatives for controlling abnormal flows:

1. If possible, increase the number of filters in operation (adjusts loading rates). Filters should have active slime growths on media,

2. Reduce or stop filter recycle or recirculation rate (adjusts DO and dilution), and/or

3. Operate filters in *PARALLEL OPERATION*[12] rather than *SERIES OPERATION*[13] (adjusts loading rates).

What can a trickling filter plant operator do to control plant effluent during abnormal flows? The following paragraphs suggest methods for you to consider when operating your plant. If you understand how your plant operates and how to adjust for abnormal conditions, you can select the best procedures.

CONDITION ONE. HIGH FLOW DUE TO STORM WATER INFLOW AND GROUNDWATER INFILTRATION

A. If collection system is clean from good maintenance or from recent storms, the high flows usually will dilute the solids and BOD and possibly reduce influent and solids loadings to below the ADWF loading.

B. If A above is true, then the influent could have a higher than normal DO.

PLANT OPERATION PLAN

1. Reduce or stop filter recycle or recirculation flows.

   a. Recycling flows to a filter are for dilution, DO increase, and to maintain the hydraulic load on the filter media.

   b. Reduction in recycle flows in some plant designs also will reduce the hydraulic loading on the secondary clarifiers, thus providing for a better removal of solids sloughed from the filter resulting from high flows.

2. Place filters in parallel operation. Since the storm water has diluted the inflow, the DO will be higher in the inflow. Loading each filter equally (half of total flow to each filter) instead of forcing the total flow across each filter will reduce sloughing and keep each filter functioning.

3. Keep influent screening equipment operating and check often, especially if collection system is a combined sanitary and storm water system which may have leaves and other debris in the flow.

4. Increase postchlorination rates to match flows in order to ensure proper disinfection.

Other adjustments that may be considered when high flows occur include:

5. Use the collection system as a storage reservoir by throttling the plant influent gate or by changing pumping elevations of the pumping start and stop cycles.

NOTE: a. This should be done only when the lowest manhole elevation in the system is known and the maximum water level can be held at least 2 feet (0.6 m) below the manhole rim elevation to prevent system overflow.

   b. Beware of pump problems developing when the wet well water level is higher than normal. High water levels reduce the head a pump must work against, which results in a higher pumped flow, which calls for an oversized pump motor.

6. Use chemicals on the influent to reduce solids and BOD load by causing chemical precipitation in the primary clarifiers.

7. Inspect the collection system for major defects and develop programs to reduce inflow and infiltration.

8. Reduce pumping of digester supernatant back to headworks in order to keep the hydraulic loading as low as possible.

9. Reduce the pumping of sludge to anaerobic digesters because sludge temperature will be lower than usual due to the lower temperatures of the storm waters.

CONDITION TWO. HIGH FLOWS DUE TO A BROKEN COLLECTION SYSTEM PIPE

High flows consist of relatively clear water from a creek or stream above broken pipe or from groundwater infiltration.

PLANT OPERATION PLAN

1. Operate plant using same procedures as outlined under CONDITION ONE.

2. Have broken pipe repaired as soon as possible.

---

[12] *Parallel Operation.* Wastewater being treated is split and a portion flows to one treatment unit while the remainder flows to another similar treatment unit. Also see SERIES OPERATION.

[13] *Series Operation.* Wastewater being treated flows through one treatment unit and then flows through another similar treatment unit. Also see PARALLEL OPERATION.

CONDITION THREE.   HIGH FLOWS DUE TO CLEARANCE OF MAIN LINE SEWER STOPPAGES

A. Surge or slug of high flows results when stoppage is cleared.

B. Influent septic, odorous, and probably has a high solids and BOD load.

PLANT OPERATION PLAN

1. Increase prechlorination rates in order to:

    a. Control odors, and
    b. Reduce influent BOD load.

2. Store released surge in collection system and allow to flow slowly into plant.

3. Place filters in series operation. Apply the initial loading to the first filter and have it perform as a roughing filter to reduce the shock load on the next filter. See Section 6.13, "Roughing Filters," for a more complete description.

4. Increase recycle rates to both filters in order to:

    a. Increase dilution of influent to filter, and
    b. Increase DO content of water applied to filter.

5. Increase postchlorination rates to maintain effective disinfection.

6. Frequently skim surface of primary clarifier to keep grease and other floatables off the filter media and out of the orifices on the distributor arms.

7. Frequently pump sludge from bottom of primary clarifier to reduce solids load to filters.

8. Stop recycle flows of supernatant or other heavy solids back to plant influent or to primary clarifier. These flows would complicate the filter problems.

Other adjustments that may be considered when a septic main line sewer stoppage is cleared include:

9. Feed chemicals into plant influent, and

10. Arrange with collection system maintenance crews to inform plant when a system stoppage is cleared and expected to hit the plant.

CONDITION FOUR.   INDUSTRIAL DISCHARGES

Influent conditions that may be expected following industrial discharges are listed below, followed by recommended plant operating plans for each condition. Your POTW's (**P**ublicly **O**wned **T**reatment **W**orks) pretreatment facility inspection program is designed to prevent toxic wastewaters from causing treatment plants to violate their NPDES discharge requirements. This section will help you respond properly if a toxic waste reaches your treatment plant.

A. High flows with normal solids and BOD loadings.

PLANT OPERATION PLAN

Operate plant using same procedures as outlined under CONDITION ONE.

B. High flows with high solids loadings.

PLANT OPERATION PLAN

Operate plant using same procedures as outlined under CONDITION THREE.

C. High flows and toxic material in plant influent. A toxic material often only can be detected if the industry notifies the plant or if the influent sampling equipment monitors for a specific chemical and sends out an alarm. Typical toxic conditions include an excessively high or low pH, or the presence of excess amounts of ammonia, heavy metals, or hydrocarbons. Unfortunately, in most plants the influent is not monitored continuously for high or low pH or the specific toxic material that causes the problem. Usually the operator never realizes that something toxic has hit the plant until a day or two later when the biological process is not functioning and the solids content in the plant effluent is high.

PLANT OPERATION PLAN

1. If a toxic chemical is known (it usually is not), a neutralizing agent may be added to the influent. For example, chlorine is used to counteract cyanide. If a high pH is detected (in excess of 9.0), lower the pH by adding an acid such as acetic or sulfuric. If the pH is below 6.0, increase the pH with caustic soda (sodium hydroxide—Na(OH)). If the pH is not properly adjusted you could produce toxic hydrogen cyanide or cyanogen chloride. Overchlorination may destroy the entire biological process. Always wear appropriate safety gear (goggles and/or face shield, impervious gloves, and protective clothing) when handling chemicals.

2. Operate the plant filters in series with a high recirculation rate in order to dilute the influent and bring the pH toward neutral. The first filter may slough and lose its biological culture from the media, but you may save the biological growths on the second filter. If the filters are operated in parallel, the toxic flow may strip the media in both filters of biological growths.

Other steps that may be considered when a toxic waste reaches a trickling filter include:

3. Store toxic waste in collection system until diluted by other wastewater and/or gradually release during high flows.

4. Locate industry responsible for toxic discharge and require industry to start a source control program. Be sure your sewer-use ordinances are enforced.

5. Restrict toxic discharges from industry. Allow releases to occur slowly, rather than in slugs or large batches. Also allow discharges to be released when plant can handle toxic wastes. Another type of restriction on very toxic agents is to allow discharges to the collection system only when tested, neutralized, and plant flows are sufficient to dilute the discharge. Under these conditions the wastewater treatment plant may handle all incoming wastewater without special procedures and also without the possibility of the plant processes becoming upset.

Some treatment plants will have more alternatives than other plants due to design considerations and equipment, but the best option is to have a plan for any abnormal condition that could occur at your plant.

6.417   Operational Problems With Upstream or Downstream Treatment Processes

When a treatment process has operational problems, you should be aware of the possible compensating adjustments or actions that may be available to maintain plant effluent quality.

1. Screening

If a comminutor or other screening equipment has failed, the operator of a trickling filter plant should frequently clean the bypass bar rack to remove as much of the debris as possible. Frequent skimming of the primary clarifiers also will help to reduce plugging of the orifices in the distributor arm.

The frequency of flushing the distributor arms should be increased to daily to keep them clean. Also, the distributor orifices should be cleaned more frequently to ensure an even application of wastewater to the media surface. The recirculation and influent feed pumps also should be checked for proper flows and discharge pressures because the pump impellers may become plugged or loaded with debris, thus reducing flows and requiring shutdown of the pumps for cleaning.

2. Grit

Large volumes of grit will seldom reach the filters unless a primary clarifier is bypassed or excessively overloaded with flow. When the primary clarifier is excessively hydraulically overloaded, the grit will be deposited in the underdrain which will allow septic dams and conditions to develop and produce odors. Grit also can increase ponding on a filter by filling the small voids between the media and reducing the downward flow of water. When this occurs the filter is taken out of service, flooded, drained, the media washed off with water from a hose under high pressure, and the underdrain cleaned.

3. Primary Clarifier

When a primary clarifier is out of service in a trickling filter plant, operate the filters in series. Use the first filter as a roughing filter to remove the larger pieces of suspended wastes. The first filter can perform satisfactorily this way for several weeks, but will require a cleanup of the filter afterward. This procedure should keep the second filter healthy and leave only one filter to clean.

4. Secondary Clarifier

When a secondary clarifier is out of service, solids in the treated effluent will increase and will accumulate in the chlorine contact basin. Operate the filters in series. Apply normal recirculation flows to the first filter and do not recirculate to the second filter. Apply the normal flow as evenly as possible to the second filter. This procedure should minimize sloughing from the media on the second filter for a while and, hopefully, until the secondary clarifier is back in service.

A simple device containing a coarse hardware-cloth screen or a similar device could be installed in the underdrain channel or outlet box to catch the sloughings. The screen would have to be removed and cleaned frequently.

5. Chlorinator

When one chlorinator is not working, bring another chlorinator on line, if one is available. If no other chlorinators are available, reduce the hydraulic load on the secondary clarifiers as much as possible to obtain the best solids removal. If you do not have a standby chlorinator, develop emergency procedures now to chlorinate your effluent when the chlorinator is not working. Some smaller plants use sodium hypochlorite as a standby disinfectant.

6.42   Troubleshooting

Table 6.2 summarizes Section 6.41 and is a quick reference to common problems and suggested ways to correct them.

TABLE 6.2   SOLUTIONS TO TRICKLING FILTER PROBLEMS

Solutions

Problems	Increase recirculation rate	Spray (jet) filter surface	Stop distributor to flush out	Rake or stir filter surface	Dose filter with chlorine	Flood filter	Let filter dry out	Check plant influent	Check ventilation	Protect outside areas from splash	Use masking agents	Check exhaust odor control system	Apply insecticides	Keep weeds down	Decrease recirculation	Parallel operation	Reduce spray (adjust nozzles)	Partially open end gates	Keep sludge pumped out of the clarifier	Replacement of media	See Section for details
Ponding	x	x	x	x	x	x*	x*													x*	6.410
Filter Flies	x				x	x*							x	x				x			6.412
Odors	x							x	x	x	x	x									6.411
Icing of the Filter															x	x	x	x			6.415
Low Dissolved Oxygen in Secondary Clarifier	x							x	x										x		
Floating Sludge in Secondary Clarifier								x											x		
High Effluent Suspended Solids															x						6.414

* Drastic measure. See Section 6.41, "Response to Abnormal Conditions."

Troubleshooting Guide—Trickling Filters

(Adapted from *PERFORMANCE EVALUATION AND TROUBLESHOOTING AT MUNICIPAL WASTEWATER TREATMENT FACILITIES,* Office of Water Program Operations, US EPA, Washington, DC.)

INDICATOR/OBSERVATION	PROBABLE CAUSE	CHECK OR MONITOR	SOLUTION
1. Filter ponding.	1a. Excessive organic loading.	1a. Visual inspection of ponded areas to determine cause. Check loading rates.	1a. Maintain filter loadings (both organic (lbs BOD/day 1,000 cu ft or kg BOD/day/ 1,000 cu m) and hydraulic load) within plant guidelines.
	1b. Accumulation of leaves and debris.	1b. Inspect surface of media for rags, rubber goods, and leaves where ponding is occurring.	1b. 1. Remove debris from filter surface. 2. Clean distributor orifices. 3. Check preliminary treatment screening equipment if large amount of rags. 4. Maintain trees and shrubs adjacent to filter to prevent leaf drop on media.
	1c. Excessive sloughing. Excessive biological growth.	1c. Growths clogging filter media.	1c. 1. Check hydraulic loading. 2. Flood filter for 24 hours to loosen and flush growth from voids when returned to service. 3. Check for equal distribution from filter nozzles, readjust orifices to maintain equal distribution onto media. 4. Jet with hose, rake, or stir media with bar to loosen growths and clean media. 5. Dose with approximately 5 mg/*L* chlorine for several hours to clean media. If done during low flows, keep the chlorine dose to a minimum.
	1d. Improper operation of primary treatment units.	1d. Excessive suspended solids, scum, BOD in primary effluent.	1d. 1. Correct operation of primary clarification units. 2. Apply chemical aids to primary influent to improve efficiency (chlorine, coagulants).
	1e. Insects, snails, moss.	1e. Visual inspection of ponded areas to determine cause.	1e. Flush filter and chlorinate filter influent to produce a chlorine residual of 0.5-1.0 mg/*L* for several hours.
	1f. Media too small or not uniform in size, broken down by extreme temperatures.	1f. Screen size of media for uniformity.	1f. Replace media.
2. Filter flies.	2a. Excessive biological growth on filters.	2a. Visual inspection.	2a. Remove excessive growth as described in 1c.
	2b. Plant grounds around filters provide breeding ground for flies.	2b. Inspect grounds.	2b. Maintain grounds so as not to provide a sanctuary for flies.

INDICATOR/OBSERVATION		PROBABLE CAUSE		CHECK OR MONITOR		SOLUTION
2.	Filter flies. (continued)	2c.	Hydraulic loading too low to wash filter of fly larvae.	2c.	Hydraulic loading should be greater than 200 GPD/sq ft (8,000 L/day/sq m).	2c. Prevent completion of fly life cycle by the following remedies: 1. Increase recirculation rate. 2. Flood filter for 24 hours at intervals frequent enough to prevent completion of life cycle. 3. Chlorinate to produce a residual of 1.0 mg/L for a few hours each week. 4. Apply with caution an approved insecticide to filter walls and areas breeding flies.
		2d.	Poor distribution of wastewater, especially along filter walls.	2d.	Check distribution with pans.	2d. 1. Unclog spray orifices or nozzles, end gate vent nozzles. 2. Change orifice plates on leading edge of distributor to slow down distributor, providing higher flushing action on media to clear voids of growth.
3.	Odors.	3a.	Anaerobic waste being applied to filters.	3a.	Monitor plant prechlorination and pre-aeration units.	3a. 1. Attempt to maintain aerobic conditions throughout the plant by prechlorination, pre-aeration, and proper primary treatment operation. 2. Increase recirculation rate on filter to improve aerobic conditions. 3. Chlorinate filter influent to reduce organic and H_2S load.
		3b.	Poor ventilation of filter.	3b.	Check filter underdrain for debris and organic material and the media for heavy biological growth.	3b. 1. Flush and clean filter underdrains and channels, remove all sediment and debris. 2. Flush vent tubes or draft tubes for open and free air passage. 3. Apply forced ventilation by mechanical fans or blowers to underdrains.
		3c.	Excessive biological growths.	3c.	Inspect media voids.	3c. 1. Increase recirculation rate to filter. 2. Flood and flush filter media to remove excessive growths.
		3d.	Poor housekeeping.	3d.	Visual inspection.	3d. Remove debris from filter media and wash down distributor splash plates and walls above media. Immediately remove slime growths caused by wastewater splash on exposed structures and walls above media.

INDICATOR/OBSERVATION	PROBABLE CAUSE	CHECK OR MONITOR	SOLUTION
4. Ice buildup on filter media.	4a. Climate.	4a. Air and wastewater temperature.	4a. 1. Decrease recirculation. 2. When used, operate two-stage filters in parallel. 3. Adjust orifices and splash plates for coarse spray. 4. Partially open dump gates at outer end of distributor arms to provide a stream rather than a spray. 5. Break up and remove ice formation. 6. Construct wind screens, covers, or canopies to reduce heat losses. 7. Add hot water or steam to the filter influent, if necessary.
	4b. Uneven distribution during freezing weather.	4b. Visual inspection.	4b. Adjust distributors for more even flow (remove debris if it has clogged orifices).
5. Uneven distribution of flow on filter surface.	5a. Clogging of distributor orifices.	5a. Check with pans. Look for ponding in some areas with concurrent drying in other areas.	5a. 1. Open end gates, flush distributor arms. 2. Clean distributor nozzles.
	5b. Inadequate hydraulic distribution on filter.	5b. Hydraulic loading on surface area. Check with pans.	5b. Maintain equal hydraulic load on all of filter surface area by adjusting orifices and recirculation rates.
	5c. Seal leaks.	5c. Seal.	5c. Replace seal.
6. Snails, moss, and roaches.	6a. Climatic conditions and geographical location.	6a. Visual inspection.	6a. 1. Chlorinate to produce residual of 0.5-1.0 mg/L. 2. Flush filter with maximum recirculation rate.
7. Increase in secondary clarifier effluent suspended solids.	7a. Excessive sloughing from filter due to seasonal change.	7a. Monitor filter effluent for SS, BOD, DO, and pH.	7a. 1. Change mode of operation. 2. Polymer addition to clarifier influent if SS uncontrollable by filter operation change.
	7b. Excessive sloughing due to organic loading.	7b. Organic loading and percent BOD removal by filter from influent to effluent.	7b. Increase clarifier underflow rate.
	7c. Excessive sloughing due to pH or toxic conditions.	7c. pH, toxic substances.	7c. 1. Increase recirculation for dilution. 2. Maintain pH between 5.5 and 9.0 and preferably between 6.5 and 8.5 by use of chemicals.
	7d. Denitrification in clarifier.	7d. Check effluent for nitrification and see if sludge floats in clumps.	7d. Increase clarifier underflow rate.
	7e. Final clarifier hydraulically overloaded.	7e. Clarifier overflow rate (should not exceed 1,200 GPD/sq ft).	7e. If due to recirculation, reduce recirculation rate during peak flow periods.
	7f. Final clarifier equipment malfunction.	7f. Check for: 1. High sludge blanket. 2. Broken sludge collection/ pumping equipment. 3. Broken baffles. 4. Uneven flows over effluent weirs.	7f. 1. Pump settled sludge more frequently. 2. Repair or replace broken equipment; adjust weirs to an equal elevation.
	7g. Temperature currents in final clarifier.	7g. Temperature profile of clarifier.	7g. Install baffles to stop short-circuiting.

This table is primarily useful as an "idea file" that will help you to avoid overlooking a possible solution to an operating problem. A review of Sections 6.01 and 6.02 will also help in deciding on a course of action. Be sure to keep in mind that trickling filters respond slowly to changes. For example, if you change the recirculation rate, the biological system could take several days to stabilize to the new conditions. Therefore, change only one thing at a time, and wait until it has leveled out to evaluate the effects of the change. Also, remember that plant units are interrelated. For example, a change in the recirculation rate (where the recirculation pattern includes a clarifier) will change the hydraulic loading on the clarifier. This will change the efficiency of the clarifier, which will then carry a different amount of solids to subsequent units. Increased solids removal in the primary or secondary clarifiers may affect loadings on the digesters downstream. If the filter recycle flows do not involve the clarifiers, the flows to clarifiers will not change and the solids removal efficiency of the clarifier should not change either.

The important thing in plant operation is to avoid problems, and the key to avoiding problems is *ALERTNESS*. Many of the common problems develop slowly and go undetected in their early stages. For instance, ponding will occur as a result of a buildup of heavy slime growths. The alert operator, who carefully inspects the filters daily, will detect the developing problem and make corrections. To the casual observer, this operator doesn't have any problems!

QUESTIONS

Write your answers in a notebook and then compare your answers with those on page 206.

6.4H What are four possible sources of abnormal conditions?

6.4I How can an operator adjust the trickling filter process when excess flows occur?

6.4J How do high flows from the clearance of a main line sewer stoppage differ from high flows caused by a broken sewer pipe?

6.4K Why are toxic industrial discharges difficult to detect until it is too late?

6.4L How are toxic industrial discharges often detected or suspected?

6.4M Why should an operator only make one change at a time in trickling filter operation and wait approximately one week before making another change when attempting to correct a trickling filter problem?

6.5 MAINTENANCE

6.50 Bearings and Seals (Figure 6.8)

The bearings in distributors may be located in the base of the center column or at the top. Both types will have a water seal at the base to prevent wastewater leakage. This is to avoid uneven distribution of the wastewater over the media, and also to protect the bearings when they are located in the base. Many older distributors used a mercury seal. Mercury should not be used because mercury ions are toxic to living organisms, including operators.

Generally, the bearings ride on removable races (tracks) in a bath of oil. The oil usually specified is a turbine oil with oxidation and corrosion inhibitors added. The manufacturer's literature or the plant O & M manual should specify what type of oil to use. If this information is not available, the representative from any major oil company can recommend a suitable oil.

Be sure to monitor the oil very carefully. The level and condition of the oil are crucial to the life of the equipment, and should be checked weekly. To check, drain out about a pint of oil into a clean container and, if the oil is clean and free of water, return it to the unit. If the oil is dirty, drain it and refill with a mixture of approximately 1/4 oil and 3/4 solvent (such as kerosene), and operate the distributor for a few minutes. Drain again, and refill with the correct oil. (Note: Drawing off some of the oil to check it is important. You can see if the oil is contaminated and verify that the oil level sensing line is not plugged.)

Water in the oil will appear at the bottom of the oil in the container. If water is found in the oil, either the sealing fluid is low or the gasket must be replaced in mechanical seals. Refer to the manufacturer's literature or the plant O & M manual for instructions.

6.51 Distributor Arms

Work on distributor orifices only after the arms have stopped moving, unless you are using a pressure hose for cleaning and stand outside the filter wall. The distributor arms should be flushed weekly by opening the end dump gates one at a time. Also clean debris off of the filter surface each day, cleaning the orifices as often as needed. When there is considerable plugging, you should install a coarse hardware-cloth or similar type screen ahead of the filters, if possible. Screens are easier to clean and good distribution is maintained over the filter media. Observe the distributor daily for smooth operation. If it becomes jumpy, seems to vibrate, or slows down with the same amount of wastewater passing through it, the bearings and races are probably damaged and will require replacement. A thorough oil check each week will probably keep this from happening. Adjust the turnbuckles occasionally on the guy rods to keep the rotating distributor arms at the proper level to provide even flow over all of the media.

The speed of rotation of the distributor should not be excessive. Rotation of the distributor is due to the reaction of the water flowing through the orifices. This is similar to the backward thrust of a water hose or the spinning of some types of lawn sprinklers. Speed is controlled by regulation of flow through the orifices. (On larger distributors, approximately 1 RPM is normal. The manufacturer's literature or plant O & M manual will state the maximum allowable speed.) If the distributor rotates too fast, it may damage the bearing races on the turntable.

To reduce the speed of rotation, provision usually is made on the front of each arm for orifices which are easily installed (Figure 6.3, page 172). The reaction of the water flowing

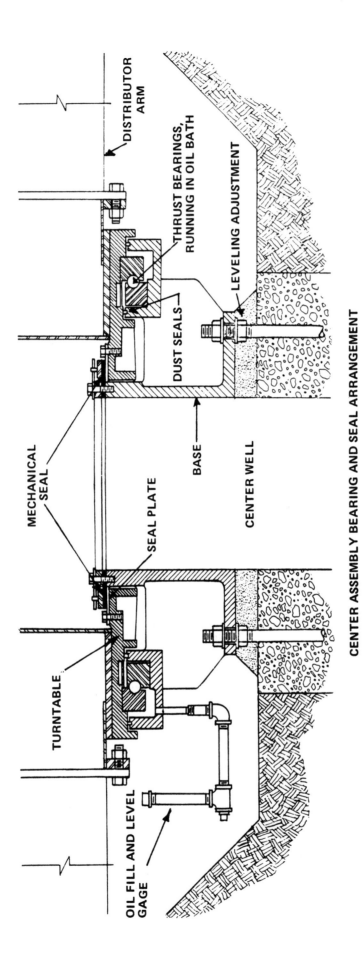

DISTRIBUTOR ARM

THRUST BEARINGS, RUNNING IN OIL BATH

LEVELING ADJUSTMENT

DUST SEALS

MECHANICAL SEAL

SEAL PLATE

BASE

TURNTABLE

CENTER WELL

OIL FILL AND LEVEL GAGE

CENTER ASSEMBLY BEARING AND SEAL ARRANGEMENT

Fig. 6.8   Trickling filter bearing and mechanical seal

through these orifices cancels some of the thrust of the regular orifices. If the speed of rotation is too slow, check for mechanical problems and, if none, increase flow to distributor.

Since most distributors appear rather large and bulky, many operators are surprised to find that they are delicately balanced. As soon as wastewater begins to flow from the orifices, the distributor arm should start to move. The fan-like pattern as the wastewater leaves the deflecting plates should be uniform. If the plates have developed a slime growth that is affecting uniform distribution, the slime should be brushed off.

6.52   Fixed Nozzles

Fixed-nozzle trickling filter beds are similar to lawn sprinkler systems because the distribution piping system is buried under the media and feeds into riser pipes spaced evenly across the filter bed. Each riser pipe is equipped with a spray head called a "nozzle" at the top of the riser pipe above the filter bed media (Figure 6.5, page 176). The nozzles are designed to handle high flows at low pressures and to pass some debris that would be in the wastewater in order to prevent plugging of the nozzle. The riser pipes and nozzles are spaced evenly just like a lawn irrigation system in order to provide equal distribution of water on the filter media surface.

Fixed nozzles should be observed frequently to determine that each is putting out a desired spray pattern and evenly covering the media.

These nozzles may become plugged or flow restricted, thus causing poor spray patterns just like a plugged or restricted orifice will on a rotating distributor arm. If a nozzle is not spraying properly, the system should be shut down. In some installations the feed distribution pipe to that row of nozzles may be shut off by closing the valve, thus stopping water flow to those nozzles in that particular row. On the faulty nozzle, the ball is removed and the rags or debris are removed from the cone and deflector. Proper cleaning may require turning the system on, blowing air through the riser pipe to remove the stoppage, and then reassembling the nozzle.

If frequent plugging is a problem, a screen may be installed ahead of the filter pumps or siphon to keep the rags and debris out of the distribution system and nozzles.

6.53   Underdrains

The underdrains are buried under the filter bed. Usually cleanouts or flusher branches are located on the head end of each line or channel for flushing to remove sludge deposits or debris from the underdrain system. If flushing will not clear the line and your agency or city's collection system maintenance section has a high velocity cleaner for cleaning sewer lines, borrow their services for an hour or so and have them clean the underdrains. You may wish to schedule this cleaning procedure every three to six months in order to keep the underdrain system open and clear. A clear underdrain system provides for a fast carry off of filter bed effluent, promotes unrestricted air flow for media ventilation, and reduces problems caused by odors, septic conditions, and ponding.

6.54   Recirculation Pumps

Refer to Volume II, Chapter 15, "Maintenance," for information on how to maintain recirculation pumps.

QUESTIONS

Write your answers in a notebook and then compare your answers with those on page 206.

6.5A   What is the purpose of the seal in a rotary distributor?

6.5B   Why should you drain some of the oil from the distributor each time it is checked?

6.5C   How would you slow down the rotational speed of a distributor?

6.5D   What maintenance should be done on trickling filter underdrains?

6.6   SAFETY

In order to work around a trickling filter safely, several precautions should be taken. *FIRST, SHUT OFF THE FLOW TO THE FILTER AND ALLOW THE DISTRIBUTOR TO STOP ROTATING BEFORE ATTEMPTING TO WORK ON IT.* On all but the very small units, the force of the rotating distributor arms is about the equivalent of a good-sized truck. *YOU JUST CAN'T REACH OUT AND STOP ONE WITHOUT ENDANGERING YOURSELF.* Serious injuries can result.

The slime growth on a filter is very slippery. *EXTREME CARE SHOULD BE TAKEN WHEN WALKING ON THE FILTER MEDIA.* Rubber boots with deeply ridged soles will help your footing. Do not carry oil in glass containers.

Most synthetic media filters are supplied with a heavy "walking surface." However, most of these walking surfaces are not adequate for providing safe access to the distributor for bearing lubrication. If there is adequate clearance between the top of the media and the bottom of the distributor, plastic grating can be placed from the edge of the filter tower to the center pier to provide a safe walking surface.

QUESTIONS

Write your answers in a notebook and then compare your answers with those on page 206.

6.6A   Why should the flow to a trickling filter be shut off before attempting to work on the filter?

6.6B   Why should an operator walk carefully on the filter media?

END OF LESSON 2 OF 3 LESSONS
on
TRICKLING FILTERS

Please answer the discussion and review questions next.

DISCUSSION AND REVIEW QUESTIONS

Chapter 6.   TRICKLING FILTERS

(Lesson 2 of 3 Lessons)

Write the answers to these questions in your notebook before continuing. The question numbering continues from Lesson 1.

9. Define the term *SHOCK LOAD*.

10. How would you attempt to identify the source or cause of an abnormal condition?

11. What is a trend chart and how can it be used?

12. What should an operator do before an abnormal condition occurs?

13. How would you correct a ponding problem?

14. Why should a ponding problem be corrected as soon as possible?

15. What action would you take to prevent odor problems from developing in a trickling filter?

16. Develop a *PLANT OPERATION PLAN* for when a surge of septic wastewater hits a treatment plant from the clearance of a sewer line stoppage.

17. Why should wastewater be kept from leaking into the bearings of the distributor base?

18. Why should the flow to a trickling filter be shut off before attempting to work on the filter?

CHAPTER 6.   TRICKLING FILTERS

(Lesson 3 of 3 Lessons)

6.7   LOADING CRITERIA

6.70   Typical Loading Rates (See Section 6.73 for metric)

STANDARD-RATE FILTER:

Media (rock)	— 6 to 8 ft depth, growth sloughs periodically
Hydraulic Loading	— 25 to 100 gal/day/sq ft or 0.02 to 0.07 GPM/sq ft
Organic (BOD) Loading	— 5 to 25 lbs BOD/day/1,000 cu ft

HIGH-RATE FILTERS: (Rock or Synthetic Media)

Media (rock)	— 3 to 5 ft depth, growth sloughs continuously
Hydraulic Loading	— 100 to 1,000 gal/day/sq ft or 0.07 to 0.7 GPM/sq ft
Organic (BOD) Loading	— 25 to 100 lbs BOD/day/1,000 cu ft
Media (synthetic)	— 15 to 30 ft depth, growth sloughs continuously
Hydraulic Loading	— 350 to 2,100 gal/day/sq ft or 0.2 to 1.5 GPM/sq ft
Organic (BOD) Loading	— 50 to 300 lbs BOD/day/1,000 cu ft

ROUGHING FILTER:

Media (synthetic)	— 15 to 30 ft depth, growth sloughs continuously
Hydraulic Loading	— 1,400 to 4,200 gal/day/sq ft or 1.0 to 3.0 GPM/sq ft
Organic (BOD) Loading	— 100 to over 300 lbs BOD/day/ 1,000 cu ft

6.71   Computing Hydraulic Loading

Hydraulic loading on a trickling filter is the amount of wastewater applied per day over the surface area of the media. This term also is called the hydraulic surface loading. In computing hydraulic loadings, several bits of information must be gathered. To figure the hydraulic loading, we must know:

1. The gallons per day applied to the filter, and
2. The surface area of the filter.

NOTE: Hydraulic loadings are expressed as:

gal/sq ft/day,[14]   or
gal/day/sq ft = GPD/sq ft[14]

Both expressions mean the same. The hydraulic rate indicates the number of gallons of wastewater per day applied to each square foot of surface area or the gallons of water applied to each square foot each day.

Suppose we have a high-rate filter that is fed by a pump rated at 2,100 GPM and the filter diameter is 100 feet.

$$\text{Hydraulic Loading, GPD/sq ft} = \frac{\text{Flow Rate, GPD}}{\text{Surface Area, sq ft}}$$

For our problem, we must obtain the flow rate in GPD and surface area[15] in square feet or ft^2.

$$\text{Flow Rate, GPD} = 2,100 \ \frac{\text{gal}}{\text{min}} \times \frac{60 \ \text{min}}{\text{hr}} \times \frac{24 \ \text{hr}}{\text{day}}$$
$$= 3,024,000 \ \text{gal/day}$$

$$\text{Surface Area, sq ft} = 0.785 \times (\text{Diameter, ft})^2$$
$$= 0.785 \times 100 \ \text{ft} \times 100 \ \text{ft}$$
$$= 7,850 \ \text{sq ft}$$

$$\text{Hydraulic Loading, GPD/sq ft} = \frac{\text{Flow Rate, GPD}}{\text{Surface Area, sq ft}}$$
$$= \frac{3,024,000 \ \text{GPD}}{7,850 \ \text{sq ft}}$$
$$= 385 \ \text{GPD/sq ft}$$

Note that in computing hydraulic loadings the flow used is the total flow applied to the filter or the plant inflow plus the recirculated flow. The recirculated flow is usually about equal to the plant inflow or up to two times the inflow.

6.72   Computing Organic (BOD) Loading

Using the same filter as in the above example of hydraulic loading, assume that the laboratory test results show that the wastewater being applied to the filter has a BOD of 100 mg/L; the filter depth is 3 feet. We need to know the pounds of BOD applied per day and the volume of the media in cubic feet.

NOTE: Organic (BOD) loadings are expressed as:

lbs BOD/1,000 cu ft/day, or
lbs BOD/day/1,000 cu ft.

---

[14] Loadings as well as test results should always be presented using the same units. Theoretically a rate should have the TIME UNIT LAST (gal/sq ft/day); however, because flows are calculated as gal/day, it is easier to understand if loadings are reported as gal/day/sq ft.
[15] Area of a Circle, sq ft = 0.785 x Diameter, ft x Diameter, ft, or
= 0.785 D^2

Both expressions mean the same. The organic loading indicates the pounds of BOD applied per day to the volume of filter media for treatment.

$$\text{Organic (BOD) Loading,} \atop \text{lbs BOD/day/1,000 cu ft} = \frac{\text{BOD Applied, lbs BOD/day}}{\text{Volume of Media in 1,000 cu ft}}$$

To solve this problem we must first calculate the BOD applied in lbs/day and volume of media in cu ft.

Volume of Media, cu ft = (Surface Area, sq ft)(Depth, ft)

= (7,850 sq ft)(3 ft)

= 23,550 cu ft

Volume of Media, in 1,000 cu ft = 23.5 (1,000 cu ft units)

= 23.5 thousand cubic feet

BOD Applied, lbs BOD/day = (Flow, MGD)(BOD, mg/L)(8.34 lbs/gal)[16]

$$= 3.024 \, \frac{\text{M gal}}{\text{day}} \times \frac{100 \text{ mg}}{\text{M mg}} \times \frac{8.34 \text{ lbs}}{\text{gal}}$$

= 2,522 lbs BOD/day

$$\text{Organic (BOD) Loading,} \atop \text{lbs BOD/day/1,000 cu ft} = \frac{\text{BOD Applied, lbs BOD/day}}{\text{Volume of Media (in 1,000 cu ft)}}$$

$$= \frac{2,522 \text{ lbs BOD/day}}{23.5(1,000 \text{ cu ft})}$$

= 107 lbs BOD/day/1,000 cu ft

NOTE: If flow is given in GPM, divide by 700 GPM/MGD to get the approximate flow in MGD.

In computing BOD loadings (organic loadings), it is standard practice to ignore the BOD of the recirculated effluent, where recirculation is used. To attempt to perform this calculation (using the recirculated load) is complicated and makes it difficult to compare your loadings and resulting effluent quality with other plants. Organic loadings may also be based on chemical oxygen demand (COD) or soluble biochemical oxygen demand (SBOD). In these cases the COD or SBOD in mg/L is used in place of the BOD, mg/L, in the formula to obtain the lbs COD/day/1,000 cu ft or lbs SBOD/day/1,000 cu ft.

If your plant must provide *NITRIFICATION*,[17] lower the organic loading to 20 to 30 pounds of BOD per day per 1,000 cubic feet of filter media. This should produce an effluent with an ammonia nitrogen level below 1 mg/L and a nitrate-nitrogen concentration of around 15 mg/L.

6.73   Typical Loading Rates (Metric)

The next three sections show typical loading rates for trickling filters in the metric system and how to calculate the hydraulic loading and organic loading using the metric system.

STANDARD-RATE FILTER:

Media (rock)	— 1.8 to 2.4 meters depth, growth sloughs periodically
Hydraulic Loading	— 1.0 to 4.0 cu m/day/sq m
Organic (BOD) Loading	— 0.08 to 0.4 kg/day/cu m

HIGH-RATE FILTERS: (Rock or Synthetic Media)

Media (rock)	— 0.9 to 1.5 meters depth, growth sloughs continuously
Hydraulic Loading	— 4.0 to 40 cu m/day/sq m
Organic (BOD) Loading	— 0.4 to 1.6 kg/day/cu m
Media (synthetic)	— 5 to 10 meters depth, growth sloughs continuously
Hydraulic Loading	— 15 to 90 cu m/day/sq m
Organic (BOD) Loading	— 0.8 to 4.8 kg/day/cu m

ROUGHING FILTER:

Media (synthetic)	— 5 to 10 meters depth, growth sloughs continuously
Hydraulic Loading	— 60 to 100 cu m/day/sq m
Organic (BOD) Loading	— 1.5 to over 4.8 kg/day/cu m

6.74   Computing Hydraulic Loading (Metric)

Suppose we have a high-rate filter that is fed by a pump rated at 0.132 cu m/sec (2,100 GPM), and the filter diameter is 30.5 m (100 feet):

$$\text{Hydraulic Loading, cu m/day/sq m} = \frac{\text{Flow Rate, cu m/day}}{\text{Surface Area, sq m}}$$

For our problem, we must obtain the flow rate in cubic meters per day and the surface area in square meters.

$$\text{Flow Rate,} \atop \text{cu m/day} = 0.132 \, \frac{\text{cu m}}{\text{sec}} \times \frac{60 \text{ sec}}{\text{min}} \times \frac{60 \text{ min}}{\text{hr}} \times \frac{24 \text{ hr}}{\text{day}}$$

= 11,405 cu m/day

---

[16] *The units of this formula can be proved by remembering that one liter weighs or equals one million milligrams.*

$$\frac{mg}{L} = \frac{mg}{1,000,000 \text{ mg}} = \frac{mg}{\text{M mg}}$$

Therefore,

$$MGD \times \frac{mg}{L} \, BOD \times 8.34 \, \frac{lbs}{gal} = \frac{\text{M gal}}{day} \times \frac{mg \, BOD}{\text{M mg}} \times \frac{lbs}{gal} = lbs \, BOD/day$$

[17] *Nitrification (NYE-truh-fuh-KAY-shun).   An aerobic process in which bacteria change the ammonia and organic nitrogen in wastewater into oxidized nitrogen (usually nitrate). The second-stage BOD is sometimes referred to as the "nitrogenous BOD" (first-stage BOD is called the "carbonaceous BOD").*

Surface Area, sq m = (0.785)(Diameter, m)^2

= 0.785 x 30.5 m x 30.5 m

= 730.2 sq m

$$\text{Hydraulic Loading,} \atop \text{cu m/day sq m} = \frac{\text{Flow Rate, cu m/day}}{\text{Surface Area, sq m}}$$

$$= \frac{11,405 \text{ cu m/day}}{730.2 \text{ sq m}}$$

= 15.6 cu m/day/sq m

or = 385 GPD/sq ft

6.75 Computing Organic (BOD) Loading (Metric)

Using the same filter as in the above example of hydraulic loading, assume that the laboratory test results show that the wastewater being applied to the filter has a BOD of 100 mg/L. The filter depth is 0.9 meter. We need to know the kilograms of BOD applied per day and the volume of the media in cubic meters.

$$\text{Organic (BOD) Loading,} \atop \begin{array}{c} \text{kg BOD/day} \\ \hline \text{cu m} \end{array} = \frac{\text{BOD Applied, kg/day}}{\text{Volume of Media in cu m}}$$

To solve this problem we must first calculate the volume of the media in cubic meters and the BOD applied in kilograms per day.

Volume of Media, cu m = (Surface Area, sq m)(Depth, m)

= (730.2 sq m)(0.9 m)

= 657.5 cu m

$$\text{BOD Applied,} \atop \text{kg BOD/day} = (\text{BOD, mg}/L)(\text{Flow, cu m/day})(1,000 \text{ kg/cu m})[18]$$

$$= \frac{100 \text{ mg}}{1,000,000 \text{ mg}} \times \frac{11,405 \text{ cu m}}{\text{day}} \times \frac{1,000 \text{ kg}}{\text{cu m}}$$

= 1,140.5 kg/day

$$\text{Organic (BOD) Loading,} \atop \begin{array}{c} \text{kg BOD/day} \\ \hline \text{cu m} \end{array} = \frac{\text{BOD Applied, kg BOD/day}}{\text{Volume of Media, cu m}}$$

$$= \frac{1,140.5 \text{ kg BOD/day}}{657.5 \text{ cu m}}$$

= 1.7 kg BOD/day/cu m

or = 107 lbs BOD/day/1,000 cu ft

QUESTIONS

Write your answers in a notebook and then compare your answers with those on pages 206 and 207.

6.7A What is hydraulic loading on a trickling filter and how is it expressed?

6.7B What information must be available to figure the hydraulic loading on a trickling filter?

6.7C What is the hydraulic loading on a trickling filter 80 feet in diameter that is receiving a flow of 3,200 GPM?

6.7D Is the filter in Problem 6.7C within the normal hydraulic loading range for high-rate filters?

6.7E What information is needed to figure the BOD loading on a trickling filter?

6.7F Compute the BOD loading on a standard-rate trickling filter with a diameter of 100 feet, media depth of 8 feet, and that is receiving 350 GPM of wastewater with a BOD of 100 mg/L.

6.7G Is the filter in Problem 6.7F loaded within normal limits for a standard-rate filter?

6.8 TRICKLING FILTER/SOLIDS CONTACT (TF/SC) PROCESS

A modification of the trickling filter process is the trickling filter/solids contact process. Final effluent from trickling filters passes through an aerated solids contact tank (similar to the activated sludge aeration tanks shown in Chapter 8, Figures 8.2 and 8.4) and then through secondary clarifiers with flocculator center wells. The sludge from the secondary clarifiers is mixed with trickling filter effluent and the mixture flows through the aeration solids contact tank.

The aerated solids contact time (detention time) is one hour or less based on total flow, including recycle or return sludge flow. The solids retention time or sludge age (see Chapter 8, Section 8.200) of the aerated solids contact tank is less than two days. The mixed liquor suspended solids (MLSS, see Chapter 11, "Activated Sludge," Volume II) in the aeration contact tank can range from 1,000 to 2,500 mg/L and have little effect on the final effluent suspended solids. For additional information, see "Full-Scale Studies of the Trickling Filter/Solids Contact Process," by R. M. Matasci, C. Kaempfer, and J. A. Heidman, Journal Water Pollution Control Federation, Volume 58, Number 11, November 1986, pages 1043 to 1049.

A survey of 30 TF/SC plants indicated that proper clarifier operation is the key for success. The process is more forgiving and less susceptible to shock loads than the activated sludge

---

[18] The density of water is 1,000 kg per cu m. Also remember that 1 liter of water weighs 1,000,000 mg.

process but it does require greater operator attention than the conventional trickling filter process. Most operators maintain a sludge blanket of less than 0.5 foot to ensure fresh (aerobic) conditions.

6.9 REVIEW OF PLANS AND SPECIFICATIONS

Plans and specifications should be reviewed by operators so they can:

1. Become familiar with a proposed plant,
2. Learn what will be constructed, and
3. Offer suggestions on how the plant can be designed for easier and more effective operation and maintenance.

When reviewing plans, carefully study those areas influencing how the plant will be operated and maintained. These areas should include:

1. Site.

   a. Access to the filter. Consider roads for maintenance equipment and walkways for personnel.

   b. Overhead clearance. Determine locations and distances to electrical power and telephone lines. Be sure they will not interfere with the boom of a crane lifting a turntable.

   c. Trees and shrubs near filters. Trees should not be planted or allowed to remain close to open filters because leaves will plug the voids in the media, cause ponding, and also prevent ventilation. Evergreen trees are recommended over trees that lose their leaves. Be sure that trees are planted where their roots can't get into plant piping.

   d. Location of hose bibs (high-pressure water faucets). Place hose bibs at convenient locations for washing down the filter and other maintenance jobs.

2. Trickling filter structure.

   a. Access (walking surface) to turntable seals and also oil drain, fill, and level plugs. Be sure sufficient space is provided for necessary maintenance work.

   b. Layout of underdrain grills, channels, and channel slopes. Access and space must be provided for flushing out solids, carrying solids away, and also proper ventilation.

   c. Location of valves and gates. Provisions must be made to allow flooding of the filter media and also dewatering of effluent control boxes and underdrain collector channels. Be sure valves are located between observation and sampling manholes and the filter in order to allow and observe flooding of the filter.

   d. Access to effluent boxes. Access and space must be available for removal of effluent box covers or grates and also maintenance of slide gates.

   e. Center column support. Support should be wide enough for timbers and jacks to be used to raise the distributor from the turntable for race maintenance.

   f. Covered trickling filters.

      (1) The operator cannot easily see if the distributor arm is moving under the cover. Some type of device that causes a light to flash when the arm passes a certain point should be placed on the end of one of the arms. Then the operator can determine the speed of rotation of the distributor arm by watching the flashing light.

      (2) If the filter is completely covered, a forced-air ventilation system is needed. If odors cause complaints from neighbors, an odor-scrubbing device will be needed also. The odor scrubbing fans should be installed so the forced-air ventilation will be in the same direction as natural air currents. These ventilation fans must be equipped with airtight seals to the drive motors to avoid corrosion problems.

      (3) Proper materials must be used to avoid corrosion of the roof structure.

   g. Walls should extend a minimum of 4 to 5 feet above the surface of the trickling filter media to prevent staining/spraying of the outside of the filter tower.

   h. Vents should be designed to prevent staining/spraying of the outside of the filter towers. They should also be designed to close if forced ventilation is used.

3. Equipment.

   a. Distributors

      (1) Adjustable orifice plates should be installed on both the leading and trailing edges of the distributor arms.

      (2) Safety stops should be installed to prevent endgate handle from catching in the media during flushing of the distributor arm.

      (3) Turnbuckles on guy rods must have sufficient thread length to make necessary adjustments.

   b. Valves

      (1) Valves must seat properly against design heads to prevent leakage back into channel during dewatering operations.

      (2) A protective coating must be applied to all gates and frames.

      (3) Stop nuts must be installed on all valve stems.

4. Safety.

   a. Guardrails must be located where necessary.

   b. Look for areas where splashing water could cause slippery surface and suggest corrective changes.

   c. Be sure switches for turning off pumps to distributor arms are located so that the distributor arm can be stopped quickly and easily when necessary.

   d. Install 115- to 120-volt receptacles at appropriate locations to permit the use of drop lights when inspecting vents and underdrains.

NOTE: Some items may not be covered on the plans, but may be discussed in the specifications.

When reading specifications, study the following items:

1. Equipment details should indicate the sizes, capacities, flow rates, pressures, horsepowers, efficiencies, and materials. Be sure protective guards and coatings are provided.

2. Performance capabilities of equipment.

3. Testing details. Will the equipment be tested at the manufacturer's factory and/or at the plant site? What are the tests supposed to accomplish?

4. Responsibility of equipment manufacturer's representative regarding:

   a. Installation inspection,

   b. Installation testing,

   c. Training of staff for equipment operation and maintenance,

   d. Assistance during start-up, and

   e. Warranty period and conditions.

5. Number of copies of prints, O & M manuals, and manufacturer's service manuals.

6. Lists of equipment spare parts (including serial and/or stock numbers) and quantity of each spare part that should be provided.

7. Safety

   a. Safety equipment provided to protect operators.

   b. Equipment has necessary approvals of safety agencies.

6.10 ARITHMETIC ASSIGNMENT

Turn to the Arithmetic Appendix at the back of this manual and read all of Section A.6, *FORCE, PRESSURE, AND HEAD.*

In Section A.13, *TYPICAL WASTEWATER TREATMENT PLANT PROBLEMS (ENGLISH SYSTEM)*, read and work the problems in Section A.133, Trickling Filters.

6.11 ADDITIONAL READING

1. *MOP 11*, Chapter 21, "Trickling Filters, Rotating Biological Contactors, and Combined Processes."*

2. *NEW YORK MANUAL*, Chapter 5, "Secondary Treatment."*

3. *TEXAS MANUAL*, Chapter 12, "Trickling Filters."*

4. *PROCESS CONTROL MANUAL FOR AEROBIC BIOLOGICAL WASTEWATER TREATMENT FACILITIES*, U.S. Environmental Protection Agency. Obtain from National Technical Information Service (NTIS), 5285 Port Royal Road, Springfield, VA 22161. Order No. PB-279474. EPA No. 430-9-77-006. Price, $84.00, plus $5.00 shipping and handling per order.

* Depends on edition.

6.12 METRIC CALCULATIONS

Refer to Sections 6.73, 6.74, and 6.75 for the solutions to all problems in this chapter using metric calculations.

6.13 ACKNOWLEDGMENT

Material in this chapter was reviewed by John F. Harrison whose contributions are greatly appreciated.

END OF LESSON 3 OF 3 LESSONS ON TRICKLING FILTERS

Please answer the discussion and review questions next.

DISCUSSION AND REVIEW QUESTIONS

Chapter 6. TRICKLING FILTERS

(Lesson 3 of 3 Lessons)

Write the answers to these questions in your notebook. The question numbering continues from Lesson 2.

A trickling filter is 70 feet in diameter and the depth of the media is 5 feet. The average daily flow is 2,450 GPM and the BOD of the influent to the filter is 100 mg/*L*. Calculate:

19. Flow in gallons per day.

20. Surface area of filter.

21. Hydraulic surface loading.

22. Volume of filter media.

23. BOD applied in pounds per day.

24. Organic loading.

Show your work.

SUGGESTED ANSWERS

Chapter 6. TRICKLING FILTERS

ANSWERS TO QUESTIONS IN LESSON 1

Answers to questions on page 177.

6.0A   Primary treatment is effective in removing (a) settleable solids, and (b) scum or floatable solids, but is not nearly as effective in removing (c) lighter suspended solids or dissolved solids which may exert a strong oxygen demand on the receiving waters.

6.0B   The purpose of secondary treatment is to convert dissolved or suspended materials into a form more readily separated from the water being treated. This process produces an overall plant removal of suspended solids and BOD of 90 percent or more.

6.0C   The distributor arms or pipes on a trickling filter rotate because of the reaction from the force of the water leaving the arms (as with a lawn sprinkler or fire hose), or by mechanical means (a motor and gears).

6.0D   The trickling filter process works by distributing settled wastewater over the filter media. Microorganisms grow on the filter media and convert colloidal and soluble oxygen-demanding substances to forms that will separate from the wastewater being treated.

6.0E   Recirculation increases the efficiency of a trickling filter by increasing the time of contact of the wastewater with the biological slime growth on the filter media, and by washing off excess growths (sloughing). Sloughing keeps the biological film in an aerobic condition and seeds the lower regions of the filter with active organisms. Sometimes recirculation is used to prevent intermittent drying of slimes on the filter media.

Answers to questions on page 179.

6.1A The three general classifications of trickling filters are standard-rate, high-rate, and roughing filters.

6.1B The principal differences between standard-rate and high-rate filters include BOD loadings, hydraulic loadings, depth of the media, recirculation, and effluent quality.

Answers to questions on page 180.

6.2A Items that should be checked before placing a trickling filter in service:

☐ Check type and amount of oil used in all oil reservoirs.

☐ Examine distributor arms for rotation and level.

☐ Inspect distributor orifices.

☐ Remove debris from underdrain system.

☐ Touch up any damage to painted surfaces.

☐ Examine valves for seating and smooth operation.

☐ Remove any trash on or in the media.

6.2B During start-up of a trickling filter heavy chlorination is necessary to reduce the health hazard and the pollutional load in the receiving waters because the slime growths have not developed on the filter media.

6.2C Items requiring daily inspection:

☐ Ponding

☐ Filter flies

☐ Odors

☐ Plugged orifices

☐ Roughness or vibration of distributor arms

☐ Leakage past the distributor turntable seal

☐ Splash beyond the filter media

☐ Cleanup of slimes not on media

6.2D If a clarifier is operated below design hydraulic loading, the solids in the clarifier will become septic and cause a poor effluent. When the hydraulic loading is too high, some solids may be washed out of the clarifier.

Answers to questions on page 181.

6.2E Before shutting down a trickling filter or any other major plant process or piece of equipment, always take a few minutes to plan what you are going to do.

6.2F The following items should be considered when planning to shut down a trickling filter:

1. Incoming flow rates.
2. How will shutdown affect rest of plant?
3. Are the necessary tools and other needed items available? and
4. Are there any other tasks that should be performed while the filter is off the line?

Answers to questions on page 183.

6.3A To determine if the distribution of wastewater over the trickling filter is even, place pans of the same size in the media at several points along the radius of the filter. Run the distributor until the pans are almost full. The amount of water in each pan should not vary from the average by more than 5 percent.

6.3B Laboratory tests used to measure the efficiency of a trickling filter include suspended solids, BOD, settleable solids, dissolved oxygen (DO), chlorine demand, clarity, pH, nutrients, and coliform count.

6.3C (1) Calculate the efficiency of a trickling filter plant if the suspended solids of the plant influent is 200 mg/L and the plant effluent suspended solids is 20 mg/L.

$$\text{BOD Efficiency, \%} = \frac{(\text{In} - \text{Out})}{\text{In}} \times 100\%$$

$$= \frac{(200 \text{ mg/}L - 20 \text{ mg/}L)}{200 \text{ mg/}L} \times 100\%$$

$$= 90\%$$

(2) What is the efficiency of the trickling filter only if the effluent suspended solids from the primary clarifier (wastewater applied to filter) is 140 mg/L and the effluent suspended solids remains at 20 mg/L?

$$\text{Trickling Filter Efficiency, \%} = \frac{(\text{In} - \text{Out})}{\text{In}} \times 100\%$$

$$= \frac{(140 \text{ mg/}L - 20 \text{ mg/}L)}{140 \text{ mg/}L} \times 100\%$$

$$= 85.7\%$$

ANSWERS TO QUESTIONS IN LESSON 2

Answers to questions on page 186.

6.4A The major consideration for daily operation of a trickling filter is to use the lowest recirculation rates that will produce good results (meet NPDES permit requirements), but not cause ponding or other problems.

6.4B Trickling filter effluent should be from 3 to 6 mg/L DO for rock media and 4 to 8 mg/L for synthetic media.

Answers to questions on page 188.

6.4C Causes of ponding include excessive organic loadings without corresponding high recirculation rate, media too small or not sufficiently uniform in size, accumulation of debris in filter spaces (voids), a high organic growth rate followed by a shock load and rapid, uncontrolled sloughing, or an excessive growth of insect larvae or snails.

6.4D To correct a ponding problem:

1. Locate cause.
2. Increase hydraulic loading by increasing recirculation.
3. Adjust distributor orifice plates so distributor will rotate more slowly and flush off some of the slime.
4. If media are nonuniform, they should be replaced.
5. Spray filter surface with a high-pressure water stream or stop distributor and allow it to flush problem area.
6. Rake or turn ponding area.
7. Dose filter with chlorine.
8. Flood filter.
9. Shut off filter for a few hours and allow growth to dry so part of it may be flushed out.

There are many possible ways to correct ponding. The approach for a particular problem should be aimed at the cause and the quickest and easiest way to correct the situation. You are not expected to have listed all the answers, but to know some of them.

6.4E   To control odor problems in a trickling filter:

1. Maintain aerobic conditions in collection system and primary treatment units.
2. Maintain ventilation in filter.
3. Increase recirculation rate if odor problems develop.
4. Keep the wastewater splash from the distributor away from exposed structures.
5. Use commercial masking agents.
6. Refer to O & M manual for proper operation of covered filter and forced-air ventilation system equipment. Also, maintain good operation and housekeeping procedures.

6.4F   Trickling filter flies can be controlled by:

1. Increasing recirculation rate.
2. Keeping orifice openings clear, including end gates of distributor arms.
3. Applying approved insecticides with caution to filter walls and to other plant structures.
4. Flooding filter for 24 hours at intervals frequent enough to prevent completion of the life cycle.
5. Dosing with about 1 mg/*L* chlorine for a few hours each week.
6. Keeping area around the filter clean, including removing weeds, cutting grass, and pruning shrubbery.

6.4G   A trickling filter should not be taken out of service during icing conditions because the quality of the effluent will be reduced and additional maintenance problems could develop.

Answers to questions on page 195.

6.4H   Four possible sources of abnormal conditions include:

1. Storm water inflow and groundwater infiltration,
2. Broken collection system pipe that permits excess inflow from groundwater or a creek or stream,
3. Clearance of a main line sewer stoppage, or
4. Industrial discharges.

6.4I   During excess flows, the operator can make the following adjustments to the trickling filter process:

1. If possible, increase the number of filters in operation (adjusts loading rates). Filters should have active slime growths on media,
2. Reduce or stop filter recycle or recirculation rate (adjusts DO and dilution), and/or
3. Operate filters in parallel operation rather than series operation (adjusts loading rates).

6.4J   High flows from the clearance of a sewer stoppage contain a septic, odorous influent with a high solids and BOD load, but flows from a broken pipe may not be septic or have the other problems.

6.4K   Toxic industrial discharges are difficult to detect because (1) the plant influent is usually not monitored for the toxic substance or condition (high or low pH), and (2) industry may not notify the operator when a toxic spill or dump occurs.

6.4L   Toxic industrial discharges are often detected or suspected when excessive amounts of solids start appearing in the plant effluent, indicating that something has upset the organisms living in the slime growths on the trickling filter.

6.4M   When attempting to correct a trickling filter problem, make only one change at a time and wait approximately one week to allow the organisms (biological process) to stabilize so you can evaluate the effectiveness of each change.

Answers to questions on page 197.

6.5A   The purpose of the seal in a rotary distributor is to prevent wastewater leakage from the center column before the wastewater is distributed over the media.

6.5B   Some oil should be drained from the distributor each time it is checked to examine the condition of the oil and to be sure the oil level sensing line is not plugged.

6.5C   The rotational speed of a distributor can be reduced by opening orifices on the front side of the distributor arms.

6.5D   Trickling filter underdrain maintenance consists of flushing the lines or channels to remove sludge deposits or debris every three to six months.

Answers to questions on page 197.

6.6A   Flow to a trickling filter should be shut off before attempting to work on a filter because the rotating arms can cause serious injury.

6.6B   Walk carefully on the filter media because the slime growths on the media are very slippery.

ANSWERS TO QUESTIONS IN LESSON 3

Answers to questions on page 201.

6.7A   Hydraulic loading on a trickling filter is the amount of wastewater applied per day over the surface area, expressed as gallons/day/square foot, or GPD/sq ft.

6.7B   Information needed to calculate hydraulic loading includes the gallons per day applied to the filter and surface area of the filter in square feet. These two items are all that are needed. However, it is necessary to convert the other data into the proper form (GPM to gal/day and surface area to sq ft).

6.7C   What is the hydraulic loading on a trickling filter 80 feet in diameter that is receiving a flow of 3,200 GPM?

GIVEN: Diameter = 80 ft
              Flow   = 3,200 GPM

REQUIRED: Hydraulic Loading

$$\text{Hydraulic Loading, GPD/sq ft} = \frac{\text{Flow Rate, GPD}}{\text{Surface Area, sq ft}}$$

Find *FLOW RATE, GPD*, and *SURFACE AREA* in sq ft.

$$\text{Flow Rate, GPD} = 3{,}200 \frac{\text{gal}}{\text{min}} \times \frac{60 \text{ min}}{\text{hr}} \times \frac{24 \text{ hr}}{\text{day}}$$

$$= 4{,}608{,}000 \text{ gal/day}$$

$$\text{Surface Area, sq ft} = 0.785 \times (\text{Diameter, ft})^2$$

$$= 0.785 \times 80 \text{ ft} \times 80 \text{ ft}$$

$$= 5{,}024 \text{ sq ft}$$

$$\text{Hydraulic Loading, GPD/sq ft} = \frac{\text{Flow Rate, GPD}}{\text{Surface Area, sq ft}}$$

$$= \frac{4,608,000 \text{ GPD}}{5,024 \text{ sq ft}}$$

$$= 917 \text{ GPD/sq ft}$$

6.7D Yes. The hydraulic loading range for high-rate filters is 100 to 1,000 GPD/sq ft.

6.7E To calculate the BOD loading on a trickling filter, the pounds of BOD applied per day and the volume of filter media in thousands of cubic feet are needed. Since BOD is reported in mg/L, the standard formula for converting mg/L to pounds per day is needed (MGD x mg/L x 8.34), and usually GPM will have to be converted to MGD (700 GPM = 1 MGD). There are many conversion factors that can be used to obtain the correct answers, and you should use the ones you are familiar with and understand.

6.7F Compute the BOD loading on a standard-rate trickling filter with a diameter of 100 feet, media depth of 8 feet, and that is receiving 350 GPM of wastewater with a BOD of 100 mg/L.

GIVEN: Diameter = 100 ft
          Depth   = 8 ft
          Flow   = 350 GPM
          BOD   = 100 mg/L

REQUIRED: Organic (BOD) Loading

$$\text{Organic (BOD) Loading, lbs BOD/day/1,000 cu ft} = \frac{\text{BOD Applied, lbs BOD/day}}{\text{Volume of Media, 1,000 cu ft}}$$

Find *SURFACE AREA*, *MEDIA VOLUME*, and *BOD APPLIED*.

Surface Area, sq ft = 0.785 x (Diameter, ft)^2
                    = 0.785 x 100 ft x 100 ft
                    = 7,850 sq ft

Volume of Media, cu ft = (Surface Area, sq ft)(Depth, ft)
                       = (7,850 sq ft)(8 ft)
                       = 62,800 cu ft
                       = 62.8(1,000 cu ft)

$$\text{Flow Rate, MGD} = \frac{350 \text{ GPM}}{700 \text{ GPM/MGD}}$$

$$= 0.5 \text{ MGD}$$

BOD Applied, lbs BOD/day = (Flow, MGD)(BOD, mg/L)(8.34 lbs/gal)

$$= 0.5 \frac{\text{M gal}}{\text{day}} \times 100 \frac{\text{mg}}{\text{M mg}} \times 8.34 \frac{\text{lbs}}{\text{gal}}$$

$$= 417 \text{ lbs BOD/day}$$

$$\text{Organic (BOD) Loading, lbs BOD/day/1,000 cu ft} = \frac{\text{BOD Applied, lbs BOD/day}}{\text{Volume of Media, 1,000 cu ft}}$$

$$= \frac{417 \text{ lbs BOD/day}}{62.8(1,000 \text{ cu ft})}$$

$$= 6.6 \text{ lbs BOD/day/1,000 cu ft}$$

6.7G Yes. The BOD loading range for standard-rate filters is 5 to 25 pounds of BOD per day per 1,000 cu ft.

APPENDIX

Monthly Data Sheet

CLEANWATER, U.S.A.
WATER POLLUTION CONTROL PLANT

OPERATOR: _____

MONTHLY RECORD ____ 20 ____

DATE	DAY	WEATHER	FLOW MGD	RAW WASTEWATER TEMP	RAW pH	SETT. SOLIDS	RAW BOD	RAW SUSP SOLIDS	PRIM. EFF. BOD	PRIM. SUSP SOLIDS	PRIM. D.O.	FINAL pH	FINAL BOD	FINAL SUSP SOLIDS	FINAL D.O.	CL2 RES.	DIGESTION RAW SLUDGE GALS/DAY	SCUM GALS/DAY	CO2 %	GAS PROD. FT3/DAY	MIXING HRS.	REMARKS
1	M	FAIR	1.200	70	7.3	8	190	208	132	95	2.3	7.6	29	21	7.0	2.1	5940	100	32	10800	4	
2	T	"	1.051	69	7.2	10	205	218	143	101	1.9	7.4	31	24	6.8	2.4	5135	110	32	11450	1	
3	W	"	1.120	70	7.3	11	220	222	153	108	2.0	7.0	23	26	5.9	3.0	5380	90	33	11220	8	
4	T	"	0.987	70	7.1	11	184	201	127	92	2.1	7.2	28	32	6.3	1.8	4765	120	33	11570	8	SLUDGE TO #1 BED – 16,000 GAL.
5	F	CLDY	1.008	68	7.0	7	232	248	162	120	2.0	7.3	31	25	8.1	1.9	5010	115	34	10990	8	
6	S	"	1.102	68	7.2	9	211	210	147	105	2.2	7.3	30	30	7.8	2.1	5240	120	32	11200	4	
7	S	FAIR	0.974	69	7.3	8	199	215	139	98	2.1	7.2	26	29	7.4	2.0	4895	110	32	12100	4	
8																						
9																						
10																						
11																						
12																						
13																						
14																						
15																						
16																						
17																						
18																						
19																						
20																						
21																						
22																						
23																						
24																						
25																						
26																						
27																						
28																						
29																						
30																						
31																						
MAX.																						
MIN.																						
AVG.			1.016	69	7.1	9	195	210	130	100	2.0	7.3	30	27	6.4	2.0	4980	100	33	11,000	4	

SUMMARY DATA

% REMOVAL	BOD	S.S
INF–PRI	30.0	52.4
INF–EFF	84.6	87.2

SLUDGE DATA

% SOLIDS – AVG.	4.4
LBS. DRY SOLIDS / DAY	1,827
% VOL. SOLIDS – AVG.	76
LBS. VOL. SOLIDS / DAY	1,388
LBS. VOL. SOLIDS/1000 FT3/DAY	27.7
GALS. SLUDGE TO BEDS	48,000
CU. YDS. CAKE REMOVED	22
FT3 GAS/LB. VOL. SOLIDS	8.0
FT3 GAS / MG FLOW	10,800

COST DATA

MAN DAYS 44 PAYROLL	$1,250
POWER PURCHASED	450
OTHER UTILITIES (GAS, H2O)	60
GASOLINE, OIL, GREASE	30
CHEMICALS AND SUPPLIES	95
MAINTENANCE	140
VEHICLE COSTS	70
OTHER	20
TOTAL	$2,105
OPERATING COST/MG TREATED	$66.83
OPERATING COST/CAPITA/MO.	$0.21

FLOW METER:
LAST 445237
1st 413749
TOTAL: 31,488 MG

ELECTRIC METER:
LAST 5196
1st 4821
MULT. 80 x 375 = 30,000 KWH

RAW SLUDGE:
LAST 657,314
1st 699,814
STROKES ____ SCUM 3100
TOTAL: 154,000 x 1.0 = 154,000 GALS

GAS METER:
LAST 724296
1st 383296
TOTAL: 341,000 FT3

CHAPTER 7

ROTATING BIOLOGICAL CONTACTORS

by

Richard Wick

TABLE OF CONTENTS

Chapter 7.   ROTATING BIOLOGICAL CONTACTORS

OBJECTIVES

Chapter 7. ROTATING BIOLOGICAL CONTACTORS

Following completion of Chapter 7, you should be able to:

1. Describe a rotating biological contactor and the purpose of each important part,

2. Start up and operate a rotating biological contactor,

3. Operate a rotating biological contactor under abnormal conditions,

4. Shut down and restart a rotating biological contactor,

5. Maintain and troubleshoot a rotating biological contactor,

6. Safely perform the duties of the operator of a rotating biological contactor,

7. Review the plans and specifications for a rotating biological contactor, and

8. Calculate the hydraulic and organic loadings on a rotating biological contactor.

WORDS

Chapter 7. ROTATING BIOLOGICAL CONTACTORS

BIODEGRADABLE (BUY-o-dee-GRADE-able)   BIODEGRADABLE

Organic matter that can be broken down by bacteria to more stable forms which will not create a nuisance or give off foul odors is considered biodegradable.

COMPOSITE (come-PAH-zit) (PROPORTIONAL) SAMPLE   COMPOSITE (PROPORTIONAL) SAMPLE

A composite sample is a collection of individual samples obtained at regular intervals, usually every one or two hours during a 24-hour time span. Each individual sample is combined with the others in proportion to the rate of flow when the sample was collected. The resulting mixture (composite sample) forms a representative sample and is analyzed to determine the average conditions during the sampling period.

GRAB SAMPLE   GRAB SAMPLE

A single sample of water collected at a particular time and place which represents the composition of the water only at that time and place.

INHIBITORY SUBSTANCES   INHIBITORY SUBSTANCES

Materials that kill or restrict the ability of organisms to treat wastes.

MPN (pronounce as separate letters)   MPN

MPN is the Most Probable Number of coliform-group organisms per unit volume of sample water. Expressed as a density or population of organisms per 100 mL of sample water.

NAMEPLATE   NAMEPLATE

A durable metal plate found on equipment which lists critical installation and operating conditions for the equipment.

NEUTRALIZATION (new-trall-i-ZAY-shun)   NEUTRALIZATION

Addition of an acid or alkali (base) to a liquid to cause the pH of the liquid to move toward a neutral pH of 7.0.

NITRIFICATION (NYE-truh-fuh-KAY-shun)   NITRIFICATION

An aerobic process in which bacteria change the ammonia and organic nitrogen in wastewater into oxidized nitrogen (usually nitrate). The second-stage BOD is sometimes referred to as the "nitrogenous BOD" (first-stage BOD is called the "carbonaceous BOD").

PYROMETER (pie-ROM-uh-ter)   PYROMETER

An apparatus used to measure high temperatures.

SOLUBLE BOD   SOLUBLE BOD

Soluble BOD is the BOD of water that has been filtered in the standard suspended solids test. The soluble BOD is a measure of food for microorganisms that is dissolved in the water being treated.

SUPERNATANT (sue-per-NAY-tent)   SUPERNATANT

Liquid removed from settled sludge. Supernatant commonly refers to the liquid between the sludge on the bottom and the scum on the surface of an anaerobic digester. This liquid is usually returned to the influent wet well or to the primary clarifier.

CHAPTER 7.   ROTATING BIOLOGICAL CONTACTORS

7.0   DESCRIPTION OF ROTATING BIOLOGICAL CONTACTORS

Rotating biological contactors (RBCs) are a secondary biological treatment process (Figure 7.1) for domestic and BIODEGRADABLE[1] industrial wastes. Biological contactors have a rotating "shaft" surrounded by plastic discs called the "media." The shaft and media are called the "drum" (Figures 7.2 and 7.3). A biological slime grows on the media when conditions are suitable. This process is very similar to a trickling filter where the biological slime grows on rock or other media and settled wastewater (primary clarifier effluent) is applied over the media. With rotating biological contactors, the biological slime grows on the surface of the plastic disc media. The slime is rotated into the settled wastewater and then into the atmosphere to provide oxygen for the organisms (Figure 7.2). The wastewater being treated either flows parallel to the rotating shaft or flows perpendicular to the shaft as it flows from stage to stage or tank to tank.

The plastic disc media are made of high-density plastic circular sheets usually 12 feet (3.6 m) in diameter. These sheets are bonded and assembled onto horizontal shafts up to 25 feet (7.5 m) in length. Spacing between the sheets provides the hollow (void) space for distribution of wastewater and air (Figures 7.3 and 7.4).

The rotating biological contactor process uses several plastic media drums. Concrete or coated steel tanks usually hold the wastewater being treated. The media rotate at about 1.5 RPM while approximately 40 percent of the media surface is immersed in the wastewater (Figure 7.4). As the drum rotates, the media pick up a thin layer of wastewater which flows over the biological slimes on the discs. Organisms living in the slimes use organic matter from the wastewater for food and dissolved oxygen from the air, thus removing wastes from the water being treated. As the attached slimes pass through the wastewater, some of the slimes are sloughed from the media as the media rotates downward into the wastewater being treated. The effluent with the sloughed slimes flows to the secondary clarifier where the slimes are removed from the effluent by settling. Figure 7.5 shows the location of a rotating biological contactor process in a wastewater treatment plant. The process is located in the same position as the trickling filter or activated sludge aeration basin. Usually the process operates on a "once-through" scheme, with no recycling of effluent or sludge, which makes it a simple process to operate.

The major parts of the process are listed in Table 7.1 along with their purposes. The concrete or steel tanks are commonly shaped to conform to the general shape of the media. This shape eliminates dead spots where solids could settle out and cause odors and septic conditions.

The rotating biological contactor process is usually divided into four different stages (Figure 7.6). Each stage is separated by a removable baffle, concrete wall or cross-tank bulkhead. Wastewater flow is either parallel or perpendicular to the shaft. Each bulkhead or baffle has an underwater orifice or hole to permit flow from one stage to the next. Each section of media between bulkheads acts as a separate stage of treatment.

Staging is used in order to maximize the effectiveness of a given amount of media surface area. Organisms on the first-stage media are exposed to high levels of BOD and reduce the BOD at a high rate. As the BOD levels decrease from stage to stage, the rate at which the organisms can remove BOD decreases and NITRIFICATION[2] starts.

Treatment plants requiring four or more shafts of media usually are arranged so that each shaft serves as an individual stage of treatment. The shafts are arranged so the flow is perpendicular to the shafts (Figure 7.6, Layout No. 3). Plants with fewer than four shafts are usually arranged with the flow parallel to the shaft (Figure 7.6, Layout No. 1).

Rotating biological contactors are covered for several reasons relating to climatic conditions:

1. Protect biological slime growths from freezing;

2. Prevent intense rains from washing off some of the slime growths;

3. Stop exposure of media to direct sunlight to prevent growth of algae;

4. Avoid exposure of media to sunlight which may cause the media to become brittle; and

5. Provide protection for operators from sun, rain, snow, or wind while maintaining equipment.

---

[1] Biodegradable (BUY-o-dee-GRADE-able).   Organic matter that can be broken down by bacteria to more stable forms which will not create a nuisance or give off foul odors is considered biodegradable.

[2] Nitrification (NYE-truh-fuh-KAY-shun).   An aerobic process in which bacteria change the ammonia and organic nitrogen in wastewater into oxidized nitrogen (usually nitrate). The second-stage BOD is sometimes referred to as the "nitrogenous BOD" (first-stage BOD is called the "carbonaceous BOD").

TREATMENT PROCESS   FUNCTION

PRELIMINARY TREATMENT

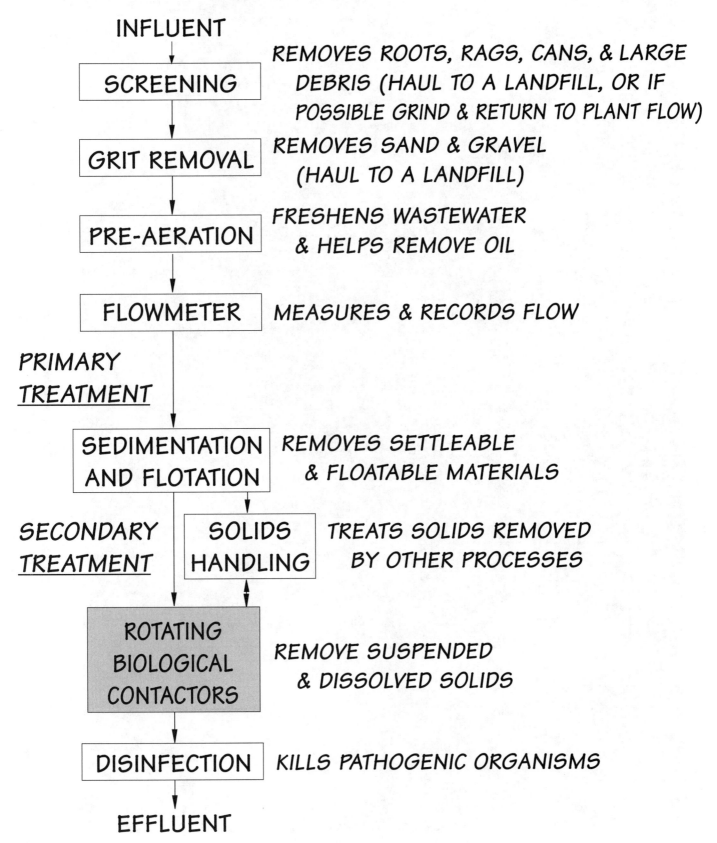

INFLUENT

SCREENING — REMOVES ROOTS, RAGS, CANS, & LARGE DEBRIS (HAUL TO A LANDFILL, OR IF POSSIBLE GRIND & RETURN TO PLANT FLOW)

GRIT REMOVAL — REMOVES SAND & GRAVEL (HAUL TO A LANDFILL)

PRE-AERATION — FRESHENS WASTEWATER & HELPS REMOVE OIL

FLOWMETER — MEASURES & RECORDS FLOW

PRIMARY TREATMENT

SEDIMENTATION AND FLOTATION — REMOVES SETTLEABLE & FLOATABLE MATERIALS

SECONDARY TREATMENT

SOLIDS HANDLING — TREATS SOLIDS REMOVED BY OTHER PROCESSES

ROTATING BIOLOGICAL CONTACTORS — REMOVE SUSPENDED & DISSOLVED SOLIDS

DISINFECTION — KILLS PATHOGENIC ORGANISMS

EFFLUENT

Fig. 7.1   Flow diagram of treatment plant

Fig. 7.2   Rotating biological contactors
(Permission of Autotrol Corporation)

Fig. 7.3   *Plastic disc media and biological contactor drum*
(Permission of Autotrol Corporation)

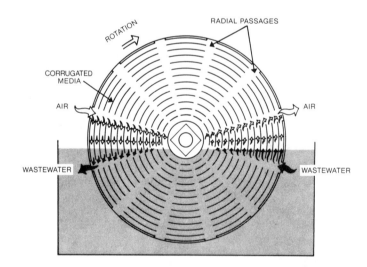

Media cross section

End-view sketch illustrates exchange
of air and wastewater

Fig. 7.4   *Sections of the plastic disc media*
(Permission of Autotrol Corporation)

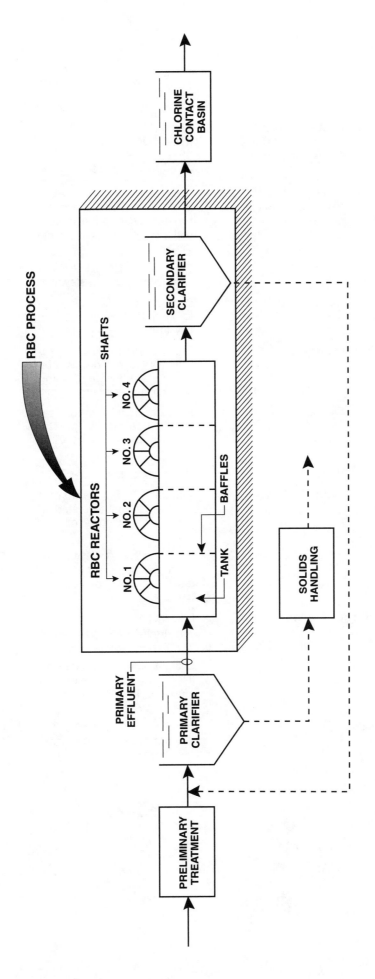

Fig. 7.5   Typical rotating biological contactor (reactor) treatment plant

TABLE 7.1   PURPOSE OF PARTS OF A ROTATING BIOLOGICAL CONTACTOR

Part	Purpose
Concrete or Steel Tank Divided Into Bays (Sections) by Baffles (Bulkheads)	Tank. Holds the wastewater being treated and allows the wastewater to come in contact with the organisms on the discs.
	Bays and baffles. Prevent short-circuiting of wastewater.
Orifice or Weir Located in Baffle	Controls flow from one stage to the next stage or from one bay to the next bay.
Rotating Media	Provide support for organisms. Rotation provides food (from wastewater being treated) and air for organisms.
Cover Over Contactor	Protects organisms from severe fluctuations in the weather, especially freezing. Also contains odors.
Drive Assembly	Rotates the media.
Influent Lines With Valves	Influent lines. Transport wastewater to be treated to the rotating biological contactor.
	Influent valves. Regulate influent to contactor and also isolate contactor for maintenance.
Effluent Lines With Valves	Effluent lines. Convey treated wastewater from the contactor to the secondary clarifier.
	Effluent valves. Regulate effluent from the contactor and also isolate contactor for maintenance.
Underdrains	Allow for removal of solids which may settle out in tank.

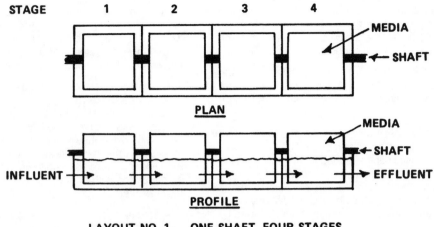

LAYOUT NO. 1   ONE SHAFT, FOUR STAGES
FLOW PARALLEL TO SHAFT

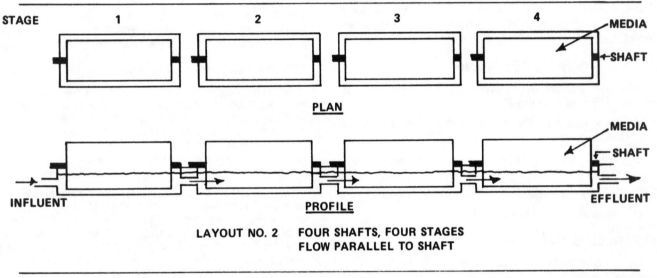

LAYOUT NO. 2   FOUR SHAFTS, FOUR STAGES
FLOW PARALLEL TO SHAFT

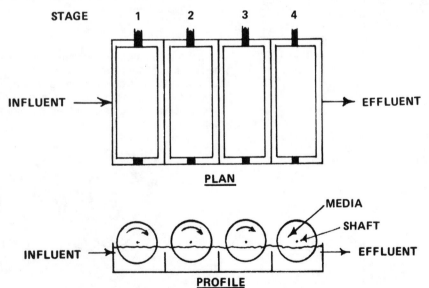

LAYOUT NO. 3   FOUR SHAFTS, FOUR STAGES
FLOW PERPENDICULAR TO SHAFT

Fig. 7.6   Possible rotating biological contactor layouts

Fiberglass covers in the shape of the media are easily removed for maintenance. In some areas, the rotating biological contactors are covered by a building. In other areas only a roof is placed over the media for protection against sunlight. The type of cover depends on climatic conditions.

Three types of drive assemblies are used to rotate the shafts supporting the media: (1) motor with chain drive (Figure 7.7), (2) motor with direct shaft drive, and (3) air drive (Figure 7.8).

The first type of drive assembly consists of a motor, belt drive, gear or speed reducer, and chain drive. The third type of drive unit consists of plastic cups attached to the outside of the media (Figure 7.8). A small air header below the edge of the media releases air into the cups. The air in the cups creates a buoyant force which then makes the shaft turn. With any of the three types of drive assemblies, the mainshaft is supported by two main bearings.

Individual units are usually provided with influent and effluent line valving to allow isolation for maintenance reasons. Usually the units are not shut down during low flow conditions because power consumption is minimal and as the flows decrease, the percent of BOD removal increases.

QUESTIONS

Write your answers in a notebook and then compare your answers with those on page 237.

7.0A   How does a rotating biological contactor (RBC) treat wastewater?

7.0B   Why is the RBC process divided into four or more stages?

7.0C   What is the purpose of a cover over the RBC unit?

7.0D   What are the three types of drive units?

7.1 PROCESS OPERATION

Performance by rotating biological contactors is affected by hydraulic loadings and temperatures below 55°F (13°C). Plants have been designed to treat flows ranging from 18,000 GPD to 50 MGD (70,000 liters/day to 200 M*L*D), however the majority of plants treat flows of less than 5 MGD (20 M*L*D). Typical operating and performance characteristics are as follows:

Characteristic	Range
Hydraulic Loading[3]	
BOD Removal	1.5 to 6 GPD/sq ft
	(60 to 240 liters per day/sq m)
Nitrogen Removal	1.5 to 1.8 GPD/sq ft
	(60 to 72 liters per day/sq m)
Organic Loading[3]	
Soluble BOD[4,5]	2.5 to 4 lbs BOD/day/1,000 sq ft
	(12 to 20 kg BOD/day/1,000 sq m)
BOD Removal	80 to 95 percent
Effluent Total BOD	10 to 30 mg/*L*
Effluent Soluble BOD	5 to 15 mg/*L*
Effluent NH₃-N	1 to 10 mg/*L*
Effluent NO₃-N	2 to 7 mg/*L*

See Section 7.5, "Loading Calculations," for procedures showing how to calculate the hydraulic and organic loadings on rotating biological contactors. Both hydraulic and organic loads to a rotating biological contactor should be evaluated. Experience indicates that the organic load (especially to the first stage) often controls the performance of the RBC.

Advantages of rotating biological contactors over trickling filters include the elimination of the rotating distributor with its problems, the elimination of the problems caused by ponding on the media, and filter flies. More efficient use of the media is achieved due to the even or uniform rotation of the media into the wastewater being treated. Another advantage of RBCs is the lack of anaerobic conditions in properly designed facilities as compared with anaerobic conditions found in the bottom of trickling filters.

A limitation of the process, as compared with trickling filters, is the lack of flexibility due to the absence of provisions for recirculation; however, in most installations recirculation is not needed, but will improve effluent quality when used. Rotating biological contactors are more sensitive to industrial wastes than trickling filters. Care must be taken to ensure that industrial organic loadings to the RBCs do not cause low dissolved oxygen conditions in the treatment system.

7.10 Pretreatment Requirements

Rotating biological contactors are usually preceded by preliminary treatment processes consisting of screening and grit removal, and by primary settling. Grit and large organic matter, if not removed, can settle beneath the drums and form sludge deposits which reduce the effective tank volume, produce septic conditions, scrape the slimes from the media, and possibly stall the unit.

Some rotating biological contactor plants have aerated flow equalization tanks between the primary clarifiers and the rotating biological contactors. Flow equalization tanks may be installed to equalize or balance highly fluctuating flows and to allow for the dilution of strong wastes and neutralization of highly acidic or alkaline wastes. These equalization tanks are capable of reducing or eliminating shock loads and providing pre-aeration to the RBC.

7.11 Start-Up

Prior to plant start-up, become familiar with and understand the contents of the plant O & M manual. If you have any questions, ask the design engineer or the manufacturer's representative. These persons should instruct the operator on the proper operation of the plant and maintenance of the equipment.

---

[3] *Hydraulic and organic loadings depend on influent flow, influent soluble BOD, effluent BOD, temperature, and surface area of plastic media. Manufacturers provide charts converting flow to hydraulic and organic loadings for their media.*

[4] *Soluble Bod.   Soluble BOD is the BOD of water that has been filtered in the standard suspended solids test. The soluble BOD is a measure of food for microorganisms that is dissolved in the water being treated.*

[5] *BOD loading for first stage or first shafts only.*

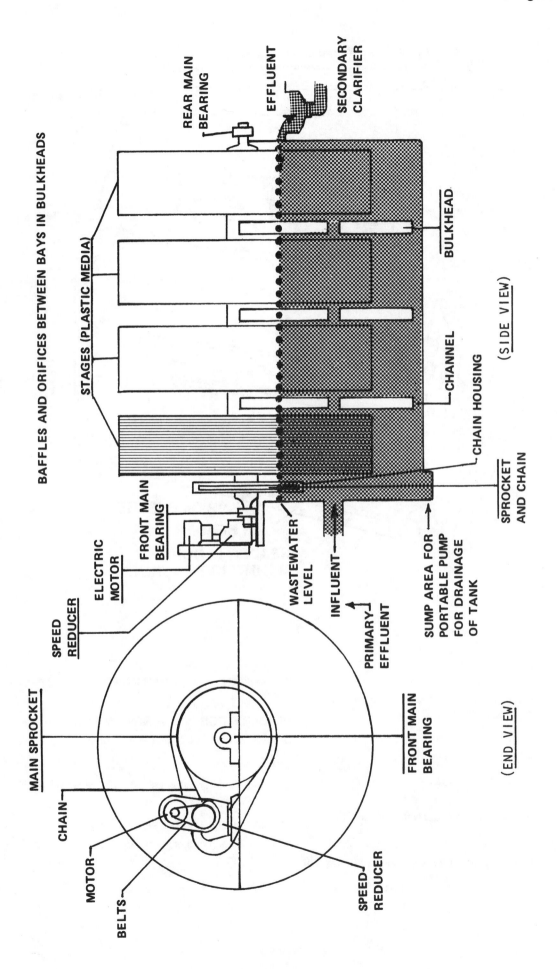

BAFFLES AND ORIFICES BETWEEN BAYS IN BULKHEADS

REAR MAIN BEARING

EFFLUENT

SECONDARY CLARIFIER

STAGES (PLASTIC MEDIA)

BULKHEAD

CHANNEL

CHAIN HOUSING

(SIDE VIEW)

SPROCKET AND CHAIN

FRONT MAIN BEARING

ELECTRIC MOTOR

SPEED REDUCER

WASTEWATER LEVEL

INFLUENT

PRIMARY EFFLUENT

SUMP AREA FOR PORTABLE PUMP FOR DRAINAGE OF TANK

MAIN SPROCKET

CHAIN

MOTOR

BELTS

SPEED REDUCER

FRONT MAIN BEARING

(END VIEW)

Fig. 7.7   Motor with chain drive unit

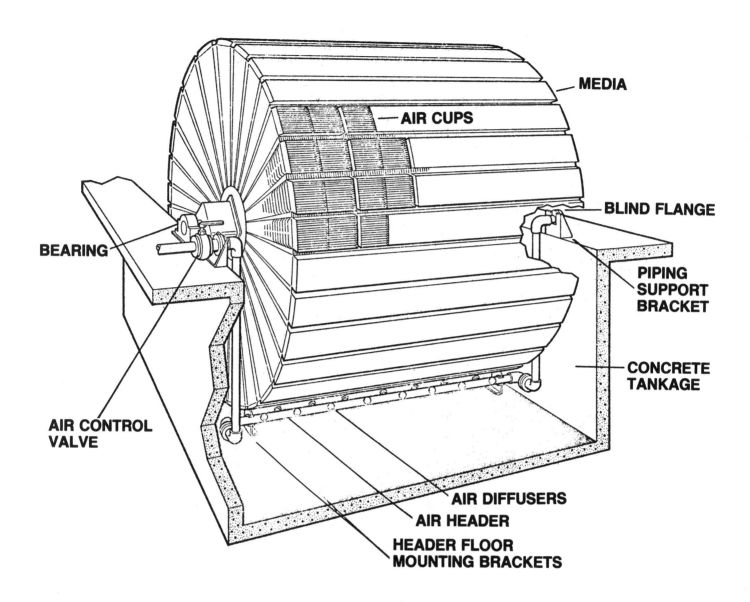

MEDIA

Media with polyethylene air cups attached to the outer perimeter capture air as it is released from the air header.

AIR CUPS

Air cups, attached to the media, capture air which in turn rotates the media.

AIR CONTROL VALVE

Butterfly control valve controls inlet air supply to each unit.

AIR HEADER

Lightweight headers that carry the air through the system run the length of the media assembly and are easily removable for cleaning.

HEADER FLOOR MOUNTING BRACKETS

Brackets secure header to the floor of the tank.

AIR DIFFUSER

Coarse-bubble air diffusers distribute air from the header into the air cups and media.

PIPING SUPPORT BRACKET

Bracket on each end of the header holds unit in place.

Fig. 7.8   Air drive unit
(Permission of Autotrol Corporation)

7.110 Pre-Start Checks for New Equipment

Before starting any equipment or allowing any wastewater to enter the treatment process, check the following items:

1. *TIGHTNESS*

    Inspect the following for tightness in accordance with manufacturer's recommendations.

    a. Anchor bolts
    b. Mounting studs
    c. Bearing caps
        Check any torque limitations.
    d. Locking collars
    e. Jacking screws
    f. Roller chain
        Be sure chain is properly aligned.
    g. Media
        Unbalanced media may cause slippage.
    h. Belts
        Use matched sets on multiple-belt drives.

2. *LUBRICATION*

    Be sure the following parts have been correctly lubricated with proper lubricants in accordance with manufacturer's recommendations.

    a. Mainshaft bearings
    b. Roller chain
    c. Speed reducer

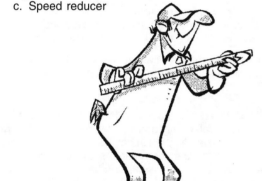

3. *CLEARANCES*

    a. Between media and tank wall.
    b. Between media and baffles or cover support beams.
    c. Between chain casing and media.
    d. Between roller chain, sprockets, and chain casing.

4. *SAFETY*

    Be sure safety guards are properly installed over chains and other moving parts.

7.111 Procedure for Starting Unit

Actual start-up procedures for a new unit should be in your plant O & M manual and provided by the manufacturer. A typical starting procedure is outlined below.

1. Switch on power, allow shaft to rotate one turn, turn off the power, lock out, and tag the main breaker. Inspect and cor-

rect if necessary during this revolution:

    a. Movement of chain casing.

    b. Unusual noises.

    c. Direction of media rotation. Where wastewater flow is parallel to the rotating media shaft, the direction of rotation is not critical. If the wastewater flow is perpendicular to the rotating media shaft, the media should be moving through the wastewater against the direction of flow (see Figure 7.6).

2. Switch on power and allow shaft to rotate for 15 minutes. Inspect the following:

    a. Chain drive sprocket alignment.

    b. Noises in bearings, chain drives, and drive package.

    c. Motor amperage. Compare with *NAMEPLATE*[6] value.

    d. Temperature of mainshaft bearing (by hand) and drive package pillow block. If too hot for the hand, use a *PYROMETER*[7] or thermometer. Temperature should not exceed 200°F (93°C).

    e. Tightness of shaft bearing cap bolts. Tighten to manufacturer's recommended torque.

    f. Determine number of revolutions per minute for drum and record for future reference.

3. Open inlet valve and allow wastewater to fill the tank (all four stages if in one tank). Open the outlet valve to allow water to flow through the tank. Make inspections listed in steps 1 and 2 again while drum is rotating. Shut off power, lock out, and tag the main breaker to make any corrections.

4. Check the relationship between the clarifier inlet and the rotating biological contactor outlet for hydraulic balance. This means that you want to be sure the tank containing the biological contactor will not overflow and cause stripping of the biomass.

5. See Section 7.20 for break-in maintenance instructions which start after eight hours of operation.

Development of biological slimes can be encouraged by regulating the flow rate and strength of the wastewater applied to nearly constant levels by the use of recirculation if available. Maintaining building temperatures at 65°F (18°C) or higher will help. The best rotating speed is one that will shear off growth at a rate that will provide a constant "hungry and reproductive" film of microorganisms exposed to the wastewater being treated.

Allow one to two weeks for an even growth of biological slimes (biomass) to develop on the surface of the media with normal strength wastewater. After start-up, a slimy growth (biomass) will appear. During the first week, excessive sloughing will occur naturally. This sloughing is normal and the sloughed material is soon replaced with a fairly uniform, shaggy, brown-to-gray appearing biomass with very few or no bare spots.

Follow the same start-up procedures whether a plant is starting at less than design flow or at full design flow. Start-up during cold weather takes longer because the organisms in the slime growth (biomass) are not as active and require more time to grow and reproduce.

---

[6] *Nameplate. A durable metal plate found on equipment which lists critical installation and operating conditions for the equipment.*
[7] *Pyrometer (pie-ROM-uh-ter). An apparatus used to measure high temperatures.*

7.12 Operation

Rotating biological contactor treatment plants are not difficult to operate and produce a good effluent provided the operator properly and regularly performs the duties of inspecting the equipment, testing the influent and effluent, observing the media, maintaining the equipment, and taking corrective action when necessary.

7.120 Inspecting Equipment

This treatment process has relatively few moving parts. There is a drive train to rotate the shaft and there are bearings upon which the shaft rotates. Check the following items when inspecting equipment.

1. Feel outer housing of shaft bearing to see if it is running hot. Use a pyrometer or thermometer if temperature is too hot for your hand. If temperature exceeds 200°F (93°C), the bearings may need to be replaced. Also check for proper lubrication and be sure the shaft is properly aligned. The longer the shaft, the more critical the alignment.

2. Listen for unusual noises in motor bearings. Locate cause of unusual noises and correct.

3. Feel motors to determine if they are running hot. If hot, determine cause and correct.

4. Look around drive train and shaft bearing for oil spills. If oil is visible, check oil levels in the speed reducers and chain drive system. Also look for damaged or worn out gaskets or seals.

5. Inspect chain drive for alignment and tightness.

6. Inspect belts for proper tension.

7. Be sure all guards over moving parts and equipment are in place and properly installed.

8. Clean up any spills, messes, or debris.

7.121 Testing Influent and Effluent

Wastewater analysis is required to monitor overall plant and process performance. Because there are few process control functions to be performed, only a minimal analysis is required to monitor and report daily performance. To determine if the rotating biological contactors are operating properly, you should measure (1) BOD, (2) suspended solids, (3) pH, and (4) dissolved oxygen (DO). Performance is best monitored by analysis of a 24-hour *COMPOSITE SAMPLE*[8] for BOD and suspended solids on a daily basis. DO and pH should be measured using *GRAB SAMPLES*[9] at specific times. Actual frequency of tests may depend on how often you need the results for plant control and also how often your NPDES permit requires you to sample and analyze the plant effluent.

DISSOLVED OXYGEN

The DO in the wastewater being treated beneath the rotating media will vary from stage to stage. A plant designed to treat primary effluent for BOD and suspended solids removal will usually have 0.5 to 1.0 mg/L DO in the first stage. The DO level should increase to 1 to 3 mg/L at the end of the first stage. A plant designed for nitrification to convert ammonia and organic nitrogen compounds to nitrate usually will have four stages and DO levels of 4 to 8 mg/L. The difference between an RBC unit designed for BOD removal and one designed for nitrification is the design flow applied per square foot of media surface area. DO in the first stage of a nitrification unit will be more than 1 mg/L DO and often as high as 2 to 3 mg/L.

EFFLUENT VALUES

Typical BOD, suspended solids, and ammonia and nitrate effluent values for rotating biological contactors depend on NPDES permit requirements and design effluent values. As flows increase, effluent values increase because a greater flow is applied to each square foot of media while the time the wastewater is in contact with the slime growths is reduced. Also, the greater the levels of BOD, suspended solids, and nitrogen in the influent, the greater the levels in the plant effluent. Figure 7.9 shows influent and effluent values for a rotating biological contactor. The influent and effluent data plotted are seven-day moving averages which smooth out daily fluctuations and reveal trends. Procedures for calculating moving averages are explained in Volume II, Chapter 18, "Analysis and Presentation of Data."

If analysis of samples reveals a decrease in process efficiency, look for three possible causes:

1. Reduced wastewater temperatures,
2. Unusual variations in flow and/or organic loadings, and
3. High or low pH values (less than 6.5 or greater than 8.5).

Once the cause of the problem has been identified, possible solutions can be considered and the problem corrected.

TEMPERATURE

Wastewater temperatures below 55°F (13°C) will result in a reduction of biological activity and in a decrease in BOD or organic material removal. Not much can be done by the operator except to wait for the temperatures to increase again. Under severe conditions, provisions can be made to heat the building, the air inside the RBC unit cover, or the RBC unit influent.

Solar heat can be used effectively to maintain temperature in buildings and enclosures without drying out the biological slime growths. Ceilings should be kept low to effectively use available heat. If existing buildings have high ceilings, large vaned fans can be mounted on the ceilings to direct heat downward.

INFLUENT VARIATIONS

When large daily influent flow and/or organic (BOD) variations occur, a reduction in process efficiency is likely to result. Before corrective steps are taken, the exact extent of the problem and resulting change in process efficiency must be determined. In most cases, when the influent flow and/or organic peak loads are less than three times the daily average values during a 24-hour period, little decrease in process efficiency will result.

---

[8] *Composite (come-PAH-zit) (Proportional) Sample.   A composite sample is a collection of individual samples obtained at regular intervals, usually every one or two hours during a 24-hour time span. Each individual sample is combined with the others in proportion to the rate of flow when the sample was collected. The resulting mixture (composite sample) forms a representative sample and is analyzed to determine the average conditions during the sampling period.*

[9] *Grab Sample.   A single sample of water collected at a particular time and place which represents the composition of the water only at that time and place.*

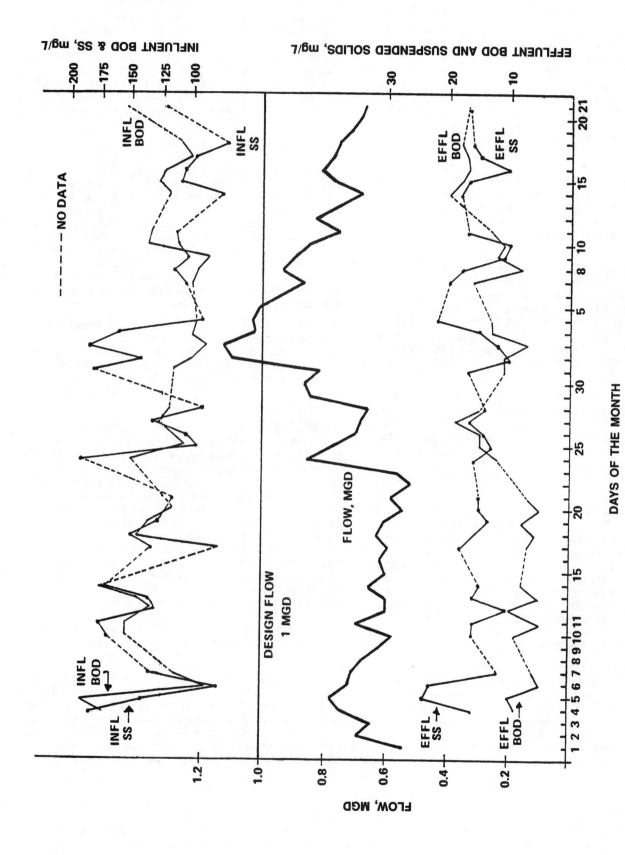

Fig. 7.9 Typical BOD and suspended solids values for an RBC unit

In treatment plants where the influent flow and/or organic loads exceed design values for a sustained period, the effluent BOD and suspended solids must be measured regularly during this period to determine if corrective action is required.

During periods of severe organic overload, the bulkhead or baffle between stages one and two may be removed. This procedure provides a larger amount of media surface area for the first stage of treatment. If the plant is continuously overloaded and the effluent violates the NPDES permit requirements, additional treatment units should be installed. A possible short-term solution to an organic overload problem might be the installation of facilities to recycle effluent; however, this would cause a greater increase of any hydraulic overload.

pH

Every wastewater has an optimum pH level for best treatability. Domestic wastewater pH varies between 6.5 and 8.5 and will have little effect on organic removal efficiency. However, if this range is exceeded at any time (due to industrial waste discharges, for example), a decrease in efficiency is likely.

To adjust the pH toward 7.0, either pre-aerate the influent or add chemicals. If the pH is too low, add sodium bicarbonate or lime. If the pH is too high, add acetic acid. The amount of chemical to be added depends on the characteristics of the water and can best be determined by adding chemicals to samples in the lab and measuring the change in pH. Always wear appropriate safety gear (goggles and/or face shield, impervious gloves, protective clothing) when handling chemicals.

When dealing with nitrification, pH and alkalinity are very critical. The pH should be kept as close as possible to a value of 8.4 when nitrifying. The alkalinity level in the raw wastewater should be maintained at a level at least 7.1 times the influent ammonia concentration to allow the reaction to go to completion without adversely affecting the microorganisms. Sodium bicarbonate can be used to increase both the alkalinity and pH.

Another cause of pH variations could be the addition of SUPERNATANT[10] from a digester. The supernatant should be tested for pH and suspended solids. Without testing the supernatant, you will not know what kind of load you're placing on the rest of the plant. Sometimes it's best to drain supernatant at low flows to the plant. Caution should be taken to avoid overloading the process. If the supernatant pH is too low, supernatant could be drawn off during high flows when these flows can be used for dilution and NEUTRALIZATION.[11]

7.122   Observing the Media

Rotating biological contactors use bacteria and other living organisms growing on the media to treat wastes. Because of this, you can use your sight and smell to identify problems. The slime growth or biomass should have a brown-to-gray color, no algae present, a shaggy appearance with a fairly uniform coverage, and very few or no bare spots. The odor should not be offensive, and certainly there should be no sulfide (rotten egg) smells.

BLACK APPEARANCE

If the appearance becomes black and odors that are not normal do occur, then this could be an indication of solids or BOD overloading. These conditions would probably be accompanied by low DO in the plant effluent. Compare previous influent suspended solids and BOD values with current test results to determine if there is an increase. To solve this problem, place another rotating biological contactor unit in service, if possible, or try to pre-aerate the influent to the RBC unit. Also review the operation of the primary clarifiers and sludge digesters to be sure they are not the source of the overload.

WHITE APPEARANCE

A white appearance on the disc surface also might be present during high loading conditions. This might be due to a type of bacteria that feeds on sulfur compounds. The overloading could result from industrial discharges containing sulfur compounds upon which certain sulfur-loving bacteria thrive and produce a white slime biomass. Corrective action consists of placing another RBC unit in service or trying to pre-aerate the influent to the unit. During periods of severe organic or sulfur overloading, remove the bulkhead or baffle between stages one and two. Prechlorination of plant influent also will control sulfur-loving bacteria.

Another cause of overloading may be sludge deposits that have been allowed to accumulate in the bottom of the bays. To remove these deposits, drain the bays, wash the sludge deposits out, and return the unit to service. Be sure the orifices in the baffles between the bays are clear.

SLOUGHING

If severe sloughing or loss of biomass occurs after the start-up period and process difficulty arises, the causes may be due to the influent wastewater containing toxic or INHIBITORY SUBSTANCES[12] that kill the organisms in the biomass or restrict their ability to treat wastes. To solve this problem, steps must be taken to eliminate the toxic substance even though this may be very difficult and costly. Biological processes will never operate properly as long as they attempt to treat toxic wastes. Until the toxic substance can be located and eliminated, loading peaks should be dampened (reduced) and a diluted, uniform concentration of the toxic substance allowed to reach the media in order to minimize harm to the biological culture. While the corrections are made at the plant, dampening may be accomplished by regulating inflow to the plant. Be careful not to flood any homes or overflow any low manholes. Toxic wastes may be diluted using plant effluent (until the toxic material reaches the effluent) or any other source of water supply.

Another problem that could cause loss of biomass is an unusual variation in flow and/or organic loading. In small communities one cause may be high flow during the day and near zero flow at night. During the day the biomass is receiving food and oxygen and starts growing; then the night flow drops to near zero—available food is reduced and nearly stops. The biomass starts sloughing off again due to lack of food.

---

[10] Supernatant (sue-per-NAY-tent).   Liquid removed from settled sludge. Supernatant commonly refers to the liquid between the sludge on the bottom and the scum on the surface of an anaerobic digester. This liquid is usually returned to the influent wet well or to the primary clarifier.

[11] Neutralization (new-trall-i-ZAY-shun).   Addition of an acid or alkali (base) to a liquid to cause the pH of the liquid to move toward a neutral pH of 7.0.

[12] Inhibitory Substances.   Materials that kill or restrict the ability of organisms to treat wastes.

Possible solutions to sloughing of the biomass due to excessive variations in plant flow and/or organic loading include throttling peak conditions and recycling from the secondary clarifier or RBC effluent during low flows. Be very careful when throttling plant inflows that low elevation homes are not flooded or that manholes do not overflow. Usually RBC units do not have provisions for any recycling from the secondary clarifier. If low flows at night are creating operation problems due to lack of organic matter, a possible solution is the installation of a pump to recirculate water from the secondary clarifier. If recirculation is provided, try to maintain a hydraulic loading rate of greater than 1.0 to 1.5 GPD/sq ft. A flow equalization tank can be used to provide fairly continuous or even flows.

7.123 Control of Snails

Snails are not a problem in RBCs used to remove carbonaceous biochemical oxygen demand (CBOD) because the growth of "bugs" (microbes) removing CBOD is high and the microbial slime (treating the wastewater) consumed by snails is quickly replaced by new growth. Snails are a problem where the RBC is expected to remove nitrogenous biochemical oxygen demand (NBOD) because snails remove slow-growing nitrifying bacteria and interfere with nitrification. Snail shells are a problem when they clog pipes and pumps.

Chlorination is commonly used to control snails on RBCs. One approach is to take an RBC train (the RBCs treating a particular slug or flow of wastewater) off line. Add chlorine to the water to a concentration of 60 to 70 mg/L, rotate the RBC in the superchlorinated solution for two to three days, and then return the RBC to service. Chlorination also can be used to control filamentous microorganisms. The superchlorinated water can be dechlorinated before it is discharged from the facility by applying sulfur dioxide. When the RBC is returned to service it will take a period of time for the biomass to recover and reach full treatment capacity.

Another approach to control snails is to increase the pH to 10. A pH of 10 will kill snails without harming the microbial growth on the RBCs. The pH can be increased by adding caustic soda, sodium hydroxide, or lime and maintaining the RBC exposure for eight hours. Operators may have to increase the pH to 10 every one to two months to control the snails.

Adjustment of the wastewater application rate at upstream trickling filters where snails lay their eggs will wash away a large proportion of the eggs and thereby help to reduce the population of snails in RBCs. Also, if secondary solids are being recycled to the head of the plant, snail eggs and snails may be recycled along with the solids. Some operators reduce this transfer of snails by sending the solids from the secondary clarifier directly to anaerobic digesters.

7.124 Operational Troubleshooting

Indicators of possible rotating biological contactor process operational problems, probable causes, and solutions are summarized in Table 7.2.

7.13 Abnormal Operation

Abnormal operating conditions may develop under the following circumstances:

1. High or low flows,
2. High or low solids loading, and
3. Power outages.

When your plant must treat high or low flows or solids (organic) loads, abnormal conditions develop as the treatment efficiency drops. For solutions to these problems, refer to Section 7.12, "Operation," and Table 7.2. One advantage of RBC units is the fact that high flows usually do not wash the slime growths off the media; consequently the organisms are present and treating the wastewater during and after the high flows.

A power outage requires the operator to take certain precautions to protect the equipment and the slime growths while no power is available. If the power is off for less than four hours, nothing needs to be done. If the power outage lasts longer than four hours, the RBC shaft needs to be turned about one-quarter of a turn every four hours. Turning prevents all the slime growth from accumulating on the bottom portion of the plastic disc media. Before attempting to turn the shaft, lock out and tag the power in case the outage ends abruptly. To turn the shaft, *REMOVE THE BELT GUARD USING EXTREME CARE.* Turn the shaft by using the spokes on the pulley or a strap wrench on the shaft. *DO NOT EXPOSE YOUR FINGERS OR HANDS TO THE PINCH POINT BETWEEN THE BELT(S) AND THE PULLEY(S).* Place a wedge-shaped block between the belts and belt pulley or sheave to hold the shaft and media in the desired location. Actually, the shaft is very delicately balanced and easy to rotate. Do not try to weld handles or brackets to the shaft to facilitate turning because this will throw the shaft off balance.

WARNING. If the shaft starts to roll back to its original position before you get the block properly inserted, do not try to stop the shaft. Let it roll back and stop. If you try to stop the shaft from rolling back, you could injure yourself and also damage the belts and the belt's pulley or sheave.

Gently spray water on the slime growth that is not submerged frequently enough to keep the biomass moist whenever the drum is not rotating.

If the power outage lasts longer than 12 hours, more than normal sloughing will occur from the media when the unit is placed back in service. When the sloughing becomes excessive, increase the sludge pumping rate from the secondary clarifier.

7.14 Shutdown and Restart

The rotating biological contactor may be stopped by turning off the power to the drive package. If the process is to be stopped for longer than four hours, follow the precautions listed in Section 7.13, "Abnormal Operation," when a power outage occurs. Do not allow one portion of the media to be submerged in the wastewater being treated for more than four hours. Occasionally spray the media not submerged to prevent the slime growth from drying out whenever the drum is not rotating.

If the tank holding the wastewater being treated must be drained, a portable sump pump may be used. A sump is usually located at the end of the unit by the motor. Pump the water either to the primary clarifier or to the inlet end of an RBC unit in operation. A trough running the full length of the tank allows the solids to be pumped out. While the tank is empty, inspect for cracks and any other damage and make necessary repairs.

Try to keep the slime growths moist to minimize sloughing and a reduction in organism activity when the process starts again. A loss in process efficiency can result if the slimes are washed off the media. *DO NOT WASH THE SLIME GROWTH OFF THE MEDIA* because you will be washing away the organisms that treat the wastewater. If the unit is to be out of service for longer than one day, the slimes may be washed off the media to prevent the development of odor problems.

TABLE 7.2 TROUBLESHOOTING GUIDE—ROTATING BIOLOGICAL CONTACTORS

(Adapted from *PERFORMANCE EVALUATION AND TROUBLESHOOTING
AT MUNICIPAL WASTEWATER TREATMENT FACILITIES*,
Office of Water Program Operations, US EPA, Washington, DC.)

INDICATOR/OBSERVATION	PROBABLE CAUSE	CHECK OR MONITOR	SOLUTION
1. Decreased treatment efficiency.	1a. Organic overload.	1a. Check peak organic loads—BOD, SS, DO, pH, temperature.	1a. 1. Improve pretreatment of plant. 2. Place another RBC in service if available. 3. Remove bulkhead between stages 1 and 2 for larger first stage. 4. Recycle effluent as a possible short-term solution.
	1b. Hydraulic overload.	1b. Check peak hydraulic loads—if less than twice the daily average, should not be the cause.	1b. 1. Flow equalization; eliminate source of excessive flow. 2. Balance flows between reactors. 3. Store peak flows in collection system, monitor possible overflows of collection system.
	1c. pH too high or too low.	1c. Desired range is 6.5 - 8.5 for secondary treatment; 8 - 8.5 for nitrification.	1c. 1. Eliminate source of undesirable pH or add acid or base to adjust pH. When nitrifying, maintain alkalinity at 7 times the influent NH_3 concentration. 2. Sodium bicarbonate can be used to increase both pH and alkalinity.
	1d. Low wastewater temperatures.	1d. Temperatures less than 55° F will reduce efficiency.	1d. 1. Cover RBC to contain heat of wastewater. 2. Heat influent to unit or building.
2. Excessive sloughing of biomass from discs.	2a. Toxic materials in influent.	2a. Determine material and its source.	2a. 1. Eliminate toxic material if possible—if not, use flow equalization to reduce variations in concentration so biomass can acclimate. 2. Recycle effluent for dilution.
	2b. Excessive pH variations.	2b. pH below 5 or above 10 can cause sloughing.	2b. Eliminate source of pH variations or maintain control of influent pH.
	2c. Unusual variation in flow and/or organic loading.	2c. Influent flow rate(s) and organic strength.	2c. Eliminate/reduce variations by throttling peak conditions and recycling from the secondary clarifier or RBC effluent during low flows.
			2d. Monitor industrial contributors for flow variations.
3. Development of white biomass over most of disc area.	3a. Septic influent or high H_2S concentrations.	3a. Influent odor.	3a. Pre-aerate wastewater or add sodium nitrate or hydrogen peroxide or place another RBC unit in service. Prechlorination of influent will also control sulfur-loving bacteria.
	3b. First stage is overloaded organically.	3b. Organic loading on first stage.	3b. 1. Improve pretreatment of plant. 2. Place another RBC in service, if available. 3. Adjust baffles between first and second stages to increase total surface area in first stage.
4. Solids accumulating in reactors.	4a. Inadequate pretreatment.	4a. Determine if solids are grit or organic.	4a. Remove solids from reactors and provide improved grit removal or primary settling.

Restart rotation by applying power to the drive unit. Before applying power, inspect the shaft and drive unit for possible interference from such items as tools or bulkheads. If slippage occurs from an unbalanced media, inspect and adjust alignment and tension.

QUESTIONS

Write your answers in a notebook and then compare your answers with those on page 237.

7.1A Why should debris, grit, and suspended solids be removed before the wastewater being treated reaches the RBC unit?

7.1B List the major items for a pre-start check.

7.1C What are the main operational duties for an operator of an RBC unit?

7.2 MAINTENANCE

Rotating biological contactors have few moving parts and require minor amounts of preventive maintenance. Chain drives, belt drives, sprockets, rotating shafts, and any other moving parts should be inspected and maintained in accordance with manufacturers' instructions or your plant's O & M manual. All exposed parts, bearing housing, shaft ends, and bolts should be painted or covered with a layer of grease to prevent rust damage. Motors, speed reducers, and all other metal parts should be painted for protection.

Maintenance also includes the repair or replacement of broken parts. A preventive maintenance program that keeps equipment properly lubricated and adjusted to help reduce wear and breakage requires less time and money than a program that waits for breakdowns to occur before taking any action. The frequency of inspection and lubrication is usually provided by manufacturer's instructions and also may be found in the plant O & M manual. The following sections indicate a typical maintenance program for a rotating biological contactor treatment process. More details can be found in a plant O & M manual.

7.20 Break-In Maintenance

AFTER 8 HOURS OF OPERATION

1. Recheck tightening torque of capscrews in all split-tapered bushings in the drive package.

2. Visually inspect hubs and capscrews for general condition and possibility of rubbing against an obstruction.

3. Inspect belt drive (drive package) and tighten as needed.

AFTER 24 HOURS OF OPERATION

1. Inspect all chain drives.

AFTER 40 HOURS OF OPERATION

1. Inspect all belt drives in drive packages.

AFTER 100 HOURS OF OPERATION

1. Change oil in speed reducer. Use manufacturer's recommended lubricants.

2. Clean magnetic drain plug in speed reducer.

3. Check all capscrews in split-tapered bushings and setscrews in drive package output sprocket and bearing for tightness.

4. Inspect belt drive of drive package.

AFTER 3 WEEKS OF OPERATION

1. Change oil in chain casing. Be sure oil level is at or above the mark on the dipstick. Use manufacturer's recommended lubricants.

7.21 Preventive Maintenance Program

Interval		Procedure
Daily	1.	Check for hot shaft and bearings. Replace bearings if temperature exceeds 200°F (93°C) and there are problems.
Daily	2.	Listen for unusual noises in shaft and bearings. Identify cause of noise and correct if necessary.
Weekly	3.	Grease the mainshaft bearings and drive bearings. Use manufacturer's recommended lubricants. Add grease slowly while shaft rotates. When grease begins to ooze from the housing, the bearings contain the correct amount of grease. Add six full strokes where bearings cannot be seen.
4 wk.	4.	Inspect all chain drives.
4 wk.	5.	Inspect mainshaft bearings and drive bearings.
4 wk.	6.	Apply a generous coating of general purpose grease to mainshaft stub ends, mainshaft bearings, and end collars.
3 mo.	7.	Change oil in chain casing. Use manufacturer's recommended lubricants. Be sure oil level is at or above the mark on the dipstick.
3 mo.	8.	Inspect belt drive.
6 mo.	9.	Change oil in speed reducer. Use manufacturer's recommended lubricants.
6 mo.	10.	Clean magnetic drain plug in speed reducer.
6 mo.	11.	Purge the grease in the double-sealed shaft seals of the speed reducer by removing the plug located 180 degrees from the grease fitting on both the input and output seal cages. Pump grease into the seal cages and then replace the plug. Use manufacturer's recommended grease.
12 mo.	12.	Grease motor bearings. Use manufacturer's recommended grease. To grease motor bearings, stop motor and remove drain plugs. Inject new grease with pressure gun until all old grease has been forced out of the bearing through the grease drain. Run motor until all excess grease has been expelled. This may require up to several hours running time for some motors. Replace drain plugs.

7.22 Housekeeping

Properly designed systems have sufficient turbulence so solids or sloughed slime growths should not settle out on the bottom of the bays. If grease balls appear on the water surface in the bays, they should be removed with a dip net or screen device.

If media comes apart, squeeze the two unbonded sections together with a pair of pliers. Take another pair of pliers and force a heated nail through the media. The heat from the nail will melt the plastic and make a plastic weld between the two sections of media.

7.23 Troubleshooting Guide

7.230 Roller Chain Drive

Trouble	Probable Cause	Corrective Action
1. Noisy Drive	1a. Moving parts rub stationary parts.	1a. Tighten and align casing and chain. Remove dirt or other interfering matter.
	1b. Chain does not fit sprockets.	1b. Replace with correct parts.
	1c. Loose chain.	1c. Maintain a taut chain at all times.
	1d. Faulty lubrication.	1d. Lubricate properly.
	1e. Misalignment or improper assembly.	1e. Correct alignment and assembly of the drive.
	1f. Worn parts.	1f. Replace worn chain or bearings. Reverse worn sprockets before replacing.
2. Rapid Wear	2a. Faulty lubrication.	2a. Lubricate properly.
	2b. Loose or misaligned parts.	2b. Align and tighten entire drive.
3. Chain Climbs Sprockets	3a. Chain does not fit sprockets.	3a. Replace chain or sprockets.
	3b. Worn-out chain or worn sprockets.	3b. Replace chain. Reverse or replace sprockets.
	3c. Loose chain.	3c. Tighten.
4. Stiff Chain	4a. Faulty lubrication.	4a. Lubricate properly.
	4b. Rust or corrosion.	4b. Clean and lubricate.
	4c. Misalignment or improper assembly.	4c. Correct alignment and assembly of the drive.
	4d. Worn-out chain or worn sprockets.	4d. Replace chain. Reverse or replace sprockets.
5. Broken Chain or Sprockets	5a. Shock or overload.	5a. Repair or replace broken parts. Add more RBC units or remove baffles between stages 1 and 2.
	5b. Wrong size chain, or chain that does not fit sprockets.	5b. Replace chain. Reverse or replace sprockets.
	5c. Rust or corrosion.	5c. Replace parts. Correct corrosive conditions.
	5d. Misalignment.	5d. Correct alignment.
	5e. Interferences.	5e. Make sure no solids interfere between chain and sprocket teeth. Loosen chain if necessary for proper clearance over sprocket teeth.

7.231   Belt Drive

Trouble	Probable Cause	Corrective Active
1. Excessive edge wear	1. Misalignment or non-rigid centers.	1. Check alignment and/or reinforcement mounting.
2. Jacket wear on pressure-face side of belt tooth*	2. Bent flange.	2. Straighten flange.
3. Excessive jacket wear between belt teeth (exposed tension members)*	3. Excessive overload and/or excessive belt tightness.	3. Reduce installation tension and/or increase drive load-carrying capacity.
4. Cracks in Neoprene backing	4a. Excessive installation tension.	4a. Reduce installation tension.
	4b. Exposure to excessively low temp. (below −30°F or −35°C).	4b. Eliminate low temperature condition or consult factory for proper belt construction.
5. Softening of Neoprene backing	5. Exposure to excessive heat (+200°F or 93°C) and/or oil.	5. Eliminate high temperature and oil condition or consult factory for proper belt construction.
6. Tensile or tooth shear failure*	6a. Small or sub-minimum diameter pulley.	6a. Increase pulley diameter.
	6b. Belt too narrow.	6b. Increase belt width.
7. Excessive pulley tooth wear (on pressure face and/or OD)*	7a. Excessive overload and/or excessive belt tightness.	7a. Reduce installation tension and/or increase drive load-carrying capacity.
	7b. Insufficient hardness of pulley material.	7b. Surface-harden pulley or use harder material.
8. Unmounting of flange	8a. Incorrect flange installation.	8a. Reinstall flange correctly.
	8b. Misalignment.	8b. Correct alignment.
9. Excessive drive noise	9a. Misalignment.	9a. Correct alignment.
	9b. Excessive installation tension.	9b. Reduce tension.
	9c. Sub-minimum pulley diameter.	9c. Increase pulley diameter.
10. Tooth shear*	10a. Less than 6 teeth in mesh (TIM).	10a. Increase TIM or use next smaller pitch.
	10b. Excessive load.	10b. Increase drive load-carrying capacity.
11. Apparent belt stretch	11. Reduction of center distance or non-rigid mounting.	11. Re-tension drive and/or reinforce mounting.
12. Cracks or premature wear at belt tooth root*	12. Improper pulley groove top radius.	12. Regroove or install new pulley.
13. Tensile break	13a. Excessive load.	13a. Increase load-carrying capacity of drive.
	13b. Sub-minimum pulley diameter.	13b. Increase pulley diameter.

* Pertains to a timing belt system only. Recent systems use a V-belt drive.

7.232   Bearings and Motors

Trouble	Probable Cause	Corrective Action
1. Shaft bearings running hot or failing. If temperature exceeds 200°F (93°C) the bearings may need to be replaced.	1a. Inadequate lubrication.	1a. 1. Lubricate bearings per manufacturer's instructions. 2. Check tightness (torque) and alignment of bearings.
	1b. Shaft misalignment.	1b. Align shaft properly. The longer the shaft, the more critical the alignment.
2. Motors running hot.	2a. Inadequate maintenance.	2a. 1. Lubricate per manufacturer's instructions. 2. Maintain correct oil level and oil viscosity in speed reducer.
	2b. Improper chain drive alignment.	2b. Align properly.

7.3   SAFETY

Any equipment with moving parts or electrical components should be considered a potential safety hazard. *ALWAYS SHUT OFF THE POWER TO UNIT, TAG THE SWITCH, AND LOCK THE MAIN BREAKER IN THE "OFF" POSITION BEFORE WORKING ON A UNIT.*

7.30   Slow-Moving Equipment

Slow-moving equipment does not appear dangerous. Unfortunately, moving parts such as the sprockets, chains, belts, and sheaves can cause serious injury by tearing and/or crushing your hands or legs.

7.31   Wiring and Connections

Wiring and connections should be inspected regularly for potential hazards such as loose connections and bare wires. Again, always shut off, tag, and lock out the main breaker before working on a unit.

7.32   Slippery Surfaces

Caution must be taken on slippery surfaces. Falls can result in serious injuries. Any spilled oil or grease must be cleaned up immediately. If covers over the media allow sufficient space for walkways, condensed moisture on surfaces will create slippery places. If the temperature of the air within the enclosure can be kept several degrees above the temperature of the wastewater, condensation is significantly reduced. This condensation cannot be avoided completely so walk carefully at all times. The application of a nonskid material on walkways can greatly reduce the potential for slips and falls.

7.33   Infections and Diseases

Precautions must be taken to prevent infections in cuts or open wounds and illnesses from waterborne diseases. After working on a unit, always wash your hands before smoking or eating. *GOOD PERSONAL HYGIENE MUST BE PRACTICED BY ALL OPERATORS AT ALL TIMES.*

7.4   REVIEW OF PLANS AND SPECIFICATIONS

When reviewing plans and specifications, be sure the following items are included in the design of rotating biological contactors.

1. Enclosure to protect biomass from freezing temperature. Enclosure constructed of suitable corrosion-resistant materials and has windows or louvered structures in sides for ventilation. Forced ventilation is not necessary.

2. Heating. A source of heat is helpful during winter operation to minimize the corrosion caused by condensation and to improve operator comfort. If the temperature of the air within the enclosure is kept several degrees above the temperature of the wastewater, condensation is significantly reduced. Keep ceilings low to effectively use available heat.

3. Recirculation. Provisions for recirculating secondary effluent and also secondary sludge will allow for flexibility and can improve performance.

QUESTIONS

Write your answers in a notebook and then compare your answers with those on page 237.

7.2A   How often should maintenance be performed on an RBC unit during start-up?

7.3A   List some possible safety hazards operators encounter when working around RBC units.

7.5   LOADING CALCULATIONS

7.50   Typical Loading Rates

Hydraulic and organic loadings on rotating biological contactors depend on influent soluble BOD, effluent total and soluble BOD requirements, and wastewater temperature. The loading rates given here are for "typical" values and values for your plant could be different.

Characteristic	Range
Hydraulic Loading	
BOD Removal	1.5 to 6 GPD/sq ft (60 to 240 liters per day/sq m)
Nitrogen Removal	1.5 to 1.8 GPD/sq ft (60 to 72 liters per day/sq m)
Organic Loading	
Soluble BOD	2.5 to 4 lbs BOD/day/1,000 sq ft (12 to 20 kg BOD/day/1,000 sq m)
Total BOD	6 to 8 lbs BOD/day/1,000 sq ft (30 to 40 kg BOD/day/1,000 sq m)

7.51   Computing Hydraulic Loading

Hydraulic loading on a rotating biological contactor is the amount (gallons) of wastewater per day that flows past the rotating media. To calculate the hydraulic loading, we must know:

1. Gallons per day treated by the rotating biological contactor, and

2. The surface area of the media in square feet.

EXAMPLE 1

A rotating biological contactor treats a flow of 3.5 MGD. The surface area of the media is 1,000,000 square feet (provided by manufacturer). What is the hydraulic loading in GPD/sq ft?

Known	Unknown
Flow, MGD = 3.5 MGD	Hydraulic Loading, GPD/sq ft
Surface Area, sq ft = 1,000,000 sq ft	

Calculate the hydraulic loading in gallons per day of wastewater per square foot of surface media.

$$\text{Hydraulic Loading, GPD/sq ft} = \frac{\text{Flow, gal/day}}{\text{Surface Area, sq ft}}$$

$$= \frac{3,500,000 \text{ GPD}}{1,000,000 \text{ sq ft}}$$

$$= 3.5 \text{ GPD/sq ft}$$

7.52 Computing Organic (BOD) Loading

Organic loadings on rotating biological contactors are based on soluble BOD. Soluble BOD is the BOD of water that has been filtered in the standard suspended solids test. If soluble BOD information is not available, soluble BOD may be estimated on the basis of the total BOD and the suspended solids as follows:

$$\text{Soluble BOD, mg/}L = \text{Total BOD, mg/}L - \text{Suspended BOD, mg/}L$$

where

Suspended BOD, mg/L = K x Suspended Solids, mg/L

Therefore,

$$\text{Soluble BOD, mg/}L = \text{Total BOD, mg/}L - (\text{K x Suspended Solids, mg/}L)$$

where

K = 0.5 to 0.7 for most domestic wastewaters.

EXAMPLE 2

The rotating biological contactor in Example 1 treats an influent with a total BOD of 200 mg/L and suspended solids of 250 mg/L. Assume a K value of 0.5 to calculate the soluble BOD. What is the organic loading in pounds of soluble BOD per day per 1,000 square feet of media surface?

Known	Unknown
Flow, MGD = 3.5 MGD	Organic Loading, lbs BOD/day/1,000 sq ft
Surface Area, sq ft = 1,000,000 sq ft	
Total BOD, mg/L = 200 mg/L	
SS, mg/L = 250 mg/L	
K = 0.5	

1. Estimate the soluble BOD treated by the rotating biological contactor.

$$\text{Soluble BOD, mg/}L = \text{Total BOD, mg/}L - (\text{K x Suspended Solids, mg/}L)$$

$$= 200 \text{ mg/}L - (0.5 \text{ x } 250 \text{ mg/}L)$$

$$= 200 \text{ mg/}L - 125 \text{ mg/}L$$

$$= 75 \text{ mg/}L$$

2. Determine the soluble BOD applied to the rotating biological contactor in pounds of soluble BOD per day.

$$\text{BOD Applied, lbs BOD/day} = (\text{Soluble BOD, mg/}L)(\text{Flow, MGD})(8.34 \text{ lbs/gal})$$

$$= (75 \text{ mg/}L)(3.5 \text{ MGD})(8.34 \text{ lbs/gal})$$

$$= 2,189 \text{ lbs Soluble BOD/day}$$

3. Calculate the organic loading in pounds of soluble BOD per day per 1,000 square feet of media surface. Since the media surface area is 1,000,000 square feet, there are 1,000 • 1,000 square feet of surface area.

$$\text{Organic BOD Loading, lbs BOD/day/1,000 sq ft} = \frac{\text{Soluble BOD Applied, lbs BOD/day}}{\text{Surface Area of Media (in 1,000 sq ft)}}$$

$$= \frac{2,189 \text{ lbs Soluble BOD/day}}{1,000 • 1,000 \text{ sq ft}}$$

$$= 2.2 \text{ lbs BOD/day/1,000 sq ft}$$

7.53 Typical Loading Rates (Metric)

The next three sections show typical loading rates for rotating biological contactors and how to calculate the hydraulic and organic loading using the metric system.

Characteristic	Range
Hydraulic Loading	
BOD Removal	0.06 to 0.24 cu m/day/sq m
Nitrogen Removal	0.06 to 0.07 cu m/day/sq m
Organic Loading	
Soluble BOD	15 to 25 gm BOD/day/sq m
Total BOD	36 to 50 gm BOD/day/sq m

7.54 Computing Hydraulic Loading (Metric)

Hydraulic loading on a rotating biological contactor is the amount (cubic meters) of wastewater per day that flows past the rotating media.

EXAMPLE 3

A rotating biological contactor treats a flow of 15,000 cubic meters per day. The surface area of the media is 100,000 square meters (provided by the manufacturer). What is the hydraulic loading in cu m/day/sq m?

Known	Unknown
Flow, cu m/day = 15,000 cu m/day	Hydraulic Loading, cu m/day/sq m
Surface Area, sq m = 100,000 sq m	

Calculate the hydraulic loading in cubic meters per day of wastewater per square meter of surface media.

$$\text{Hydraulic Loading, cu m/day/sq m} = \frac{\text{Flow, cu m/day}}{\text{Surface Area, sq m}}$$

$$= \frac{15,000 \text{ cu m/day}}{100,000 \text{ sq m}}$$

$$= 0.15 \text{ cu m/day/sq m}$$

7.55 Computing Organic (BOD) Loading (Metric)

EXAMPLE 4

The rotating biological contactor in Example 3 treats an influent with a soluble BOD of 75 mg/L. What is the organic loading in grams of soluble BOD per day per square meter of media surface?

Known	Unknown
Flow, cu m/day = 15,000 cu m/day	Organic Loading, gm BOD/day/sq m
Surface Area, sq m = 100,000 sq m	
Soluble BOD, mg/L = 75 mg/L	

1. Determine the soluble BOD applied to the rotating biological contactor in grams of soluble BOD per day.

$$
\begin{aligned}
\text{BOD Applied,} \\
\text{gm/day}
\end{aligned}
= (\text{BOD, mg/}L)(\text{Flow, cu m/day})(1{,}000\ \text{kg/cu m})(1{,}000\ \text{gm/kg})
$$

$$
= \frac{75\ \text{mg}}{1{,}000{,}000\ \text{mg}} \times \frac{15{,}000\ \text{cu m}}{\text{day}} \times \frac{1{,}000\ \text{kg}}{\text{cu m}} \times \frac{1{,}000\ \text{gm}}{\text{kg}}
$$

$$
= 1{,}125{,}000\ \text{gm BOD/day}
$$

2. Calculate the organic loading in grams of soluble BOD per day per square meter of media surface.

$$
\begin{aligned}
\text{Organic BOD Loading,} \\
\text{gm BOD/day/sq m}
\end{aligned}
= \frac{\text{Soluble BOD Applied, gm BOD/day}}{\text{Surface Area of Media, sq m}}
$$

$$
= \frac{1{,}125{,}000\ \text{gm Soluble BOD/day}}{100{,}000\ \text{sq m}}
$$

$$
= 11.2\ \text{gm BOD/day/sq m}
$$

QUESTIONS

Write your answers in a notebook and then compare your answers with those on page 237.

A rotating biological contactor treats a flow of 2.0 MGD with an influent soluble BOD of 100 mg/L. The total surface area of the media is 500,000 square feet.

7.5A What is the hydraulic loading in gallons per day per square foot of media surface?

7.5B What is the organic loading in pounds of soluble BOD per day per 1,000 square feet of surface area?

7.5C If the first stage surface area is 250,000 square feet, what is the soluble BOD loading on the first stage only?

7.6 ARITHMETIC ASSIGNMENT

Turn to the Arithmetic Appendix at the back of this manual and read all of Section A.7, *VELOCITY AND FLOW RATE*.

In Section A.13, *TYPICAL WASTEWATER TREATMENT PLANT PROBLEMS (ENGLISH SYSTEM)*, read and work the problems in Section A.134, Rotating Biological Contactors.

7.7 ACKNOWLEDGMENTS

Jerry Greene, Autotrol Corporation, reviewed this section and provided helpful suggestions, photographs, drawings, and procedures. John R. Harrison and Mark Lambert also reviewed this chapter and their contributions are sincerely appreciated.

7.8 ADDITIONAL READING

1. *MOP 11*, Chapter 21, "Trickling Filters, Rotating Biological Contactors, and Combined Processes."*

2. *NEW YORK MANUAL*, Chapter 5, "Secondary Treatment."*

* Depends on edition.

7.9 METRIC CALCULATIONS

Refer to Section 7.5, "Loading Calculations," for metric calculations.

END OF LESSON ON ROTATING BIOLOGICAL CONTACTORS

Please answer the discussion and review questions next.

DISCUSSION AND REVIEW QUESTIONS

Chapter 7. ROTATING BIOLOGICAL CONTACTORS

Write the answers to these questions in your notebook. The purpose of these questions is to indicate to you how well you understand the material in the chapter.

1. Describe the rotating biological contactor (RBC) process and discuss how it works.

2. What RBC equipment would you inspect on a regular basis and how would you do it for an RBC unit?

3. What water quality indicators would you test for in the effluent from an RBC treatment plant?

4. How do the slime growths (biomass) on the plastic media look under (a) normal conditions, and (b) abnormal conditions?

5. What factors can cause a decrease in the process efficiency of an RBC unit and how can these problems be corrected?

SUGGESTED ANSWERS

Chapter 7.   ROTATING BIOLOGICAL CONTACTORS

Answers to questions on page 222.

7.0A   A rotating biological contactor treats wastewater by allowing the slime growths to develop on the surface of plastic media. The slime growths contain organisms that remove the organic materials from the wastewater.

7.0B   The RBC process is divided into four or more stages to increase the effectiveness of a given amount of media surface area. Organisms on the first-stage media are exposed to high levels of BOD and reduce the BOD at a high rate. As the BOD levels decrease from stage to stage, the rate at which the organisms can remove BOD decreases and nitrification starts.

7.0C   The purpose of a cover over the RBC unit is to protect organisms from severe changes in weather, especially freezing. Other reasons include preventing heavy rains from washing off slime and protecting operators from weather while maintaining equipment. Covers also contain odors and prevent the growth of algae on the media.

7.0D   The three types of drive units are:

1. Motor with chain drive,
2. Motor with direct shaft drive, and
3. Air drive.

Answers to questions on page 231.

7.1A   Debris, grit, and suspended solids should be removed to prevent sludge deposits forming beneath the media. These sludge deposits can reduce the effective tank volume, produce septic conditions, scrape the slimes from the media, and possibly stall the unit.

7.1B   The major items on a pre-start check include

1. Tightness of bolts and parts,
2. Lubrication of equipment,
3. Clearances for moving parts, and
4. Proper installation of safety guards.

7.1C   The main operational duties for an operator of an RBC unit include

1. Inspecting equipment,
2. Testing influent and effluent,
3. Observing the media,
4. Maintaining the equipment, and
5. Correcting any problems.

Answers to questions on page 234.

7.2A   During start-up, maintenance should be performed on an RBC unit:

1. After 8 hours of operation,
2. After 24 hours of operation,
3. After 40 hours of operation,
4. After 100 hours of operation, and
5. After 3 weeks of operation.

7.3A   Possible safety hazards to operators working around RBC units include:

1. Moving parts and equipment,
2. Electrical power,
3. Bare electrical wires and loose connections,
4. Slippery surfaces,
5. Infections in cuts and open wounds, and
6. Illnesses from waterborne diseases.

Answers to questions on page 236.

7.5A , 7.5B, and 7.5C

A rotating biological contactor treats a flow of 2.0 MGD with an influent soluble BOD of 100 mg/L. The total surface area of the media is 500,000 square feet.

Known		Unknown	
Flow, MGD	= 2.0 MGD	7.5A	Hydraulic Loading, GPD/sq ft
Soluble BOD, mg/L	= 100 mg/L	7.5B	Organic Loading, lbs BOD/day/1,000 sq ft
Surface Area, sq ft	= 500,000 sq ft		
Surface Area, sq ft (First Stage)	= 250,000 sq ft	7.5C	First Stage BOD Loading, lbs BOD/day/1,000 sq ft

7.5A   Calculate the hydraulic loading in gallons per day of wastewater per square foot of media surface.

$$\text{Hydraulic Loading, GPD/sq ft} = \frac{\text{Flow, gal/day}}{\text{Surface Area, sq ft}}$$

$$= \frac{2,000,000 \text{ gal/day}}{500,000 \text{ sq ft}}$$

$$= 4 \text{ GPD/sq ft}$$

7.5B   1. Determine the soluble BOD applied to the rotating biological contactor in pounds of soluble BOD per day.

$$\text{BOD Applied, lbs BOD/day} = (\text{Soluble BOD, mg/}L)(\text{Flow, MGD})(8.34 \text{ lbs/gal})$$

$$= (100 \text{ mg/}L)(2.0 \text{ MGD})(8.34 \text{ lbs/gal})$$

$$= 1,668 \text{ lbs Soluble BOD/day}$$

2. Calculate the organic loading in pounds of soluble BOD per day per 1,000 square feet of media surface.

$$\text{Organic BOD Loading, lbs BOD/day/1,000 sq ft} = \frac{\text{Soluble BOD Applied, lbs BOD/day}}{\text{Surface Area of Media (in 1,000 sq ft)}}$$

$$= \frac{1,668 \text{ lbs Soluble BOD/day}}{500 \cdot 1,000 \text{ sq ft}}$$

$$= 3.3 \text{ lbs BOD/day/1,000 sq ft}$$

7.5C   Calculate the organic loading on the first stage in pounds of soluble BOD per day per 1,000 square feet of first stage surface area.

$$\text{Organic BOD Loading, lbs BOD/day/1,000 sq ft} = \frac{\text{Soluble BOD Applied, lbs BOD/day}}{\text{First Stage Surface Area (in 1,000 sq ft)}}$$

$$= \frac{1,668 \text{ lbs Soluble BOD/day}}{250 \cdot 1,000 \text{ sq ft}}$$

$$= 6.6 \text{ lbs BOD/day/1,000 sq ft}$$

NOTE: Soluble BOD loading on first stage should be less than 4 lbs BOD/day/1,000 sq ft.

CHAPTER 8

ACTIVATED SLUDGE

Volume I. Package Plants and Oxidation Ditches

Volume II, Chapter 11
Operation of Conventional Activated Sludge Plants

ADVANCED WASTE TREATMENT, Chapter 2
Pure Oxygen Plants and Operational Control Options

by

John Brady

Revised by

Ross H. Gudgel

TABLE OF CONTENTS

Chapter 8. ACTIVATED SLUDGE

PACKAGE PLANTS AND OXIDATION DITCHES

LESSON 3

OBJECTIVES

Chapter 8.   ACTIVATED SLUDGE

PACKAGE PLANTS AND OXIDATION DITCHES

The activated sludge process is a very important wastewater treatment process. For this reason, the chapters on activated sludge have been divided into three parts and will be presented in three separate manuals.

I. Package Plants and Oxidation Ditches (Volume I)

II. Operation of Conventional Activated Sludge Plants (Volume II)

III. *(ADVANCED WASTE TREATMENT)* Pure Oxygen Plants and Operational Control Options

If you are the operator of a package plant or oxidation ditch, Volume I will provide you with the information you need to know to operate your plant. Volume II and *ADVANCED WASTE TREATMENT* will help you better understand the activated sludge plant or a modification. Volume I will help you understand the activated sludge process and Volume II will tell you how to operate your plant. *ADVANCED WASTE TREATMENT* will explain to you alternative means of operational control that may work very well for your plant. If you operate a pure oxygen plant, *ADVANCED WASTE TREATMENT* will tell you what you need to know to operate the pure oxygen system. All three parts contain information important to the proper operation of your plant. *ADVANCED WASTE TREATMENT* also contains information helpful to operators using the activated sludge process to treat special wastes such as industrial wastes.

The following objectives apply to the treatment plants covered in each of the three volumes. After completion of the appropriate part on activated sludge, you should be able to:

1. Explain the principles of the activated sludge process and the factors that influence and control the process,

2. Inspect a new activated sludge facility for proper installation,

3. Place a new activated sludge process into service,

4. Schedule and conduct operation and maintenance duties,

5. Collect samples, interpret lab results, and make appropriate adjustments in treatment processes,

6. Recognize factors that indicate an activated sludge process is not performing properly, identify the source of the problem, and take corrective action,

7. Conduct your duties in a safe fashion,

8. Determine aerator loadings and understand the application of different loading guidelines,

9. Keep records for an activated sludge plant,

10. Identify the common modifications of the activated sludge process,

11. Review plans and specifications for an activated sludge plant,

12. Describe each of the process stages used to treat wastewater in a sequencing batch reactor (SBR),

13. Place a new sequencing batch reactor in service,

14. Collect and analyze samples and make appropriate process adjustments during start-up and normal operation,

15. Safely operate and maintain a sequencing batch reactor, and

16. Review plans and specifications for a sequencing batch reactor.

WORDS

Chapter 8.   ACTIVATED SLUDGE

PACKAGE PLANTS AND OXIDATION DITCHES

ABSORPTION (ab-SORP-shun)   ABSORPTION

The taking in or soaking up of one substance into the body of another by molecular or chemical action (as tree roots absorb dissolved nutrients in the soil).

ACTIVATED SLUDGE (ACK-ta-VATE-ed sluj)   ACTIVATED SLUDGE

Sludge particles produced in raw or settled wastewater (primary effluent) by the growth of organisms (including zoogleal bacteria) in aeration tanks in the presence of dissolved oxygen. The term "activated" comes from the fact that the particles are teeming with bacteria, fungi, and protozoa. Activated sludge is different from primary sludge in that the sludge particles contain many living organisms which can feed on the incoming wastewater.

ACTIVATED SLUDGE (ACK-ta-VATE-ed sluj) PROCESS   ACTIVATED SLUDGE PROCESS

A biological wastewater treatment process which speeds up the decomposition of wastes in the wastewater being treated. Activated sludge is added to wastewater and the mixture (mixed liquor) is aerated and agitated. After some time in the aeration tank, the activated sludge is allowed to settle out by sedimentation and is disposed of (wasted) or reused (returned to the aeration tank) as needed. The remaining wastewater then undergoes more treatment.

ADSORPTION (add-SORP-shun)   ADSORPTION

The gathering of a gas, liquid, or dissolved substance on the surface or interface zone of another material.

AERATION (air-A-shun) LIQUOR   AERATION LIQUOR

Mixed liquor. The contents of the aeration tank including living organisms and material carried into the tank by either untreated wastewater or primary effluent.

AERATION (air-A-shun) TANK   AERATION TANK

The tank where raw or settled wastewater is mixed with return sludge and aerated. The same as aeration bay, aerator, or reactor.

AEROBES   AEROBES

Bacteria that must have molecular (dissolved) oxygen (DO) to survive. Aerobes are aerobic bacteria.

AEROBIC (AIR-O-bick) DIGESTION   AEROBIC DIGESTION

The breakdown of wastes by microorganisms in the presence of dissolved oxygen. This digestion process may be used to treat only waste activated sludge, or trickling filter sludge and primary (raw) sludge, or waste sludge from activated sludge treatment plants designed without primary settling. The sludge to be treated is placed in a large aerated tank where aerobic microorganisms decompose the organic matter in the sludge. This is an extension of the activated sludge process.

AGGLOMERATION (a-GLOM-er-A-shun)   AGGLOMERATION

The growing or coming together of small scattered particles into larger flocs or particles which settle rapidly. Also see FLOC.

AIR LIFT PUMP   AIR LIFT PUMP

A special type of pump. This device consists of a vertical riser pipe submerged in the wastewater or sludge to be pumped. Compressed air is injected into a tail piece at the bottom of the pipe. Fine air bubbles mix with the wastewater or sludge to form a mixture lighter than the surrounding water which causes the mixture to rise in the discharge pipe to the outlet. An air-lift pump works like the center stand in a percolator coffee pot.

ALIQUOT (AL-li-kwot)   ALIQUOT

Representative portion of a sample. Often an equally divided portion of a sample.

ANAEROBES   ANAEROBES

Bacteria that do not need molecular (dissolved) oxygen (DO) to survive.

ANOXIC (an-OX-ick)   ANOXIC

A condition in which atmospheric or dissolved molecular oxygen is *NOT* present in the aquatic (water) environment and nitrate is present. Oxygen deficient or lacking sufficient oxygen. The term is similar to anaerobic.

BACTERIAL (back-TEAR-e-al) CULTURE   BACTERIAL CULTURE

In the case of activated sludge, the bacterial culture refers to the group of bacteria classified as AEROBES, and FACULTATIVE organisms, which covers a wide range of organisms. Most treatment processes in the United States grow facultative organisms which use the carbonaceous (carbon compounds) BOD. Facultative organisms can live when oxygen resources are low. When "nitrification" is required, the nitrifying organisms are OBLIGATE AEROBES (require oxygen) and must have at least 0.5 mg/L of dissolved oxygen throughout the whole system to function properly.

BATCH PROCESS   BATCH PROCESS

A treatment process in which a tank or reactor is filled, the wastewater (or other solution) is treated or a chemical solution is prepared, and the tank is emptied. The tank may then be filled and the process repeated. Batch processes are also used to cleanse, stabilize or condition chemical solutions for use in industrial manufacturing and treatment processes.

BIOMASS (BUY-o-MASS)   BIOMASS

A mass or clump of organic material consisting of living organisms feeding on the wastes in wastewater, dead organisms and other debris. Also see ZOOGLEAL MASS.

BULKING (BULK-ing)   BULKING

Clouds of billowing sludge that occur throughout secondary clarifiers and sludge thickeners when the sludge does not settle properly. In the activated sludge process, bulking is usually caused by filamentous bacteria or bound water.

CATHODIC (ca-THOD-ick) PROTECTION   CATHODIC PROTECTION

An electrical system for prevention of rust, corrosion, and pitting of steel and iron surfaces in contact with water, wastewater or soil. A low-voltage current is made to flow through a liquid (water) or a soil in contact with the metal in such a manner that the external electromotive force renders the metal structure cathodic. This concentrates corrosion on auxiliary anodic parts which are deliberately allowed to corrode instead of letting the structure corrode.

COAGULATION (co-AGG-you-LAY-shun)   COAGULATION

The clumping together of very fine particles into larger particles (floc) caused by the use of chemicals (coagulants). The chemicals neutralize the electrical charges of the fine particles, allowing them to come closer and form larger clumps. This clumping together makes it easier to separate the solids from the water by settling, skimming, draining or filtering.

COMPOSITE (come-PAH-zit) (PROPORTIONAL) SAMPLE   COMPOSITE (PROPORTIONAL) SAMPLE

A composite sample is a collection of individual samples obtained at regular intervals, usually every one or two hours during a 24-hour time span. Each individual sample is combined with the others in proportion to the rate of flow when the sample was collected. The resulting mixture (composite sample) forms a representative sample and is analyzed to determine the average conditions during the sampling period.

CONING (CONE-ing)   CONING

Development of a cone-shaped flow of liquid, like a whirlpool, through sludge. This can occur in a sludge hopper during sludge withdrawal when the sludge becomes too thick. Part of the sludge remains in place while liquid rather than sludge flows out of the hopper. Also called coring.

CONTACT STABILIZATION   CONTACT STABILIZATION

Contact stabilization is a modification of the conventional activated sludge process. In contact stabilization, two aeration tanks are used. One tank is for separate reaeration of the return sludge for at least four hours before it is permitted to flow into the other aeration tank to be mixed with the primary effluent requiring treatment. The process may also occur in one long tank.

DENITRIFICATION (dee-NYE-truh-fuh-KAY-shun) (ACTIVATED SLUDGE)   DENITRIFICATION

(1) The anoxic biological reduction of nitrate nitrogen to nitrogen gas.

(2) The removal of some nitrogen from a system.

(3) An anoxic process that occurs when nitrite or nitrate ions are reduced to nitrogen gas and nitrogen bubbles are formed as a result of this process. The bubbles attach to the biological floc in the activated sludge process and float the floc to the surface of the secondary clarifiers. This condition is often the cause of rising sludge observed in secondary clarifiers or gravity thickeners. Also see NITRIFICATION.

DIFFUSED-AIR AERATION   DIFFUSED-AIR AERATION

A diffused-air activated sludge plant takes air, compresses it, and then discharges the air below the water surface of the aerator through some type of air diffusion device.

DIFFUSER   DIFFUSER

A device (porous plate, tube, bag) used to break the air stream from the blower system into fine bubbles in an aeration tank or reactor.

DISSOLVED OXYGEN   DISSOLVED OXYGEN

Molecular (atmospheric) oxygen dissolved in water or wastewater, usually abbreviated DO.

ENDOGENOUS (en-DODGE-en-us) RESPIRATION   ENDOGENOUS RESPIRATION

A situation where living organisms oxidize some of their own cellular mass instead of new organic matter they adsorb or absorb from their environment.

F/M RATIO   F/M RATIO

See FOOD/MICROORGANISM RATIO.

FACULTATIVE (FACK-ul-TAY-tive)   FACULTATIVE

Facultative bacteria can use either dissolved molecular oxygen or oxygen obtained from food materials such as sulfate or nitrate ions. In other words, facultative bacteria can live under aerobic, anoxic, or anaerobic conditions.

FILAMENTOUS (FILL-a-MEN-tuss) ORGANISMS   FILAMENTOUS ORGANISMS

Organisms that grow in a thread or filamentous form. Common types are *Thiothrix* and *Actinomycetes.* A common cause of sludge bulking in the activated sludge process.

FLIGHTS   FLIGHTS

Scraper boards, made from redwood or other rot-resistant woods or plastic, used to collect and move settled sludge or floating scum.

FLOC   FLOC

Clumps of bacteria and particles or coagulants and impurities that have come together and formed a cluster. Found in aeration tanks, secondary clarifiers and chemical precipitation processes.

FOOD/MICROORGANISM RATIO   FOOD/MICROORGANISM RATIO

Food to microorganism ratio. A measure of food provided to bacteria in an aeration tank.

$$\frac{\text{Food}}{\text{Microorganisms}} = \frac{\text{BOD, lbs/day}}{\text{MLVSS, lbs}}$$

$$= \frac{\text{Flow, MGD} \times \text{BOD, mg/}L \times 8.34 \text{ lbs/gal}}{\text{Volume, MG} \times \text{MLVSS, mg/}L \times 8.34 \text{ lbs/gal}}$$

or by calculator math system

$$= \text{Flow, MGD} \times \text{BOD, mg/}L \div \text{Volume, MG} \div \text{MLVSS, mg/}L$$

$$\text{or metric} \quad = \frac{\text{BOD, kg/day}}{\text{MLVSS, kg}}$$

$$= \frac{\text{Flow, M}L\text{/day} \times \text{BOD, mg/}L \times 1 \text{ kg/M mg}}{\text{Volume, M}L \times \text{MLVSS, mg/}L \times 1 \text{ kg/M mg}}$$

Commonly abbreviated F/M Ratio.

HEADER   HEADER

A large pipe to which the ends of a series of smaller pipes are connected. Also called a MANIFOLD.

LINEAL (LIN-e-al)   LINEAL

The length in one direction of a line. For example, a board 12 feet long has 12 lineal feet in its length.

MANIFOLD   MANIFOLD

A large pipe to which the ends of a series of smaller pipes are connected. Also called a HEADER.

MEAN CELL RESIDENCE TIME (MCRT)   MEAN CELL RESIDENCE TIME (MCRT)

An expression of the average time that a microorganism will spend in the activated sludge process.

$$\text{MCRT, days} = \frac{\text{Total Suspended Solids in Activated Sludge Process, lbs}}{\text{Total Suspended Solids Removed From Process, lbs/day}}$$

$$\text{or metric} = \frac{\text{Total Suspended Solids in Activated Sludge Process, kg}}{\text{Total Suspended Solids Removed From Process, kg/day}}$$

NOTE: Operators at different plants calculate the Total Suspended Solids (TSS) in the Activated Sludge Process, lbs (kg), by three different methods.

1. TSS in the Aeration Basin or Reactor Zone, lbs (kg),
2. TSS in Aeration Basin and Secondary Clarifier, lbs (kg), or
3. TSS in Aeration Basin and Secondary Clarifier Sludge Blanket, lbs (kg).

These three different methods make it difficult to compare MCRTs in days among different plants unless everyone uses the same method.

MECHANICAL AERATION   MECHANICAL AERATION

The use of machinery to mix air and water so that oxygen can be absorbed into the water. Some examples are: paddle wheels, mixers, or rotating brushes to agitate the surface of an aeration tank; pumps to create fountains; and pumps to discharge water down a series of steps forming falls or cascades.

MICROORGANISMS (MY-crow-OR-gan-IS-zums)   MICROORGANISMS

Very small organisms that can be seen only through a microscope. Some microorganisms use the wastes in wastewater for food and thus remove or alter much of the undesirable matter.

MIXED LIQUOR   MIXED LIQUOR

When the activated sludge in an aeration tank is mixed with primary effluent or the raw wastewater and return sludge, this mixture is then referred to as mixed liquor as long as it is in the aeration tank. Mixed liquor also may refer to the contents of mixed aerobic or anaerobic digesters.

MIXED LIQUOR SUSPENDED SOLIDS (MLSS)   MIXED LIQUOR SUSPENDED SOLIDS (MLSS)

Suspended solids in the mixed liquor of an aeration tank.

MIXED LIQUOR VOLATILE SUSPENDED SOLIDS
  (MLVSS)

MIXED LIQUOR VOLATILE SUSPENDED SOLIDS
(MLVSS)

The organic or volatile suspended solids in the mixed liquor of an aeration tank. This volatile portion is used as a measure or indication of the microorganisms present.

NITRIFICATION (NYE-truh-fuh-KAY-shun)   NITRIFICATION

An aerobic process in which bacteria change the ammonia and organic nitrogen in wastewater into oxidized nitrogen (usually nitrate). The second-stage BOD is sometimes referred to as the "nitrogenous BOD" (first-stage BOD is called the "carbonaceous BOD"). Also see DENITRIFICATION.

OBLIGATE AEROBES   OBLIGATE AEROBES

Bacteria that must have atmospheric or dissolved molecular oxygen to live and reproduce.

OXIDATION (ox-uh-DAY-shun)   OXIDATION

Oxidation is the addition of oxygen, removal of hydrogen, or the removal of electrons from an element or compound. In wastewater treatment, organic matter is oxidized to more stable substances. The opposite of REDUCTION.

POLYELECTROLYTE (POLY-ee-LECK-tro-lite)   POLYELECTROLYTE

A high-molecular-weight substance that is formed by either a natural or synthetic process. Natural polyelectrolytes may be of biological origin or derived from starch products, cellulose derivatives, and alignates. Synthetic polyelectrolytes consist of simple substances that have been made into complex, high-molecular-weight substances. Often called a POLYMER.

POLYMER (POLY-mer)   POLYMER

A long chain molecule formed by the union of many monomers (molecules of lower molecular weight). Polymers are used with other chemical coagulants to aid in binding small suspended particles to larger chemical flocs for their removal from water.

PROTOZOA (pro-toe-ZOE-ah)   PROTOZOA

A group of motile microscopic organisms (usually single-celled and aerobic) that sometimes cluster into colonies and generally consume bacteria as an energy source.

REDUCTION (re-DUCK-shun) REDUCTION

Reduction is the addition of hydrogen, removal of oxygen, or the addition of electrons to an element or compound. Under anaerobic conditions (no dissolved oxygen present), sulfur compounds are reduced to odor-producing hydrogen sulfide (H_2S) and other compounds. The opposite of OXIDATION.

RISING SLUDGE RISING SLUDGE

Rising sludge occurs in the secondary clarifiers of activated sludge plants when the sludge settles to the bottom of the clarifier, is compacted, and then starts to rise to the surface, usually as a result of denitrification, or anaerobic biological activity that produces carbon dioxide and/or methane.

SECCHI (SECK-key) DISC SECCHI DISC

A flat, white disc lowered into the water by a rope until it is just barely visible. At this point, the depth of the disc from the water surface is the recorded Secchi disc transparency.

SEIZING SEIZING

Seizing occurs when an engine overheats and a component expands to the point where the engine will not run. Also called freezing.

SEPTIC (SEP-tick) SEPTIC

A condition produced by anaerobic bacteria. If severe, the sludge produces hydrogen sulfide, turns black, gives off foul odors, contains little or no dissolved oxygen, and the wastewater has a high oxygen demand.

SHOCK LOAD (ACTIVATED SLUDGE) SHOCK LOAD

The arrival at a plant of a waste which is toxic to organisms in sufficient quantity or strength to cause operating problems. Possible problems include odors and bulking sludge which will result in a high loss of solids from the secondary clarifiers into the plant effluent and a biological process upset that may require several days to a week to recover. Organic or hydraulic overloads also can cause a shock load.

SLUDGE AGE SLUDGE AGE

A measure of the length of time a particle of suspended solids has been retained in the activated sludge process.

$$\text{Sludge Age, days} = \frac{\text{Suspended Solids Under Aeration, lbs or kg}}{\text{Suspended Solids Added, lbs/day or kg/day}}$$

SLUDGE DENSITY INDEX (SDI) SLUDGE DENSITY INDEX (SDI)

This calculation is used in a way similar to the Sludge Volume Index (SVI) to indicate the settleability of a sludge in a secondary clarifier or effluent. The weight in grams of one milliliter of sludge after settling for 30 minutes. SDI = 100/SVI. Also see SLUDGE VOLUME INDEX (SVI).

SLUDGE VOLUME INDEX (SVI) SLUDGE VOLUME INDEX (SVI)

This is a calculation which indicates the tendency of activated sludge solids (aerated solids) to thicken or to become concentrated during the sedimentation/thickening process. SVI is calculated in the following manner: (1) allow a mixed liquor sample from the aeration basin to settle for 30 minutes; (2) determine the suspended solids concentration for a sample of the same mixed liquor; (3) calculate SVI by dividing the measured (or observed) wet volume (mL/L) of the settled sludge by the dry weight concentration of MLSS in grams/L.

$$\text{SVI, m}L\text{/gm} = \frac{\text{Settled Sludge Volume/Sample Volume, m}L\text{/}L}{\text{Suspended Solids Concentration, mg/}L} \times \frac{1{,}000 \text{ mg}}{\text{gram}}$$

SOLIDS CONCENTRATION SOLIDS CONCENTRATION

The solids in the aeration tank which carry microorganisms that feed on wastewater. Expressed as milligrams per liter of Mixed Liquor Volatile Suspended Solids (MLVSS, mg/L).

STABILIZED WASTE STABILIZED WASTE

A waste that has been treated or decomposed to the extent that, if discharged or released, its rate and state of decomposition would be such that the waste would not cause a nuisance or odors in the receiving water.

STEP-FEED AERATION STEP-FEED AERATION

Step-feed aeration is a modification of the conventional activated sludge process. In step aeration, primary effluent enters the aeration tank at several points along the length of the tank, rather than all of the primary effluent entering at the beginning or head of the tank and flowing through the entire tank in a plug flow mode.

SUPERNATANT (sue-per-NAY-tent)   SUPERNATANT

Liquid removed from settled sludge. Supernatant commonly refers to the liquid between the sludge on the bottom and the scum on the surface of an anaerobic digester. This liquid is usually returned to the influent wet well or to the primary clarifier.

TOC (pronounce as separate letters)   TOC

Total **O**rganic **C**arbon. TOC measures the amount of organic carbon in water.

TURBIDITY METER   TURBIDITY METER

An instrument for measuring and comparing the turbidity of liquids by passing light through them and determining how much light is reflected by the particles in the liquid. The normal measuring range is 0 to 100 and is expressed as Nephelometric Turbidity Units (NTUs).

VOLUTE (vol-LOOT)   VOLUTE

The spiral-shaped casing which surrounds a pump, blower, or turbine impeller and collects the liquid or gas discharged by the impeller.

ZOOGLEAL (ZOE-glee-al) MASS   ZOOGLEAL MASS

Jelly-like masses of bacteria found in both the trickling filter and activated sludge processes. These masses may be formed for or function as the protection against predators and for storage of food supplies. Also see BIOMASS.

CHAPTER 8.   ACTIVATED SLUDGE

PACKAGE PLANTS AND OXIDATION DITCHES

(Lesson 1 of 3 Lessons)

8.0   THE ACTIVATED SLUDGE PROCESS

8.00   Wastewater Treatment by Activated Sludge

When wastewater enters an activated sludge plant, the preliminary treatment processes (Chapter 4) remove the coarse or heavy solids (grit) and other debris, such as roots, rags, and boards. Primary clarifiers (Chapter 5) remove much of the floatable and settleable material. Normally the activated sludge process treats settled wastewater, but in some plants the raw wastewater flows from the preliminary treatment processes directly to the activated sludge process.

8.01   Definitions

ACTIVATED SLUDGE (Figure 8.1). Activated sludge consists of sludge particles produced in raw or settled wastewater (primary effluent) by the growth of organisms (including zoogleal bacteria) in aeration tanks in the presence of dissolved oxygen. The term "activated" comes from the fact that the particles are teeming with bacteria, fungi, and protozoa.

ACTIVATED SLUDGE PROCESS (Figure 8.2). The activated sludge process is a biological wastewater treatment process that uses *MICROORGANISMS*[1] to speed up decomposition of wastes. When activated sludge is added to wastewater, the microorganisms feed and grow on waste particles in the wastewater. As the organisms grow and reproduce, more and more waste is removed, leaving the wastewater partially cleaned. To function efficiently, the mass of organisms *(SOLIDS CONCENTRATION*[2]*)* needs a steady balance of food *(FOOD/MICROORGANISM RATIO*[3]*)* and oxygen.

8.02   Process Description

Secondary treatment in the form of the activated sludge process (Figures 8.3 and 8.4) is aimed at *OXIDATION*[4] and removal of soluble or finely divided suspended materials that were not removed by previous treatment. Aerobic organisms do this in a few hours as wastewater flows through an aeration tank. The organisms *STABILIZE*[5] soluble or finely suspended solids by partial oxidation forming carbon dioxide, water, and sulfate and nitrate compounds. The remaining solids are changed to a form that can be settled and removed as sludge during sedimentation.

After the aeration period, the wastewater is routed to a secondary settling tank for a liquid-organism (water-solids) separation. Settled organisms in the final clarifier are in a deteriorating condition due to lack of oxygen and food and should be returned to the aeration tank as quickly as possible. The remaining clarifier effluent is usually chlorinated and discharged from the plant.

Conversion of dissolved and suspended material to settleable solids is the main objective of high-rate activated sludge processes, while low-rate processes stress oxidation. The oxidation may be by chemical or biological processes. In the acti-

---

[1] *Microorganisms (MY-crow-OR-gan-IS-zums).   Very small organisms that can be seen only through a microscope. Some microorganisms use the wastes in wastewater for food and thus remove or alter much of the undesirable matter.*

[2] *Solids Concentration.   The solids in the aeration tank which carry microorganisms that feed on wastewater. Expressed as milligrams per liter of Mixed Liquor Volatile Suspended Solids (MLVSS, mg/L).*

[3] *Food/Microorganism Ratio.   Food to microorganism ratio. A measure of food provided to bacteria in an aeration tank.*

$$\frac{Food}{Microorganisms} = \frac{BOD, \ lbs/day}{MLVSS, \ lbs}$$

$$= \frac{Flow, \ MGD \times BOD, \ mg/L \times 8.34 \ lbs/gal}{Volume, \ MG \times MLVSS, \ mg/L \times 8.34 \ lbs/gal}$$

$$or \ metric \quad = \frac{BOD, \ kg/day}{MLVSS, \ kg}$$

$$= \frac{Flow, \ ML/day \times BOD, \ mg/L \times 1 \ kg/M \ mg}{Volume, \ ML \times MLVSS, \ mg/L \times 1 \ kg/M \ mg}$$

Commonly abbreviated F/M Ratio.

[4] *Oxidation (ox-uh-DAY-shun).   Oxidation is the addition of oxygen, removal of hydrogen, or the removal of electrons from an element or compound. In wastewater treatment, organic matter is oxidized to more stable substances. The opposite of reduction.*

[5] *Stabilized Waste.   A waste that has been treated or decomposed to the extent that, if discharged or released, its rate and state of decomposition would be such that the waste would not cause a nuisance or odors in the receiving water.*

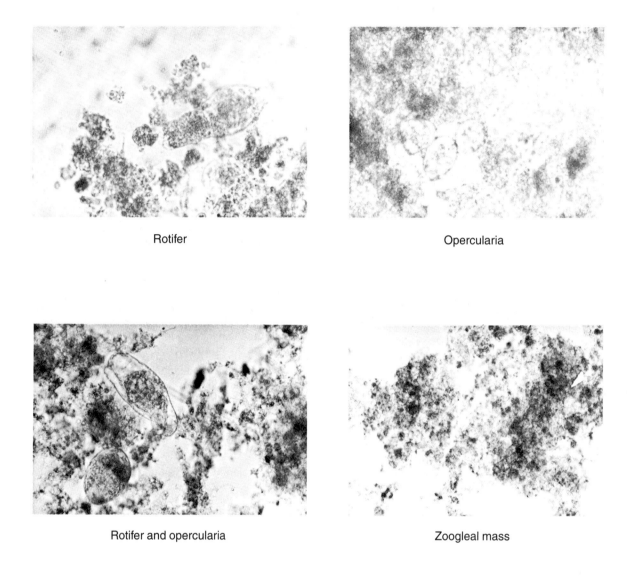

Rotifer

Opercularia

Rotifer and opercularia

Zoogleal mass

Fig. 8.1   Microorganisms in activated sludge

For information on the use of microorganisms to operate the activated sludge process, see Chapter 2, "Activated Sludge," in *ADVANCED WASTE TREATMENT.* Also see Section 11.10, "Microbiology for Activated Sludge," by Paul V. Bohlier in Chapter 11 of *OPERATION OF WASTEWATER TREATMENT PLANTS,* Volume II.

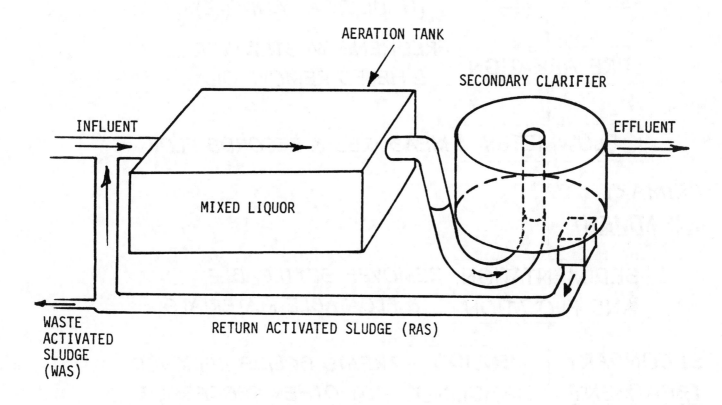

Fig. 8.2   Activated sludge process

TREATMENT PROCESS   FUNCTION

PRELIMINARY TREATMENT

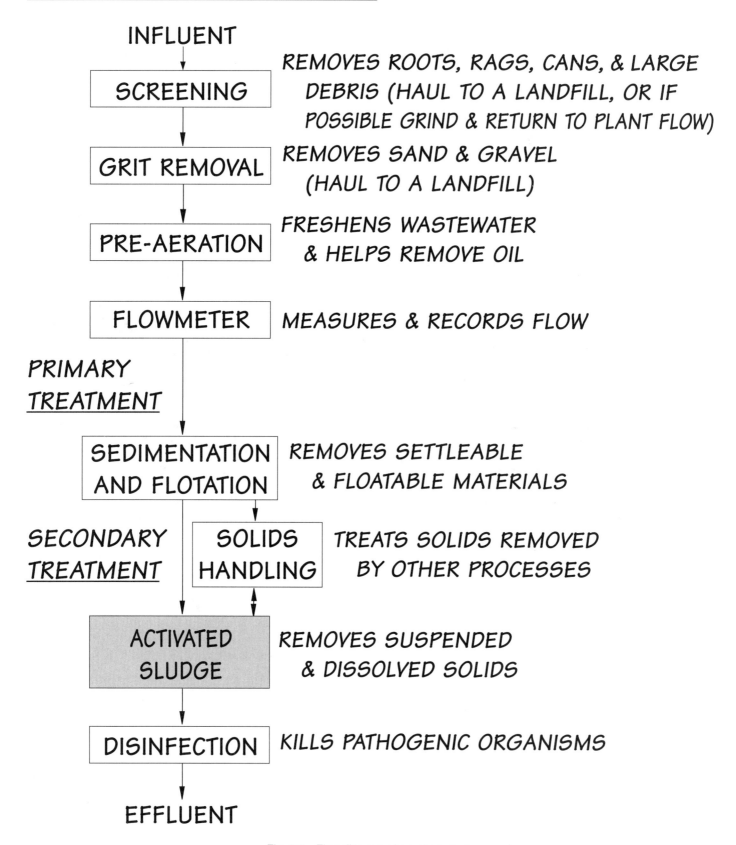

INFLUENT

SCREENING — REMOVES ROOTS, RAGS, CANS, & LARGE DEBRIS (HAUL TO A LANDFILL, OR IF POSSIBLE GRIND & RETURN TO PLANT FLOW)

GRIT REMOVAL — REMOVES SAND & GRAVEL (HAUL TO A LANDFILL)

PRE-AERATION — FRESHENS WASTEWATER & HELPS REMOVE OIL

FLOWMETER — MEASURES & RECORDS FLOW

PRIMARY TREATMENT

SEDIMENTATION AND FLOTATION — REMOVES SETTLEABLE & FLOATABLE MATERIALS

SECONDARY TREATMENT

SOLIDS HANDLING — TREATS SOLIDS REMOVED BY OTHER PROCESSES

ACTIVATED SLUDGE — REMOVES SUSPENDED & DISSOLVED SOLIDS

DISINFECTION — KILLS PATHOGENIC ORGANISMS

EFFLUENT

Fig. 8.3   Flow diagram of a typical plant

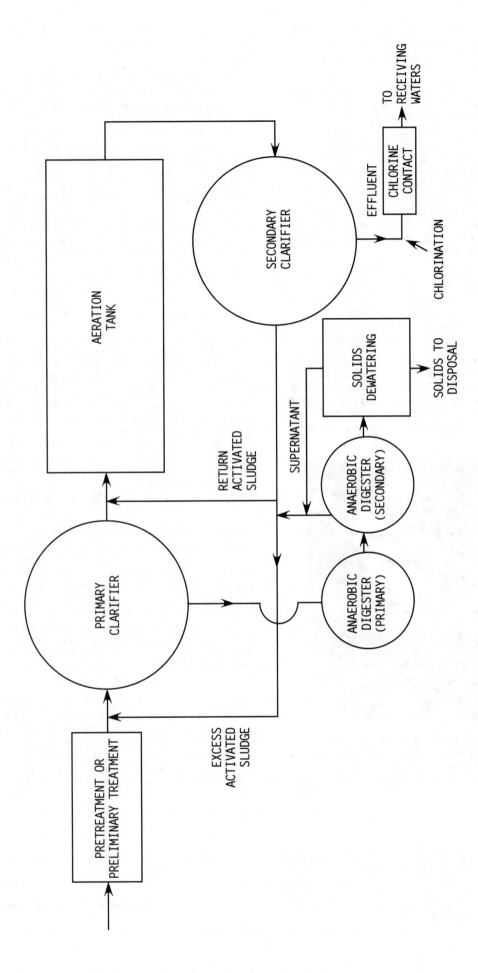

Fig. 8.4   Plan layout of a typical activated sludge plant

vated sludge process, the biochemical oxidation carried out by living organisms is stressed. The same organisms also are effective in conversion of substances to settleable solids if the plant is operated properly.

When wastewater enters the aeration tanks, it is mixed with the activated sludge to form a mixture of sludge, carrier water, and influent solids. These solids come mainly from the discharges from homes, factories, and businesses. The activated sludge that is added contains many different types of helpful living organisms that were grown during previous contact with wastewater. These organisms are the workers in the treatment process. They use the incoming wastes for food and as a source of energy for their life processes and for the reproduction of more organisms. These organisms will use more food contained in the wastewater in treating the wastes. The activated sludge also forms a lacy network or floc mass that entraps many materials not used as food.

Some organisms (workers) will require a long time to use the available food in the wastewater at a given waste concentration. Many organisms will compete with each other in the use of available food (waste) to shorten the time factor and increase the portion of waste stabilized. The ratio of food to organisms is a primary control in the activated sludge process.

Organisms tend to increase with waste (food) load and time spent in the aeration tank. Under favorable conditions the operator will remove (sludge wasting) the excess organisms to maintain the required number of workers for effective waste treatment. Therefore, removal of organisms from the treatment process (sludge wasting) is a very important control technique.

Oxygen, usually supplied from air, is needed by the living organisms as they oxidize wastes to obtain energy for growth. Insufficient oxygen will slow down aerobic organisms, make *FACULTATIVE* [6] organisms work less efficiently, and favor pro-

duction of foul-smelling intermediate products of decomposition and incomplete reactions.

An increase in organisms in an aeration tank will require greater amounts of oxygen. More food in the influent encourages more organism activity and more oxidation; consequently, more oxygen is required in the aeration tank. An excess of oxygen is required for complete waste stabilization. Therefore, the dissolved oxygen (DO) content in the aeration tank is an essential control test. Some minimum level of oxygen must be maintained to favor the desired type of organism activity to achieve the necessary treatment efficiency. If the DO in the aeration tank is too low, *FILAMENTOUS ORGANISMS* [7] will thrive and the sludge *FLOC* [8] will not settle in the secondary clarifier. Also, if the DO is too high, pinpoint floc will develop and not be removed in the secondary clarifier. Therefore, the proper DO level must be maintained so solids will settle properly and the plant effluent will be clear.

Flows must be distributed evenly among two or more similar treatment units. If your plant is equipped with a splitter box or a series of boxes, it will be necessary to periodically check and estimate whether the flow is being split as intended.

Activated sludge solids concentrations in the aerator and the secondary clarifier should be determined by the operator for process control purposes. Solids are in a deteriorating condition as long as they remain in the secondary clarifier. Depth of sludge blanket in the secondary clarifier and concentrations of solids in the aerator are very important for successful wastewater treatment. Centrifuge tests will give a quick estimate of solids concentrations and locations in the units. Precise solids tests should be made periodically for comparison with centrifuge solids tests. Before any changes are made in the mode of operation, precise solids measurements should be obtained. Settleability tests show the degree and volume of solids settling that may be obtained in a secondary clarifier; however, visual plant checks show what is actually happening.

Primary clarifiers are designed to remove material that settles to the bottom or floats to the top. Activated sludge helps this process along by collecting and *AGGLOMERATING* [9] the tiny particles in the primary effluent or raw wastewater so that they will settle better. If for some reason the organisms fail to make this change in the soluble solids, then the secondary clarifier effluent quality will not be satisfactory. For the activated sludge process to work properly, the operator must control the number of organisms, the dissolved oxygen in the aeration tanks, and the treatment time. When these factors are under proper control, the organisms will convert soluble solids and agglomerate the fine particles into a floc mass.

A floc mass is made up of millions of organisms (10^{12} to 10^{18}/100 mL in a good activated sludge), including bacteria, fungi, yeast, protozoa, and worms. When a floc mass is returned to the aerator from the final clarifier, the organisms grow as a result of taking food from the inflowing wastewater. The surface of the floc mass is irregular and promotes the transfer of wastewater pollutants into the solids by means of

---

[6] *Facultative (FACK-ul-TAY-tive).   Facultative bacteria can use either dissolved molecular oxygen or oxygen obtained from food materials such as sulfate or nitrate ions. In other words, facultative bacteria can live under aerobic, anoxic, or anaerobic conditions.*

[7] *Filamentous (FILL-a-MEN-tuss) Organisms.   Organisms that grow in a thread or filamentous form. Common types are Thiothrix and Actinomycetes. A common cause of sludge bulking in the activated sludge process.*

[8] *Floc.   Clumps of bacteria and particles or coagulants and impurities that have come together and formed a cluster. Found in aeration tanks, secondary clarifiers and chemical precipitation processes.*

[9] *Agglomeration (a-GLOM-er-A-shun).   The growing or coming together of small scattered particles into larger flocs or particles which settle rapidly. Also see FLOC.*

mechanical entrapment, absorption, adsorption, or adhesion. Many substances not used as food also are transferred to the floc mass, thus improving the quality of the plant effluent.

Material taken into the floc mass is partially oxidized to form cell mass and oxidation products. Ash or inorganic material (silt and sand) taken in by the floc mass increase the density of the mass. Mixing the contents of the aerator causes the floc masses to bump into each other and form larger clumps. Eventually these masses become heavy enough to settle to the bottom of the secondary clarifier where they can be removed easily. This sludge now contains most of the organisms and waste material that had been mixed in the wastewater.

The next step in the activated sludge process is removal of sludge from the secondary clarifier. Some of the material is converted and released to the atmosphere in the form of stripped gases (carbon dioxide or volatile gases not converted and released in the aeration tank). That leaves water and sludge solids. A certain amount of the solids (waste activated sludge) will be returned to the aerator to treat incoming wastewater. The operator must pump these solids to the aerator. The rest of the waste activated sludge must be removed and disposed of so that it does not continue in the plant flow. After the sludge solids have been removed from the final clarifier, the treated wastewater moves to advanced waste treatment processes and/or the disinfection process.

The successful operation of an activated sludge plant requires the operator to be aware of the many factors influencing the process and to check them repeatedly. To keep the organisms working in the activated sludge, you *MUST* provide a suitable environment. High concentrations of acids, bases, and other toxic substances are undesirable and may kill the working organisms. Uneven flows of wastewater may cause overfeeding, starvation, and other problems that upset the activated sludge process. Failure to supply enough oxygen can cause an unfavorable environment which results in decreased organism activity.

While successful operation of an activated sludge plant involves an understanding of many factors, actual control of the process as outlined in this section is relatively simple. *CONTROL CONSISTS OF MAINTAINING THE PROPER SOLIDS (FLOC MASS) CONCENTRATION IN THE AERATOR FOR THE WASTE (FOOD) INFLOW BY ADJUSTING THE WASTE SLUDGE PUMPING RATE AND REGULATING THE OXYGEN SUPPLY TO MAINTAIN A SATISFACTORY LEVEL OF DISSOLVED OXYGEN IN THE PROCESS.*

QUESTIONS

Write your answers in a notebook and then compare your answers with those on page 284.

8.0A   What is the purpose of the activated sludge process in treating wastewater?

8.0B   What is a stabilized waste?

8.0C   Why is air added to the aeration tank in the activated sludge process?

8.0D   What happens to the air requirement in the aeration tank when the strength (BOD) of the incoming wastewater increases?

8.0E   What factors could cause an unsuitable environment for the activated sludge process in an aeration tank?

8.1   REQUIREMENTS FOR CONTROL

Effective control of the activated sludge process depends on the operator's ability to interpret and adjust several interrelated factors. Some of these factors are:

1. Effluent quality requirements.

2. Wastewater flow, concentration, and characteristics of the wastewater received.

3. Amount of activated sludge (containing the working organisms) to be maintained in the process relative to inflow.

4. Amount of oxygen required to stabilize wastewater oxygen demands and to maintain a satisfactory level of dissolved oxygen to meet organism requirements.

5. Equal division of plant flow and waste load between duplicate treatment units (two or more clarifiers or aeration tanks).

6. Transfer of the pollutional material (food) from the wastewater to the floc mass (solids or workers) and separation of the solids from the treated wastewater.

7. Effective control and disposal of in-plant residues (solids, scums, and supernatants) to accomplish ultimate disposal in a nonpolluting manner.

8. Provisions for maintaining a suitable environment for the work force of living organisms treating the wastes to keep them healthy and happy.

Effluent quality requirements may be stated by your regulatory agency in terms of percentage removal of wastes. Current regulations frequently specify allowable quantities of wastes that may be discharged. These quantities are based upon flow and concentrations of significant items such as solids, oxygen demand, coliform bacteria, nitrogen, and oil as specified by your regulatory agencies in your NPDES permit.

The effluent quality requirements in your NPDES permit usually determine what kind of activated sludge operation you can use and how tightly you must control the process. For example, if an effluent containing 50 mg/*L* of suspended solids and BOD (refers to five-day BOD) is satisfactory, a high-rate activated sludge process will probably meet your needs. If the limit is 10 mg/*L*, the high-rate process would not be suitable. If a high degree of treatment is required, very close process control and additional treatment after the activated sludge process may be needed. Today secondary treatment plants are expected to remove 85 percent of the BOD and provide an effluent with a 30-day average BOD of less than 30 mg/*L*.

The treatment plant operator has little control over the make-up or amount of influent coming into the treatment plant. However, municipal ordinances may limit or forbid dumping substances into the collection system if they could seriously damage your treatment facilities or create safety hazards. Even with these laws to protect the collection system, some type of inspection program may be necessary (Chapter 5, "Industrial Wastewaters," in Volume I of the *INDUSTRIAL WASTE TREATMENT* manual). You may have to work out some additional plans for disposal, pretreatment, or controlled discharge to be sure that substances harmful to your treatment plant are diluted before they enter the plant.

QUESTIONS

Write your answers in a notebook and then compare your answers with those on pages 284 and 285.

8.1A   What two different ways may effluent quality requirements be stated by regulatory agencies?

8.1B   How might harmful industrial waste discharges be regulated to protect an activated sludge process?

Please answer the discussion and review questions next.

DISCUSSION AND REVIEW QUESTIONS

Chapter 8.   ACTIVATED SLUDGE

(Lesson 1 of 3 Lessons)

At the end of each lesson in this chapter you will find some discussion and review questions. The purpose of these questions is to indicate to you how well you understand the material in the lesson. Write the answers to these questions in your notebook before continuing.

1. Some activated sludge plants do not have a primary clarifier. True or False?

2. Define activated sludge.

3. Sketch the activated sludge process.

4. Why should activated sludge in the final clarifier be returned to the aeration tank as soon as possible?

5. Define facultative bacteria.

6. How can the operator control the activated sludge process?

CHAPTER 8.   ACTIVATED SLUDGE

PACKAGE PLANTS AND OXIDATION DITCHES

(Lesson 2 of 3 Lessons)

8.2   PACKAGE PLANTS (EXTENDED AERATION)

8.20   Purpose of Package Plants

8.200   Use of Package Plants

You may be assigned to operate a small extended aeration plant (Figure 8.5). You may not have laboratory facilities and may not be supplied materials for simple tests, such as dissolved oxygen (DO), pH, or settleable solids. Fortunately, this type of plant is usually under a light load and/or sized for a long retention of the solids (extended aeration), and the *SLUDGE AGE* [10] may be ten to twenty days (lower ages are associated with higher temperatures). The operation of these plants is similar to the operation of any other activated sludge plant. A high-quality effluent requires attention, understanding, and good plant operation.

This type of plant comes in many sizes, but basically there are just two compartments or tanks made from one large tank. The larger compartment is used for aeration, and the smaller one for clarification and settling. Usually provisions are not made for a primary settling compartment. The plant may be mechanically aerated or a small air compressor may provide air through diffusers. The settling tank is usually a double-hopper tank equipped with a pump or *AIR LIFT PUMP* [11] system for the return of sludge from the hoppers of the settling tank back to the aeration tank. Some plants have a third compartment that is used for *AEROBIC DIGESTION* [12] (Figure 8.6).

8.201   Types of Package Plant Treatment Processes
(Figure 8.7)

The most common types of treatment processes are the extended aeration type, the contact stabilization type, and the complete mix type. These processes are essentially modifica-

tions of the conventional activated sludge process and describe the structural arrangement of the aeration tank as well as the various arrangements of process streams that are used to provide process flexibility. Realistically, almost all package plants are of the extended aeration type. They have long solids retention times, high mixed liquor suspended solids, and low food/microorganism ratios.

1. Extended Aeration

    Extended aeration is similar to a conventional activated sludge process except that the bugs (organisms) are retained in the aeration tank longer and do not get as much food. The bugs get less food because there are more of them to feed. Mixed liquor suspended solids concentrations are from 2,000 to 6,000 mg/*L*. In addition to the bugs consuming the incoming food, they also consume any stored food in the dead bugs. The new products are carbon dioxide, water, and a biologically inert residue. Extended aeration does not produce as much waste sludge as other processes; however, wasting still is necessary to maintain proper control of the process.

2. Contact Stabilization

    Contact stabilization is similar to conventional activated sludge except that the capture of the waste material and the digestion of that material by the bugs (organisms) is done in different aeration tanks. The bugs can "adsorb" the waste material on the cell wall in only fifteen to thirty minutes, but it takes several hours to "absorb" the material through the cell wall.

    In conventional activated sludge, adsorption and absorption are done in one tank; therefore, the wastewater has to remain there for a longer time. In both cases, the bugs flow

---

[10] *Sludge Age.   A measure of the length of time a particle of suspended solids has been retained in the activated sludge process.*

$$\text{Sludge Age, days} = \frac{\text{Suspended Solids Under Aeration, lbs or kg}}{\text{Suspended Solids Added, lbs/day or kg/day}}$$

[11] *Air Lift Pump.   A special type of pump. This device consists of a vertical riser pipe submerged in the wastewater or sludge to be pumped. Compressed air is injected into a tail piece at the bottom of the pipe. Fine air bubbles mix with the wastewater or sludge to form a mixture lighter than the surrounding water which causes the mixture to rise in the discharge pipe to the outlet. An air-lift pump works like the center stand in a percolator coffee pot.*

[12] *Aerobic (AIR-O-bick) Digestion.   The breakdown of wastes by microorganisms in the presence of dissolved oxygen. This digestion process may be used to treat only waste activated sludge, or trickling filter sludge and primary (raw) sludge, or waste sludge from activated sludge treatment plants designed without primary settling. The sludge to be treated is placed in a large aerated tank where aerobic microorganisms decompose the organic matter in the sludge. This is an extension of the activated sludge process.*

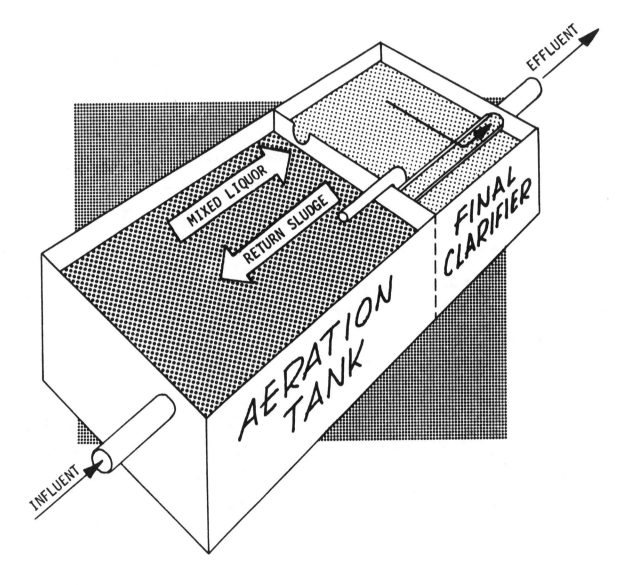

Fig. 8.5   Package plant (two compartments)

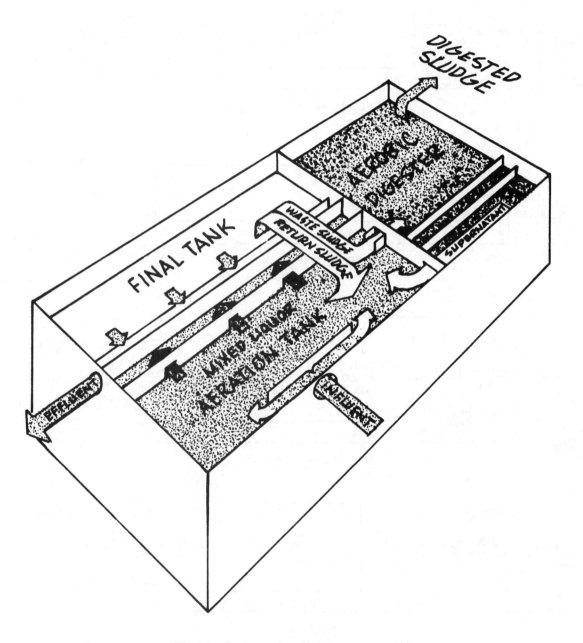

Fig. 8.6   Package plant (three compartments)

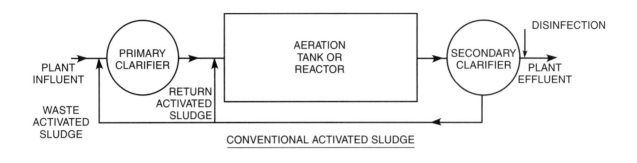

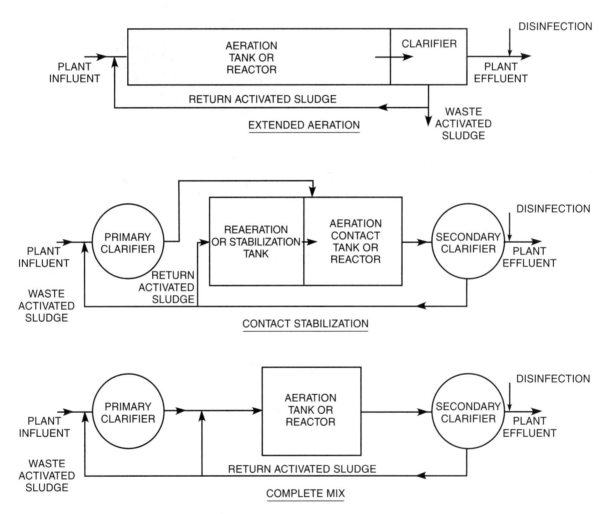

NOTE: ALL FOUR PLANTS HAVE AERATION HEADERS RUNNING THE <u>LENGTH</u> OF THE AERATION TANK.
COMPLETE MIX ALSO HAS AERATION HEADERS RUNNING <u>ACROSS</u> THE AERATION TANK
AT REGULAR INTERVALS.

Fig. 8.7 Types of package aeration plants

to the clarifier to be separated from the wastewater, but in contact stabilization the settled bugs still have to digest their food. Another aeration tank called a stabilization or re-aeration tank is provided separately for this step. Here the bugs digest the food and then are returned hungry to the original aeration tank (contact tank) ready to eat more food. The mixed liquor suspended solids (MLSS) concentration of the contact tank should be maintained around 1,500 to 2,000 mg/L. If the MLSS gets too high, the sludge that the microorganisms form is disposed of or "wasted" as in conventional activated sludge.

In most package plants the adsorption/oxidation is achieved in one tank.

3. Complete Mix

In an ideal complete-mix activated sludge plant, the contents of the tank are completely mixed (the MLSS are uniformly mixed throughout the entire aeration tank). To ensure that this is achieved, special arrangements are often employed to *UNIFORMLY DISTRIBUTE THE INFLUENT AND WITHDRAW THE EFFLUENT FROM THE AERATION TANK.* Attention to the tank shape and to intensive mixing is important. There are some means that the operator may use to evaluate the degree to which a particular process operates in the complete-mix mode. First and foremost, the entire contents of the tank should be as uniform as possible. This can be confirmed by measurements of DO and suspended solids. If the tank is thoroughly mixed, these measurements should be nearly uniform. The settleability of the complete-mix sludges is generally well within the accepted range of normal operation. The MLSS in the aeration tank ranges from 2,000 to 5,000 mg/L.

Most package plants have one influent line and one influent port. The aeration compartment has one aeration header and the tank contents are completely mixed in a very short time. A limitation of the complete-mix type is that the process may be more susceptible to short-circuiting.

8.202 Aeration Methods

Two methods are commonly used to supply oxygen from the air to the bacteria—*MECHANICAL AERATION* and *DIFFUSED AERATION.* Both methods are mechanical processes with the difference being whether the mechanisms are at or in the aerator or at a remote location.

Mechanical aeration devices agitate the water surface in the aerator to cause spray and waves by paddle wheels (Figure 8.8), mixers, rotating brushes (Figure 8.11, page 271), or some other method of splashing water into the air or air into the water so that oxygen can be absorbed.

Mechanical aerators in the tank tend to be lower in installation and maintenance costs. Usually they are more versatile in terms of mixing, production of surface area of bubbles, and oxygen transfer per unit of applied power.

Diffused air systems use a device called a diffuser (Figure 8.9) to break up the air stream from the blower system into fine bubbles in the mixed liquor. The smaller the bubble, the greater the oxygen transfer due to the greater surface area of rising air bubbles surrounded by water. Unfortunately, fine bubbles will tend to regroup into larger bubbles while rising unless they are broken up by suitable mixing energy and turbulence.

QUESTIONS

Write your answers in a notebook and then compare your answers with those on page 285.

8.2A How many compartments are there in a typical package extended aeration plant? What is the purpose of each compartment?

8.2B What are some common characteristics of all package aeration plants?

8.21 Pre-Start Check-Out

If the package plant is being installed, a hole is dug large enough to accommodate the plant. Usually the plant is brought to the site by truck and placed in the excavation (hole) by a crane. The inlet sewer and discharge lines are connected and the plant is ready for service after power has been connected.

While the tank is empty, check the following:

1. Tank must be level from one end to the other.

2. If the tank is constructed of metal, is *CATHODIC PROTECTION*[13] required?

3. Condition of paint on exterior and interior.

4. Removal of all rocks and debris from both compartments.

5. If equipped with comminutor or grinder, check lubrication, clearance of cutters, and operation.

---

[13] *Cathodic (ca-THOD-ick) Protection.   An electrical system for prevention of rust, corrosion, and pitting of steel and iron surfaces in contact with water, wastewater or soil. A low-voltage current is made to flow through a liquid (water) or a soil in contact with the metal in such a manner that the external electromotive force renders the metal structure cathodic. This concentrates corrosion on auxiliary anodic parts which are deliberately allowed to corrode instead of letting the structure corrode.*

Fig. 8.8 Mechanical aeration device
(Permission of INFILCO INC.)

Fig. 8.9   Air diffuser
(Courtesy Paul Hallbach, National Training Center, Water Quality Office/EPA)

NOTE: Diffuser does not appear to be level. Most of the air is leaving from the higher left side which has a lower head of water over it.

6. Check aeration device:

   a. Lubrication
   b. Direction of rotation
   c. Mechanical aeration—proper agitator depth
   d. Compressor, if diffused air
      (1) Air filter and oil bath
      (2) Air header and valves
      (3) Air lift tubes and valves
      (4) Diffusers installed
      (5) Swing header lifts easily and moves freely

7. Record and file the following data:

   a. Plant model and serial number
   b. Two copies of plant operation and maintenance manual
   c. Nameplate data from equipment
      (1) Comminutor
      (2) Comminutor motor
      (3) Aeration motor
      (4) Compressor or agitator
      (5) Amperage of running equipment
      (6) Oils and greases specified for each unit

8. Check influent gate or valve for proper operation.

8.22 Starting the Plant

(See Volume II, Chapter 11, for detailed instructions.) The aerator compressor should be started and air introduced to the diffusers prior to the introduction of wastewater. If wastewater enters the tank before air is coming out of the diffusers, the diffusers could become plugged. If the plant is the diffused-air type with air lifts for return sludge, the air line valve to the air lifts will have to be closed until the settling compartment is filled. Otherwise all the air will attempt to go to the empty compartment and none will go to the diffusers. Once the settling compartment is filled from the overflow from the aeration tank, the air lift valves may be opened. They will have to be adjusted to return a constant stream of water and solids to the aeration tank. This adjustment is usually two to three turns open on the air valve to each air lift.

There may be a buildup of foam during the first week or so of start-up. A one-inch (25-mm) water hose with a lawn sprinkler may be used to keep it under control until sufficient mixed liquor solids are obtained.

Try to build up the solids or mixed liquor suspended solids (MLSS) as quickly as possible during start-up. This can be achieved by not wasting sludge until the desired level of MLSS is achieved.

8.23 Operation of Aeration Equipment

Aeration equipment should be operated continuously. In a diffused-air system, the operator controls air flow to the diffuser with the header control valve. This valve forces excess air to the air lifts in the settling compartment. Good treatment rarely results from interrupted operation and should not be attempted. You can judge how well the aeration equipment is working by the appearance of the water in the settling compartment and the effluent that goes over the weir. If the water is murky or cloudy and the aeration compartment has a rotten egg odor (H_2S), not enough air is being supplied. The air supplied or aeration rate should be increased. If the water is clear in the settling compartment, the aeration rate is probably sufficient. Try to maintain a DO level of around 2 mg/L throughout the aeration tank if you have a DO probe or lab equipment to measure the DO. Try to measure the DO at different locations in the aeration tank as well as from top to bottom.

QUESTIONS

Write your answers in a notebook and then compare your answers with those on page 285.

8.2C Why should package plants have cathodic protection?

8.2D Why should the valve to the air lift pumps be closed until the settling compartment is filled?

8.2E What would you do if the water in the aeration compartment was murky or cloudy and the aeration compartment had a rotten egg odor (H_2S)?

8.24 Wasting Sludge

Many older package plants did not come equipped with facilities for wasting sludge. The reason was that the extended aeration system was considered capable of stabilizing the influent suspended solids to a level sufficient for a tolerable release to the receiving waters. Experience has shown that this concept was wrong and that some sludge must be wasted routinely to achieve optimum plant performance. The operator must waste a portion of the plant solids content out of the system periodically. For best results, in terms of effluent quality, waste about five percent of the solids each week during summer operation to prevent excessive solids "burping."

To waste the excess activated sludge, turn off the return pumps or air lifts for one hour and continue to let the rest of the plant function. After one hour of not returning sludge, use a portable pump to pump about five percent of the waste solids from the settling compartment to a sand or soil drying bed. The amount of solids pumped is determined by measuring the depth of the sludge blanket and then lowering it five percent. Record the pumping time and weekly waste solids for this time period if results are satisfactory.

If your package activated sludge plant does not have an aerobic digester, applying waste activated sludge to drying beds may cause odor problems. If odors from waste sludge drying beds are a problem, consider the following solutions:

1. Waste the excess activated sludge into an aerated holding tank. This tank can be pumped out and the sludge disposed of in an approved sanitary landfill. If aerated long enough, the sludge could be applied to drying beds.

2. The excess or waste activated sludge can be removed by a septic tank pumper and disposed of in an approved sanitary landfill.

3. Arrange for disposal of the excess activated sludge at a nearby treatment plant.

Annually, check the bottom of the hoppers for rocks, sticks, and grit deposits. Also check the tail pieces of the air lifts to be sure that they are clear of rags and rubber goods and in proper working condition.

Frequency and amount of wasting may be revised after several months of operation by examining:

1. The amount of carryover of solids in the effluent.

2. The depth to which the solids settle in the aeration compartment when the aeration device is off (should be greater than one-third of the distance from top to bottom).

3. The appearance of floc and foam in the aeration compartment as to color, settleability, foam makeup, and excess solids on water surface of the tank.

4. Results of laboratory testing (Section 8.26).

A white, fluffy foam indicates low solids content in the aerator while a brown, leathery foam suggests high solids concentrations. If you notice high effluent solids levels at the same time each day, the solids loading may be too great for the final clarifier. Excessive solids indicate the mixed liquor suspended solids concentration is too high for the flows and more solids should be wasted.

8.25 Operation

8.250 Normal Operation

Package activated sludge plants should be checked every day. Each visit should include the following:

1. Check the appearance of aeration and final clarification compartments.

2. Check aeration unit for proper operation and lubrication.

3. Check return sludge line for proper operation. If the air lift is not flowing properly, briefly close the outlet valve which forces the air to go down and out the tail piece. This will blow it out and clear any obstructions. Reopen the discharge valve and adjust to desired return sludge flow.

4. Check comminuting device for lubrication and operation.

5. Hose down aeration tank and final compartment.

6. Brush weirs when necessary.

7. Skim off grease and other floating material such as plastic and rubber goods.

8. Check plant discharge for proper appearances, grease, or material of wastewater origin that is not desirable.

8.251 Abnormal Operation

Remember that changing conditions or abnormal conditions can upset the microorganisms in the aeration tank. As the temperature changes from season to season, the activity of the organisms speeds up or slows down. Also the flows and waste (food as measured by BOD and suspended solids) in the plant influent change seasonally. All of these factors require the operator to gradually adjust aeration rates, return sludge rates, and wasting rates. Abnormal conditions may consist of high flows or solids concentrations as a result of storms or weekend loads. These problems require the op-

erator and the plant to be prepared and to do the best possible job with available facilities.

Toxic wastes such as pesticides, detergents, solvents, or high or low pH levels can upset or kill the microorganisms in the aeration tank. Plant effluent usually does not deteriorate until after the toxic substance has passed through the plant. To correct problems from toxic substances, try to locate the source and prevent future discharges. If the microorganisms in the aeration tank have been killed, try to build up the microorganisms as if you were working with a new plant.

An outstanding example of a toxic substance added by operators is the use of chlorine for odor control (prechlorination). Chlorine is a toxicant and should not be allowed to enter the activated sludge process in uncontrolled amounts because it is not selective with respect to type of organisms damaged. Although chlorine is used in larger activated sludge plants for the control of bulking sludge, its use in package plants is not recommended. It may kill the organisms that you should be retaining as workers. Chlorine is effective in disinfecting the plant effluent *AFTER* treatment by the activated sludge process.

8.252 Troubleshooting

When problems develop in the activated sludge process, try to identify the problem, the cause of the problem, and select the best possible solution. Remember that the activated sludge process is a biological process and may require from three days to a week or longer to show any response to the proper corrective action. Allow seven or more days for the process to stabilize after making a change in the treatment process.

1. Solids in the effluent.

   a. If effluent appears turbid (muddy or cloudy), the return activated sludge pumping rate is out of balance. Try increasing the return sludge rate. Also consider the possible presence of something toxic to the microorganisms or a hydraulic overload washing out some of the solids.

   b. If the activated sludge is not settling in the clarifier (sludge bulking), several possible factors could be causing this problem. Look for too low a solids level in the system, low dissolved oxygen concentrations in the aeration tank, strong, stale *SEPTIC*[14] influent, high grease levels in influent, or alkaline wastes from a laundry.

   c. If the solids level is too high in the sludge compartment of the secondary clarifier, solids will appear in the effluent. Try increasing the return sludge pumping rate.

   d. If odors are present and the aeration tank mixed liquor appears black as compared with the usual brown color, try increasing aeration rates and look for septic dead spots.

   e. If light-colored floating sludge solids are observed on the clarifier surface, try reducing the aeration rates. Try to maintain the dissolved oxygen at around 2 mg/L throughout the entire aeration tank.

2. Odors

   a. If the effluent is turbid and the aeration tank mixed liquor appears black as compared with the usual brown color, try increasing aeration rates and look for septic dead spots.

---

[14] *Septic (SEP-tick). A condition produced by anaerobic bacteria. If severe, the sludge produces hydrogen sulfide, turns black, gives off foul odors, contains little or no dissolved oxygen, and the wastewater has a high oxygen demand.*

b. If clumps of black solids appear on the clarifier surface, try increasing the return sludge rate. Also be sure the sludge return line is not plugged and that there are no septic dead spots around the edges or elsewhere in the clarifier.

c. Examine method of wasting and disposing of waste activated sludge to be sure this is not the source of the odors.

d. Poor housekeeping could result in odors. Do not allow solids to accumulate or debris removed from wastewater to sit around the plant in open containers.

3. Foaming/Frothing

Foaming is usually caused by too low a solids level while frothing is caused by too long a solids retention time.

a. If too much activated sludge was wasted, reduce wasting rate.

b. If overaeration caused excessive foaming, reduce aeration rates.

c. If plant is recovering from overload or septic conditions, allow time for recovery.

Foaming can be controlled by water sprays or commercially available defoaming agents until the cause is corrected by reducing or stopping wasting and building up solids levels in the aeration tank.

To learn more about the operation of an activated sludge process under both normal and abnormal conditions, you may refer to Volume II, Chapter 11, "Activated Sludge." There you will also find a troubleshooting guide for activated sludge plants.

8.253   Shutdown

Shutdown procedures depend on whether the plant is being shut down because of operational problems, for maintenance and repairs, or for the off-season such as would be the case in a resort area. Activated sludge microorganisms (like people) die quickly from lack of oxygen if the aeration system is out of service for a short time period. Whenever the tank must be drained, try to determine the groundwater level. A high groundwater level can float a tank and cause considerable damage to structures and pipes. Diffusers should be clean before the tank is returned to service. If the package plant is shut down during the off-season, "moth ball" the equipment to prevent damage from weather and moisture. Exact procedures will depend on location and climate.

8.254   Operational Strategy

This section provides a brief summary of the basic concepts in the operation and control of the activated sludge process as it relates to both package plants and oxidation ditches. The following list outlines items that you the operator must consider in the day-to-day operation of your treatment plant.

1. Do the influent flow characteristics vary significantly during the year? Is your activated sludge process operation adequate to provide suitable treatment for these variations?

2. Is adequate pretreatment and collection system monitoring being practiced to avoid downstream mechanical or process failure?

3. Are routine solids tests performed (centrifuge, settleability, depth of blanket, visual observations) with results plotted on graphs to assist you in determining if a change in the process mode or operation is necessary?

4. Is suitable aeration time and mixing being provided to allow adequate oxidation, conversion, and floc formation of the solids?

5. Is adequate sludge wasting being practiced to properly maintain a favorable food to microorganism balance throughout the system?

6. If an increase or decrease in organisms results, is the oxygen level adjusted accordingly to maintain proper solids settling and production of a clear final effluent?

7. Is the return sludge flow rate such that it allows for a high concentration of solids which will reduce the amount of water returned to the aerator?

8. Do you visit your plant on a regular basis to observe process conditions, check equipment for proper operation, lubricate and maintain equipment, and clean process tanks and related equipment?

9. When a problem develops in your activated sludge process, do you refer to Section 8.252, "Troubleshooting"?

10. Do you avoid injuries by avoiding hazards and following safe procedures?

11. Before leaving your plant for the day, do you make a final and detailed check of the equipment for proper operation? Do you ensure that flow rates are set properly, that flow gates are set for possible storm and/or high flow conditions, that timer-controlled equipment and equipment alarms are set properly, and that equipment is stored and buildings and gates are properly locked?

8.26   Laboratory Testing

Testing for solids condition may be accomplished by the settling test. Using a quart jar, take a sample from the aeration compartment after the aeration device has been operating for ten to fifteen minutes; fill the jar to the top. Let the jar stand and watch the floc form and settle to the bottom of the jar. At the end of thirty minutes, the jar should be approximately half full of the settled solids, or slightly less, and have a chocolate brown color with clear water above it. The solids should appear granular. If the solids do not settle half way and the water above them is cloudy or murky in appearance, a longer aeration period, more air, or solids wasting is needed. If the solids settle to less than a quarter of the jar's depth and the water above the solids is murky or cloudy, no wasting of solids should be done and the solids level in the aerator should be allowed to increase.

If the solids settle to the bottom of the jar with a clear liquor on top and stay down one hour and come up in two hours, this is an indication of good operation. Solids should never be allowed to remain in the settling compartment longer than two hours. If the solids should rise in one hour, this is an indication usually of too much air, or too many solids. Make a slight adjustment to reduce the air to the aeration compartment or increase the return sludge rate.

Another possible cause of solids rising in one hour could be that there are not enough solids under aeration. When this happens the sludge will rise because of the high respiration (breathing) rate of the overtaxed organisms and the rapid DO depletion caused by these organisms in the settling compartment. Under these circumstances you would want to try to increase or build up the mixed liquor suspended solids. To identify the cause of problems and select the proper solution, you must keep good records and observe what is happening in your plant.

The final clarifier should be equipped with a scum baffle. A properly operated plant will produce some light, oxidized floc that will float to the surface of the settling compartment. A scum baffle will prevent this flow from leaving the compartment in the plant effluent. The better the treatment, the more likely scum froth will develop, unless the unit is septic.

QUESTIONS

Write your answers in a notebook and then compare your answers with those on page 285.

8.2F   If it becomes necessary to waste sludge, how much and when should it be wasted?

8.2G   How frequently should a package plant be visually checked by an operator?

8.2H   What should you do if you take a sample in a jar from the aeration compartment and after 30 minutes

    a. solids do not settle to the bottom half of the jar?
    b. solids settle to the bottom and then float to the top?

8.27 Safety

Operators of wastewater treatment plants have one of the worst safety records in the United States. If you are the operator of a small package plant, you must be extremely careful. Frequently you may be the only person at the plant and there will be no one nearby to come to your rescue or help you if you are hurt or seriously injured. *THEREFORE, PRACTICE SAFETY, AVOID HAZARDOUS CONDITIONS, AND USE SAFE PROCEDURES.*

You can be killed by toxic gases (hydrogen sulfide), electric shock, and drowning. Slippery surfaces, falls, and attempting to lift heavy objects can cause serious injury. Cuts and bruises can lead to infection and pathogens (disease-causing bacteria) in wastewater can make you sick. Chlorine and dangerous lab chemicals can blind you or seriously burn your skin. Whenever you must enter a confined space additional personnel will be required. All work in confined spaces must be done in full compliance with your local safety regulatory agency's requirements. Do electrical troubleshooting *ONLY* if you are qualified and authorized to do so. If you must work alone doing routine operation and maintenance, phone your office at regular intervals and have someone check on you if you fail to report. You can avoid injuries if you avoid hazards and follow safe procedures.

8.28 Maintenance

Maintenance of equipment in package plants should follow the manufacturer's instructions. Items requiring attention include:

1. Plant cleanliness. Wash down tank walls, weirs, and channels to reduce the collection of odor-causing materials;

2. Aeration equipment:

    a. Air blowers and air diffusion units, and

    b. Mechanical aerators;

3. Air lift pumps;

4. Scum skimmer;

5. Sludge scrapers;

6. Froth spray system;

7. Weirs, gates, and valves; and

8. Raw wastewater pumps.

8.29 Additional Reading

1. *PACKAGE TREATMENT PLANTS—OPERATIONS MANUAL*, U.S. Environmental Protection Agency. Obtain from National Technical Information Service (NTIS), 5285 Port Royal Road, Springfield, VA 22161. Order No. PB-279444. EPA No. 430-9-77-005. Price, $39.50, plus $5.00 shipping and handling per order. *NOTE:* This is an outstanding publication and if you are the operator of a package activated sludge plant, you should obtain this Operations Manual.

2. *OPERATOR'S POCKET GUIDE TO ACTIVATED SLUDGE—PART II: PROCESS CONTROL AND TROUBLESHOOTING.* Obtain from Arcadis, Inc., Attn: Pocket Guide Department, 11490 Westheimer, Suite 600, Houston, TX 70777. Price, $3.50, plus $2.00 shipping and handling. (Quantity discounts available.)

QUESTIONS

Write your answers in a notebook and then compare your answers with those on page 285.

8.2I   How can operators avoid being injured?

8.2J   What safety precautions should be taken if you must work alone?

END OF LESSON 2 OF 3 LESSONS ON ACTIVATED SLUDGE

Please answer the discussion and review questions next.

DISCUSSION AND REVIEW QUESTIONS

Chapter 8.   ACTIVATED SLUDGE

(Lesson 2 of 3 Lessons)

Write the answers to these questions in your notebook before continuing. The question numbering continues from Lesson 1.

7. What is a temporary way to control foaming in the aeration compartment of a package plant?

8. How would you operate the aeration device in a package plant?

9. How would you waste sludge in a package plant?

10. Briefly explain what would cause package aeration plant problems such as solids in the effluent, odors, and foaming.

CHAPTER 8.   ACTIVATED SLUDGE

PACKAGE PLANTS AND OXIDATION DITCHES

(Lesson 3 of 3 Lessons)

8.3   OXIDATION DITCHES

8.30   Use of Oxidation Ditches

8.300   Flow Path for Oxidation Ditches

The oxidation ditch (Figure 8.10 and Table 8.1) is a modified form of the activated sludge process and is usually operated in the extended aeration mode.

The main parts of the oxidation ditch are the aeration basin which generally consists of two channels placed side by side and connected at the ends to produce one continuous loop of the wastewater flow, a brush rotor assembly (Figure 8.11), settling tank, return sludge pump, and excess sludge handling facilities.

There is usually no primary settling tank or grit removal system used in this process. Inorganic solids such as sand, silt, and cinders are captured in the oxidation ditch and removed during sludge wasting or cleaning operations. The raw wastewater passes directly through a bar screen to the ditch. The bar screen is necessary for the protection of the mechanical equipment such as rotor and pumps. Comminutors or barminutors may be installed after the bar screen or instead of a bar screen. The oxidation ditch forms the aeration basin and here the raw wastewater is mixed with previously formed active organisms. The rotor is the aeration device that entrains (dissolves) the necessary oxygen into the liquid for microbial life and keeps the contents of the ditch mixed and moving. The velocity of the liquid in the ditch must be maintained to prevent settling of solids, normally 1.0 to 1.5 feet per second (0.3 to 0.45 m/sec). The ends of the ditch are well rounded to prevent eddying and dead areas, and the outside edges of the curves are given erosion protection.

The mixed liquor flows from the ditch to a clarifier for separation. The clarified water passes over the effluent weir and is chlorinated. Plant effluent is discharged to either a receiving stream, percolation ditches, or a subsurface disposal or leaching system. The settled sludge is removed from the bottom of the clarifier by a pump and is returned to the ditch or wasted. Scum that floats to the surface of the clarifier is removed and either returned to the oxidation ditch for further treatment or disposed of by burial.

Since the oxidation ditch is operated as a closed system, the amount of volatile suspended solids will gradually increase. It will periodically become necessary to remove some sludge from the process. Wasting of sludge lowers the MLSS (Mixed Liquor Suspended Solids) concentration in the ditch and keeps the microorganisms more active. Control of sludge concentration by wasting of excess sludge is one of the reasons for the high reductions possible by this process. Excess sludge may be dried directly on sludge drying beds or stored

TABLE 8.1   PURPOSES OF PARTS IN AN OXIDATION DITCH

Part	Purpose
Oxidation Ditch	Provides detention time where activated sludge microorganisms treat wastewater.
Rotor	Causes surface aeration which transfers oxygen from air to water for respiration by microorganisms. Also keeps the contents of the ditch mixed and moving.
Level Control Weir	Regulates how deep rotors sit in the flow of wastewater. This affects the amount of oxygen dissolved in or transferred to the water being treated. Overflow goes to final settling tank.
Final Settling Tank	Allows activated sludge to be separated from water being treated. Clear effluent leaves plant and settled activated sludge is either returned to oxidation ditch or wasted.
Return Sludge Pump	Returns settled activated sludge from final settling tank to oxidation ditch or to excess sludge handling facilities.
Excess Sludge Handling Facilities	Treat waste activated sludge for ultimate disposal.

in a holding tank or in sludge lagoons for later disposal to larger treatment plants or approved sanitary landfills.

The basic process design results in simple, easy operation. A high mixed liquor suspended solids (MLSS) concentration, usually between 2,000 to 6,000 mg/L (some plants carry 6,000 to 8,000 mg/L MLSS), is carried in the ditch and the plant may be capable of handling shock and peak loads without upsetting plant operation. There is no foam problem after solids buildup, as experienced with other types of activated sludge plants. Cold weather operation has less effect on plant efficiency than other processes because of the large number of microorganisms in the ditch.

Operating plants in the United States are achieving BOD removals of about 90 percent and as high as 98 percent.

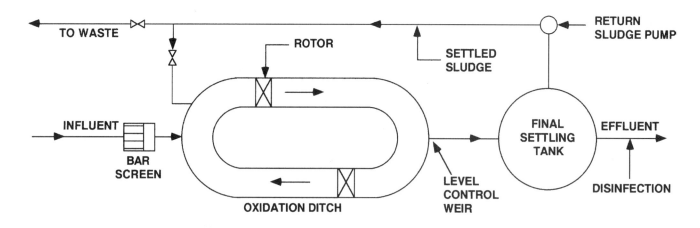

Fig. 8.10   Oxidation ditch plant

(Source: "Oxidation Ditch" prepared by William L. Berk for the
New England Regional Wastewater Institute, South Portland, Maine 04106. August 1970)

8.301   Description of Oxidation Ditches

1. Plant flow	0.2 to 20.0 MGD (750 to 75,000 cu m/day)
2. BOD loading	10 to 50 lbs BOD/day per 1,000 cubic feet of ditch (200 to 800 kg/day per 1,000 cu m)
3. F/M ratio[15]	0.03 to 0.1 lb BOD/day/lb MLVSS
4. MLSS concentration	2,000 to 6,000 mg/L
5. Sludge age	20 to 35 days
6. Ditch detention time	3 to 24 hours
7. Minimum velocity	1.0 fps (0.3 mps)
8. DO levels	0.5 to 3.0 mg/L
9. Liquid level	3.0 to 7.0 feet (1 to 2 m)

See Section 8.35, "Operational Guidelines," for procedures to calculate these items. Use of the F/M ratio and sludge age to control the activated sludge process will be discussed in Volume II, Chapter 11.

8.31   Safety

Lost time, injury, and even death are the results of not being concerned with applying the rules of safety to all activities involved in operating and maintaining a plant. Practicing "safety" is not just knowing what to do; it is a lifestyle. Personnel must not only acquire this "lifestyle," but must also know what to do

if an accident occurs. Refer to Volume II, Chapter 14, "Plant Safety," for more information.

Some safety precautions that should be observed at all times when working in a treatment plant include:

1. Wear safety shoes with steel toes, shanks, and soles that retard slipping. Cork-inserted composition soles provide the best traction for all-around use.

2. Slippery algal growths should be scrubbed off and washed away whenever they appear.

3. Keep all areas clear of spilled oil or grease. Use soap and water, not gasoline or solvents, for cleaning.

4. Wear gloves when working with equipment or while in direct contact with wastewater.

5. Do not leave tools, equipment, and materials where they could create a safety hazard.

6. Adequate lighting should be provided for night work and in areas with limited existing lighting.

7. Ice conditions in winter require spiked shoes and the sanding of icy areas if the ice cannot be thawed away with wash water.

8. Remove only sections of handrails, deck plates, or grating necessary for the immediate job. Removed sections should be properly stored out of the way and properly secured against falling into tanks. The area should be barricaded to prevent unauthorized personnel entry and possible injury.

9. Do not walk on top of the oxidation ditch sidewalls; you may slip and fall into the ditch.

---

[15] *F/M Ratio.   Food to microorganism ratio. A measure of food provided to bacteria in an aeration tank.*

$$\frac{Food}{Microorganisms} = \frac{BOD,\ lbs/day\ or\ kg/day}{MLVSS,\ lbs\ or\ kg}$$

$$= \frac{Flow,\ MGD \times BOD,\ mg/L \times 8.34\ lbs/gal}{(Ditch\ Volume,\ MG \times SS,\ mg/L \times Volatile\ Portion,\ decimal \times 8.34\ lbs/gal)}$$

Brush rotor

Brush rotor

Outboard bearing assembly

Brush rotor drive motor and
gear reducer assembly

Fig. 8.11 Brush rotor

QUESTIONS

Write your answers in a notebook and then compare your answers with those on page 285.

8.3A   Why are the ends of oxidation ditches well rounded?

8.3B   List the major components of an oxidation ditch treatment process.

8.3C   Why should operators avoid walking on top of oxidation ditch sidewalls?

8.32   Start-Up

There are two primary objectives of start-up. One is to make certain that all mechanical equipment is operating properly. The second is to develop a proper microbial floc (activated sludge) in the oxidation ditch. Floc development is essential for the plant to succeed in reducing the quantities of polluting materials in the raw wastewater.

The start-up procedures presented here should be used along with the manufacturer's start-up procedures for all components of the plant. The plant operator, the contractor, engineer, and equipment manufacturer's representative should all be present at the start-up.

During start-up of the plant, some construction may still be in progress. Special care should be given to ensure that all safety procedures are followed at all times.

8.320   Pre-Start Inspection

The inlet structure should be checked for debris and all debris cleaned from the structure prior to start-up.

The ditch structure should be cleaned of all debris prior to start-up. Check the walkways to ensure there is no debris that can later fall into the channel. Inspect the influent and effluent lines to be sure that they are free of debris.

If you are preparing to start a rotor that is a new installation, one that has not been in operation for some time, or one that has just been overhauled, a thorough inspection check should be made prior to starting the rotor to prevent damage to the rotor and injury to personnel.

Rotor checklist.

1. The ON/OFF switch is OFF and the main breaker is locked out and tagged.

2. Motor secured to gear reducer.

3. Gear reducer assembly secured to the mounting platform.

4. Rotor cylinder shaft secured to reducer coupling bolts.

5. Rotor blades and teeth secured to the cylinder.

6. Driver, rotor, bearings, and stand(s) properly aligned.

7. Rotor turns with a reasonable pull on the cylinder.

8. Rotor anti-rotation screws properly adjusted.

9. Proper oil type and quantity in the reducer.

10. All bearings greased with the proper lubricant.

11. All lube oil line fittings tight.

12. Gear reducer housing air vent "open."

13. All bolts tight.

14. All tools and foreign material are clear of the rotor assembly.

15. Safety guards over moving parts properly installed and secure.

Following completion of the pre-start checks, the rotor assembly is ready to start. Be sure all personnel are clear of the rotor assembly. Turn the main power breaker ON. The ON/OFF switch should now be positioned to ON and the rotor will start.

If the rotor assembly is part of a new oxidation ditch installation, the following additional steps should be taken:

1. Start the rotor(s) with the rotor(s) out of the water.

2. Check and record the motor amperage and voltage on each phase.

3. Check the rotation of the rotor and run the rotor for at least one hour (see "bump start" below).

4. Recheck the rotor support bearing(s) and drive alignment and realign if necessary according to manufacturer's installation drawings.

5. Tighten all nuts, bolts, and set screws to prepare the rotor for regular operation.

Check and record the items noted under Section 8.330, "Normal Operation, Records, Mechanical."

NOTE: When starting a new or recently overhauled rotor assembly, a "bump start" is recommended *BEFORE* a routine full start and run is attempted.

A "bump start" allows operation of the rotor assembly for 2 to 3 seconds and is accomplished by briefly positioning the ON/OFF switch to ON and immediately returning it to the OFF position.

This short run-time will allow you to determine if the rotor assembly operates freely and properly, operates in the proper rotation, and will avoid extensive damage to the rotor assembly if it is not properly installed.

Make sure the adjustable weir operates freely and does not bind. Set the weir at the proper elevation in accordance with O & M manual or manufacturer's instructions.

The clarifier structure and piping should be inspected and cleared of all debris. All control gates and valves should be checked for smooth operation and proper seating. Refer to Chapter 5, "Sedimentation and Flotation," for proper clarifier start-up procedures.

The return sludge and waste sludge systems should be examined for leakage and all valves should be operated one complete cycle and set for normal operation. Pumps should be manually operated with liquid in them to check proper operation. One pump should be operated and checked for vibration, excess noise, overheating, and the amperage reading record. The same procedure should then be repeated for the second pump with the first pump shut off. If your plant has waste sludge treatment facilities (holding tanks, lagoons, or drying beds), operate the necessary valves to allow pumping to this part of the plant. Operate each pump manually and note the discharge. Shut both pumps off and return the valves to the normal flow position for returning the settled sludge back to the oxidation ditch.

8.321 Plant Start-Up

Start to fill the ditch with water or wastewater. If possible, add a water tank or two full of healthy seed activated sludge from a nearby plant. Divert all wastewater to be treated into the ditch. If the ditch was initially filled with wastewater and one or two days were required to fill the ditch during hot weather, odor problems could develop. Start the rotors when the water level reaches the bottom of the rotor blades.

During plant start-up, a dark gray color of the developing MLSS may be seen. A dark gray color usually indicates a lack of bacterial buildup in the mixed liquor. If this condition continues for more than several days, check your return sludge system to see that it is operating properly.

Do not start discharging to the clarifier when the water reaches the rotors. Allow the wastewater to continue to fill the ditch and to be treated. When the water level approaches the maximum submergence of the rotors and the peak motor load, then start allowing some of the water to be discharged to the clarifier. During start-up, an unstable clarifier effluent may result due to the inadequate biological treatment. As this effluent is generally the discharged product (as final effluent), chlorination is to be used to reduce health hazards on the receiving waters during this time. State regulatory agencies should be contacted to ensure that the receiving waters will not be harmed as a result of heavily chlorinating the plant effluent.

During the period of start-up, wastewater testing procedures should be initiated as soon as possible. The actual flow rates should be recorded and also the incoming BOD and COD levels.

Building up of the MLSS concentration is the most important activity during the start-up process. At least three and possibly up to fifteen days are required to build up the MLSS concentration. In the event the actual MLSS concentration cannot be determined on a daily basis, you should at least record daily the results of the 30-minute sludge settleability tests. See Section 8.26, "Laboratory Testing." During this period, maintain the highest possible return activated sludge rate.

Dissolved oxygen (DO) should be measured at a sampling location approximately 15 feet (4.5 m) upstream from the rotor. Until the desired MLSS concentration is reached and the 30-minute sludge settleability volume reaches 20 percent, a minimum DO concentration of 2.0 mg/L should be kept in the ditch. Following this period, a lesser DO concentration may be desired, but never less than 0.5 mg/L at a point 15 feet (4.5 m) upstream of the rotors.

Following start-up, when the plant has stabilized, the solids should settle rapidly in the clarifier leaving a clear, odorless and stable effluent. The solids should look like particles, golden to rich dark brown in color, with sharply defined edges.

You should not expect immediate results from the start-up procedures. Plant start-up takes time, sometimes over a month if a "seed" activated sludge from another treatment plant is not available. Also, some conditions may occur during start-up that would, under normal conditions, indicate a poorly operating process such as light foaming in the ditch or a cloudy supernatant in the settleable solids tests. These conditions should only be temporary if the information in this section is applied properly.

QUESTIONS

Write your answers in a notebook and then compare your answers with those on page 285.

8.3D What are the two primary objectives of start-up?

8.3E What items or structures should be inspected during the pre-start inspection?

8.3F During plant start-up, when should the rotors be started?

8.3G When an oxidation ditch is operating properly, how should the activated sludge solids appear?

8.33 Operation

8.330 Normal Operation

Process controls and operation of an oxidation ditch are similar to the activated sludge process. To obtain maximum performance efficiency, the following control methods must be maintained:

1. Proper food supply (measured as BOD or COD and no toxicants) for the microorganisms;

2. Proper DO levels in the oxidation ditch;

3. Proper ditch environment (no toxicants, and sufficient microorganisms to treat the wastes);

4. Proper ditch detention time to treat the wastes by control of the adjustable weir; and

5. Proper water/solids separation in the clarifier.

PROPER FOOD SUPPLY FOR THE MICROORGANISMS

Influent flows and waste characteristics are subject to limited control by the operator. Municipal ordinances may prohibit discharge to the collection system of materials damaging to treatment structures or human safety. Control over wastes dumped into the collection system requires a pretreatment facility inspection program to ensure compliance. Alternate means of disposal, pretreatment, or controlled discharge of significantly damaging wastes may be required in order to permit dilution to an acceptable level by the time the waste arrives at the treatment plant. Refer to Chapter 4, "Preventing and Minimizing Wastes at the Source," in INDUSTRIAL WASTE TREATMENT, Volume I, for waste discharge control methods.

PROPER DO LEVELS

Proper operation of the process depends on the rotor assembly supplying the right amount of oxygen to the waste flow in the ditch. For the best operation, a DO concentration of 0.5 to 2.0 mg/L should be maintained just upstream (15 feet or 4.5 m) of the rotors. Overoxygenation wastes power and excessive DO levels can cause a pinpoint floc to form that does not settle and is lost over the weir in the settling tank. Control of rotor oxygenation is achieved by adjusting the ditch outlet level control weir.

The level or elevation of the rotors is fixed but the deeper the rotors sit in the water, the greater the transfer of oxygen from the air to the water (greater DO). The ditch outlet level control weir regulates the level of water in the oxidation ditch.

PROPER ENVIRONMENT

The oxidation ditch process with its long-term aeration basin is designed to carry MLSS concentrations of 2,000 to 6,000 mg/L. This provides a large organism mass in the system.

Performance of the ditch and ditch environment can be evaluated by conducting a few simple tests and general observations. The color and characteristics of the floc in the ditch as well as the clarity of the effluent should be observed and recorded daily. Typical tests are settleable solids, DO upstream of the rotor, pH, and residual chlorine in the plant effluent. These test procedures are outlined in Volume II, Chapter 16, "Laboratory Procedures and Chemistry." Laboratory tests such as BOD, COD, suspended solids, volatile solids, total solids, and microscopic examinations should be performed periodically by the plant operator or an outside laboratory. The results will aid you in determining the actual operating efficiency and performance of the process.

Oxidation ditch solids are controlled by regulating the return sludge rate and waste sludge rate. Remember that solids continue to deteriorate as long as they remain in the clarifier. Adjust the return sludge rate to return the microorganisms in a healthy condition from the final settling tank to the oxidation ditch. If dark solids appear in the settling tank, either the return sludge rate should be increased (solids remaining too long in clarifier) or the DO levels are too low in the oxidation ditch.

Adjusting the waste sludge rate regulates the solids concentration (number of microorganisms) in the oxidation ditch. If the surface of the ditch has a white, crisp foam, reduce the sludge wasting rate. Some plants have operated successfully with an MLSS of 6,000 to 8,000 mg/L and very little wasting of solids. If the surface has a thick, dark foam, increase the wasting rate. Waste activated sludge may be removed from the ditch by pumping to a sludge holding tank, to sludge drying beds, to sludge lagoons, or to a tank truck. Ultimate disposal may be to larger treatment plants or to approved sanitary landfills.

Remember that this is a biological treatment process and several days may be required before the process responds to operation changes. Make your changes slowly, be patient, and observe and record the results. Allow seven or more days for the process to stabilize after making a change. For additional information on the regulation of the process, see Section 8.26, "Laboratory Testing," which discusses the use of the settling test for operational control. For more details on controlling the activated sludge process, see Volume II, Chapter 11.

PROPER TREATMENT TIME AND FLOW VELOCITIES

Treatment time is directly related to the flow of wastewater and is controlled by an adjustable weir. Velocities in the ditch should be maintained at 1.0 to 1.5 feet per second (0.3 to 0.45 m/sec) to prevent the deposition of floc. With this in mind, the ditch contents should travel the complete circuit of the ditch, or from rotor to rotor, every 3 to 6 minutes. If the rotors are operated by time clocks (30 minutes off and 30 minutes on, for example), the velocities in the ditch must be sufficient to resuspend any settled material.

PROPER WATER/SOLIDS SEPARATION

MLSS that have entered and settled in the secondary clarifier are continuously removed from the clarifier as return sludge, by pump, for return to the oxidation ditch. Usually all sludge formed by the process and settled in the clarifier is returned to the ditch, except when wasting sludge. Scum that is captured on the surface of the clarifier also is removed from the clarifier and either returned to the oxidation ditch for further treatment or disposed of by burial.

OBSERVATIONS

Some aspects of the operation of your oxidation ditch plant can be controlled and adjusted with the help of some general observations. General observations of the plant are important to help you determine whether or not your oxidation ditch is operating as intended. These observations include color of the mixed liquor in the ditch, odor at the plant site, and clarity of the ditch and sedimentation tank surfaces.

COLOR. You should note the color of the mixed liquor in the ditch daily. A properly operating oxidation ditch plant mixed liquor should have a medium to rich dark brown color. If the MLSS, following proper start-up, changes color from a dark brown to a light brown and the MLSS appear to be thinner than before, the sludge waste rate may be too high which may cause the plant to lose efficiency in removing waste materials. By decreasing sludge waste rates before the color lightens too much, you can ensure that the plant effluent quality will not deteriorate due to low MLSS concentrations.

If the MLSS becomes black, the ditch is not receiving enough oxygen and has gone "anaerobic." The oxygen output of the rotors must be increased to eliminate the black color and return the process to normal aerobic operation. This is done by increasing the submergence level of the rotor.

ODOR. When the oxidation ditch plant is operating properly, there will be little or no odor. Odor, if detected, should have an earthy smell. If an odor other than this is present, you should check and determine the cause. Odor similar to rotten eggs indicates that the ditch may be going anaerobic, requiring more oxygen or a higher ditch velocity to prevent deposition of solids. The color of the MLSS could be black if this were the case.

Odor may also be a sign of poor housekeeping. Grease and solids buildup on the edge of the ditch or sedimentation tank will go anaerobic and cause odors. With an oxidation ditch, odors are much more often caused by poor housekeeping than poor operation.

CLARITY. In a properly operating oxidation ditch a "layer" of "clear" water or supernatant is usually visible a few feet upstream from the rotor. The depth of this relatively clear water may vary from almost nothing to as much as two or more inches (5 cm) above the mixed liquor. The clarity will depend on the ditch velocity and the settling characteristics of the activated sludge solids.

Two other good indications of a properly operating oxidation ditch are the clarity of the settling tank water surface and the oxidation ditch surface free of foam buildup. Foam buildup in the ditch (normally not enough to be a nuisance) is usually caused by an insufficient MLSS concentration. Most frequently foam buildup is only seen during plant start-up and will gradually disappear.

Clarity of the effluent from the secondary clarifier discharged over the weirs is the best indication of plant performance. A very clear effluent shows that the plant is achieving excellent pollutant removals. A cloudy effluent often indicates a problem with the plant operation.

RECORDS

Daily, or as scheduled, the items listed below should be checked and a record made of these checks. Accurate records are essential and invaluable in evaluating rotor efficiency and in establishing normal operating conditions.

Check:

ITEM	CONDITION
Mechanical	
1. Motor	High or uneven amperage
2. Motor	High temperature
3. Motor	Unusual noise
4. Motor	Operating hours
5. Gear reducer	Unusual bearing or gear noise
6. Gear reducer	Proper oil level (use sight glass or dipstick to check)
7. Gear reducer housing	Air vent pipe "open"
8. Outboard bearing	Unusual bearing noise
9. Rotor	Unusual noise or vibration
10. Rotor	Remove any debris caught in rotor blades such as rags, weeds, or plastic goods.
Operation	
11. DO concentrations	Use portable DO probe or perform Winkler lab test. DO should be taken approximately 15 feet (4.5 m) *UPSTREAM* of the rotor(s). Maintain 0.5 to 2.0 mg/L at this point.
12. Ditch velocity	1.0 to 1.5 fps (0.3 to 0.45 mps)
13. Ditch water level	Read staff gage in ditch. Convert this reading to brush aerator submergence.
14. Ditch surface condition	Neither white foam nor a heavy brown scum
15. Mixed Liquor Suspended Solids concentration (MLSS)	2,000 to 5,000 mg/L

FINAL PLANT SURVEY

Before leaving for the day, one final inspection should be made around the plant. The following questions may help in seeing that you have left the plant in a condition so that it will be operating properly the next time it is attended.

1. Are any pieces of equipment operating poorly? Will they need to be checked before the next scheduled day of operator attendance (hot bearings, loose belts)?

2. Are return sludge rates set at the correct level?

3. Are flowmeters clean and operating?

4. Are inlet gates set properly in case of high flows?

5. Has the rotor level of submergence been set properly?

6. If some equipment is time-clock controlled, are the time clocks set?

7. If remote alarms are used to warn someone about equipment failures, are these set properly?

8. Is equipment stored and locked so as to prevent vandalism?

9. Are outside lights on or set to come on?

PERFORMANCE

Figure 8.12 shows the performance of an actual oxidation ditch.

QUESTIONS

Write your answers in a notebook and then compare your answers with those on page 286.

8.3H How is the DO in the oxidation ditch regulated?

8.3I What should be the velocity in the ditch?

8.3J What observations should be made daily to help indicate the performance of the oxidation ditch?

8.331 Abnormal Operation

In making the routine checks that are noted under Section 8.330, "Normal Operation, Records, Mechanical," you may occasionally find some conditions to be abnormal. Serious damage to the rotor assembly may result if the abnormal conditions are not corrected as soon as possible.

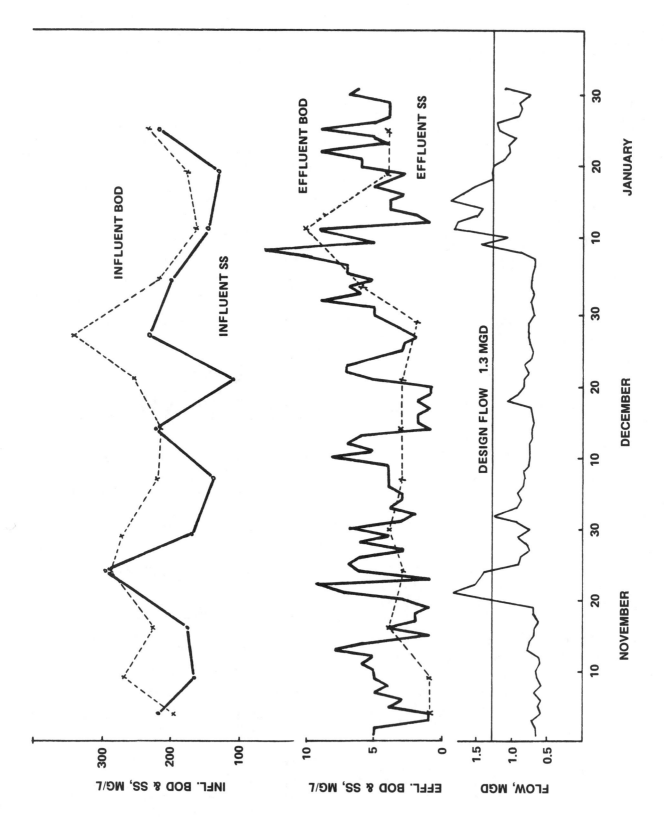

Fig. 8.12 Actual performance of an oxidation ditch

Rotor assembly operation is *ESSENTIAL* for the efficient operation of the oxidation ditch. Loss of rotor-generated mixing and air entrainment to the mixed liquor in the ditch for an extended period of time will turn your oxidation ditch activated sludge process into an upset, bulking activated sludge process.

Listed in Table 8.2 are some abnormal conditions, possible causes, and your response to the conditions that will aid you in the safe and efficient operation of the rotor.

Environmental factors that affect the wastewater treatment process include temperature and precipitation. The wastewater temperature affects the activity of the microorganisms. During cold weather this reduced activity might lower the efficiency of the treatment system. Besides biological effects of temperature, the flocculation and sedimentation of the mixed liquor solids are not as effective at lower temperatures.

Ice buildup will hinder or stop altogether the proper operation of mechanical parts such as rotors and sludge scraper mechanisms. Chunks of ice may develop that float in the ditch and eventually enter the area of the rotor assembly. Unless adequate safeguards are provided, serious damage to the rotor *WILL* result. Some of the safeguards that *MUST* be considered are:

1. The oxidation ditch in cold weather areas *SHOULD BE OPERATED FOR THE MAXIMUM DETENTION TIME PRACTICAL* in order to conserve as much heat in the wastewater as possible. This action will inhibit ice chunk formation. Set the effluent control weir at the highest possible level.

2. The splashing and/or spraying action produced by the rotor will allow ice to form on the rotor assembly. Serious consideration must be given to covering the whole rotor assembly with a structure made of wood, fiberglass, or other suitable material that will bridge over the ditch. Installations of this type generally are not heated.

NOTE: If it is necessary to shut the rotor off for maintenance, *ICE WILL FORM ON THE ROTOR.* Prior to restarting the rotor, hose it off with water to thaw the ice. Otherwise the ice could cause vibrations that can damage the rotor.

CAUTION: Ice conditions in winter require spiked shoes and sanding icy areas if the ice cannot be thawed away with wash water.

In some treatment plants, very heavy rains or snow melts cause flows to the treatment plant to exceed three to four times design flow. This is generally accompanied by a weaker wastewater in terms of BOD and COD due to the dilution effect of the storm water. These hydraulic overloads may exceed the capacity of the clarifier to settle sludge solids properly. When this occurs, extremely high BOD, COD, and suspended solids concentrations will be discharged to the receiving stream in the final effluent and possible process upsets can occur if corrective action is not taken. Chemicals such as alum, ferric chloride, and polymers can be added to the final settling tank to assist in settling solids during abnormal conditions. When chemicals such as alum are used, the volume of return sludge is increased and the pH of the sludge may be reduced.

Another method of preventing hydraulic overloads from causing this discharge of high BOD, COD, and suspended solids concentrations is to shut down one or more rotor assemblies in the ditch. This will allow the ditch to act as a large settling tank, keeping the MLSS from flowing into the clarifier where they could be washed out. When the plant flow de-

TABLE 8.2   ABNORMAL BRUSH ROTOR OPERATION

Item	Abnormal Condition	Possible Cause	Operator Response
Motor	Motor "off"	Ambient temperature in switchboard panel or room too high.	Reduce temperature with fans.
		Degree of rotor submergence results in excessive amperage draw.	Adjust rotor submergence.
		Motor shorted or burned out.	Check for overloading. Repair motor.
	Motor "hot" to the touch	Dry bearings.	Lubricate bearings.
		Excessive grease.	Remove grease and grease properly.
		Worn bearings.	Replace bearings.
		Excessive amperage draw (motor load).	Adjust rotor submergence.
Gear Reducer	Grinding, chipping, whirring, or whining noise	Low oil level.	Drain and/or refill. Check for leakage and make repairs.
		Gear drive part(s) partially or totally worn.	Replace worn parts.
Bearings	Grinding or bumping noise	Insufficient bearing grease.	Grease bearing(s) routinely.
		Worn bearing(s).	Replace bearing(s).

creases to more normal levels, the rotor(s) can then be restarted to resume normal operation.

Other than the above corrective actions, without plant modifications there is not much the operator can do to offset the changes in treatment efficiency caused by temperature changes and high flows during abnormal conditions. However, ice buildup can be controlled by frequent observation and removal during cold periods. For persistent cold weather problems, construction of a *LIGHTWEIGHT* building over the ditch and clarifier may be more economical in the long run than fighting ice. The normal wastewater temperature will generally supply sufficient heat inside a building to prevent ice from forming.

Usually it is necessary to vary the amount of MLSS in the ditch as seasons change. Because the microorganisms are not as active in winter at low temperatures, the MLSS will need to be higher in the winter than in the summer if complete nitrification is desired.

8.332 Shutdown

Shutdown of the oxidation ditch may be necessary for emergency repairs. Try to schedule any shutdowns so the activated sludge microorganisms will be without air for the shortest possible time period. Problems such as odors and a loss of the microorganism culture can start within two hours after the rotors have been shut off. The microorganism culture may start to deteriorate within 10 minutes, but can recover quickly if the downtime does not exceed four hours. If possible, try to keep one rotor operating at all times.

To prevent injury to personnel, shut down the rotor assembly whenever any maintenance function is performed.

1. Turn the ON/OFF switch to OFF.

2. Turn the main power breaker OFF.

3. Lock out and tag the main power breaker in the OFF position.

Maintenance work may now be performed.

If the oxidation ditch treats seasonal loads and is shut down during the off-season, protect equipment from the weather and moisture.

8.333 Troubleshooting

Refer to Section 8.252, "Troubleshooting." The problems of package aeration plants and the solutions are very similar to those of oxidation ditches.

If floatables appear in the final settling tank, examine the baffle around the level control weir. Since oxidation ditches do not have primary clarifiers, plastic goods and other floatables can be a problem if the baffle is not properly adjusted.

8.334 Operational Strategy

Refer to Section 8.254, "Operational Strategy." The items considered in the operational strategy for package aeration plants are the same as those considered for oxidation ditches.

8.34 Maintenance

8.340 Housekeeping

A general cleanup each day at the plant is important. This not only gives you a more pleasant place to work, but also helps your plant perform better.

Daily cleanup usually includes removing and burying debris that may have accumulated on the bar screen; removing grease and scum from the surface of the clarifier; and washing or brushing down the ditch and clarifier weirs and walls.

8.341 Equipment Maintenance

Regularly scheduled equipment maintenance must be performed. As boring as some equipment manufacturers' instruction manuals may be, *READ THEM!* Learn what they say. You should check each piece of equipment daily to see that it is functioning properly. You may have very few mechanical devices in your oxidation ditch plant, but they are all important. The rotors and pumps should be inspected to see that they are operating properly. If pumps are clogged, the obstructions should be removed. Listen for unusual noises. Check for loose bolts. Uncovering a mechanical problem in its early stages could prevent a costly repair or replacement at a later date.

Lubrication should also be performed with a fixed operating schedule and properly recorded. Follow the lubrication and maintenance instructions furnished with each piece of equipment. If you are unable to find the instructions, write to the manufacturer for a new set. Make sure that the proper lubricants are used. Overlubrication is wasteful and reduces the effectiveness of lubricant seals and may cause overheating of bearings or gears.

Painting should be done periodically. In addition to beautifying your plant, it gives a good protective coating on all iron and/or metal surfaces and will prolong the life of the metal.

NEVER PAINT OVER equipment identification tags! You will regret it if you do.

You should make it your business to know the manufacturer of every piece of equipment in your plant. Knowing the manufacturer and how to quickly contact the manufacturer may save important time and money when equipment breakdowns occur.

1. Motors (See Volume II, Chapter 15, "Maintenance," for additional details.)

    Motors should be greased after about 2,000 hours of operation or as often as conditions and/or the manufacturer recommends. The motor must be stopped when greasing begins.

    Remove the filler and drain plugs, free the grease holes of any hardened grease, add new grease through the filler hole until it starts to come out of the drain hole. Start the motor and let it run for about 15 minutes to expel any excess grease. Stop the motor and install the filler and drain plugs.

    Rotor assembly motors are generally very much exposed to a high degree of moisture. For this reason the motor should be checked at least yearly by an electrician to be sure all parts are in good working condition.

2. Gear Reducer

    Generally all new oil-lubricated equipment has a "break-in" period of about 400 hours. After this time the oil should be drained from the gear reducer, the unit flushed and new oil added. This procedure removes fine metal particles that have worn off the internal components as a result of the initial close tolerances as the equipment was "breaking-in." If large quantities of fine metal particles are found after the "break-in" period, the manufacturer should be consulted.

A high-quality, turbine-type oil is normally used in the gear reducer assembly. Frequency of oil change is after about 1,400 hours of operation under normal service conditions.

3. Bearings

a. Gear Reducer Bearings

These bearings are generally greased twice per week *WHILE THE ROTOR ASSEMBLY IS IN OPERATION* to ensure proper distribution of the lubricant.

Extension tubes or pipes are usually attached to the bearing cap. A grease fitting is then installed on the other end of the tube or pipe. This allows for safety when greasing the bearings while the rotor is operating by extending the grease fitting away from the hazardous area of the operating rotor. This grease fitting modification may be used in other equipment applications where lubrication of running equipment would present a safety hazard.

b. Rotor Intermediate and Outboard Bearings

These bearings are generally lubricated *DAILY WHILE THE ROTOR ASSEMBLY IS IN OPERATION.*

In most bearing lubrication applications, overlubrication is wasteful and reduces the effectiveness of lubricant seals and may cause overheating of bearings. Rotor intermediate and outboard bearings *ARE GENERALLY THE EXCEPTION.* As a rule, these bearings cannot be overlubricated.

Some rotor bearings are protected by neoprene seals which retain lubricant and keep out moisture. The rotor shaft may have "slingers" near the bearing and a detachable shield may cover the bearing. Be sure you do not overlubricate and destroy the seals.

4. Lubrication

Lubrication of this equipment in locations that have extreme weather variations is critical. Oil and grease must be changed, using the proper type and grade for the expected weather conditions, as determined by the equipment manufacturer.

NOTE: The proper type and grade of oil is a necessity. If it is too thin or too thick it will interfere with proper functioning of bearings and gears.

Bearing grease changes may be accomplished by flushing the grease housing with a 30 to 90 weight oil. The oil is added in the same manner as the grease.

QUESTIONS

Write your answers in a notebook and then compare your answers with those on page 286.

8.3K What problems can be caused by ice during cold weather?

8.3L Why are more microorganisms (higher MLSS) needed in the ditch during cold weather than during warm weather?

8.3M Why is a general cleanup each day at the plant important?

8.35 Operational Guidelines

8.350 English System

Section 8.301, "Description of Oxidation Ditches," lists operational guidelines for an oxidation ditch plant. This section outlines the procedures for you to follow to calculate the guidelines for your oxidation ditch. The first step is to draw a sketch of your oxidation ditch, to obtain the important dimensions, and to record the appropriate flow and wastewater characteristics.

1. Oxidation ditch dimensions.

   a. Length, ft = 200 ft

   b. Bottom, ft = 8 ft

   c. Height, ft = 8 ft

   d. Depth, ft = 4 ft

   e. Radius, ft = 28 ft

   f. Slope = 2

2. Flow and waste characteristics.

   a. Flow, MGD = 0.2 MGD

   b. BOD, mg/L = 200 mg/L

   c. Influent SS, mg/L = 200 mg/L

   d. MLSS, mg/L = 4,000 mg/L

   e. Volatile Matter in MLSS = 70%

3. Calculate oxidation ditch volume in cubic feet and million gallons.

   a. Determine cross-sectional area.

   Find average width of water in ditch, W in feet.

   $$W, ft = B + \frac{D}{S}$$

   $$= 8 ft + \frac{4 ft}{2}$$

   $$= 8 ft + 2 ft$$

   $$= 10 ft$$

   Area, sq ft = W D

   $$= 10 ft \times 4 ft$$

   $$= 40 sq ft$$

   b. Determine the centerline length of the ditch. Find the length around the two ends.

   Ends L, ft = $2 \pi R$

   $$= 2 \times 3.14 \times 28 ft$$

   $$= 176 ft$$

   Total L, ft = Ends L, ft + 2 L, ft

   $$= 176 ft + (2 \times 200 ft)$$

   $$= 576 ft$$

   c. Calculate ditch volume.

   Vol, cu ft = Length, ft x Area, sq ft

   $$= 576 ft \times 40 sq ft$$

   $$= 23,040 cu ft$$

   $$= 23.04 \bullet (1,000 cu ft)(\text{read as } 23.04\text{ one thousand cubic feet})$$

   Vol, gal = 23,040 cu ft x 7.48 gal/cu ft

   $$= 172,339 gal$$

   $$= 0.172 MG$$

4. Calculate the BOD loading in pounds BOD per day and pounds BOD per day per 1,000 cu ft of ditch volume.

   BOD Loading, lbs BOD/day = Flow, MGD x BOD, mg/L x 8.34 lbs gal

   $$= 0.2 MGD \times 200 mg/L \times 8.34 lbs/gal*$$

   $$= 334 lbs BOD/day$$

   $$\text{BOD Loading, lbs BOD/day/1,000 cu ft} = \frac{BOD, lbs BOD/day}{Ditch Vol, 1,000 cu ft}$$

   $$= \frac{334 lbs BOD/day}{23.04 \bullet (1,000 cu ft)}$$

   $$= 14 lbs BOD/day/1,000 cu ft$$

* Remember that 1 liter = 1,000,000 mg and that

$$\frac{lbs}{day} = \frac{Mil Gal}{day} \times \frac{mg}{Mil mg} \times \frac{lbs}{gal} = \frac{lbs}{day}$$

5. Determine the Food/Microorganism Ratio.

   a. Microorganisms are measured as pounds of Mixed Liquor Volatile Suspended Solids under aeration in the oxidation ditch. VM means Volatile Matter.

   MLVSS, lbs = Vol, MG x MLSS, mg/L x VM x 8.34 lbs/gal

   $$= 0.172 MG \times 4,000 mg/L \times 0.70 \times 8.34 lbs/gal$$

   $$= 4,017 lbs MLVSS$$

   b. Calculate F/M Ratio.

   $$\frac{F}{M}, lbs BOD/day/lb MLVSS = \frac{BOD, lbs BOD/day}{MLVSS, lbs}$$

   $$= \frac{334 lbs BOD/day}{4,017 lbs MLVSS}$$

   $$= 0.08 lb BOD/day/lb MLVSS$$

6. Determine the Sludge Age, days.

   a. Calculate pounds of solids under aeration.

   Aeration Solids, lbs = Vol, MG x MLSS, mg/L x 8.34 lbs/gal

   $$= 0.172 MG \times 4,000 mg/L \times 8.34 lbs/gal$$

   $$= 5,738 lbs$$

   b. Calculate solids fed to ditch, lbs/day.

   Solids Added, lbs/day = Flow, MGD x Infl SS, mg/L x 8.34 lbs/gal

   $$= 0.2 MGD \times 200 mg/L \times 8.34 lbs/gal$$

   $$= 334 lbs/day$$

   c. Calculate Sludge Age, days.

   $$Sludge Age, days = \frac{Solids Under Aeration, lbs}{Solids Added, lbs/day}$$

   $$= \frac{5,738 lbs}{334 lbs/day}$$

   $$= 17 days$$

7. Calculate ditch detention time, hours.

   $$Detention Time, hours = \frac{Ditch Volume, MG \times 24 hr/day}{Flow, MGD}$$

   $$= \frac{0.172 MG \times 24 hr/day}{0.2 MGD}$$

   $$= 20.6 hours$$

8.351 Metric System

This section will show you how to calculate operational criteria using the metric system. To start you must collect the same basic information.

1. Oxidation ditch dimensions.

   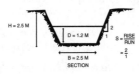

   a. Length, m = 60 meters

   b. Bottom, m = 2.5 meters

   c. Height, m = 2.5 meters

   d. Depth, m = 1.2 meters

   e. Radius, m = 9 meters

   f. Slope   = 2

2. Flow and waste characteristics.

   a. Flow, cu m/day   = 750 cu m/day

   b. BOD, mg/L   = 200 mg/L

   c. Influent SS, mg/L = 200 mg/L

   d. MLSS, mg/L   = 4,000 mg/L

   e. Volatile Matter = 70%
      in MLSS

3. Calculate oxidation ditch volume in cubic meters.

   a. Determine cross-sectional area.

      Find average width of water in ditch, W in meters.

      $$W, m = B + \frac{D}{S}$$

      $$= 2.5\ m + \frac{1.2\ m}{2}$$

      $$= 3.1\ m$$

      $$Area,\ sq\ m = W\ D$$

      $$= 3.1\ m \times 1.2\ m$$

      $$= 3.72\ sq\ m$$

   b. Determine the centerline length of the ditch. Find the length around the two ends.

      $$Ends\ L,\ m = 2\ \pi\ R$$

      $$= 2 \times 3.14 \times 9\ m$$

      $$= 56.5\ m$$

      $$Total\ L,\ m = Ends\ L,\ m + 2\ L,\ m$$

      $$= 56.5\ m + (2 \times 60\ m)$$

      $$= 176.5\ m$$

   c. Calculate ditch volume.

      $$Vol,\ cu\ m = Length,\ m \times Area,\ sq\ m$$

      $$= 176.5\ m \times 3.72\ sq\ m$$

      $$= 657\ cu\ m$$

      $$= 0.657 \cdot (1,000\ cu\ m)(read\ as\ 0.657\ one\ thousand\ cubic\ meters)$$

4. Calculate the BOD loading in kg BOD per day and kg BOD per day per 1,000 cu m of ditch volume.

   $$BOD\ Loading,\ kg\ BOD/day = Flow, \frac{cu\ m}{day} \times BOD\ \frac{mg}{L} \times \frac{1\ kg}{1,000,000\ mg} \times \frac{1,000\ L}{1\ cu\ m}$$

   $$= 750\ \frac{cu\ m}{day} \times 200\ \frac{mg}{L} \times \frac{1\ kg}{1,000,000\ mg} \times \frac{1,000\ L}{1\ cu\ m}$$

   $$= 150\ kg\ BOD/day$$

   $$BOD\ Loading,\ kg\ BOD/day/1,000\ cu\ m = \frac{BOD,\ kg\ BOD/day}{Ditch\ Volume,\ 1,000\ cu\ m}$$

   $$= \frac{150\ kg\ BOD/day}{0.657 \cdot (1,000\ cu\ m)}$$

   $$= 228\ kg\ BOD/day/1,000\ cu\ m$$

5. Determine the Food/Microorganism Ratio.

   a. Calculate kilograms of MLVSS under aeration in the oxidation ditch.

      $$MLVSS,\ kg = Vol,\ cu\ m \times MLSS, \frac{mg}{L} \times VM \times \frac{1\ kg}{1,000,000\ mg} \times \frac{1,000\ L}{1\ cu\ m}$$

      $$= 657\ cu\ m \times 4,000\ \frac{mg}{L} \times 0.70 \times \frac{1\ kg}{1,000,000\ mg} \times \frac{1,000\ L}{1\ cu\ m}$$

      $$= 1,840\ kg\ MLVSS$$

   b. Calculate F/M Ratio.

      $$\frac{F}{M},\ kg\ BOD/day/kg\ MLVSS = \frac{BOD,\ kg/day}{MLVSS,\ kg}$$

      $$= \frac{150\ kg\ BOD/day}{1,840\ kg\ MLVSS}$$

      $$= 0.08\ kg\ BOD/day/kg\ MLVSS$$

6. Determine the Sludge Age, days.

   a. Calculate kilograms of solids under aeration.

      $$Aeration\ Solids,\ kg = Vol,\ cu\ m \times MLSS,\ mg/L \times \frac{1\ kg}{1,000,000\ mg} \times \frac{1,000\ L}{1\ cu\ m}$$

      $$= 657\ cu\ m \times 4,000\ mg/L \times \frac{1\ kg}{1,000,000\ mg} \times \frac{1,000\ L}{1\ cu\ m}$$

      $$= 2,628\ kg$$

   b. Calculate solids fed to ditch, kg/day.

      $$Solids\ Added,\ kg/day = Flow,\ cu\ m/day \times Infl\ SS,\ mg/L \times \frac{1\ kg}{1,000,000\ mg} \times \frac{1,000\ L}{1\ cu\ m}$$

      $$= 750\ cu\ m/day \times 200\ mg/L \times \frac{1\ kg}{1,000,000\ mg} \times \frac{1,000\ L}{1\ cu\ m}$$

      $$= 150\ kg/day$$

c. Calculate Sludge Age, days.

$$\text{Sludge Age, days} = \frac{\text{Solids Under Aeration, kg}}{\text{Solids Added, kg/day}}$$

$$= \frac{2,628 \text{ kg}}{150 \text{ kg/day}}$$

$$= 17.5 \text{ days}$$

7. Calculate ditch detention time, hours.

$$\text{Detention Time, hours} = \frac{\text{Ditch Volume, cu m x 24 hr/day}}{\text{Flow, cu m/day}}$$

$$= \frac{657 \text{ cu m x 24 hr/day}}{750 \text{ cu m/day}}$$

$$= 21 \text{ hours}$$

QUESTION

Write your answers in a notebook and then compare your answers with those on page 286.

8.3N   Determine the BOD loading on an oxidation ditch in:

1. Pounds of BOD/day, and

2. Kilograms of BOD/day.

The inflow is 0.8 MGD and the influent BOD is 250 mg/L.

1 MGD = 3,785 cu m/day

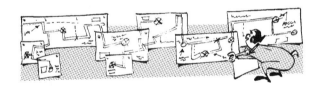

8.4   REVIEW OF PLANS AND SPECIFICATIONS

As an operator you can be very helpful to design engineers in pointing out some design features that would make your job easier. This section indicates some of the items that you should look for when reviewing plans and specifications for expansion of existing facilities or construction of new package plants and oxidation ditches.

8.40   Package Plants

1. Is the plant designed for implementation of modifications to the activated sludge process (adequate flexibility to accommodate future treatment requirements)?

2. Are adequate standby units (equipment) designed into the system?

3. Are ladders, railings, and walkways provided to allow safe, easy access to equipment, pipes, and valves for normal operation, routine maintenance, or repair?

4. Are adequate remote and local controls provided for the mechanical equipment?

5. Are the equipment and related instrumentation designed to operate at low flows and loads common in the early stages of plant operation?

6. Are adequate dewatering systems provided to permit rapid servicing of submerged equipment?

7. Is the chlorination facility flexible enough to allow for pre-chlorination and are adequate control devices provided?

8. Is the treatment plant's total connected horsepower adequate to allow operation of all equipment in parallel?

9. Are flow equalization facilities provided to handle high flows during wet weather or industrial discharges?

10. Is adequate support equipment provided to allow easy and safe removal of aeration diffusers?

11. Is a laboratory provided with appropriate equipment for conducting at least minimum process control tests?

12. Are adequate sludge drying beds provided with consideration given to wet weather conditions?

13. Is a sludge transfer pump provided to transfer sludge to the drying beds?

14. Are the beds designed for easy removal of dried sludge and are there provisions for proper disposal of the sludge?

15. Is there a provision for waste or digested sludge liquid storage in place of or in addition to sludge drying beds?

16. Is standby or auxiliary power provided?

8.41   Oxidation Ditches

1. Influent and return activated sludge should enter an oxidation ditch just upstream of a rotor assembly to afford immediate mixing with mixed liquor in the channel.

2. Effluent should exit the oxidation ditch upstream of the rotor and far enough upstream from the injection of the influent and return activated sludge to prevent short circuiting.

3. Water level in the aeration channel should be controlled by an adjustable weir. In calculating weir height or "set point," use maximum raw flow plus maximum recirculated flow to prevent excessive rotor immersions.

4. Walkways with railings must be provided across the aeration channel to provide access to the rotor for maintenance. The normal location is upstream of the rotor. Location should be such to prevent spray from the rotor on the walkway. Approved flotation devices should be provided at strategic locations if accidental entry into the channel is possible.

5. Horizontal baffles, placed within 15 feet (4.5 m) downstream of the rotor, should be used on all basins with water depths over 6 feet (1.8 m) to provide proper mixing of the entire depth of the basin.

6. In a single oxidation ditch or aeration channel, the rotor drive assembly should be on the outboard side for ease of access.

7. The ditch should be constructed with some type of lining. Consideration should be given to the most economical means of lining available in the particular plant location. As typical liners, consideration could be given to gunite or shotcrete, poured concrete or earthen walls, asphalt, or precast concrete or tile.

8. All drive and gear assemblies should be elevated out of the water and safe and easy access should be provided for maintenance.

9. Standby or auxiliary power must be provided to operate critical equipment.

10. Floating aerators and related equipment should be provided in case of rotor failure in order to maintain adequate treatment.

11. Secondary clarifier.

   a. Surface Rate: 600 gal/day/sq ft (24 cu m/day/sq m) based upon plant design flow. Where wide variations in plant hydraulic loading are expected, care should be taken to limit maximum instantaneous surface rate to 1,200 gal/day/sq ft (49 cu m/day/sq m).

   b. Solids Loading: Normal oxidation ditch operation and surface loading rates will prevent excessive solids loading to the final clarifier. However, if complete nitrification is required 12 months per year, the maximum instantaneous solids loading must not exceed 45 lbs/day/sq ft (200 kg/day/sq m).

   c. Detention Time: Three hours based on plant design flow. In no case, however, should the final clarifier have a side water depth less than eight feet (2.4 m).

12. Effluent disposal.

   a. Receiving Stream: The discharge pipe should be located where flood water will not flow back into the plant if there is a power outage or pump failure. A tide gate or check valve can be installed in the effluent line to prevent backflows. Be sure the outlet in the receiving waters is submerged to reduce foam and scum problems.

   b. Percolation Ditch: Design should be based on a percolation rate of gallons per *LINEAL* [16] foot of ditch. Provisions must be made to prevent blowing sand and/or dirt from entering the ditch.

13. Winter climates.

   a. Provisions should be made to maximize detention times in order to conserve as much heat as possible in the wastewater.

   b. Is it practical to cover the oxidation ditch? If not, consider a large and a small ditch or a single ditch that can be modified to use only half its total capacity.

   c. Be sure that all equipment requiring normal maintenance is housed and/or heated. This will extend the useful life of the equipment and facilitate service work. Changing a gear box, repairing a pipe, or installing electrical components becomes a major task at very low temperatures.

   d. The rotor assembly should be provided with a lightweight cover to prevent rotor icing.

   e. Equip the clarifier with a lightweight tarpaulin to keep heavy snow out of the clarifier and to reduce freezing problems.

   f. A subsurface disposal pipeline and percolation field should be provided for effluent disposal in winter months. A percolation ditch will freeze over in the winter.

QUESTIONS

Write your answers in a notebook and then compare your answers with those on page 286.

8.4A   What provisions should be made to permit rapid servicing of submerged equipment?

8.4B   What items would you check when reviewing the location of a walkway intended to provide access to the rotor for maintenance?

8.5   ARITHMETIC ASSIGNMENT

Turn to the Arithmetic Appendix at the back of this manual and read all of Section A.8, *PUMPS*. Also work the example problems and check the arithmetic using your calculator.

In Section A.13, *TYPICAL WASTEWATER TREATMENT PLANT PROBLEMS (ENGLISH SYSTEM)*, read and work the problems in Section A.135, Activated Sludge.

8.6   METRIC CALCULATIONS

Refer to Section 8.35, "Operational Guidelines," for metric calculations.

END OF LESSON 3 OF 3 LESSONS ON ACTIVATED SLUDGE VOL. 1

Please answer the discussion and review questions next.

---

[16] *Lineal (LIN-e-al).   The length in one direction of a line. For example, a board 12 feet long has 12 lineal feet in its length.*

DISCUSSION AND REVIEW QUESTIONS

Chapter 8.   ACTIVATED SLUDGE

(Lesson 3 of 3 Lessons)

Write the answers to these questions in your notebook. The question numbering continues from Lesson 2.

11. What happens to inorganic solids such as grit, sand, and silt that enter an oxidation ditch plant?

12. During plant start-up, what does a dark gray color in the mixed liquor indicate and what would you do if this condition persists for more than several days?

13. During plant start-up, when should water be discharged to the clarifier?

14. How can an operator determine if the sludge wasting rate should be increased or decreased?

15. What would you do if solids were in the effluent of the final settling tank during high flows caused by storms or during high influent solids levels caused by the cleaning of the collection system sewers?

SUGGESTED ANSWERS

Chapter 8.   ACTIVATED SLUDGE

PACKAGE PLANTS AND OXIDATION DITCHES

ANSWERS TO QUESTIONS IN LESSON 1

Answers to questions on page 255.

8.0A   The purpose of the activated sludge process in treating wastewater is to oxidize and remove soluble or finely divided suspended materials that were not removed by previous treatment.

8.0B   A stabilized waste is a waste that has been treated or decomposed to the extent that, if discharged or released, its rate and state of decomposition would be such that the waste would not cause a nuisance or odors.

8.0C   Air is added to the aeration tank in the activated sludge process to provide oxygen to sustain the living organisms as they oxidize wastes to obtain energy for growth. The application of air also encourages mixing in the aerator.

8.0D   Air requirements increase when the strength (BOD) of the incoming wastewater increases because more food (wastes) encourages biological activity (reproduction and respiration).

8.0E   Factors that could cause an unsuitable environment for the activated sludge process in an aeration tank include:

1. High concentrations of acids, bases, and other toxic substances.
2. Uneven flows of wastewater that cause overfeeding or starvation.
3. Failure to supply enough oxygen.

Answers to questions on page 256.

8.1A   Effluent quality requirements may be stated by regulatory agencies in terms of percentage removal of wastes or allowable quantities of wastes that may be discharged.

8.1B Harmful industrial waste discharges may be regulated by preventing discharge to the collection system, requiring pretreatment, or controlling the discharge in order to protect an activated sludge process.

ANSWERS TO QUESTIONS IN LESSON 2

Answers to questions on page 261.

8.2A A typical package extended aeration plant may have either two or three compartments. The purpose of each compartment is:

1. Aeration. Aerate and mix waste to be treated with activated sludge.
2. Clarification and Settling. Allow activated sludge to be separated from water being treated. Clarified effluent leaves plant and settled activated sludge is returned to aeration tank to treat more wastewater.
3. Aerobic digestion (optional). Treatment of waste activated sludge.

8.2B Common characteristics of all package aeration plants include long solids retention times, high mixed liquor suspended solids, and low food/microorganism ratios.

Answers to questions on page 264.

8.2C Package plants should have cathodic protection to prevent rust, corrosion, and pitting of steel and iron surfaces in contact with water, wastewater, or soil.

8.2D The valve to the air lift pumps must be closed until the settling compartment is filled or all of the air will attempt to flow out the air lifts and no air will flow out the diffusers.

8.2E If the water in the aeration compartment is murky or cloudy and the aeration compartment has a rotten egg odor (H_2S), you should increase the aeration rate.

Answers to questions on page 267.

8.2F For best results, in terms of effluent quality, waste about five percent of the solids each week during warm weather operation.

8.2G A package plant should be visually checked by an operator every day.

8.2H a. If the solids do not settle in the jar, the aeration rate should be increased and the MLSS concentration checked for proper level.

b. If the solids settle and then float to the surface, the aeration rate should be reduced a little each day until the solids settle properly. Also reduce the wasting rate to increase the MLSS content.

Answers to questions on page 267.

8.2I Operators can avoid being injured on the job by practicing safety, by avoiding hazardous conditions, and by using safe procedures.

8.2J If you must work alone, phone your office at regular intervals and have someone check on you if you fail to report.

ANSWERS TO QUESTIONS IN LESSON 3

Answers to questions on page 272.

8.3A The ends of oxidation ditches are well rounded to prevent eddying and dead areas.

8.3B The major components of an oxidation ditch treatment process include:

1. Oxidation ditch,
2. Rotor,
3. Level control weir,
4. Final settling tank,
5. Return sludge pump, and
6. Excess sludge handling facilities.

8.3C Operators should avoid walking on top of oxidation ditch sidewalls to avoid slipping and falling into the ditch.

Answers to questions on page 273.

8.3D The two primary objectives of start-up are to:

1. Make certain that all mechanical equipment is operating properly, and
2. Develop a proper microbial floc (activated sludge) in the oxidation ditch.

8.3E The following items or structures should be inspected during the pre-start inspection.

1. Inlet structure,
2. Oxidation ditch structure,
3. Rotor,
4. Adjustable weir,
5. Clarifier structure and piping, and
6. Return sludge and waste sludge systems.

8.3F During plant start-up, start the rotors when the water level in the ditch reaches the bottom of the rotor blades.

8.3G When an oxidation ditch is operating properly, the activated sludge solids should look like particles, golden to rich dark brown in color, with sharply defined edges.

Answers to questions on page 275.

8.3H   The DO in the oxidation ditch is regulated by raising or lowering the ditch outlet level control weir. This weir controls the level of water in the ditch which in turn regulates the degree of submergence of the rotors. The greater the submergence, the more oxygen is transferred from the air.

8.3I   The ditch velocity should be maintained between 1.0 and 1.5 fps (0.3 to 0.45 mps).

8.3J   Daily observations of ditch color, odors, lack of foam on the aerator surface, and settling tank clarity help indicate the performance of the oxidation ditch.

Answers to questions on page 279.

8.3K   During cold weather ice buildup will hinder or stop altogether the proper operation of mechanical parts such as rotors and sludge scraper mechanisms.

8.3L   More microorganisms are needed to treat the same amount of wastes during cold weather than during warm weather because the colder the water, the less active the microorganisms.

8.3M   A general cleanup each day provides a more pleasant place to work and also improves plant performance.

Answers to question on page 282.

8.3N   Determine the BOD loading on an oxidation ditch in (1) pounds of BOD/day, and (2) kilograms of BOD/day. The inflow is 0.8 MGD and the influent BOD is 250 mg/L. 1 MGD = 3,785 cu m/day.

Known	Unknown
Flow, MGD = 0.8 MGD	1. BOD Loading, lbs BOD/day
BOD, mg/L = 250 mg/L	2. BOD Loading, kg BOD/day
1 MGD = 3,785 cu m/day	

1. Determine BOD loading, lbs BOD/day.

$$\text{BOD Loading, lbs BOD/day} = \text{Flow, MGD} \times \text{BOD, mg/}L \times 8.34 \text{ lbs/gal}$$

$$= 0.8 \text{ MGD} \times 250 \text{ mg/}L \times 8.34 \text{ lbs/gal}$$

$$= 1,668 \text{ lbs BOD/day}$$

2. Determine BOD loading, kg BOD/day.

a. Convert flow from MGD to cubic meters per day.

$$\text{Flow, cu m/day} = \text{Flow, MGD} \times \frac{3,785 \text{ cu m/day}}{1 \text{ MG}}$$

$$= \frac{0.8 \text{ MGD} \times 3,785 \text{ cu m/day}}{1 \text{ MGD}}$$

$$= 3,028 \text{ cu m/day}$$

b. Calculate BOD loading, kg BOD/day.

$$\text{BOD Loading, kg BOD/day} = \text{Flow, cu m/day} \times \text{BOD, mg/}L \times \frac{1 \text{ kg}}{1,000,000 \text{ mg}} \times \frac{1,000 \text{ } L}{1 \text{ cu m}}$$

$$= 3,028 \text{ cu m/day} \times 250 \text{ mg/}L \times \frac{1 \text{ kg}}{1,000,000 \text{ mg}} \times \frac{1,000 \text{ } L}{1 \text{ cu m}}$$

$$= 757 \text{ kg BOD/day}$$

Answers to questions on page 283.

8.4A   Adequate dewatering systems should be provided to permit rapid servicing of submerged equipment.

8.4B   When reviewing the location of a walkway intended to provide access to the rotor for maintenance, check:

1. Distance from walkway to rotor,
2. Normal location is upstream from rotor, and
3. Spray from the rotor will not fall on the walkway.

END OF ANSWERS TO QUESTIONS IN LESSON 3

CHAPTER 9

WASTEWATER STABILIZATION PONDS

by

A. Hiatt

TABLE OF CONTENTS

Chapter 9.   WASTEWATER STABILIZATION PONDS

OBJECTIVES

Chapter 9.   WASTEWATER STABILIZATION PONDS

Following completion of Chapter 9, you should be able to:

1. Explain how wastewater stabilization ponds work and what factors influence and control pond treatment processes,

2. Place a new pond into operation,

3. Schedule and conduct normal and abnormal operational and maintenance duties,

4. Collect samples, interpret lab results, and make appropriate adjustments in pond operation,

5. Recognize factors that indicate a pond is not performing properly, identify the source of the problem, and take corrective action,

6. Develop a pond operating strategy,

7. Conduct your duties in a safe fashion,

8. Determine pond loadings,

9. Identify the different types of ponds,

10. Keep records for a waste treatment pond facility, and

11. Review plans and specifications for new ponds.

WORDS

Chapter 9. WASTEWATER STABILIZATION PONDS

ADVANCED WASTE TREATMENT ADVANCED WASTE TREATMENT

Any process of water renovation that upgrades treated wastewater to meet specific reuse requirements. May include general cleanup of water or removal of specific parts of wastes insufficiently removed by conventional treatment processes. Typical processes include chemical treatment and pressure filtration. Also called TERTIARY TREATMENT.

ALGAE (AL-gee) ALGAE

Microscopic plants which contain chlorophyll and live floating or suspended in water. They also may be attached to structures, rocks, or other submerged surfaces. Algae produce oxygen during sunlight hours and use oxygen during the night hours. Their biological activities appreciably affect the pH, alkalinity, and dissolved oxygen of the water.

ALGAL (AL-gull) BLOOM ALGAL BLOOM

Sudden, massive growths of microscopic and macroscopic plant life, such as green or blue-green algae, which can, under the proper conditions, develop in lakes, reservoirs, and also in ponds.

BIOFLOCCULATION (BUY-o-flock-u-LAY-shun) BIOFLOCCULATION

The clumping together of fine, dispersed organic particles by the action of certain bacteria and algae. This results in faster and more complete settling of the organic solids in wastewater.

CHEMICAL OXYGEN DEMAND (COD) CHEMICAL OXYGEN DEMAND (COD)

A measure of the oxygen-consuming capacity of organic matter present in wastewater. COD is expressed as the amount of oxygen consumed from a chemical oxidant in mg/L during a specific test. Results are not necessarily related to the biochemical oxygen demand (BOD) because the chemical oxidant may react with substances that bacteria do not stabilize.

COMPOSITE (come-PAH-zit) (PROPORTIONAL) SAMPLE COMPOSITE (PROPORTIONAL) SAMPLE

A composite sample is a collection of individual samples obtained at regular intervals, usually every one or two hours during a 24-hour time span. Each individual sample is combined with the others in proportion to the rate of flow when the sample was collected. The resulting mixture (composite sample) forms a representative sample and is analyzed to determine the average conditions during the sampling period.

DUCKWEED DUCKWEED

A small, green, cloverleaf-shaped floating plant, about one-quarter inch (6 mm) across which appears as a grainy "scum" on the surface of a pond.

FACULTATIVE (FACK-ul-TAY-tive) POND FACULTATIVE POND

The most common type of pond in current use. The upper portion (supernatant) is aerobic, while the bottom layer is anaerobic. Algae supply most of the oxygen to the supernatant.

FREE OXYGEN FREE OXYGEN

Molecular oxygen available for respiration by organisms. Molecular oxygen is the oxygen molecule, O_2, that is not combined with another element to form a compound.

GRAB SAMPLE GRAB SAMPLE

A single sample of water collected at a particular time and place which represents the composition of the water only at that time and place.

MEDIAN MEDIAN

The middle measurement or value. When several measurements are ranked by magnitude (largest to smallest), half of the measurements will be larger and half will be smaller.

MOLECULAR OXYGEN MOLECULAR OXYGEN

The oxygen molecule, O_2, that is not combined with another element to form a compound.

OVERTURN OVERTURN

The almost spontaneous mixing of all layers of water in a reservoir or lake when the water temperature becomes similar from top to bottom. This may occur in the fall/winter when the surface waters cool to the same temperature as the bottom waters and also in the spring when the surface waters warm after the ice melts. This is also called "turnover."

PARALLEL OPERATION PARALLEL OPERATION

Wastewater being treated is split and a portion flows to one treatment unit while the remainder flows to another similar treatment unit. Also see SERIES OPERATION.

PERCOLATION (PURR-co-LAY-shun) PERCOLATION

The movement or flow of water through soil or rocks.

pH (pronounce as separate letters) pH

pH is an expression of the intensity of the basic or acidic condition of a liquid. Mathematically, pH is the logarithm (base 10) of the reciprocal of the hydrogen ion activity.

$$pH = Log \frac{1}{[H^+]}$$

The pH may range from 0 to 14, where 0 is most acidic, 14 most basic, and 7 neutral. Natural waters usually have a pH between 6.5 and 8.5.

PHOTOSYNTHESIS (foe-toe-SIN-thuh-sis) PHOTOSYNTHESIS

A process in which organisms, with the aid of chlorophyll (green plant enzyme), convert carbon dioxide and inorganic substances into oxygen and additional plant material, using sunlight for energy. All green plants grow by this process.

POPULATION EQUIVALENT POPULATION EQUIVALENT

A means of expressing the strength of organic material in wastewater. In a domestic wastewater system, microorganisms use up about 0.2 pound of oxygen per day for each person using the system (as measured by the standard BOD test). May also be expressed as flow (100 gallons per day per person) or suspended solids (0.2 lb SS/day/person).

$$\text{Population Equivalent, persons} = \frac{\text{Flow, MGD x BOD, mg/}L\text{ x 8.34 lbs/gal}}{0.2 \text{ lb BOD/day/person}}$$

RIPRAP RIPRAP

Broken stones, boulders, or other materials placed compactly or irregularly on levees or dikes for the protection of earth surfaces against the erosive action of waves.

SERIES OPERATION SERIES OPERATION

Wastewater being treated flows through one treatment unit and then flows through another similar treatment unit. Also see PARALLEL OPERATION.

SPLASH PAD SPLASH PAD

A structure made of concrete or other durable material to protect bare soil from erosion by splashing or falling water.

STABILIZED WASTE STABILIZED WASTE

A waste that has been treated or decomposed to the extent that, if discharged or released, its rate and state of decomposition would be such that the waste would not cause a nuisance or odors in the receiving water.

STOP LOG STOP LOG

A log or board in an outlet box or device used to control the water level in ponds and also the flow from one pond to another pond or system.

TERTIARY (TER-she-AIR-ee) TREATMENT TERTIARY TREATMENT

Any process of water renovation that upgrades treated wastewater to meet specific reuse requirements. May include general cleanup of water or removal of specific parts of wastes insufficiently removed by conventional treatment processes. Typical processes include chemical treatment and pressure filtration. Also called ADVANCED WASTE TREATMENT.

TOXIC (TOX-ick)   TOXIC
A substance which is poisonous to a living organism.

TOXICITY (tox-IS-it-tee)   TOXICITY
The relative degree of being poisonous or toxic. A condition which may exist in wastes and will inhibit or destroy the growth or function of certain organisms.

CHAPTER 9.   WASTEWATER STABILIZATION PONDS

USED FOR TREATMENT OF WASTEWATER AND OTHER WASTES

(Lesson 1 of 3 Lessons)

9.0   USE OF PONDS

Shallow ponds (3 to 5 feet or 1 to 1.5 meters deep) are often used to treat wastewater and other wastes instead of, or in addition to, conventional waste treatment processes. (See Figures 9.1 and 9.2 for typical plant layouts and Table 9.1 for purpose of pond parts.) When discharged into ponds, wastes are treated or *STABILIZED*[1] by several natural processes acting at the same time. Heavy solids settle to the bottom where they are decomposed by bacteria. Lighter suspended material is broken down by bacteria in suspension. Some wastewater is disposed of by evaporation from the pond surface.

Dissolved nutrient materials, such as nitrogen and phosphorus, are used by green *ALGAE*,[2] which are actually microscopic plants floating and living in the water. The algae use carbon dioxide (CO_2) and bicarbonate to build body protoplasm. In so growing, they need nitrogen and phosphorus in their metabolism much as land plants do. Like land plants, they release oxygen and some carbon dioxide as waste products.

Ponds can serve as very effective treatment facilities. Extensive studies of their performance have led to a better understanding of the natural processes by which ponds treat wastes. Information is provided here on the natural processes and ways operators can regulate pond processes for efficient waste treatment.

TABLE 9.1   PURPOSE OF POND PARTS

Part	Purpose
Flowmeter	Measures and records flows into pond.
Bar Screen	Removes coarse material from pond influent.
Pond Inlets	Distribute influent in pond.
Pond Depth and Outlet Control	Regulates outflow from pond and depth of water in pond. Allows pond to be drained for cleaning and inspection.
Outlet Baffle	Prevents scum and other surface debris from flowing to next pond or receiving waters.
Dike or Levee	Separates ponds and holds wastewater being treated in ponds.
Transfer Line	Conveys wastewater from one pond to another.
Recirculation Line	Returns pond effluent rich in algae and oxygen from second pond to first pond for seeding, dilution, and process control.
Chlorination	Applies chlorine to treated wastewater for disinfection purposes.
Chlorine Contact Basin	Provides contact time for chlorine to disinfect pond effluent.
Effluent Line	Conveys treated wastewater to receiving waters, to point of reuse (irrigation), or to land disposal site.

9.1   HISTORY OF PONDS IN WASTE TREATMENT

The first wastewater collection systems in the ancient Orient and in ancient Europe discharged wastewater into nearby bodies of water. These systems accomplished their intended purpose until overloading, as in modern systems, made them objectionable.

---

[1] *Stabilized Waste.   A waste that has been treated or decomposed to the extent that, if discharged or released, its rate and state of decomposition would be such that the waste would not cause a nuisance or odors in the receiving water.*

[2] *Algae (AL-gee).   Microscopic plants which contain chlorophyll and live floating or suspended in water. They also may be attached to structures, rocks, or other submerged surfaces. Algae produce oxygen during sunlight hours and use oxygen during the night hours. Their biological activities appreciably affect the pH, alkalinity, and dissolved oxygen of the water.*

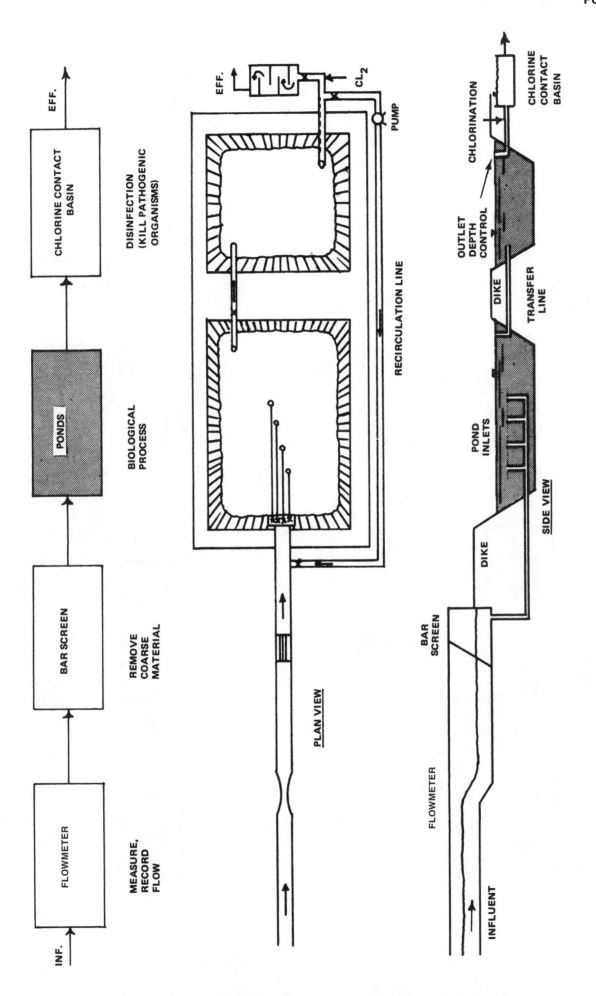

Fig. 9.1 Typical plant; ponds only

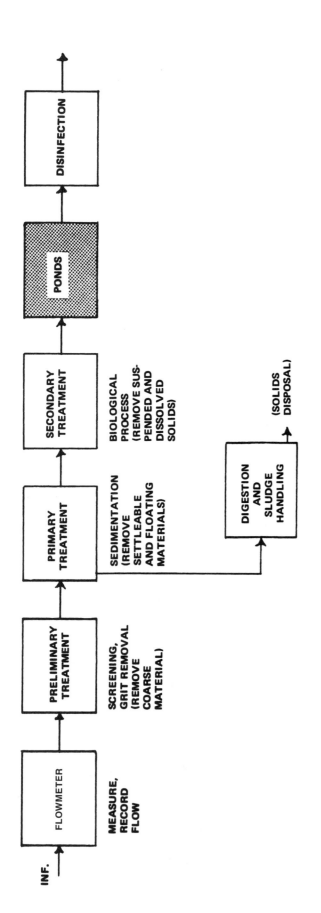

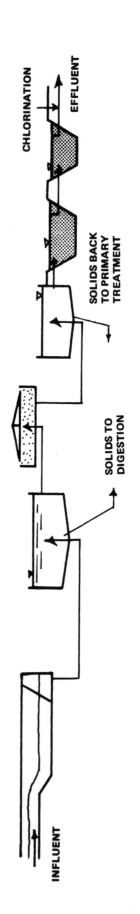

Fig. 9.2   Typical plant; ponds after secondary treatment

In ancient times, ponds and lakes were purposefully fertilized with organic wastes to encourage the growth of algae which, in turn, greatly increased the production of fish due to the food supply provided by the algae. This practice still continues and is a recognized art in Germany.

The first ponds constructed in the United States were built for the purpose of keeping wastewaters from flowing into places where they would be objectionable. Once built, these ponds performed a treatment process that finally became recognized as such.

The tendency over the years has been to equate pond treatment efficiency with the absence of odors. Actually, the opposite is true as the greatest organic load destroyed per unit of area (high treatment efficiency) may be accompanied by objectionable odors.

Since 1958, engineers have designed and constructed a great number of ponds using research by qualified biological consultants, current scientific knowledge of ponding, and the experience of past successes and failures. When operated in a knowledgeable and purposeful manner, these ponds have successfully performed a variety of functions.

As a complete process, the ponding of wastewater offers many advantages for smaller installations. This is true provided that land is not costly and the location is isolated from residential, commercial, and recreational areas. The advantages are that a pond:

1. Does not require expensive equipment;

2. Does not require highly trained operating personnel;

3. Is economical to construct;

4. Provides treatment that is equal or superior to some conventional processes;

5. Is a satisfactory method of treating wastewater on a temporary basis;

6. Is adaptable to changing loads;

7. Is adaptable to land application;

8. Consumes little energy;

9. Serves as a wildlife habitat;

10. Has an increased potential design life;

11. Has few sludge handling and disposal problems; and

12. Is probably the most trouble-free of any treatment process when used correctly, *PROVIDED A CONSISTENTLY HIGH-QUALITY EFFLUENT IS NOT REQUIRED.*

The limitations are that a pond:

1. May produce odors;

2. Requires a large area of land;

3. Treats wastes inconsistently depending on climatic conditions;

4. May contaminate groundwaters; and

5. May have high suspended solids levels in the effluent.

QUESTIONS

Write your answers in a notebook and then compare your answers with those on page 331.

9.1A If a pond is giving off objectionable odors, are the wastes being effectively treated? Explain your answer.

9.1B Discuss the limitations of ponds.

9.2 POND CLASSIFICATIONS AND APPLICATIONS

Ponding of raw wastewater, as a complete treatment process, is used to treat the wastes of single families as well as large cities up to the size of the city of Melbourne, Australia. Currently, a portion of Melbourne's wastewater is disposed of on a 28,000 acre farm. During the dry season, most of the treatment is accomplished by broad irrigation and grass filtration (flowing through grasslands) while during the wet season most of the 130 MGD flow is handled by ponds and discharged to receiving waters. Ponds designed to receive wastes with no prior treatment are often referred to as "raw wastewater (sewage) lagoons" or "stabilization ponds" (Figure 9.3). This requires sizable areas of land.

Ponds are quite commonly used in series (one pond following another) after a primary wastewater treatment plant to provide additional clarification, BOD removal, and disinfection. These ponds are sometimes called "oxidation ponds." Ponds are sometimes used in series after a trickling filter plant, thus giving a form of *TERTIARY TREATMENT*.[3] These are sometimes called "polishing ponds." Ponds placed in series with each other can provide a high quality effluent which is acceptable for discharge into many watercourses. If the detention time is long enough, many ponds can meet fecal coliform standards.

A great many variations in ponds are possible due to differences in depth, operating conditions, and loadings. A bold line of distinction among different types of ponds is often impossible. Current literature generally uses three broad pond classifications: *AEROBIC*, *ANAEROBIC*, and *FACULTATIVE*.

AEROBIC ponds are characterized by having dissolved oxygen distributed throughout their contents practically all of the time. They usually require an additional source of oxygen to supplement the rather minimal amount that can be diffused from the atmosphere at the water surface. The additional source of oxygen may be supplied by algae during daylight hours, by mechanical agitation of the surface, or by compressors bubbling air through the pond.

ANAEROBIC ponds, as the name implies, usually are without any dissolved oxygen throughout their entire depth. Treatment depends on fermentation of the sludge at the pond bottom. This process can be quite odorous under certain conditions, but it is highly efficient in destroying organic wastes. Anaerobic ponds are mainly used for processing industrial wastes, although some domestic waste ponds become anaerobic when they are badly overloaded.

FACULTATIVE (FACK-ul-TAY-tive) ponds are the most common type in current use. The upper portion (supernatant) of these ponds is aerobic, while the bottom layer is anaerobic. Algae supply most of the oxygen to the supernatant. Facultative

---

[3] Tertiary (TER-she-AIR-ee) Treatment. *Any process of water renovation that upgrades treated wastewater to meet specific reuse requirements. May include general cleanup of water or removal of specific parts of wastes insufficiently removed by conventional treatment processes. Typical processes include chemical treatment and pressure filtration. Also called advanced waste treatment.*

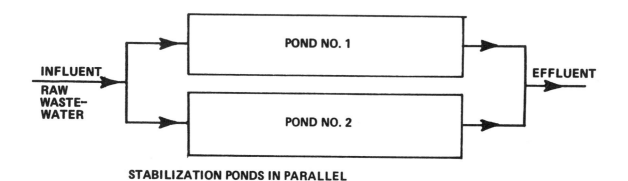

STABILIZATION PONDS IN SERIES

STABILIZATION PONDS IN PARALLEL

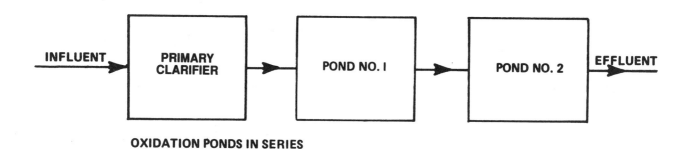

OXIDATION PONDS IN SERIES

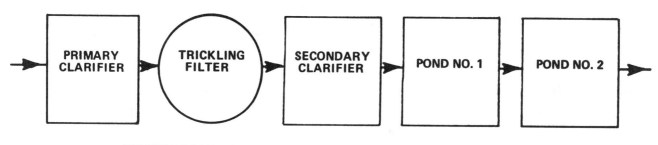

POLISHING PONDS IN SERIES

Fig. 9.3   Pond classifications

ponds are most common because it is almost impossible to maintain completely aerobic or anaerobic conditions all the time at all depths of the pond.

Pond uses may be classified according to detention time. A pond with a detention time of less than three days will perform in ways similar to a sedimentation or settling tank. Some growth of algae will occur in the pond, but it will not have a major effect on the treatment of the wastewater.

Abundant growth of algae will be observed in ponds with detention periods from three to around twenty days, but large amounts of algae will be found in the pond effluent. In some effluents, the stored organic material may be greater than the amount in the influent. Detention times in this range merely allow the organic material to change form and delay problems until the algae settle out in the receiving waters. Effluent BODs may show considerable reductions from influent BODs, but this is because BOD is a rate estimate (oxygen used during a five-day period). The rate of oxygen use is temporarily slowed down, but will increase when anaerobic decomposition of settled dead algae cells starts.

Longer detention periods in ponds provide time for the sedimentation of algae. Usually, this will occur in facultative ponds having anaerobic conditions on the bottom and aerobic conditions on the surface. Combined aerobic-anaerobic treatment provided by long detention periods produces definite stabilization of the pond influent during treatment in the pond.

Controlled-discharge ponds are facultative ponds with long detention times of up to 180 days or longer. These ponds may discharge effluent only once (fall) or twice (fall and spring) a year.

QUESTIONS

Write your answers in a notebook and then compare your answers with those on page 331.

9.2A   What is the difference between raw wastewater (sewage) lagoons, oxidation ponds, and polishing ponds?

9.2B   What is the difference between the terms "aerobic," "anaerobic," and "facultative" when applied to ponds?

9.2C   How does the use of a pond vary depending on detention time?

9.3   EXPLANATION OF TREATMENT PROCESS

As mentioned in the previous section, waste disposal ponds also are classified according to their dissolved oxygen content. Oxygen in an aerobic pond is distributed throughout the entire depth practically all the time. An anaerobic pond is predominantly without oxygen most of the time because oxygen requirements are much greater than the oxygen supply. In a facultative pond, the upper portion is aerobic most of the time, whereas the bottom layer is predominantly anaerobic.

In aerobic ponds or in the aerobic layer of facultative ponds, organic matter contained in the wastewater is first converted to carbon dioxide and ammonia, and finally, to algae in the presence of sunlight. Algae are simple one- or many-celled microscopic plants which are essential to the successful operation of both aerobic and facultative ponds.

By using sunlight through *PHOTOSYNTHESIS*,[4] the algae use the carbon dioxide in the water to produce *FREE OXYGEN*,[5] making it available to the aerobic bacteria that inhabit the pond. Each pound of algae in a healthy pond is capable of producing 1.6 pounds of oxygen on a normal summer day. Algae live on carbon dioxide and other nutrients in the wastewater. At night when light is no longer available for photosynthesis, algae use up the oxygen by respiration and produce carbon dioxide. The alternate use and production of oxygen and carbon dioxide can result in diurnal (daily) variations of both *pH*[6] and dissolved oxygen. During the day algae use carbon dioxide, which raises the pH, while at night they produce carbon dioxide and the pH is lowered. Algae are found in the soil, water, and air; they occur naturally in a pond without seeding and multiply greatly under favorable conditions. Figure 9.4 illustrates the role of algae in treating wastes in a pond.

In anaerobic ponds or in the anaerobic layer of facultative ponds, the organic matter is first converted by a group of organisms called the "acid producers" to carbon dioxide, nitrogen, and organic acids. In an established pond, at the same time, a group called the "methane fermenters" breaks down the acids and other products of the first group to form methane

---

[4] *Photosynthesis (foe-toe-SIN-thuh-sis).   A process in which organisms, with the aid of chlorophyll (green plant enzyme), convert carbon dioxide and inorganic substances into oxygen and additional plant material, using sunlight for energy. All green plants grow by this process.*

$$CO_2 + H_2O \xrightarrow{\text{Light}} CH_2O_x + O_2$$

[5] *Free Oxygen.   Molecular oxygen available for respiration by organisms. Molecular oxygen is the oxygen molecule, O_2, that is not combined with another element to form a compound.*

[6] *pH (pronounce as separate letters).   pH is an expression of the intensity of the basic or acidic condition of a liquid. Mathematically, pH is the logarithm (base 10) of the reciprocal of the hydrogen ion activity.*

$$pH = Log \frac{1}{[H^+]}$$

The pH may range from 0 to 14, where 0 is most acidic, 14 most basic, and 7 neutral. Natural waters usually have a pH between 6.5 and 8.5.

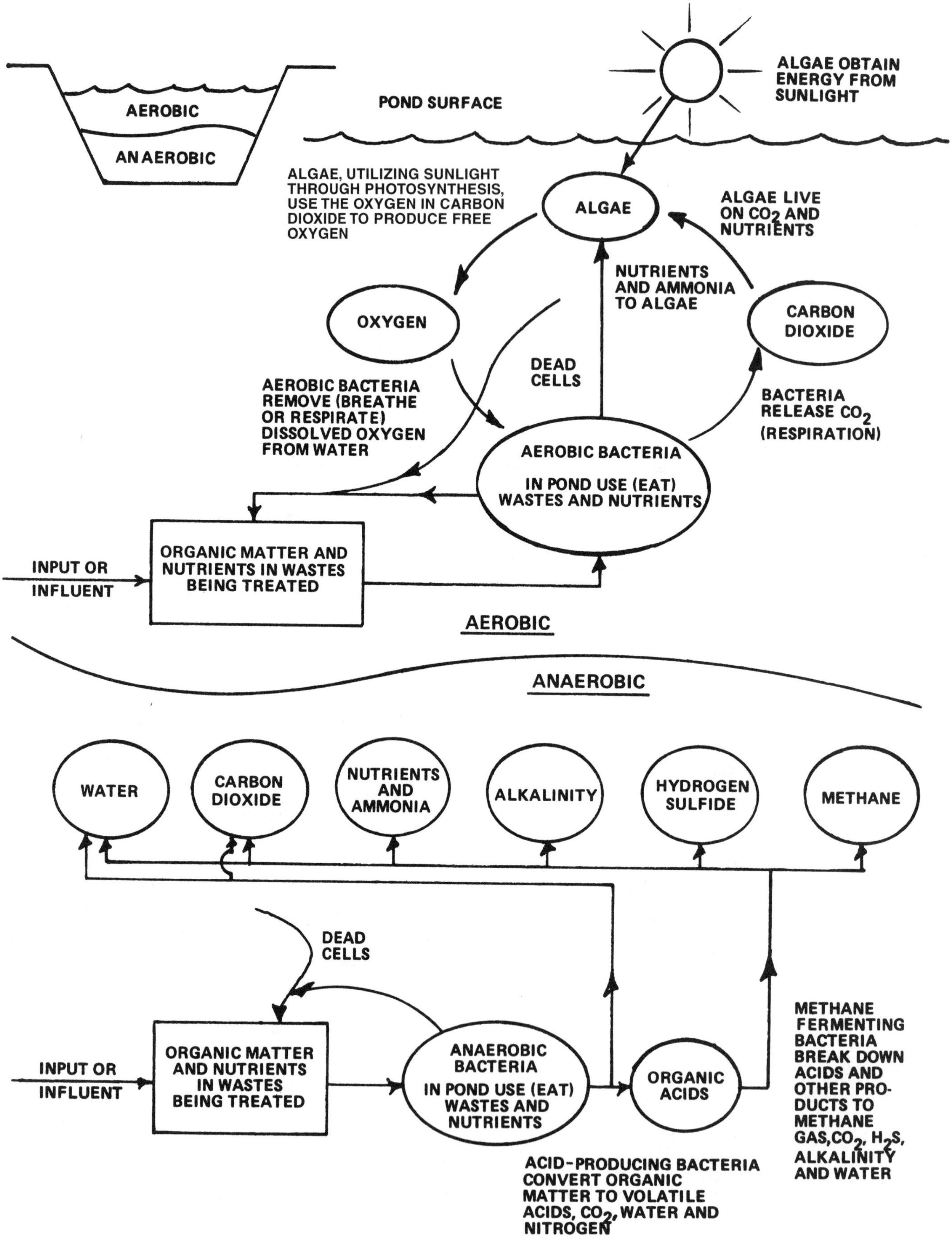

Fig. 9.4   Process of decomposition in aerobic and anaerobic layers of a pond

gas and alkalinity. Water is another end product of organic decomposition. This process is described in Figure 9.4.

In a successful facultative pond, the processes characteristic of aerobic ponds occur in the surface layers, while those similar to anaerobic ponds occur in its bottom layers.

During certain periods, sludge decomposition in the anaerobic zone is interrupted and it begins to accumulate. If sludge accumulation occurs and decomposition does not set in, it is probably due to a lack of the right bacteria, low pH, presence of substances that slow or stop the process, or a low temperature. Under these circumstances, the acid production will continue at a slower rate, but the rate of gas (methane) production slows down considerably.

Sludge storage in ponds is continuous with small amounts stored during warm weather and larger amounts when it is cold. During low temperatures, the bacteria cannot multiply fast enough to handle the waste. When warm weather comes, the "acid producers" begin to decompose the accumulated sludge deposits built up during the winter. If the organic acid production is too great, a lowered pH will occur with the possibilities of an upset pond and resulting hydrogen sulfide odors.

Hydrogen sulfide ordinarily is not a problem in properly designed and operated ponds because it dissociates (divides) into hydrogen and hydrosulfide ions at high pH and may form insoluble metallic sulfides or sulfates. This high degree of dissociation and the formation of insoluble metallic sulfides are the reasons that ponds having a pH above 8.5 do not emit odors, even when hydrogen sulfide is present in relatively large amounts. An exception occurs in northern climates during the spring when the pH is low and the pond is just getting started; then hydrogen sulfide odors can be a problem.

All of the organic matter that finds its way to the bottom of a stabilization pond through the various processes of sludge decomposition is subject to *METHANE FERMENTATION*, provided that proper conditions exist or become established. In order for methane fermentation to exist, an abundance of organic matter must be deposited and continually converted to organic acids. An abundant population of methane bacteria must be present. They require a pH level of from 6.5 to 7.5 within the sludge, alkalinity of several hundred mg/L to buffer (neutralize) the organic acids (volatile acid/alkalinity relationship), and suitable temperatures. Once methane fermentation is established, it accounts for a considerable amount of the organic load removal.

QUESTIONS

Write your answers in a notebook and then compare your answers with those on page 331.

9.3A   How is oxygen produced by algae?

9.3B   Where do the algae found in a pond come from?

9.3C   What happens to unstable organic matter in a pond?

9.4 POND PERFORMANCE

The treatment efficiencies that can be expected from ponds vary more than most other treatment devices. Some of the many variables are:

1. Physical Factors

    a. type of soil
    b. surface area
    c. depth
    d. wind action
    e. sunlight
    f. temperature
    g. short-circuiting
    h. inflow variations

2. Chemical Factors

    a. organic material
    b. pH
    c. solids
    d. concentration and nature of waste

3. Biological Factors

    a. type of bacteria
    b. type and quantity of algae
    c. activity of organisms
    d. nutrient deficiencies
    e. toxic concentrations

The performance expected from a pond depends largely on its design. The design, of course, is determined by the waste discharge requirements or the water quality standards to be met in the receiving waters. Overall treatment efficiency may be about the same as primary treatment (only settling of solids), or it may be equivalent to the best secondary biological treatment plants. Some ponds, usually those located in hot, arid areas, have been designed to take advantage of *PERCOLATION*[7] and high evaporation rates so that there is no discharge.

Depending on design, ponds can be expected to provide BOD removals from 50 to 90 percent. Facultative ponds, under normal design loads with 50 to 60 days' detention time, will usually remove approximately 90 to 95 percent of the coliform bacteria and 70 to 80 percent of the BOD load approximately 80 percent of the time. Controlled-discharge ponds with 180-day detention times can produce BOD removals from 85 to 95 percent, total suspended solids removals from 85 to 95 percent, and fecal coliform reductions up to 99 percent.

Physical sedimentation by itself has been found to remove approximately 90 percent of the suspended solids in three days. About 80 percent of the dissolved organic solids can be removed by biologic action in ten days. However, in a pond with a healthy algae and bacteria population, a phenomenon known as *BIOFLOCCULATION*[8] can occur. This will remove approximately 85 percent of both suspended and dissolved solids within hours. Bioflocculation is accelerated by increased temperature, wave action, and high dissolved oxygen content.

---

[7] Percolation (PURR-co-LAY-shun).   The movement or flow of water through soil or rocks.

[8] Bioflocculation (BUY-o-flock-u-LAY-shun).   The clumping together of fine, dispersed organic particles by the action of certain bacteria and algae. This results in faster and more complete settling of the organic solids in wastewater.

Pond detention times are sometimes specified by regulatory agencies to ensure adequate treatment and removal of pathogenic bacteria. Many agencies specify effluent or receiving water quality standards in terms of *MEDIAN*[9] and maximum MPN (Most Probable Number) values that should not be exceeded. In critical water use areas, chlorination or other means of disinfection can be used to further reduce the coliform level (see Chapter 10, "Disinfection and Chlorination").

A pond is generally regarded as not fulfilling its function when it creates a visual or odor nuisance, or leaves a high BOD, solids, grease, or coliform-group bacteria concentration in the discharge (unless it was designed to be anaerobic in the first stages and aerobic in later ponds for final treatment).

QUESTIONS

Write your answers in a notebook and then compare your answers with those on page 331.

9.4A   What biological factors influence the treatment efficiency of a pond?

9.4B   What is bioflocculation?

9.4C   What factors indicate that a pond is not fulfilling its function (operating properly)?

END OF LESSON 1 OF 3 LESSONS
on
WASTEWATER STABILIZATION PONDS

Please answer the discussion and review questions next.

DISCUSSION AND REVIEW QUESTIONS

Chapter 9.   WASTEWATER STABILIZATION PONDS

(Lesson 1 of 3 Lessons)

At the end of each lesson in this chapter you will find some discussion and review questions. The purpose of these questions is to indicate to you how well you understand the material in the lesson. Write the answers to these questions in your notebook before continuing.

1. When wastewater flows through different treatment processes in a plant, where might ponds be located?

2. Why are most ponds "facultative ponds"?

3. Where does the oxygen come from that is produced by algae in a pond?

4. What is photosynthesis?

5. What are the three types of factors that may influence pond performance?

---

[9] *Median.   The middle measurement or value. When several measurements are ranked by magnitude (largest to smallest), half of the measurements will be larger and half will be smaller.*

CHAPTER 9.   WASTEWATER STABILIZATION PONDS

(Lesson 2 of 3 Lessons)

9.5   STARTING THE POND

One of the most critical periods of a pond's life is the time that it is first placed in operation. If at all possible, at least one foot (0.3 m) of water should be in the pond before wastes are introduced. The water should be turned into the pond in advance to prevent odors developing from waste solids exposed to the atmosphere. Thus a source of water should be available when starting a pond.

A good practice is to start ponds during the warmer part of the year because a shallow starting depth allows the contents of the pond to cool too rapidly if nights are cold. Generally speaking, the warmer the pond contents, the more efficient the treatment processes.

ALGAL BLOOMS[10] normally will appear from seven to twelve days after wastes are introduced into a pond, but it generally takes at least sixty days to establish a thriving biological community. A definite green color is evidence that a flourishing algae population has been established. After this length of time has elapsed, bacterial decomposition of bottom solids will usually become established. This is generally evidenced by bubbles coming to the surface near the pond inlet where most of the sludge deposits occur. Although the bottom is anaerobic, travel of the gas through the aerobic surface layers generally prevents odor release.

Wastes should be discharged to the pond intermittently during the first few weeks with constant monitoring of the pH. The pH in the pond should be kept above 7.5 if possible. Initially the pH of the bottom sludge will be below 7 due to the digestion of the sludge by acid-producing bacteria. If the pH

starts to drop, discharge to the pond should be diverted to another pond or diluted with makeup water (water from another source) if another pond is not available until the pH recovers. A high pH is essential to encourage a balanced anaerobic fermentation (bacterial decomposition) of bottom sludge. This high pH also is indicative of high algal activity since removal of the carbon dioxide from the water in algal metabolism tends to keep the pH high. A continuing low pH indicates acid production which will cause odors. Soda ash (sodium carbonate) may be added to the influent to a pond to increase the pH.

QUESTIONS

Write your answers in a notebook and then compare your answers with those on pages 331 and 332.

9.5A   Why should at least one foot of fresh water cover the pond bottom before wastes are introduced?

9.5B   Why should ponds be started during the warmer part of the year if at all possible?

9.5C   What does a definite green color in a pond indicate?

9.5D   When bubbles are observed coming to the pond surface near the inlet, what is happening in the pond?

9.6   DAILY OPERATION AND MAINTENANCE

BECAUSE PONDS ARE RELATIVELY SIMPLE TO OPERATE, THEY ARE PROBABLY NEGLECTED MORE THAN ANY OTHER TYPE OF WASTEWATER TREATMENT PROCESS. Many of the complaints that arise regarding ponds are the result of neglect or poor housekeeping. Following are listed the day-to-day operational and maintenance duties that will help to ensure peak treatment efficiency and to present your plant to its neighbors as a well-run waste treatment facility. If problems develop in a pond, refer to this section and Section 9.10, "Review of Plans and Specifications," as troubleshooting guides.

9.60   Scum Control

Scum accumulation is a common characteristic of ponds and is usually greatest in the spring when the water warms and vigorous biological activity resumes. Ordinarily, wind action will break up scum accumulations and cause them to settle; however, in the absence of wind or in sheltered areas, other means must be used. If scum is not broken up, it will dry on top and become crusted. Not only is the scum more difficult to break up then, but a species of blue-green algae is apt to become established on the scum. This can give rise to dis-

---

[10] *Algal (Al-gull) Bloom.   Sudden, massive growths of microscopic and macroscopic plant life, such as green or blue-green algae, which can, under the proper conditions, develop in lakes, reservoirs, and also in ponds.*

agreeable odors. If scum is allowed to accumulate, it can reach proportions where it cuts off a significant amount of sunlight from the pond. When this happens the production of oxygen by algae is reduced and odor problems can result. Rafts of scum cause a very unsightly appearance in ponds and can quite likely become a source of botulism that will have a devastating effect on any waterfowl and shore birds that may be attracted to the facility.

Many methods of breaking up scum have been used, including agitation with garden rakes from the shore, jets of water from pumps or tank trucks, and the use of outboard motors on boats in large ponds. Outboard motors should be of the air-cooled type to avoid plugging the cooling system with algae and scum. Scum is broken up most easily if it is attended to promptly.

9.61 Odor Control

Eventually, odors probably will come from a wastewater treatment plant no matter what kind of process is used. Most odors are caused by overloading (see Section 9.109 to determine pond loading) or poor housekeeping practices and can be remedied by taking corrective measures. If a pond is overloaded, stop loading and divert influent to other ponds, if available, until the odor problem stops. Then gradually start loading the pond again. Once a pond develops odor problems, it is more apt to cause trouble than other ponds.

There are times, such as when unexpected shutdowns occur, that plant processes may be upset and cause odors. For these unexpected occurrences, it is strongly advised that a careful plan for emergency odor control be available. Odors usually occur during the spring warm-up in colder climates because biological activity has been reduced during cold weather. When the water warms, microorganisms become active, use up all of the available dissolved oxygen, and odors are produced under these anaerobic conditions.

There are several suggested ways to reduce odors in ponds. These ways include recirculation from aerobic units, the use of floating aerators, and heavy chlorination. Recirculation from an aerobic pond to the inlet of an anaerobic pond (1 part recycle flow to 6 parts influent flow) will reduce or eliminate odors. Usually, floating aeration and chlorination equipment are too expensive to have standing idle waiting for an odor problem to develop. Odor-masking chemicals also have been promoted for this purpose and have some uses for concentrated sources of specific odors. However, in almost all

cases, process procedures of the type mentioned previously are preferable. In any event, it is poor procedure to wait until the emergency arises to plan for odor control. Often several days are needed to receive delivery of materials or chemicals if they are required. Try to have possible alternate methods of control ready to go in case they are needed.

In some areas, sodium nitrate has been added to ponds as a source of oxygen for microorganisms rather than sulfate compounds, thus preventing odors. To be effective, sodium nitrate must be dispersed throughout the water in the pond. Once mixed in the pond, it acts very quickly because many common organisms (facultative groups) may use the oxygen in nitrate compounds instead of dissolved oxygen. The amount of sodium nitrate depends on the oxygen demand. Some operators dosed their ponds successfully between 0.3 and 0.4 mg/L of sodium nitrate. Liquid sodium hypochlorite or chlorine solution is a faster acting solution, but not necessarily the best chemical because it will interfere with biological stabilization of the wastes.

A different odor-control strategy will be needed if the ponds have already frozen over in the winter but are expected to develop odor problems when the ice thaws in the spring and the water in the pond turns over. In this situation, some operators spread sodium nitrate on the ice surface before thaw so that the chemical will be available to supply oxygen when needed.

9.62 Weed and Insect Control

Weed control is an essential part of good housekeeping and is not a formidable task with modern herbicides and soil sterilants. Weeds around the edge are most objectionable because they allow a sheltered area for mosquito breeding and scum accumulation. In most average ponds, there has been little need for mosquito control when edges are kept free of weed growth. Weeds also can hinder pond circulation. Aquatic weeds, such as tules, will grow in depths shallower than three feet (1 m), so an operating pond level of at least this depth is necessary. Tules may emerge singly or be well scattered, but should be removed promptly by hand as they will quickly multiply from the root system. One of the best methods for control of undesirable vegetation is achieved by a daily practice of close inspection and immediate removal of the young plants (including roots).

Suspended vegetation, such as duckweed, usually will not flourish if the pond is exposed to a clean sweep of the wind. Dike vegetation control is aided by regular mowing and use of a cover grass that will crowd out undesirable growth. Because emergent weed growth will occur only when sunlight is able to reach the pond bottom, the single best preventive measure against emergent growth is to maintain a water depth of at least three feet (1 m). Due to greater water clarity, the amount of sunlight reaching the bottom will be greater in secondary or final ponds than in primary ponds. Because shallow water promotes growth, there will likely always be a battle to keep emergent weeds from becoming well established around lagoon banks.

Whenever emergent or suspended weeds are being pulled from the lagoon, such protective gear as waterproof gloves, boots, and goggles should be worn to reduce the chance of infection from pathogens that may be present in the water. Pulled weeds should be buried to prevent odor and insect problems. Although most stabilization ponds are no deeper than five feet (1.7 m), there is still sufficient depth to drown a person, especially if that person gets caught in a sticky clay liner. Using the buddy system and approved flotation devices will greatly increase the safety level when performing any pond maintenance, especially when using a boat or mowing the dike. Control measures include:

Emergent Weeds

1. Keep the water level above three feet.

2. Pull out new (first-year) growth by hand.

3. Drown the weeds by raising the water level.

4. Lower the water level, cut the weeds or burn them with a gas burner, and raise the water level (cut and drown).

5. Use herbicides as a last resort.

6. Use *RIPRAP*[11] (large broken stones or concrete) along the bank. If riprap is used and growth continues, herbicides will be the only alternative.

7. Install a pond liner.

Suspended Vegetation

1. Keep the pond exposed to a clean sweep of the wind.

2. A few ducks may be used to eat light growth of duckweed.

3. Small ponds may be skimmed with rakes or boards. This may have to be repeated.

4. Excessive growth can be mechanically harvested.

5. Use herbicides as a last resort.

Dike Vegetation

1. Mow regularly during the growing season. Dike slopes may be cut using sickle bars or weed-eater equipment. When using heavy equipment, mowers designed especially for cutting slopes are preferable. Any tractor used on the dike should have a low center of gravity. The tractor must be provided with an approved roll-over protective structure and seat belts if within certain use/weight classifications. Check with your local safety regulatory agency for applicable rules.

2. Seed or reseed slopes with desirable grasses that will form a thick and somewhat impenetrable mat.

3. Use herbicides as a last resort.

Herbicides are a last resort in vegetation control for ponds, not only because of the obvious hazards facing the operator, but also because of dangers presented to the biological growth in the pond and receiving stream (if discharging).

Care must be taken to follow mixing, application, and storage directions exactly. Safety for the operator and for any other persons who might come in contact with the herbicide is of utmost importance. Proper protective gear and warnings (verbal and written) are vital. Also, desirable vegetation could be killed by herbicide drift on a windy day or by other improper applications.

Availability, formulation, trade names, and federal clearance of state registration for certain herbicides may change. Do not use stored herbicides without checking on their current status of approval.

Some herbicides are effective only on certain plants, and herbicides are most effective at specific stages in the life of the plant. Cattails, for instance, are usually sprayed when the shoots are two to three feet tall and cattails are developing.

Perhaps the most important factor in the selection of the proper herbicide is site compatibility. In addition to the other important instructions on the product label, a comprehensive listing of approved application sites is usually provided. This list should be checked to verify that the product is the best choice for each specific application. Herbicides that are effective on many types of undesirable lagoon vegetation and that are approved for both pond and river (discharging lagoon) applications are often the logical choice.

Material in this section was adapted from an article entitled "Control of Undersirable Lagoon Vegetation," by Doug Matheny; it appeared in the Oklahoma Water Pollution Control Association's *DRIP*, Spring 1988.

Duckweed consists of tiny green plants that float on the surface of a pond. Duckweed stops sunlight penetration and hinders surface aeration, thus reducing dissolved oxygen in the pond. Rakes, water sprays or pushing a board with a boat can be used to move the duckweed into a corner of a pond where it can be physically removed from the pond. Another approach is to install a surface discharge in the downwind corner of a pond. The prevailing winds will blow the duckweed into the corner and the duckweed will be removed by the surface discharge. The duckweed can be captured by a wire mesh bucket placed in the effluent line after the surface outlet. The bucket will need to be removed and cleaned as the duckweed accumulates.

Mosquitoes will breed in sheltered areas of standing water where there is vegetation or scum to which the egg rafts of the female mosquito can become attached. These egg rafts are fragile and will not withstand the action of disturbed water surfaces such as caused by wind action or normal currents. Keeping the water edge clear of vegetation and keeping any scum broken up will normally give adequate control. Shallow, isolated pools left by the receding pond level should be drained or sprayed with a larvacide. Mosquitofish (*Gambusia*) are an effective means of controlling mosquitoes. A stocking density of 2,500 fish per acre has been reported. However, stocking levels will depend on the local situation and environmental conditions, as well as the desire to sustain the fish population.

Any of several minute shrimp-like animals may infest the pond from time to time during the warmer months of the year (March-November). These microcrustaceans live on algae and at times will appear in such numbers as to almost clear the pond of algae. During the more severe infestations, there will be a sharp drop in the dissolved oxygen of the pond, ac-

---

[11] *Riprap. Broken stones, boulders, or other materials placed compactly or irregularly on levees or dikes for the protection of earth surfaces against the erosive action of waves.*

companied by a lowered pH. This is a temporary condition because the microcrustaceans will outrun the algae supply, and there will be a mass die-off of insects which will be followed by a rapid greening up of the pond again. When the algae concentration in a pond is low under these conditions, ponds operated on a batch basis may find this a good time for release of water due to low suspended solids values.

Ordinarily there should be no great concern about these infestations because they soon balance themselves; however, in the case of a heavily loaded pond, a sustained low dissolved oxygen content may give rise to obnoxious odors. In that event, any of several commercial sprays can be used to control the shrimp-like animals.

Chironomid midges are often produced in wastewater ponds in sufficient numbers to be serious nuisances to nearby residential areas, farm workers, recreation sites, and industrial plants. When emerging in large numbers, they may also create traffic hazards. At present the only satisfactory control is through the use of insecticides. Control measures are time consuming and may be difficult, particularly if there is a discharge to a receiving stream. If possible, lower the level in the ponds enough to contain a day's inflow before applying an insecticide. Holding the insecticide for at least one day will kill more insects and reduce the effect of the insecticide on receiving waters. For better results, insecticides should be applied on a calm day and any recirculation pumps should be stopped.

CAUTION. Before attempting to apply any insecticide or pesticide, contact your local official in charge of approving pesticide applications. This person can tell you which chemicals may be applied, the conditions of application, and safe procedures.

9.63   Levee Maintenance

Levee slope erosion caused by wave action or surface runoff from precipitation is probably the most serious maintenance problem. If allowed to continue, it will result in a narrowing of the levee crown which will make accessibility with maintenance equipment most difficult.

If the levee slope is composed of easily erodible material, one long-range solution is the use of bank protection such as stone riprap or broken concrete rubble. Good bank-protection materials include small pieces of broken street materials, curbs, gutters, and also bricks and other suitable materials from building demolition. Also a semi-porous plastic sheet has been used with riprap. This sheet allows the two-way movement of air and water, but prevents the movement of soils. The sheet also discourages weed growth and digging by crayfish (crawdads). Geosynthetics are an effective means of levee erosion control.

Portions of the pond levee or dike not exposed to wave action should be planted with a low-growing spreading grass to prevent erosion by surface runoff. Native grasses may naturally seed the levees, or local highway departments may be consulted for suitable grasses to control erosion. If necessary, grass may have to be mowed to prevent it from becoming too high. Do not allow grazing animals to control vegetation because they may damage the levees near the waterline and possibly complicate erosion problems.

Plants or grasses with long roots, such as willows and alfalfa, should not be allowed to grow on levees because they may damage the levees and possibly cause levee failure and costly repair. Burrowing animals such as muskrats, badgers, squirrels, and gophers also may cause levees to fail. Remove these animals from levees as soon as possible and repair their burrowed holes immediately.

Levee tops should be crowned so that rainwater will drain over the side in a sheet flow. Otherwise the water may flow a considerable distance along the levee crown and gather enough flow to cause erosion when it finally spills over the side and down the slope.

If the levees are to be used as roadways during wet weather, they should be paved or well graveled.

If seepage or leakage from the ponds appears on the outside of levees, ask your engineer or a qualified person to investigate and solve this problem before further damage occurs to the levee.

9.64   Headworks and Screening

Be sure to clean the bar screen as frequently as necessary. The screen should be inspected at least once or twice a day with more frequent visits during storm periods. Screenings should be disposed of daily in a sanitary manner, such as by burial, to avoid odors and fly breeding. A trench dug by a backhoe near the bar screen provides a convenient location for disposal by burial. Another method of disposal is to place screenings in garbage cans and request that your local garbage service dispose of the screenings at a sanitary landfill disposal site.

Many pond installations have grit chambers at the headworks to protect raw wastewater lift pumps or prevent plugging of the influent lines. There are many types of grit removal equipment. Grit removed by the various types of mechanical equipment or by manual means will usually contain small amounts of organic matter and should therefore be disposed of in a sanitary manner. Disposal by burial is the most common method.

QUESTIONS

Write your answers in a notebook and then compare your answers with those on page 332.

9.6A   Why should scum not be allowed to accumulate on the surface of a pond?

9.6B   How can scum accumulations be broken up?

9.6C   What are the causes of odors from a pond?

9.6D   What precaution would you take to be prepared for an odor problem that might develop?

9.6E   Why are weeds objectionable in and around ponds?

9.6F   How can weeds be controlled and removed in and around ponds?

9.6G   Why should insects be controlled?

9.6H   Why should a pond be lowered before an insecticide is applied?

9.65 Some Operating Hints

1. Anaerobic ponds should be covered and isolated for odor control and followed by aerobic ponds. Floating polystyrene planks can be used to cover anaerobic ponds and can be painted white for protection from the sun. These will help to confine odors and heat and tend to make the anaerobic ponds more efficient.

2. Placing *PONDS IN SERIES*[12] tends to cause the first pond to become overloaded and may never allow it to recover; the overload may be carried to the next pond in series. Feeding *PONDS IN PARALLEL*[12] allows the distribution of the incoming load evenly between units. Whether ponds are operated in series or in parallel should depend on the loading situation and NPDES permit requirements.

   When operating ponds in series, the accumulation of solids in the first pond may become a serious problem after a long period of use. Periodically the flow should be routed around the first pond. This pond should then be drained and the solids removed and buried.

3. A large amount of recirculation, say 25 to 100 percent, can be very helpful. This allows the algae and other aerobic organisms to become thoroughly mixed with incoming raw wastewater. At the same time, good oxygen transfer can be attained by passing the incoming water over a deck and down steps or other type of aerator. This procedure can cause heat loss, however.

4. Heavy chlorination at the recirculation point can assist in odor control, but will probably interfere with treatment.

5. As with any treatment process, it is necessary to measure the important water quality indicators (DO and solids) at frequent, regular intervals and plot them so that you have some idea of the direction the process is taking in time to take corrective action when necessary.

6. When solids start floating to the surface of a pond during the spring or fall *OVERTURN*,[13] the pond may have to be taken out of service and cleaned. Measurement of the sludge depth on the bottom of a pond also will indicate when a pond should be cleaned. Usually ponds are cleaned when the wet sludge is over one foot (0.3 m) deep.

7. Before applying insecticides or herbicides, be sure to check with appropriate authorities regarding the long-term effects of the pesticide you plan to use. Do not apply pesticides that may be toxic to organisms in the receiving waters.

8. Algae in effluent. Unfortunately there is little an operator can do by changing pond operating procedures to effectively remove algae from a pond effluent. The best approach is to operate ponds in series and to draw off the effluent below the surface by use of a good baffling arrangement.

   If the algae must be removed from the effluent of a pond, additional facilities may be designed and constructed (see Section 9.11, "Eliminating Algae From Pond Effluents"). Algae removal processes include microscreening, slow sand filtration, dissolved air flotation, and algae harvesting. In final ponds that are operated in series with periodic discharges, alum may be added in doses of less than 20 mg/L to improve effluent quality before discharge.

9.66 Abnormal Operation

Abnormal operation occurs when ponds are overloaded because the BOD loads are too high. Excessive BOD levels can occur when influent loads exceed design capacity due to population increases, industrial growth, or industrial dumps. Under these conditions new facilities must be constructed or the BOD loading must be reduced at the source. Repeated wide fluctuations in BOD loads over short time periods will also interfere with pond performance and create a nearly constant state of abnormal operation.

Another type of overloading can occur when too much flow is diverted to one pond. This can happen when an operator accidentally feeds one pond more than the other or when a pipe opening is blocked by rags, solids, or grit due to low pipe velocities, and thus too much flow is diverted to another pond. When this happens and the overloaded pond starts producing odors, take the pond out of service and divert flows to the other ponds until the overloaded pond recovers. Hopefully ponds in service will not become overloaded. Also be sure to remove any rags, solids, or grit that caused the overloading and inspect the other pipes to prevent this problem from happening again in the other ponds.

Usually ponds do not become overloaded during storms and periods of high runoff because there is not a significant increase in the BOD loading on the ponds.

Large amounts of brown or black scum on the surface of a pond is an indication that the pond is overloaded. Scum on the surface of a pond often leads to odor problems. The best way to control scum is to take corrective action as soon as possible (see Section 9.60, "Scum Control").

---

[12]

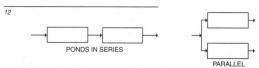

[13] *Overturn. The almost spontaneous mixing of all layers of water in a reservoir or lake when the water temperature becomes similar from top to bottom. This may occur in the fall/winter when the surface waters cool to the same temperature as the bottom waters and also in the spring when the surface waters warm after the ice melts. This is also called "turnover."*

During winter conditions the pond can become covered with ice and snow. Sunlight is no longer available to the algae and oxygen cannot enter the water from the atmosphere. Without dissolved oxygen available for aerobic decomposition, anaerobic decomposition of the solids occurs. Anaerobic decomposition takes place slowly because of the low temperatures. By keeping the pond surface at a high level, a longer detention time will be obtained and heat losses will be minimized. During the period of ice cover, odorous gases formed by anaerobic decomposition accumulate under the ice and are dissolved into the wastewater being treated.

Some odors may be observed in the spring just after the ice cover breaks up because the pond is still in an anaerobic state and some of these dissolved gases are being released. Melting of ice in the spring provides dilution water with a high oxygen content, thus the ponds usually become facultative in a few days after breakup of the ice if they are not organically (BOD) overloaded.

9.67 Batch Operation

Some ponds do not discharge continuously. These ponds may discharge only once (fall) or twice (fall and spring) a year. Discharges should be made only when necessary and, if possible, during the nonrecreational season when flows are high in the receiving waters.

If your pond is allowed to discharge intermittently (controlled discharge), you must work closely with your pollution control agency and be sure that you are in compliance with the National Pollutant Discharge Elimination System (NPDES) permit. Before and during the discharge, samples should be collected from the pond being emptied and from the receiving waters both upstream and downstream from the point of discharge. Samples are usually analyzed for DO, BOD, pH, total suspended solids, and coliform-group bacteria.

Ponds should not be emptied too quickly. If a pond is emptied too quickly, the wet side slope along the shoreline may slide into the pond. Usually ponds are emptied in two weeks or less depending on how much water is to be discharged. Normally 1.0 to 1.5 feet (0.3 to 0.45 m) of water is left in the bottom of the pond.

9.68 Controlled Discharge

A common operational modification to aerated and facultative lagoons is the controlled-discharge pond mode in which pond discharge is prohibited during the winter months in cold climates and/or during the peak algal growth periods in all climates. In this approach, each cell in the system is isolated and then discharged sequentially. Sufficient storage capacity is provided in the lagoon system to allow wastewater storage during winter months, peak algal growth periods, or receiving stream low-flow periods. In the Great Lakes states of Michigan, Minnesota, and Wisconsin, as well as in the province of Ontario, Canada, many pond systems are designed to discharge in the spring and/or fall when water quality effects are minimized. As a secondary benefit, operational costs are lower than for a continuous-discharge lagoon because of reduction in laboratory monitoring requirements and the need for less operator control.

A similar modification is called a hydrograph controlled-release lagoon (HCRL). Water is retained in the pond until flow volume and conditions in the receiving stream are adequate for discharge, thus eliminating the need for costly additional treatment. An HCRL system has three principal components: a stream-flow monitoring system, a storage cell, and an effluent discharge system. The stream-flow monitoring system measures the flow rate in the stream and transmits the data to the effluent discharge system. The effluent discharge system consists of a controller and a discharge structure. The controller operates a discharge device, such as a motor-driven sluice gate, through which wastewater is discharged from the storage lagoon; however, these tasks can be manually performed.

Key considerations for HCRLs include:

- An effluent release model keyed to flow (as measured by depth) in the receiving stream is needed,

- Outlets should be capable of drawing from different depths to ensure best quality, and

- Storage cells should be sized by use of water-balance equations.

9.69 Shutting Down a Pond

Ponds may be shut down for short periods of time without any problems developing. For example, if flow to a pond must be stopped to repair a pipe or a valve, no precautionary procedures are necessary. If a pond is full and received no flows for a long period of time, start up the pond with caution and gradually increase the load. If the full load is applied immediately, the pond may become overloaded because the microorganism population in the pond is low and insufficient to treat the load.

Stop all flow to a pond when emptying it to remove bottom deposits, repair inlet or outlet structures, or repair levees. Drain by use of discharge valves or pump water from one pond to other ponds. Try to feed the other ponds equally to prevent them from becoming overloaded. Frequently there is a time lag between the overloading of a pond and the development of problems. Therefore, watch the other ponds and lab results closely for any signs (odors, low pH, low DO, drop in alkalinity, loss of green color) of potential problems developing.

9.610 Operating Strategy

In order to prevent ponds from developing odors or discharging an effluent in violation of the NPDES permit requirements, develop a plan to keep the ponds operating as intended.

A. Maintain constant water elevations in the ponds.

If the NPDES permit allows the discharge of a pond into receiving waters, keep a constant water level to help maintain constant loadings. When the water surface elevation starts to drop, look for the following possible causes:

1. Discharge valve open too far or a *STOP LOG* [14] is missing;

2. Levees leaking due to animal burrows, cracks, soil settlement, or erosion; and

3. Inlet lines plugged or restricted and causing wastewater to back up into the collection system.

When the water surface starts to rise, look for:

1. Discharge valve closed or lines plugged, and

2. Sources of infiltration.

NOTE: Under some conditions you may not want to maintain constant water levels in your ponds. For example, you may allow the water surface to fluctuate to:

    1. Control shoreline aquatic vegetation,

    2. Control mosquito breeding and burrowing rodents,

    3. Handle fluctuating inflows, and

    4. Regulate discharge (continuous, intermittent, or seasonal).

B. Distribute inflow equally to ponds.

All ponds designed to receive the flow should receive the same hydraulic and organic (BOD) loadings.

C. Keep pond levees or dikes in good condition.

Proper maintenance of pond levees can be a time-saving activity. Regularly inspect levees for leaks and erosion and correct any problems before they become serious. If erosion is a problem at the waterline, install riprap. Do not allow weeds to grow along the waterline and keep weeds on the levee mowed. If insect larvae are observed on the pond surface, spray with an appropriate insecticide before problems develop.

D. Observe and test pond condition.

Daily visual observations can reveal if a pond is treating the wastewater properly. The pond should be a deep green color indicating a healthy algae population. Scum and floating weeds should be removed to allow sunlight to reach the algae in the pond.

Once or twice a week, tests should be conducted to determine pond dissolved oxygen level, pH, and temperature. Effluent dissolved oxygen should be measured at this time also. Other effluent tests should be conducted at least weekly and include BOD, suspended solids, dissolved solids, coliform-group bacteria, and chlorine residual. If ponds are operated on a batch or controlled discharge basis, these effluent tests will have to be determined only during periods of discharge.

During warm summer months algae populations tend to be high and may cause high suspended solids concentrations in the effluent. An advantage of ponds in arid regions is that this is also a period of high evaporation rates. Under these conditions effluent flows may drop to almost zero or may be stopped. In the fall and winter when the weather is cool and sunlight is reduced, the algae population in ponds and thus the suspended solids are reduced. This situation could allow ponds to meet effluent requirements during this period.

If test results reveal that certain water quality indicators (such as DO, BOD, pH, or suspended solids) are tending to move in the wrong direction, try to identify the cause and take corrective action. Remember that ponds are a biological process and that changes resulting from corrective action may not occur until a week or so after changes were made.

E. Troubleshooting.

The "Troubleshooting Guide" on pages 310 and 311 provides you with step-by-step procedures to follow when problems develop in a pond.

QUESTIONS

Write your answers in a notebook and then compare your answers with those on page 332.

9.6I   Why are the contents of ponds recirculated?

9.6J   When do ponds that are operated on a batch basis discharge?

9.6K   What factors would you consider when developing an operating strategy for ponds?

9.7 SURFACE AERATORS (Figures 9.5 and 9.6 and Table 9.2)

Surface aerators have been used in two types of applications:

1. To provide additional air for ponds during the night, during cold weather, or for overloaded ponds; and

2. To provide a mechanical aeration device for ponds operated as aerated lagoons. Aerated lagoons operate like activated sludge aeration tanks without returning any settled activated sludge.

In both cases, the aerators are operated by time clocks with established ON/OFF cycles. Laboratory tests on the dissolved oxygen in a pond indicate the time period for ON and OFF cycles to maintain aerobic conditions in the surface layers of the pond. Adjustments in the ON/OFF cycles are necessary when changes occur in the quantity and quality of the influent and seasonal weather conditions. Some experi-

---

[14] *Stop Log. A log or board in an outlet box or device used to control the water level in ponds and also the flow from one pond to another pond or system.*

Troubleshooting Guide—Ponds

(Adapted from *PERFORMANCE EVALUATION AND TROUBLESHOOTING AT
MUNICIPAL WASTEWATER TREATMENT FACILITIES*,
Office of Water Program Operations, US EPA, Washington, DC.)

INDICATOR/OBSERVATION	PROBABLE CAUSE	CHECK OR MONITOR	SOLUTION
1. Poor quality effluent.	1a. Mixing/agitation equipment failure.	1a. Monitor surface aerators, rotors, or aeration equipment and DO of pond water.	1a. 1. Restart out of service mixing or aeration units. 2. Increase operating time. 3. Increase recycle flow from effluent to influent. 4. Provide additional mixing (small boat and outboard motor for one hour for every two hours of daylight, or at least three times a day).
	1b. Organic overload.	1b. Monitor influent laboratory data: BOD, SS, DO, pH, temperature.	1b. 1. Increase or start recirculation of effluent to influent. 2. Mix pond contents hourly by surface aerators or outboard motor on small boat at least three times per day. 3. Increase run cycle on surface aeration equipment to maintain at least 1.0 mg/L DO. 4. Add chemicals to help reduce pond load by prechlorination for BOD reduction, add sodium nitrate or hydrogen peroxide for oxygen input.
	1c. Excessive turbidity from scum mats.	1c. Floating mats of scum or sludge on surface and corners of pond.	1c. 1. Break up scum mats a. water sprays b. poles or rakes c. mixing equipment d. outboard motor 2. Check pond influent for excess grease or scum. 3. If seasonal temperature change, rising sludge from pond may have to be broken up daily by mixing methods in item 1c.1. above.
	1d. Blockage of light by excessive plant growth (tules, reeds, and/or grasses) near dikes.	1d. Visual inspection for weed growth in and near ponds.	1d. 1. Remove plant growth. 2. Schedule regular herbicide applications to levees and dikes as a last resort.
	1e. Low temperature.	1e. Monitor air and pond water temperature.	1e. 1. Freezing weather, raise pond levels to increase water depth. 2. Reduce recirculation rates. 3. If possible, operate ponds in series.
	1f. Toxic material in influent.	1f. Color change, low DO, low pH for no apparent reason.	1f. 1. Sample and try to identify toxic material. 2. Increase recirculation from effluent to influent. 3. Increase surface aeration or pond mixing times. 4. Implement new or enforce existing sewer-use ordinances. 5. See 1b. above.
	1g. Loss of pond volume caused by sludge accumulation.	1g. Sludge depth.	1g. Remove sludge.

INDICATOR/OBSERVATION	PROBABLE CAUSE	CHECK OR MONITOR	SOLUTION
2. Low dissolved oxygen in ponds.	2a. Low algal growth.	2a. Visual inspection, pond not green, low DO during daylight hours, none from sun-up until noon. Odor.	2a. 1. Recirculate last pond effluent to inlet of first pond. 2. Same as 1b.
	2b. Excessive scum accumulation.	2b. Pond surface conditions.	2b. 1. Break up and sink resuspended scum. 2. Skim off grease balls and scum. Dispose of in landfill.
3. Odors.	3a. Anaerobic conditions or spring and fall turnovers.	3a. Monitor pond loading, BOD, SS, pH, DO, and temperature.	3a. 1. Limit organic load by diverting influent flows to several ponds. 2. Same as 1b.
	3b. Hydrogen sulfide in pond influent.	3b. Monitor total dissolved H_2S.	3b. 1. Check collection system. 2. Pretreat with chlorine or pre-aeration. 3. Aerate plant influent in small pond at least 30 minutes before applying to other ponds.
4. Inability to maintain sufficient liquid.	4a. Leakage.	4a. 1. Seepage around dikes. 2. Sampling wells located on the outside perimeter of the pond(s).	4a. 1. Apply bentonite clay to the pond water to seal leak. 2. If sampling well analysis indicates contamination, the lagoon/pond site may require lining.
	4b. Excessive evaporation or percolation.	4b. Detention time in pond is probably too long.	4b. Divert land drainage or stream flow into pond.
5. Insect generation.	5a. Layers of scum and excessive plant growth in sheltered portions of the pond.	5a. Visual inspection. Mosquitoes.	5a. 1. Weed and scum removal. 2. Application of approved insecticides or larvacides.
	5b. Shallow pools of standing water outside pond.	5b. Visual inspection.	5b. Cut vegetation outside pond and fill in potholes that collect standing water nearby.
6. Levee erosion.	6a. Windy conditions.	6a. Visual inspection.	6a. 1. Riprap levee. 2. Construct a wind barrier around pond.
	6b. Excessive surface aerator operating time.	6b. Aerator operating time.	6b. Reduce aerator operating time if DO levels allow.
7. Excessive weeds and tule growth.	7a. Pond too shallow.	7a. Visual inspection for weeds in the area.	7a. Deepen all pond areas to at least 3 feet.
	7b. Inadequate maintenance program to control vegetation.	7b. Maintenance program.	7b. 1. Correct program deficiency. 2. Install pond lining. 3. Herbicide program as a last resort.
	7c. Poor circulation.	7c. Visual inspection of flow characteristics.	7c. Fluctuate pond level.
8. Animals burrowing into the dikes.	8. Burrowing animals (gophers, squirrels, crayfish).	8. Visual inspection.	8. 1. Alter pond level several times in rapid succession. 2. Remove animals as soon as possible. 3. Provide riprap with semiporous sheet on levee slopes.
9. Groundwater contamination.	9. Leakage through bottom and/or sides of pond.	9. Seepage around pond dikes.	9. Apply bentonite clay to pond water to seal leak.

enced operators have correlated their lab test results to pond appearance and regulate the ON/OFF cycles using the following rule: "If the pond has foam on the surface, reduce the operating time of the aerator; and if there is no evidence of foam on the pond surface, increase the operating time of the aerator." If there is a trace of foam on the surface, the operating time is satisfactory.

Surface aerators may be either stationary or floating. Maintenance of surface aerators should be conducted in accordance with manufacturer's recommendations. Always turn off, tag, and lock out electric current when repairing surface aerators. Special precautions may be necessary to handle problems with icing and winter maintenance. Overhead guy wires have been used to prevent aerators from turning over when iced up. The operator should be aware that if a power failure occurs, even for a short time, the pond surface will freeze rapidly and possibly damage mechanical aerators. If an aeration system becomes plugged due to deposits of carbonate, try reducing the aeration or adding carbon dioxide gas.

Another method of aerating ponds is by the use of air compressors connected to plastic tubes placed across the pond bottom. Holes are drilled in the plastic tubes which serve as diffusers to disperse air in the pond. With this method of aeration, diffusers can become plugged and create maintenance problems.

Aeration tubing requires an effective maintenance program. The type of maintenance depends on the type of aeration system and problems encountered.

1. Gas cleaning. Some aeration tubing systems require cleaning on a weekly basis with anhydrous hydrogen chloride gas. The gas removes deposits on the tubing slits and destroys biological slime that may grow in the tubing. The deposits are caused by the precipitation of carbonate and bicarbonate compounds.

   Anhydrous hydrogen chloride gas is extremely dangerous. Be cautious when using it. If you have gas connections on the lagoon embankment, be sure to stand upwind from the HCl tank. High concentrations of hydrogen chloride gas are highly corrosive to eyes, skin, and mucous membranes. In systems using central manifolds, you must be cautious of HCl gas accumulating in low areas (HCl gas is 1.3 times heavier than air). Also, the monel needle must be closed prior to opening the cylinder valve. Two gas manifolds have ruptured in Vermont due to excess pressures in the manifolds as a result of the monel needle being open

TABLE 9.2   PURPOSE OF AERATOR PARTS

Part	Purpose
PLATFORM AERATOR COMPONENTS	
Aerator	Introduces oxygen into pond.
Electric Motor	Provides energy to drive aerator.
Drive Reduction Gear Box	Converts torque from motor to drive impeller.
Draft Tube	Conveys bottom contents of pond to surface for aeration.
Discharge Guide	Regulates spray patterns for oxygen transfer to water.
Jacking Screw	Adjusts aerator impeller level in water to regulate oxygen transfer and motor loading amps.
FLOATING AERATOR COMPONENTS (additional)	
Pontoons or Floats	Provide platform for motor.
Guy Wires	Maintain aerator positions in ponds and are anchored to pond levee.
Power Cable	Conveys power to motor.
Impeller (aerator)	Pumps water into air to be aerated.

too far. The installation of a pressure regulator between the monel needle and the cylinder valve will minimize the potential of excess pressure developing. Remember, HCl reacts with base metals to form a flammable gas (hydrogen). Therefore, all materials used in the manifold, header, transport, and regulation systems must be compatible with HCl. Begin at the last header in the lagoon and work your way forward. Apply the gas from one side with the opposite header shut off and then reverse the procedure. If the potential for exposure to hydrogen chloride exists, you should use appropriate, approved breathing protection, cold insulating gloves, protective clothing, and a face shield for eye protection.

2. High-pressure air purging. Some aerated lagoon facilities are capable of applying 90 psi (620 kPa) of compressed air to the system just after the blowers and the air manifold. The air manifold valves and blowers should be off. This is not the pressure level that will reach the lagoon tubing because of the friction losses in the tubing system. High-pressure cleaning will greatly deform the slits in the tubing and break off deposits farther back in the tubing. High-pressure air may be applied for three to five minutes once a year.

3. Slit deformation. The aeration tubing slits may be allowed to deform by shutting off the air once a day for about 20 minutes. The weight of the water above the tubing will cause the slits to be deformed inward, thus breaking up the deposits.

4. Removal of tubing and cleaning. A very thorough but time-consuming job is the drawdown of the lagoon and removal of the tubing for cleaning. The tubing can be flexed by hand or with special equipment to break up deposits and the inner surfaces of the tubing may be blown out with high-pressure air. You may avoid draining the lagoon to clean the tubing by using a boat, retrieving one end of a line of tubing, and manually flexing each section of tubing.

Fig. 9.5   Surface aerator
(Permission of EIMCO)

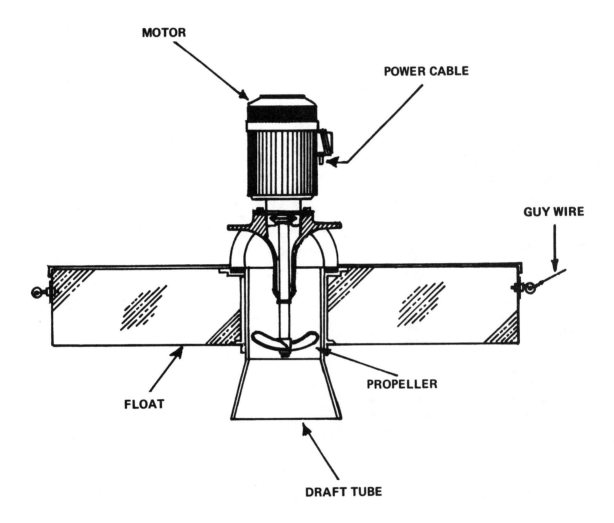

Fig. 9.6   Floating surface aerator
(Permission of Aqua-Jet)

5. Blowouts. Blowouts of aeration tubing and boils caused by slit enlargements must be repaired. The damaged area must be cut out and a special coupling inserted and clamped in place. The lagoon water level may have to be lowered so the line can be hooked and drawn to the surface for repair.

Acknowledgment

Material in this section was provided by Jon Jewett.

9.8 SAMPLING AND ANALYSIS

9.80 Importance

Probably the most important sampling that can be accomplished easily by any operator is routine pH and dissolved oxygen analysis. A good practice is to take pH, temperature, and dissolved oxygen tests several times a week, and occasionally during the night. These values should be recorded because they will serve as a valuable record of performance. The time of day should be varied occasionally for the tests so that the operator becomes familiar with the pond's characteristics at various times of the day. Usually the pH and dissolved oxygen will be lowest just at sunrise. Both will get progressively higher as the day goes on, reaching their highest level in late afternoon.

Be very careful to avoid getting any atmospheric oxygen into the sample taken to measure dissolved oxygen. This is most necessary when samples are taken in the early morning or if the dissolved oxygen in the pond is low from overloading. If possible, measure the dissolved oxygen with an electric meter and probe, being careful not to allow the membrane on the end of the probe to be exposed to the atmosphere during actual DO measurement of the water sample.

Ponds often have clearly developed individuality, each being a biological community that is unique. Apparently identical adjacent ponds receiving the same influent in the same amount often have a different pH and a different dissolved oxygen content at any given time. One pond may generate considerable scum while its neighbor doesn't have any scum. For this reason, *EACH* pond should be given routine pH and dissolved oxygen tests. Such testing may indicate an unequal loading because of the internal clogging of influent or distribution lines that might not be apparent from visual inspection. Tests also may indicate differences or problems that are being created by a buildup of solids or solids recycling.

When an operator becomes familiar with operating a pond, the results of some of the chemical tests can be related to visual observations. A deep green sparkling color generally indicates a high pH and a satisfactory dissolved oxygen content. A dull green color or lack of color generally indicates a declining pH and a lowered dissolved oxygen content. A gray color indicates the pond is being overloaded or not working properly.

9.81 Frequency and Location of Lab Samples

The frequency of testing and expected ranges of test results vary considerably from pond to pond, but you should establish those ranges within which your pond functions properly. Test results will also vary during the hours of the day. Table 9.3 summarizes the typical tests, locations, and frequency of sampling.

TABLE 9.3   FREQUENCY AND LOCATION OF LAB SAMPLES

Test	Frequency[a]	Location	Common Range
pH[b]	Weekly	Pond	7.5+
Dissolved Oxygen (DO)[b]	Weekly	Pond Effluent	4 - 12 mg/L 4 - 12 mg/L
Temperature	Weekly	Pond	
BOD[c]	Weekly	Influent Effluent	100 - 300 mg/L 20 - 50 mg/L
Coliform-Group Bacteria	Weekly	Effluent	MPN >24,000/100 mL (unchlorinated)
Chlorine Residual	Daily	Effluent	0.5 - 2.0 mg/L
Suspended Solids[d]	Weekly	Influent Effluent	100 - 350 mg/L 40 - 80 mg/L
Dissolved Solids	Weekly	Influent	250 - 800 mg/L

[a] Tests may be less frequent for ponds with long detention times (greater than 100 days).
[b] pH values above 9.0 and DO levels over 15 mg/L are not uncommon.
[c] Contact your regulatory agency to determine whether effluent samples should be filtered to remove algae before testing. If the samples must be filtered, the agency will recommend the proper procedures.
[d] Effluent suspended solids consist of algae, microorganisms, and other suspended matter.

Samples should always be collected from the same point or location. Raw wastewater samples for pond influent tests may be collected either at the wet well of the influent pump station or at the inlet control structure. Samples of pond effluent should be collected from the outlet control structure or from a well-mixed point in the outfall channel. Pond samples may be taken from the four corners of the pond. The samples should be collected from a point eight feet (2.5 m) out from the water's edge and one foot (0.3 m) below the water surface. Be careful; try not to stir up material from the pond bottom. Do not collect pond samples during or immediately after high winds or storms because solids will be stirred up after such activity.

BOD should be measured on a weekly basis. Samples should be taken during the day at low flow, medium flow, and high flow. The average of these three tests will give a reasonable indication of the organic load of the wastewater being treated. If it is suspected that the BOD varies sharply during the day or from day to day, or if unusual circumstances exist, the sampling frequency should be increased to obtain a clear definition of the variations. If the pond DO level is supersaturated (see Volume II, Chapter 16, section on DO tests), the sample must be aerated to remove the excess oxygen before the BOD test is performed. A typical data sheet for a plant consisting mainly of ponds is provided in the Appendix at the end of this chapter. Figure 9.7 shows the influent and effluent BOD and suspended solids values from an actual pond.

A *GRAB SAMPLE*[15] is a single sample. Grab samples are used to measure temperature, pH, dissolved oxygen, and chlorine residual. These tests must be performed immediately after the sample is collected in order to obtain accurate re-

[15] Grab Sample.   A single sample of water collected at a particular time and place which represents the composition of the water only at that time and place.

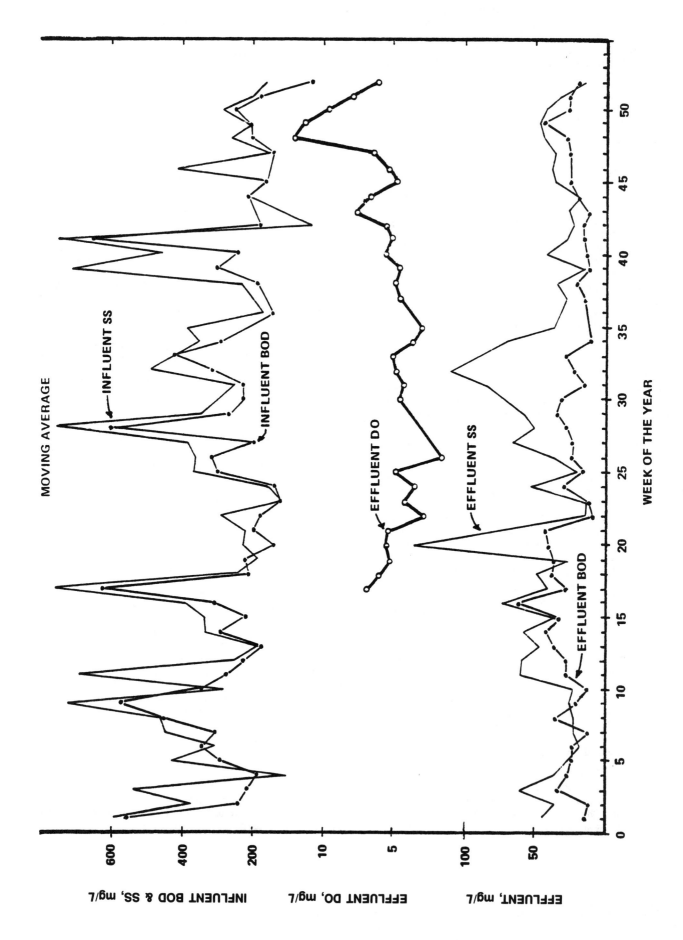

Fig. 9.7   Pond influent and effluent BOD and suspended solids values and effluent DO values

sults. *COMPOSITE SAMPLES*[16] of pond influent or effluent are collected by gathering individual samples at regular intervals over a selected period of time. The individual samples are then mixed together in proportion to the flow at the time of sampling. Pond samples may be composited by mixing equal portions from the four corners of the pond. Composite samples should be placed in a refrigerator or ice chest as soon as possible after they are collected. BOD and suspended solids are measured using composite samples.

Tests of pH, DO, and temperature are important indicators of the condition of the pond, whereas BOD, coliform, and solids tests measure the efficiency of the pond in treating wastes. BOD is also used to calculate the loading on the pond.

In order to estimate the organic loading on the pond, the operator must have some knowledge of the biochemical oxygen demand (BOD) of the waste and the approximate average daily flow. Influent BOD and solids will vary with time of day, day of week, and season, but a pond is a good equalizer if not overloaded.

Tests for pond alkalinity can provide helpful information to an operator. After you have determined "normal" alkalinity levels for your pond, a sudden change in alkalinity of 10 to 20 mg/L or more may indicate that a problem is developing. A change in alkalinity may be a warning that the pH of the pond could change in a day or two if corrective action is not taken. If a trend in changes in alkalinity continues in one direction for two or three days, the cause of this change should be identified. For useful results, the alkalinity test should be performed every day.

9.82 Expected Treatment Efficiencies[17]

Table 9.4 is provided as a guide to indicate probable removal efficiencies of typical ponds.

TABLE 9.4 EXPECTED RANGES OF REMOVAL BY PONDS

Test	Detention Time	Expected Removal
BOD		50 to 90%
BOD (facultative pond)	50 to 60 days	70 to 80%[18]
Coliform Bacteria (facultative pond)	50 to 60 days	90 to 95%
Suspended Solids	After 3 days	90%
Dissolved Organic Solids	After 10 days	80%

The calculation of pond BOD removal efficiency is figured in terms of the percentage of BOD removed.

EXAMPLE

The influent BOD to a series of ponds is 300 mg/L, and the effluent BOD is 60 mg/L. What is the efficiency of BOD removal?

$$\text{BOD Removal, \%} = \frac{(\text{In} - \text{Out})}{\text{In}} \times 100\%$$

$$= \frac{(300 \text{ mg/}L - 60 \text{ mg/}L)}{300 \text{ mg/}L} \times 100\%$$

$$= \frac{240 \text{ mg/}L}{300 \text{ mg/}L} \times 100\%$$

$$= 0.80 \times 100\%$$

$$= 80\%$$

9.83 Response to Poor Pond Performance

See Section 9.6, "Daily Operation and Maintenance," especially Section 9.65, "Some Operating Hints."

QUESTIONS

Write your answers in a notebook and then compare your answers with those on page 332.

9.7A Surface aerators have been used in what two types of applications in ponds?

9.8A If the pH and dissolved oxygen are dropping dangerously low in a pond, how can this situation be corrected?

9.8B Why should the pH, temperature, and dissolved oxygen be measured in a pond?

9.8C If the color of a pond is dull green, gray, or colorless, what is happening in the pond?

9.8D Influent BOD to a series of ponds is 200 mg/L. If the BOD in the effluent of the last pond is 40 mg/L, what is the BOD removal efficiency?

---

[16] Composite (come-PAH-zit) (Proportional) Sample. *A composite sample is a collection of individual samples obtained at regular intervals, usually every one or two hours during a 24-hour time span. Each individual sample is combined with the others in proportion to the rate of flow when the sample was collected. The resulting mixture (composite sample) forms a representative sample and is analyzed to determine the average conditions during the sampling period.*

[17] *Waste Removal, %* $= \dfrac{(\text{In} - \text{Out})}{\text{In}} \times 100\%$

[18] *Expected removal approximately 80 percent of the time with poorer removals during the remainder of the time.*

9.9 SAFETY

Even though a pond has little mechanical equipment, there still are hazards. Catwalks should have guardrails and nonskid walking surfaces. Headworks and any enclosed appurtenances should be well ventilated to prevent dangerous gas accumulations.

```
                    ┌──────────────┐
                    │   WARNING    │
┌───────────────────┴──────────────┴────────────────────┐
│                                                        │
│  AN OPERATOR SHOULD ALWAYS BE ACCOMPA-                 │
│  NIED BY A HELPER WHEN PERFORMING ANY                  │
│  TASK THAT IS DANGEROUS SINCE POND LOCA-               │
│  TIONS ARE USUALLY QUITE ISOLATED. IMMEDI-             │
│  ATE AID MIGHT BE NEEDED TO PREVENT SERI-              │
│  OUS INJURY OR LOSS OF LIFE.                           │
│                                                        │
└────────────────────────────────────────────────────────┘
```

Be very careful when removing debris from channels and ponds. Do not attempt to lift too much. Make certain you have secure footing so you won't slip and fall. Never stand in a boat or lean over too far to one side, for you could fall into the pond and also possibly tip over the boat. Always wear a life jacket when in a boat.

Electrical wires and electrical equipment are always a source of potential danger. Exercise caution when cutting weeds or removing vegetation such as trees next to electrical wires. Electrical wires in damp areas can be especially dangerous. Be careful when spraying weeds around electrical wires and equipment because the spray could act as a conductor. Always turn off, tag, and lock out electric current when repairing surface aerators and other equipment operated by electricity.

When applying pesticides or herbicides, be sure they are approved by the appropriate officials for your specific use and *FOLLOW THE DIRECTIONS EXACTLY.* This includes following the directions for using the proper procedures for mixing or preparing the solution, applying the solution, disposing of any excess solution and the containers, and cleaning up before you go home. Protective clothing, gloves, and respiratory protection may be required depending on the product, product strength, application method, and wind conditions. Not only can you kill the target pest or weed, but carelessness can harm nearby grasses, plants, trees, fish, birds, and even people, including yourself. Also, failure to follow directions may result in harm to the algae and microorganisms in the pond.

Tetanus and typhoid are ever-present dangers when working around wastewater. Adequate precautions should be observed by applying first-aid procedures to all cuts and scrapes and always washing before eating or smoking.

Fences should surround ponds to keep unauthorized persons and animals out of the pond area. They should be located in such a manner that they will not interfere with mechanical or hand maintenance of levee slopes.

QUESTIONS

Write your answers in a notebook and then compare your answers with those on page 332.

9.9A   What safety devices should be provided on catwalks over ponds?

9.9B   Why should an operator be accompanied by a helper when performing any dangerous task?

END OF LESSON 2 OF 3 LESSONS
on
WASTEWATER STABILIZATION PONDS

Please answer the discussion and review questions next.

DISCUSSION AND REVIEW QUESTIONS

Chapter 9.   WASTEWATER STABILIZATION PONDS

(Lesson 2 of 3 Lessons)

Write the answers to these questions in your notebook before continuing. The question numbering continues from Lesson 1.

6. Why should water be introduced into a new pond before any wastewater?

7. Why is good housekeeping an important factor in operating a properly functioning pond?

8. Why may chlorine compounds or chlorine solution not be the best method of odor control in a pond?

9. What precautions should be taken when applying an insecticide?

10. What lab tests measure the condition of a pond?

11. Estimate the BOD removal efficiency of a series of ponds if the influent BOD is 250 mg/L, and the effluent BOD is 50 mg/L.

12. Why should fences be placed around ponds?

CHAPTER 9.   WASTEWATER STABILIZATION PONDS

(Lesson 3 of 3 Lessons)

9.10   REVIEW OF PLANS AND SPECIFICATIONS

A careful review of the plans and specifications of a proposed pond can provide the operator an opportunity to suggest design improvements and changes before construction which will allow the operator and the ponds to do a better job. Guidelines for reviewing plans and specifications are provided in this section. If you are having trouble with an existing pond or ponds, check the items discussed in this section for help in locating possible sources of problems.

9.100   Location

The general considerations for the location of other types of wastewater treatment plants also apply to the location of ponds. Isolation should be as great as can be economically provided. Ponds should be isolated to prevent nuisances such as odors and insects associated with ponds from disturbing residential, commercial, and recreational neighbors, as well as to prevent possible traffic hazards from insects. Attention to the direction of prevailing winds with due regard for present and projected downwind residential, commercial, and recreational development is of utmost importance.

Winds can have both favorable and unfavorable impacts on ponds. Winds are desirable in terms of blowing surface scum and weeds to one side of the pond where they can be removed. Also winds can be helpful by mixing the contents of a pond, such as DO, algae, and incoming wastes. An undesirable aspect of winds is the creation of waves which can erode the pond levee. Both of these factors should be considered when selecting the location of the ponds and the arrangement or length of the ponds. If high winds are to be expected in the area where ponds will be constructed, try to arrange the ponds so the winds will blow across the short width of the pond rather than the length in order to reduce levee erosion caused by waves.

9.101   Chemistry of Waste

Before the design of any pond is undertaken, it should be determined whether the waste to be treated contains any *TOXIC*[19] constituents that may interfere with growth of algae or bacteria. Certain wastes, such as dairy products and wine products, are difficult to treat because of their low pH. Any processing waste must be carefully investigated before one can be certain that it can be successfully treated by ponding. Some process wastes contain powerful fungicides and disinfectants that may have a great inhibitive effect on the biological activity in a pond. Other wastes may have nutrient deficiencies that could inhibit the growth of desirable types of algae.

In addition, some natural water supplies have a high sulfur content or other chemicals that limit the possibility of desired sludge decomposition.

9.102   Headworks and Screening

A headworks with a bar screen is desirable to remove rags, bones, and other large objects that might lodge in pipes or control structures.

A trash shredder is a luxury that may not be warranted. Any material that gets past an adequate bar screen will, in all probability, not harm the influent pump. Any fecal matter will be pulverized when going through the pump.

9.103   Flow Measuring Devices

An influent measuring device should be installed to give a direct reading on the daily volume of wastes that are introduced into the ponds. This information, along with a BOD measurement of the influent, is required to estimate the organic loading on the pond. Comparison of influent and effluent flow rates is necessary for estimating percolation and evaporation losses.

A measuring device provides basic data for prediction of future plant expansion needs or for detecting unauthorized or abnormal flows. Reliable, well-kept records on flow volume help justify budgets and greatly assist an engineer's design of a plant expansion or new installation.

9.104   Inlet and Outlet Structures

Inlet structures should be simple, foolproof, and constructed of standard manufactured articles so that replacement parts are readily available. Telescoping friction-fit tubes (see Figure 9.8) for regulating spill or discharge height should be avoided because a biological growth may become attached and prevent the tubes from telescoping if they are not cleaned regularly. Occasional dosages of hypochlorite solution can effectively discourage growths. Also, the formation of ice can prevent adjust-

---

[19] *Toxic (TOX-ick).   A substance which is poisonous to a living organism.*

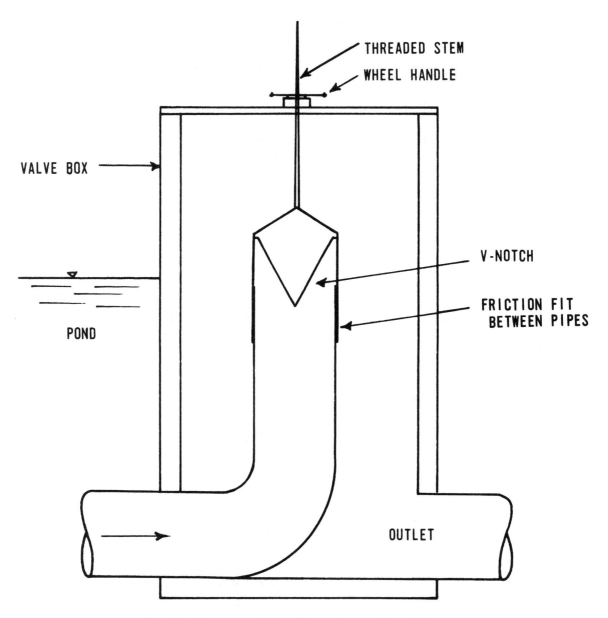

Fig. 9.8 Telescoping friction-fit tubes for regulating discharge. These tubes must be exercised regularly to prevent becoming stuck.

ment of the tubes. If freezing is a problem, place a polyethylene floating ring sprayed with urea around the friction tube of the telescopic valve to prevent freeze-ups. This device will act as a floating baffle to keep scum and floating debris from clogging up the tube and entering the effluent.

Figure 9.1 (page 295) shows four inlets to one pond. With this type of installation, usually only one inlet valve is open at a time. If all the valves were open, velocities in the pipes might become too low and solids would settle out in the pipes. To overcome low velocities, close all but one valve or recycle pond effluent back to the inlet. When all inlet valves are open, the load is more evenly distributed throughout the pond, but requires sufficient recycle flow to maintain the desired velocities in the inlet lines.

A submerged inlet will minimize the occurrence of floating material and will help conserve the heat of the pond by introducing the warmer wastewater into the depths of the pond. Warm wastewater introduced at the bottom of a cold water mass will channel to the surface and spread unless it is promptly and vigorously mixed with cold water. Warm wastewater spilled onto the surface of the pond will spread out in a thin layer on the surface and not contribute to the warmth of the lower regions of the pond where heat is needed for bacterial decomposition. Inlet and outlet structures should be so located in relation to each other to minimize possible short-circuiting.

Outlet structures (Figures 9.9 and 9.10) should consist of a baffled and submerged pipe inlet to prevent scum and other floating surface material from leaving the pond. The actual level of the pond and rate of outflow can be controlled by the use of flash boards in the outlet structure. A row boat may be used when access to the outlet baffle shown in Figure 9.9 is required.

Valves that have stems extending into the stream flow should be avoided. Stringy material and rags will collect, form an obstruction, and may render the valve inoperative.

Free overfalls (Figure 9.11) at the outlet should be avoided to minimize release of odors, foaming, and gas entrapment which may hamper pipe flows. Free overfalls should be converted to submerged outfalls if they are causing nuisances and other problems.

If a pond has a surface outlet, floating material can be kept out of the effluent by building a simple baffle around the outlet. The baffle can be constructed of wood or other suitable material. This baffle should be securely supported or anchored.

9.105 Levees

The selection of the steepness of the levee slope must depend on several variables. A steep slope erodes quicker from wave wash unless the levee material is of a rocky nature or else protected by riprap. However, a steep slope minimizes waterline weed growth. Equipment operation and the performance of routine maintenance are more difficult on steep slopes. A gentle slope will erode the least from wave wash. Also, it is easier to operate equipment and easier to perform routine maintenance on a gentle slope. However, waterline weed growth will have a much greater opportunity to flourish.

Ensure that provisions are made to adequately compact and/or seal the levee banks to prevent leaking. Pipes passing through levees should be as close to horizontal as possible to reduce the possibility of leaks. Also cut-off walls should be in-

stalled when pipes pass through levees to prevent leakage around the outside of the pipes. Proper compaction and sealing is necessary around pipes, SPLASH PADS,[20] cleanouts, valves, inlet and outlet control structures, and also recirculation, transfer, and drain lines. Once a leak develops, stopping it can be very difficult.

The top of the levee should be at least ten feet (3.1 m) wide to allow for maintenance vehicles. Provisions should be made for a rounded or sloping top to allow for drainage. Pave or gravel the levee surface if it will be used as a roadway during wet weather.

9.106 Pond Depths

The operational depth of ponds deserves considerable attention. Depending upon conditions, ponds of less than three feet (1 m) of depth may be completely aerobic if there are no solids on the bottom (unlikely) because of the depth of sunlight penetration. This means that the treatment of wastes is accomplished essentially by converting the wastes to algal cell material. Ponds of this shallow depth with short detention times are apt to be irregular in performance because algal blooms will increase to such proportions that a mass die-off will occur. When this happens, all algae sink to the bottom and thereby add to the organic load. Such conditions could lead to the creation of an anaerobic pond. The bottoms of shallow ponds will become anaerobic when solids collect on the bottom and after sunset.

Discharges from shallow, aerobic ponds contain large amounts of algae. To operate efficiently, these ponds should have some means of removing the algae grown in the pond before the effluent is discharged to the receiving waters. If the algae are not removed from the effluent, the organic matter in the wastewater is not removed or treated and the problem is merely transferred to some downstream pool. See Section 9.11, "Eliminating Algae From Pond Effluents," for a discussion of possible methods of algae removal.

An observed phenomenon of lightly loaded, shallow secondary ponds and tertiary ponds is that they are apt to become infested with filamentous algae and mosses that not only limit the penetration of sunlight into the pond, but hamper circulation of the pond's contents and clog inlet and outlet structures. When the loading is increased, this condition improves because these algae and mosses require relatively clean water (low nutrients) for their environment.

Pond depths of four feet (1.2 m) or more allow a greater conservation of heat from the incoming wastes. This encourages biological activity as the ratio between pond volume and pond area is more favorable. In facultative ponds, depths over four feet (1.2 m) provide physical storage for dissolved oxygen accumulated during the day. The stored dissolved oxygen carries over through the night when no oxygen is released by the algae, unless floating algae and poor circulation keep all the oxygen near the surface. This physical storage of DO is very important during the colder months when nights are long.

A pond operating depth of at least three feet (1 m) is recommended to prevent tule and cattail growth. Weeds that emerge along the shoreline can be effectively controlled by spraying with any of several products available. Ponds designed for depths less than three feet (1 m) should be lined to prevent troublesome weed growth.

---

[20] Splash Pad.   A structure made of concrete or other durable material to protect bare soil from erosion by splashing or falling water.

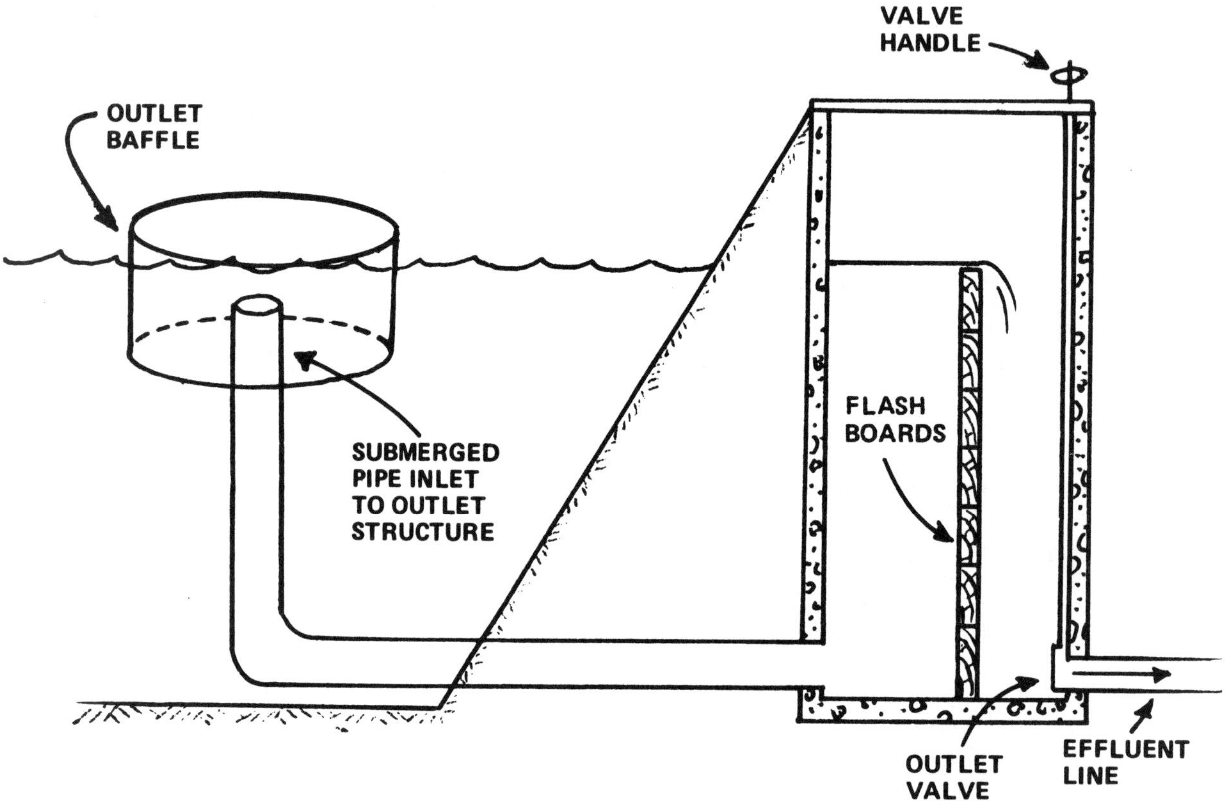

Fig. 9.9   Pond outlet structure

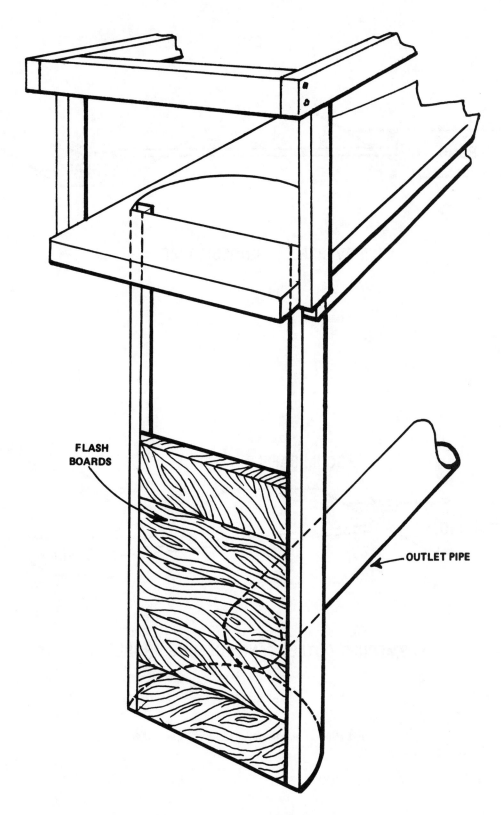

Fig. 9.10   Pond outlet control structure

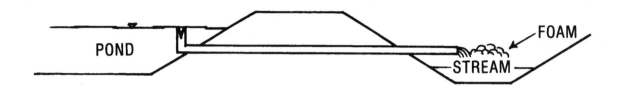

FREE OVERFALL -- UNDESIRABLE

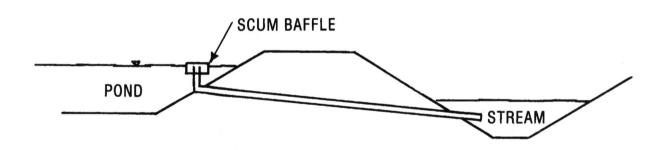

SUBMERGED OUTLET -- NO FOAMING PROBLEMS

Fig. 9.11   Free overfall and submerged outlet

9.107 Fencing and Signs

The pond area must be surrounded by a fence capable of keeping livestock out and discouraging trespassing. A gate wide enough to allow mowing equipment and other maintenance vehicles to enter the pond facility should be provided. All access gates should have locks.

Signs should be posted along the fence around the ponds to indicate the nature of the facility and to forbid trespassing. The signs should not be more than 300 feet apart (90 m), but the spacing should also comply with local penal codes governing trespass.

9.108 Surface Aerators

Provisions must be made for easy access and sufficient space for maintenance and repair of fixed aerators. Alternate anchor points should be installed in order to move floating aerators. Also, be sure the electrical cables are long enough to permit easy movement of the aerator and large enough to handle anticipated loads.

QUESTIONS

Write your answers in a notebook and then compare your answers with those on pages 332 and 333.

9.10A   Why are some wastes not easily treated by ponds?

9.10B   Why should the influent to a pond be metered?

9.10C   Why should the inlet to a pond be submerged?

9.10D   Why should the outlet be submerged?

9.10E   Why should free overfalls be avoided?

9.10F   How could problems created by a surface outlet be reduced or corrected?

9.10G   What is the minimum recommended pond operating depth?

9.10H   Why are shallow ponds apt to be irregular in performance?

9.109 Pond Loading (English and Metric Calculations)

The waste loading on a pond is generally spoken of in relation to its area, and may be stated in several different ways:

1. lbs of BOD per day per acre = lbs BOD/day/acre (This is called "organic loading.");

2. inches (or feet) of depth added per day (This is called "hydraulic loading" or "overflow rate."); or

3. persons (or population served) per acre (This is called "population loading.").

Detention time is related directly to pond hydraulic loading, which is actually the rate of inflow of wastewater. It may be expressed as million gallons per day (MGD), or as the number of acre-inches per day or acre-feet per day (one acre-foot covers one acre to a depth of one foot or twelve inches and is equal to 43,560 cubic feet). We must know the pond volume in order to determine detention time; this is most easily computed on an acre-foot basis.

A. Detention Time

$$\text{Detention (in days)} = \frac{\text{Pond Volume, ac-ft}}{\text{Influent Rate, ac-ft/day}}$$

This equation does not take into consideration water that may be lost through evaporation or percolation. Detention time may vary from 30 to 120 days, depending on the treatment requirements to be met. Controlled discharge ponds may have minimum detention times of 180 days. Ponds whose discharges are disposed of by land application may have 210-day minimum detention times.

B. Population Loading

Loading calculated on a population-served basis is expressed simply as:

$$\text{No. of Persons per Acre} = \frac{\text{Population Served, persons}}{\text{Surface Area of Pond, ac}}$$

The population loading may vary from 50 to 500 persons per acre, depending on many local factors.

C. Hydraulic Loading

The hydraulic loading or overflow rate is expressed as:

$$\text{Inches per Day} = \frac{\text{Inflow, ac-in per day}}{\text{Average Pond Area, ac}}$$

The hydraulic loading may vary from half an inch to several inches per day, depending on the organic load of the influent.

NOTE: If the wastewater inflow rate is known in million gallons per day (MGD), it can be converted to an equivalent number of acre-inches per day as follows:

Inflow, acre-inches per day = Inflow, MGD x 36.8[21]

If the pond detention time is known, the hydraulic loading can also be calculated, as follows:

$$\text{Inches per Day} = \frac{\text{Depth of Pond, in}}{\text{Detention Time, days}}$$

---

[21] $1 \text{ MGD} = \frac{1,000,000 \text{ gal}}{\text{day}} \times \frac{1 \text{ cu ft}}{7.48 \text{ gal}} \times \frac{1 \text{ ac}}{43,560 \text{ sq ft}} \times \frac{12 \text{ in}}{1 \text{ ft}} = 36.8 \frac{\text{ac-in}}{\text{day}}$

D. Organic Loading

The organic loading is expressed as:

$$\text{Organic Load, lbs BOD per Day per Acre} = \frac{(\text{BOD, mg/}L)(\text{Flow, MGD})(8.34 \text{ lbs/gal})^{22}}{\text{Pond Area, ac}}$$

Typical organic loadings may range from 10 to 50 lbs BOD per day per acre. Recirculation will help an overloaded pond.

EXAMPLE CALCULATIONS

NOTE TO OPERATORS: If you have difficulty following the work shown in Example 1 below, you should refer to Section 9.16, "Ponds—Math Supplement," page 330, for further details. If you have no trouble, continue with the lesson.

EXAMPLE 1

Use of the pond loading formulas can be illustrated by examining a typical situation. The following data should be obtained so that all the calculations can be performed:

Essential Data

1. Depth of Pond = 4 feet

2. Width of Pond

   Bottom = 412 feet
   Water Surface = 428 feet
   Average Width = 420 feet

3. Length of Pond

   Bottom = 667 feet
   Water Surface = 683 feet
   Average Length = 675 feet

4. Side Slopes

   (2 ft horizontal
   to 1 ft vertical) = 2:1

5. Influent Flow = 0.2 million gallons per day
   = 0.2 MGD
   = 200,000 gallons per day

6. Influent BOD = 200 mg/L
   = 200 lbs BOD per million lbs of wastewater

7. Population = 2,000 persons

To calculate the loading criteria, first determine the pond area and volume.

I. POND AREA, ACRES

$$\text{Avg Pond Area, ac} = \frac{(\text{Avg Width, ft})(\text{Avg Length, ft})}{43,560 \text{ sq ft/ac}}$$

$$= \frac{(420 \text{ ft})(675 \text{ ft})}{43,560 \text{ sq ft/ac}}$$

$$= 6.51 \text{ ac}$$

II. POND VOLUME, ACRE-FEET

$$\text{Volume, ac-ft} = (\text{Avg Area, ac})(\text{Depth, ft})$$

$$= (6.51 \text{ ac})(4 \text{ ft})$$

$$= 26.04 \text{ ac-ft (say 26 ac-ft)}$$

Convert flow rate from gallons per day to ac-ft/day.

$$\text{Flow Rate, ac-ft/day} = \frac{200,000 \text{ gal}}{\text{day}} \times \frac{\text{cu ft}}{7.48 \text{ gal}} \times \frac{\text{ac}}{43,560 \text{ sq ft}}$$

$$= 0.61 \text{ ac-ft/day}$$

III. LOADING CRITERIA

NOTE TO OPERATORS: Details for calculations in the remainder of Example 1 have not been given. If you have trouble, go to the end of this lesson and study Section 9.16, "Ponds—Math Supplement," for details, and try to apply them to this section.

1. *DETENTION TIME*

$$\text{Detention Time, days} = \frac{\text{Pond Volume, ac-ft}}{\text{Flow Rate, ac-ft/day}}$$

$$= \frac{26 \text{ ac-ft}}{0.61 \text{ ac-ft/day}}$$

$$= 42.6 \text{ days}$$

2. *POPULATION LOADING*

$$\text{Number of Persons per Acre} = \frac{\text{Population Served by Sewer System, persons}}{\text{Surface Pond Area, ac}}$$

$$= \frac{(2,000 \text{ Persons})(43,560 \text{ sq ft/ac})}{(428 \text{ ft})(683 \text{ ft})}$$

$$= 298 \text{ persons/ac}$$

NOTE: If there is a significant waste flow from industry mixed in with the domestic waste, an adjustment must be made to take the industrial waste into consideration. This is usually done by analyzing the industrial waste and converting it to a *POPULATION EQUIVALENT.*[23]

---

[22] Recall lbs/day = (Conc, mg/M mg)(M gal/day)(8.34 lbs/gal)

[23] Population Equivalent.   A means of expressing the strength of organic material in wastewater. In a domestic wastewater system, microorganisms use up about 0.2 pound of oxygen per day for each person using the system (as measured by the standard BOD test). May also be expressed as flow (100 gallons per day per person) or suspended solids (0.2 lb SS/day/person).

$$\text{Population Equivalent, persons} = \frac{\text{Flow, MGD} \times \text{BOD, mg/L} \times 8.34 \text{ lbs/gal}}{0.2 \text{ lb BOD/day/person}}$$

3. HYDRAULIC LOADING (OVERFLOW RATE)

$$\text{Inches per Day} = \frac{\text{Depth of Pond, in}}{\text{Detention Time, days}}$$

$$= \frac{(\text{Depth, 4 ft})(12 \text{ in/ft})}{42.6 \text{ days}}$$

$$= 1.13 \text{ in/day}$$

4. ORGANIC LOADING

$$\frac{\text{Organic Load,}}{\text{lbs BOD/day/ac}} = \frac{(\text{BOD, mg/}L)(\text{Flow, MGD})(8.34 \text{ lbs/gal})}{\text{Area, ac}}$$

$$= \frac{200 \text{ lbs}}{\text{M lbs}} \times \frac{0.2 \text{ M gal}}{\text{day}} \times \frac{8.34 \text{ lbs}}{\text{gal}} \times \frac{1}{6.5 \text{ ac}}$$

$$= 51 \text{ lbs BOD/day/ac}$$

EXAMPLE 2

NOTE TO OPERATORS: Details for calculations in Example 2 have not been given. If you have trouble, go back and study the procedures for Example 1 and try to apply them to Example 2.

Suppose that a small wastewater treatment plant must be completely shut down for major repairs that will require several months of work. Enough vacant land is near the plant to enable 16 acres of temporary ponds to be constructed as raw wastewater (sewage) lagoons. Determine if this is feasible, given the following data:

Influent Rate	= 1 MGD
Influent BOD	= 150 mg/L
	= 150 lbs BOD per million lbs of wastewater
Pond Area	= 16 acres
Average Operating Depth	$= 42 \text{ inches} = \frac{42 \text{ in}}{12 \text{ in/ft}} = 3.5 \text{ ft}$

Assume that at least a 60-day detention period (average time the wastewater must take to flow through the pond for disinfection) is desired for bacterial die-off.

Assume that the organic loading (BOD) should not exceed 50 lbs per day per acre.

Calculate what the *WASTE DETENTION TIME* would be in the pond:

One acre-foot = 325,829 gallons

Pond Volume, ac-ft = Pond Area, ac x Pond Depth, ft

$$= 16 \text{ ac} \times 3.5 \text{ ft}$$

$$= 56 \text{ ac-ft}$$

Influent Flow Rate = 1,000,000 gal per day

$$= \frac{1,000,000 \text{ GPD}}{325,829 \text{ gal/ac-ft}}$$

$$= 3.07 \text{ ac-ft per day}$$

$$\text{Detention Time} = \frac{56 \text{ ac-ft}}{3.07 \text{ ac-ft per day}}$$

$$= 18.2 \text{ days}$$

Thus, the detention time would not be sufficient to satisfy requirements. Increasing the depth to 5 feet would help.

Calculate the organic loading:

$$\frac{\text{Organic Loading,}}{\text{lbs BOD/day}} = (\text{BOD, mg/}L)(\text{Flow, MGD})(8.34 \text{ lbs/gal})$$

$$= (150 \text{ mg/}L)(1 \text{ MGD})(8.34 \text{ lbs/gal})$$

$$= 1,251 \text{ lbs BOD per day}$$

The organic loading per acre of pond would be:

$$\frac{\text{Organic Loading,}}{\text{lbs BOD/day/ac}} = \frac{\text{Loading, lbs BOD/day}}{\text{Area, ac}}$$

$$= \frac{1,251 \text{ lbs BOD per day}}{16 \text{ ac}}$$

$$= 78.2 \text{ lbs BOD/day/ac}$$

Therefore, the organic loading would exceed the desired maximum of 50 lbs BOD/day/acre.

QUESTION

Write your answers in a notebook and then compare your answers with those on page 333.

9.10I   Given a pond receiving a flow of 2.0 MGD from 20,000 people. Influent BOD is 180 mg/L. Pond area is 24 acres, and the average operating depth is four feet. Determine the detention time, organic loading, population loading, and hydraulic loading.

EXAMPLE METRIC CALCULATIONS

EXAMPLE 1

Use of the pond loading formulas can be illustrated by examining a typical situation. The following data should be obtained so that all calculations can be performed:

Essential Data

1. Depth of Pond   = 2.0 meters

2. Width of Pond

Bottom	= 125 meters
Water Surface	= 133 meters
Average Width	= 129 meters

3. Length of Pond

Bottom	= 200 meters
Water Surface	= 208 meters
Average Length	= 204 meters

4. Side Slopes

(2 m horizontal to 1 m vertical)   = 2:1

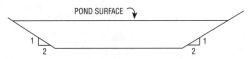

5. Influent Flow   = 800 cubic meters per day

6. Influent BOD   = 200 mg/L

7. Population   = 2,000 persons

To calculate the loading criteria, first determine the pond area and volume.

I.   POND AREA, SQUARE METERS

Pond Area, sq m = (Avg Width, m)(Avg Length, m)

= (129 m)(204 m)

= 26,316 sq m

II.   POND VOLUME, CUBIC METERS

Volume, cu m = (Avg Area, sq m)(Depth, m)

= (26,316 sq m)(2 m)

= 52,632 cu m

III.   LOADING CRITERIA

1.   *DETENTION TIME*

$$\text{Detention Time, days} = \frac{\text{Pond Volume, cu m}}{\text{Flow Rate, cu m/day}}$$

$$= \frac{52,632 \text{ cu m}}{800 \text{ cu m/day}}$$

$$= 66 \text{ days}$$

2.   *POPULATION LOADING*

$$\begin{matrix}\text{Number of}\\\text{Persons per}\\\text{square meter}\end{matrix} = \frac{\text{Population Served by Sewer System, persons}}{\text{Surface Pond Area, sq m}}$$

$$= \frac{2,000 \text{ persons}}{(133 \text{ m})(208 \text{ m})}$$

$$= 0.072 \text{ persons/sq m}$$

NOTE: If there is a significant waste flow from industry mixed in with the domestic waste, an adjustment must be made to take the industrial waste into consideration. This is usually done by analyzing the industrial waste and converting it to a population equivalent.

3.   *HYDRAULIC LOADING (OVERFLOW RATE)*

$$\begin{matrix}\text{Centimeters}\\\text{per Day}\end{matrix} = \frac{\text{Depth of Pond, cm}}{\text{Detention Time, days}}$$

$$= \frac{(\text{Depth, 2 m})(100 \text{ cm/m})}{66 \text{ days}}$$

$$= 3.03 \text{ cm/day}$$

4.   *ORGANIC LOADING*

$$\begin{matrix}\text{Organic Load,}\\\text{gm BOD/}\\\text{day/sq m}\end{matrix} = \frac{(\text{BOD, mg/}L)(\text{Flow, cu m/day})(1 \text{ gm/1,000 mg})(1,000 \text{ }L\text{/cu m})}{\text{Area, sq m}}$$

$$= \frac{200 \text{ mg/}L \times 800 \text{ cu m/day} \times 1 \text{ gm/1,000 mg} \times 1,000 \text{ }L\text{/cu m}}{26,316 \text{ sq m}}$$

$$= 6.1 \text{ gm BOD/day/sq m}$$

EXAMPLE 2

Suppose that a small wastewater treatment plant must be completely shut down for major repairs that will require several months of work. Enough vacant land is near the plant to enable 65,000 square meters of temporary ponds to be constructed as raw wastewater (sewage) lagoons. Determine if this is feasible, given the following data:

Influent Rate	= 4,000 cu m/day
Influent BOD	= 150 mg/L
Pond Area	= 65,000 square meters
Average Operating Depth	= 1.2 meters

Assume that at least a 60-day detention period (average time the wastewater must take to flow through the pond for disinfection) is desired for bacterial die-off.

Assume that the organic loading (BOD) should not exceed 5 grams per day per square meter.

Calculate what the *WASTE DETENTION TIME* would be in the pond.

$$\text{Pond Volume, cu m} = \text{Pond Area, sq m} \times \text{Pond Depth, m}$$

$$= 65,000 \text{ sq m} \times 1.2 \text{ m}$$

$$= 78,000 \text{ cu m}$$

$$\text{Detention Time, days} = \frac{\text{Pond Volume, cu m}}{\text{Influent Flow, cu m/day}}$$

$$= \frac{78,000 \text{ cu m}}{4,000 \text{ cu m/day}}$$

$$= 19.5 \text{ days}$$

Thus, the detention time would not be sufficient to satisfy requirements. Increasing the depth to 3.5 meters would help.

Calculate the organic loading:

$$\begin{matrix}\text{Organic Loading,}\\\text{gm BOD/day}\end{matrix} = (\text{BOD, mg/}L)(\text{Flow, cu m/day})(1 \text{ gm/1,000 mg})(1,000 \text{ }L\text{/cu m})$$

$$= (150 \text{ mg/}L)(4,000 \text{ cu m/day})(1 \text{ gm/1,000 mg})(1,000 \text{ }L\text{/cu m})$$

$$= 600,000 \text{ gm BOD per day}$$

The organic loading per square meter of pond area would be:

$$\begin{matrix}\text{Organic Loading,}\\\text{gm BOD/day/sq m}\end{matrix} = \frac{\text{Loading, gm BOD/day}}{\text{Area, sq m}}$$

$$= \frac{600,000 \text{ gm BOD/day}}{65,000 \text{ sq m}}$$

$$= 9.2 \text{ gm BOD/day/sq m}$$

Therefore, the organic loading would exceed the desired maximum of 5 gm BOD/day/sq m.

9.11 ELIMINATING ALGAE FROM POND EFFLUENTS

Algae are usually present in the effluent from ponds with continuous discharges. Algae can create undesirable impacts on the receiving waters in terms of a loss of aesthetic values, increased turbidity, suspended solids and biochemical oxygen demand, and also the development of nuisance conditions. In most areas, the algae in the effluent increase the suspended solids concentration to the point that the NPDES effluent limitation on total suspended solids is exceeded, thus resulting in a permit violation.

Researchers have attempted to develop cost-effective treatment processes for removing algae from pond effluents. These efforts have included the use of centrifuges, chemical coagulation, filtration, microstraining, magnetic separation, and ultrafiltration. Although some of these processes have been very effective, none of these techniques have achieved a high level of acceptance due to the increased treatment costs associated with the processes.

Duckweed Systems

The Lemna Duckweed System is a patented process that uses aquatic duckweed plants for wastewater treatment. This system is used effectively as a polishing pond after a conventional wastewater treatment pond. The duckweed covering the polishing pond's surface prevents the penetration of sunlight and causes the algae to die and settle out of the wastewater being treated. Plastic grids (approximately 10 ft by 10 ft square) are placed on the surface of the pond to prevent the wind from blowing all the duckweed to one side of the pond. The population of duckweed within each grid reproduces and must be harvested on a regular basis for the system to be effective.

Duckweed plants are capable of removing phosphorus and nitrogen from the water. With sufficient detention time (greater than 30 days) and intensive harvesting, significant nutrient removal may be possible. Duckweed needs water temperatures of 50°F (10°C) or greater to be effective. If the water temperatures drop below 50°F, duckweed will recover when the water temperatures increase to 50°F.

The key considerations for duckweed systems include:

- A 30-day retention time, shallow pond depth, and frequent harvesting (every 1 to 3 days at peak season) if nutrient removal is desired,

- A plan must be in place for disposing of the harvested plants (that is, processing on site, transportation, and ultimate disposal),

- Post-aeration of the effluent may be necessary to meet dissolved oxygen (DO) requirements, and

- Capability to draw effluent from several levels and thereby avoid high algal concentrations near the water surface during discharge.

Land Treatment

Another approach to improving effluent quality from waste stabilization ponds is to apply the pond effluent to an overland-flow treatment system followed by a constructed wetland wastewater treatment system.

Overland-flow systems consist of grass planted on slowly permeable soil and slopes from 0 to 8 percent. The slope lengths range from 100 to 200 feet (30 to 60 meters). The application rate is expressed in inches (centimeters) per day and depends on treatment requirements and local conditions. Application periods usually range from 6 to 12 hours per day. BOD removals are in the 75 to 90 percent range and suspended solids in the 50 to 90 percent range. For additional information on overland-flow treatment systems, see *ADVANCED WASTE TREATMENT*, Chapter 8, "Wastewater Reclamation and Recycling," in this series of operator training manuals, available from the Office of Water Programs, California State University, Sacramento.

Constructed Wetlands

If higher levels of effluent quality are desired, effluent from an overland-flow system can be treated by constructed wetlands. Wetlands systems often consist of two cells and they can be of either the free surface flow (FSF) type or the submerged flow (SF) type. The vegetation in the wetland system may consist mainly of cattails. In cold winter months, the cattails may die and the organic debris is partially burned and the remaining unburned material is removed to prevent plugging of the drains. For additional information on constructed wetlands, see *SMALL WASTEWATER SYSTEM OPERATION AND MAINTENANCE*, Volume II, Chapter 13, "Alternative Wastewater Treatment, Disposal, and Reuse Methods," in this series of operator training manuals, available from the Office of Water Programs, California State University, Sacramento.

Straw

Research in England has proven that blue-green algae can be controlled by the active agent produced during the decomposition of barley straw. Wheat straw works, though not as well as barley straw. Straw should be applied to the final wastewater stabilization pond twice a year, once in the autumn and once again in the spring before algal growth starts. Recommended applications are about one-third ounce of straw per 1.33 cubic yards of water in the pond (a 3-foot-deep pond that is 20 feet by 50 feet would need about 2 pounds of straw, more if the water in the pond changes quickly). A dose of 5 grams of straw per cubic meter is also suggested. Water quality indicators should be monitored to ensure desired performance of the straw.

QUESTIONS

Write your answers in a notebook and then compare your answers with those on page 333.

9.11A What are the undesirable effects of algae on the receiving waters?

9.11B How does the Lemna Duckweed System eliminate algae from pond effluents?

9.11C What types of land treatment systems are used to improve effluent quality from waste stabilization ponds?

9.12 ARITHMETIC ASSIGNMENT

Turn to the Arithmetic Appendix at the back of this manual and read Section A.9, *STEPS IN SOLVING PROBLEMS*, and A.136, Ponds.

9.13 ACKNOWLEDGMENT

Liberal use has been made of the many papers presented by Professor W. J. Oswald of the University of California at Berkeley on the subject of the treatment of wastes by ponding.

9.14   ADDITIONAL READING

1. *MOP 11*, Chapter 23, "Natural Biological Processes."*

2. *NEW YORK MANUAL*, page 5-43, "Stabilization Ponds."*

3. *TEXAS MANUAL*, Chapter 16, "Stabilization Ponds."*

4. *STABILIZATION POND OPERATION AND MAINTE-NANCE*. Obtain from Minnesota Pollution Control Agency, 520 Lafayette Road, St. Paul, MN 55155-4194. Price, $15.00.

5. *OPERATIONS MANUAL—STABILIZATION PONDS*, U.S. Environmental Protection Agency. Obtain from National Technical Information Service (NTIS), 5285 Port Royal Road, Springfield, VA 22161. Order No. PB-279443. EPA No. 430-9-77-012. Price, $39.50, plus $5.00 shipping and handling per order.

* Depends on edition.

9.15   METRIC CALCULATIONS

Refer to Section 9.109, "Pond Loading (English and Metric Calculations)."

END OF LESSON 3 of 3 LESSONS
on
WASTEWATER STABILIZATION PONDS

Please answer the discussion and review questions next. (You may skip Section 9.16 unless you would like to see the arithmetic calculations for Example 1 in this chapter.)

9.16   PONDS—MATH SUPPLEMENT (Details of Example Calculations)

References: See the Appendix, "How to Solve Wastewater Treatment Plant Arithmetic Problems," at the end of this manual.

Solution: Example 1

I. Pond Area, sq ft = (Average Width, ft)(Average Length, ft)

A. Calculate average width.

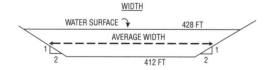

$$\text{Average Width, ft} = \frac{\text{Water Surface Width, ft} + \text{Bottom Width, ft}}{2}$$

$$= \frac{428 \text{ ft} + 412 \text{ ft}}{2}$$

$$\begin{array}{r} 428 \\ + 412 \\ \hline 840 \end{array}$$

$$= \frac{840 \text{ ft}}{2}$$

$$\begin{array}{r} 420 \\ 2\overline{)840} \\ 8 \\ \hline 04 \\ 4 \\ \hline 00 \end{array}$$

$$= 420 \text{ ft}$$

B. Calculate average length.

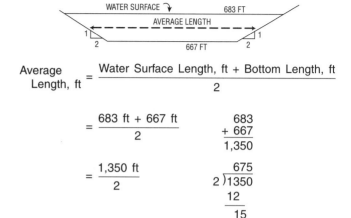

$$\text{Average Length, ft} = \frac{\text{Water Surface Length, ft} + \text{Bottom Length, ft}}{2}$$

$$= \frac{683 \text{ ft} + 667 \text{ ft}}{2}$$

$$\begin{array}{r} 683 \\ + 667 \\ \hline 1,350 \end{array}$$

$$= \frac{1,350 \text{ ft}}{2}$$

$$\begin{array}{r} 675 \\ 2\overline{)1350} \\ 12 \\ \hline 15 \\ 14 \\ \hline 10 \\ 10 \\ \hline 0 \end{array}$$

$$= 675 \text{ ft}$$

C. Calculate pond area.

Area, sq ft = (Average Width, ft)(Average Length, ft)

$$= (420 \text{ ft})(675 \text{ ft})$$

$$= 283,500 \text{ sq ft}$$

$$\begin{array}{r} 675 \\ \times 420 \\ \hline 000 \\ 13\ 50 \\ 270\ 0 \\ \hline 283,500 \end{array}$$

Units: When we multiply ft by ft, we obtain square feet or ft^2.

$$\text{Area, acres} = \frac{\text{Area, sq ft}}{43,560 \text{ sq ft/ac}}$$

$$= \frac{283,500 \text{ sq ft}}{43,560 \text{ sq ft/ac}}$$

$$= 6.5 \text{ ac}$$

$$\begin{array}{r} 6.508 \\ 43560\overline{)283500.} \\ 261360 \\ \hline 22140\ 0 \\ 21780\ 0 \\ \hline 360\ 00 \\ 000\ 00 \\ \hline 360\ 000 \\ 348\ 480 \\ \hline 11\ 520 \end{array}$$

Units: The sq ft on top (numerator) and bottom (denominator) cancel out, and the /acre on the bottom shifts to the top (numerator).

Our result is 6.508 acres, but we will round off our answer to the nearest tenth (0.1), or 6.5. This is sufficient accuracy.

II. Calculate pond volume.

$$\text{Pond Volume, ac-ft} = (\text{Area, ac})(\text{Depth, ft})$$

$$= (6.51 \text{ ac})(4 \text{ ft})$$

$$= 26.04 \text{ ac-ft}$$

$$\begin{array}{r} 6.51 \\ \times 4 \\ \hline 26.04 \end{array}$$

DISCUSSION AND REVIEW QUESTIONS

Chapter 9.   WASTEWATER STABILIZATION PONDS

(Lesson 3 of 3 Lessons)

Write the answers to these questions in your notebook. The question numbering continues from Lesson 2.

13. Why is it desirable for a pond to be isolated from neighbors?

14. How can scum be prevented from leaving a pond?

15. How can erosion of levee slopes be controlled?

A pond receives an inflow of 0.01 MGD from 100 people. The pond is 150 feet long, 150 feet wide, and 4 feet deep. Influent BOD is 200 mg/L. Determine the following loading criteria:

16. Detention time in days;

17. Population loading in persons per acre;

18. Hydraulic loading in inches per day; and

19. Organic loading in pounds of BOD applied per day per acre.

SUGGESTED ANSWERS

Chapter 9.   WASTEWATER STABILIZATION PONDS

ANSWERS TO QUESTIONS IN LESSON 1

Answers to questions on page 297.

9.1A   Probably. When the greatest organic load per unit of area (high treatment efficiency) is destroyed, objectionable odors may develop.

9.1B   The limitations of a pond are that it may:

1. Produce odors,
2. Require a large amount of land,
3. Treat wastes inconsistently depending on climatic conditions,
4. Contaminate groundwaters, or
5. Leave high levels of suspended solids in the effluent.

Answers to questions on page 299.

9.2A   The difference between raw wastewater (sewage) lagoons, oxidation ponds, and polishing ponds is the amount of treatment wastewater receives before reaching the pond. Wastewater receiving no treatment flows directly into a raw wastewater (sewage) lagoon. A pond located after a primary clarifier or sedimentation tank is called an "oxidation pond," and a "polishing pond" is placed after a trickling filter or activated sludge plant.

9.2B   AEROBIC ponds have DO distributed throughout the pond; ANAEROBIC ponds do not contain any DO. Most ponds are facultative and have aerobic (have DO) conditions on the surface and are anaerobic (no DO) on the bottom.

9.2C   The use of a pond will vary depending on the detention period. Ponds with detention times less than three days will act like sedimentation tanks. In ponds with a detention period from three to twenty days, the organic material in the influent will be converted to algae, and high concentrations of algae will be found in the effluent. Ponds with longer detention periods provide time for sedimentation of algae and a better effluent.

Answers to questions on page 301.

9.3A   Algae produce oxygen from the oxygen in the carbon dioxide molecule (CO_2) through photosynthesis.

9.3B   Algae simply appear in a pond on their own without seeding. They are found in soil, water, and air and multiply under favorable conditions.

9.3C   Organic matter in a pond is converted to carbon dioxide and ammonia and, finally, in the presence of sunlight, to algae. The organic matter in anaerobic bottom sections is first converted by a group of organisms called "acid producers" to carbon dioxide, nitrogen, and organic acids. Next, a group called the "methane fermenters" breaks down the acids and other products of the first group to form methane gas. Another end product of organic reduction is water.

Answers to questions on page 302.

9.4A   Biological factors influencing the treatment efficiency of a pond include the type of bacteria, type and quantity of algae, activity of organisms, nutrient deficiencies, and toxic concentrations.

9.4B   Bioflocculation is the clumping together of fine, dispersed organic particles into settleable solids by the action of certain bacteria and algae.

9.4C   A pond is not functioning properly when it creates a visual or odor nuisance, or leaves a high BOD, solids, grease, or coliform-group bacteria concentration in the effluent unless it was designed to be anaerobic in the first stages and aerobic in later ponds for final treatment.

ANSWERS TO QUESTIONS IN LESSON 2

Answers to questions on page 303.

9.5A   At least one foot of water should cover the pond bottom before wastes are introduced to prevent decomposing solids from being exposed and causing odor problems.

9.5B Ponds should be started during the warmer months because higher temperatures are associated with efficient treatment processes.

9.5C A definite green color in a pond indicates a flourishing algae population and is a good sign.

9.5D When bubbles are observed coming to the pond surface near the inlet, this indicates that the solids that settled to the bottom are being decomposed anaerobically by bacterial action.

Answers to questions on page 306.

9.6A Scum should not be allowed to accumulate on the surface of a pond because it is unsightly, may prevent sunlight from reaching the algae, and an odor-producing species of algae may develop on the scum. Also scum can become a source of botulism.

9.6B Scum accumulations may be broken up with rakes, jets of water, or by use of outboard motors.

9.6C Most odors are caused in ponds by overloading or poor housekeeping.

9.6D To prepare for an odor problem, a careful plan for emergency odor control must be developed. For example, an odor control chemical should be available before an odor problem develops. Sodium nitrate or a floating aerator will help control odors and improve treatment of the wastewater.

9.6E Weeds are objectionable in and around ponds because they provide a shelter for the breeding of mosquitoes and scum accumulation and also hinder pond circulation.

9.6F Weeds may be controlled by removing them manually or by using herbicides and soil sterilants.

9.6G Insects should be controlled because they may, in sufficient numbers, be a serious nuisance to nearby residential areas, farm workers, recreation sites, industrial plants, and drivers on highways.

9.6H A pond should be lowered before the application of an insecticide to improve the destruction of insects and reduce the effect of the insecticide on the receiving waters by holding the wastewater at least one day. Lowering of the pond also will dry up weeds and insects.

Answers to questions on page 309.

9.6I The contents of ponds are recirculated to allow algae and other aerobic organisms to become thoroughly mixed with incoming raw wastewater.

9.6J Ponds that are operated on a batch basis may discharge only once (fall) or twice (fall and spring) a year.

9.6K To develop an operating strategy for ponds, consider the following factors:

1. Maintain constant water elevations in the ponds;
2. Distribute inflow equally to ponds;
3. Keep pond levees or dikes in good condition; and
4. Observe and test pond condition.

Answers to questions on page 317.

9.7A Surface aerators have been used (1) to provide additional air for ponds during the night, during cold weather, and for overloaded ponds, and (2) to provide air for ponds operated as aerated lagoons.

9.8A When the pH and dissolved oxygen drop dangerously low, the loading should be reduced or stopped. Recirculating water from a healthy pond to the problem pond should help the situation. Recirculation from outlet to inlet areas is beneficial for seeding, DO, and mixing.

9.8B pH, temperature, and dissolved oxygen should be measured to provide a record of pond performance and to indicate the status (health) of the pond and whether corrective action is or may be necessary. DO may be expected to be low in the morning and increase with sunlight hours.

9.8C When a pond turns dull green, gray, or colorless, generally the pH and dissolved oxygen have dropped too low. This condition may be caused by overloading or lack of circulation.

9.8D Influent BOD to a series of ponds is 200 mg/L. If the BOD in the effluent of the last pond is 40 mg/L, what is the BOD removal efficiency?

$$\text{BOD Removal, \%} = \frac{(\text{In} - \text{Out})}{\text{In}} \times 100\%$$

$$= \frac{(200 \text{ mg}/L - 40 \text{ mg}/L)}{200 \text{ mg}/L} \times 100\%$$

$$= \frac{160 \text{ mg}/L}{200 \text{ mg}/L} \times 100\%$$

$$= 0.80 \times 100\%$$

$$= 80\%$$

Answers to questions on page 318.

9.9A Catwalks over ponds should have guardrails and non-skid walking surfaces.

9.9B An operator should be accompanied by a helper when performing any dangerous task because immediate aid might be needed to prevent serious injury or loss of life.

ANSWERS TO QUESTIONS IN LESSON 3

Answers to questions on page 325.

9.10A Some wastes are not easily treated by ponds because they contain substances with interfering concentrations that hinder algal or bacterial growth.

9.10B The influent to a pond should be metered to justify budgets, indicate unexpected fluctuations in flows that may cause upsets, and provide data for future expansion when necessary.

9.10C The inlet to a pond should be submerged to distribute the heat of the influent as much as possible and to minimize the occurrence of floating material.

9.10D The outlet of a pond should be submerged to prevent the discharge of floating material.

9.10E Free overfalls should be avoided to minimize odors, foaming, and gas entrapment which may hamper the flow of water in pipes. They are generally controlled with pipes at the outfall.

9.10F The discharge of floating material over a surface outlet may be corrected by constructing a baffle around the outlet.

9.10G   The minimum recommended pond operating depth is three feet. At shallower depths, aquatic weeds become a nuisance and pond performance is apt to be irregular.

9.10H   Any shallow pond is apt to be irregular in performance because the algae grow, die, and settle to the bottom, thus creating a new organic load. Algae *PRODUCE* and *STORE* organic matter that must be stabilized later. Objectionable "burps" of unstable material are common in pond effluents.

Answers to question on page 327.

9.10I   Given: Flow   = 2.0 MGD
               Population   = 20,000 people
               Influent BOD   = 180 mg/L
               Pond Area   = 24 acres
               Average Depth   = 4 feet

        Find: Detention Time
               Organic Loading
               Population Loading
               Hydraulic Loading

Determine pond volume in acre-feet.

Volume, ac-ft = (Avg Area, ac)(Depth, ft)

= (24 ac)(4 ft)

= 96 ac-ft

Convert flow rate from MGD to ac-ft per day.

$$\text{Flow, ac-ft/day} = \frac{2{,}000{,}000 \text{ gal}}{\text{day}} \times \frac{\text{cu ft}}{7.48 \text{ gal}} \times \frac{\text{acre}}{43{,}560 \text{ sq ft}}$$

= 6.1 ac-ft/day

Determine detention time in days.

$$\text{Detention Time, days} = \frac{\text{Pond Volume, ac-ft}}{\text{Flow Rate, ac-ft/day}}$$

$$= \frac{96 \text{ ac-ft}}{6.1 \text{ ac-ft/day}}$$

= 15.7 days

```
        15.7
6.1) 96.0
     61
     350
     305
     450
     427
```

Calculate organic loading in pounds of BOD per day per acre.

$$\text{Organic Loading, lbs BOD/day/ac} = \frac{(\text{BOD, mg/}L)(\text{Flow, MGD})(8.34 \text{ lbs/gal})}{\text{Area, ac}}$$

$$= \frac{180 \text{ lbs}}{\text{M lbs}} \times \frac{2.0 \text{ MG}}{\text{day}} \times \frac{8.34 \text{ lbs}}{\text{gal}} \times \frac{1}{24 \text{ ac}}$$

= 125 lbs BOD/day/ac

Estimate the population loading in persons per acre.

$$\text{Population Loading, persons/ac} = \frac{\text{Population, persons}}{\text{Surface Area, ac}}$$

$$= \frac{20{,}000 \text{ persons}}{24 \text{ ac}}$$

= 833 persons/ac

Calculate the hydraulic loading in inches per day.

$$\text{Hydraulic Loading, in/day} = \frac{\text{Pond Depth, in}}{\text{Detention Time, days}}$$

$$= \frac{(4 \text{ ft})(12 \text{ in/ft})}{15.7 \text{ days}}$$

$$= \frac{48}{15.7}$$

= 3.06 in/day

```
          3.057
15.7) 48.0
      47.1
       900
       785
      1150
      1099
```

Answers to questions on page 329.

9.11A   Algae can create undesirable effects on the receiving waters in terms of a loss of aesthetic values, increased turbidity, suspended solids, and biochemical oxygen demand; algae can also cause the development of nuisance conditions.

9.11B   The Lemna Duckweed System eliminates algae from pond effluents by covering the pond's surface with duckweed, thus preventing the penetration of sunlight and causing algae to die and settle out of the wastewater being treated.

9.11C   Land treatment systems used to improve effluent quality from stabilization ponds consist of applying the effluent to an overland-flow treatment system followed by a constructed wetland wastewater treatment system.

APPENDIX

Monthly Data Sheet

CLEANWATER, U.S.A.
WATER POLLUTION CONTROL PLANT

MONTHLY RECORD _____ 20____   OPERATOR: _____

DATE	DAY	WEATHER	TIME OF VISIT	FLOW (MGD) EFFLUENT	TEMP °F NO.1 POND	TEMP °F NO.2 POND	pH NO.1 POND	pH NO.2 POND	D.O. NO.1 POND	D.O. NO.2 POND	B.O.D. INFLUENT	B.O.D. EFFLUENT	CL2 FEED RATE LB/DAY	CL2 RESIDUAL MG/L	E-COLI M.P.N./100	MIXING HRS NO.1 POND	MIXING HRS NO.2 POND	REMARKS
1	T	OVERCAST	9:30 A.M.	0.025	68	70	7.6	7.9	0.8	1.4	183	32	35	3.2	62	-0-	-0-	EFFL. SS = 85 mg/l
2	W	CLEAR	2:00 P.M.	0.038	76	79	7.7	8.4	1.3	8.7			35	1.4				
3	T																	
4	F																	
5	S																	
6	S																	
7	M	CLEAR	10:00 A.M.	0.026	70	74	7.6	8.1	1.2	7.6	158	29	35					EFFL. SS = 53 mg/l
8	T	CLEAR	10:30 A.M.	0.027	69	73	7.6	8.1	1.1	7.8								
9	W	CLEAR	3:15 P.M.	0.039	77	80	7.6	8.3	1.5	11.2			35	0.8	2400			
10	T	HOT																
11	F																	
12	S																	
13	S																	
14	M	CLEAR	9:30 A.M.	0.023	70	75	7.8	8.5	1.9	7.6	162	33	40	4.2	700	-0-	-0-	INCREASED CL2 FEED RATE TO 40#/DAY — EFFL. SS = 37 mg/l
15	T																	
16	W	CLEAR	1:30 P.M.	0.031	75	79	7.8	8.4	1.7	10.6			40	2.5				
17	T																	
18	F	CLEAR	3:30 P.M.	0.037	79	85	7.8	8.6	2.1	14.5			40	2.8	23			
19	S																	
20	S																	
21	M	CLEAR	9:00 A.M.	0.020	71	76	7.6	8.4	1.0	7.8			40	5.6				DECLARED HOLIDAY - NO PLANT CHECK
22	T																	
23	W	CLEAR	10:30 A.M.	0.028	72	78	7.7	8.4	1.0	8.1	178	30	40	5.0	6	-0-	-0-	EFFL. SS = 48 mg/l
24	T																	
25	F	CLEAR	2:15 P.M.	0.035	80	87	7.8	8.5	1.3	11.4			40	3.8				
26	S																	
27	S																	
28	M	CLEAR	10:00 A.M.	0.026	73	79	7.7	8.5	1.1	8.6	168	35	40	4.7		-0-	-0-	CL2 FEED RATE BACK TO 35#/DAY
29	T																	
30	W	CLEAR	4:15 P.M.	0.038	81	88	7.8	8.5	1.8	15.3			35	1.1	6			EFFL. SS = 57 mg/l
31	T																	
MAX				0.039	81	88	7.8	8.6	2.1	15.3	183	35	40	5.6	2400	-0-	-0-	
MIN				0.020	68	70	7.6	7.9	0.8	7.6	162	29	35	0.8	6	-0-	-0-	
AVG	CLEAR			0.030	73	78	7.7	8.3	1.3	9.2	169	31	37.2	2.7	62	-0-	-0-	

FLOW METER:
LAST 232602
1st 231672
TOTAL 0.930 MG

ELECTRIC METER:
LAST 5978
1st 5817
MULT. 40 x 161 = 6440 KWH

SUMMARY DATA

% REMOVAL B.O.D.	81.6 %
LBS. B.O.D. / ACRE / DAY	52.8
DETENTION TIME — DAYS	86.7

PONDS OPERATED IN SERIES ALL MONTH.

MIXERS NOT OPERATED DURING MONTH.

COST DATA

MAN DAYS @ 2.5 HRS = 1 DAY — PAYROLL	245.52
POWER PURCHASED	69.60
OTHER UTILITIES (GAS, H2O)	NONE
GASOLINE, OIL, GREASE	NONE
CHEMICALS & SUPPLIES	138.60
MAINTENANCE	NONE
VEHICLE COSTS	57.40
OTHER	NONE
TOTAL	$ 511.12
OPERATING COST / CAPITA / MO	$ 1.68
COST / 1000 GAL. TREATED	$ 0.55

CHAPTER 10

DISINFECTION AND CHLORINATION

by

Leonard W. Hom

(With a special section by J. L. Beals)

Revised by

Tom Ikesaki

TABLE OF CONTENTS

Chapter 10.   DISINFECTION AND CHLORINATION

OBJECTIVES

Chapter 10.   DISINFECTION AND CHLORINATION

Following completion of Chapter 10, you should be able to:

1. Explain the principles of wastewater disinfection with chlorine,

2. Control the chlorination process to obtain the desired effluent disinfection,

3. Handle chlorine safely,

4. Detect chlorine leaks and take appropriate corrective action,

5. Inspect new chlorination facilities for proper installation,

6. Schedule and conduct chlorination operation and maintenance duties,

7. Recognize factors that indicate the chlorination process is not performing properly, identify the source of the problem, and take corrective action,

8. Conduct your duties in a safe fashion,

9. Determine chlorine dosages,

10. Explain applications and limitations of uses of chlorine other than for disinfection,

11. Keep records of chlorination operation, and

12. Safely operate and maintain a sulfur dioxide dechlorination system.

WORDS

Chapter 10.   DISINFECTION AND CHLORINATION

AIR GAP

An open vertical drop, or vertical empty space, between a drinking (potable) water supply and the point of use in a wastewater treatment plant. This gap prevents the contamination of drinking water by backsiphonage because there is no way wastewater can reach the drinking water supply.

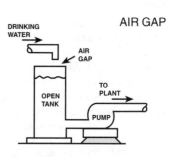

AIR PADDING

Pumping dry air (dew point –40°F) into a container to assist with the withdrawal of a liquid or to force a liquified gas such as chlorine or sulfur dioxide out of a container.

AMBIENT (AM-bee-ent) TEMPERATURE

Temperature of the surroundings.

AMPEROMETRIC (am-PURR-o-MET-rick)

A method of measurement that records electric current flowing or generated, rather than recording voltage. Amperometric titration is a means of measuring concentrations of certain substances in water.

BACTERIA (back-TEAR-e-ah)

Bacteria are living organisms, microscopic in size, which usually consist of a single cell. Most bacteria use organic matter for their food and produce waste products as a result of their life processes.

BIODEGRADATION (BUY-o-deh-grah-DAY-shun)

The breakdown of organic matter by bacteria to more stable forms which will not create a nuisance or give off foul odors.

BREAKOUT OF CHLORINE

A point at which chlorine leaves solution as a gas because the chlorine feed rate is too high. The solution is saturated and cannot dissolve any more chlorine. The maximum strength a chlorine solution can attain is approximately 3,500 mg/*L*. Beyond this concentration molecular chlorine is present which will break out of solution and cause "off-gassing" at the point of application.

BREAKPOINT CHLORINATION

Addition of chlorine to water or wastewater until the chlorine demand has been satisfied. At this point, further additions of chlorine result in a residual that is directly proportional to the amount of chlorine added beyond the breakpoint.

CHLORAMINES (KLOR-uh-means)   CHLORAMINES

Compounds formed by the reaction of hypochlorous acid (or aqueous chlorine) with ammonia.

CHLORINATION (KLOR-uh-NAY-shun)   CHLORINATION

The application of chlorine to water or wastewater, generally for the purpose of disinfection, but frequently for accomplishing other biological or chemical results.

CHLORINE CONTACT CHAMBER   CHLORINE CONTACT CHAMBER

A baffled basin that provides sufficient detention time of chlorine contact with wastewater for disinfection to occur. The minimum contact time is usually 30 minutes. (Also commonly referred to as basin or tank.)

CHLORINE DEMAND   CHLORINE DEMAND

Chlorine demand is the difference between the amount of chlorine added to wastewater and the amount of residual chlorine remaining after a given contact time. Chlorine demand may change with dosage, time, temperature, pH, and nature and amount of the impurities in the water.

Chlorine Demand, mg/L = Chlorine Applied, mg/L − Chlorine Residual, mg/L

CHLORINE REQUIREMENT   CHLORINE REQUIREMENT

The amount of chlorine which is needed for a particular purpose. Some reasons for adding chlorine are reducing the number of coliform bacteria (Most Probable Number), obtaining a particular chlorine residual, or oxidizing some substance in the water. In each case a definite dosage of chlorine will be necessary. This dosage is the chlorine requirement.

CHLORORGANIC (klor-or-GAN-ick)   CHLORORGANIC

Organic compounds combined with chlorine. These compounds generally originate from, or are associated with, living or dead organic materials.

COLIFORM (COAL-i-form)   COLIFORM

One type of bacteria. The presence of coliform-group bacteria is an indication of possible pathogenic bacterial contamination. The human intestinal tract is one of the main habitats of coliform bacteria. They may also be found in the intestinal tracts of warm-blooded animals, and in plants, soil, air, and the aquatic environment. Fecal coliforms are those coliforms found in the feces of various warm-blooded animals; whereas the term "coliform" also includes other environmental sources.

COLORIMETRIC MEASUREMENT   COLORIMETRIC MEASUREMENT

A means of measuring unknown chemical concentrations in water by *MEASURING A SAMPLE'S COLOR INTENSITY.* The specific color of the sample, developed by addition of chemical reagents, is measured with a photoelectric colorimeter or is compared with "color standards" using, or corresponding with, known concentrations of the chemical.

COMBINED AVAILABLE CHLORINE   COMBINED AVAILABLE CHLORINE

The total chlorine, present as chloramine or other derivatives, that is present in a water and is still available for disinfection and for oxidation of organic matter. The combined chlorine compounds are more stable than free chlorine forms, but they are somewhat slower in disinfection action.

COMBINED AVAILABLE CHLORINE RESIDUAL   COMBINED AVAILABLE CHLORINE RESIDUAL

The concentration of residual chlorine that is combined with ammonia, organic nitrogen, or both in water as a chloramine (or other chloro derivative) and yet is still available to oxidize organic matter and help kill bacteria.

COMBINED CHLORINE   COMBINED CHLORINE

The sum of the chlorine species composed of free chlorine and ammonia, including monochloramine, dichloramine, and trichloramine (nitrogen trichloride). Dichloramine is the strongest disinfectant of these chlorine species, but it has less oxidative capacity than free chlorine.

COMBINED RESIDUAL CHLORINATION   COMBINED RESIDUAL CHLORINATION

The application of chlorine to water or wastewater to produce a combined available chlorine residual. The residual may consist of chlorine compounds formed by the reaction of chlorine with natural or added ammonia (NH_3) or with certain organic nitrogen compounds.

DPD (pronounce as separate letters) METHOD   DPD METHOD

A method of measuring the chlorine residual in water. The residual may be determined by either titrating or comparing a developed color with color standards. DPD stands for N,N-diethyl-p-phenylene-diamine.

DECHLORINATION (dee-KLOR-uh-NAY-shun) DECHLORINATION

The removal of chlorine from the effluent of a treatment plant. Chlorine needs to be removed because chlorine is toxic to fish and other aquatic life.

DEGRADATION (deh-gruh-DAY-shun) DEGRADATION

The conversion or breakdown of a substance to simpler compounds. For example, the degradation of organic matter to carbon dioxide and water.

DEW POINT DEW POINT

The temperature to which air with a given quantity of water vapor must be cooled to cause condensation of the vapor in the air.

DISINFECTION (dis-in-FECT-shun) DISINFECTION

The process designed to kill or inactivate most microorganisms in wastewater, including essentially all pathogenic (disease-causing) bacteria. There are several ways to disinfect, with chlorination being the most frequently used in water and wastewater treatment plants. Compare with STERILIZATION.

EDUCTOR (e-DUCK-ter) EDUCTOR

A hydraulic device used to create a negative pressure (suction) by forcing a liquid through a restriction, such as a Venturi. An eductor or aspirator (the hydraulic device) may be used in the laboratory in place of a vacuum pump. As an injector, it is used to produce vacuum for chlorinators. Sometimes used instead of a suction pump.

ELECTRON ELECTRON

(1) A very small, negatively charged particle which is practically weightless. According to the electron theory, all electrical and electronic effects are caused either by the movement of electrons from place to place or because there is an excess or lack of electrons at a particular place.

(2) The part of an atom that determines its chemical properties.

ENTERIC ENTERIC

Of intestinal origin, especially applied to wastes or bacteria.

ENZYMES (EN-zimes) ENZYMES

Organic substances (produced by living organisms) which cause or speed up chemical reactions. Organic catalysts and/or biochemical catalysts.

FILAMENTOUS (FILL-a-MEN-tuss) ORGANISMS FILAMENTOUS ORGANISMS

Organisms that grow in a thread or filamentous form. Common types are *Thiothrix* and *Actinomycetes*. A common cause of sludge bulking in the activated sludge process.

FREE AVAILABLE CHLORINE FREE AVAILABLE CHLORINE

The amount of chlorine available in water. This chlorine may be in the form of dissolved gas (Cl_2), hypochlorous acid (HOCl), or hypochlorite ion (OCl$^-$), but does not include chlorine combined with an amine (ammonia or nitrogen) or other organic compound.

FREE AVAILABLE RESIDUAL CHLORINE FREE AVAILABLE RESIDUAL CHLORINE

That portion of the total available residual chlorine remaining in water or wastewater at the end of a specified contact period. Residual chlorine will react chemically and biologically as hypochlorous acid (HOCl) or hypochlorite ion (OCl$^-$). This does not include chlorine that has combined with ammonia, nitrogen, or other compounds.

FREE CHLORINE FREE CHLORINE

Free chlorine is chlorine (Cl_2) in a liquid or gaseous form. Free chlorine combines with water to form hypochlorous (HOCl) and hydrochloric (HCl) acids. In wastewater free chlorine usually combines with an amine (ammonia or nitrogen) or other organic compounds to form combined chlorine compounds.

FREE RESIDUAL CHLORINATION FREE RESIDUAL CHLORINATION

The application of chlorine or chlorine compounds to water or wastewater to produce a free available chlorine residual directly or through the destruction of ammonia (NH_3) or certain organic nitrogenous compounds.

HEPATITIS (HEP-uh-TIE-tis) HEPATITIS

Hepatitis is an inflammation of the liver caused by an acute viral infection. Yellow jaundice is one symptom of hepatitis.

HYPOCHLORINATION (HI-poe-KLOR-uh-NAY-shun)   HYPOCHLORINATION

The application of hypochlorite compounds to water or wastewater for the purpose of disinfection.

HYPOCHLORINATORS (HI-poe-KLOR-uh-NAY-tors)   HYPOCHLORINATORS

Chlorine pumps, chemical feed pumps or devices used to dispense chlorine solutions made from hypochlorites such as bleach (sodium hypochlorite) or calcium hypochlorite into the water being treated.

HYPOCHLORITE (HI-poe-KLOR-ite)   HYPOCHLORITE

Chemical compounds containing available chlorine; used for disinfection. They are available as liquids (bleach) or solids (powder, granules, and pellets) in barrels, drums, and cans. Salts of hypochlorous acid.

MPN (pronounce as separate letters)   MPN

MPN is the **M**ost **P**robable **N**umber of coliform-group organisms per unit volume of sample water. Expressed as a density or population of organisms per 100 mL of sample water.

MOTILE (MO-till)   MOTILE

Capable of self-propelled movement. A term that is sometimes used to distinguish between certain types of organisms found in water.

NITROGENOUS (nye-TRAH-jen-us)   NITROGENOUS

A term used to describe chemical compounds (usually organic) containing nitrogen in combined forms. Proteins and nitrate are nitrogenous compounds.

NOMOGRAM (NOME-o-gram)   NOMOGRAM

A chart or diagram containing three or more scales used to solve problems with three or more variables instead of using mathematical formulas.

ORTHOTOLIDINE (or-tho-TOL-uh-dine)   ORTHOTOLIDINE

Orthotolidine is a colorimetric indicator of chlorine residual. If chlorine is present, a yellow-colored compound is produced. This reagent is no longer approved for tests of effluent chlorine residual.

OXIDIZING AGENT   OXIDIZING AGENT

Any substance, such as oxygen (O_2) or chlorine (Cl_2), that will readily add (take on) electrons. When oxygen or chlorine is added to wastewater, organic substances are oxidized. These oxidized organic substances are more stable and less likely to give off odors or to contain disease-causing bacteria. The opposite is a REDUCING AGENT.

PARASITIC (PAIR-a-SIT-tick) BACTERIA   PARASITIC BACTERIA

Parasitic bacteria are those bacteria which normally live off another living organism, known as the "host."

PATHOGENIC (PATH-o-JEN-ick) ORGANISMS   PATHOGENIC ORGANISMS

Bacteria, viruses, cysts, or protozoa which can cause disease (giardiasis, cryptosporidiosis, typhoid, cholera, dysentery) in a host (such as a person). There are many types of organisms which do *NOT* cause disease and which are *NOT* called pathogenic. Many beneficial bacteria are found in wastewater treatment processes actively cleaning up organic wastes.

POSTCHLORINATION   POSTCHLORINATION

The addition of chlorine to the plant discharge or effluent, *FOLLOWING* plant treatment, for disinfection purposes.

POTABLE (POE-tuh-bull) WATER   POTABLE WATER

Water that does not contain objectionable pollution, contamination, minerals, or infective agents and is considered satisfactory for drinking.

PRECHLORINATION   PRECHLORINATION

The addition of chlorine in the collection system serving the plant or at the headworks of the plant *PRIOR TO* other treatment processes mainly for odor and corrosion control. Also applied to aid disinfection, to reduce plant BOD load, to aid in settling, to control foaming in Imhoff units and to help remove oil.

REAGENT (re-A-gent)   REAGENT

A pure chemical substance that is used to make new products or is used in chemical tests to measure, detect, or examine other substances.

REDUCING AGENT

Any substance, such as base metal (iron) or the sulfide ion (S^{2-}), that will readily donate (give up) electrons. The opposite is an OXIDIZING AGENT.

RELIQUEFACTION (re-LICK-we-FACK-shun)

The return of a gas to the liquid state; for example, a condensation of chlorine gas to return it to its liquid form by cooling.

RESIDUAL CHLORINE

The concentration of chlorine present in water after the chlorine demand has been satisfied. The concentration is expressed in terms of the total chlorine residual, which includes both the free and combined or chemically bound chlorine residuals.

ROTAMETER (RODE-uh-ME-ter)

A device used to measure the flow rate of gases and liquids. The gas or liquid being measured flows vertically up a tapered, calibrated tube. Inside the tube is a small ball or bullet-shaped float (it may rotate) that rises or falls depending on the flow rate. The flow rate may be read on a scale behind or on the tube by looking at the middle of the ball or at the widest part or top of the float.

SACRIFICIAL ANODE

An easily corroded material deliberately installed in a pipe or tank. The intent of such an installation is to give up (sacrifice) this anode to corrosion while the water supply facilities remain relatively corrosion free.

SAPROPHYTIC (SAP-row-FIT-ick) ORGANISMS

Organisms living on dead or decaying organic matter. They help natural decomposition of the organic solids in wastewater.

SEPTICITY (sep-TIS-it-tee)

Septicity is the condition in which organic matter decomposes to form foul-smelling products associated with the absence of free oxygen. If severe, the wastewater produces hydrogen sulfide, turns black, gives off foul odors, contains little or no dissolved oxygen, and the wastewater has a high oxygen demand.

STERILIZATION (STARE-uh-luh-ZAY-shun)

The removal or destruction of all microorganisms, including pathogenic and other bacteria, vegetative forms and spores. Compare with DISINFECTION.

TITRATE (TIE-trate)

To *TITRATE* a sample, a chemical solution of known strength is added drop by drop until a certain color change, precipitate, or pH change in the sample is observed (end point). Titration is the process of adding the chemical reagent in small increments (0.1 – 1.0 milliliter) until completion of the reaction, as signaled by the end point.

TOTAL CHLORINE

The total concentration of chlorine in water, including the combined chlorine (such as inorganic and organic chloramines) and the free available chlorine.

TOTAL CHLORINE RESIDUAL

The total amount of chlorine residual (value for residual chlorine, including both free chlorine and chemically bound chlorine) present in a water sample after a given contact time.

CHAPTER 10.   DISINFECTION AND CHLORINATION

(Lesson 1 of 5 Lessons)

10.0   NEED FOR DISINFECTION

10.00   Effectiveness in Microorganism Removal by Various Treatment Processes

Homes, hospitals, and industrial facilities all discharge liquid and solid waste materials into the wastewater collection system. Diseases from human discharges may be transmitted by wastewater. Typical disease-causing microorganisms include bacteria, viruses, and parasites. These microorganisms are commonly referred to as *PATHOGENIC* (disease-causing) *ORGANISMS*.[1] These microorganisms can cause the following types of illnesses:

Bacteria-caused:	Internal Parasite-caused:
Salmonellosis	Amoebic Dysentery
Shigellosis	Ascaris (giant roundworm)
Typhoid Fever	Giardiasis
Cholera	Cryptosporidiosis
Paratyphoid	**Virus-caused:**
Bacillary Dysentery	Polio
Anthrax	Infectious Hepatitis

NOTE: To date there is no record of an operator becoming infected by the HIV virus or developing AIDS due to on-the-job activities.

Disease-producing microorganisms are potentially present in all wastewaters. These microorganisms must be removed or killed before treated wastewater can be discharged to the receiving waters. The purpose of disinfection is to destroy pathogenic microorganisms and thus prevent the spread of waterborne diseases.

Wastewater treatment processes usually remove some of the pathogenic microorganisms through the following processes:

1. Physical removal through sedimentation and filtration,

2. Natural die-away or die-off of microorganisms in an unfavorable environment, and

3. Destruction by chemicals introduced for treatment purposes.

Effectiveness of treatment processes in removing microorganisms depends on the type of treatment processes and the hydraulic and organic loadings. Typical destruction/removals by various processes are summarized below:

Treatment Process	Microorganism Removal, %
Screening	10-20
Grit Channel	10-25
Primary Sedimentation	25-75
Chemical Precipitation	40-80
Trickling Filters	90-95
Activated Sludge	90-98
Chlorination	98-99

Although pathogenic microorganisms are reduced in number by the various treatment processes listed and by natural die-away or die-off in unfavorable environments, many microorganisms still remain. To ensure that essentially all pathogenic microorganisms are destroyed in the effluents of wastewater treatment plants, disinfection is practiced. Since chlorine is the most widely used chemical for disinfection, this chapter will deal primarily with the principles and practices of chlorine disinfection and dechlorination (Figure 10.1). (The Appendix at the end of this chapter describes another method of disinfection that is increasingly being used in wastewater treatment plants: disinfection using ultraviolet (UV) light systems.)

10.01   Disinfection

Two terms you should understand are "disinfection" and "sterilization." *DISINFECTION* is the destruction of all pathogenic microorganisms, while *STERILIZATION* is the destruction of all microorganisms.

The main objective of disinfection is to prevent the spread of disease by protecting:

1. Public water supplies,
2. Receiving waters used for recreational purposes, and
3. Shellfish growing areas.

Disinfection is effective because pathogenic microorganisms are more sensitive to destruction by chlorination than nonpathogens. Many nonpathogenic microorganisms are classified as *SAPROPHYTES*[2] and are essential to wastewater treatment processes. Chlorination for disinfection purposes results in killing essentially all of the pathogens in the

---

[1] *Pathogenic (Path-o-jen-ick) Organisms.   Bacteria, viruses, cysts, or protozoa which can cause disease (giardiasis, cryptosporidiosis, typhoid, cholera, dysentery) in a host (such as a person). There are many types of organisms which do NOT cause disease and which are NOT called pathogenic. Many beneficial bacteria are found in wastewater treatment processes actively cleaning up organic wastes.*

[2] *Saprophytic (SAP-row-FIT-ick) Organisms.   Organisms living on dead or decaying organic matter. They help natural decomposition of the organic solids in wastewater.*

TREATMENT PROCESS   FUNCTION

PRELIMINARY TREATMENT

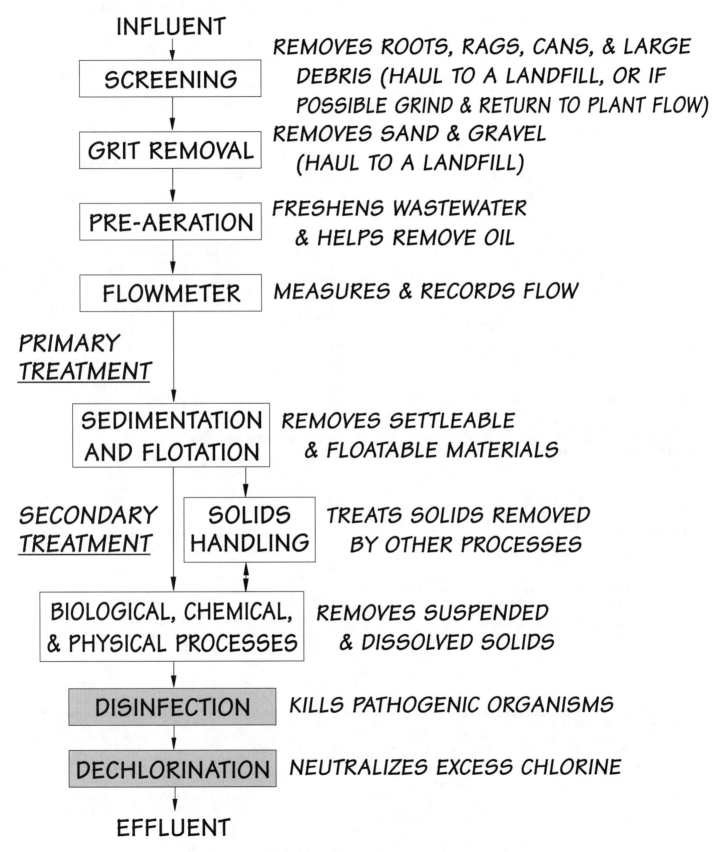

INFLUENT

SCREENING — REMOVES ROOTS, RAGS, CANS, & LARGE DEBRIS (HAUL TO A LANDFILL, OR IF POSSIBLE GRIND & RETURN TO PLANT FLOW)

GRIT REMOVAL — REMOVES SAND & GRAVEL (HAUL TO A LANDFILL)

PRE-AERATION — FRESHENS WASTEWATER & HELPS REMOVE OIL

FLOWMETER — MEASURES & RECORDS FLOW

PRIMARY TREATMENT

SEDIMENTATION AND FLOTATION — REMOVES SETTLEABLE & FLOATABLE MATERIALS

SECONDARY TREATMENT

SOLIDS HANDLING — TREATS SOLIDS REMOVED BY OTHER PROCESSES

BIOLOGICAL, CHEMICAL, & PHYSICAL PROCESSES — REMOVES SUSPENDED & DISSOLVED SOLIDS

DISINFECTION — KILLS PATHOGENIC ORGANISMS

DECHLORINATION — NEUTRALIZES EXCESS CHLORINE

EFFLUENT

Fig. 10.1   Typical flow diagram of wastewater treatment plant

plant effluent. No attempt is made to *STERILIZE* wastewater because it is unnecessary and impractical. Many other sensitive organisms in contact with chlorine are destroyed too. Chlorine is a nonselective killer. Organisms are affected by chlorine on the basis of their sensitivity and growth rate and also on the concentration of chlorine and exposure time.

One of the main uses of chlorine in wastewater treatment is for disinfection. Chlorine is relatively easy to obtain and cheap to manufacture. Even at relatively low dosages, chlorine is extremely effective.

Today people are living more intimately with wastewater than ever before. Wastewater treatment plant effluent may be used for irrigating lawns, parks, cemeteries, freeway planting, golf courses, college campuses, athletic fields, and other public areas. Recreational lakes used for boating, swimming,

water skiing, fishing, and other water sports are frequently made up partially or, in a few cases, solely of treated effluents. As public contact has increased and diluting waters have decreased or become of poor quality, it has become obvious that more consideration must be given to disinfection practices.

QUESTIONS

Write your answers in a notebook and then compare your answers with those on page 418.

10.0A   What is the purpose of disinfection? Why is this important?

10.0B   How are pathogenic bacteria destroyed or removed from water?

10.0C   Why are wastes not sterilized?

10.0D   Why is chlorine used for disinfection?

10.02   Reaction of Chlorine in Wastewater

Chlorine is applied to wastewater as *FREE CHLORINE*[3] (Cl_2), as *HYPOCHLORITE*[4] ion (OCl^-), or as chlorine dioxide (ClO_2). In either the free chlorine or hypochlorite ion form, chlorine is an extremely active chemical and acts as a potent *OXIDIZING AGENT.*[5] In the chlorine dioxide form, chlorine is not as potent an oxidizing agent in the pH ranges commonly found in most wastewaters. Since chlorine is very reactive, it is often used up by side reactions before disinfection takes place. These side reactions can be with such substances as organic material, hydrogen sulfide, phenols, thiosulfate, and ferrous iron. These side reactions occur first and use up a major portion of the chlorine necessary to meet the *CHLORINE DEMAND*[6][7] for a wastewater.

10.020   Free Chlorine (Cl_2)

Free chlorine combines with water to form hypochlorous and hydrochloric acids:

$$\text{Chlorine} + \text{Water} \rightleftharpoons \text{Hypochlorous Acid} + \text{Hydrochloric Acid}$$

$$Cl_2 + H_2O \rightleftharpoons HOCl + H^+ + Cl^-$$

In solutions that are dilute and have a pH above 4, the formation of HOCl (hypochlorous acid) is most complete and leaves little Cl_2 existing. The hypochlorous acid is a weak acid and is very poorly dissociated (broken up) at pH levels below pH 6. Thus any free chlorine or hypochlorite (OCl^-) added to water will immediately form either HOCl or OCl^- and what will be formed is controlled by the pH value of the water. This is extremely important since HOCl and OCl^- differ in disinfection ability. HOCl has a 40 to 80 times greater disinfection potential than OCl^-. Normally in wastewater with a pH of 7.3 (depends on temperature), 50 percent of the chlorine present will be in the form of HOCl and 50 percent in the form of OCl^-. The higher the pH level, the greater the percent of OCl^-.

10.021   Hypochlorite (OCl^-)

When hypochlorite compounds are used in wastewater, they are usually in the form of sodium hypochlorite (NaOCl). Calcium hypochlorite ($Ca(OCl)_2$) normally is not used because of its cost, calcium sludge-forming characteristics, and explosive nature.

The use of hypochlorite in wastewater follows a reaction similar to that of chlorine gas.

$$\text{Sodium Hypochlorite} + \text{Water} \rightarrow \text{Sodium Hydroxide} + \text{Hypochlorous Acid} + \text{Hypochlorite Ion} + \text{Hydrogen Ion}$$

$$2\ NaOCl + 2\ H_2O \rightarrow 2\ NaOH + HOCl + OCl^- + H^+$$

The difference between chlorine gas and hypochlorite compounds is in the side reactions formed. The reaction of chlorine gas tends to decrease the pH, which favors the HOCl (hy-

---

[3] *Free Chlorine.*   Free chlorine is chlorine (Cl_2) in a liquid or gaseous form. Free chlorine combines with water to form hypochlorous (HOCl) and hydrochloric (HCl) acids. In wastewater free chlorine usually combines with an amine (ammonia or nitrogen) or other organic compounds to form combined chlorine compounds.

[4] *Hypochlorite (HI-poe-KLOR-ite).*   Chemical compounds containing available chlorine; used for disinfection. They are available as liquids (bleach) or solids (powder, granules, and pellets) in barrels, drums, and cans. Salts of hypochlorous acid.

[5] *Oxidizing Agent.*   Any substance, such as oxygen (O_2) or chlorine (Cl_2), that will readily add (take on) electrons. When oxygen or chlorine is added to wastewater, organic substances are oxidized. These oxidized organic substances are more stable and less likely to give off odors or to contain disease-causing bacteria. The opposite is a reducing agent.

[6] *Chlorine Demand.*   Chlorine demand is the difference between the amount of chlorine added to wastewater and the amount of *RESIDUAL CHLORINE*[7] remaining after a given contact time. Chlorine demand may change with dosage, time, temperature, pH, and nature and amount of the impurities in the water. Chlorine Demand, mg/L = Chlorine Applied, mg/L – Chlorine Residual, mg/L.

[7] *Residual Chlorine.*   The concentration of chlorine present in water after the chlorine demand has been satisfied. The concentration is expressed in terms of the total chlorine residual, which includes both the free and combined or chemically bound chlorine residuals.

pochlorous acid) formation, while the hypochlorite increases the pH with the formation of hydroxyl ions (OH⁻) by the formation of sodium hydroxide. At a high pH of around 10, the hypochlorous acid (HOCl) dissociates.

$$\text{Hypochlorous Acid} \rightleftharpoons \text{Hydrogen Ion} + \text{Hypochlorite Ion}$$

$$HOCl \rightleftharpoons H^+ + OCl^-$$

This pH condition lasts only a brief time at the contact of the hypochlorite solution and the wastewater to be treated. Since the hypochlorite ion (OCl⁻) is a relatively ineffective disinfectant, the sodium hypochlorite solution should be as dilute as possible. This is extremely important in wastewater disinfection.

10.022   Chlorine Dioxide (ClO₂)

Chlorine dioxide reacts in the following manner:

$$\text{Chlorine Dioxide} + \text{Water} \rightarrow \text{Chlorate Ion} + \text{Chlorite Ion} + \text{Hydrogen Ions}$$

$$2\ ClO_2 + H_2O \rightarrow ClO_3^- + ClO_2^- + 2\ H^+.$$

The oxidizing capability of chlorine dioxide is not all used in wastewater treatment because reactions with the substances (*REDUCING AGENTS*[8]) in wastewater only cause the reduction of chlorine dioxide to chlorite.

$$\text{Chlorine Dioxide} + \text{Electron} = \text{Chlorite Ion}$$

$$ClO_2 + e^- = ClO_2^-$$

Therefore, chlorine dioxide is not as reactive as chlorine and chlorine is a better oxidant. In waters with a pH above 8.5, chlorine dioxide is a very effective disinfectant. Chlorine dioxide usually does not react with ammonia.

Chlorine dioxide is very unstable and must be generated at the plant site. Generally chlorine dioxide is prepared by the injection of sodium chlorite into the chlorine solution line from a chlorinator.

$$\text{Sodium Chlorite} + \text{Chlorine} \rightarrow \text{Sodium Chloride} + \text{Chlorine Dioxide}$$

$$2\ NaClO_2 + Cl_2 \rightarrow 2\ NaCl + 2\ ClO_2$$

Chlorine dioxide may find more applications in the water treatment field than in wastewater treatment because chlorine dioxide is less likely than chlorine to produce cancer-causing compounds and to cause tastes and odors in water.

10.03   Reactions of Chlorine Solutions With Impurities in Wastewater

Since wastewater contains a great number of complex substances and chemicals, many of these substances have a serious effect on wastewater chlorination. A few of the major impurities will be discussed in the following sections.

10.030   Inorganic Reducing Materials

Chlorine reacts rapidly with many reducing agents and more slowly with others. These reactions complicate the use of chlorine for disinfecting purposes. One of the most widely

known inorganic reducing materials is hydrogen sulfide. Hydrogen sulfide reacts with chlorine to form sulfuric acid (sulfate) or elemental sulfur depending on the concentration of hydrogen sulfide, pH, and temperature.

1. $\text{Hydrogen Sulfide} + \text{Free Chlorine} + \text{Water} \rightarrow \text{Sulfuric Acid} + \text{Hydrochloric Acid}$

$$H_2S + 4\ Cl_2 + 4\ H_2O \rightarrow H_2SO_4 + 8\ HCl$$

2. $\text{Water} + \text{Free Chlorine} \rightarrow \text{Hypochlorous Acid} + \text{Hydrogen Sulfide} \rightarrow \text{Elemental Sulfur} + \text{Hydrochloric Acid} + \text{Water}$

$$H_2O + Cl_2 \rightarrow HOCl + H_2S \rightarrow S^0 + HCl + H_2O$$
$$\downarrow$$

One part of hydrogen sulfide (H₂S) takes about 8.5 parts of chlorine in equation 1 and 2.2 parts of chlorine in equation 2. Since these reactions occur before a chlorine residual occurs, the demand caused by the inorganic salts must be satisfied first before any disinfection can take place. Ferrous iron, manganese, and nitrite are examples of other inorganic reducing agents that react with chlorine.

10.031   Reaction With Ammonia (NH₃)

Since ammonia is present in all domestic wastewaters, the reaction of ammonia with chlorine is of great significance. When chlorine is added to waters containing ammonia, the ammonia reacts with hypochlorous acid (HOCl) to form monochloramine, dichloramine, and trichloramine. The formation of these chloramines depends on the pH of the solution and the initial chlorine-ammonia ratio.

$$\text{Ammonia} + \text{Hypochlorous Acid} \rightarrow \text{Chloramine} + \text{Water}$$

$NH_3 + HOCl \rightarrow NH_2Cl + H_2O$	monochloramine
$NH_2Cl + HOCl \rightarrow NHCl_2 + H_2O$	dichloramine
$NHCl_2 + HOCl \rightarrow NCl_3 + H_2O$	trichloramine

In general at the pH levels that are usually found in wastewater (pH 6.5 to 7.5), monochloramine and dichloramine exist together. At pH levels below 5.5, dichloramine exists by itself. Below pH 4.0, trichloramine is the only compound found.

The mono- and dichloramine forms have definite disinfection powers and are of interest in the measurement of chlorine residuals. Dichloramine has a more effective disinfecting power than monochloramine.

$$NH_3 + HOCl \rightarrow \_\_Cl + H_2O$$
$$NH_3 + 2HO\_\_ \rightarrow \_\_Cl_2 + 2H_2O$$
$$NH_3 + 3H\_\_Cl \rightarrow NCl_3 + 3H_2O$$

---

[8] *Reducing Agent.   Any substance, such as base metal (iron) or the sulfide ion (S²⁻), that will readily donate (give up) electrons. The opposite is an oxidizing agent.*

If enough chlorine is added to react with the inorganic compounds and *NITROGENOUS*[9] compounds, then this chlorine will react with organic matter to produce *CHLORORGANIC*[10] compounds or other combined forms of chlorine, which have slight disinfecting action. Then, if enough chlorine is added to react with all the above compounds, any additional chlorine will exist as *FREE AVAILABLE CHLORINE*[11] which has the highest disinfecting action (Figure 10.2). This situation rarely exists in wastewater that contains nitrogenous compounds. The term *BREAKPOINT CHLORINATION*[12] refers to the breakpoint shown on Figure 10.2.

The exact mechanism of this disinfection action is not fully known. In some theories, chlorine is considered to exert a direct action against the bacterial cell, thus destroying it. Another theory is that the toxic character of chlorine inactivates the *ENZYMES*[13] which the living microorganisms need to "digest" their food supply. As a result, the organisms die of starvation. From the point of view of wastewater treatment, the mechanism of the action of chlorine is much less important than its effects as a disinfecting agent.

The demand by inorganic and organic materials is referred to as "chlorine demand." The chlorine that remains in combined forms having disinfecting properties plus any free chlorine is referred to as "residual chlorine." The sum of the chlorine demand and the chlorine residual is the chlorine dose.

Chlorine Dose = Chlorine Demand + Chlorine Residual

where

Chlorine Residual = Combined Chlorine Forms + Free Chlorine

10.04   Hypochlorination[14] of Wastewater

Hypochlorination is usually not the most effective method of disinfection. Higher costs and the disinfecting deficiencies of hypochlorination make chlorination with liquid or gaseous chlorine the most effective method in most cases. Some cities recently have switched to hypochlorination for safety reasons because chlorine gas is not involved. As previously described, when the pH is increased the hypochlorite ion (OCl^-) formation is increased and is less efficient than the hypochlorous acid ($HOCl$). However, when the hypochlorite solution is added to wastewater, the solution becomes diluted with the pH usually approaching that of the wastewater.

Hypochlorination will raise the pH of the wastewater being treated. The rise in pH will decrease the effectiveness of the hypochlorite, thereby requiring a higher dosage. This, in turn, will increase the pH even more.

10.05   Factors Influencing Disinfection

Both *CHLORINE ADDITION* and *CONTACT TIME* are essential for effective killing or inactivation of pathogenic micro-

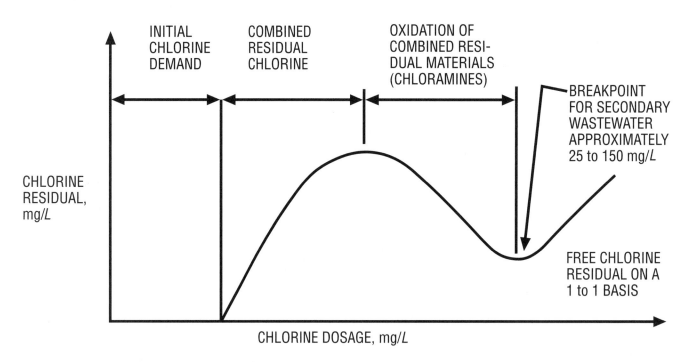

Fig. 10.2   Breakpoint chlorination curve

---

[9] *Nitrogenous (nye-TRAH-jen-us).   A term used to describe chemical compounds (usually organic) containing nitrogen in combined forms. Proteins and nitrate are nitrogenous compounds.*

[10] *Chlororganic (klor-or-GAN-ick).   Organic compounds combined with chlorine. These compounds generally originate from, or are associated with, living or dead organic materials.*

[11] *Free Available Chlorine.   The amount of chlorine available in water. This chlorine may be in the form of dissolved gas (Cl_2), hypochlorous acid ($HOCl$), or hypochlorite ion (OCl^-), but does not include chlorine combined with an amine (ammonia or nitrogen) or other organic compound.*

[12] *Breakpoint Chlorination.   Addition of chlorine to water or wastewater until the chlorine demand has been satisfied. At this point, further additions of chlorine result in a residual that is directly proportional to the amount of chlorine added beyond the breakpoint.*

[13] *Enzymes (EN-zimes).   Organic substances (produced by living organisms) which cause or speed up chemical reactions. Organic catalysts and/ or biochemical catalysts.*

[14] *Hypochlorination (HI-poe-KLOR-uh-NAY-shun).   The application of hypochlorite compounds to water or wastewater for the purpose of disinfection.*

organisms. Experimental determination of the best combination of combined residual chlorine and contact time is necessary to ensure both proper chlorine dose and minimum use of chlorine. Changes in pH affect the disinfection ability of chlorine and the operator must reexamine the best combination of chlorine addition and contact time when the pH fluctuates. Critical factors influencing disinfection are summarized as follows:

1. Injection point and method of mixing to get disinfectant in contact with wastewater being disinfected.

2. Design (shape) of contact chambers. Contact chambers are designed in various sizes and shapes. Rectangular contact chambers often allow short-circuiting and consequently reduced contact times. Baffles often are installed to increase mixing action, to obtain better distribution of disinfectant, and to reduce short-circuiting which in turn increases contact time. Pipelines have proved to be good contact chambers.

3. Contact time. With good initial mixing, the longer the contact time, the better the disinfection. Most chlorine contact basins are designed to provide a contact time of 30 minutes. In general, extending the chlorine contact time is more effective than increasing the chlorine dose to improve disinfection.

4. Effectiveness of upstream treatment processes. The lower the suspended solids and dissolved organic content of the wastewater, the better the disinfection.

5. Temperature. The higher the temperature, the more rapid the rate of disinfection.

6. Dose rate and type of chemical. Normally the higher the dose rate, the quicker the disinfection rate. The form or type of chemical also influences the disinfection rate.

7. pH. The lower the pH, the better the disinfection.

8. Numbers and types of organisms. The greater the organisms' concentrations, the longer the time required for disinfection. Bacteria cells are killed quickly and easily, but bacterial spores are extremely resistant.

10.06 Chlorine Requirements

The quantity of chlorine-demanding substances differs from plant to plant, so that the amount of chlorine that has to be added to ensure proper disinfection also differs. The amount of chlorine required to satisfy the chlorine-demanding substances is called the *CHLORINE DEMAND*. This demand is equal to the chlorine dose minus the chlorine residual.

Chlorine Demand = Chlorine Dose – Chlorine Residual

The chlorine residual is determined by one of several laboratory tests. The method of choice must be one of those approved by state and federal water pollution control agencies, otherwise the results will not be recognized. These tests are discussed in the laboratory section of Volume II (Chapter 16). The amount of residual that one should maintain is determined by the desired microorganism population. The microorganism population is usually specified by state or federal NPDES permit requirements. The microorganism population usually is estimated by determining the *MPN*[15] of *COLIFORM*[16] group organisms present. This determination does not test for individual pathogenic microorganisms, but uses the coliform group of organisms as the indicator organism. Coliform organisms usually are found in most wastewaters. See Chapter 16 for the laboratory test to determine the MPN of coliforms present.

Calculations to determine the chlorine dosage and chlorine demand are illustrated in the following example problem.

EXAMPLE

A chlorinator is set to feed 50 pounds of chlorine per 24 hours; the wastewater flow is at a rate of 0.85 MGD; and the chlorine as measured by the chlorine residual test after thirty minutes of contact time is 0.5 mg/*L*. Find the chlorine dosage and chlorine demand in mg/*L*.

$$\text{Chlorine Feed or Dose, mg/}L = \frac{\text{Pounds of Chlorine}}{\text{Million Pounds of Water}}$$

$$= \frac{50 \text{ lbs Chlorine/day}}{(0.85 \text{ MG/day})(8.34 \text{ lbs/gal})}$$

$$= \frac{59 \text{ lbs Chlorine/MG}}{8.34 \text{ lbs/gal}}$$

$$= 7.1 \text{ lbs Chlorine/million pounds water}$$

$$= 7.1 \text{ ppm (Parts Per Million parts)}$$

$$= 7.1 \text{ mg/}L$$

$$\text{Chlorine Demand, mg/}L = \text{Chlorine Dose, mg/}L - \text{Chlorine Residual, mg/}L$$

$$= 7.1 \text{ mg/}L - 0.5 \text{ mg/}L$$

$$= 6.6 \text{ mg/}L$$

---

[15] *MPN (pronounce as separate letters). MPN is the **M**ost **P**robable **N**umber of coliform-group organisms per unit volume of sample water. Expressed as a density or population of organisms per 100 mL of sample water.*

[16] *Coliform (COAL-i-form). One type of bacteria. The presence of coliform-group bacteria is an indication of possible pathogenic bacterial contamination. The human intestinal tract is one of the main habitats of coliform bacteria. They may also be found in the intestinal tracts of warm-blooded animals, and in plants, soil, air, and the aquatic environment. Fecal coliforms are those coliforms found in the feces of various warm-blooded animals; whereas the term "coliform" also includes other environmental sources.*

QUESTIONS

Write your answers in a notebook and then compare your answers with those on page 418.

10.0E   What does chlorine produce when it reacts with organic matter and with nitrogenous compounds?

10.0F   How is the chlorine dosage determined?

10.0G   How much chlorine must be added to wastewater to produce disinfecting action?

10.0H   How is the chlorine demand determined?

10.0I   Calculate the chlorine demand of treated domestic wastewater if:

Flow Rate = 1.2 MGD
Chlorinator = 70 lbs of chlorine per 24 hours
Residual   = 0.4 mg/L after thirty minutes

The objective of disinfection is the destruction or inactivation of pathogenic bacteria, and the ultimate measure of the effectiveness is the bacteriological result. The measurement of residual chlorine does supply a tool for practical control. If the residual chlorine value commonly effective in most wastewater treatment plants does not yield satisfactory bacteriological kills in a particular plant, the residual chlorine that does must be determined and used as a control in that plant. In other words, the 0.5 mg/L residual chlorine, while generally effective, is not a rigid standard but a guide that may be changed to meet local requirements.

One special case would be the use of chlorine in the effluent from a plant serving a tuberculosis hospital. Studies have shown that a residual of at least 2.0 mg/L should be maintained in the effluent from this type of institution, and that detention time should be at least two hours at the average rate of flow instead of the thirty minutes normally used as the basis for plant design. Two-stage chlorination may be particularly effective in this case.

The following list of chlorine dosages provides a reasonable guideline to produce chlorine residual adequate for applications indicated for domestic wastewaters. Individual plants may require higher or lower dosages, depending on the type and amount of suspended and dissolved organic compounds in the chlorinated sample.

APPLICATION	DOSAGE RANGE, mg/L
Collection Systems	
Slime Control	1-15
Corrosion Control	2-9[a]
Odor Control	2-9[a]
Treatment	
Grease Removal	1-10
BOD Reduction	0.5-2[b]
Inorganic Compounds	2-12
Filter Ponding	1-10
Filter Flies	0.1-0.5
Activated Sludge	
Bulking	1-10
Digester	
Supernatant	20-150
Foaming	2-20
Disinfection	
Raw Wastewater	10-30
Primary Clarifier Effluent	5-20
Chemical Precipitation	2-6
Trickling Filter Effluent[c]	3-20
Activated Sludge Effluent[c]	2-8
Advanced Waste Treatment	1-5

[a] per mg/L H_2S
[b] per mg/L BOD destroyed
[c] after secondary clarification

QUESTIONS

Write your answers in a notebook and then compare your answers with those on page 418.

10.0J   What is the objective of disinfection?

10.0K   How is the effectiveness of the chlorination process for a particular plant determined?

END OF LESSON 1 OF 5 LESSONS ON DISINFECTION AND CHLORINATION

Please answer the discussion and review questions next.

DISCUSSION AND REVIEW QUESTIONS

Chapter 10.   DISINFECTION AND CHLORINATION

(Lesson 1 of 5 Lessons)

At the end of each lesson in this chapter you will find some discussion and review questions. The purpose of these questions is to indicate to you how well you understand the material in the lesson. Write the answers to these questions in your notebook before continuing.

1. Why must wastewaters be disinfected?

2. Why is chlorination used to disinfect wastewater?

3. What constituents in wastewater are mainly responsible for the chlorine demand?

4. To improve disinfection, which is more effective—increasing the chlorine dose, or extending the chlorine contact time?

5. How do suspended and dissolved organic compounds in an effluent affect disinfection?

CHAPTER 10.   DISINFECTION AND CHLORINATION

(Lesson 2 of 5 Lessons)

10.1   POINTS OF CHLORINE APPLICATION
(Figure 10.3)

10.10   Collection System Chlorination

One of the primary benefits of up-sewer chlorination is to prevent the deterioration of structures. Other benefits include odor and *SEPTICITY*[17] control, and possibly BOD reduction to decrease the load imposed on the wastewater treatment processes. In some instances, the maximum benefit may result from a single application of chlorine at a point on the main intercepting sewer before the junction of all feeder sewer lines.

In others, several applications at more than one point on the main intercepting sewer or at the upper ends of the feeder lines may prove most effective. Due to high costs, *CHLORINATION SHOULD BE CONSIDERED AS A TEMPORARY OR EMERGENCY MEASURE IN MOST CASES*, with emphasis being placed on proper design of the system. Aeration also is effective in controlling septic conditions in collection systems. Although many problems result from improper design or design for future capacity requirements, the need for hydrogen sulfide protection exists under the best of conditions in some locations.

10.11   Prechlorination

"Prechlorination" is defined as the addition of chlorine to wastewater at the entrance to the treatment plant, ahead of settling units and prior to the addition of other chemicals. In addition to its application for aiding disinfection and odor control at this point, prechlorination is applied to reduce plant BOD load, as an aid to settling, to control foaming in Imhoff units, and to help remove oil. Current trends are away from prechlorination to up-sewer aeration or other chemical treatment for control of odors.

10.12   Plant Chlorination

Chlorine is added to wastewater during treatment by other processes, and the specific point of application is related to the results desired. The purpose of plant chlorination may be for control and prevention of odors, corrosion, sludge bulking, digester foaming, filter ponding, filter flies, and as an aid in sludge thickening. Here again, chlorination should be an emergency measure. Extreme care must be exercised when applying chlorine because it can interfere with or inhibit biological treatment processes.

10.13   Chlorination Before Filtration

More stringent discharge requirements are causing many agencies to provide filtration for solids removal before discharge (Chapter 4, *ADVANCED WASTE TREATMENT* manual). The better designs provide a means of chlorinating the water before application to the filters to kill algae and other large biological organisms. Prechlorination tends to prevent the development of biological growths that might cause short-circuiting or excessive backwashing in the filter media. Post chlorination of the effluent from the filter would be in addition to this application.

10.14   Postchlorination

"Postchlorination" is defined as the addition of chlorine to municipal or industrial wastewater following other treatment processes. This point of application should be before a *CHLORINE CONTACT CHAMBER*[18] and after the final settling unit in the treatment plant. The most effective place for chlorine application for disinfection is after treatment and on a well-clarified effluent. Postchlorination is used primarily for disinfection. As a result of chlorination for disinfection, some reduction in BOD may be observed; however, chlorination is rarely practiced solely for the purpose of BOD reduction.

---

[17] *Septicity (sep-TIS-it-tee).   Septicity is the condition in which organic matter decomposes to form foul-smelling products associated with the absence of free oxygen. If severe, the wastewater produces hydrogen sulfide, turns black, gives off foul odors, contains little or no dissolved oxygen, and the wastewater has a high oxygen demand.*

[18] *Chlorine Contact Chamber.   A baffled basin that provides sufficient detention time of chlorine contact with wastewater for disinfection to occur. The minimum contact time is usually 30 minutes. (Also commonly referred to as basin or tank.)*

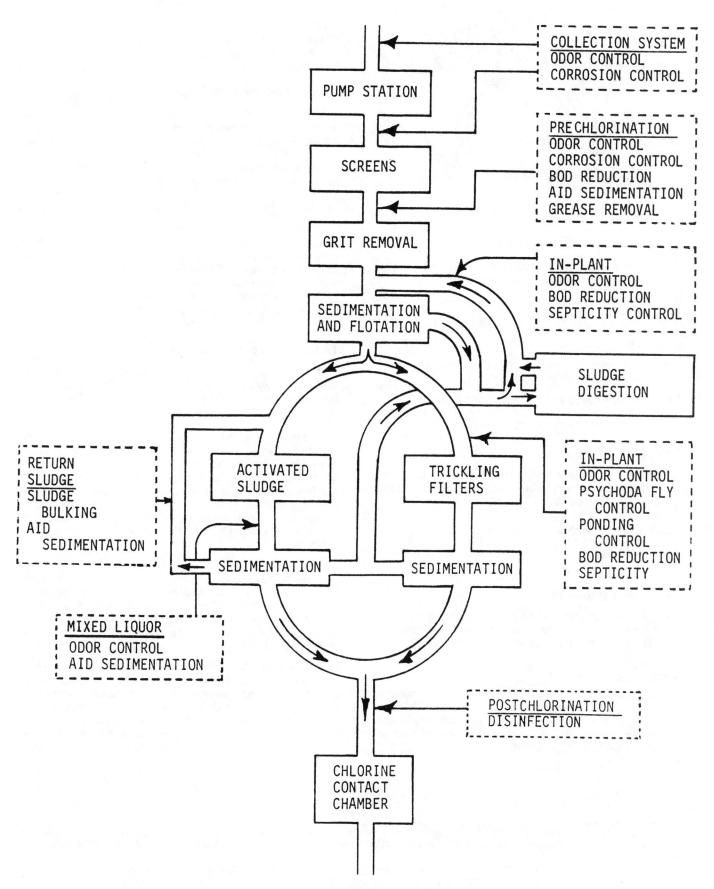

Fig. 10.3   Points of chlorine application

QUESTIONS

Write your answers in a notebook and then compare your answers with those on pages 418 and 419.

10.1A   What is the purpose of up-sewer chlorination?

10.1B   Where should chlorine be applied in sewers?

10.1C   What are the main reasons for prechlorination?

10.1D   Why might chlorine be added to wastewater during treatment by other processes?

10.1E   What is the objective of postchlorination?

10.2 CHLORINATION PROCESS CONTROL

10.20 Chlorinator Control

The control of chlorine flow to points of application is accomplished by six basic methods and a seventh method that combines two of the basic six.

10.200 Manual Control

Feed-rate adjustment and starting and stopping of equipment is done by hand.

10.201 Start-Stop Control

Feed-rate adjustment is done by hand; starting and stopping (by interrupting injector water supply) is controlled by starting of wastewater pump, flow switch, and level switch.

10.202 Step-Rate Control

Chlorinator feed rate is varied according to the number of wastewater pumps in service. As each pump starts, a preset quantity of chlorine is added to the flow of chlorine existing at starting time. This system can be applied conveniently with installations using up to eight pumps.

10.203 Timed-Program Control

Chlorine feed rate is varied on a timed step-rate basis regulated to correspond to the times of flow changes or by using a time-pattern transmitter that uses a revolving cam cut to match a flow pattern.

10.204 Flow-Proportional Control

Chlorinator feed rate is controlled by a system that converts wastewater flow information into a chlorinator control valve setting. This can be accomplished by a variety of flow metering equipment, including all process control instrumentation presently available and nearly all metering equipment now in use on wastewater systems.

10.205 Chlorine Residual Control

Chlorine feed rate is controlled to a desired chlorine residual (usually *COMBINED AVAILABLE CHLORINE*[19]) level. After mixing and reaction time (about five minutes maximum), a wastewater sample is *TITRATED*[20] by an *AMPEROMETRIC*[21] analyzer-recorder (or indicator). As the residual chlorine level varies above or below the desired (set point) level, the chlorinator is caused to change its feed rate to bring the chlorine residual back to the desired level.

10.206 Compound Loop Control

Any "automatic" control system (step-rate, timed-program, flow proportional, or residual) can be used in two ways: (1) by positioning the feed-rate valve; or (2) by varying the vacuum differential across the feed-rate valve. For instance, a flow-proportional (or step-rate, or timed-program) control system may position the feed-rate valve, and a residual-control system may vary the vacuum differential across the feed-rate valve. Thus, changes in flow cause changes in feed-rate valve position, but changes in chlorine demand may occur without any flow change. When this happens, the residual analyzer detects a change in chlorine residual and, by varying the vacuum differential across the feed-rate valve, causes the chlorinator to change rates to meet the desired chlorine residual level.

Various combinations of compound loop control can be devised. Generally speaking, the part of the system requiring the fastest response should be applied to valve positioning (since it responds faster). If flow changes are rapid, flow control should be by valve position. If flow and demand change rates are nearly the same, the magnitude of change may dictate the selection of control.

The selection of control methods should be based on treatment costs and treatment results (required or desired). A waste discharger must normally meet a disinfection standard. A small treatment plant might do this with a compound loop control system costing several thousand dollars, but may save less than one hundred dollars a year in chlorine consumed. In this case the expense would not be justified. A manual system,

---

[19] *Combined Available Chlorine. The total chlorine, present as chloramine or other derivatives, that is present in a water and is still available for disinfection and for oxidation of organic matter. The combined chlorine compounds are more stable than free chlorine forms, but they are somewhat slower in disinfection action.*

[20] *Titrate (TIE-trate). To TITRATE a sample, a chemical solution of known strength is added drop by drop until a certain color change, precipitate, or pH change in the sample is observed (end point). Titration is the process of adding the chemical reagent in small increments (0.1 – 1.0 milliliter) until completion of the reaction, as signaled by the end point.*

[21] *Amperometric (am-PURR-o-MET-rick). A method of measurement that records electric current flowing or generated, rather than recording voltage. Amperometric titration is a means of measuring concentrations of certain substances in water.*

which would meet the maximum requirements and overchlorinate at minimum requirement periods, might be used. It is not unheard of for a plant to have maximum chlorine residual requirements because of irrigation and/or marine life tolerances. In these cases, the uncontrolled or haphazard application of chlorine cannot be considered, no matter how large the added cost.

A specified chlorine residual level may be required at some point downstream from the best residual control sample point. In this case, a residual analyzer should be used to monitor and record residuals at this point. It may also be used to change the control set point of the controlling residual analyzer.

Ultimate control of dosage for disinfection rests on the results desired, that is, the bacterial level or concentration acceptable or permissible at the point of discharge. Determination of chlorine requirements according to the current edition of *STANDARD METHODS FOR THE EXAMINATION OF WATER AND WASTEWATER* is the best method of control. You must remember that the chlorine requirement or chlorine dose will vary with wastewater flow, time of contact, temperature, pH, and major waste constituents such as hydrogen sulfide, ammonia, and amount of dead and living organic matter.

The chlorine requirements for various flow rates and contact times can be determined on either a plant or a laboratory scale. If the determination is made on a laboratory scale, you should expect the plant requirements to be somewhat higher. This is to be expected since the mixing and actual contact times can be more carefully controlled in the laboratory. It is preferable that the determinations be made by both methods and the results compared. If the chlorine requirement as determined by full plant tests is significantly greater than that determined by laboratory testing, a wastage of chlorine is indicated. The two major causes of a large discrepancy between laboratory and plant results are poor mixing at the point of chlorine injection and short-circuiting in the contact chamber. Either problem can usually be solved at a relatively small expense as compared to the savings that can be achieved by the reduced use of chlorine.

QUESTIONS

Write your answers in a notebook and then compare your answers with those on page 419.

10.2A   How can chlorine gas feed be controlled?

10.2B   Define amperometric titration.

10.2C   Control of chlorine dosage depends on the bacterial _____ desired.

10.21 Chlorination Control Nomogram [22]

Determination of the chlorine residual may give information that confirms the previous choice of dosage or indicates that the dosage should be adjusted. The contact period must be specified since longer contact periods increase chlorine uptake.

Since feed rate is expressed in pounds per day, the rate setting of the feeder must be calculated from determination of the chlorine required and the flow. The simplest means of making the calculation is by a chlorination control nomogram taken from the *WPCF MANUAL OF PRACTICE NO. 11* (MOP 11), 1968 (Figure 10.4). The following sequence is a guide to using this nomogram:

1. Lay a straightedge (ruler) on the point on Line A representing flow, and on the point on Line B representing *CHLORINE REQUIRED*,[23] and read the point on Line C which shows the setting for the chlorine feeder.

2. For any value in excess of maximum indicated on Scales A, B, or C, introduce proper factor of ten or multiple thereof. The application of a factor of ten will be presented later in Example 2.

3. Greatest accuracy will be obtained when the angle of the straightedge approaches a right angle with Line B. Multipliers of ten may be applied to aid in accomplishing this objective.

4. If straightedge does not cross all three scales, introduce necessary factors of ten and move straightedge to point where all three scales will be crossed.

Let's try some examples using Figure 10.4. Assume the given chlorine dosage was selected on the basis of preliminary tests as capable of producing the desired results.

EXAMPLE 1

Given: Maximum Flow Rate = 0.5 MGD
        Chlorine Dosage = 1.0 mg/*L*

Procedure: Place one end of straightedge (ruler) on 0.5 MGD (Line A) and draw a line through the point on Line B representing a chlorine dosage of 1.0 mg/*L*. Extend the line to Line C and read the point indicating the chlorine feed rate.

Answer: Chlorine Feed Rate = 4.2 lbs per 24 hours

Check:

Chlorine Feed Rate, lbs/day = (Max Flow, MGD)(Dosage, mg/*L*)(8.34 lbs/gal)

$$= 0.5\ \frac{MG}{day}\ \times\ 1.0\ \frac{mg^{24}}{M\ mg}\ \times\ 8.34\ \frac{lbs}{gal}$$

$$= 4.17, \text{ say } 4.2 \text{ lbs per day}$$

---

[22] *Nomogram (NOME-o-gram). A chart or diagram containing three or more scales used to solve problems with three or more variables instead of using mathematical formulas.*

[23] *Chlorine Requirement.   The amount of chlorine which is needed for a particular purpose. Some reasons for adding chlorine are reducing the number of coliform bacteria (Most Probable Number), obtaining a particular chlorine residual, or oxidizing some substance in the water. In each case a definite dosage of chlorine will be necessary. This dosage is the chlorine requirement.*

[24] *Recall concentrations in mg/L are the same as mg/M mg.*

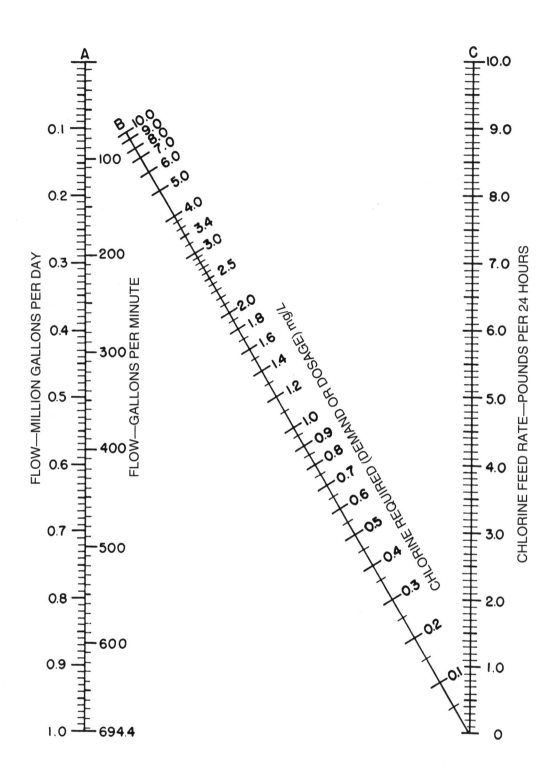

Fig. 10.4 Chlorination control nomogram
(Source: *WPCF MOP No. 11*, 1968)

EXAMPLE 2

Given: Maximum Flow Rate = 5.0 MGD

         Chlorine Dosage = 1.0 mg/L

Procedure: 5.0 MGD is off our scale on Line A. Reduce flow by a factor of ten to 0.5 MGD, or 5.0 MGD/10 = 0.5 MGD. The problem is the same as Example 1, and the chart gives a chlorine feed rate of 4.2 lbs per 24 hr. The flow rate is actually ten times 0.5 MGD (10 x 0.5 MGD = 5 MGD). Therefore the required chlorine feed rate is 10 x 4.2 lbs per 24 hours (10 x 4.2 = 42) or 42 lbs per 24 hours.

NOTE: On a cold day, a 150-lb cylinder may not be adequate to provide this feed rate because the lines may freeze.

EXAMPLE 3

Given: Maximum Flow Rate = 0.4 MGD

         Chlorine Dosage = 5.0 mg/L

Procedure: A line through a flow of 0.4 MGD (Line A) and 5.0 mg/L (Line B) misses Line C. The chlorine dosage should be reduced by a factor of ten to be able to use the nomogram.

$$\text{Chlorine Dosage} = \frac{5.0\ \text{mg}/L}{10} = 0.5\ \text{mg}/L$$

$$\begin{array}{ll}\text{Chlorine} \\ \text{Feed Rate}\end{array} = \begin{array}{l}1.7\ \text{lbs per 24 hours} \\ \text{(from chart)}\end{array}$$

$$\begin{array}{ll}\text{Actual} \\ \text{Feed Rate}\end{array} = (10)(1.7\ \text{lbs per 24 hr})$$

$$= 17\ \text{lbs per 24 hr}$$

The chlorine feed rate must be ten times the rate from the chart because the chlorine required was reduced by a factor of ten to use the chart.

The results from the chart should be checked using the mathematical calculations used in Example 1 to avoid errors. Note that the chlorine requirement should take into consideration the chlorine demand so that a desired residual is obtained after a given contact period. As discussed in Section 10.06, chlorine requirements vary with the wastewater characteristics, concentration, flow, and temperature. Adjustment of chlorine feeder rates to meet all these variations is the ultimate goal of good operation practice. More frequent adjustments are usually required for primary effluent than for secondary effluent.

Suggested schedule for adjusting manual feeder rates:

1. In large plants—at least every hour,

2. In medium-sized plants—every two to four hours, and

3. In small plants—every eight hours.

Other methods of chlorinator control have been described in Section 10.20.

10.22 Hypochlorinator[25] Feed Rate and Control

Chlorine for disinfection and other purposes is provided in some plants by the use of hypochlorite compounds. The useful chlorine concentration in hypochlorite compounds is lower than the concentration in pure gaseous or liquid chlorine and is, therefore, generally more expensive per unit of chlorine. However, the safety, storage, and application equipment costs and the training costs for hypochlorite usage can be much lower in smaller plants and, therefore, hypochlorite use is often cost effective when all costs are considered.

The amount of chlorine delivered depends on the type of hypochlorite. For example, HTH (high test hypochlorite) contains approximately 65 percent chlorine by weight, and chlorinated lime contains 34 percent.

Manufacturers of hypochlorite compounds define available chlorine as the amount of gaseous chlorine required to make the equivalent hypochlorite chlorine. If you prepare a hypochlorite solution for disinfection and immediately measure the chlorine residual, you will find the chlorine residual about half of the expected value based on the manufacturer's amount of available chlorine. When hypochlorite is mixed with water, approximately half of the chlorine forms hydrochloric acid (HCl) and the other half forms hypochlorite (OCl$^-$), the chlorine residual that you measured.

EXAMPLE 4

A wastewater requires a chlorine feed rate of 17 lbs per day. How many pounds of chlorinated lime will be required to provide the needed chlorine? Assume each pound of chlorinated lime contains 0.34 lb of available chlorine.

$$\begin{array}{ll}\text{Chlorinated Lime} \\ \text{Feed Rate, lbs/day}\end{array} = \frac{\text{Chlorine Required, lbs/day}}{\text{Portion of Chlorine in lbs of Hypochlorite}}$$

$$= \frac{17\ \text{lbs/day}}{0.34\ \text{lb}}$$

$$= 50\ \text{lbs/day}$$

Hypochlorinators can be installed on either small systems or large systems. Usually the larger systems use liquid chlorine because of lower costs. However, some very large cities use hypochlorite because it is safer. Treatment plants serving these cities are located in highly populated areas where escaping chlorine gas could threaten the lives of many people.

The items in feed systems of hypochlorinators and their construction are discussed below.

1. Some type of storage system for the solution. The storage container is made of corrosion-resistant materials such as plastic or other similar materials.

2. Solution piping which is usually fiberglass or PVC (polyvinyl chloride).

3. Diffusers that are made of the same material as the piping systems.

4. Valves and *EDUCTORS*[26] that are made of PVC.

5. Pumps that are made of corrosion-resistant materials. Epoxy-lined systems have been successful.

---

[25] *Hypochlorinators (HI-poe-KLOR-uh-NAY-tors).   Chlorine pumps, chemical feed pumps or devices used to dispense chlorine solutions made from hypochlorites such as bleach (sodium hypochlorite) or calcium hypochlorite into the water being treated.*

[26] *Eductor (e-DUCK-ter).   A hydraulic device used to create a negative pressure (suction) by forcing a liquid through a restriction, such as a Venturi. An eductor or aspirator (the hydraulic device) may be used in the laboratory in place of a vacuum pump. As an injector, it is used to produce vacuum for chlorinators. Sometimes used instead of a suction pump.*

6. Flowmeters. These meters can be constructed using a Hastelloy C straight-through metal tube *ROTAMETER*[27] with the float position determined magnetically and the flow rate transmitted either electrically or pneumatically.

7. Chlorine residual analyzers of the amperometric type are commonly installed.

8. Automatic controls that consist of a hypochlorite flow controller, recorder and totalizer, a ratio control station, and the necessary electronic signal converters.

The feed system can usually be operated automatically or by manual control. The operation of the hypochlorite system will usually cost twice as much as a liquid chlorine system. Maintenance of the hypochlorinator system requires more operator-hours than do the liquid chlorine systems.

QUESTIONS

Write your answers in a notebook and then compare your answers with those on page 419.

10.2D   How is the rate of dosage for a chlorinator determined?

10.2E   Determine the chlorine feed rate, pounds per 24 hours, if you are treating:

1. A flow of 0.5 MGD and the chlorine required is 1.0 mg/*L*.
2. Flow = 0.8 MGD and chlorine required = 4.0 mg/*L*.
3. Flow = 6 MGD and chlorine required = 25 mg/*L*.

10.2F   How frequently should feeder rates be adjusted with manual controls? Why?

10.2G   How many pounds of HTH (high test hypochlorite) should be used per day by a hypochlorinator dosing a flow of 0.55 MGD at 12 mg/*L* of chlorine?

10.23   Chlorine Solution Discharge Lines, Diffusers, and Mixing

10.230   *Solution Discharge Lines*

Solution discharge lines are made from a variety of materials depending upon the requirements of service. Two primary requirements are that the line must be resistant to the corro-sive effects of chlorine solution and of adequate size to carry the required flows. Additional considerations are pressure conditions, flexibility (if required), resistance to external corrosion and stresses when underground or passing through structures, ease and tightness of connections, and the adaptability to field fabrication or alteration.

Development of plastics has contributed greatly to chemical solution transmission. Polyvinyl chloride (PVC) pipe and black polyethylene flexible tubing have all but eliminated the use of rubber hose. Both are generally less expensive and both outlast rubber in normal service. The use of hose is almost exclusively limited to applications where flexibility is required or where extremely high back pressures exist.

PVC and polyethylene can be field fabricated and altered. PVC should be Schedule 80 to limit its tendency to partially collapse under vacuum conditions. Schedule 80 PVC may be threaded and assembled with ordinary pipe tools or may be installed using solvent-welded fittings.

Rubber-lined steel pipe has been used for many years where resistance to external stresses is required. It cannot be field fabricated or altered and is thus somewhat restricted in application. PVC lining of steel pipe has not yet become economically competitive, but other plastics which can readily compete with rubber lining and are adaptable to field fabrication and alteration have been developed.

Never use neoprene hose to carry chlorine solutions because it will become hard and brittle in a short time.

10.231   *Chlorine Solution Diffusers*

Chlorine diffusers are normally constructed of the same materials used for solution lines. Their design is an extremely important part of a chlorination program. This importance is almost completely related to the mixing of the chlorine solution with the wastewater being treated; however, strength and flexibility also must be given consideration. In most *CIRCULAR, FILLED* conduits flowing at 0.25 ft/sec (0.08 m/sec) or greater, a solution injected at the center of the pipeline will mix with the entire flow after flowing ten pipe diameters downstream. Mixing in open channels can be accomplished by the use of a hydraulic jump (Figure 10.5) or by sizing diffuser orifices so that a high velocity (about 16 ft/sec or 4.8 m/sec) is attained at the diffuser discharge. This accomplishes two things: (1) introducing a pressure drop to get equal discharge from each orifice, and (2) imparting sufficient energy to the surrounding wastewater to complete the mixing. Generally speaking, a diffuser should be supplied for each two to three feet (0.6 to 1 meter) of channel depth. Diffusers should be placed across the width of the channel rather than in the direction of flow. Mixing also can be achieved by the use of high-speed mechanical mixers specifically designed for this purpose.

Fig. 10.5   Hydraulic jump

---

[27] *Rotameter (RODE-uh-ME-ter).   A device used to measure the flow rate of gases and liquids. The gas or liquid being measured flows vertically up a tapered, calibrated tube. Inside the tube is a small ball or bullet-shaped float (it may rotate) that rises or falls depending on the flow rate. The flow rate may be read on a scale behind or on the tube by looking at the middle of the ball or at the widest part or top of the float.*

10.232 Mixing

Mixing as well as the speed of mixing are extremely important ahead of a chlorine contact tank or a residual sampling point. Since a contact tank is usually designed for low velocity, little mixing occurs after wastewater enters it. Mixing must be achieved before the contact tank is entered. The same is true for a chlorine residual sampling point. If good mixing does not occur somewhere upstream from the sampling point, erratic results will be obtained by the residual chlorine analyzing system.

QUESTIONS

Write your answers in a notebook and then compare your answers with those on page 419.

10.2H   Chlorine solution discharge lines may be made of _____ or _____.

10.2I   Why does little mixing of the chlorine solution with wastewater occur in chlorine contact basins?

10.24   Measurement of Chlorine Residual

Refer to Volume II, Chapter 16, "Laboratory Procedures and Chemistry," for procedures to measure chlorine residual. Amperometric titration provides for the most convenient and the most repeatable results; however, the equipment costs more than equipment for the other methods. The starch-iodide test can be used, but will have limited success in turbid or muddy waters. DPD[28] tests can be used and are less expensive than the other methods, but this also is a colorimetric method. The orthotolidine method is no longer a test that is recognized by federal agencies. Figure 10.6 shows a typical automatic chlorine residual analyzer.

ORP (Oxidation-Reduction Potential) probes are being used to optimize chlorination processes in wastewater treatment plants. ORP (also called the redox potential) is a direct measure of the effectiveness of a chlorine residual in disinfecting a plant effluent. Chlorine forms that are toxic to microorganisms (including coliforms) are missing one or more electrons in their molecular structure. They satisfy their need for electrons by taking electrons from any organic substances or microorganisms present in the water being treated. When microorganisms lose electrons they become inactivated and can no longer transmit a disease or reproduce.

The ability of chlorine to take electrons (the electrical attraction or electrical potential) is the ORP and is measurable in millivolts. The strength of the millivoltage (or the redox measurement) is directly proportional to the oxidative disinfection strength of the chlorine in the treatment system. The higher the concentration of chlorine disinfectant, the higher the measured ORP voltage. Conversely, the higher the concentration of organics (chlorine demanding substances), the lower the measured ORP voltage. The redox sensing unit (ORP probe) measures the voltage present in the wastestream being treated and thus provides a direct measure of the disinfecting power of the disinfectant present in the wastestream.

In a typical installation, a High Resolution Redox (HRR) chlorine controller monitors the chlorine residual using a redox (ORP) probe suspended in the chlorine contact chamber approximately 6.5 minutes downstream from the chlorine injection point. The controller converts the redox signal to a 4 to 20 milliamp (mA) signal that automatically adjusts the chlorine feed rate from the chlorinator. A second High Resolution Redox (HRR) probe, which is suspended approximately four minutes downstream from the point of sulfur dioxide (SO_2) (used to dechlorinate the wastestream) injection, sends a signal to a second controller that adjusts the sulfur dioxide feed from the sulfonator.

The High Resolution Redox (HRR) units control the chlorination and dechlorination chemical feed rates according to actual demand in the treatment processes. These HRR units automatically treat the wastestream with the chlorine and sulfur dioxide dosages required to maintain chemical residuals in the ideal ranges, regardless of changes in the chemical demand or wastestream (water) flow.

Maintenance for the chlorine ORP probe consists of cleaning the unit's sensor once a month. To ensure that the sulfur dioxide controller is highly responsive at all times, the sulfur dioxide sensor should be cleaned three times a week.

For additional information, see "Optimizing Chlorination/ Dechlorination at a Wastewater Treatment Plant," *PUBLIC WORKS*, January, 1995, page 33.

QUESTIONS

Write your answers in a notebook and then compare your answers with those on page 419.

10.2J   What does an ORP probe measure in a disinfection system?

10.2K   What happens to a microorganism when it loses an electron?

10.2L   What maintenance is required on ORP probes?

10.25   Start-Up of Chlorinators

Before starting any chlorination system, read the plant Operation and Maintenance Manual and the manufacturer's literature to become familiar with the equipment. Review the plans or drawings of the facility. Determine what equipment, pipelines, pumps, tanks, and valves are to be placed into service or are in service. The current status of the entire chlorination system must be known before starting or stopping any portion of the system. During emergencies you must act quickly and may not have time to check out each of the steps outlined above, but you still must follow established procedures.

---

[28] DPD (pronounce as separate letters) Method. A method of measuring the chlorine residual in water. The residual may be determined by either titrating or comparing a developed color with color standards. DPD stands for N,N-diethyl-p-phenylene-diamine.

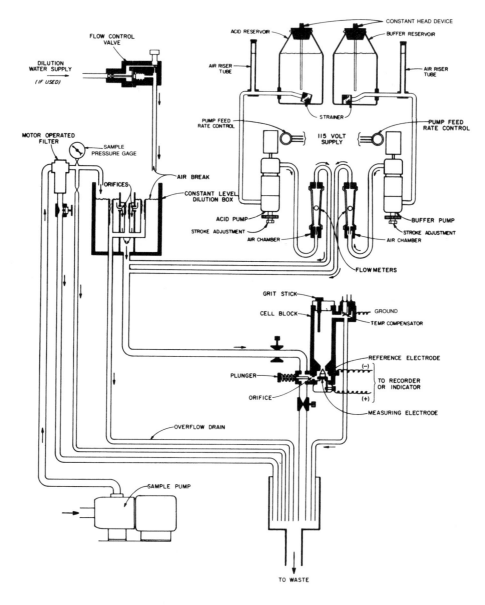

Fig. 10.6   Automatic chlorine residual analyzer

(Permission of Wallace & Tiernan Division, Penwalt Corporation)

Procedures for start-up, operation, shutdown, and trouble-shooting are outlined in this section and are intended to be typical procedures for all types of chlorinators. For specific directions, see the manufacturer's literature and the plant O and M manual.

10.250 Gas Chlorinators

Start-up procedures for chlorinators using chlorine gas from containers are outlined in this section.

1. Be sure chlorine gas valve at the chlorinator is closed. This valve should have been closed since the chlorinator is out of service.

2. All chlorine valves on the supply line should have been closed during shutdown. Be sure they are still closed. If any valves are required to be open for any reason, this exception should be indicated by a tag on the valve.

3. Inspect all tubing, manifold, and valve connections for potential leaks and be sure all joints are properly gasketed.

4. Check chlorine solution distribution lines to be sure that system is properly valved to deliver chlorine solution to desired point of application.

5. Open chlorine metering orifice slightly by adjusting the chlorine feed rate control.

6. Start the injector water supply system. Usually the source of water is plant effluent (with a minimum of suspended solids) or potable water (after an *AIR GAP*[29] or air-break system) supply. Injector water is pumped at an appropriate flow rate and the flow through the injector creates sufficient vacuum in the injector to draw chlorine. Chlorine is absorbed and mixed in the water at the injector. This chlorine solution is conveyed to the point of application.

7. Examine injector water supply system.

   a. Note reading of injector supply pressure gage. If reading is abnormal (different from usual reading), try to identify cause and correct. Injectors should operate at an inlet pressure greater than 50 psi.

   b. Note reading of injector vacuum gage. If system does not have a vacuum gage, have one installed. If the vacuum reading is less than normal, the machine may function at a lower feed rate, but will be unable to deliver at rated capacity.

8. Inspect chlorinator vacuum lines for leaks.

9. Crack open the chlorine container valve and allow gas to enter the line. Inspect all joints for leaks by placing an ammonia-soaked rag near each joint. A polyethylene "squeeze bottle" filled with ammonia water[30] to dispense ammonia vapor may also be used. Care must be taken to avoid spraying ammonia water on any leak or touching the cloth to any metal. The formation of a white cloud or vapor will indicate a chlorine leak. Start with the valve at the chlorine container, move down the line and check all joints between this valve and the next one downstream. If the downstream valve passes the ammonia test, open the valve and continue to the next valve. If there are no leaks to the chlorinator, open the cylinder valve approximately one (1) complete turn to obtain maximum discharge and continue with the start-up procedure.

10. Inspect the chlorinator.

   a. Chlorine gas pressure at the chlorinator should be between 20 and 30 psi (1.4 to 2.1 kg/sq cm). Control the chlorine pressure by using the pressure regulating valve.

   b. Operate chlorinator at complete range of feed rates.

   c. Check operation on manual and automatic settings.

11. Chlorinator is ready for use. Log in the time the system is placed into operation and the application point.

10.251 Liquid Chlorine Chlorinators

Start-up procedures for chlorinators using liquid chlorine from containers are outlined in this section. In most plants liquid chlorine is delivered in one-ton containers. However, liquid chlorine may be delivered in railroad tank cars to very large plants.

1. Inspect all joints, valves, manifolds, and tubing connections in chlorination system, including application lines for proper fit and for leaks. Make sure that all joints have gaskets.

2. If chlorination system has been broken open or exposed to the atmosphere, verify that the system is dry. Usually once a system has been dried out, it is never opened again to the atmosphere. However, if moisture in the air enters the system or by any other means, it readily mixes with chlorine and forms hydrochloric acid which will corrode the pipes, valves, joints, and fittings.

---

[29] *Air Gap.   An open vertical drop, or vertical empty space, between a drinking (potable) water supply and the point of use in a wastewater treatment plant. This gap prevents the contamination of drinking water by backsiphonage because there is no way wastewater can reach the drinking water supply. See Chapter 15, Section 15.7, paragraph 18, "Air Gap Separation Systems," in Volume II.*

[30] *Use a concentrated ammonia solution containing 28 to 30 percent ammonia as NH_3 (this is the same as 58 percent ammonium hydroxide, NH_4OH, or commercial 26° Baumé).*

To verify that the system is dry, determine the *DEW POINT*[31] (must be lower than −40°F or −40°C). If not dry, turn the evaporator on, pass dry air through the evaporator, and force this air through the system. If this step is omitted and moisture is in the system, serious corrosion damage can result and the entire system may have to be repaired.

3. Start up the evaporators. Fill the water bath and adjust the device according to the manufacturer's directions.

4. Turn on the heaters on the evaporators.

5. Wait until the temperature of the evaporators reaches 180°F (82°C). This may take over an hour on large units.

6. Inspect and close all valves on the chlorine supply line. One chlorine valve on the evaporator (the inlet valve) should be open to allow for expansion of the chlorine if heated.

7. Open the chlorine metering orifice slightly by adjusting the chlorine feed rate knob. This is to prevent damage to the rotameter.

8. Start the injector water supply system.

9. Examine injector water supply system.

   a. Note reading of injector water supply pressure gage. If gage reading is abnormal (different from usual reading), try to identify the cause and correct it. Injectors should operate at an inlet pressure greater than 50 psi.

   b. Note reading of injector vacuum gage. If system does not have a vacuum gage, have one installed. If the vacuum reading is less than normal, the machine may function at a lower feed rate, but will be unable to deliver at the maximum rated capacity.

10. Inspect chlorinator vacuum lines for leaks.

11. On the system connected to the gas side of the supply container, crack open the *CHLORINE GAS LINE* at the chlorine container. All liquid chlorine systems should be checked by using gas because of the danger of leaks (gas is less dangerous). Inspect the joints between this valve and the next one downstream. If this valve passes the ammonia leak test, continue to the next valve down the line. Follow this procedure until the evaporator is reached. Before allowing chlorine to enter the evaporator and the chlorinator, make sure that all valves between the evaporator and the chlorinator are open. Heat in the evaporator will expand the gas, and if the system is closed, excessive pressure can develop. Chlorine should never be trapped in a line, evaporator, or chlorinator. After the gas test reconnect to liquid side and retest at connection and pigtail valve.

12. If no problems develop, the supply line to the evaporator can be put in service by opening the valve 1½ to 2 turns.

13. Check the operation of the chlorinator.

   a. Operate over complete range of chlorine feed rates.

   b. Check operation on manual and automatic settings.

14. If the chlorinator is operating properly, close the gas valve on the container, wait a few minutes to evacuate pigtail connector, close valve from pigtail to manifold, disconnect pigtail, replace gas valve cap, remove liquid valve cap, clean valve face and threads, install new gasket to connect pigtail, crack valve open and then close. Check for leak at new connection. If no leak, open liquid valve two turns (usually one turn open is sufficient).

15. After admitting liquid chlorine to the system, wait until the temperature of the evaporator again reaches 180°F (82°C) and full working pressure (100 psi or 7 kg/sq cm). Inspect the evaporator by looking for leaks around pipe joints, unions, and valves.

16. The system is ready for normal operation.

---

[31] *Dew Point Test.   Dew point is the temperature to which air with a given quantity of water vapor must be cooled to cause condensation of the vapor in the air. One way to measure the dew point is with a special dew point apparatus. This apparatus consists of a small box. The gas or air being tested enters the box on one side and leaves on the opposite side. One of the other sides has an observation window. A polished cup is inserted firmly in the top.*

   Pass a sample of air or gas being tested through this apparatus. Adjust the flow so it can be felt against wetted lips, but not readily felt by the hand. Pour acetone into the cup. Allow the sample to pass through the cup for about five minutes. Add small amounts of crushed dry ice to the acetone. Stir continuously with a thermometer. Carefully add dry ice to the acetone as necessary to slowly lower the temperature. When dew or moisture first appears on the outside polished surface of the cup, read the temperature from the thermometer. This temperature is the DEW POINT. The lower the dew point, the less moisture in the air. The amount of moisture in the air can be determined from a dew point temperature chart or table provided by manufacturers of dew point measuring equipment.

QUESTIONS

Write your answers in a notebook and then compare your answers with those on page 419.

10.2M   What would you do before attempting to start any chlorination system?

10.2N   How is the chlorination system inspected for leaks?

10.26 Normal and Abnormal Operation

Normal operation of the chlorination process requires regular observation of facilities and a regular preventive maintenance program. When something abnormal is observed or discovered, corrective action must be taken. This section outlines normal operation procedures and also responses to abnormal conditions.

10.260 Container Storage Area

Daily

1. Inspect building or area for ease of access by authorized personnel to perform routine and emergency duties.

2. Be sure fan and ventilation equipment are operating properly.

3. Read scales, charts, or meters at the same time every day to determine use of chlorine and any other chemicals. Notify plant superintendent when chlorine supply is low.

4. Look at least once per shift for chlorine and chemical leaks.

5. Try to maintain temperature of storage area below temperature of chlorinator room.

6. Determine manifold pressure before and after chlorine pressure regulating valve by reading upstream and downstream pressure gages.

7. Be sure all chlorine containers are properly secured.

Weekly

1. Clean building or storage area.

2. Check operation of chlorine leak detector alarm.

Monthly

1. Exercise all valves, including flex connector's auxiliary, manifold, filter bypass, and switchover valves.

2. Inspect all flex connectors and replace any that have been kinked or flattened.

3. Inspect hoisting equipment.

   a. Cables: frayed or cut.

   b. Container hoisting beam and hooks: cracked or bent.

   c. Controls: operate properly; do not stick or respond sluggishly; cords not frayed; safety chains or cables in place.

   d. Hoist travels on monorail easily and smoothly.

4. Examine building ventilation.

   a. Ducts and louvers: clean and operate freely.

   b. Fans and blowers: operate properly; guards in place; equipment properly lubricated.

5. Perform preventive maintenance as scheduled.

   a. Lubricate equipment (monthly to quarterly).

   b. Repack valves and regulators (6 to 12 months).

   c. Clean and replace valve seats and stems (annually).

   d. Clean filters and replace glass wool. (*CAUTION:* Glass wool soaked with liquid chlorine or chlorine impurities may burn your skin or give off sufficient chlorine gas to be dangerous.) (annually)

   e. Paint equipment (annually).

   f. Inspect condition and parts of all repair kits and safety equipment (monthly).

10.261 Evaporators

Evaporators are used to convert liquid chlorine to gaseous chlorine for use by gas chlorinators.

Daily

1. Check evaporator water bath to be sure water level is at midpoint of sight glass.

2. Be sure water bath temperature is between 160 and 195°F (71 to 91°C). Low alarm should sound at 160°F (71°C) and high alarm should sound at 200°F (93°C).

3. Determine chlorine inlet pressure to evaporator. Pressure should be same pressure as on supply manifold from containers (20 to 100 psi or 1.4 to 7 kg/sq cm).

4. Determine chlorine outlet temperature from evaporator. Typical range is 90 to 105°F (32 to 41°C). High alarm should sound at 110°F (43°C). At low temperatures the chlorine pressure-reducing valve (CPRV) will close due to the low temperature in the water bath.

5. Check CPRV operation.

6. If evaporator is equipped with water bath recirculation pump at back of evaporator, determine if pump is operating properly.

7. Look for leaks and repair any discovered.

Abnormal Evaporator Conditions

1. Evaporator water level low. Water level not visible in sight glass!

   Troubleshooting Measurements

   a. Determine actual level of water.

   b. Measure temperature of water.

   c. Check temperature and pressure of chlorine in evaporator system and supply lines back to containers and chlorinators for possible overpressure of system (pressure should not exceed 100 psi or 7 kg/sq cm).

Corrective Action

a. If chlorine pressure on system is near or over 100 psi (7 kg/sq cm), close supply container valves to stop chlorine addition to the system, increase feed rate of chlorinator to use chlorine in the system, and drop the pressure back down to a safe range. *NOTE:* Alarm system is usually set to sound at 110 psi (7.7 kg/sq cm). If system pressure was at or over 110 psi (7.7 kg/sq cm), inspect alarm circuit to determine cause of alarm failure.

b. If temperature of water bath is at an abnormal level, find the cause. Water bath levels are usually set in the following sequence:

(1) 160°F or 71°C: low temperature alarm.

(2) 185 to 195°F or 85 to 91°C: normal operating range.

(3) 185°F or 85°C: actuates chlorine pressure reducing valve (CPRV) to open position.

(4) 200°F or 93°C: high temperature alarm.

If water temperature is above 200°F (93°C), alarm should have sounded. Open control switch on evaporator heaters to stop current flow to heating elements. If temperature is in normal range, return to correcting the original problem of a low water level.

c. Evaporator water levels are controlled by a solenoid valve. First check to see if drain valve is fully closed and then use the following steps.

(1) Override solenoid valve and fill water bath to proper level.

(2) If water cannot be added, take evaporator out of service:

(a) Switch to another evaporator, or

(b) Switch supply to gas side of containers. May have to connect more containers to supply sufficient chlorine gas.

(3) Repair valve and actuating sensor.

NOTE: The installation of a manual bypass on the water line will allow continued operation of the evaporator in the event of solenoid failure and would facilitate solenoid valve replacement with the unit on line.

2. Low water temperature in evaporator.

a. Check chemical (chlorine) flow-through rate. Rate may have exceeded unit's capacity and may require two evaporators to be on line to handle chemical feed rate.

b. Inspect immersion heaters for proper operation. First examine control panel for thermal overload on breaker. Most evaporators are equipped with two to three heating elements. An inspection of the electrical system will indicate if the breakers are shorted or open and will locate the problem. Replace any heating elements that have failed.

c. If no spare evaporators are available, operate from the valves on the chlorine gas supply. If necessary, reduce the chlorine feed rate to keep the chlorination system working properly.

3. No chlorine gas flow to the chlorinator.

a. Inspect chlorine pressure-reducing valve downstream from evaporator and determine if valve is in the open or closed position.

(1) Valve may be closed due to low water temperature in evaporator (less than 185°F or 85°C).

(2) Valve may be closed due to loss of vacuum on system or loss of continuity of electrical control circuits which may have been caused by a momentary power drop. Correct problem and reset valve.

(3) Valve may be out of adjustment and restricting gas flow through the valve due to a low pressure setting.

b. Inspect supply containers and manifold. Possible sources of lack of chlorine gas flow to chlorinators include:

(1) Containers empty.

(2) Container chlorine supply lines incorrectly connected to gas instead of liquid side of containers. High flow rates of gas will freeze chlorine, thus frosting valves and flex connectors.

(3) Chlorine manifold filters plugged. Check pressure upstream and downstream from filter. Pressure drop should not exceed 10 psi (0.7 kg/sq cm). Frost on manifold will indicate a restriction in the filters.

(4) Inspect manifold and system for closed valves. Most systems operate properly with all chlorine valves at only *ONE TURN OPEN* position.

Monthly

1. Exercise all valves, including inlet, outlet, chlorine pressure reducing (CPRV), water, drain, and fill valves.

2. Inspect evaporator cathode protection meter (if so equipped). Cathodic protection protects the metal water tank (on *SULFONATORS*[32]) and piping from corrosion due to electrolysis. Electrolysis is the flow of electrical current and is the reverse of metal plating. In electrolysis, the flow away of certain compounds from the metal causes corrosion and holes in a short time. This type of corrosion is controlled by either a *SACRIFICIAL ANODE*[33] made of magnesium and zinc or by applied small electrical currents to suppress or reverse the normal corroding current flow.

---

[32] *Sulfonators (see Section 10.8, "Dechlorination") are similar to chlorinators.*

[33] *Sacrificial Anode.   An easily corroded material deliberately installed in a pipe or tank. The intent of such an installation is to give up (sacrifice) this anode to corrosion while the water supply facilities remain relatively corrosion free.*

3. Check setting of CPRV (chlorine pressure-reducing valve) in order to maintain desired pressure of chlorine gas to chlorinators.

4. Inspect heating and ventilation equipment in chlorinator area. Maintain a higher temperature in the chlorinator area than in the chlorine storage area.

5. Perform scheduled routine preventive maintenance.

   a. Drain and flush water bath.

   b. Clean evaporator tank.

   c. Check heater elements.

   d. Repack gasket and reseat pressure-reducing valves (annually).

   e. Replace anodes (annually).

   f. Paint system (annually).

10.262 Chlorinators, Including Injectors

Daily

1. Check injector water supply pressure. Pressures will range from 40 to 90 psi (2.8 to 6.3 kg/sq cm) depending on system.

2. Determine injector vacuum. Values will range from 15 to 25 inches (38 to 64 cm) of mercury.

3. Check chlorinator vacuum. Values will range from 5 to 10 inches (13 to 25 cm) of mercury.

4. Determine chlorinator chlorine supply pressure. Values will range from 20 to 40 psi (1.4 to 2.8 kg/sq cm).

5. Read chlorinator feed rate on rotameter tube. Is feed rate at required level? Record rotameter reading and time.

6. Examine and record mode of control.

   a. Manual

   b. Automatic (single input)

   c. Automatic (dual input)

7. Measure chlorine residual at application point.

8. Inspect system for chlorine leaks.

9. Inspect auxiliary components.

   a. Flow signal input. Does chlorinator feed rate change when flow changes? Chlorinator response is normally checked by biasing (adjusting) flow signal which may drive dosage control unit on chlorinator to full open or closed position. When switch is released, chlorinator will return to previous feed rate. During this operation the unit should have responded smoothly through the change. If the response was not smooth, look for mechanical problems of binding, lubrication, or vacuum leaks.

   b. If chlorinator also is controlled by a residual analyzer, be sure the analyzer is working properly. Check the following items.

      (1) Actual chlorine residual is properly indicated.
      (2) Recorder alarm set point.
      (3) Recorder control set point.
      (4) Sample water flow.
      (5) Sample water flow to cell block after dilution with fresh water.
      (6) Adequate flow of dilution water.
      (7) Filter system and drain.

      (8) Cell block.

         (a) Buffer pump and solution feed correct. Run pH test of cell block.

         (b) Check amount of grit in cell block and add more grit if amount is low. Grit is used to keep the electrode free of slimes and chemical scales in order to provide quick and accurate readings.

         (c) Cell block hydraulics purger.

      (9) Run comparison tests of chlorine residuals. Do tests match with analyzer output readings?

      (10) If residual analyzer samples two streams, start other stream flow and compare tested residual of that stream with analyzer output readings. Standardize analyzer output readings against tested residuals. Enter changes and corrections in log.

      (11) Change recorder chart daily. Date the chart for recordkeeping purposes.

      (12) Check recorder output signal controlling chlorinator for control responses on feed rate. Correct feed rates through ratio controller.

Weekly

1. Put chlorinator on manual control. Operate feed rate adjustment through full range from zero to full scale (250, 500, 1,000, 2,000, 4,000, 6,000, 8,000, or 10,000 pounds/day). At each end of scale check:

   a. Chlorinator vacuum.

   b. Injector vacuum.

   c. Solution line pressure.

   d. Chlorine pressure at chlorinator.

      If any of the readings do not produce normal set points, make proper adjustments.

      (1) Injector should produce necessary vacuum at chlorinator (5 to 10 inches or 13 to 25 centimeters of mercury).

      (2) Adjust CPRV to obtain sufficient pressure and chemical feed for full-rate operation of chlorinator. This CPRV is on the main chlorine supply line and is usually located near or in the chlorine container area.

2. If unit performs properly through complete range of feed rates, return unit to automatic control. If any problems develop, locate source and correct.

3. Clean chlorine residual analyzer, including the following items:

   a. Clean filters.

   b. Clean sample line.

   c. Clean hydraulic dilution wells and baffles.

   d. Flush discharge hoses and pipes.

   e. Clean and flush cell block.

   f. Fill buffer reservoirs.

   g. Check buffer pump and feed rate.

   h. Wipe machine clean and keep it clean.

Monthly

1. Exercise all chlorine valves.

2. Inspect heaters and room ventilation equipment.

3. Check chlorinator vent line to outside of structure for any obstructions that could prevent free access to the atmosphere. Bugs and wasps like vent lines for nests. If this becomes a problem install a fine wire mesh over the open end of the vent line.

4. Inspect unit for vacuum leaks.

5. Clean rotameter sight glass.

6. Inspect all drain lines and hoses.

7. Perform scheduled routine maintenance.

   a. Repack seat and stem of valves.

   b. Inspect tubing and fittings for leaks. Wash and dry thoroughly before reassembling.

   c. Inspect control system.

      (1) Electrical and electronics
      (2) Pneumatics
      (3) Lubrication
      (4) Calibration of total system

   d. Chlorine analyzer.

      (1) Lubrication of chart drives, filter drives, and pumps.
      (2) Clean and flush all piping and hoses, filters, tubing, cell blocks, and hydraulic chambers.
      (3) Replace electrodes in cell block.
      (4) Replace buffer pumps and system solenoids. Clean acid and iodide reservoirs.
      (5) Calibrate unit with known standards.
      (6) Repaint unit.

   e. Inspect safety equipment, including self-contained breathing equipment and repair kits.

Annually

1. Disassemble, clean, and regasket chlorinator.

Possible Abnormal Conditions

1. Chlorine leak in chlorinator.

   Shut off gas flow to chlorinator. Leave injector on line. Place another chlorinator on line, if available, until repairs are completed. Allow the chlorinator to operate under a vacuum with zero psi showing on the chlorine pressure gage for three to five minutes to remove chlorine gas. If the system must be opened to repair the leak, the vacuum on the chlorinator may minimize the potential for any remaining chlorine gas to be released to the atmosphere. If effective evacuation of chlorine gas from the unit cannot be ensured, appropriate respiratory protection must be used. Repair the leak.

2. Gas pressure too low, less than 20 psi (1.4 kg/sq cm). Alarm indicated. Check chlorine supply.

   a. Empty containers, switch to standby units.

   b. Evaporator shut down. See Section 10.261, "Evaporators."

   c. Inspect manifold for closed valves or restricted filters. Correct by switching to another manifold, cleaning replacing filters, or setting valves and controls to proper position.

3. Injector vacuum too low.

   a. Adjust injector to achieve required vacuum.

   b. Inspect injector to achieve required vacuum.

      (1) Pump off: start pump.
      (2) Strainers dirty: clean strainers.
      (3) Pump worn out and will not deliver appropriate flow and pressure to injector: use other unit and/or repair or replace pump.
      (4) Inspect valves in system. Place valves in proper position.

   c. Inspect solution line discharge downstream from injector. Check for the following items:

      (1) Valve closed or partially closed.
      (2) Line broken or restriction reducing flow or increasing back pressure.
      (3) Diffuser plugged, thus restricting flow and creating a higher back pressure on discharge line and injector. Clean diffuser and flush pipe.

NOTE: The installation of a pressure gage on the discharge side of the injector or on the suction line would alert the operator to an abnormal system back pressure.

4. Low chlorine residual. Alarm indicator is on from chlorine residual analyzer.

   Determine actual chlorine residual and compare with residual reading from chlorine analyzer. If residual analyzer is indicating a different chlorine residual, recalibrate analyzer and readjust. If chlorine residual analyzer is correct and chlorine residual is low, check the following items:

   a. Sample pump

      (1) Operation, flow, and pressure.
      (2) Sample lines clean and free of solids or algae that could create a chlorine demand.
      (3) Strainer dirty and restricting flow, thus preventing adequate pressure (15 to 20 psi or 1.0 to 1.4 kg/sq cm) at analyzer.

   b. Control system if chlorinator is on automatic control. If chlorine feed rate remains too low, take chlorinator off of automatic control and switch to manual control. Set chlorinator to proper feed rate as determined by previous adequate feed rates (see Section 10.28, "Operational Strategy").

c. Chlorine demand higher than the amount one chlorinator can supply. Place additional chlorinator on line.

5. High chlorine usage. Effluent appears low in solids, yet coliform counts in effluent are high. High chlorine demand may be caused by either nitrite lock or breakpoint chlorination problems. Nitrite lock is a condition that occurs when nitrification is not complete.

To determine whether the high chlorine demand is probably caused by nitrite lock or breakpoint chlorination, perform the following test:

1. Collect a sample of secondary effluent before chlorination in a 250-mL beaker.

2. Pour 100 mL of sample into another 250-mL beaker.

3. Add several potassium iodide crystals to sample beaker and mix until they dissolve.

4. Add 5 mL glacial acetic acid or white distilled vinegar. If a yellow, orange, or red color appears, nitrite is present.

5. If color produced is not obvious, add a squirt of starch solution. If a blue color appears, nitrite is present.

6. If no color appears, the plant is probably experiencing breakpoint chlorination.

Breakpoint chlorination occurs when the chlorine to ammonia ratio exceeds approximately 8:1. When this ratio is exceeded, the chlorine will oxidize the ammonia to nitrate or other end products.

The options for the operator are twofold: (1) increase the chlorine concentration until breakpoint is reached, and (2) decrease the chlorine concentration until the chlorine:ammonia-N ratio is below 5:1. The latter option is what is often recommended but it is a tough call for operators since reducing chlorine feed when you don't have a residual seems nonsensical.

Nitrite lock occurs when nitrification is not complete: ammonia is oxidized to nitrite, but nitrite is not oxidized to nitrate. This can occur when the Nitrobacter population has not become established, such as during start-up of a nitrification system, or when Nitrobacter is inhibited. Free chlorine will react with nitrite in a 5:1 ratio of Cl_2 to NO_2-N. Chloramines do not readily react with nitrite.

There are several controls that can be used to prevent or offset the effects of nitrite lock. NOTE: It is important that corrective actions be taken at the onset of nitrite lock so Nitrobacter is not wasted from the system.

1. Reduce the level of nitrification—This is a viable option only if nitrification is not required.

2. Add supplemental alkalinity—If the pH of the secondary effluent is below 6.8, supplemental alkalinity can be added to prevent the low pH from inhibiting Nitrobacter.

3. Increase the ammonia concentration above the nitrite concentration—The ammonia concentration can be increased by introducing a non-nitrified effluent, a high ammonia in-plant recycle stream, or commercially available ammonia. If the ammonia concentration is higher than the nitrite concentration, chloramines will form that do not readily react with nitrite.

Nitrite lock can also be caused by low dissolved oxygen, extreme temperatures, and other forms of toxicity (even self-induced such as chlorine for filamentous control or ammonia from in-plant wastestreams).

Acknowledgment

Information in this section was provided by Woodie Muirhead. His contribution is greatly appreciated.

QUESTIONS

Write your answers in a notebook and then compare your answers with those on page 419.

10.2O Normal operation of a chlorinator includes daily inspection of what facilities or areas?

10.2P What is the purpose of evaporators?

10.2Q What abnormal conditions could be encountered when operating an evaporator?

10.2R How can you determine if a chlorine residual analyzer is working properly?

10.2S What are possible chlorinator abnormal conditions?

10.27 Shutdown of a Chlorinator
(operating from container gas connection)

10.270 Short-Term Shutdown

The following is a typical procedure for shutting down a chlorinator for a time period of less than one week.

1. Close chlorine container gas outlet valve.

2. Allow chlorine gas to completely evacuate the system through the injector. Chlorine gas pressure gages will fall to zero psi on the manifold and the chlorinator.

3. Close chlorinator gas discharge valve. The chlorinator may remain in this condition indefinitely and is ready to be placed back in service by reopening the chlorinator discharge valve and chlorine container gas outlet valve. After these valves have been reopened, inspect for chlorine leaks throughout the chlorination system.

4. Shut down injector if dedicated to the chlorinator being removed from service.

5. Disconnect the chlorine container from the system or routinely check the pressure in the chlorinator because the chlorine container valve may leak. Also the chlorinator inlet valve may be closed.

10.271 Long-Term Shutdown

1. Perform steps one, two, three, and four above for short-term shutdown.

2. Turn off chlorinator power switch, lock out, and tag.

3. Secure chlorinator gas manifold and chlorinator valve in closed position.

10.28 Operational Strategy

The operator should strive to keep the chlorination system well maintained and operational. There will be occasions when pieces of equipment will be out of service for maintenance and repair. These activities should be scheduled during the times of the year when demands on the chlorination system are low, such as the winter months. Plants may be required to provide chlorination/dechlorination services for:

1. Disinfection: postchlorination of effluent.

2. Process Control: activated sludge, RBCs, and trickling filter treatment processes.

3. Odor Control: usually seasonal during the warmer months from May through October.

4. Dechlorination: usually seasonal during July through November. (See Section 10.8, "Dechlorination." Chlorination and dechlorination facilities are similar.)

Under these circumstances, maintenance of chlorination facilities requiring disassembly, cleaning, repairing, and reassembling should be done during the winter months of November through April. If a chlorination unit fails, there will be standby capabilities for at least postchlorination to disinfect the plant effluent. Even if all systems fail, the operator is still expected to meet discharge disinfection requirements. A supply of HTH or other dry hypochlorite compound should be available for emergencies of this nature because dry hypochlorite compounds can be fed manually while repairs are being made. A table of dosages or calculation procedures and a calibrated measuring container should be available so the proper dosage of chemical can be applied.

Usually wastewater treatment plants are designed and equipped with standby or backup capabilities so at least the effluent can be disinfected. In this case, two areas could cause trouble:

1. Run out of chlorine.

This problem is strictly the operator's problem. You are the only person who maintains the chlorine supply inventory. You must request more chlorine with sufficient lead time for the vendor to acquire and deliver the chlorine. Therefore, your plant should never run out of chlorine or any other chemical.

Care must also be exercised that the connected containers do not empty and that sufficient containers are connected so that the maximum dosage rates can be achieved without exceeding maximum withdrawal rates from the connected containers. Weighing scales or automatic switchover equipment should always be provided. The latter is highly desirable particularly when the plant is unattended part of the day. Even in this case, it is important to note when the switchover has occurred so that the empty containers can be replaced with full ones before the second set empties.

2. Failure of automatic control equipment.

Assume you are operating a chlorination system on "auto" control to meet NPDES permit coliform requirements. A 4.5-mg/L chlorine residual must be maintained at the outlet of the chlorine contact basin to meet this requirement at your plant. If the plant operators have recorded the past chlorinator feed rates every two hours during daily routine checks, a history of feed rates at various times during the day will have been established for "auto" control. With this information available, manual operation only requires setting the feed rate to correspond to the same hour and flow conditions. Daily flows at most treatment plants are consistent (except during storm flows) where feed rates are known for every day of the year. By reviewing data for the same day for three previous weeks,

the information below can be obtained and calculations can be made for intermediate hours (1000 and 1200 hours). To calculate these suggested manual feed rates, determine the average feed rate between the 0900 and 1100 average feed rates and also between the 1100 and 1300 average feed rates. See Volume II, Chapter 18, Section 18.4, "Average or Arithmetic Mean," and Section A.35, "Data Analysis," in the Appendix at the end of Volume II.

Time	Last Week	Two Weeks Ago	Three Weeks Ago	Average	Suggested Manual Feed Rates
0900	1,055 lbs/day	1,065 lbs/day	1,060 lbs/day	1,060 lbs/day	1,060 lbs/day
1000	—			—	1,500 lbs/day
1100	1,900 lbs/day	1,900 lbs/day	1,800 lbs/day	1,900 lbs/day	1,900 lbs/day
1200	—			—	1,900 lbs/day
1300	1,800 lbs/day	1,850 lbs/day	1,850 lbs/day	1,850 lbs/day	1,850 lbs/day

By recording readings of chlorine feed rates, it is very simple to manually program equipment and maintain high plant performance during instrumentation or equipment failures. In this example, note that the suggested feed rate is adjusted manually every hour in an attempt to provide adequate disinfection at all times.

10.29 Troubleshooting

A regular program of scheduled preventive maintenance is essential to keep chlorinators functioning properly. Even well-maintained equipment, however, may break down or not operate as intended. Table 10.1 lists several problems you may encounter with chlorinators or the chlorine supply system and suggests steps you can take to troubleshoot and correct each problem.

QUESTIONS

Write your answers in a notebook and then compare your answers with those on page 419.

10.2T   List the steps for a short-term shutdown of a chlorinator.

10.2U   What are two areas that could hinder your ability to meet NPDES permit coliform requirements when your plant is equipped with standby or backup capabilities?

END OF LESSON 2 OF 5 LESSONS ON DISINFECTION AND CHLORINATION

Please answer the discussion and review questions next.

TABLE 10.1   TROUBLESHOOTING GUIDE—CHLORINATION

(Adapted from *PERFORMANCE EVALUATION AND TROUBLESHOOTING AT MUNICIPAL WASTEWATER TREATMENT FACILITIES*, Office of Water Program Operations, US EPA, Washington, DC.)

INDICATOR/OBSERVATION	PROBABLE CAUSE	CHECK OR MONITOR	SOLUTION
1. Low chlorine gas pressure at chlorinator.	1a. Insufficient number of cylinders connected to system. Supply valve closed or partly closed.	1a. Reduce feed rate and note if pressure rises appreciably after short period of time. If so, 1a. is the cause.	1a. Connect enough cylinders to the system so that chlorine feed rate does not exceed the withdrawal rate from the cylinders. Icing or very cold conditions can be noted at the cylinder/container valve if inadequate supply is the problem.
	1b. Stoppage or flow restriction between cylinders and chlorinators. Gas pressure reducing valve closed/malfunctioning.	1b. Reduce feed rate and note if icing and cooling effect on supply lines continues.	1b. Valve out, evacuate line then disassemble chlorine header system to point where cooling begins, locate stoppage and clean with solvent.
2. No chlorine gas pressure at chlorinator.	2a. Chlorine cylinders empty, not connected to system, or supply valve closed.	2a. Visual inspection of system gages.	2a. 1. Replace empty cylinders. 2. Connect cylinders. 3. Open supply valve.
	2b. Plugged or damaged pressure-reducing valve.	2b. Inspect valve. High chlorine pressure upstream of valve, low pressure downstream.	2b. Repair the reducing valve after shutting off supply valves, evacuating gas in the header system.
3. Chlorinator will not feed chlorine.	3a. No chlorine supply.	3a. Check chlorine supply and pressure gages.	3a. 1. Restore chlorine supply to chlorinator. 2. Check chlorine pressure-reducing valve (CPRV) on evaporator on chlorine supply header.
	3b. 1. Inadequate injector vacuum.	3b. 1. Check injector supply pump for proper output.	3b. 1. Start injector supply pump, obtain proper output in flow and psi.
	2. Injector diaphragm ruptured.	2. Injector diaphragm.	2. Replace injector diaphragm, adjust injector to obtain proper vacuum for operating chlorination system.

TABLE 10.1   TROUBLESHOOTING GUIDE—CHLORINATION (continued)

INDICATOR/OBSERVATION	PROBABLE CAUSE	CHECK OR MONITOR	SOLUTION
3. Chlorinator will not feed chlorine. (continued)	3c. Air leak in chlorinator.	3c. Check chlorinator components for secureness and proper connections.	3c. Retighten connections, replace faulty diaphragms, or ruptured tubing, defective seals, or O-rings.
	3d. Plugged diffuser.	3d. Check back pressure on chlorine water supply to contact basin.	3d. Clean diffuser.
	3e. V-notch orifice of chlorinator (chlorine) out of adjustment/ disengaged.	3e. Travel of V-notch orifice.	3e. Adjust or reconnect orifice, lubricate stem.
4. Chlorine gas escaping from chlorine pressure-reducing valve (CPRV).	4a. Main diaphragm of CPRV ruptured due to improper assembly or fatigue.	4a. Place ammonia bottle near termination of CPRV vent line to confirm leak.	4a. Disassemble valve and diaphragm; reassemble correctly.
5. Inability to maintain chlorine feed rate without icing of chlorine system.	5a. Insufficient supply.	5a. Check chlorine supply header pressure gage.	5a. Add more cylinders to meet chlorine feed demand.
	5b. Insufficient evaporator capacity.	5b. Check evaporator temperature and in/out pressure.	5b. 1. Place another evaporator in service. 2. If operating pre-post-chlorination system with separate units—place pre system on gas from cylinder supply, leave post system on evaporators. 3. Clean water bath on evaporator, ensure evaporator power is ON, ensure heater element is functioning properly, ensure solenoid is not malfunctioning and allowing water to circulate through evaporator.
	5c. 1. CPRV dirty (supply manifold). 2. Restriction in line. 3. Withdrawal rate too high.	5c. 1. Check chlorine pressure downstream of CPRV. 2. Chlorine system supply line pressures. 3. Feed rate.	5c. 1. Clean CPRV. 2. Locate and remove restriction. 3. Lower withdrawal rate or place more chlorine containers on line.
6. Chlorine evaporator system unable to maintain water bath temperature sufficient to keep external CPRV open.	6a. Heating element malfunction.	6a. Evaporator water bath temperature.	6a. Remove and replace heating element.
	6b. Solenoid valve malfunction.	6b. Solenoid valve.	6b. Repair/replace defective solenoid.
7. Wide variation in chlorine residual produced in effluent.	7a. Variation in chlorine demand.	7a. Analyze chlorine demand of plant effluent to determine demand during various flow periods.	7a. Program postchlorination feed rates during day to meet chlorine demand and supply desired chlorine residual to meet disinfection requirements.

TABLE 10.1 TROUBLESHOOTING GUIDE—CHLORINATION (continued)

INDICATOR/OBSERVATION	PROBABLE CAUSE	CHECK OR MONITOR	SOLUTION
7. Wide variation in chlorine residual produced in effluent. (continued)	7b. Chlorine contact basin.	7b. 1. Determine detention time for various portions of day.	7b. 1. Maintain minimum of thirty minutes of chlorine contact time with effluent in basin.
		2. Dye test basin at peak flows.	2. Baffle basin or mix to prevent short-circuiting.
		3. Solids deposit in basin.	3. Clean contact basin to avoid solids resuspending during peak flows and increasing chlorine demand.
		4. Sample location.	4. Sample other locations for best application point.
	7c. Chlorine diffuser.	7c. Chlorine diffuser for blockage, damage, and proper location for even chlorine dispersion.	7c. 1. Clean diffuser orifices.
			2. Replace broken or damaged parts on diffuser.
			3. Change location or style of diffuser for better mixing of chlorine with effluent.
	7d. Inadequate feed rate adjustment of post-chlorinators.	7d. Monitor effluent for chlorine residual.	7d. Reprogram chlorine feed rates to meet demand conditions.
	7e. Flow proportioning chlorine control devices not working properly.	7e. 1. Check flowmeter output.	7e. 1. Recalibrate flowmeter to correctly measure plant flow.
		2. Check flowmeter proportioning output device.	2. Maintain equipment so that flowmeter reading is correctly transmitting to chlorinator controller equivalent readings by mechanical cam, air, or electronic signals.
		3. Check chlorinator controller for in/out response.	3. Adjust chlorinator controller so that chlorinator follows plant flow signal.
	7f. Malfunction of auto control.	7f. Verify that chlorinator feed rate is what is required at given flow.	7f. Repair control system. May require manufacturer's field service personnel to perform repairs.
8. Chlorine residual analyzer recorder controller does not control chlorine residual properly.	8a. Electrodes fouled.	8a. Visual inspection.	8a. Clean electrodes.
	8b. Insufficient potassium iodide being added for amount of residual being measured.	8b. Potassium iodide dosage.	8b. Adjust potassium iodide feed to correspond with residual being measured.
	8c. Buffer additive system malfunctioning.	8c. See if pH of sample going through cell is maintained.	8c. Repair buffer additive system.

TABLE 10.1   TROUBLESHOOTING GUIDE—CHLORINATION (continued)

INDICATOR/OBSERVATION	PROBABLE CAUSE	CHECK OR MONITOR	SOLUTION
8. Chlorine residual analyzer recorder controller does not control chlorine residual properly. (continued)	8d. Malfunctioning of analyzer cell.	8d. Disconnect analyzer cell and apply a simulated signal to recorder mechanism.	8d. Call authorized service personnel to repair electrical components.
	8e. Poor mixing of chlorine at point of application.	8e. Set chlorine feed rate at constant dosage and analyze a series of grab samples for consistency.	8e. Install mixing device to cause turbulence at point of application.
	8f. Chlorinator rotameter tube range is improper size.	8f. Check tube range to see if it gives too small or too large an incremental change in feed rate.	8f. Replace rotameter tube with a proper range of feed rates.
	8g. Loop time too long.	8g. Check loop time.	8g. Reduce loop time by doing the following: 1. Move injector closer to point of application 2. Increase velocity in sample line to analyzer cell. 3. Move cell closer to sample point. 4. Move sample point closer to point of application.
9. Coliform count fails to meet required standards for disinfection.	9a. Chlorine residual too low.	9a. Chlorine residual.	9a. 1. Increase chlorine feed rate. 2. Increase chlorine contact time.
	9b. Inadequate chlorine residual control.	9b. Continuously record residual in effluent.	9b. Use chlorine residual analyzer to monitor and control the chlorine dosage automatically.
	9c. Inadequate chlorination equipment capacity.	9c. Check capacity of equipment.	9c. Replace equipment as necessary to provide treatment based on maximum flow through plant.
	9d. Solids buildup in contact chamber.	9d. Visual inspection.	9d. Clean contact chamber to reduce solids buildup.
	9e. Short-circuiting in contact chamber.	9e. Contact time.	9e. 1. Install baffling in contact chamber. 2. Install mixing device in contact chamber.
	9f. Coliform regrowth in piping/sample station.	9f. Effluent coliform sampling station.	9f. Modify system as needed to provide adequate chlorine residual to prevent regrowth.
	9g. High chlorine demand, effluent appears low in solids.	9g. Nitrite lock. or 9g. Breakpoint chlorination.	9g. 1. Reduce level of nitrification. 2. Add supplemental alkalinity. 9g. 1. Increase chlorine until breakpoint is reached. 2. Decrease chlorine until chlorine:ammonia-N ratio is below 5.1.

TABLE 10.1   TROUBLESHOOTING GUIDE—CHLORINATION (continued)

INDICATOR/OBSERVATION	PROBABLE CAUSE	CHECK OR MONITOR	SOLUTION
10. Chlorine residual too high in plant effluent to meet requirements.	10a. Chlorine feed rate too high.	10a. Chlorine residual.	10a. Reduce chlorine feed rate.
	10b. Malfunctioning chlorine residual control.	10b. Operation of residual control system/analyzer.	10b. Repair/calibrate chlorine feed/residual control loops.
11. *BREAKOUT* (breakaway) *OF CHLORINE*[34]	11a. Overfeeding chlorine.	11a. Chlorine feed rate.	11a. Decrease chlorine feed rate to minimize breakout and maximize application efficiency.
	11b. Insufficient injector water flow.	11b. Injector flow or supply/pump system.	11b. Adjust injector to maximum water flow. Place additional pumps on line to provide maximum capacity for dissolving chlorine.
	11c. Excess mixing.	11c. Chlorine mixer speed.	11c. Decrease mixer speed if variable speed, if not, turn unit off. Excess mixing may cause release of chlorine under breakout conditions.
	11d. Inadequate mixing.	11d. Chlorine mixer speed.	11d. Increase mixer speed to encourage dispersion of chlorine solution and increase dissolution capacity of flow stream.
	11e. Inadequate diffuser submergence.	11e. Depth of diffuser at application point.	11e. Lower diffuser to increase flow level over diffuser discharge point and provide additional contact between flow and chlorine solution.
	11f. Inadequate diffuser size.	11f. Diffuser size/design.	11f. Install a larger diffuser to disperse chlorine solution over a larger flow area.

CHLORINE SUPPLY SYSTEM (must be performed by trained and qualified operators)

INDICATOR/OBSERVATION	PROBABLE CAUSE	CHECK OR MONITOR	SOLUTION
12. Discoloration at joint, cadmium plating gone. Green or reddish colored deposits.	12. Chlorine leak.	12. Check suspect area with fumes of ammonia solution.	12. Disassemble and repair/replace defective joint as quickly as possible.
13. Small drop of liquid on joint.	13. Chlorine leak/condensation.*	13. Check suspect area with fumes of ammonia solution.	13. If chlorine leak confirmed, repair/replace defective joint as quickly as possible.*
14. Corrosion at gas gages or pressure switches.	14. Chlorine leak/condensation.*	14. Check suspect area with fumes of ammonia solution.	14. If chlorine leak confirmed, repair/replace defective gages/pressure switches as quickly as possible.*

* Condensation may be caused by line restrictions or improper temperature differential between the storage and chlorine equipment area; these possibilities should be investigated if no chlorine leaks are found.

[34] Breakout of Chlorine.   A point at which chlorine leaves solution as a gas because the chlorine feed rate is too high. The solution is saturated and cannot dissolve any more chlorine. The maximum strength a chlorine solution can attain is approximately 3,500 mg/L. Beyond this concentration molecular chlorine is present which will break out of solution and cause "off-gassing" at the point of application.

DISCUSSION AND REVIEW QUESTIONS

Chapter 10.   DISINFECTION AND CHLORINATION

(Lesson 2 of 5 Lessons)

Write the answers to these questions in your notebook before continuing. The question numbering continues from Lesson 1.

6. Where might chlorine be applied in the treatment of wastewater?

7. Where is chlorine usually applied for disinfection purposes?

8. What is the ultimate control of chlorine dosage for disinfection?

9. Calculate the chlorinator setting (lbs per 24 hours) to treat a waste with a chlorine demand of 12 mg/*L*, when a chlorine residual of 2 mg/*L* is desired, if the flow is 1 MGD.

10. Determine the chlorine feed rate for a flow of 0.75 MGD and a chlorine dosage of 18 mg/*L*.

11. Why must the chlorine solution be well mixed with the wastewater?

12. What precautions would you take before starting any chlorination system?

13. When should the maintenance and repair of chlorination systems be scheduled?

CHAPTER 10.   DISINFECTION AND CHLORINATION

(Lesson 3 of 5 Lessons)

10.3   CHLORINE SAFETY PROGRAM

Every good safety program begins with cooperation between the employee and the employer. The employee must take an active part in the overall program. The employee must be responsible and should take all necessary steps to prevent accidents. This begins with the attitude that as good an effort as possible must be made by everyone. Safety is everyone's problem. The employer also must take an active part by supporting safety programs. There must be funding to purchase equipment and to enforce safety regulations required by OSHA and industrial safety programs. The following items should be included in all safety programs.

1. Establishment of a safety program.

2. Written rules and specific safety procedures.

3. Periodic hands-on training using safety equipment:

    a. Leak-detection equipment
    b. Respiratory protection equipment
    c. Atmospheric monitoring devices
    d. Repair kits

4. Establishment of emergency procedures for chlorine leaks and first aid.

5. Establishment of a maintenance and calibration program for safety devices and equipment.

All persons handling chlorine should be thoroughly aware of its hazardous properties. Personnel should know the location and use of the various pieces of protective equipment and be instructed in safety procedures. In addition, an emergency procedure should be established and each individual should be instructed how to follow the procedures. An emergency checklist also should be developed and available. For additional information on this topic, see the Water Environment Federation's Manual of Practice No. 1, *SAFETY AND HEALTH IN WASTEWATER SYSTEMS*, and The Chlorine Institute's *CHLORINE MANUAL*, 6th edition.[35] Also see Volume II, Chapter 14, "Plant Safety."

10.30   Chlorine Hazards

Chlorine is a gas, 2.5 times heavier than air, extremely toxic, and corrosive in moist atmospheres. Dry chlorine gas can be safely handled in steel containers and piping, but with moisture must be handled in corrosion-resisting materials such as silver, glass, Teflon, and certain other plastics. Chlorine gas at container pressure should never be piped in silver, glass, Teflon, or any other plastic material. Even in dry atmospheres, the gas is very irritating to the mucous membranes of the nose, to the throat, and to the lungs; a very small percentage in the air causes severe coughing. Heavy exposure can be fatal. (See Table 10.2.)

> **WARNING**
>
> WHEN ENTERING A ROOM THAT MAY CONTAIN CHLORINE GAS, OPEN THE DOOR SLIGHTLY AND CHECK FOR THE SMELL OF CHLORINE. NEVER GO INTO A ROOM CONTAINING CHLORINE GAS WITH HARMFUL CONCENTRATIONS IN THE AIR WITHOUT A SELF-CONTAINED AIR SUPPLY, PROTECTIVE CLOTHING AND HELP STANDING BY. HELP MAY BE OBTAINED FROM YOUR CHLORINE SUPPLIER AND YOUR LOCAL FIRE DEPARTMENT.

10.31   Why Chlorine Must Be Handled With Care

You must always remember that chlorine is a hazardous chemical and must be handled with respect. Concentrations of chlorine gas in excess of 1,000 ppm (0.1% by volume in air) may be fatal after a few breaths.

TABLE 10.2   PHYSIOLOGICAL RESPONSE TO CONCENTRATIONS OF CHLORINE GAS[36,37]

Effect	Parts of Chlorine Gas Per Million Parts of Air By Volume (ppm)
Slight symptoms after several hours' exposure	1[a]
Detectable odor	0.08 – 0.4
60-minute inhalation without serious effects	4
Noxiousness	5
Throat irritation	15
Coughing	30
Dangerous from one-half to one hour	40
Death after a few deep breaths	1,000

[a] OSHA regulations specify that exposure to chlorine shall at *NO* time exceed 1 ppm.

---

[35] Write to: Water Environment Federation (WEF), Publications Order Department, 601 Wythe Street, Alexandria, VA 22314-1994. Order No. MO2001. Price to members, $26.74; nonmembers, $36.74; price includes cost of shipping and handling. The Chlorine Institute, Inc., 1300 Wilson Boulevard, Arlington, VA 22209. Pamphlet 1. Price to members, $15.00; nonmembers, $30.00; plus 10 percent of order total for shipping and handling.

[36] Adapted from data in U.S. Bureau of Mines TECHNICAL PAPER 248 (1955).

[37] The maximum **P**ermissible **E**xposure **L**imit (PEL) is 0.5 ppm (8-hour weighted average). The IDLH level is 10 ppm. IDLH is the **I**mmediately **D**angerous to **L**ife or **H**ealth concentration. This concentration represents a maximum level from which one could escape within 30 minutes without any escape-impairing symptoms or any irreversible health effects.

Because the characteristic sharp odor of chlorine is noticeable even when the amount in the air is small, it is usually possible to get out of the gas area before serious harm is suffered. This feature makes chlorine less hazardous than gases such as carbon monoxide, which is odorless, and hydrogen sulfide, which impairs your sense of smell in a short time.

Inhaling chlorine causes general restlessness, panic, severe irritation of the throat, sneezing, and production of much saliva. These symptoms are followed by coughing, retching and vomiting, and difficulty in breathing. Chlorine is particularly irritating to persons suffering from asthma and certain types of chronic bronchitis. Liquid chlorine causes severe irritation and blistering on contact with the skin.

10.32   Protect Yourself From Chlorine

Every person working with chlorine should know the proper ways to handle it, should be trained in the use of appropriate respiratory protective devices and methods of detecting hazards, and should know what to do in case of emergencies.

WARNING

CANISTER-TYPE "GAS MASKS" ARE USUALLY **INADEQUATE** AND **INEFFECTIVE** IN SITUATIONS WHERE CHLORINE LEAKS OCCUR AND ARE THEREFORE NOT RECOMMENDED FOR USE UNDER ANY CIRCUMSTANCES. **SELF-CONTAINED AIR OR SUPPLIED-AIR TYPES OF BREATHING APPARATUS ARE RECOMMENDED.** CHEMICAL CARTRIDGE RESPIRATORS (Figure 10.7) MAY BE USED FOR ESCAPE ONLY.

Self-contained air supply and demand-breathing equipment must fit properly and be used properly. Pressure-demand and rebreather kits may be safer. Pressure-demand units use more air from the air bottle which reduces the time a person may work on a leak. There are certain physical constraints when using respiratory protection. Confirm requirements with your local safety regulatory agency.

Before entering an area with a chlorine leak, wear protective clothing. A chemical suit will prevent chlorine from contacting the sweat on your body and forming hydrochloric acid. Chemical suits are very cumbersome, but should be worn when the chlorine concentration is high. A great deal of practice is required to perform effectively while wearing a chemical suit.

The best protection that one can have when dealing with chlorine is to respect it. Each individual should practice rules of safe handling and good *PREVENTIVE MAINTENANCE.*

PREVENTION IS THE BEST EMERGENCY TOOL YOU HAVE.

Plan ahead.

1. Have your fire department and other available emergency response agencies tour the area so that they know where the facilities are located. Give them a clearly marked map indicating the location of the chlorine storage area, chlorinators, and emergency equipment.

2. Have regularly scheduled practice sessions in the use of respiratory protective devices, chemical suits, and chlorine repair kits. Involve all personnel who may respond to a chlorine leak.

3. Have a supply of ammonia available to detect chlorine leaks.

4. Write emergency procedures:

    Prepare a *CHLORINE EMERGENCY LIST* of names or companies and phone numbers of persons to call during an emergency and ensure that all involved personnel are trained in notification procedures. This list should be posted at plant telephones and should include:

    a. Fire department,
    b. Chlorine emergency personnel,
    c. Chlorine supplier, and
    d. Police department.

5. Follow established procedures during all emergencies.

    a. Never work alone during chlorine emergencies.

    b. Obtain help immediately and quickly repair the problem. *PROBLEMS DO NOT GET BETTER.*

    c. Only authorized and properly trained persons with adequate equipment should be allowed in the danger area to correct the problem.

    d. If you are caught in a chlorine atmosphere without appropriate respiratory protection, shallow breathing is safer than breathing deeply. Recovery depends upon the duration and amount of chlorine inhaled, so it is important to keep that amount as small as possible.

Mouthbit chemical cartridge respirators are used for escape only

Half-facepiece chemical cartridge respirators are suitable for work masks within their approval limits, or for escape

Fig. 10.7 *Chemical cartridge respirators*
(Courtesy of PPG Industries, Inc., Chemical Division)

e. If you discover a chlorine leak, leave the area immediately unless it is a very minor leak. Small leaks can be found by using a rag soaked with ammonia. A polyethylene "squeeze bottle" filled with ammonia water to dispense ammonia vapor may also be used. Care must be taken to avoid spraying ammonia water on any leak or touching the ammonia-soaked cloth to any metal. A white gas will form near the leak so it can be located and corrected.

f. Notify your police department that you need help if it becomes necessary to stop traffic on roads and to evacuate persons in the vicinity of the chlorine leak.

6. Develop emergency evacuation procedures for use during a serious chlorine leak. Coordinate these procedures with your police department and other officials. Be sure that all facility personnel are thoroughly trained in any evacuation procedures that are developed.

7. Post emergency procedures in all operating areas.

8. Inspect equipment and routinely make any necessary repairs.

9. At least twice weekly, inspect areas where chlorine is stored and where chlorinators are located. Remove all obstructions from the areas.

10. Schedule routine maintenance on *ALL* chlorine equipment at least once every six months or more frequently.

11. Have health appraisal for employees on chlorine emergency duty. All those who have heart and/or respiratory problems should not be allowed on emergency teams. There may be other physical constraints. Contact your local safety regulatory agency for details.

Remember:

Small amounts of chlorine cause large problems. Leaks never get better.

10.33 First-Aid Measures

Mild Cases

Whenever you have a mild case of chlorine exposure, which does happen from time to time around chlorination equipment, you should first leave the contaminated area. Move slowly, breathe lightly without exertion, remain calm, keep warm, and resist coughing. Notify other operators and have them repair the leak immediately.

If clothing has been contaminated, remove as soon as possible. Otherwise the clothing will continue to give off chlorine gas which will irritate the body even after leaving the contaminated area. Immediately wash off area affected by chlorine. Shower and put on clean clothes.

If victim has slight throat irritation, immediate relief can be accomplished by drinking milk. Drinking spirits of peppermint also will help reduce throat irritation. A mild stimulant such as hot coffee or hot tea is often used to control coughing.

Extreme Cases

1. Follow established emergency procedures.

2. Always use proper safety equipment. Do not enter area without a self-contained breathing apparatus.

3. Remove patient from affected area immediately. Call a physician and begin appropriate treatment immediately.

4. First aid:

a. Remove contaminated clothes to prevent clothing giving off chlorine gas which will irritate the body.

b. Keep patient warm and cover with blankets if necessary.

c. Place patient in a comfortable position on back.

d. If breathing is difficult, administer oxygen if equipment and trained personnel are available.

e. If breathing appears to have stopped, begin CPR (cardiopulmonary resuscitation) immediately. Mouth-to-mouth resuscitation or any of the other approved methods may be used.

f. *EYES!*

If even a small amount of chlorine gets into your eyes, they must be flushed immediately with large amounts of lukewarm water so that all traces of chlorine are flushed from the eyes (at least 15 minutes). Hold the eyelids apart gently but firmly to ensure complete flushing of all eye and lid tissues.

QUESTIONS

Write your answers in a notebook and then compare your answers with those on page 420.

10.3A   What are the hazards of chlorine gas?

10.3B   What type of breathing apparatus is recommended when repairing a chlorine leak?

10.3C   What first-aid measures should be taken if a person comes in contact with chlorine?

10.4 CHLORINE HANDLING

10.40 Chlorine Containers

10.400 Cylinders

Cylinders containing 100 to 150 pounds (45 to 68 kg) of chlorine are convenient for very small treatment plants with capacities less than 0.5 MGD (1,890 cu m/day). These cylinders are usually of seamless steel construction (Figure 10.8).

A fusible plug is placed in the valve below the valve seat (Figure 10.9). This plug is a safety device. The fusible metal softens or melts at 158 to 165°F (70 to 74°C) to prevent build-up of excessive pressures and the possibility of rupture due to a fire or high surrounding temperatures.

Cylinders will not explode and can be handled safely.

The following are procedures for handling chlorine cylinders.

1. Move cylinders with a properly balanced hand truck with clamp supports that fasten at least two-thirds of the way up the cylinder.

2. 100- and 150-pound (45- and 68-kg) cylinders can be rolled in a vertical position. Lifting of these cylinders should be avoided except with approved equipment. Use a lifting clamp, cradle, or carrier. Never lift with chains, rope slings, or magnetic hoists.

3. Protective hood or cap should always be replaced when moving a cylinder.

4. Cylinders must be kept away from direct heat (steam pipes or radiators).

Chlorine Cylinder

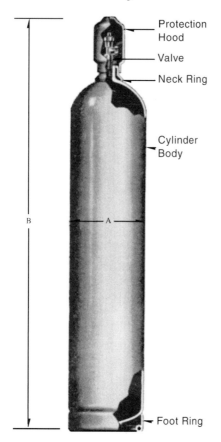

Protection Hood

Valve

Neck Ring

Cylinder Body

Foot Ring

Net Cylinder Contents	Approx. Tare, Lbs.*	Dimensions, Inches	
		A	B
100 Lbs.	73	8¼	54½
150 Lbs.	92	10¼	54½

* Stamped tare weight on cylinder shoulder does not include valve protection hood.

Fig. 10.8   Chlorine cylinder
(Courtesy of PPG Industries, Inc., Chemical Division)

5. Cylinders are stored in an upright position.

6. Cylinders must be firmly secured to an immovable object. Use a safety chain or clamp to prevent either empty or full cylinders from falling over or being knocked down.

7. Store empty cylinders separately from full cylinders. *NOTE:* Never store chlorine cylinders near turpentine, ether, anhydrous ammonia, finely divided metals, hydrocarbons, or other materials that are flammable in air or react violently with chlorine.

10.401   Ton Tanks

Ton tanks are of welded construction and have a loaded weight of as much as 3,700 pounds (1,680 kg). They are about 80 inches (200 cm) in length and 30 inches (75 cm) in outside diameter. The ends of the tanks are crimped inward to provide a substantial grip for lifting clamps (Figure 10.10).

The following are some characteristics of and procedures for handling ton tanks.

Most ton tanks have eight openings for fusible plugs and valves (Figures 10.10 and 10.11). Generally, two operating valves are located on one end near the center. There are six or eight fusible metal safety plugs, three or four on each end. These are designed to melt within the same temperature range as the safety plug in the cylinder valve.

> **WARNING**
>
> IT IS VERY IMPORTANT THAT FUSIBLE PLUGS SHOULD NOT BE TAMPERED WITH UNDER ANY CIRCUMSTANCES AND THAT THE TANK SHOULD NOT BE HEATED. ONCE THIS PLUG OPENS, ALL OF THE CHLORINE IN THE TANK WILL BE RELEASED.

Ton tanks are shipped by rail in multi-unit tank cars. They also may be transported by truck or semi-trailer.

Ton tanks should be handled with a suitable lift clamp in conjunction with a hoist or crane of at least two-ton capacity (Figure 10.10).

Ton tanks should be stored and used on their sides, above the floor or ground, on steel or concrete supports. They should not be stacked more than one high.

Ton tanks should be placed on trunnions (pivoting mounts) that are equipped with rollers so that the withdrawal valves may be positioned one above the other. The upper valve will discharge chlorine gas, and the lower valve will discharge liquid chlorine (see Figure 10.10). The ability to rotate tanks is also a safety feature. In case of a liquid leak the container can be rolled so that the leaking chlorine escapes as a gas rather than as a liquid. Trunnion rollers should not exceed 3½ inches (9 cm) in diameter so that the containers will not rotate too easily and be turned out of position. Roller shafts should be equipped with a zerk-type lubrication fitting. Roller bearings are not advised because of the ease with which they rotate. Locking devices should be used when these rollers are used to prevent ton tanks from rolling while connected.

10.402   Chlorine Tank Cars

Chlorine tank cars are of 16-, 30-, 55-, 85-, or 90-ton capacity. All have four-inch (10-cm) cork board insulation protected by a steel jacket. The dome of the standard car contains four angle valves plus a safety valve. The safety valve has a relief pressure of either 225 psi or 375 psi depending on the test pressure of the tank car. The test pressure of car is identified by the last three digits of the classification number located near the right end of the car (for example, 300 psi or 500 psi). The two angle valves located on the axis line of the tank are equipped for discharging liquid chlorine. The two angle valves at right angles to the axis of the tank deliver gaseous chlorine (Figure 10.12).

Unloading of tank cars should be performed by trained personnel in accordance with Interstate Commerce Commission (ICC) regulations.

In most situations, chlorine is withdrawn from tank cars as a liquid and then passed through chlorine evaporators. Sometimes DRY air is passed into the tank car through one of the

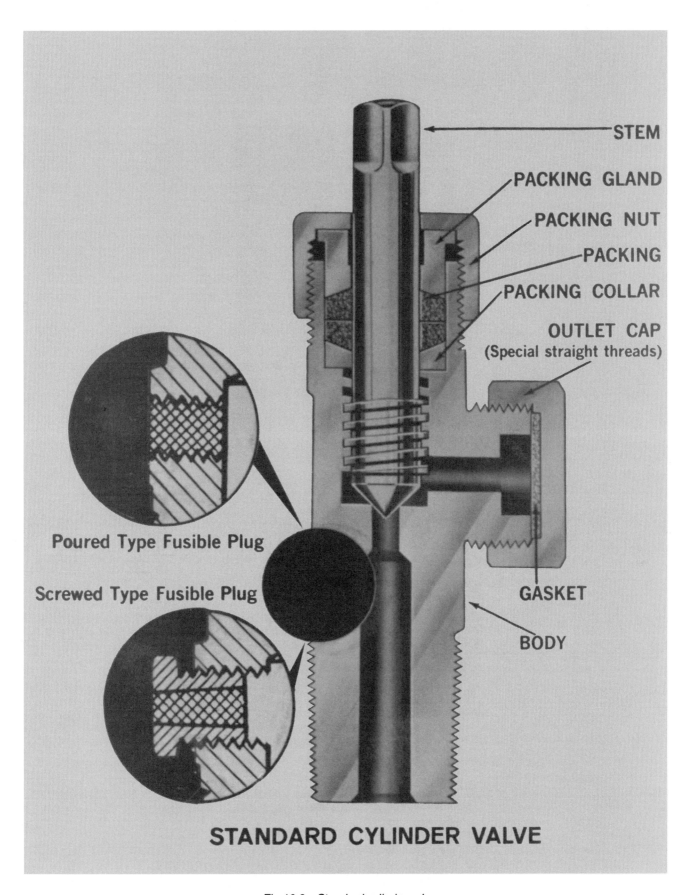

STEM

PACKING GLAND

PACKING NUT

PACKING

PACKING COLLAR

OUTLET CAP
(Special straight threads)

GASKET

BODY

Poured Type Fusible Plug

Screwed Type Fusible Plug

STANDARD CYLINDER VALVE

Fig.10.9   Standard cylinder valve
(Reproduced with permission of the Dow Chemical Company (1959, 1966))

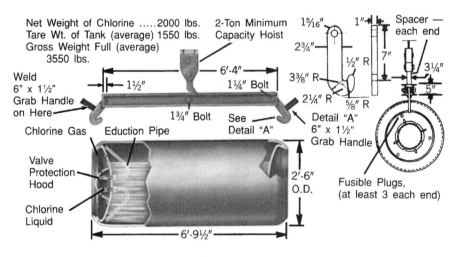

Net Weight of Chlorine .....2000 lbs.
Tare Wt. of Tank (average) 1550 lbs.
Gross Weight Full (average)
3550 lbs.

2-Ton Minimum Capacity Hoist

1⁵⁄₁₆″

Spacer — each end

Weld 6" x 1½" Grab Handle on Here

1½″   6'-4″   1¼″ Bolt

2¾″

3⅜″ R

1″   7″

½″ R

3¼″

5″

1¾″ Bolt   See Detail "A"

2¼″ R   ⁵⁄₈″ R

Detail "A" 6" x 1½" Grab Handle

Chlorine Gas   Eduction Pipe

Valve Protection Hood

Chlorine Liquid

2'-6″ O.D.

6'-9½″

Fusible Plugs, (at least 3 each end)

Fig. 10.10   Ton tank lifting beam
(Courtesy of PPG Industries)

NOTE: Grab handles recommended by Denis Unreiner, Medicine Hat, Alberta, Canada

Ton Tank Valve   100 and 150 lb Cylinder Valve

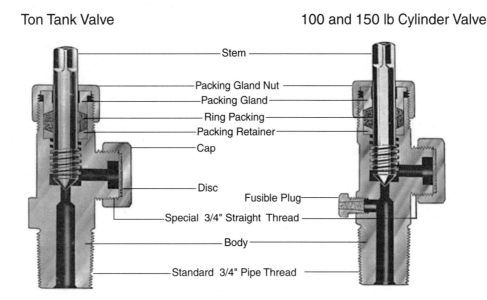

- Stem
- Packing Gland Nut
- Packing Gland
- Ring Packing
- Packing Retainer
- Cap
- Disc
- Fusible Plug
- Special 3/4" Straight Thread
- Body
- Standard 3/4" Pipe Thread

Fig. 10.11   Comparison of ton tank valve with cylinder valve
(Courtesy of PPG Industries, Inc., Chemical Division)

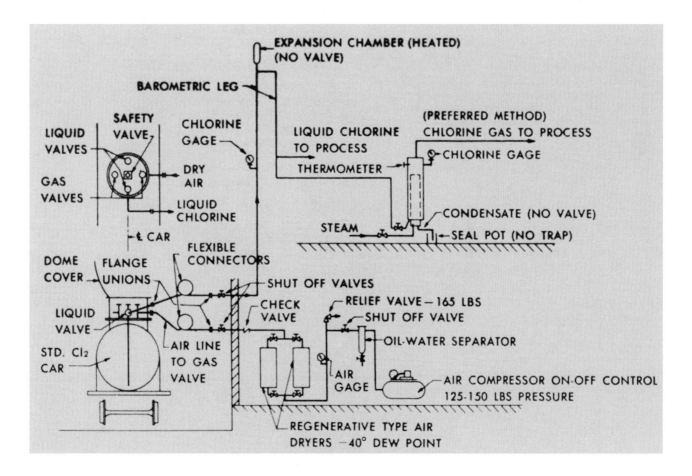

Fig. 10.12   Typical chlorine tank car unloading arrangement
(Reproduced with permission of the Dow Chemical Company (1959, 1966))

gas valves to assist in liquid withdrawal. This practice is referred to as *AIR PADDING*.[38]

QUESTIONS

Write your answers in a notebook and then compare your answers with those on page 420.

10.4A   How may chlorine be delivered to a plant?

10.4B   What is the purpose of the fusible plug?

10.41   Removing Chlorine From Containers

10.410   Connections

The outlet threads on container valves are specialized threads; they are not ordinary tapered pipe threads. Use only the fittings and gaskets furnished by your chlorine supplier or chlorinator equipment manufacturer when making connections to chlorine containers. Do not try to use regular pipe thread fittings. Whenever you make a new connection, always use a new gasket. The outlet threads on container valves should always be inspected before being connected to the chlorine system. Containers with outlet threads that are badly worn, cross threaded, or corroded should be rejected and returned to the supplier. The connecting nut on the chlorine system should also be inspected and replaced if it develops any of these problems. Since the threads on the cylinder connection may be worn, yoke-type connectors (Figure 10.13) are recommended.

Flexible ³⁄₈-inch 2,000-pound (psi) (0.95-cm, 140-kg/sq cm) annealed (toughened) copper tubing is recommended for connections between chlorine containers and stationary piping. Care should be taken to prevent sharp bends in the tubing because this will weaken it and eventually the tubing will start leaking. Many operators recommend use of a sling to hold the tubing when disconnecting it from an empty cylinder to prevent the tubing from flopping around and getting kinked or getting dirt inside it. Cap or plug the connectors and/or valves when they are not connected to prevent entry of dirt/debris and moisture.

A shutoff valve is needed after the container valve or at the beginning of stationary piping to simplify changing of containers.

Whenever changing a chlorine container, use new washers or gaskets. Always purchase washers or gaskets from your chlorine supplier in order to obtain the proper material.

10.411   Valves

Do not use wrenches longer than six inches (15 cm), pipe wrenches, or wrenches with an extension on container valves. If you do, you could exert too much force and break the valve. Use only a square end open or box wrench (see Figure 10.13) (this wrench can be obtained from your chlorine supplier). To unseat the valve, strike the end of the wrench with the heel of your hand to rotate the valve stem in a counter-clockwise direction. Then open slowly. One complete turn permits maximum discharge. Do not force the valve beyond the full open position or you may strip the internal valve stem threads. If the valve is too tight to open in this manner, loosen the packing gland nut (Figure 10.11) slightly to free the stem. Open the valve, then retighten the packing nut. If you are uncertain how to loosen the nut, you should return the container to the supplier. Do not use organic lubricants on valves and lines because further chemical reactions can block the lines.

10.412   Ton Tanks

One-ton tanks (Figure 10.10) must be *PLACED ON THEIR SIDES WITH THE VALVES IN A VERTICAL POSITION*. Connect the flexible tubing to the *TOP VALVE* to remove chlorine gas from a tank. The *BOTTOM VALVE* is used to remove liquid chlorine and is used only with a chlorine evaporator. The valves are similar to those on the smaller chlorine cylinders (fusible plugs are not located at valves on ton containers) and must be handled with the same care.

10.413   Railroad Tank Cars

Unloading of tank cars must be performed by trained personnel in accordance with established safety procedures and practices. Figure 10.14 shows the layout of a facility for unloading a railroad tank car. This section lists the procedures to follow to safely receive, connect, and disconnect railroad tank cars. Your facility may be slightly different. This procedure can be made more specific for your facility by numbering each valve in the procedures.

Fig. 10.13   Yoke and adapter-type connectors

(Permission of The Chlorine Institute, Inc.)

---

[38] *Air Padding.   Pumping dry air (dew point –40°F) into a container to assist with the withdrawal of a liquid or to force a liquified gas such as chlorine or sulfur dioxide out of a container.*

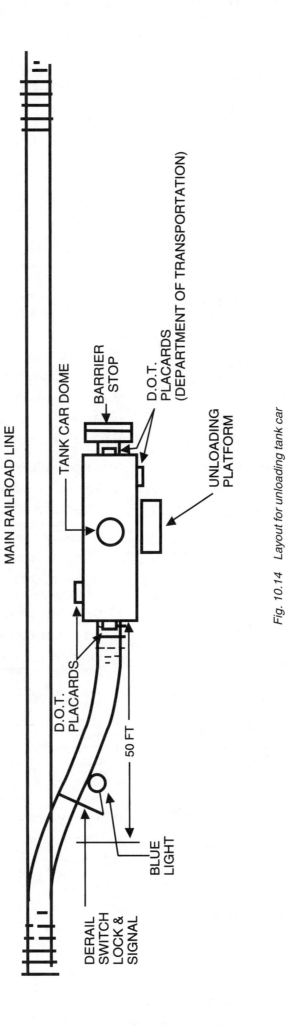

Fig. 10.14   Layout for unloading tank car

Receiving Tank Cars

1. Inspect and refuse delivery if car is damaged or has corrosion that would make it unsafe. Once the tank car has been uncoupled, the railroad is relieved of all responsibility. Check that the invoice numbers match with the numbers on the car.

2. Spot the tank car so the unloading platform is aligned with the car catwalk.

3. Verify that the railroad has set the brakes.

4. Set two wheel chocks, one before and one after, on the rail nearest the chlorine tower (Figure 10.15).

5. Place warning light and sign one car length from tank car (Figure 10.15).

6. Close and lock derailing mechanism (Figure 10.14).

7. Record on log sheet the date and time the tank car arrived and was accepted or rejected on treatment plant property.

8. Sign name on log sheet.

Connecting Tank Cars

1. Material needed before leaving office:

   a. Keys to tool locker on chlorine tower.
   b. Electrical safety tag.
   c. A buddy; no one is to work alone on the chlorine tower.

2. Verify that system valving is appropriate. Develop a valve numbering system and required valve positions for your installation.

3. Tag pad air compressor so it will not be turned off during purging operations.

4. Check emergency eye wash and deluge shower for proper operation.

5. Put on escape respirator.

6. Lower and lock down drawbridge.

7. Check that valves on the unloading tower are properly positioned.

8. Open pad air valve if pad air pressure is higher than chlorine pressure.

9. Open tank car lid and verify that all car valves are closed. Check valve area with ammonia to locate any leaks.

10. Remove plug from tank car valve farthest from the chlorine tower and install gage assembly. Check plug area with ammonia before complete removal because the tank car valve may not be seated properly. *NOTE:* Use a suitable pipe dope on male threads only. Tape should not be used for these connections. Tape fragments have caused leakage from angle valves and have prevented excess flow valves from operating when needed.

11. Record tank car pressure. If pressure is below desired level, then padding the tank car will be required.

12. Install flex air hose to connect pad air piping to chlorine piping.

13. Open valves to initiate purging of chlorine flex line. Valves should be opened only slightly until the chlorine flex line is completely connected.

14. Remove plug from tank car valve and connect chlorine flex line. *USE A NEW LEAD WASHER IN THE AMMONIA COUPLING FOR THE CHLORINE LINE.*

15. After no leaks can be heard from connections, close valves.

16. Remove flex air hose to break pad air/chlorine piping connection.

17. Open valve on tank car. *TEST WITH AMMONIA GAS.*

18. Open valves to process (chlorinator/evaporator area) slowly. Too rapid opening may permit a sudden surge of chlorine and cause the excess flow valve (Figure 10.16) in the eductor pipe to close, thus stopping flow. Usually a sharp metallic click will be heard when the valve closes. If the valve closes, pressure above and below the ball must be equalized to reopen it.

19. Open valves. This will connect chlorine unloading tower to chlorine building.

20. Store all gear and lock tool cabinet.

21. Remove compressor tag.

22. Record date and time job was completed on log sheet.

23. Sign name on log sheet.

24. Record date and time tank car was placed in service.

25. Record date and time tank car was taken out of service.

Disconnecting Tank Cars

1. Materials needed before leaving office:

   a. Keys to tool locker on chlorine tower.
   b. Electrical safety tag.
   c. A buddy; no one is to work alone on the chlorine tower.

2. Verify valves are closed.

3. Tag pad air compressor so it will not be turned off during purging operations.

4. Check eye wash and shower for operation.

5. Put on escape respirator.

6. Close valve that disconnects the chlorine tower from process (chlorinator/evaporator area).

7. Verify that pad air pressure is higher than liquid chlorine line.

8. Open pad air supply valve to chlorine tower.

9. Install flex air hose to connect pad air piping to chlorine piping.

10. Open valves that will allow the air to force the chlorine back into the tank car.

11. Wait five minutes.

12. Close valve on chlorine tank car.

13. Close valves and then crack open slightly. Loosen chlorine coupling and test with ammonia gas until no chlorine gas is escaping.

14. Remove chlorine flex line.

15. Reassemble and cap chlorine flex line and verify that no air is escaping.

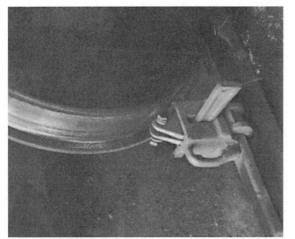

Wheel chock for tank cars.

Tank car stop.

Tank car derail and stop sign. It stands
between tracks.

Fig. 10.15   Tank car safety equipment
(Courtesy of PPG Industries, Inc.)

Fig. 10.16   Excess flow valve
(Courtesy of PPG Industries, Inc.)

16. Close tank car valve farthest from the chlorine tower.

17. Remove gage assembly slowly and test with ammonia gas. Put cap on gage assembly.

18. Put plugs back in chlorine tank car valves and secure lid.

DON'T LEAVE VALVE HANDLE IN CHLORINE TANK CAR

19. Close pad air to chlorine valves.

20. Remove flex air hose to break pad air/chlorine piping connection.

21. Store all gear and lock tool cabinet.

22. Raise and lock drawbridge.

23. Change Department of Transportation (D.O.T.) placards to read "danger chlorine empty."

24. Place warning light and sign at the end of the spur.

25. Place wheel chocks at the end of the spur.

26. Unlock and open derailing mechanism.

27. Remove compressor tag.

28. Call railroad and notify them to pick up empty tank car.

29. Record name of railroad official contacted.

30. Record date and time of disconnect on log sheet.

31. Sign Bill of Lading (shipping receipt).

32. Sign name on log sheet to indicate that job is completed.

10.42   Chlorine Leaks

Chlorine leaks must be taken care of immediately or they will become worse.

Corrective measures should be undertaken only by trained operators wearing proper safety equipment. A team of operators should be trained to safely repair chlorine leaks.

All other persons should leave the danger area until conditions are safe again.

If the leak is large, all persons in the adjacent areas should be warned and evacuated. Obtain assistance from off-site emergency response agencies. You must always consider your neighbors . . . PEOPLE, animals, and plants.

1. *BEFORE ANY NEW SYSTEM IS PUT INTO SERVICE*, it should be cleaned, dried, and tested for leaks. Pipelines may be cleaned and dried by flushing and steaming from the high end to allow condensate and foreign materials to drain out, or by the use of commercially available cleaning solvents compatible with chlorine. After the empty line is heated thoroughly, dry air may be blown through the line until it is dry. After drying, the system may be tested for tightness with 150 psi (10.5 kg/sq cm) dry air. Leaks may be detected by application of soapy water to the outside of joints. Small quantities of chlorine gas may now be introduced into the line, the test pressure built up with air, and the system tested for leaks with ammonia. Whenever a new system is tested for leaks, at least one chlorinator should be on the line to withdraw chlorine from the system in case of a leak. The same is true in case of an emergency leak at any installation. If a chlorinator is not running, at least one or more should be started. Preferably, all available chlorinators should be put on the line.

2. *TO FIND A CHLORINE LEAK*, tie a rag on a stick, *DIP THE RAG*[39] in a strong ammonia solution and hold the rag near the suspected points. A polyethylene "squeeze bottle" filled with ammonia water to dispense ammonia vapor may also be used. Care must be taken to avoid spraying ammonia water on any leak or touching the cloth to any metal. White fumes will indicate the exact location of the leak. Location of leaks by this method may not be possible for large leaks that diffuse the gas over large areas. Do not use an ammonia spray bottle because the entire room could turn white if it is full of chlorine gas.

3. *IF THE LEAK IS IN THE EQUIPMENT* in which chlorine is being used, close the valves on the chlorine container at once. Repairs should not be attempted while the equipment is in service. All chlorine piping and equipment that is to be repaired by welding should be flushed with water or steam. Before returning equipment to use, it *MUST* be cleaned, dried, and tested as previously described.

4. *IF THE LEAK IS IN A CHLORINE CYLINDER OR CONTAINER*, use the proper emergency repair kit (Figure 10.17) supplied by most chlorine suppliers. These kits can be used to stop most leaks in a chlorine cylinder or container and can usually be delivered to a plant within a few hours if one is not already at the site of the leak. *IT IS*

---

[39] *A one-inch (2.5-cm) paint brush may be used instead of a rag.*

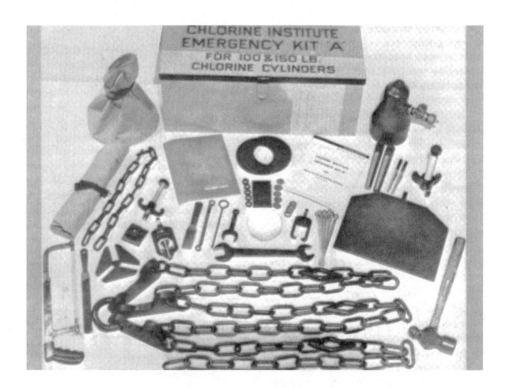

Emergency Kit "A" for Chlorine Cylinders

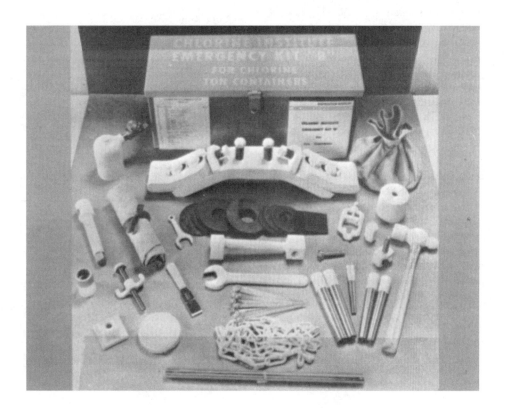

Emergency Kit "B" for Chlorine Ton Containers

Fig. 10.17   Chlorine Institute emergency repair kits
(Permission of The Chlorine Institute, Inc.)

ADVISABLE TO HAVE EMERGENCY REPAIR KITS AVAILABLE AT YOUR PLANT AT ALL TIMES AND TO TRAIN PERSONNEL IN THEIR USE. Respiratory protective equipment should be located outside chlorine storage areas. The repair kits may be located within chlorine storage areas because during an emergency requiring their use, you will already be wearing approved respiratory protection and thus they will be accessible.

5. If chlorine is escaping as a liquid from a cylinder or a ton tank, turn the container so that the leaking side is on top. In this position, the chlorine will escape only as a gas, and the amount that escapes will be only 1/15 as much as if the liquid chlorine were leaking. Keeping the chlorinators running also will reduce the amount of chlorine gas leaking out of a container. Increase the feed rate to cool the supply tanks as much as possible.

6. *FOR SITUATIONS IN WHICH A PROLONGED OR UN-STOPPABLE LEAK* is encountered, emergency disposal of chlorine should be provided. Chlorine may be absorbed in solutions of caustic soda, soda ash, or agitated hydrated lime slurries (Table 10.3). Chlorine should be passed into the solution through an iron pipe or a properly weighted rubber hose to keep it immersed in the absorption solution. If fixed piping exists between chlorination and absorption units, a barometric loop or vacuum breaking device must be installed to prevent back siphonage of the absorption chemical into the chlorine system and subsequent chemical reactions. The container should not be immersed because the leaks will be aggravated due to the corrosive effect, and the container may float when partially empty. In some cases it may be advisable to move the container to an isolated area. Discuss the details of such precautions with your chlorine supplier.

7. *NEVER PUT WATER ON A CHLORINE LEAK.* A mixture of water and chlorine will increase the rate of corrosion of the container and make the leak larger. Besides, water may warm the chlorine, thus increasing the pressure and forcing the chlorine to escape faster.

8. *LEAKS AROUND VALVE STEMS* can often be stopped by closing the valve or tightening the packing gland nut. Tighten the nut or stem by turning it clockwise.

9. *LEAKS AT THE VALVE DISCHARGE OUTLET* can often be stopped by replacing the gasket or adapter connection.

10. *LEAKS AT FUSIBLE PLUGS AND CYLINDER VALVES* usually require special handling and emergency equipment. Call your chlorine supplier immediately and obtain an emergency repair kit for this purpose if you do not have a kit readily available.

11. *PINHOLE LEAKS* in the walls of cylinder and ton tanks can be stopped by using a clamping pressure saddle with a turnbuckle available in repair kits. This is only a temporary measure, and the container must be emptied as soon as possible.

If a repair kit is not available, use your ingenuity. One operator stopped a pinhole leak temporarily until a repair kit arrived by placing several folded layers of neoprene packing over a leak, a piece of scrap steel plate over the packing, wrapping a chain around the cylinder and steel plate, and applying leverage pressure with a crowbar.

12. *A LEAKING CONTAINER* must not be shipped. If the container leaks or if the valves do not work properly, keep the container until you receive instructions from your chlorine supplier for returning it. If a chlorine leak develops in transit, keep the vehicle moving until it reaches an open area.

13. Do not accept delivery of containers showing evidence of leaking, stripped threads, or abuse of any kind.

14. If a chlorine container develops a leak, be sure your supplier does not charge you for the unused chlorine.

15. Chlorine leaks may be detected by chlorine gas detection devices. Alarm systems may be connected to these devices. Be sure to follow the manufacturer's recommendations regarding frequency of checking and testing detection devices and alarm systems.

TABLE 10.3   CHLORINE ABSORPTION SOLUTIONS*

Absorption Solution		Alkali (lbs)	Water (gal)
Caustic Soda (100%)	a	125	40
	b	188	60
	c	2,500	800
Soda Ash	a	300	100
	b	450	150
	c	6,000	2,000
Hydrated Lime**	a	125	125
	b	188	188
	c	2,500	2,500

> Chlorine Container Size (lbs net):
> a = 100, b = 150, c = 2,000

* Source: The Chlorine Institute, Inc.
** Hydrated lime solution must be continuously and vigorously agitated while chlorine is to be absorbed.

QUESTION

Write your answer in a notebook and then compare your answer with the one on page 420.

10.4C   How would you look for a chlorine leak?

END OF LESSON 3 OF 5 LESSONS
on
DISINFECTION AND CHLORINATION

Please answer the discussion and review questions next.

DISCUSSION AND REVIEW QUESTIONS

Chapter 10. DISINFECTION AND CHLORINATION

(Lesson 3 of 5 Lessons)

Write the answers to these questions in your notebook before continuing. The question numbering continues from Lesson 2.

14. What type of breathing apparatus should be worn when entering an area in which chlorine gas is present?

15. How could your police department assist you in the event of a serious chlorine leak?

16. Why should clothing be removed from a person who has been in an area contaminated with liquid or gaseous chlorine?

17. Why should chlorine containers and cylinders be stored where they won't be heated?

18. Why are slings used to hold chlorine tubing when changing chlorine cylinders?

19. Why should water never be poured on a chlorine leak?

20. How would you attempt to repair a pinhole leak in a chlorine cylinder?

CHAPTER 10.   DISINFECTION AND CHLORINATION

(Lesson 4 of 5 Lessons)

10.5   CHLORINATION EQUIPMENT AND MAINTENANCE
by J. L. Beals

10.50   Chlorinators

Chlorine usually is delivered by vacuum-solution feed chlorinators (Figures 10.18 and 10.19). The chlorine gas is controlled, metered, introduced into a stream of injector water, and then conducted as a solution to the point of application.

The primary advantage of vacuum operation is safety. If a failure or breakage occurs in the vacuum system, the chlorinator either stops the flow of chlorine into the equipment or allows air to enter the vacuum system rather than allowing chlorine to escape into the surrounding atmosphere. In case the chlorine inlet shutoff fails, a vent valve discharges the incoming gas to the outside of the chlorinator building.

The operating vacuum is provided by a hydraulic injector. The injector operating water absorbs the chlorine gas, and the resultant chlorine solution is conveyed to a chlorine diffuser through corrosion-resistant conduit.

A vacuum chlorinator also includes a vacuum regulating valve to dampen fluctuations and give smooth operation. A vacuum relief prevents excessive vacuum within the equipment.

A typical vacuum control chlorinator is shown in Figures 10.18 and 10.19 and the purposes of the parts are listed in Table 10.4. Chlorine gas flows from a chlorine container to the gas inlet (see Figure 10.19). After entering the chlorinator the gas passes through a rotameter which indicates the rate of gas flow. The rate is controlled by a V-notch variable orifice. The gas next passes through a spring-loaded pressure regulating valve which maintains the proper operating pressure. The gas then moves to the injector where it is dissolved in water and leaves the chlorinator as a chlorine solution (HOCl) ready for application.

10.51   Evaporators

Chlorine evaporators are installed in treatment plants where large quantities of chlorine are used. An evaporator (Figure 10.20) is a hot water heater surrounding a steel tank. Water is usually heated by electricity. Heat in the water is transferred to the liquid chlorine in an inner steel tank. Water bath heaters are used to provide an even distribution of heat around the center tank to eliminate the problem of hot spots on the inner

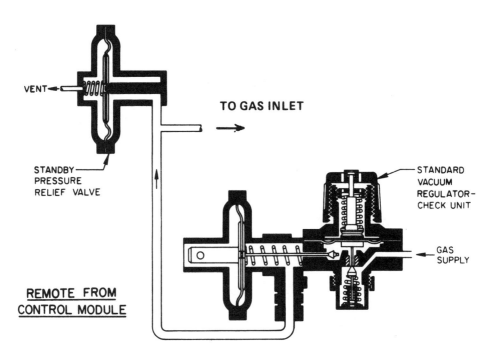

Fig. 10.18   Chlorinator gas pressure controls

(Permission of Wallace & Tiernan Division, Penwalt Corporation)

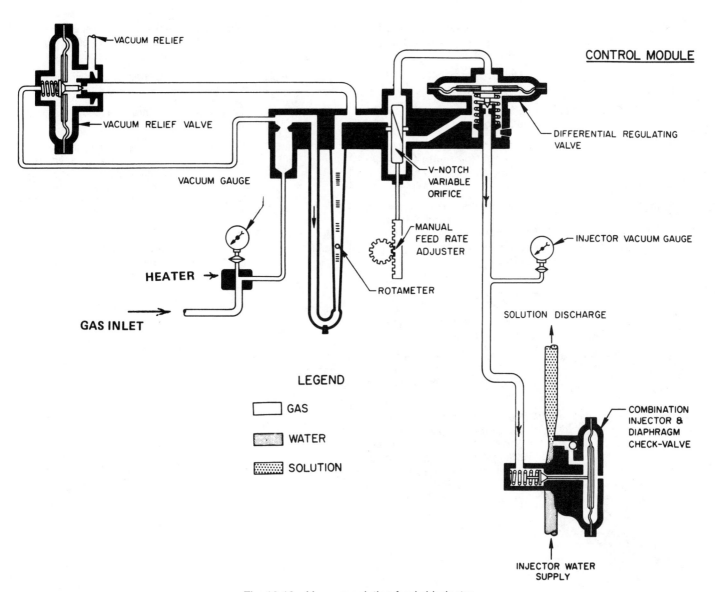

CONTROL MODULE

VACUUM RELIEF

VACUUM RELIEF VALVE

VACUUM GAUGE

DIFFERENTIAL REGULATING VALVE

V-NOTCH VARIABLE ORIFICE

HEATER →

GAS INLET

MANUAL FEED RATE ADJUSTER

ROTAMETER

INJECTOR VACUUM GAUGE

SOLUTION DISCHARGE

LEGEND

GAS

WATER

SOLUTION

COMBINATION INJECTOR & DIAPHRAGM CHECK-VALVE

INJECTOR WATER SUPPLY

Fig. 10.19 Vacuum-solution feed chlorinator
(Permission of Wallace & Tiernan Division, Penwalt Corporation)

TABLE 10.4   CHLORINATOR PARTS AND PURPOSE
(Figures 10.18 and 10.19)

Part	Purpose	Part	Purpose
Pressure Gage (Not shown on Figure 10.18)	Indicates chlorine gas pressure at chlorinator system from chlorine manifold and supply (20 psi minimum and 40 psi maximum or 2.4 kg/sq cm minimum and 4.8 kg/sq cm maximum).	V-notch Plug and Variable Orifice	Control chlorine feed rate by regulating flow of chlorine gas. A wide V-notch in the plug allows high feed rates through the orifice and a small V-notch in the plug provides low feed rates.
Gas Supply	Provides source of chlorine gas from containers to chlorinator system.	Vacuum Relief Valve	Relieves excess vacuum by allowing air to enter system and reduce vacuum.
Vacuum Regulator-Check Unit	Maintains a constant vacuum on chlorinator.	Vacuum Relief	Provides source of air to reduce excess vacuum.
Standby Pressure Relief	Relieves excess gas pressure on chlorinator.	Injector Vacuum Gage	Indicates vacuum at the injector.
Vent	Discharges any excess chlorine gas (pressure) to atmosphere outside of chlorination building.	Diaphragm Check Valve	Regulates chlorinator vacuum which in turn adjusts chlorinator feed rate. Receives signal from chlorine feed rate controls and then adjusts feed rate by regulating vacuum.
Gas Inlet	Allows entrance of chlorine gas to chlorinator. Gas flows from chlorine container through supply line and gas manifold to inlet.	Manual Feed Rate Adjuster	Regulates chlorine feed rate manually. Most chlorination systems have automatic feed rate controls with a manual override.
Heater	Prevents reliquefaction of chlorine gas.	Injector Water Supply	Provides source of water for chlorine solution. Must provide sufficient pressure and volume to operate injector.
Vacuum Gage	Indicates vacuum on chlorinator system.		
Rotameter Tube and Float	Indicate chlorinator feed rate by reading top of float or center of ball for rate marked on tube.	Injector	Mixes or injects chlorine gas into water supply. Creates sufficient vacuum to operate chlorinator and to pull metered amount of chlorine gas.
Differential Regulating (Reducing) Valve	Regulates (reduces) chlorinator chlorine gas pressure. Serves to maintain a constant differential across the orifice in order to obtain repeatable chlorine gas flow rates at a given V-notch orifice opening regardless of fluctuations in the injector vacuum.	Solution Discharge	Discharges solution mixture of chlorine and water.

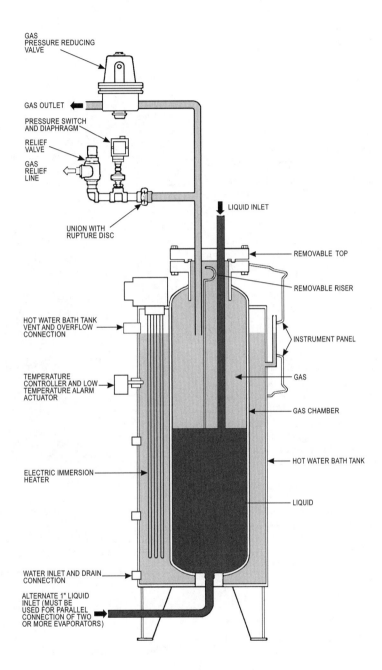

Fig. 10.20 Evaporator

(Permission of Wallace & Tiernan Division, Penwalt Corporation)

tank. Elimination of hot spots makes the evaporator easier to control and reduces the danger of overheating the chlorine and causing pressurization of chlorine by expansion.

Liquid chlorine containers are connected to the chlorine system through the liquid valve. When the liquid chlorine flowing from the container reaches the evaporator, the liquid chlorine vaporizes. The temperature of the chlorine gas is around 110 to 120°F (43 to 49°C). Chlorine gas flows from the evaporator to the gas manifold. Chlorine gas manifolds have gas filters to remove small particles from the gas.

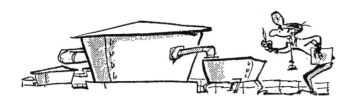

When the temperature of the water in the evaporator water jacket reaches a preset level, valves operate (open and close) to allow chlorine gas to flow to the chlorinators. If the water temperature falls below the set point, a valve closes to prevent the carryover of liquid chlorine to the chlorinator.

The level of liquid chlorine in the evaporator tank is automatically regulated by pressure from the chlorination system demand for chlorine. When the chlorine demand is high, pressure at the liquid chlorine supply containers exceeds the evaporator gas pressure and liquid chlorine flows into the inner tank. As the chlorine demand decreases, the evaporator's inner-tank chlorine gas pressure increases from the vaporization of the liquid chlorine. This increased pressure reduces the liquid chlorine flow into the evaporator. Equilibrium usually is obtained when the liquid flow rate and rate of vaporization are equal.

When chlorine gas leaves the evaporator and passes through the gas pressure-reducing valve, the chlorine gas enters the inlet block on the chlorinator. The chlorinator meters the chlorine gas at the desired dosage rate to the injector. A heater at the chlorinator inlet block vaporizes any liquid droplets that may have carried over from the evaporator. Try to keep the chlorinator room 10°F (6°C) warmer than the storage or evaporator room to prevent *RELIQUEFACTION*[40] of chlorine gas back to liquid chlorine.

10.52   Hypochlorinators (hi-poe-KLOR-uh-NAY-tors)

Hypochlorinators are chlorine pumps or devices used to feed chlorine solutions made from hypochlorites such as bleach (sodium hypochlorite) or calcium hypochlorite. Hypochlorite compounds are available as liquids or various forms of solids (powder, pellets), and in a variety of containers or in bulk. Hypochlorination systems consist of a water meter and a diaphragm metering pump. The pump feeds a hypochlorite solution in proportion to the wastewater flow.

10.53   Chlorine Dioxide Facility (Figure 10.21)

Most existing chlorination units may be used to produce chlorine dioxide. In addition to the existing chlorination system, a diaphragm pump, solution tank, mixer, chlorine dioxide generating tower, and electrical controls are needed. The diaphragm pump and piping must be made of corrosion-resistant materials because of the corrosive nature of chlorine dioxide. Usually PVC or polyethylene pipe is used.

Special consideration must be given when handling sodium chlorite. Sodium chlorite is usually supplied as a salt and is very combustible around organic compounds. Whenever spills occur, sodium chlorite must be neutralized with anhydrous sodium sulfite. Combustible materials (including gloves) should not be worn when handling sodium chlorite. If sodium chlorite comes in contact with clothing, the clothes should be removed immediately and soaked in water to remove all traces of sodium chlorite or they should be burned immediately. Due to the hazards of safety handling sodium chlorite, chlorine dioxide has not been widely used to treat wastewater.

QUESTIONS

Write your answers in a notebook and then compare your answers with those on page 420.

10.5A   How is chlorine delivered (fed) to the point of application?

10.5B   Why has chlorine dioxide not been widely used to treat wastewater?

10.54   Installation and Maintenance

The following are some features of importance when working with chlorine facilities. Also examine these items when reviewing plans and specifications.

1. Chlorinators should be located as near the point of application as possible.

---

[40] *Reliquefaction (re-LICK-we-FACK-shun).   The return of a gas to the liquid state; for example, a condensation of chlorine gas to return it to its liquid form by cooling.*

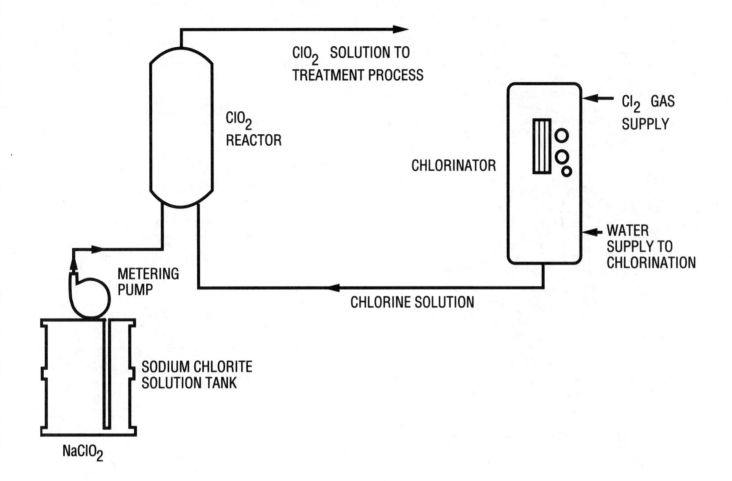

Fig. 10.21   Chlorine dioxide facility

Source: *AN ASSESSMENT OF OZONE AND CHLORINE DIOXIDE TECHNOLOGIES FOR TREATMENT OF MUNICIPAL WATER SUPPLIES, EXECUTIVE SUMMARY.* U.S. Environmental Protection Agency, Cincinnati, Ohio 45268, EPA-600/8-78-018, October 1978.

2. If possible, there should be a separate room for chlorinators and chlorine container storage (above ground) to prevent chlorine gas leaks from damaging equipment and harming personnel. There should be no access to this room from a room containing equipment or where personnel work.

3. Ample working space around the equipment and storage space for spare parts should be provided.

4. There should be an ample supply of water to operate the chlorinator at required capacity under maximum pressure conditions at the chlorinator injector discharge.

5. The building should be adequately heated. The temperature of the chlorine cylinder and chlorinator should be above 50°F (10°C). Line heaters may be used to keep chlorine piping and chlorinator at higher temperatures to prevent condensing of gas into liquid in the pipelines and chlorinator. In general, a temperature difference of 5 to 10°F (3 to 6°C) is recommended. The maximum temperature at which a chlorine cylinder is stored should not exceed 100°F (43°C).

6. It is not advisable to draw more than 40 pounds of chlorine from any one 100- to 150-pound (45- to 68-kg) cylinder in a 24-hour period because of the danger of freezing and slowing down the chlorine flow. With ton containers, the limit of chlorine gas is about 8 pounds of chlorine per day per °F (6.4 kg/°C) ambient temperature. When evaporators are provided, these limitations do not apply.

7. There should be adequate light.

8. There must be adequate ventilation. Exhaust ventilation should be taken from a point within twelve inches (30 centimeters) of the floor because chlorine is heavier than air. Mechanical ventilation must be at a rate of not less than one cubic foot of air per minute per square foot (0.005 cubic meter of air per second per square meter) of floor area of the storage area. Normally, ventilation from chlorine storage rooms is discharged to the atmosphere, but when a chlorine leak occurs the ventilated air containing the chlorine should be routed to a treatment system to remove the chlorine. A caustic scrubbing system can be used to treat air containing chlorine from a leak. The treatment system should be designed to reduce the maximum discharge concentration of chlorine to one-half the IDLH (**I**mmediately **D**angerous to **L**ife or **H**ealth) at the point of discharge to the atmosphere. The IDLH for chlorine is 10 ppm. A secondary standby source of power is required for the chlorine detection, alarm, ventilation, and treatment systems.

9. Adequate measuring and controlling of chlorine dosage is required. Scales and recorders indicating loss in weight are desirable as a continuous check and as a record of the continuity of chlorination. *RECORD* weights daily.

10. There should be continuity of chlorination. When chlorination is practiced for disinfection, it is needed continuously for the protection of downstream water users. Therefore, arrange that chlorination will function for 1,440 minutes per day. To secure continuous chlorination, the chlorine gas lines from cylinders should feed to the manifold so that the cylinders can be removed without interrupting the feed of gas. Duplicate units with automatic cylinder switchover should be provided. Hypochlorinators are sometimes used during emergencies.

11. For additional information on chlorinator maintenance, refer to Chapter 15, "Maintenance," in Volume II.

QUESTIONS

Write your answers in a notebook and then compare your answers with those on page 420.

10.5C Why should chlorinators be in a separate room?

10.5D Why is room temperature important for proper chlorinator operation?

10.5E Why should not more than 40 pounds of chlorine per day be drawn from any one cylinder?

10.5F How can chlorinator rooms be ventilated?

10.5G Why is adequate ventilation important in a chlorinator room?

10.5H How can chlorination rates be checked against the chlorinator setting?

10.5I Why should chlorination be continuous?

10.5J How can continuous chlorination be achieved?

10.55 Installation Requirements

Portions of these paragraphs are from Section A of Wallace & Tiernan Catalog Sheet Nos. 5.110 and 5.111. They are reproduced with the permission of Wallace & Tiernan Division, Penwalt Corporation.

10.550 Piping, Valves, and Manifolds

After you have determined (a) the availability of various types of chlorine containers and selected the type most suited to your particular needs, (b) the inventory required and the space needed, (c) the method of handling the containers, and (d) the type of weighing scales to be used, the final step in regard to chlorine supply is to plan the required piping.

CONNECTIONS AT CONTAINERS: Standard practice is to connect an auxiliary tank valve (either union- or yoke-type) to the container valve to facilitate changing containers, to minimize the release of gas when containers are changed, and to serve as a shutoff valve in the event the container valve is defective. From this auxiliary valve a flexible connection is used to connect to the manifold or, in the case of small installations, directly to the chlorinator.

CONNECTIONS AT CHLORINATOR: In general, small chlorinators are equipped to receive a flexible connection directly from the chlorine container and no other piping is necessary. Larger chlorinators may use a flexible connection from a manifold, if located close to the container, or may have piping from the manifold to the chlorinator where the distance is greater.

CONNECTIONS AT EVAPORATOR: Where evaporators are used, the piping from the manifold to the evaporator carries liquid chlorine. The piping from the evaporator to the chlorinator carries chlorine gas. Evaporators normally are furnished with all necessary intermediate valves and fittings.

PIPING—MATERIALS AND JOINTS: Best practice calls for the use of seamless carbon steel (Schedule 80) pipe for conducting chlorine gas or liquid and fittings that are 3,000-lb (1,360-kg) forged steel. Except in unusual cases, the size will be ¾ or 1 inch (1.9 to 2.5 cm). In most installations, it will be found practical to use threaded joints. These joints should be

put together using Teflon tape as a joint lubricant. Unions of the flanged, ammonia type, two-bolt oval are recommended. For pipe sizes larger than one inch (2.5 cm) in diameter, a four-bolt oval should be used. From the standpoint of maintenance, line valves should be kept to a minimum. Insulation is required only in those unusual cases where it is necessary to prevent chlorine gas lines from becoming chilled or liquid lines from becoming overheated.

PIPING—CHLORINE GAS: It is important to observe the correct temperature conditions in conducting chlorine gas from the location of the containers to the point of use. To avoid difficulty with reliquefaction of chlorine, piping and control equipment should be at a higher temperature than that of the chlorine containers. In general, a difference of 5 to 10°F (3 to 6°C) is recommended. It is preferable to run chlorine gas lines overhead through relatively warm areas rather than along the floor or through basement areas where lower temperatures may be encountered.

When it is not possible to secure suitable temperature conditions, the use of an external chlorine pressure-reducing valve near the containers is recommended.

The use of a chlorine pressure-reducing valve is also recommended in those localities where severe temperature changes are likely to be encountered during a 24-hour period.

PIPING—LIQUID CHLORINE: In the case of liquid chlorine, it is important to avoid conditions that will encourage vaporization. Thus, it is important to keep liquid chlorine lines as cool as, or cooler than, the containers. Avoid running liquid chlorine lines through overheated areas where gasification is likely to take place.

Valves in liquid chlorine lines should be kept to a minimum. This is particularly important to avoid situations where it is easy to close two valves in a line thus trapping liquid which, upon an increase in temperature, may develop dangerous pressures.

The use of an expansion chamber is recommended where traps occur in the line or where it is necessary to run lines a considerable distance. As the name implies, the chamber provides an area for expansion in the event that valves at both ends of the line are closed. The expansion chamber should have a capacity of at least 20 percent of the line volume or section of the line that it protects. The chamber is typically equipped with a rupture disc, pressure gage, and alarm switch.

VALVES AND MANIFOLDS: Chlorine valves consist of the following: (a) auxiliary tank valves for use at the container, (b) header valves for use on or in conjunction with manifolds, (c) line valves for insertion in liquid and gas lines for shutoff purposes, and (d) pressure-reducing valves to reduce the pressure in gas lines where necessary. Manifolds are assemblies designed to receive the flexible connections from the container, generally provide a shutoff valve, and include the means of connecting to the chlorinator piping. They are available in types and sizes to accommodate any required number of containers and may be mounted in any convenient manner.

10.551 Chlorinator Injector Water Supply

The injector operating water supply serves to produce the vacuum under which vacuum chlorinators function and to dissolve the chlorine and discharge it as a solution at the point of application. The quantity of water required and the minimum pressure at which it is supplied depend upon:

1. Maximum chlorinator feed rate, and

2. Back pressure at injector discharge (pressure at point of chlorine application, plus friction loss in solution line, and plus or minus elevation differences between injector and point of application).

Water quantity and pressure can vary from one to two GPM (4.5 to 11 cu m/day) at 15 psi (1 kg/sq cm) [for 10 lbs/day (4.5 kg/day) at 0 back pressure] and up to 360 GPM (1,960 cu m/day) at 60 psi (4 kg/sq cm) [8,000 lbs/day (3,630 kg/day) at 20 psi (1.4 kg/sq cm) back pressure]. In some extremely high back pressure situations, injector water up to 300 psi (21 kg/sq cm) may be required. These conditions do not occur often in wastewater treatment installations, and back pressures exceeding 20 psi (1.4 kg/sq cm) (except in force main applications) are uncommon.

Plant effluent is used frequently as injector operating water. When this is the case, a pump is usually required to provide the required quantity and pressure. If a pump is used exclusively for injector operation, it should be designed for 25 to 50 percent over capacity to allow for wear. Injector water is often supplied from a service water system also providing water for other purposes. If a potable water supply is the only source of injector water, precautions must be taken to ensure against direct cross-connections between wastewater and potable water to prevent contamination of the potable water. (Consult your local public health authority.)

Injector water requirements vary so widely depending upon make, model, capacity, and back pressure that it is advisable to consult the chlorinator manufacturer if a new system is to be installed or if an existing system must be altered and any of these operating conditions are to be changed.

10.56   Review of Plans and Specifications

See the previous two sections, 10.54 and 10.55, for items that should be considered when reviewing plans and specifications.

QUESTIONS

Write your answers in a notebook and then compare your answers with those on page 420.

10.5K   What is the best piping material for conducting chlorine gas or liquid?

10.5L   Plant _____ is used frequently as the chlorinator injector water supply.

10.6   OTHER USES OF CHLORINE

10.60   Odor Control

Chlorination is used to inhibit the growth of odor-producing bacteria and to destroy hydrogen sulfide (H_2S), the most common odor nuisance, which has the smell of rotten eggs. Hydrogen sulfide, in addition to creating an odor nuisance, can be an explosion hazard when mixed with air in certain concentrations. Breathing H_2S can impair your ability to smell, and too much will paralyze your respiratory center, causing death in severe cases. It also can cause corrosion of metals and concrete, being particularly damaging to electrical equipment even in low concentrations.

Hydrogen sulfide may be found in significant quantities in any collection and treatment system where sufficient time is allowed for its development. It may be expected to be present most often in new systems where flows are extremely low in comparison with design capacity, and particularly in lift stations where pump operating cycles may be at a low frequency. Collection systems that serve large areas often allow time for H_2S development even when operating at design capacity.

The purpose of this section is not to discuss the reasons for odor production, but rather their elimination or control by chlorination; however, the correction of an odor problem will usual-

ly require a decision being made between system modification and treatment. Sometimes both may be required. Choices of this type often hinge on the costs involved, and it will frequently be found that modifications to major system components are far more costly than treatment. When this is the case, chlorination is usually the most economical solution. Other solutions include the use of air or ozone.

Sulfide compounds and gases develop whenever given time to do so. The rate of sulfide production increases with temperature (about 7 percent on the average with each 1°C increase in wastewater temperature).

Hydrogen sulfide odors can be inactivated by chlorination at levels well below the chlorine demand point. This is commonly referred to as "sub-residual chlorination." The reason that this is true is based on the fact that the $Cl_2 + H_2S$ reaction precedes most other chlorine-consuming reactions. Since it is known that bacterial kills occur at sub-residual levels, it is logical that odor-producing bacteria can be reduced in numbers without satisfying the chlorine demand. This can be accomplished without significantly interfering with organisms beneficial to the treatment processes.

The quantities of chlorine required to accomplish control of odors vary widely from plant to plant and at any given plant fluctuate over a broad range. Hydrogen sulfide is generally found in higher concentrations when flows are low. For this reason it is usually not economical to chlorinate for odor control in direct proportion to flow. To establish a basis for treatment, tests should be run over periods that include all the various conditions that could possibly affect odor production.

When the requirements are known, the primary concern is to apply chlorine at the proper location. The best locations are generally up-sewer ahead of the plant influent structures, and up-sewer ahead of lift stations. This is done to allow mixing and reaction time before the waste reaches a point of agitation.

Sometimes force mains empty into the gravity sections of a collection system several hours after pumping. If odor problems result, a treatment point should be placed upstream at a point where the sewer is still under pressure and flowing full; thus treatment can be completed before odors are released to the atmosphere.

Hydrogen sulfide should not be considered merely an odor nuisance. Always keep in mind that it can create an explosion hazard, it can paralyze your respiratory center, and it should always be considered a source of corrosion. For these reasons, odor masking agents should not be used except possibly as additional treatment for odors not eliminated by chlorination. Excessive use of masking agents could prevent detection of a serious problem condition.

10.61   Protection of Structures

The destruction of hydrogen sulfide in wastewater also reduces the production of sulfuric acid that is highly corrosive to sewer systems and structures. This is particularly significant where temperatures are high and time of travel in the sewer system is unusually long. The treatment is similar to that for odor control: chlorination sufficient to prevent hydrogen sulfide formation or to destroy hydrogen sulfide that has been produced (about 2 mg/L chlorine per mg/L of hydrogen sulfide). Sulfide problems also may be corrected by oxygenation in sewers. The choice between oxygenation and chlorination will usually depend on the costs and safety risks involved. Chlorination is effective and cheaper than oxygenation, but chlorine leaks can be a serious hazard to the public.

10.62 Aid to Treatment

Among its many uses, chlorine improves treatment efficiency in the following ways.

10.620 *Sedimentation and Grease Removal*

Prechlorination at the inlet of a settling tank improves clarification by improving settling rate, reducing septicity of raw wastewater, and increasing grease removal. Maximum grease removal is achieved when chlorination in the range of one to ten mg/L is combined with aeration ("aerochlorination"). This is an expensive procedure, and some studies have indicated that benefits are minimal. Generally, grease removal in this manner is considered a beneficial side effect or "bonus" reaction to chlorine which is essentially applied for other reasons. Excess chlorination ahead of secondary processes can inhibit the bacterial action critical to the process and decrease sedimentation efficiency.

10.621 *Trickling Filters*

Continuous chlorination at the filter influent controls slime growths and destroys filter fly larvae (Psychoda). Generally the chlorine is applied to produce a residual of 0.5 mg/L (continuous) at the orifices or nozzles. Caution should be used because some filter growth may be severely damaged by excessive chlorination. Suspended solids will increase in a trickling filter effluent after chlorination for filter fly control. Also, it will be difficult to evaluate filter performance on the basis of BOD removals because chlorine can interfere with the BOD test. As a general statement, it would be well to look closely at loadings, operation, and general adequacy of the process when filter fly chlorination is continuously necessary, because continuous chlorination may be an expensive alternative for adequate design and operation.

10.622 *Activated Sludge*

Chlorination of return sludge in the range of 1 to 10 mg/L reduces bulking of activated sludge that is caused by overloading. The point of application should be where the return sludge will be in contact with the chlorine solution for about one minute before the sludge is mixed with the incoming settled wastewater. Chlorine is also commonly used to control *FILAMENTOUS ORGANISMS.*[41] Again, chlorine used in this manner is an expensive alternative for adequate design and operation. The main effort should be directed toward process improvement, considering chlorination mainly as an emergency solution. Never forget that chlorine is toxic to organisms that are needed to treat the incoming wastes.

10.623 *Reduction of BOD*

The reduction of BOD by chlorination has been explored by numerous investigators. These individuals have proposed four types of reactions to explain the reduction of BOD level.

1. Direct oxidation. Chlorine is an oxidizing agent and directly oxidizes the wastes instead of the organisms.

2. Formation with nitrogen compounds of bactericidal chloramines by substitution of chlorine for hydrogen.

3. Formation with carbon compounds of substances that are no longer decomposable by substitution of chlorine for hydrogen.

4. Addition of chlorine to unsaturated compounds to form non-decomposable substances.

These investigators have demonstrated that chlorination of raw wastewater can produce the following effects.

1. Reduce BOD by at least 2 mg/L for each mg/L of chlorine absorbed up to the point at which a residual is produced.

2. The reduction is increased with increasing chlorine dosages. This addition is not without limits.

3. The reduction seems to be permanent.

QUESTIONS

Write your answers in a notebook and then compare your answers with those on page 420.

10.6A How can hydrogen sulfide odors be controlled? Why?

10.6B How can sulfuric acid damage to structures be minimized or eliminated? Why?

10.7 ACKNOWLEDGMENTS

Portions of the information contained in this chapter were taken in part from Chapter 17, "Disinfection and Chlorination," Water Pollution Control Federation *MANUAL OF PRACTICE NO. 11*; and from Chapter 7, "Chlorination of Sewage," *MANUAL OF INSTRUCTION FOR SEWAGE TREATMENT PLANT OPERATORS* (New York Manual). Both publications are excellent references for additional study. J. L. Beals provided many helpful comments.

END OF LESSON 4 OF 5 LESSONS ON DISINFECTION AND CHLORINATION

Please answer the discussion and review questions next.

---

[41] *Filamentous (FILL-a-MEN-tuss) Organisms. Organisms that grow in a thread or filamentous form. Common types are Thiothrix and Actinomycetes. A common cause of sludge bulking in the activated sludge process.*

DISCUSSION AND REVIEW QUESTIONS

Chapter 10.   DISINFECTION AND CHLORINATION

(Lesson 4 of 5 Lessons)

Write the answers to these questions in your notebook before continuing. The question numbering continues from Lesson 3.

21. How is the rate of chlorine gas flow in a chlorinator controlled?

22. How often should the weights of chlorine cylinders be recorded?

23. Why must chlorination be continuous?

24. Why should the temperature of chlorine piping and control equipment be higher than the temperature of the chlorine containers?

25. Why must direct cross-connections between a public water supply and the injector water supply be avoided?

26. Why should hydrogen sulfide production be controlled?

27. What is "sub-residual chlorination"?

CHAPTER 10.   DISINFECTION AND CHLORINATION

(Lesson 5 of 5 Lessons)

10.8   DECHLORINATION[42]

10.80   Need for Dechlorination

Receiving waters such as streams, rivers, and lakes provide habitat for fish and numerous other types of aquatic organisms. The need for protection of this environment from toxic substances (such as chlorine) has prompted regulatory agencies to require that no measurable chlorine residual be allowed to enter receiving waters in the effluents from wastewater treatment plants. Removal of chlorine from treatment plant effluents is called "dechlorination."

Dechlorination is the physical or chemical removal of all traces of residual chlorine remaining after the disinfection process and prior to the discharge of the effluent to the receiving waters. This is commonly accomplished by the use of sulfur compounds such as sulfur dioxide, sodium sulfite, or sodium metabisulfite. Activated carbon has been used, but was found to be extremely expensive in large applications.

Dechlorination may be achieved by the following treatment processes:

1. Long detention periods. Prolonged detention periods provide sufficient time for dissipation of residual chlorine.

2. Aeration. Bubbling air through the water with a chlorine residual in the last portion of long, narrow chlorine contact basins will remove a chlorine residual.

3. Sunlight. Chlorine may be destroyed by sunlight. This is accomplished by spreading the chlorinated effluent in a thin layer and exposing it to sunlight.

4. Activated carbon. Residual chlorine can be removed from water by adsorption on activated carbon.

5. Chemical reactions. Sulfur dioxide (SO_2) is frequently used because it reacts instantaneously with chlorine on approximately a one-to-one basis (1 mg/L SO_2 will react with and remove 1 mg/L chlorine residual). Other chemicals include sodium sulfite (Na_2SO_3), sodium bisulfite ($NaHSO_3$), sodium metabisulfite ($Na_2S_2O_5$), and sodium thiosulfate ($Na_2S_2O_3$).

While high chlorine residuals (2.5 to 12.0 mg/L) often are required to meet the MPN coliform requirements set by the public health agencies, fish and wildlife agencies have required dechlorination to protect aquatic life in receiving waters below the plant discharge. Aquatic life, such as salmon, trout, and similar fish, can only tolerate trace (0.01 mg/L) amounts of chlorine.

Sulfur dioxide is the most popular chemical method used for dechlorination to date. The reason for the popularity of sulfur dioxide is that it uses existing chlorination equipment and makes extensive training of operators unnecessary. This section will discuss the use of sulfur dioxide for dechlorination and only briefly mention other methods (sodium sulfite (Na_2SO_3) used in tablet dechlorination units).

10.81   Sulfur Dioxide (SO_2)

10.810   Properties

Sulfur dioxide is a colorless gas with a characteristic pungent (sharp, biting) odor. SO_2 may be cooled and compressed to a liquid. When the gas is compressed to a liquid, a colorless liquid is formed. As with chlorine, when sulfur dioxide is in a closed container the liquid and gas normally are in equilibrium. The pressure within the container bears a definite relation to the container's ambient temperature. This relationship is very similar to chlorine.

Sulfur dioxide is neither flammable nor explosive in either form, gas or liquid. Dry gaseous sulfur dioxide is not corrosive to most metals; however, in the presence of moisture it forms sulfuric acid (H_2SO_4) and is extremely corrosive. Due to this corrosive action, similar materials and equipment are used for the storage and application of both sulfur dioxide and chlorine. The sulfonator's diaphragms are manufactured of special materials to handle sulfur dioxide rather than chlorine.

Sulfur dioxide gas is more soluble in water than chlorine. Approximately one pound per gallon can be absorbed at 60°F (16°C). As the temperature increases, sulfur dioxide's solubility in water decreases. When dissolved in water, sulfur dioxide forms a weak solution of sulfurous acid (H_2SO_3).

The density of sulfur dioxide is very similar to chlorine; so much so, that it is possible to use a chlorine rotameter to measure the flow of sulfur dioxide gas without much difficulty. When using the chlorine rotameter, multiply the chlorine reading by 0.95 to obtain the pounds per day of sulfur dioxide used.

10.811   Chemical Reaction of Sulfur Dioxide With Wastewater

The chemical reaction of dechlorination results in the conversion of all active positive chlorine ions to the nonactive negative chloride ions. The reaction of sulfur dioxide (SO_2) with chlorine (HOCl) is as follows:

$$SO_2 + H_2O \rightarrow HSO_3^- + H^+$$

$$HOCl + HSO_3^- \rightarrow Cl^- + SO_4^{-2} + 2H^+$$

$$SO_2 + HOCl + H_2O \rightarrow Cl^- + SO_4^{-2} + 3H^+$$

The formation of sulfuric acid (H_2SO_4) and hydrochloric acid (HCl) from this reaction is not harmful because of the small

---

[42] Dechlorination (dee-KLOR-uh-NAY-shun).   The removal of chlorine from the effluent of a treatment plant. Chlorine needs to be removed because chlorine is toxic to fish and other aquatic life.

amount of acid produced. The pH of the effluent is not changed significantly unless the alkalinity is very low.

With combined chloramine,

$$NH_2Cl + H_2SO_3 + H_2O \rightarrow NH_4HSO_4 + HCl.$$

Similar reactions are formed with dichloramine and nitrogen trichloride. If some organic materials are present, the reaction rate may change so that an excess of sulfur dioxide may have to be applied. Unwarranted excess sulfur dioxide dosages should be avoided, not only because it is wasteful, but it may also result in dissolved oxygen reduction with a corresponding increase in BOD and COD, and a drop in the pH in the effluent discharged to receiving waters. The chemical reaction between chlorine and sulfur dioxide is approximately one to one. For example, a chlorine residual of 4 mg/L would require a sulfur dioxide dose of 4 mg/L. The chemical reaction occurs almost instantaneously.

Where it may be desirable not to use sulfur dioxide for safety reasons (use of a liquid or dry tablet rather than toxic SO_2 gas), it may be useful to substitute sodium sulfite (Na_2SO_3) or sodium metabisulfite ($Na_2S_2O_5$). The reaction then becomes:

$$Na_2SO_3 + Cl_2 + H_2O \rightarrow Na_2SO_4 + 2\ HCl.$$

When using sodium sulfite, the reaction requires 1.78 pounds of pure sodium sulfite per pound of chlorine. The speed of reaction is similar to that of sulfur dioxide. Both sodium sulfite and sodium metabisulfite require liquid storage tanks and feed pumps, but evaporators are not needed. Sodium sulfite is also available as a dry chemical in tablet form.

10.812   Application Point

The key control guidelines for the effective use of sulfur dioxide or sodium sulfite for dechlorination are (1) proper dosage based on precise monitoring of combined chlorine residual, and (2) adequate mixing at the point of application. The typical application point is just prior to the discharge of the effluent to the receiving waters. This allows time for maximum disinfection of the effluent with chlorine before dechlorination. The point of dechlorination application should be where the flow is turbulent and short-circuiting should not exist. Since the dechlorination reaction requires a relatively short time period, contact basins are not needed. Often it is not feasible to have the point of application at the remote location where the effluent is discharged to the receiving waters. Since the prime consideration is the removal of chlorine residual, this removal can be accomplished at the plant site when necessary.

QUESTIONS

Write your answers in a notebook and then compare your answers with those on page 420.

10.8A   Why are the effluents from some treatment plants dechlorinated?

10.8B   List the treatment processes that may be used to dechlorinate a plant effluent.

10.8C   What happens when sulfur dioxide gas comes in contact with moisture?

10.8D   The reaction of sulfur dioxide (SO_2) with chlorine produces sulfuric acid (H_2SO_4) and hydrochloric acid (HCl). Will these reactions cause a drop in the effluent pH?

10.82   Sulfur Dioxide Hazards

10.820   Exposure Responses to Sulfur Dioxide

Sulfur dioxide is extremely hazardous and must be handled with caution. Exercise extreme caution when working with sulfur dioxide, just like you would when handling chlorine. *REMEMBER:* Never work alone on a sulfur dioxide leak, use the "buddy system" and SCBAs. As with chlorine leaks be prepared and request emergency assistance.

Sulfur dioxide has a very strong, pungent odor. When you smell sulfur dioxide, notify your supervisor and get help. If qualified and authorized to do so, locate and repair the leak.

If you inhale sulfur dioxide gas, sulfurous acid will form on the moist mucous membranes in your body and cause severe irritation or more serious harm. The greater the exposure, the more serious the damage to your body. Exposure to high doses of sulfur dioxide can cause death due to lack of oxygen, chemical bronchopneumonia with severe bronchiolitis may be fatal several days later. In the event sulfur dioxide is inhaled, remove the victim to fresh air, use CPR (cardiopulmonary resuscitation), if necessary, and contact a doctor. Table 10.5 summarizes the impacts of the various concentrations of sulfur dioxide on the human body.

Sulfur dioxide gas is heavier than air and, therefore, will settle in low areas. Due to its low vapor pressure, the liquid changes quickly to gas when liberated, and this gas also will settle in low areas or confined spaces.

Sodium sulfite causes irritation to eyes, skin, and respiratory system on contact. Ingestion may irritate your gastrointestinal tract. Large doses may cause violent colic and diarrhea. Always wash your hands thoroughly after handling.

Suitable safety equipment, similar to that used in case of chlorine leaks, must be available whenever the potential exists for contact with sulfur dioxide or sodium sulfite. A self-contained air supply, repair kits, and chemical suits must be readily available.

Liquid sulfur dioxide has additional hazards that are associated with any compressed or liquified gas:

1. Containers burst or safety devices activate if the liquid is overpressurized or excessively heated (do not allow containers stored outside to be exposed to direct summer sunlight),

2. Violent chemical reactions result if water is drawn back into the chemical (sulfur dioxide) in the container, and

3. Body tissue freezes when in contact with a liquified gas.

TABLE 10.5   IMPACT OF SULFUR DIOXIDE ON THE HUMAN BODY[a]

Impact	Concentration
Lowest concentration detectable by odor	3-5 ppm
Lowest concentration immediately irritating to throat	8-12 ppm
Lowest concentration immediately irritating to eyes	20 ppm
Lowest concentration causing coughing	20 ppm
Maximum allowable concentration for 8-hr exposure	10 ppm
Maximum allowable concentration for 1-hr exposure	50-100 ppm
Tolerable briefly	150 ppm
Immediately dangerous concentration	400-500 ppm

[a] OSHA 8-hour TWA (Time Weighted Average) is 2 ppm and the 15-minute STEL (Short Term Exposure Limit) is 5 ppm.

10.821   Detection of Leaks

The pungent odor of sulfur dioxide can be detected whenever there is a leak somewhere in the system. The location of even the smallest leak may be readily found by the use of ammonia vapor dispensed from an aspirator or squeeze bottle in the area where a leak is suspected. If the room is full of gas, an aspirator bottle will not work because the entire room could become filled with white fumes. Leaks also are detected by the use of an ammonia swab, prepared by soaking a cloth with ammonia solution. When the ammonia vapor passes the leak, a dense white fume forms. This procedure is exactly the same as that used for detecting chlorine leaks.

NEVER USE SOAPY WATER TO LOOK FOR A LEAK. WATER COMBINES WITH SULFUR DIOXIDE TO FORM SULFURIC ACID WHICH IS VERY CORROSIVE AND WILL MAKE ANY LEAK WORSE.

10.822   What to Do in Case of Leaks

The possibility of having a leak in the system is always present; therefore, be prepared for any emergency. Leaks occurring because of broken lines, broken sight glass, or leaking joints must be handled as rapidly as possible. When these emergencies occur, only authorized employees should attempt to stop a leak. If there is any question regarding the size of the leak, a self-contained breathing apparatus must be worn. When working on any leak, the buddy system must be used. *NEVER WORK ALONE ON LEAKS.* When serious leaks occur, the source of the sulfur dioxide should be shut off by closing container valves before attempting to solve the problem.

If the leak is a minor one, the supply valve near the leak can be turned off and the leak repaired. If the leak is near or at the valve, always turn the supply off at the source, such as a cylinder or tank car.

If the leak occurs in the cylinder, isolate the cylinder from the system, rotate or turn container so gas only escapes and use the chlorine emergency kit that is described in the chlorine sections of this chapter. When the emergency equipment is installed, the container should be returned to the supplier as quickly as possible. The supplier must handle the container. Operators should not attempt to transport a damaged container. Regulations governing the movement of hazardous materials over roadways, waterways, rail, and air are issued by the U.S. Department of Transportation (D.O.T.). Check if in doubt regarding the rules and regulations. Otherwise empty the container as quickly as possible.

Ton tank leaks occur mainly at the valves and can be repaired with emergency kits. Leaky tank car valves (gas, liquid, and safety) are the only tank car leaks that can be repaired.

10.823   Safety With Sulfur Dioxide

All employees who must work around sulfur dioxide should have a regular medical checkup. No one with breathing problems, heart disorders, or similar disabilities should attempt to correct leaks. There may be other physical constraints for individuals who may use supplied air breathing apparatus. Check with your local safety regulatory agency for details.

In order to be considered qualified, operators must meet the following conditions:

1. Trained in the use of emergency equipment and procedures,

2. Good physical and mental health, and

3. Understand that the leak emergency must be stopped by a team consisting of at least two individuals. Container repair kits usually require two persons to properly repair the leak. If repairs are not performed properly and the repair device is not properly secured, pressure from within the container can blow off the repair device or temporary gaskets and the container will continue to leak.

Establish a safety training program to ensure that every operator who works with or may be exposed to sulfur dioxide knows how to handle it safely. Refer to Section 14.8, "How to Develop Safety Training Programs," in Volume II of this manual for complete details on how to set up and run an effective safety program.

A safety program should include the following items:

1. Establish a safety program.

   a. Develop emergency procedures with police and fire departments.

   b. List chemical suppliers' emergency phone numbers.

   c. List doctor's phone numbers.

   d. List fire department's and police department's phone numbers.

   e. List ambulance service's phone number.

f. Establish procedure on how to make an emergency phone call.

   (1) Give the location (address) where the accident happened, and

   (2) Instruct where the accident took place in the plant and/or where the injured party can be located. Dispatch someone to the plant entry gate to direct emergency personnel to the site of the emergency, if possible.

2. Prepare written rules to cover the above points.

3. Participate in periodic training, including hands-on use of safety equipment and use of safety procedures.

   a. Leak repair equipment

   b. Respiratory protective equipment

   c. Use and maintenance of detection equipment

4. Conduct a maintenance program for safety equipment.

10.824 First-Aid Response

Importance of First Aid

Whenever anyone is overcome by sulfur dioxide, remove the person from the contaminated area and wash the affected parts of the body with large amounts of water. If the clothing is affected, the clothes should be removed and washed thoroughly. If the individual is burned, even slightly, get medical attention from a doctor.

YOU ARE NOT A DOCTOR AND SHOULD NOT APPLY ANY MEDICATION.

Keep the individual warm and in a reclining position with head and shoulders slightly elevated. Keep the individual quiet and urge the person to resist coughing if possible.

Asphyxiation

Usually cases of asphyxiation are rare due to the pungent odor of sulfur dioxide or sodium sulfite. If the individual has stopped breathing, remove the person from the contaminated area and start CPR (cardiopulmonary resuscitation) immediately. Have someone call the fire department; fire department personnel are experts and have the necessary equipment to handle this type of emergency. Get the individual to a doctor as soon as possible.

Eyes

If sulfur dioxide or sodium sulfite gets into the eyes, wash eyes immediately with large amounts of water from an eye wash or running water hose. Keep the eyelids open while washing and wash for at least 15 minutes. Do not give medication. Transport to a doctor as soon as possible.

Skin

If sulfur dioxide or sodium sulfite gets on the skin, wash off immediately with large amounts of water. Remove any clothing that has been contaminated. If a burn has occurred, transport the injured person to a doctor for treatment and care.

Ingestion

If sodium sulfite is ingested, drink water, induce vomiting, and seek medical attention immediately.

10.825 Emergency Safety Equipment

Respiratory Protection

All wastewater treatment plants should have a self-contained pressure-demand air breathing apparatus for use with chlorine leaks. These units are equally suitable for use during sulfur dioxide leaks. These emergency breathing units must not be stored in the same room where chlorine and sulfur dioxide are stored, in rooms that these gases are routed through or in chemical feed rooms where leaks may occur. These units and repair kits must be readily accessible and near where needed, but away from an area where a leak may occur.

CANISTER MASKS ARE NOT ADEQUATE FOR SEVERE CHLORINE OR SULFUR DIOXIDE LEAKS. THIS TYPE OF EQUIPMENT SHOULD NOT BE NEAR THE PLANT BECAUSE IT MAY LEAD TO SERIOUS ACCIDENTS DUE TO LACK OF OXYGEN OR EXCESSIVE AMOUNTS OF TOXIC GASEOUS CHEMICALS. The recommended type of breathing apparatus is the unit that has its own air or oxygen supply. There are two types of self-contained breathing apparatus. One contains thirty minutes of compressed air and is the same as those used by the fire departments. Many of these units are adaptable to an air line system equipped with a 150-psi compressed breathable air cylinder that affords approximately four hours of air supply. Pressure-demand units are considered safer than the demand-breathing type. The other uses a canister to manufacture oxygen and is completely self-contained. This device is sometimes called a "rebreather kit."

Container Emergency Kits (designed to temporarily stop gaseous chemical leaks)

The following container emergency kits are available for specified applications.

1. 100-lb and 150-lb Cylinders (Kit A)

   The CHLORINE INSTITUTE EMERGENCY KIT A (Figure 10.17, page 389) contains equipment to stop leaks at the valve, fusible plug, and the tank itself.

2. Ton Tanks (Kit B)

   The CHLORINE INSTITUTE EMERGENCY KIT B contains equipment to stop leaks at the valve, fusible plug, and tank.

3. Tank Cars and Tank Trucks (Kit C)

   The CHLORINE INSTITUTE EMERGENCY KIT C contains the equipment to stop leaks only at the valve and the safety valve. *NOTE:* There are variations in the valve arrangement and clearances on some sulfur dioxide tank cars when compared with chlorine tank cars. Check your repair KIT C to ensure adequate equipment *BEFORE* an emergency arises. Modifications and/or fabrication of equipment may be required.

QUESTIONS

Write your answers in a notebook and then compare your answers with those on page 420.

10.8E   What happens when you inhale sulfur dioxide gas?

10.8F   How can a sulfur dioxide leak be detected?

10.83 Sulfur Dioxide Supply System

10.830 Sulfur Dioxide Containers

Sulfur dioxide containers and handling facilities are the same as for chlorine. Review Figures 10.8, 10.9, 10.10, 10.11, and 10.12 for pictures and drawings of chlorine containers and handling facilities; and Table 10.4 for a description of the purpose of each part. Reread Section 10.4, "Chlorine Handling," for a review of the different types of containers and facilities.

There are two important differences between the use of chlorine and sulfur dioxide:

1. Withdrawal rates of sulfur dioxide from containers are slightly lower than the rates for chlorine, and

2. Valves in sulfur dioxide systems should be made of 316 stainless steel with Teflon seats.

10.831 Supply Piping

The piping system should be the same as that required for chlorine. Sulfur dioxide usually is not corrosive to most metals when dry. In reality, you don't find completely dry sulfur dioxide gas. There is always some trace of moisture. When sulfur dioxide becomes moist, it is very corrosive, and most metals cannot stand the corrosive action. The best material to use from the source of supply to the sulfonators is Schedule 80 seamless carbon steel pipe with 3,000-pound (1,360-kg) forged steel fittings. Avoid the use of bushings by using reducing fittings instead. All unions should be of the ammonia type with lead gasket joints. Never use ground joint unions.

In some installations, plastic material has been used to connect the sulfonators to the source of supply. Plastic material is also ideally suited to carry the sulfur dioxide solution to the point of application.

The piping system carrying sulfur dioxide gas should be heated to room temperature and kept at 80 to 90°F (27 to 32°C) to prevent reliquefaction (gas changing back to liquid again).

10.832 Valves

All material used in the valving system should be approved by The Chlorine Institute. There have been problems using bronze bodies and monel stems and seats. Better results have been obtained using 316 stainless steel with Teflon seats. Plastic valves such as PVC have been used on solution lines with good results.

10.833 Sodium Sulfite Supply and Safety

Sodium sulfite tablets used for dechlorination are available in 45-pound plastic buckets. The tablets must be kept dry in a tightly closed container when not in use. Store tablets in a cool, dry place, away from acids or oxidizers. Flush away small spills of dust or powder with water. In case of a spill, shovel up the material for disposal in accordance with directions on the Material Safety Data Sheet (MSDS). In case of fire, flood with water. Do not reuse empty containers. To maintain shelf life and tablet strength, store at a temperature of 85°F (30°C).

QUESTIONS

Write your answers in a notebook and then compare your answers with those on pages 420 and 421.

10.8G   How can cylinders be kept from falling over or being knocked down?

10.8H   Why should the piping system carrying sulfur dioxide gas be heated to room temperature?

10.84 Sulfonation System[43]

10.840 Evaporator

The evaporator is nothing more than a heating system designed to increase the temperature of the liquid sulfur dioxide to the point where it will become a gas.

Liquid sulfur dioxide is piped from the source containers to the sulfur dioxide evaporator. This evaporator is a tank immersed in a constant-temperature hot water bath. The water in the bath is heated to approximately 180°F (82°C). This heat converts the liquid to gas. Liquid sulfur dioxide enters the evaporator at the bottom and sulfur dioxide gas leaves from the top. Due to the construction of the equipment, the gas and liquid are almost at equilibrium.

10.841 Sulfonator

The sulfonator is very similar to a chlorinator, except the orifice and rotameter are different and the housings and diaphragms are manufactured to handle sulfur dioxide rather than chlorine. Actually a sulfonator is a sulfur dioxide gas-metering

---

[43] Sulfonation systems consist of essentially the same equipment as used in chlorination systems. Review Figures 10.18, 10.19, and 10.20 for drawings of chlorination equipment used in sulfonation systems.

device. To achieve accuracy with safety while in use, a sulfonator should have:

1. Indicating meter (rotameter),

2. Sulfur dioxide metering orifice (V-notch),

3. Manual or automatic feed rate adjuster,

4. Vacuum differential-regulating valve,

5. Pressure-vacuum relief valve, and

6. Injector.

The sulfur dioxide feeding system is found in various sizes. These sizes refer to the maximum amount of chemical that can be fed through the control system in pounds per day. The most common of these are:

SULFUR DIOXIDE FEED SYSTEM SIZES

Size, lbs/day	Size, kg/day
50	23
100	45
200	90
250	115
400	180
2,000	900
8,000	3,600

For any size, there are a variety of sizes of rotameters that can be used. Direct mounted sulfonators on 150-pound cylinders are the most common for small systems.

10.842 Injector

The heart of any system is the injector. This injector is merely an aspirator that creates a vacuum in the low pressure area of the barrel. The vacuum is regulated by an orifice opening and the water flowing through the throat of the aspirator. The vacuum of the injector allows the chemical to flow from the storage supply through the sulfonator, the metering device, and into the injector. At the injector, the sulfur dioxide gas is dissolved in water to form sulfurous acid. This will now be referred to as the "sulfur dioxide solution." This solution flows to the point of application.

An injector system design list includes:

1. Injector water pressure gage,

2. Injector vacuum gage for remote injector installations,

3. Injector vacuum line shutoff valve at the remote injector location,

4. Sulfur dioxide solution pressure gage located immediately downstream of the injector to indicate injector back pressure (this is not required on fixed throat injector systems),

5. Injector water pressure switch for low water pressure alarm, and

6. Injector water flowmeters for multiple sulfonator systems.

10.843 Tablet Dechlorination Unit (Figure 10.22)

Tablets of sodium sulfite are placed in tubes of the dechlorination unit. The capacity of each unit is based on the flow and the sodium sulfite requirements. Models generally are designed to treat twice their rated flow capacity for short-term peak flows.

Dechlorination units are installed at the effluent from the chlorine contact tank or anywhere along the discharge line to the receiving waters. The unit must be installed level and should completely drain at low flow or no flow conditions. The outlet of the tablet feeder should allow treated wastewater to

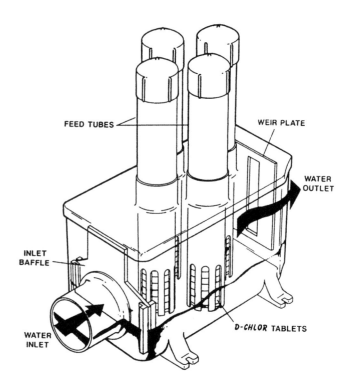

Fig. 10.22   Tablet dechlorination unit
(Permission of ELTECH International Corporation)

free fall so as not to cause flow to back up into the unit. Following dechlorination the effluent can be tested for chlorine residual using the starch-iodide method or the DPD method using standard test kits.

Before the dechlorination unit is put into operation, the system must be adjusted to provide the required sodium sulfite dosage. This is accomplished by selecting a weir of the proper size and determining the number of tubes to be filled with dechlorination tablets in accordance with the manufacturer's recommendations.

When the proper weir plate is in place and the feed tubes are charged and correctly positioned, the system is ready for operation. Admit the entire flow and then operation is continuous and automatic. After the system has been on line for approximately one hour, samples should be taken at the inlet and outlet of the dechlorination system. The inlet sample should be analyzed for residual chlorine and the outlet sample for residual chlorine and residual sulfite.

Take several samples from the inlet and outlet of the system during the first few days of operation. Fine tuning of the system can be done with the objective of minimizing residual sulfite while maintaining 100 percent dechlorination. Minimizing residual sulfite will minimize tablet use and also minimize effects on the dissolved oxygen and pH of the effluent.

If the results of the chlorine or sulfite analyses indicate that the system is not providing sufficient chemical, the necessary adjustments can be made by changing the weir size or tube configuration. For details, please refer to the later paragraphs in this section on troubleshooting.

Dechlorination tablet feeders require very little maintenance. Refilling the feed tubes and an occasional on-stream cleaning of the tablet feeder are the required maintenance. Refilling the tubes is done on a schedule based on plant flow and the weir size.

Occasional cleaning of the tablet feeder to remove accumulated residues may be required. Usually once every 6 to 12 weeks is sufficient. Solids accumulated around the feed tubes are removed by pulling the tubes a few inches off the bottom of the tablet feeder and raising the weir plate to permit the water to rush out the exit end of the tablet feeder. This action, if repeated a few times, will flush out most solids. Fibrous materials may adhere occasionally to the bottom of the feed tubes. Dislodging these with a rod or gloved hand, combined with the shearing force of water flowing through the tablet feeder, is sufficient to remove these materials. Once a year the feed tubes may require removal of internal scale buildup by simply scraping the inside surfaces of the tubes.

Troubleshooting dechlorination tablet feeders consists mainly of determining and correcting the cause of delivery of incorrect amounts of sulfite. Some of the potential problems and their remedies are outlined in the following paragraphs.

Insufficient dose or incomplete removal of chlorine. To correct this problem, try using an additional feed tube or a smaller size weir. Also, if the feed tubes are not touching the bottom of the tablet feeder, an insufficient dose will result. Check the feed tubes for contact with the bottom of the tablet feeder.

If improperly loaded, the tablets can jam, causing the stack to remain suspended above the water level. This causes low dosage levels. Check to be sure that all the tablets are flat in the stack. Should there be a tablet jam, a hard rap on the tube will loosen the bound tablets.

A gross hydraulic overload will cause a lower than desired sulfite dose. Most tablet feeders have been designed to accommodate short-term overload flows of two times the design capacity. This may be exceeded temporarily by abnormal conditions such as heavy rains and nothing should be changed if this is the case. However, if the overload into the plant becomes permanent, the additional flow must be compensated by using a larger weir and/or more feed tubes to maintain the desired sulfite level. The necessary changes can be determined by recalculating the daily flow and finding the new required weir size and tube complement.

Overdose. Sodium sulfite overdose is due to excessive consumption of tablets. Excessive overdose of tablets could result in an undesired reduction of dissolved oxygen or an effect on pH. The overdose can be corrected by removing one or more feed tubes and/or using a larger weir.

1/4-inch and 1/2-inch (0.6- and 1.2-cm) thick PVC spacer disks are also available with the tablet feeders to reduce chemical dose. The spacers are placed in the bottom of the feed tubes prior to loading the tubes with tablets. The spacers raise the entire stack of tablets off the floor of the tablet feeder causing fewer tablets to be contacted by water during normal operation. A lower chemical dose will result.

Plugging. If the flow through the weir is restricted, the water level will rise and cause backup in the tablet feeder. The most common obstacles are leaves and sticks or other solids. Remove these immediately whenever found.

Under most conditions of flow, there is sufficient shearing action by the water to keep the slots in the feed tubes from becoming obstructed. If the slots become obstructed, they may be cleared as described in the above maintenance paragraphs or by removing the feed tubes from the tablet feeder and clearing the obstructing materials.

Acknowledgment

Material in this section was obtained from ELTECH International Corporation's "Dechlorination Systems Instruction Manual." Permission to reproduce this material is appreciated.

QUESTIONS

Write your answers in a notebook and then compare your answers with those on page 421.

10.8I   What is the purpose of the sulfur dioxide evaporator?

10.8J   What happens at the injector of a sulfonation system?

10.85 Sulfonator Controls

10.850 Sulfonator Feed-Rate Control

The control of sulfur dioxide flow to point of application in plant effluent is accomplished very much like controlling chlorine flow. Control of the sulfur dioxide feed rate (dosage) to the plant effluent to remove chlorine residual depends on:

1. Chlorine residual, mg/L,

2. Plant flow rate, MGD, and

3. Amount (if any) of chlorine to be remaining after addition of sulfur dioxide.

10.851 Control Facilities

Most modern installations do not use manual control of equipment for normal operation, but from time to time, equipment failures occur and make the use of manual controls necessary. Switching from automatic to manual control is usually a matter of turning a set screw or thumb screw to release a spring or some similar mechanism (the procedure depends on the type of sulfonator). Operating under this mode or with smaller direct mounted sulfonators requires the operator to change the dose rate manually every time there is a flow change or a chlorine residual change. This is accomplished by turning the control knob to adjust to the proper dose rate of sulfur dioxide.

To determine the manual setting on a sulfonator, follow the steps outlined in the example calculation. Initially a safety factor of 3.0 mg/L of sulfur dioxide more than the chlorine residual is applied. As experience is gained, the 3.0 mg/L excess may be gradually reduced to the level actually needed.

EXAMPLE

A plant with a 2-MGD flow has an effluent chlorine residual of 4.5 mg/L. Sulfur dioxide is being applied at 1.0 mg/L more than the chlorine residual. Determine the sulfonator feed rate in pounds of sulfur dioxide per day.

Known		Unknown
Flow, MGD	= 2 MGD	Sulfonator Feed Rate in pounds SO$_2$/day
Effl Cl Res, mg/L	= 4.5 mg/L	
SO$_2$ Dose, mg/L	= Cl Res, mg/L + 1.0 mg/L	

Determine the sulfonator feed rate.

$$\begin{aligned}\text{Feed Rate, lbs/day} &= (\text{Flow, MGD})(\text{Dose, mg/}L)(8.34 \text{ lbs/gal}) \\ &= (2 \text{ MGD})(4.5 \text{ mg/}L + 1.0 \text{ mg/}L)(8.34 \text{ lbs/gal}) \\ &= 92 \text{ lbs/day or 42 kg/day}\end{aligned}$$

If plant inflow or chlorine residual changes, repeat calculations and readjust sulfonator feed rate.

Refer to Section 10.20, "Chlorinator Control," for a review of the various types of control facilities.

10.852 Selection of Method of Control

The method used in the operation is usually selected by the design engineer on the basis of a combination of cost and desired effect. A plant that is under strict requirements may use the compound loop system and could also, if funds were short, use a manually paced instrument. The cost must be determined to be reasonable before equipment is selected and installed. If operators are not at the plant 24 hours a day, then one of the automatic control modes should be considered. Even if the operator is present, automatic controls are still the best system.

10.86 Determination of Residual Sulfur Dioxide in Wastewater

Residual sulfur dioxide in plant effluent must be measured to be sure the sulfonator is not overdosing and wasting sulfur

dioxide. A residual of 0.5 mg/L or less is satisfactory. Generally, aquatic life can withstand sulfur dioxide concentrations below 20 mg/L; however, this level in an effluent is obviously wasteful. The method for testing residual sulfur dioxide is similar to that for measuring chlorine residual. This technique involves using the amperometric titration procedure. The method used is the back titration approach. The procedure is as follows:

1. Place a 200-mL sample of wastewater in the titrator.

2. Start the agitator.

3. Add 5 mL of 0.00564N phenylarsene oxide (PAO) solution to the sample and mix.

4. Add 4 mL pH 4.0 buffer solution (or enough to attain a pH of between 3.5 and 4.2) to sample and mix.

5. Adjust microammeter pointer so that it reads about 20 on the scale.

6. Add 0.0282N iodine solution in small increments from a 1-mL pipet.

7. The 5 mL of PAO solution will take all of the one mL of iodine solution. Refill the pipet and add the iodine solution in small increments. The end point will be a deflection to the right of the microammeter and will remain to the right.

8. Note the total amount of iodine solution used. This amount should be greater than 1 mL if there is to be any residual sulfur dioxide. The calculation will be:

$$\text{Residual Sulfur Dioxide, mg/}L = (\text{mL Iodine Solution} - 1.0)(4.5)$$

This calculation assumes a 200-mL sample, 5 mL of 0.00564N PAO solution, and an iodine solution that is 0.0282N.

The excess iodine measures the sulfur dioxide present in the sample. In the absence of an amperometric titrator, starch and a color change from clear to blue may be substituted for the end point.

QUESTIONS

Write your answers in a notebook and then compare your answers with those on page 421.

10.8K   Control of the sulfur dioxide feed rate (dosage) to the plant effluent to remove chlorine residual depends on what factors?

10.8L   A treatment plant with a 1.5-MGD flow has an effluent chlorine residual of 3.5 mg/L. Apply sulfur dioxide at a dose rate of 1.0 mg/L more than the chlorine residual. Determine the sulfonator feed rate in pounds of sulfur dioxide per day.

10.8M   Why is the residual sulfur dioxide measured in the plant effluent?

10.87 Operation of Sulfonation Process

10.870 Start-Up of a New System

Before any new system is started, the operators should study the piping system so they know where the shutoff valves are located. A safety program should be discussed, designed, and implemented.

The system should be cleaned, dried, and tested for leaks. Pipelines can be cleaned and dried by flushing with a cleaning solvent, steaming the line with super hot (dry) steam from the high (elevation) side of the system, and allowing the condensate and foreign material to drain out. Heat the entire line and blow dry air from one end of the line to the other. Purge line with nitrogen gas. Test the dry air for any moisture by running *DEW POINT TEST* [44] using padding air compressor air supply.

After drying the system for the first time, test the system for leaks. Pressurize the system with air to 150 psi (10.5 kg/sq cm), making sure the air is dry. Maintain the pressure for 24 hours. Some drop in pressure may occur as a result of hot compressor air cooling in the system. If there is a pressure drop due to a leak, inspect for leaks at the joints and valves by using soapy water. After this procedure indicates that the system has passed the test, small amounts of sulfur dioxide may be used. When this is done for the first time, be certain that appropriate respiratory protective devices are handy. At the first sign of odor, close the supply system. Recheck the system for leaks with ammonia. Remember: A sulfonator should be used during this check-out; otherwise there will be no place for the sulfur dioxide to go. If a leak develops, the sulfonator should be set at the highest possible feed rate in order to drain (empty) the system of sulfur dioxide so the leak may be repaired.

10.871 Start-Up of Gas Sulfonators

Start-up procedures for sulfonators using sulfur dioxide gas from containers are outlined in this section.

1. Be sure sulfur dioxide gas valve at the sulfonator is closed. This valve should have been closed already since the sulfonator is out of service.

2. All sulfur dioxide valves on the supply line should have been closed during shutdown. Be sure they are still closed. If any valves are required to be open for any reason, this exception should be indicated by a tag on the valve.

3. Inspect all tubing, manifold, and valve connections for potential leaks and be sure all joints are properly gasketed.

4. Check sulfur dioxide solution distribution lines to be sure that system is properly valved to deliver sulfur dioxide solution to desired point of application.

5. Open sulfur dioxide metering orifice slightly by adjusting sulfur dioxide feed rate control.

6. Start the injector water supply system. Usually the source of water is plant effluent (with a minimum of suspended solids) or potable water (after an air-gap system) supply. The supply water is pumped at an appropriate flow rate and pressure through the injector which creates sufficient vacuum in the injector to draw sulfur dioxide gas. Sulfur dioxide is absorbed and mixed in the water at the injector. The sulfur dioxide solution then is conveyed to the point of application in the plant effluent.

7. Examine injector water supply system.

   a. Note reading of injector supply pressure gage. If reading is abnormal (different from usual reading), try to identify cause and correct.

   b. Note reading of injector vacuum gage. If the vacuum reading is less than normal, the machine may function at a lower feed rate but will be unable to deliver at rated capacity.

8. Inspect sulfonator vacuum lines for leaks.

9. Crack open the sulfur dioxide container valve and allow gas to enter the line. Inspect all joints for leaks by placing an ammonia-soaked rag near each joint. A polyethylene "squeeze bottle" filled with ammonia water to dispense ammonia vapor may also be used. Care must be taken to avoid spraying ammonia water on or touching the cloth to any metal. A white cloud will reveal the location of a leak. Start with the valve at the sulfur dioxide container, move down the line and check all joints between this valve and the next one downstream. If the downstream valve passes the ammonia test, open the valve and continue to the next valve. If there are no leaks to the sulfonator, continue with the start-up procedure.

10. Inspect the sulfonator.

   a. Sulfur dioxide gas pressure at the sulfonator should be between 20 and 30 psi (1.4 to 2.1 kg/sq cm). Control the gas pressure by adjusting the pressure-reducing valve on the sulfur dioxide supply line.

   b. Operate sulfonator at complete range of feed rates.

   c. Check operation on manual and automatic settings.

   d. Use ammonia water to check all connections on the sulfonator.

   e. If there are no leaks, open the sulfur dioxide container valve one full turn.

11. Sulfonator is ready for use. Log in the time sulfonator was placed into service.

---

[44] *Dew Point Test. Dew point is the temperature to which air with a given quantity of water vapor must be cooled to cause condensation of the vapor in the air. One way to measure the dew point is with a special dew point apparatus. This apparatus consists of a small box. The gas or air being tested enters the box on one side and leaves on the opposite side. One of the other sides has an observation window. A polished cup is inserted firmly in the top.*

Pass a sample of air or gas being tested through the apparatus. Adjust the flow so it can be felt against wetted lips, but not readily felt by the hand. Pour acetone into the cup. Allow the sample to pass through the cup for about five minutes. Add small amounts of crushed dry ice to the acetone. Stir continuously with a thermometer. Carefully add dry ice to the acetone as necessary to slowly lower the temperature. When dew or moisture first appears on the outside polished surface of the cup, read the temperature from the thermometer. This temperature is the DEW POINT. The lower the dew point, the less moisture in the air. The amount of moisture in the air can be determined from a dew point temperature chart or table provided by manufacturers of dew point measuring equipment.

10.872   Troubleshooting the Gas Sulfonator System

Operating Symptom	Probable Cause	Remedy
Injector vacuum reading low	Hydraulic system	Check injector water supply system
	Flow restricted	Adjust injector orifice
	Low pressure	Close throat
	High pressure	Open throat
	Back pressure	Change injector and/or increase water supply to injector
	Vacuum leak in sulfonator	Locate and stop leak. Usually leak is caused by unsealed rotameter tube, leaky gasket, or ruptured diaphragm.
	Low flow of water	Increase pump output
Leaking joints	Missing gasket/loose connection	Repair joint
Sulfonator will not reach maximum output	Faulty injector (no vacuum)	Repair injector system
	Restriction in supply (no SO_2)	Find restriction in supply system
	Faulty sulfonator	Check for vacuum leaks
	Leaks	Repair leaks
	Wrong orifice	Install proper orifice
	V-notch variable orifice out of adjustment or disengaged from manual/auto feed rate adjuster	Adjust or reconnect orifice
Sulfonator will feed OK at maximum output, but will not control at low rates	Vacuum-regulating valve	Repair diaphragm
	If equipped with SPRV (Sulfonator Pressure-Reducing Valve)	Check valve capsule
		Clean SPRV cartridge, SPRV diaphragm, and SPRV gaskets
Sulfonator does not feed	Supply	Renew SO_2 supply
	Piping	Open valve
		Clean filter
Variable vacuum control, formerly working well, now will not go below 30% feed. Signal OK	SPRV	Clean SPRV
Variable vacuum control reaches full feed, but will not go below 50% feed. SPRV OK	Signal vacuum too high	Repair hole in diaphragm
		Clean dirty filter disks
		Clean converter nozzle
Variable vacuum control won't go to full feed. Gas pressure OK. SPRV OK	Plugged restrictor	Clean restrictor
	Air leak in signal	Repair air leak
Freezing of rotameter	Rate too high	Lower rate
	Restriction in rotameter orifice	Clean piping

10.873   *Start-Up of Liquid Sulfonators*

Start-up procedures for sulfonators using liquid sulfur dioxide from containers are outlined in this section.

1. Inspect all joints, valves, manifolds, and tubing connections in sulfonation system, including application lines, for proper fit and for leaks. Make sure that all joints have gaskets.

2. If sulfonation system has been broken open or exposed to the atmosphere, verify that the system is dry. Usually once a system has been dried out, it is never opened again to the atmosphere. However, if moisture enters the system in the air or by any other means, it readily mixes with sulfur dioxide and forms sulfuric acid which will corrode the pipes, valves, joints, and fittings.

   To verify that the system is dry, determine the dew point (must be lower than −40°F or −40°C). If not dry, turn the evaporator on, pass dry air through the evaporator and force this air through the system. If this step is omitted and moisture is in the system, serious corrosion damage can result and the entire system may have to be repaired.

3. Start up the evaporators. Fill the water bath and adjust the device according to the manufacturer's directions.

4. Turn on evaporator heaters. Wait until the temperature of the evaporator reaches 180°F (82°C) before proceeding to next step.

5. Inspect and close all valves on the sulfur dioxide supply line.

6. Open the sulfur dioxide metering orifice slightly. This is to prevent damage to the rotameter.

7. Start the injector water supply system.

8. Examine injector water supply system.

   a. Note reading of injector water supply pressure gage. If reading is abnormal (different from usual reading), try to identify cause and correct. Injectors should operate at greater than 50 psi (3.5 kg/sq cm) inlet pressure.

   b. Note reading of injector vacuum gage. If the vacuum reading is less than normal, the machine may function at a lower feed rate but will be unable to deliver at rated capacity.

9. Inspect sulfonator vacuum lines for leaks.

10. Close all valves on the supply line.

11. Crack open the *GAS LINE* at the sulfur dioxide container. All liquid sulfur dioxide systems should be checked out using gas because of the danger of leaks and also gas is less dangerous. Inspect the joints between this valve and the next one downstream. If this valve passes the ammonia leak test, continue to the next valve down the line. Follow this procedure until the evaporator is reached. Before allowing sulfur dioxide to enter the evaporator and the sulfonator, make sure that all valves between the evaporator and the sulfonator are open. Heat in the evaporator will expand the gas and, if the system is closed, excessive pressure can develop. Sulfur dioxide should never be trapped in a line, evaporator, or sulfonator, because heat could expand the gas to the point where pressure levels are dangerous.

12. If no problems develop, the supply line to the evaporator can be put in service by opening the valve 1 ½ to 2 turns.

13. Check the operation of the sulfonator.

    a. Operate over complete range of sulfur dioxide feed rates.

    b. Check operation on manual and automatic settings.

14. If the sulfonator is operating properly, close the gas line from the sulfur dioxide container and slowly open the liquid sulfur dioxide control valve.

15. After admitting liquid sulfur dioxide to the system, wait until the temperature of the evaporator again reaches 180°F (82°C). Inspect the evaporator.

16. The system is ready for normal operation.

10.874 Troubleshooting the Liquid Sulfonator System

Operating Symptom	Probable Cause	Remedy
Loss of SO_2 pressure at the sulfonator	Plugged SO_2 filter on supply line	Switch to another sulfonator and clean filter
	Gas pressure-reducing valve (PRV) between evaporator and sulfonator closed/malfunctioning.	Adjust, repair, or replace PRV
	SO_2 supply out	Change SO_2 supply
	Supply valve closed	Check valve and system
Liquid through rotameter	Rate too high	Lower rate or use another sulfonator
	Reliquefaction	Piping exposed to cold air; heat piping or room
	Water bath temperature off	Adjust temperature setting
	Defective evaporator	Put another evaporator on line. Repair evaporator
Loss of vacuum	Plugged diffuser, high back pressure	Repair diffuser
	Injector	Repair injector
	Loss of water pressure	Check booster pump
		Check water supply
High vacuum	Supply of SO_2	Check supply
		Check for closed valves
		Lower feed rates
Misting	Defective evaporator	Check evaporator
		Put another evaporator on line
Low temperature alarm on evaporator	Water bath temperature off	Set too low for rate being passed
		Defective water bath
		Switch to gas feed from supply
		Water solenoid valve stuck open, repair/replace

QUESTIONS

Write your answers in a notebook and then compare your answers with those on page 421.

10.8N   What is the probable cause of joints leaking sulfur dioxide gas?

10.8O   Why should sulfur dioxide never be trapped in a line between the evaporator and the sulfonator?

10.875   Normal and Abnormal Operation

The following gages should be checked routinely for proper operation.

Evaporator

1. Water level: Indicates the level of water in the water bath. The level should be in the center of the sight glass.

2. Water temperature: Normal operating range is 180 to 195°F (82 to 91°C).

3. Gas temperature: Normal operating range is 90 to 105°F (32 to 41°C). As impurities are deposited on the evaporator wall, the gas will show a drop in temperature. With experience, the operator should be able to determine when the evaporator needs cleaning.

4. Gas pressure: Should read the same pressure as the supply cylinder or tank.

5. Cathodic protection: The purpose of cathodic protection is to protect the water bath from corrosion damage. The meter should normally read in the 50 to 200 range.

Sulfonator

1. Rotameter: The rotameter indicates the dosage rate of sulfur dioxide and can be set manually or automatically; it should read the same as the established feed rate.

2. Gas pressure: This pressure is normally 20 to 30 psi (1.4 to 2.1 kg/sq cm).

3. Injector suction: Check to see if the unit fluctuates a lot. If it does, the injector water supply should be inspected for the cause.

Procedures and equipment for operating and maintaining chlorination and sulfonation systems are very similar. However, you also should be aware of the differences.

1. Sulfonator control valve diaphragms are made from different material to handle sulfur dioxide, but they may be used for chlorine also. The reverse is not true. Chlorinator control valve diaphragms cannot be used for sulfur dioxide.

2. Chlorinators used as sulfonators cannot deliver the full rated capacity of sulfur dioxide. For example, a chlorinator rated to deliver 2,000 pounds (909 kg) of chlorine per day can only deliver approximately 1,900 pounds (864 kg) of sulfur dioxide per day. A chlorinator rated at 10,000 pounds (4,545 kg) of chlorine per day can deliver only 8,000 pounds (3,636 kg) of sulfur dioxide per day.

3. Sulfur dioxide gas pressures from sulfur dioxide containers are lower than chlorine gas pressures at the same temperature. Sulfur dioxide does not vaporize at the same rate as chlorine at the same temperature. Therefore, sulfur dioxide containers are occasionally padded on the gas side with nitrogen to force liquid sulfur dioxide from a container to the evaporator. Consequently reliquefaction sometimes is a problem in the supply lines between the evaporator and the sulfonator.

For additional information on normal and abnormal operation of a sulfonator, see Section 10.26, "Normal and Abnormal Operation."

10.876   Operational Strategy

For information on the operational strategy for sulfonators, refer to the section on chlorinators, Section 10.28, "Operational Strategy."

10.877   Troubleshooting Sulfonation System

Operating Symptom	Probable Cause	Remedy
Residual chlorine at outfall	Improper mixing	Check for stratification of SO_2
	Broken diffuser	Repair diffuser
	Feed rate too low	Increase feed rate
	Increased plant inflow	Increase SO_2 feed rate
	Chlorine demand increase	Increase SO_2 feed rate or decrease chlorine feed rate if appropriate

10.878   Sulfonation System Shutdown Procedures

Sulfonator system

1. Shut off the sulfur dioxide supply. If the downtime is for a brief period, the supply can be shut off at the valve near the sulfonator. If the downtime is for a day or more, it is better to shut the supply valve at the source. This allows all the sulfur dioxide to be removed from the system.

2. If the equipment is to be dismantled, wait until the sulfur dioxide supply pressure gage reads zero. Then, remove the flexible connection at the source while still running the equipment. Attach the dry air to this connection. This will ensure that all traces of sulfur dioxide are evacuated.

3. After you are sure that all traces of sulfur dioxide are gone, the injector may be turned off. This will secure the installation. The dry air supply also should be turned off at this time.

4. Secure the open end by putting a plug on the flexible connection end. This will prevent moisture from entering the piping.

5. Begin repairs.

Evaporator

If the system is to be shut down for an extended period, it is a good idea to secure the evaporator. In order to do this, shut off the supply to the evaporators while maintaining a vacuum on the unit with a sulfonator. Once the pressure drops to zero the supply line can be disconnected and the dry air line attached to the inlet of the evaporator. This will evacuate the evaporator of residual sulfur dioxide through the sulfonator. After you are sure the sulfur dioxide is gone, isolate the evaporator. Flush the water bath with cold water to remove any foreign material. Never leave any sulfur dioxide trapped in the equipment, especially between valves.

10.88   Maintenance of the Sulfur Dioxide Systems

10.880   Supply Area

1. The area should be kept clean and free of unused objects.

2. All lifting devices such as hand trucks and hoists should be properly maintained. A maintenance program should be established.

3. Ventilation system should be periodically inspected for proper operation. Be sure that the fan is running when the switch is in the "ON" position.

4. A record should be kept of all maintenance and repairs.

10.881   Piping

1. Inspect piping periodically, and, if any discoloration appears, the piping should be replaced and tested. Repair any leaks discovered during inspection.

2. All joints should be tested periodically.

3. All fittings, when taken apart, should be checked for wear. Those that are worn must be changed. Proper gaskets should be available for use. Use new gaskets.

4. Whenever joints are opened, they should be plugged immediately. This should be done to prevent moisture from getting into the system and causing serious damage.

5. The flexible connection should be properly stored and dried before each use. Change the flexible connection periodically and throw away the old one.

6. A record should be kept of all maintenance and repairs.

10.882   Evaporator

1. The evaporator should be cleaned every six months. If the supply of sulfur dioxide is dirty, clean the evaporator more frequently.

2. After the evaporator has been cleaned in place for approximately 10 to 12 cleanings, it should be completely taken apart and cleaned. This should be done every five years on a mandatory basis regardless of the cleaning schedule.

3. The manufacturer's cleaning procedure should be followed.

4. New gaskets should be used.

5. A record of maintenance and repairs should be kept. *NEVER USE OLD, WORN PARTS.*

10.883   Sulfonators

1. Sulfonators should be cleaned every year, or more frequently if necessary.

2. Manufacturer's cleaning procedure should be followed.

3. Never use old, worn parts.

QUESTIONS

Write your answers in a notebook and then compare your answers with those on page 421.

10.8P   List the gages that should be checked routinely for proper operation of an evaporator.

10.8Q   What are the major differences between sulfonation and chlorination procedures and equipment?

10.8R   What areas of the sulfur dioxide system should be included in the maintenance program?

10.9   ARITHMETIC ASSIGNMENT

Turn to the Arithmetic Appendix at the back of this manual and read Sections A.137, Chlorination; A.138, Chemical Doses; and A.139, Blueprint Reading. Check all of the arithmetic in these sections using your electronic pocket calculator. You should be able to get the same answers.

10.10   ADDITIONAL READING

1. *MOP 11*, Chapter 13, "Odor Control," and Chapter 26, "Effluent Disinfection."*

2. *NEW YORK MANUAL*, Chapter 6, "Disinfection."*

3. *TEXAS MANUAL*, Chapter 21, "Chlorination."*

4. *CHLORINE MANUAL*, Sixth Edition. Obtain from the Chlorine Institute, Inc., 1300 Wilson Boulevard, Arlington, VA 22209. Pamphlet 1. Price to members, $15.00; nonmembers, $30.00; plus 10 percent of order total for shipping and handling.

5. *SAFETY AND HEALTH IN WASTEWATER SYSTEMS* (MOP 1). Obtain from Water Environment Federation (WEF), Publications Order Department, 601 Wythe Street, Alexandria, VA 22314-1994. Order No. MO2001. Price to members, $26.74; nonmembers, $36.74; price includes cost of shipping and handling.

* Depends on edition.

6. *CHLORINE SAFE HANDLING BOOKLET* (free to students) and *CHLORINE MANUAL* (available only to PPG customers). Order from PPG Industries, Inc., Chemicals Group, One PPG Place, Pittsburgh, PA 15272.

7. *STANDARD METHODS FOR THE EXAMINATION OF WATER AND WASTEWATER*, Greenberg, A. E., Clesceri, L. S., Eaton, A. D., 20th Edition, 1998. Obtain from Water Environment Federation (WEF), Publications Order Department, 601 Wythe Street, Alexandria, VA 22314-1994. Order No. S82010. Price to members, $164.75; nonmembers, $209.75; price includes cost of shipping and handling. Indicate your member association when ordering.

Videos on chlorine safety also are available from The Chlorine Institute.

10.11 METRIC CALCULATIONS

This section contains the solutions to all problems in this chapter using metric calculations.

10.110 Conversion Factors

lbs/day x 0.454	= kg/day
kg/day x 2.205	= lbs/day
MGD x 3,785	= cu m/day
cu m/day x 0.000264	= MGD
1,000 L	= 1 cu m

10.111 Problem Solutions

1. A chlorinator is set to feed 25 kilograms of chlorine per 24 hours; the wastewater flow is at a rate of 3,200 cu m/day, and the chlorine as measured by the chlorine residual test after 30 minutes of contact time is 0.5 mg/L. Find the chlorine dosage and chlorine demand in mg/L.

Known	Unknown
Chlorine Feed, kg/day = 25 kg/day	1. Chlorine Dose, mg/L
Flow, cu m/day = 3,200 cu m/day	2. Chlorine Demand, mg/L
Chlorine Residual, mg/L = 0.5 mg/L	

1. Calculate the chlorine dose in mg/L.

$$\text{Chlorine Feed or Dose, mg/}L = \frac{\text{Chlorine Feed, kg/day} \times 1,000 \text{ gm/kg} \times 1,000 \text{ mg/gm}}{\text{Flow, cu m/day} \times 1,000\ L/\text{cu m}}$$

$$= \frac{25 \text{ kg/day} \times 1,000 \text{ gm/kg} \times 1,000 \text{ mg/gm}}{3,200 \text{ cu m/day} \times 1,000\ L/\text{cu m}}$$

$$= 7.8 \text{ mg/}L$$

2. Determine the chlorine demand in mg/L.

$$\text{Chlorine Demand, mg/}L = \text{Chlorine Dose, mg/}L - \text{Chlorine Residual, mg/}L$$

$$= 7.8 \text{ mg/}L - 0.5 \text{ mg/}L$$

$$= 7.3 \text{ mg/}L$$

3. Calculate the chlorinator setting (kg per 24 hours) to treat a waste with a chlorine demand of 10 mg/L, when a chlorine residual of 1 mg/L is desired. The flow is 2,500 cu m per day.

Known	Unknown
Chlorine Demand, mg/L = 10 mg/L	Chlorinator Setting, kg/24 hr
Chlorine Residual, mg/L = 1 mg/L	
Flow, cu m/day = 2,500 cu m/day	

Calculate the chlorinator setting in kg per 24 hours.

$$\text{Chlorine Dose, mg/}L = \text{Chlorine Demand, mg/}L + \text{Chlorine Residual, mg/}L$$

$$= 10 \text{ mg/}L + 1 \text{ mg/}L$$

$$= 11 \text{ mg/}L$$

$$\text{Chlorinator Setting, kg/day} = \text{Chlorine Dose, mg/}L \times \text{Flow } \frac{\text{cu m}}{\text{day}} \times \frac{1,000\ L}{\text{cu m}} \times \frac{1 \text{ kg}}{1,000,000 \text{ mg}}$$

$$= 11 \frac{\text{mg}}{L} \times 2,500 \frac{\text{cu m}}{\text{day}} \times \frac{1,000\ L}{\text{cu m}} \times \frac{1 \text{ kg}}{1,000,000 \text{ mg}}$$

$$= 27.5 \text{ kg/day}$$

$$= 27.5 \text{ kg/24 hours}$$

4. A plant with a flow of 8,000 cu m per day has an effluent chlorine residual of 4.5 mg/L. Sulfur dioxide is being applied at 1.0 mg/L more than the chlorine residual. Determine the sulfonator feed rate in kilograms of sulfur dioxide per day.

Known	Unknown
Flow, cu m/day = 8,000 cu m/day	Sulfonator Feed Rate in kilograms SO_2 per day
Effl Cl Res, mg/L = 4.5 mg/L	
SO_2 Dose, mg/L = Cl Res, mg/L + 1.0 mg/L	

Determine sulfonator feed rate in kilograms of SO_2 per day.

$$\text{Feed Rate, kg/day} = \text{Flow, } \frac{\text{cu m}}{\text{day}} \times \text{Dose, } \frac{\text{mg}}{L} \times \frac{1,000\ L}{1 \text{ cu m}} \times \frac{1 \text{ kg}}{1,000,000 \text{ mg}}$$

$$= 8,000 \frac{\text{cu m}}{\text{day}} \times \left(4.5 \frac{\text{mg}}{L} + 1.0 \frac{\text{mg}}{L}\right) \times \frac{1,000\ L}{1 \text{ cu m}} \times \frac{1 \text{ kg}}{1,000,000 \text{ mg}}$$

$$= 44 \text{ kg/day}$$

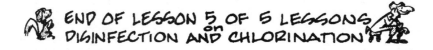

END OF LESSON 5 OF 5 LESSONS on DISINFECTION AND CHLORINATION

Please answer the discussion and review questions next.

DISCUSSION AND REVIEW QUESTIONS

Chapter 10.   DISINFECTION AND CHLORINATION

(Lesson 5 of 5 Lessons)

Write the answers to these questions in your notebook. The question numbering continues from Lesson 4.

28. Where is the typical application point for sulfur dioxide to dechlorinate a plant effluent?

29. What should you do when you smell or discover a sulfur dioxide leak?

30. What are some of the probable causes of a chlorine residual in the outfall of a plant with a dechlorination system?

SUGGESTED ANSWERS

Chapter 10.   DISINFECTION AND CHLORINATION

ANSWERS TO QUESTIONS IN LESSON 1

Answers to questions on page 348.

10.0A   The purpose of disinfection is to destroy pathogenic organisms. This is important to prevent the spread of waterborne diseases.

10.0B   Pathogenic bacteria are destroyed or removed from water by (1) physical removal through sedimentation or filtration, (2) natural die-away in an unfavorable environment, and (3) destruction through chemical treatment.

10.0C   Sterilization of wastes is impractical and unnecessary and may be detrimental to other treatment processes that are dependent on the activity of nonpathogenic saprophytes.

10.0D   Chlorine is used for disinfection because it meets the general requirements of disinfection so well, and because it has been found to be the most economically useful and available chemical for disinfection.

Answers to questions on page 352.

10.0E   Chlorine reacts with organic matter to form chlororganic compounds, and with nitrogenous compounds to form chloramines.

10.0F $\dfrac{\text{Chlorine}}{\text{Dose}} = \dfrac{\text{Chlorine}}{\text{Demand}} + \dfrac{\text{Chlorine}}{\text{Residual}}$

10.0G   To produce an effective disinfecting action, sufficient chlorine must be added after the chlorine demand is satisfied to produce a chlorine residual likely to persist through the contact period.

10.0H $\dfrac{\text{Chlorine}}{\text{Demand}} = \dfrac{\text{Chlorine}}{\text{Dose}} - \dfrac{\text{Chlorine}}{\text{Residual}}$

10.0I   Calculate the chlorine demand of treated domestic wastewater if:

Flow Rate   = 1.2 MGD
Chlorinator = 70 lbs of chlorine per 24 hours
Residual   = 0.4 mg/L after thirty minutes

$$\text{Chlorine Dose, mg/}L = \frac{70 \text{ lbs/day}}{(1.2 \text{ MG/day})(8.34 \text{ lbs/G})} = 7.0 \frac{\text{lbs}}{\text{M lbs}}$$

$$= 7.0 \text{ mg/}L$$

$$\text{Chlorine Demand, mg/}L = \text{Chlorine Dose} - \text{Chlorine Residual}$$

$$= 7.0 \text{ mg/}L - 0.4 \text{ mg/}L$$

$$= 6.6 \text{ mg/}L$$

Answers to questions on page 352.

10.0J   The objective of disinfection is the destruction or inactivation of pathogenic bacteria, and the ultimate measure of the effectiveness is the bacteriological result.

10.0K   The ultimate measure of chlorination effectiveness is the bacteriological result. The residual chlorine that yields satisfactory bacteriological results in a particular plant must be determined and used as a control in that plant.

ANSWERS TO QUESTIONS IN LESSON 2

Answers to questions on page 356.

10.1A   The purpose of up-sewer chlorination is to control odors and septicity, prevent deterioration of structures, and decrease BOD load.

10.1B   Chlorine should be applied in sewers where odor and H_2S control is necessary. These locations may be at several points in the main intercepting sewer or at the upper ends of feeder lines.

10.1C   Prechlorination provides partial disinfection and odor control. Prechlorination may also be used to reduce BOD loads, aid in settling, control foaming in Imhoff units, and to help remove oil.

10.1D   Plant chlorination provides control of odors, corrosion, sludge bulking, digester foaming, filter ponding, filter flies, or sludge thickening. Be careful that chlorination does not interfere with biological treatment processes.

10.1E Postchlorination is used primarily for disinfection.

Answers to questions on page 357.

10.2A Chlorine gas feed can be controlled by manual, start-stop, step-rate, timed-program, flow-proportional, chlorine residual, and compound loop controls.

10.2B "Amperometric titration" is a means of measuring concentrations of certain substances in water, such as the chlorine residual.

10.2C Control of chlorine dosage depends on the bacterial LEVEL or CONCENTRATION desired.

Answers to questions on page 360.

10.2D Feed in pounds per day can be calculated from the chlorine requirement and the rate of flow by the use of a chlorination control nomogram.

10.2E Determine the chlorine feed rate, pounds per 24 hours, if you are treating:

1. A flow of 0.5 MGD and the chlorine required is 1.0 mg/L.
2. Flow = 0.8 MGD and chlorine required = 4.0 mg/L.
3. Flow = 6 MGD and chlorine required = 25 mg/L.

1. 4.2 lbs per 24 hr
2. 27 lbs per 24 hr
3. 1,251 lbs per 24 hr

10.2F Manually controlled feeder rates should be controlled on the following general schedule:
Large plants: at least every hour
Medium-sized plants: every 2 to 4 hours
Small plants: every 8 hours

In small plants chlorine costs are relatively low in comparison to labor costs necessary for frequent adjustment.

10.2G How many pounds of HTH (high test hypochlorite) should be used per day by a hypochlorinator dosing a flow of 0.55 MGD at 12 mg/L of chlorine?

Find chlorine dosage in pounds per day.

$$\text{Chlorine Dosage, lbs/day} = \text{Chlorine Dose, mg/L} \times \text{Flow, MGD} \times 8.34 \text{ lbs/gal}$$

$$= 12 \text{ mg/L} \times 0.55 \text{ MGD} \times 8.34 \text{ lbs/gal}$$

$$= 55 \text{ lbs/day}$$

$$\text{HTH Feed Rate, lbs/day} = \frac{\text{Chlorine Required, lbs/day}}{\text{Portion of Chlorine in pounds of HTH}}$$

$$= \frac{55 \text{ lbs/day}}{0.65 \text{ lb}}$$

$$= 85 \text{ lbs/day}$$

Answers to questions on page 361.

10.2H Chlorine solution discharge lines may be made of polyvinyl chloride (PVC) or black polyethylene flexible tubing. Rubber-lined steel pipe has been used, but rubber hose is rarely used today.

10.2I Little mixing of chlorine solution with wastewater occurs in chlorine contact basins because of the low flow velocities in a basin.

Answers to questions on page 361.

10.2J In a disinfection system ORP is a direct measure of the effectiveness of a chlorine residual in disinfecting a plant effluent.

10.2K When a microorganism loses an electron, it becomes inactivated and can no longer transmit a disease or reproduce.

10.2L Maintenance for the chlorine ORP probe consists of cleaning the unit's sensor once a month. The sulfur dioxide probe's sensor should be cleaned three times a week to ensure a highly responsive system.

Answers to questions on page 365.

10.2M Before attempting to start any chlorination system, read the plant Operation and Maintenance Manual and the manufacturer's literature to become familiar with the equipment. Review the plans or drawings of the facility. Determine what equipment, pipelines, pumps, tanks, and valves are to be placed into service or are in service. The current status of the entire system must be known before starting or stopping any portion of the system.

10.2N Inspect the chlorination system for leaks by placing an ammonia-soaked rag near each joint and valve. A white cloud or vapor will reveal a chlorine leak. A polyethylene "squeeze bottle" filled with ammonia water to dispense ammonia vapor may also be used.

Answers to questions on page 369.

10.2O Normal operation of a chlorinator includes daily inspection of container storage area, evaporators, and chlorinators, including injectors.

10.2P Evaporators are used to convert liquid chlorine to gaseous chlorine for use by gas chlorinators.

10.2Q Abnormal conditions that could be encountered when operating an evaporator include (1) too low a water level, (2) low water temperatures, and (3) no chlorine gas flow to chlorinator.

10.2R The chlorine residual analyzer can be tested by measuring the chlorine residual and comparing this result with actual residual indicated by analyzer.

10.2S Possible chlorinator abnormal conditions include (1) chlorine leaks, (2) chlorine gas pressure too low, (3) injector vacuum too low, and (4) chlorine residual low due to a low feed rate.

Answers to questions on page 370.

10.2T Perform the following steps for the short-term shutdown of a chlorinator:

1. Close chlorine container gas outlet valve,
2. Allow chlorine gas to completely evacuate the system through the injector,
3. Close chlorinator gas discharge valve,
4. Shut down injector if dedicated to the chlorinator, and
5. Disconnect the chlorine container from the system or routinely check the pressure in the chlorinator.

10.2U Two areas that could hinder an operator's ability to meet NPDES permit coliform requirements when the plant is equipped with standby or backup capabilities are (1) running out of chlorine, and (2) failure of automatic control equipment.

ANSWERS TO QUESTIONS IN LESSON 3

Answers to questions on page 379.

10.3A   Chlorine gas is extremely toxic and corrosive in moist atmospheres.

10.3B   A properly fitting self-contained air or oxygen supply type of breathing apparatus, pressure-demand, or rebreather kits are recommended when repairing a chlorine leak.

10.3C   First-aid measures depend on the severity of the contact. Remove the victim from the gas area and keep the victim warm and quiet. Call a doctor and fire department immediately. Keep the patient breathing.

Answers to questions on page 384.

10.4A   Chlorine may be delivered to a plant in 100- or 150-pound cylinders, ton containers, or tank cars.

10.4B   Fusible plugs are provided as safety devices to prevent the building up of excessive pressures and the possibility of rupture due to a fire or high surrounding temperatures.

Answer to question on page 390.

10.4C   To look for a chlorine leak, wear a self-contained pressure-demand breathing apparatus to enter the area if the leak is severe. A rag or paint brush dipped in a strong solution of ammonia water and moved around the area will locate the leak. If the room is not full of chlorine gas, a polyethylene squeeze bottle filled with ammonia water to dispense ammonia vapor may also be used.

ANSWERS TO QUESTIONS IN LESSON 4

Answers to questions on page 396.

10.5A   Chlorine is normally delivered (fed) to the point of application as a solution feed (under vacuum); however, in some cases it is fed as a direct feed (under pressure).

10.5B   Due to the safety hazards of handling sodium chlorite, chlorine dioxide has not been widely used to treat wastewater.

Answers to questions on page 398.

10.5C   Chlorinators should be in a separate room because chlorine gas leaks can damage equipment and are hazardous to personnel.

10.5D   Room temperature is important for proper chlorinator operation to prevent condensing of gas into liquid in the pipelines and chlorinator.

10.5E   Not more than 40 pounds of chlorine per day should be drawn from a chlorine cylinder because of the danger of freezing and reduction in chlorine flow.

10.5F   Chlorinator rooms can be ventilated using forced ventilation with the outlet near the floor because chlorine is heavier than air.

10.5G   Adequate ventilation is important in a chlorinator room to remove any leaking chlorine gas.

10.5H   Chlorination rates can be checked by use of scales and recorders to measure weight loss from the container.

10.5I   Disinfection by chlorination must be continuous for the protection of downstream water users.

10.5J   Continuous chlorination can be achieved by the use of a cylinder feed manifold so cylinders can be removed without interrupting the feed of gas and also through provision of duplicate units or emergency hypochlorinators.

Answers to questions on page 400.

10.5K   The best piping material for conducting chlorine gas or liquid is seamless carbon steel (Schedule 80).

10.5L   Plant *EFFLUENT* is used frequently as the chlorinator injector water supply.

Answers to questions on page 401.

10.6A   Hydrogen sulfide odors can be controlled through chlorination by the reaction with sulfide and the delay of decomposition and stabilization. Sulfide should be controlled because it smells like rotten eggs, is a source of corrosion, can paralyze your respiratory tract, and can form explosive mixtures with air.

10.6B   Sulfuric acid damage to structures can be minimized or eliminated by chlorination or oxygenation which destroys hydrogen sulfide. Hydrogen sulfide produces sulfuric acid which damages sewer systems and structures.

ANSWERS TO QUESTIONS IN LESSON 5

Answers to questions on page 404.

10.8A   The effluents from some treatment plants are dechlorinated to protect fish and other aquatic organisms from toxic chlorine residuals.

10.8B   Dechlorination processes include:

1. Long detention periods,
2. Aeration,
3. Sunlight,
4. Activated carbon, and
5. Chemical reactions, including sulfur dioxide.

10.8C   When sulfur dioxide gas comes in contact with moisture, very corrosive sulfuric acid (H_2SO_4) is formed.

10.8D   No, sulfuric acid (H_2SO_4) and hydrochloric acid (HCl) will not lower the pH of the effluent significantly because of the small amount of acid produced, unless the alkalinity is very low.

Answers to questions on page 407.

10.8E   When you inhale sulfur dioxide gas, sulfurous acid will form on the moist mucous membranes in your body and cause severe irritation or more serious harm. Exposure to high doses of sulfur dioxide can cause death.

10.8F   A sulfur dioxide leak can be detected by soaking a cloth with ammonia solution and holding the cloth near suspected areas. A dense white fume will form near the leak. An aspirator or squeeze bottle may be used if the leak is small.

Answers to questions on page 407.

10.8G   Use a safety chain or clamp to prevent either empty or full cylinders from falling over or being knocked down.

10.8H The piping system carrying sulfur dioxide gas should be heated to room temperature and kept at 80 to 90°F (27 to 32°C) to prevent reliquefaction (gas changing back to liquid again).

Answers to questions on page 409.

10.8I The purpose of the sulfur dioxide evaporator is to heat liquid sulfur dioxide until it is converted to a gas.

10.8J At the injector, the sulfur dioxide gas is dissolved in water to form sulfurous acid. This sulfur dioxide solution flows to the point of application.

Answers to questions on page 410.

10.8K Control of the sulfur dioxide feed rate (dosage) to the plant effluent to remove chlorine residual depends on:
1. Chlorine residual, mg/L,
2. Plant flow rate, MGD, and
3. Amount (if any) of chlorine to be remaining after addition of sulfur dioxide.

10.8L A treatment plant with a 1.5-MGD flow has an effluent chlorine residual of 3.5 mg/L. Apply sulfur dioxide at a dose rate of 1.0 mg/L more than the chlorine residual. Determine the sulfonator feed rate in pounds of sulfur dioxide per day.

Known	**Unknown**
Flow, MGD = 1.5 MGD	Sulfonator Feed Rate in pounds SO$_2$/day
Effl Cl Res, mg/L = 3.5 mg/L	
SO$_2$ Dose, mg/L = Cl Res, mg/L + 1.0 mg/L	

Determine the feed rate.

Feed Rate, lbs/day = (Flow, MGD)(Dose, mg/L)(8.34 lbs/gal)

= (1.5 MGD)(3.5 mg/L + 1.0 mg/L)(8.34 lbs/gal)

= 56 lbs/day or 26 kg/day

10.8M Residual sulfur dioxide is measured in the plant effluent to be sure the sulfonator is not overdosing and wasting sulfur dioxide.

Answers to questions on page 415.

10.8N The probable cause of joints leaking sulfur dioxide gas is a missing gasket or a loose connection.

10.8O Sulfur dioxide should never be trapped in a line between the evaporator and the sulfonator because heat in the evaporator will expand the gas and the pressure could reach dangerous levels.

Answers to questions on page 416.

10.8P Gages that should be checked routinely for proper operation of an evaporator include (1) water level, (2) water temperature, (3) gas temperature, (4) gas pressure, and (5) cathodic protection.

10.8Q Major differences between sulfonation and chlorination procedures and equipment include (1) sulfonator control valve diaphragms are made of different material than chlorinators, (2) chlorinators used as sulfonators cannot deliver full rated capacity of sulfur dioxide, and (3) sulfur dioxide gas pressures from sulfur dioxide containers are lower than chlorine gas pressures at the same temperature.

10.8R Areas that should be included in the maintenance program for the sulfur dioxide system include (1) supply area, (2) piping, (3) evaporator, and (4) sulfonator.

APPENDIX
DISINFECTION USING ULTRAVIOLET (UV) SYSTEMS

Just beyond the visible light spectrum there is a band of electromagnetic radiation which we commonly refer to as ultraviolet (UV) light. When ultraviolet radiation is absorbed by the cells of microorganisms, it damages the genetic material in such a way that the organisms are no longer able to grow or reproduce, thus ultimately killing them. This ability of UV radiation to disinfect water has been understood for almost a century, but technological difficulties and high energy costs prevented widespread use of UV systems for disinfection. Today, however, with growing concern about the safety aspects of handling chlorine and the possible health effects of chlorination by-products, UV disinfection is gaining in popularity. Technological advances are being made and several manufacturers now produce UV disinfection systems for water and wastewater applications. As operating experience with installed systems increases, UV disinfection may become a practical alternative to the use of chlorination at wastewater treatment plants.

TYPES OF UV SYSTEMS

The usual source of the UV radiation for disinfection systems is from low pressure mercury vapor UV lamps that have been made into multi-lamp assemblies, as shown in Figure A-10.1. Each lamp is protected by a quartz sleeve and each has watertight electrical connections. The lamp assemblies are mounted in a rack (or racks) and these racks are immersed in the flowing water. The racks may be mounted either within an enclosed vessel or in an open channel (Figure A-10.2). Most of the UV installations in North America are of the open channel configuration.

When UV lamps are installed in open channels, they are typically placed either horizontal and parallel to the flow (Figure A-10.3) or vertical and perpendicular to the flow. In the horizontal and parallel-to-flow open channel lamp configuration, the lamps are arranged into horizontal modules of evenly spaced lamps. The number of lamps per module establishes

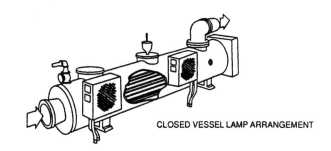

CLOSED VESSEL LAMP ARRANGEMENT

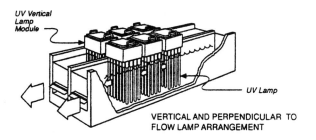

UV Vertical Lamp Module

UV Lamp

VERTICAL AND PERPENDICULAR TO FLOW LAMP ARRANGEMENT

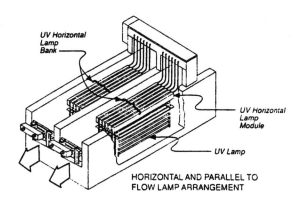

UV Horizontal Lamp Bank

UV Horizontal Lamp Module

UV Lamp

HORIZONTAL AND PARALLEL TO FLOW LAMP ARRANGEMENT

Fig. A-10.2   Typical UV lamp configurations
(Source: "Ultraviolet Disinfection," by CH2M Hill, reproduced with permission of CH2M Hill)

Fig. A-10.1   UV lamp assembly
(Reproduced with permission of Fischer & Porter (Canada) Limited)

the water depth in the channel. For example, 16 lamps could be stacked 3 inches apart to provide disinfection for water flowing through a 48-inch deep open channel.

Each horizontal lamp module has a stainless-steel frame. Each module is fitted with a waterproof wiring connector to the power distribution center. The connectors allow each module to be disconnected and removed from the channel separately for maintenance. The horizontal lamp modules are arranged in a support rack to form a lamp bank that covers the width of the UV channel and there may be several such lamp banks along the channel. The number of UV banks per channel is determined by the required UV dosage to achieve the target effluent quality.

In the vertical and perpendicular-to-flow lamp configuration, rows of lamps are grouped together into vertical modules.

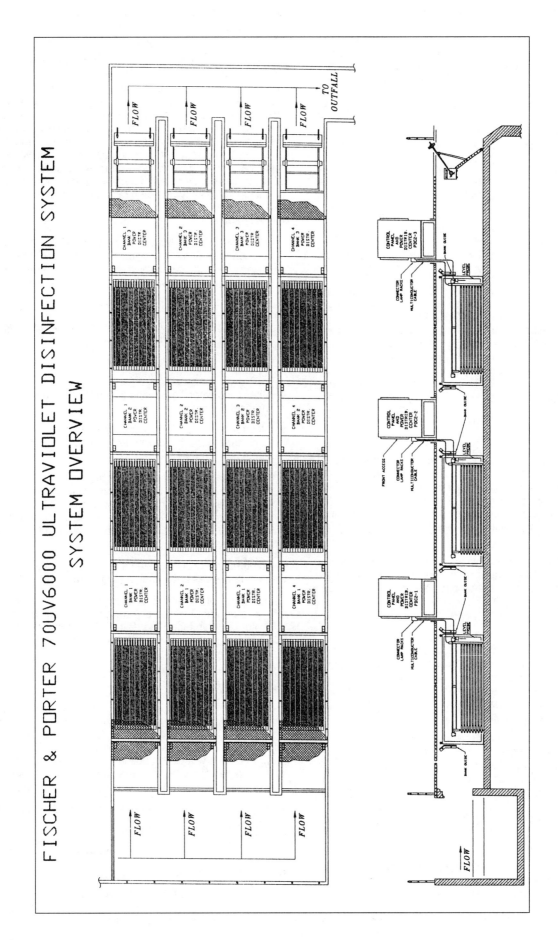

Fig. A-10.3   Horizontal, in-channel UV lamp installation
(Reproduced with permission of Fischer & Porter (Canada) Limited)

Each vertical lamp module has a stainless-steel support frame and can be removed individually from the channel for cleaning or inspection. The electrical wirings for the lamps are located within the frame above the water level. Each individual lamp can be removed from the top of the frame without removing the entire module from the channel. The length of the UV lamps establishes the depth of water in the channel. One or more vertical modules are installed to cover the width of the channel. As with the horizontal arrangement, the number of vertical lamp modules per channel will depend on the UV dosage needed to achieve the desired effluent quality.

When it is necessary to maintain pressure within the wastewater transmission system, UV lamps can be installed in a closed pressure vessel, as shown at the top of Figure A-10.2.

Another type of UV system, the thin film type, uses a chamber with many lamps spaced one-quarter inch (6 mm) apart. This system has been used in the wastewater industry for a 9-MGD (0.35-cu m/sec) secondary treatment plant that has been operating for several years.

Operators may also occasionally encounter a Teflon tube UV disinfection system, although this design is not in common use. Water flows in a thin-walled Teflon tube past a series of UV lamps. UV light penetrates the Teflon tube and is absorbed by the fluid. The advantage to this system is that water never comes in contact with the lamps. However, scale does eventually build up on the pipe walls and must be removed, or the Teflon tube must be replaced. This type of system has generally been replaced by the quartz sleeve systems described earlier.

SAFETY

WARNING: The light from a UV lamp can cause serious burns to your eyes and skin. *ALWAYS* take precautions to protect them. *NEVER* look into the uncovered parts of the UV chamber without proper protective glasses. Do not plug a UV unit into an electrical outlet or switch a unit on without having the UV lamps properly secured in the UV water chamber and the box closed.

UV lamps contain mercury vapor, a hazardous substance that will be released if a lamp is broken. Handle UV lamps with care and be prepared with the proper equipment to clean up any spills.

OPERATION

The operation of ultraviolet disinfection systems requires very little operator attention. To prevent short-circuiting and ensure that all microorganisms receive sufficient exposure to the UV radiation, the water level over the lamps must be maintained at the appropriate level. Water levels in channels can be controlled by weirs or automatic control gates.

Lamp output declines with use so the operator must monitor the output intensity and replace lamps that no longer meet design standards, as well as any lamps that simply burn out. Lamp intensity monitors can be installed to assist the operator in monitoring the level of light output. Lamp failure indicators connected to the main UV control panel will alert the operator when a lamp burns out and requires replacement. In addition, computerized systems are available to monitor and record the age (burn time) of each lamp.

Care must be taken not to exceed the maximum design turbidity levels and flow velocities when using this type of equipment. Suspended particles will shield microorganisms from the UV light and thus protect them from its destructive effects. Flows should be somewhat turbulent to ensure complete exposure of all organisms to the UV light, but flow velocity must be controlled so that the wastewater is exposed to UV radiation long enough for the desired level of disinfection to occur.

Since ultraviolet rays leave no chemical residual like chlorine does, bacteriological tests must be made frequently to ensure that adequate disinfection is being achieved by the ultraviolet system. In addition, the lack of residual disinfectant means that no protection is provided against recontamination after the treated water has left the disinfection facility. When the treated water is exposed to visible light, the microorganisms can be reactivated. Microorganisms that have not been killed have the ability to heal themselves when exposed to sunlight. The solution to this problem is to design UV systems with a high efficiency for killing microorganisms.

MAINTENANCE

A UV system is capable of continuous use if the simple maintenance routine is performed at regular intervals. By checking the following items regularly, the operator of a UV system can determine when maintenance is needed.

1. Check UV monitor for significant reduction in lamp output,

2. Monitor process for major changes in normal flow conditions such as incoming water quality,

3. Check for fouling of quartz sleeves and UV intensity monitor probes,

4. Check indicator light display to ensure that all of the UV lamps are energized,

5. Monitor elapsed time meter, microbiological results and lamp log sheet to determine when UV lamps require replacement, and

6. Check quartz sleeves for discoloration. This effect of UV radiation on the quartz is called solarization. Excessive solarization is an indication that a sleeve is close to the end of its useful service life. Solarization reduces the ability of the sleeves to transmit the necessary amount of UV radiation to the process.

Maintenance on UV systems requires two tasks: cleaning the quartz sleeves and changing the lamps.

Algae and other attached biological growths may form on the walls and floor of the UV channel. This slime can slough off, potentially hindering the disinfection process. If this condition occurs, the UV channel should be dewatered and hosed out to remove accumulated algae and slimes.

QUARTZ SLEEVE FOULING

Fouling of the quartz sleeves occurs when cations such as calcium, iron, or aluminum ions attach to protein and colloidal matter that crystallizes on the quartz sleeves. As this coating builds up on the sleeves, the intensity of the UV light decreases to the point where the buildup has to be removed for the system to remain effective. The rate at which fouling of the quartz sleeves occurs depends on several factors including:

1. Types of treatment processes prior to UV disinfection,

2. Quality of water being treated,

3. Chemicals used in the treatment processes,

4. Length of time that the lamps are submerged, and

5. Velocity of the water through the UV system. Very low or stagnant flows are especially likely to permit the settling of solids and the resulting fouling problems.

SLEEVE CLEANING

How often quartz sleeves need to be cleaned will depend on the quality of the wastewater being treated and the treatment chemicals used prior to disinfection. Dipping the UV modules for five minutes in a suitable cleaning solution will completely remove scale that has deposited on the quartz sleeves. Cleaning is best done using an inorganic acid solution with a pH between 2 and 3. The two most suitable cleaning solutions are nitric acid in strengths to approximately 50 percent concentration and a 5 percent or 10 percent solution of phosphoric acid. To clean the system while still continuing to disinfect normal flows, single modules can be removed from the channel, cleaned, and then reinstalled. The other modules remaining on line while one is being cleaned should still be able to provide for continuous disinfection.

In-channel cleaning of UV lamps is another option, but it has some disadvantages. A back-up channel is required and a much greater volume of acid solution is needed. Also, additional equipment and storage tanks for chemicals are required. Precautions must be taken to prevent damage to concrete channels from the acid cleaning solution. Epoxy coatings normally used to protect concrete from acid attack are not used in UV disinfection systems because the epoxy tends to break down under high UV-light intensities.

The type and complexity of the cleaning system will depend on the size of the system and the required frequency of cleaning. Table A-10.1 offers some guidelines.

TABLE A-10.1 RECOMMENDED UV LAMP CLEANING METHODS

Peak Flow, MGD	Location of Lamps	Type of Cleaning	Work Hours per MGD
<5	out of channel	manual wipe	1
5-20	out of channel	immersion in cleaning tank	0.5
>20	out of channel	remove bank of lamps	0.25
>20	in channel	isolate a channel	0.25

LAMP MAINTENANCE

The lamps are the only components that have to be changed on a regular basis. Their service life can be from 7,500 hours to 20,000 hours. This considerable variation can be attributed to three factors:

1. *LEVEL OF SUSPENDED SOLIDS IN THE WATER TO BE DISINFECTED AND THE FECAL COLIFORM LEVEL TO BE ACHIEVED.* Better-quality effluents or less-stringent fecal coliform standards require smaller UV doses. Since lamps lose intensity with age, the smaller the UV dose required, the greater the drop in lamp output that can be tolerated.

2. *FREQUENCY OF ON/OFF CYCLES.* High cycling rates contribute to lamp electrode failure, the most common cause of lamp failure. Limiting the number of ON/OFF cycles to a maximum of 4 per 24 hours can considerably prolong lamp life.

3. *OPERATING TEMPERATURE OF THE LAMP ELECTRODES.* System temperature usually depends on system conditions. Systems with both lamp electrodes operating at the same temperature (both electrodes submerged in the water) operate up to three times longer than systems where the two electrodes operate at different temperatures. This can occur in systems with lamps protruding through a bulkhead where only one electrode is immersed in the water and the other electrode is surrounded by air if the air temperature is routinely higher than the water temperature.

The largest drop in lamp output occurs during the first 7,500 hours. This decrease is between 30 and 40 percent. Thereafter the annual decrease in lamp output (usually 5 to 10 percent) is caused by the decreased volume of gases within the lamps and by a compositional change of the quartz (solarization) which makes it more opaque to UV light.

DISPOSAL OF USED LAMPS

Contact your appropriate regulatory agency to determine the proper way to dispose of used UV lamps. *DO NOT THROW USED LAMPS IN A GARBAGE CAN TO GET RID OF THEM* because of the hazardous mercury in the lamps.

ACKNOWLEDGMENTS

The authors wish to thank CH2M Hill, Trojan Technologies, Inc., London, Ontario, and Fischer & Porter (Canada) Limited for providing information about UV systems and for permitting the use of illustrations and portions of other proprietary materials.

CONGRATULATIONS

You've worked hard and completed a very difficult program.

APPENDIX

OPERATION OF WASTEWATER TREATMENT PLANTS

(VOLUME I)

Final Examination and Suggested Answers

How to Solve Wastewater Treatment Plant Arithmetic Problems

Abbreviations

Wastewater Words

Subject Index

FINAL EXAMINATION

VOLUME I

This final examination was prepared *TO HELP YOU RE-VIEW* the material in Volume I. The questions are divided into four types:

1. True-False,

2. Multiple Choice,

3. Short Answer, and

4. Problems.

To work this examination:

1. Write the answer to each question in your notebook,

2. After you have worked a group of questions (you decide how many), check your answers with the suggested answers at the end of this exam, and

3. If you missed a question and don't understand why, reread the material in the manual.

You may wish to use this examination for review purposes when preparing for civil service and certification examinations.

Since you have already completed this course, you do not have to send your answers to California State University, Sacramento.

True-False

1. A treatment plant operator may work for an industry.

   1. True
   2. False

2. Every operator has a responsibility to ensure that treatment plants are safe places to work and visit.

   1. True
   2. False

3. Floatable solids are easy to measure.

   1. True
   2. False

4. An industry discharging into municipal collection and treatment systems must have an NPDES permit.

   1. True
   2. False

5. Always wash your hands thoroughly before eating or smoking.

   1. True
   2. False

6. Electrical power must always be tagged and locked out before working on equipment.

   1. True
   2. False

7. Barminutors and comminutors are installed for different purposes.

   1. True
   2. False

8. The effluent from primary clarifiers is usually clearer than effluent from secondary clarifiers.

   1. True
   2. False

9. To start a clarifier, fill the tank with water and turn on the sludge collector mechanism.

   1. True
   2. False

10. In a large plant the number of secondary clarifiers on line may change with increases and decreases in seasonal flows or condition of the solids in the secondary system.

    1. True
    2. False

11. Secondary clarifiers allow for liquid-solids separation in secondary biological treatment processes.

    1. True
    2. False

12. Primary clarifier effluent is the same as trickling filter influent in some treatment plants.

    1. True
    2. False

13. Rotating biological contactors treat wastewater by a process similar to the trickling filter process.

    1. True
    2. False

14. Rotating biological contactors usually treat wastewater as a single-stage process.

    1. True
    2. False

15. More food (waste) in the influent to an aeration tank will require a decrease in the oxygen supply to the tank.

    1. True
    2. False

16. Most package plants are of the extended aeration type of activated sludge process.

    1. True
    2. False

17. Oxidation ditches are usually operated in the step-feed mode of the activated sludge process.

    1. True
    2. False

18. In very cold climates the oxidation ditch should be left un-covered to reduce ice problems.

    1. True
    2. False

19. Some ponds have been designed to take advantage of percolation and high evaporation rates so that there is no discharge.

    1. True
    2. False

20. A waste treatment pond is a biological process.

    1. True
    2. False

Multiple Choice (Select all correct answers.)

1. Organic wastes are in the effluents from which of the following industries?

    1. Dairy
    2. Fruit packing
    3. Gravel washing
    4. Paper
    5. Petroleum

2. Solids are commonly classified as

    1. Inorganic and dissolved.
    2. Inorganic and suspended.
    3. Inorganic and organic.
    4. Organic and dissolved.
    5. Organic and suspended.

3. Hydrogen sulfide gas may

    1. Cause odor problems.
    2. Create a toxic atmosphere.
    3. Damage concrete in plant.
    4. Make wastes more difficult to treat.
    5. Produce an explosive and flammable condition.

4. The purpose of primary sedimentation is to remove

    1. Pathogenic bacteria.
    2. Roots, rags, cans, and large debris.
    3. Sand and gravel.
    4. Settleable and floatable materials.
    5. Suspended and dissolved solids.

5. The main operational factors for a barminutor include

    1. Amount of debris in wastewater.
    2. Head loss through the unit.
    3. Location of unit with respect to grit channel.
    4. Number of units in service.
    5. Removal of floatables.

6. When starting or placing a comminutor in service, which of the following items would you perform?

    1. Adjust cutter blades if necessary.
    2. Check appearance and sound of comminutor.
    3. Check for proper positioning of inlet and outlet gates.
    4. Inspect for proper lubrication and oil leaks.
    5. Look for frayed cables.

7. Toxic wastes entering a treatment plant may be detected by observing the

    1. Changes in color of incoming wastewater.
    2. Movements of pointer on the toxic waste recorder.
    3. Recordings of high or low influent pH.
    4. Bulking of sludge in the clarifier.
    5. Smell of odors.

8. Which of the following factors influence the settleability of solids in a clarifier?

    1. Detention time
    2. Flow velocity and/or turbulence
    3. Movement of sludge scrapers
    4. Short-circuiting
    5. Temperature

9. Suspended solids in the effluent from a trickling filter plant may be caused by

    1. Flotation of solids in the primary clarifier.
    2. Heavy sloughing from the filters.
    3. Precipitation of solids in the secondary clarifier.
    4. Shock loading on the trickling filter.
    5. Short-circuiting through the secondary clarifier.

10. When operating a trickling filter, the operator should

    1. Adjust the process to obtain the best possible results for the least cost.
    2. Bubble oxygen up through the filter.
    3. Maintain aerobic conditions in the filter.
    4. Rotate the distributor as fast as possible to better spray settled wastewater over the media.
    5. Use the lowest recirculation rates that will yield good results to conserve power.

11. Which of the following items should be checked before starting a rotating biological contactor process?

    1. Biomass
    2. Clearances
    3. Lubrication
    4. Safety
    5. Tightness

12. What could be the cause of the slime on the media of a rotating biological contactor appearing shaggy with a brown-to-gray color?

    1. BOD overloading
    2. Fluctuating influent flows
    3. High pH values
    4. Proper operation
    5. Toxic substances in influent

13. An operator must be aware of which of the following safety hazards when working around rotating biological contactors?

    1. Infections in cuts or open wounds
    2. Loose electrical connections and bare wires
    3. Slippery surfaces caused by water, oil, or grease
    4. Slow-moving parts or equipment
    5. Waterborne diseases

14. How is excess activated sludge wasted and disposed of from package plants?

    1. Aerated in a holding tank and then disposed of in an approved sanitary landfill.
    2. At a nearby treatment plant.
    3. By incineration.
    4. Removal by a septic tank pumper.
    5. Treated by gravity thickening and anaerobic digestion.

15. Maintenance of equipment in package aeration plants includes

    1. Adjusting aeration equipment.
    2. Changing oil in speed reducer.
    3. Inspecting air lift pump.
    4. Regulating scum skimmer.
    5. Washing tank walls and channels.

16. Which of the following factors could cause a demand for more oxygen (increase in aeration rates) in an aeration tank?

    1. Increase in food (BOD) in aeration tank influent
    2. Increase in inert or inorganic wastes
    3. Increase in microorganisms
    4. Increase in pH
    5. Increase in toxic substances

17. Pond performance depends on

    1. Lack of short-circuiting.
    2. pH.
    3. Sunlight.
    4. Surface area.
    5. Type and quantity of algae.

18. An increase in plant effluent coliform level could be caused by

    1. An increase in effluent BOD.
    2. Low chlorine residual.
    3. Mixing problems.
    4. Short-circuiting in contact chamber.
    5. Solids accumulation in contact chamber.

19. Low sulfonator injector vacuum reading could be caused by

    1. High back pressure.
    2. Low flow of injector water.
    3. Missing gasket.
    4. Restricted injector flow.
    5. Wrong orifice.

20. Chlorine may be applied for hydrogen sulfide control in the

    1. Aeration tank.
    2. Collection lines.
    3. Plant effluent.
    4. Plant headworks.
    5. Trickling filter.

Short Answer

1. Define the following terms:

    a. Aerobic bacteria
    b. Biochemical oxygen demand
    c. Coliform bacteria
    d. Septic

2. Why should digester gas and air not be allowed to mix?

3. How can explosive conditions develop around bar screens and racks?

4. How can an operator regulate the flow velocity through a grit channel?

5. What is the purpose of a cyclone grit separator?

6. Define alkalinity.

7. List five water quality indicators used to measure clarifier efficiencies.

8. What areas should be studied when reviewing plans and specifications for a treatment plant?

9. What items should be checked daily when operating a trickling filter?

10. How would you attempt to avoid or correct a ponding problem on a trickling filter?

11. Define the following terms:

    a. Grab sample, and
    b. Inhibitory substances.

12. How can the development of biological slimes be encouraged when starting a new rotating biological contactor?

13. How is the activated sludge process controlled?

14. What items should the operator control to maintain the proper environment in an oxidation ditch?

15. List five advantages of ponds.

16. Why are weeds objectionable in and around ponds?

17. Why should the inlet to a pond be submerged?

18. Define the term "air gap."

19. How is the chlorine demand determined?

20. Where is chlorine usually applied for disinfection purposes?

Problems

1. Eight cubic feet of screenings were removed by a plant that treats a flow of 2 MGD during a 24-hour period. How many cubic feet of screenings were removed per MG of flow?

2. Estimate the flow velocity in a grit channel if a stick travels 36 feet in 30 seconds.

3. A circular clarifier with a diameter of 50 feet and a depth of 10 feet treats a flow of 2 MGD. Determine:

    a. Detention time, hours, and
    b. Surface loading, GPD/sq ft.

4. A trickling filter 80 feet in diameter and 4 feet deep treats a flow of 2.4 MGD with a BOD of 120 mg/L. Determine:

    a. Hydraulic loading, GPD/sq ft, and
    b. Organic loading, lbs BOD/day/1,000 cu ft.

5. A rotating biological contactor treats a flow of 2 MGD with a soluble BOD of 80 mg/L. The media surface area is 500,000 sq ft. Determine:

    a. Hydraulic loading, GPD/sq ft, and
    b. Organic loading, lbs BOD/day/1,000 sq ft.

6. An oxidation ditch has a volume of 20,000 cubic feet. The inflow is 0.2 MGD with a BOD of 200 mg/L and suspended solids of 210 mg/L. The mixed liquor suspended solids concentration is 3,500 mg/L with a volatile matter content of 70 percent. Determine the BOD loading in pounds per day per 1,000 cubic feet, F/M ratio, sludge age, and detention time.

7. A waste treatment pond treats a flow of 0.1 MGD with a BOD of 180 mg/L. The area of the pond is 3 acres. What is the organic loading on the pond?

8. Calculate the setting on a chlorinator (pounds of chlorine per 24 hours) that treats a flow of 2 MGD. The chlorine demand is 9 mg/L and the desired residual is 1 mg/L.

SUGGESTED ANSWERS
FOR FINAL EXAMINATION

VOLUME I

True-False

1. True — Treatment plant operators work for industry, sanitation districts, and cities.

2. True — Everyone has a responsibility to be sure that treatment plants are a safe place to work and visit.

3. False — Floatable solids are very difficult to measure.

4. False — An industry discharging into municipal collection and treatment systems need not obtain an NPDES permit but must meet certain specified pretreatment standards.

5. True — Always wash your hands thoroughly before eating or smoking.

6. True — Electrical power must always be tagged and locked out before working on equipment.

7. False — Barminutors and comminutors are installed for the *SAME* purpose, to shred solids and leave them in wastewater.

8. False — The effluent from secondary clarifiers is usually clearer than effluent from primary clarifiers.

9. False — To start a clarifier make sure everything is OK, including the collector mechanism, and then fill the tank with water.

10. True — In a large plant the number of secondary clarifiers on line should change with changes in flows or the condition of the solids.

11. True — Secondary clarifiers allow for liquid-solids separation in secondary biological treatment processes.

12. True — Primary clarifier effluent and trickling filter influent can be the same in some plants.

13. True — Rotating biological contactors treat wastewater by a process similar to the trickling filter process.

14. False — The rotating biological contactor process is usually divided into four stages.

15. False — More food (waste) in the influent to an aeration tank will require an increase in the oxygen supply to the tank.

16. True — Most package plants are of the extended aeration type of activated sludge process.

17. False — Oxidation ditches are usually operated in the extended aeration mode.

18. False — The oxidation ditch should be covered in very cold climates to reduce ice problems.

19. True — Some ponds have been designed so that there is no discharge.

20. True — A waste treatment pond is a biological process.

Multiple Choice

1. 1, 2, 4, 5 — Gravel washing produces inorganic wastes while the other four industries produce mainly organic wastes.

2. 3 — Solids may be classified as either (1) inorganic and organic, or (2) dissolved and suspended.

3. 1, 2, 3, 4, 5 — Hydrogen sulfide gas is very dangerous and could cause all of the problems listed.

4. 4 — The purpose of primary sedimentation is to remove settleable and floatable materials.

5. 1, 2, 4 — Location of unit and removal of floatables are important, but not main operational factors.

6. 1, 2, 3, 4 — Comminutors do not have cables, they are found on barminutors.

7. 1, 3, 4, 5 — Toxic waste recorders are not available today because of the differences in the many toxic wastes that might enter a plant.

8. 1, 2, 4, 5 — Movement of sludge scrapers does not influence settleability of solids.

9. 2, 4, 5 — Solids in the effluent from a trickling filter plant may be caused by sloughing, shock loads, and short-circuiting.

10. 1, 3, 5 — For good trickling filter operation, maintain aerobic conditions in the filter and adjust the process and recirculation rates to produce good results at the least cost.

11. 2, 3, 4, 5 — Check all items but 1, biomass, which won't develop until after the rotating biological contactor has been in operation.

12. 4 — Proper operation is the cause of the slime appearing shaggy with a brown-to-gray color.

13. 1, 2, 3, 4, 5 — All of the items listed are potential safety hazards around rotating biological contactors.

14. 1, 2, 4 — Incineration and gravity thickening usually are not used to treat waste activated sludge from package plants.

15. 1, 3, 4, 5   All items are part of equipment maintenance in a package aeration plant except that speed reducers are used with rotating biological contactors.

16. 1, 3   An increase in food and an increase in microorganisms will cause a demand for more oxygen.

17. 1, 2, 3, 4, 5   Pond performance depends on all five factors.

18. 1, 2, 3, 4, 5   An increase in plant effluent coliform level could be caused by any or all of the five factors listed.

19. 1, 2, 4   Injector vacuum reading low could be caused by high back pressure, low flow or restricted injector flow.

20. 2, 4   Chlorine may be applied for hydrogen sulfide control in the collection lines and plant headworks.

Short Answer

1. Define the following terms:

   a. Aerobic bacteria. Bacteria which will live and reproduce only in an environment containing oxygen which is available for their respiration (breathing), such as atmospheric oxygen or oxygen dissolved in water.

   b. Biochemical oxygen demand. The rate at which organisms use the oxygen in water or wastewater while stabilizing decomposable organic matter under aerobic conditions.

   c. Coliform bacteria. One type of bacteria. The presence of coliform-group bacteria is an indication of possible pathogenic bacterial contamination.

   d. Septic. A condition produced by anaerobic bacteria. If severe, the sludge produces hydrogen sulfide, turns black, gives off foul odors, contains little or no dissolved oxygen, and the wastewater has a high oxygen demand.

2. Do not allow digester gas and air to mix because this mixture is extremely explosive.

3. Explosive conditions can develop around bar screens and racks because of the possibility of explosive materials and gases from industrial discharges and the accumulation of hydrogen sulfide.

4. Flow velocities in grit channels may be regulated by the number of units in service and by using bricks or cinder blocks to change the cross-sectional shape or area.

5. The purpose of a cyclone grit separator is to separate grit from wastewater and organic material.

6. Alkalinity is the capacity of water or wastewater to neutralize acids.

7. Clarifier efficiencies are measured by the settleable solids, suspended solids, total solids, biochemical oxygen demand, and coliform bacteria tests.

8. When reviewing plans and specifications for a treatment plant, carefully study those areas influencing how the plant will be operated and maintained. Also, look carefully for potential safety hazards.

9. Items that should be checked daily when operating a trickling filter include:

   1. Any indication of ponding,
   2. Filter flies,
   3. Odors,
   4. Plugged orifices,
   5. Roughness or vibration of the distributor arms, and
   6. Leakage past the seal.

10. Ponding problems on trickling filters may be avoided or corrected by

    1. Maintaining the proper organic and hydraulic loadings,
    2. Being sure the media is of proper size, and
    3. Preventing accumulation of fibers or trash in the filter voids.

11. a. Grab Sample. A single sample of water collected at a particular time and place which represents the composition of the water only at that time and place.

    b. Inhibitory Substances. Materials that kill or restrict the ability of organisms to treat wastes.

12. Development of biological slimes can be encouraged by regulating the flow rate and strength of the wastewater applied to nearly constant levels by the use of recirculation if available. Maintaining building temperatures at 65°F (18°C) or higher will help. The best rotating speed is one that will shear off growth at a rate that will provide a constant "hungry and reproductive" film of microorganisms exposed to the wastewater being treated.

13. Control of the activated sludge process consists of maintaining the proper solids (floc mass) concentration in the aerator for the waste (food) inflow by adjusting the waste sludge pumping rate and regulating the oxygen supply to maintain a satisfactory level of dissolved oxygen in the process.

14. To maintain the proper environment in an oxidation ditch, the operator should

    1. Observe the clearness of the effluent and the color and characteristics of the floc in the ditch,
    2. Control oxidation ditch solids by regulating the return sludge rate and waste sludge rate,
    3. Prevent toxic substances from reaching the ditch, and
    4. Maintain the proper level of dissolved oxygen.

15. Five advantages of ponds include:

    1. Expensive equipment not required,
    2. Highly trained operators not essential,
    3. Economical to construct,
    4. Little energy consumed, and
    5. No sludge handling and disposal problems.

16. Weeds are objectionable in and around ponds because they provide a shelter for the breeding of mosquitoes and scum accumulation and also hinder pond circulation.

17. The inlet to a pond should be submerged to mix the influent with pond water to distribute the heat of the influent water as much as possible, to minimize the occurrence of floating material, and to control odors.

18. Air Gap. An open vertical drop, or vertical empty space, between a drinking (potable) water supply and the point of use in a wastewater treatment plant.

19. Chlorine Demand = Chlorine Dose − Chlorine Residual.

20. Chlorine is usually applied after the other treatment processes for disinfection purposes.

Problems

1.

Known	**Unknown**
Screenings, cu ft = 8 cu ft	Screenings, cu ft/MG
Flow, MGD = 2 MGD	
Time, hr = 24 hr	
or = 1 day	

Calculate the screenings removed in cubic feet of screenings per million gallons of flow.

$$\text{Screenings, } \frac{\text{cu ft}}{\text{MG}} = \frac{\text{Screenings, cu ft}}{\text{Flow, MGD} \times \text{Time, day}}$$

$$= \frac{8 \text{ cu ft}}{2 \text{ MG/day} \times 1 \text{ day}}$$

$$= 4 \text{ cu ft/MG}$$

2.

Known	**Unknown**
Distance, ft = 36 ft	Velocity, ft/sec
Time, sec = 30 sec	

Estimate the flow velocity in the grit channel.

$$\text{Velocity, ft/sec} = \frac{\text{Distance Traveled, ft}}{\text{Time, sec}}$$

$$= \frac{36 \text{ ft}}{30 \text{ sec}}$$

$$= 1.2 \text{ ft/sec}$$

3.

Known	**Unknown**
Diameter, ft = 50 ft	a. Detention Time, hr
Depth, ft = 10 ft	b. Surface Loading, GPD/sq ft
Flow, MGD = 2 MGD	

a. Estimate the detention time in hours.

$$\frac{\text{Detention}}{\text{Time, hr}} = \frac{\text{Tank Vol, cu ft} \times 7.5 \text{ gal/cu ft} \times 24 \text{ hr/day}}{\text{Flow, gal/day}}$$

$$= \frac{(0.785)(50 \text{ ft})^2 (10 \text{ ft}) \times 7.5 \text{ gal/cu ft} \times 24 \text{ hr/day}}{2,000,000 \text{ gal/day}}$$

$$= 1.8 \text{ hours}$$

b. Calculate the surface loading in gallons per day per surface square foot.

$$\frac{\text{Surface Loading,}}{\text{GPD/sq ft}} = \frac{\text{Flow, GPD}}{\text{Area, sq ft}}$$

$$= \frac{2,000,000 \text{ GPD}}{(0.785)(50 \text{ ft})^2}$$

$$= 1,020 \text{ GPD/sq ft}$$

4.

Known	**Unknown**
Diameter, ft = 80 ft	a. Hydraulic Loading, GPD/sq ft
Depth, ft = 4 ft	b. Organic Loading, lbs BOD/day/1,000 cu ft
Flow, MGD = 2.4 MGD	
BOD, mg/L = 120 mg/L	

a. Calculate the hydraulic loading in gallons per day per square foot of filter surface area.

$$\frac{\text{Hydraulic Loading,}}{\text{GPD/sq ft}} = \frac{\text{Flow, GPD}}{\text{Surface Area, sq ft}}$$

$$= \frac{2,400,000 \text{ GPD}}{(0.785)(80 \text{ ft})^2}$$

$$= \frac{2,400,000 \text{ GPD}}{5,024 \text{ sq ft}}$$

$$= 478 \text{ GPD/sq ft}$$

b. Estimate the organic loading in pounds of BOD per day per 1,000 cubic feet of media.

$$\frac{\text{Organic Loading,}}{\frac{\text{lbs BOD/day}}{1,000 \text{ cu ft}}} = \frac{\text{BOD, lbs/day}}{\text{Volume of Media (in 1,000 cu ft)}}$$

$$= \frac{(\text{Flow, MGD})(\text{BOD, mg/L})(8.34 \text{ lbs/gal})}{(0.785)(\text{Diameter, ft})^2(\text{Depth, ft})}$$

$$= \frac{(2.4 \text{ MGD})(120 \text{ mg/L})(8.34 \text{ lbs/gal})}{(0.785)(80 \text{ ft})^2(4 \text{ ft})}$$

$$= \frac{2,402 \text{ lbs BOD/day}}{20,096 \text{ cu ft}}$$

$$= \frac{2,402 \text{ lbs BOD/day}}{20.096 (1,000 \text{ cu ft})}$$

$$= 120 \text{ lbs BOD/day/1,000 cu ft}$$

5.

Known	**Unknown**
Flow, MGD = 2 MGD	a. Hydraulic Loading, GPD/sq ft
Soluble BOD, mg/L = 80 mg/L	b. Organic Loading, lbs BOD/day/ 1,000 sq ft
Surface Area, sq ft = 500,000 sq ft	

a. Calculate the hydraulic loading in gallons per day per square foot of media surface area.

$$\frac{\text{Hydraulic Loading,}}{\text{GPD/sq ft}} = \frac{\text{Flow, GPD}}{\text{Surface Area, sq ft}}$$

$$= \frac{2,000,000 \text{ GPD}}{500,000 \text{ sq ft}}$$

$$= 4 \text{ GPD/sq ft}$$

b. Estimate the organic loading in pounds of BOD per day per 1,000 square feet of media surface area.

$$\frac{\text{Organic Loading,}}{\text{lbs BOD/day}} = \frac{\text{BOD, lbs/day}}{\text{Surface Area (in 1,000 sq ft)}}$$

$$= \frac{\text{(Flow, MGD)(BOD, mg/}L\text{)(8.34 lbs/gal)}}{\text{Surface Area (in 1,000 sq ft)}}$$

$$= \frac{\text{(2 MGD)(80 mg/}L\text{)(8.34 lbs/gal)}}{\text{500 (1,000 sq ft)}}$$

$$= 2.7 \text{ lbs BOD/day/1,000 cu ft}$$

6.

Known		**Unknown**
Ditch Volume, cu ft	= 20,000 cu ft	a. BOD Loading, lbs/day/1,000 cu ft
or	= 0.15 MG	b. F/M Ratio
Flow, MGD	= 0.2 MGD	c. Sludge Age, days
BOD, mg/L	= 200 mg/L	d. Detention Time, hr
SS, mg/L	= 210 mg/L	
MLSS, mg/L	= 3,500 mg/L	
VM, %	= 70%	

a. Calculate the BOD loading in pounds of BOD per day per 1,000 cubic feet of ditch.

$$\frac{\text{BOD Loading,}}{\text{lbs BOD/day}} = \frac{\text{BOD, lbs/day}}{\text{Ditch Volume, 1,000 cu ft}}$$

$$= \frac{\text{(Flow, MGD)(BOD, mg/}L\text{)(8.34 lbs/gal)}}{\text{Volume (in 1,000 cu ft)}}$$

$$= \frac{\text{(0.2 MGD)(200 mg/}L\text{)(8.34 lbs/gal)}}{20,000 \text{ cu ft}}$$

$$= \frac{333.6 \text{ lbs BOD/day}}{20.0 \text{ (1,000 cu ft)}}$$

$$= 16.7 \text{ lbs BOD/day/1,000 cu ft}$$

b. Determine the Food/Microorganism Ratio.

$$\frac{F}{M} = \frac{\text{BOD, lbs/day}}{\text{MLVSS, lbs}}$$

$$= \frac{\text{(Flow, MGD)(BOD, mg/}L\text{)(8.34 lbs/gal)}}{\text{(Vol, MG)(MLSS, mg/}L\text{)(VM)(8.34 lbs/gal)}}$$

$$= \frac{\text{(0.2 MGD)(200 mg/}L\text{)(8.34 lbs/gal)}}{\text{(0.15 MG)(3,500 mg/}L\text{)(0.70)(8.34 lbs/gal)}}$$

$$= \frac{333.6 \text{ lbs BOD/day}}{3,065 \text{ lbs MLVSS}}$$

$$= 0.11 \text{ lb BOD/day/lb MLVSS}$$

c. Estimate the sludge age in days.

$$\frac{\text{Sludge Age,}}{\text{days}} = \frac{\text{Solids Under Aeration, lbs}}{\text{Solids Added, lbs/day}}$$

$$= \frac{\text{(Vol, MG)(MLSS, mg/}L\text{)(8.34 lbs/gal)}}{\text{(Flow, MGD)(Infl SS, mg/}L\text{)(8.34 lbs/gal)}}$$

$$= \frac{\text{(0.15 MG)(3,500 mg/}L\text{)(8.34 lbs/gal)}}{\text{(0.2 MGD)(210 mg/}L\text{)(8.34 lbs/gal)}}$$

$$= \frac{4,379 \text{ lbs}}{350 \text{ lbs/day}}$$

$$= 12.5 \text{ days}$$

d. Calculate the detention time in hours.

$$\frac{\text{Detention}}{\text{Time, hours}} = \frac{\text{(Ditch Volume, MG)(24 hr/day)}}{\text{Flow, MGD}}$$

$$= \frac{\text{(0.15 MG)(24 hr/day)}}{0.2 \text{ MGD}}$$

$$= 18 \text{ hours}$$

7.

Known	**Unknown**
Flow, MGD = 0.1 MGD	Organic Loading, lbs BOD/day/ac
BOD, mg/L = 180 mg/L	
Area, acres = 3 acres	

Calculate the organic (BOD) loading on the pond in pounds per day per acre.

$$\frac{\text{Organic Loading,}}{\text{lbs BOD/day/ac}} = \frac{\text{(Flow, MGD)(BOD, mg/}L\text{)(8.34 lbs/gal)}}{\text{Area, acres}}$$

$$= \frac{\text{(0.1 MGD)(180 mg/}L\text{)(8.34 lbs/gal)}}{3 \text{ acres}}$$

$$= \frac{150 \text{ lbs BOD/day}}{3 \text{ acres}}$$

$$= 50 \text{ lbs BOD/day/ac}$$

8.

Known	**Unknown**
Flow, MGD = 2 MGD	Chlorinator Setting, lbs/24 hours
Cl Demand, mg/L = 9 mg/L	
Cl Residual, mg/L = 1 mg/L	

a. Estimate the chlorine dose in mg/L.

$$\text{Cl Dose, mg/}L = \text{Cl Demand, mg/}L + \text{Cl Residual, mg/}L$$

$$= 9 \text{ mg/}L + 1 \text{ mg/}L$$

$$= 10 \text{ mg/}L$$

b. Calculate the chlorinator setting in pounds of chlorine per 24 hours.

$$\text{Setting, lbs/24 hr} = \text{(Flow, MGD)(Dose, mg/}L\text{)(8.34 lbs/gal)}$$

$$= \text{(2 MGD)(10 mg/}L\text{)(8.34 lbs/gal)}$$

$$= 167 \text{ lbs/day}$$

or

$$= 170 \text{ lbs/24 hours}$$

APPENDIX

HOW TO SOLVE WASTEWATER TREATMENT PLANT ARITHMETIC PROBLEMS

(VOLUME I)

TABLE OF CONTENTS

HOW TO SOLVE WASTEWATER TREATMENT PLANT ARITHMETIC PROBLEMS

OBJECTIVES

**HOW TO SOLVE WASTEWATER TREATMENT PLANT
ARITHMETIC PROBLEMS**

Following completion of this Appendix, you should be able to:

1. Add, subtract, multiply, and divide,

2. List from memory basic conversion factors and formulas, and

3. Solve wastewater treatment plant arithmetic problems.

HOW TO SOLVE WASTEWATER TREATMENT PLANT ARITHMETIC PROBLEMS

(VOLUME I)

A.0 HOW TO STUDY THIS APPENDIX

This appendix may be worked early in your training program to help you gain the greatest benefit from your efforts. Whether to start this appendix early or wait until later is your decision. The chapters in this manual were written in a manner requiring very little background in arithmetic. You may wish to concentrate your efforts on the chapters and refer to this appendix when you need help. Some operators prefer to complete this appendix early so they will not have to worry about how to do the arithmetic when they are studying the chapters. You may try to work this appendix early or refer to it while studying the other chapters.

The intent of this appendix is to provide you with a quick review of the addition, subtraction, multiplication, and division needed to work the arithmetic problems in this manual. This appendix is not intended to be a math textbook. There are no fractions because you don't need fractions to work the problems in this manual. Some operators will be able to skip over the review of addition, subtraction, multiplication, and division. Others may need more help in these areas. If you need help in solving problems, read Section A.9, "Steps in Solving Problems." Basic arithmetic textbooks are available at every local library or bookstore and should be referred to if needed. Most instructional or operating manuals for pocket electronic calculators contain sufficient information on how to add, subtract, multiply, and divide.

After you have worked a problem involving your job, you should check your calculations, examine your answer to see if it appears reasonable and, if possible, have another operator check your work before making any decisions or changes.

A.1 BASIC ARITHMETIC

In this section we provide you with basic arithmetic problems involving addition, subtraction, multiplication, and division. You may work the problems "by hand" if you wish, but we recommend you use an electronic pocket calculator. The operating or instructional manual for your calculator should outline the step-by-step procedures to follow. All calculators use similar procedures, but most of them are slightly different from others.

We will start with very basic, simple problems. Try working the problems and then comparing your answers with the given answers. If you can work these problems, you should be able to work the more difficult problems in the text of this training manual by using the same procedures.

A.10 Addition

2	6.2	16.7	6.12	43
3	8.5	38.9	38.39	39
5	14.7	55.6	44.51	34
				38
				39
2.12	0.12	63	120	37
9.80	2.0	32	60	29
11.92	2.12	95	180	259
4	23	16.2	45.98	70
7	79	43.5	28.09	50
2	31	67.8	114.00	40
13	133	127.5	188.07	80
				240

A.11 Subtraction

7	12	25	78	83
− 5	− 3	− 5	− 30	− 69
2	9	20	48	14
61	485	4.3	3.5	123
− 37	− 296	− 0.8	− 0.7	− 109
24	189	3.5	2.8	14
8.6	11.92	27.32	3.574	75.132
− 8.22	− 3.70	− 12.96	− 0.042	− 49.876
0.38	8.22	14.36	3.532	25.256

A.12 Multiplication

(3)(2)*	= 6	(4)(7)	= 28
(10)(5)	= 50	(10)(1.3)	= 13
(2)(22.99)	= 45.98	(6)(19.5)	= 117
(16)(17.1)	= 273.6	(50)(20,000)	= 1,000,000
(40)(2.31)	= 92.4	(80)(0.433)	= 34.64
(40)(20)(6)	= 4,800		
(4,800)(7.48)	= 35,904		
(1.6)(2.3)(8.34)	= 30.6912		
(0.001)(200)(8.34)	= 1.668		
(0.785)(7.48)(60)	= 352.308		
(12,000)(500)(60)(24)	= 8,640,000,000 or 8.64×10^9		
(4)(1,000)(1,000)(454)	= 1,816,000,000 or 1.816×10^9		

NOTE: The term, $\times 10^9$, means that the number is multiplied by 10^9 or 1,000,000,000. Therefore $8.64 \times 10^9 = 8.64 \times 1,000,000,000 = 8,640,000,000$.

---

* (3)(2) is the same as 3 x 2 = 6.

A.13 Division

$$\frac{6}{3} = 2 \qquad\qquad \frac{48}{12} = 4$$

$$\frac{50}{25} = 2 \qquad\qquad \frac{300}{20} = 15$$

$$\frac{20}{7.1} = 2.8 \qquad\qquad \frac{11,400}{188} = 60.6$$

$$\frac{1,000,000}{17.5} = 57,143 \qquad\qquad \frac{861,429}{30,000} = 28.7$$

$$\frac{4,000,000}{74,880} = 53.4 \qquad\qquad \frac{1.67}{8.34} = 0.20$$

$$\frac{80}{2.31} = 34.6 \qquad\qquad \frac{62}{454} = 0.137$$

$$\frac{250}{17.1} = 14.6 \qquad\qquad \frac{4,000,000}{14.6} = 273,973$$

NOTE: When we divide $1/3 = 0.3333$, we get a long row of 3s. Instead of the row of 3s, we "round off" our answer so $1/3 = 0.33$. For a discussion of rounding off numbers, see Section A.95, "Significant Figures."

A.14 Rules for Solving Equations

Most of the arithmetic problems we work in the wastewater treatment field require us to plug numbers into formulas and calculate the answer. There are a few basic rules that apply to solving formulas. These rules are:

1. Work from left to right.

2. Do all the multiplication and division above the line (in the numerator) and below the line (in the denominator); then do the addition and subtraction above and below the line.

3. Perform the division (divide the numerator by the denominator).

Parentheses ( ) are used in formulas to identify separate parts of a problem. A fourth rule tells us how to handle numbers within parentheses.

4. Work the arithmetic within the parentheses before working outside the parentheses. Use the same order stated in rules 1, 2, and 3: work left to right, above and below the line, then divide the top number by the bottom number.

Let's look at an example problem to see how these rules apply. This year one of the responsibilities of the operators at our plant is to paint both sides of the wooden fence across the front of the facility. The fence is 145 feet long and 9 feet high. The steel access gate, which does not need painting, measures 14 feet wide by 9 feet high. Each gallon of paint will cover 150 square feet of surface area. How many gallons of paint should be purchased?

STEP 1: Identify the correct formula.

$$\text{Paint Req, gal} = \frac{\text{Total Area, sq ft}}{\text{Coverage, sq ft/gal}}$$

or

$$\text{Paint Req, gal} = \frac{(\text{Fence L, ft} \times \text{H, ft} \times \text{No. Sides}) - (\text{Gate L, ft} \times \text{H, ft} \times \text{No. Sides})}{\text{Coverage, sq ft/gal}}$$

STEP 2: Plug numbers into the formula.

$$\text{Paint Req, gal} = \frac{(145 \text{ ft} \times 9 \text{ ft} \times 2) - (14 \text{ ft} \times 9 \text{ ft} \times 2)}{150 \text{ sq ft/gal}}$$

STEP 3: Work the multiplication within parentheses.

$$\text{Paint Req, gal} = \frac{(2,610 \text{ sq ft}) - (252 \text{ sq ft})}{150 \text{ sq ft/gal}}$$

STEP 4: Work the subtraction above the line.

$$\text{Paint Req, gal} = \frac{2,358 \text{ sq ft}}{150 \text{ sq ft/gal}}$$

STEP 5: Divide the numerator by the denominator.

Paint Req, gal = 15.72 gal
or 16 gallons of paint will be needed

Instructions for your electronic calculator can provide you with the detailed procedures for working the practice problems below.

$$\frac{(3)(4)}{2} = 6 \qquad\qquad \frac{64}{(8)(4)} = 2$$

$$\frac{(2 + 3)(4)}{5} = 4 \qquad\qquad \frac{54}{(4 + 2)(3)} = 3$$

$$\frac{(7 - 2)(8)}{4} = 10 \qquad\qquad \frac{48}{(8 - 3)(4)} = 2.4$$

$$\frac{(15,000)(7.48)(24)}{(1.4)(1,000,000)} = 1.9$$

$$\frac{(225 - 25)(100)}{225} = 88.9$$

$$\frac{12}{(0.432)(8.34)} = 3.3$$

$$\frac{(1,800)(0.5)(8.34)}{(110)(2.0)(8.34)} = 4.1$$

A.15 Actual Problems

Let's look at the last four problems in the previous Section A.14, "Rules for Solving Equations," as they might be encountered by an operator.

1. A rectangular sedimentation basin treats a flow of 1.4 MGD. The volume is 15,000 cubic feet. Estimate the detention time.

Known	Unknown
Basin Volume, cu ft = 15,000 cu ft	Detention Time, hours
Flow, MGD = 1.4 MGD	

Calculate the detention time in hours.

$$\text{Detention Time, hr} = \frac{(\text{Basin Volume, cu ft})(7.48 \text{ gal/cu ft})(24 \text{ hr/day})}{(\text{Flow, MGD})(1,000,000/M)}$$

$$= \frac{(15,000 \text{ cu ft})(7.48 \text{ gal/cu ft})(24 \text{ hr/day})}{(1.4 \text{ MGD})(1,000,000/M)}$$

$$= \frac{(15,000)(7.48)(24)}{(1.4)(1,000,000)}$$

$$= 1.9 \text{ hours}$$

2. The influent BOD to an activated sludge plant is 225 mg/L and the effluent BOD is 25 mg/L. What is the BOD removal efficiency of the plant?

Known	Unknown
Influent BOD, mg/L = 225 mg/L	Plant Efficiency, %
Effluent BOD, mg/L = 25 mg/L	

Calculate the efficiency of the plant in removing BOD.

$$\text{Plant BOD Efficiency, \%} = (\frac{\text{In} - \text{Out}}{\text{In}})(100\%)$$

$$= (\frac{225 \text{ mg}/L - 25 \text{ mg}/L}{225 \text{ mg}/L})(100\%)$$

$$= (\frac{225 - 25}{225})(100\%)$$

$$= 88.9\%$$

3. A chlorinator is set to feed twelve pounds of chlorine per day to a flow of 300 gallons per minute (0.432 million gallons per day). What is the chlorine dose in milligrams per liter?

Known	Unknown
Chlorinator Feed, lbs/day = 12 lbs/day	Chlorine Dose, mg/L
Flow, MGD = 0.432 MGD	

Determine the chlorine dose in milligrams per liter.

$$\text{Chlorine Dose, mg}/L = \frac{\text{Chlorinator Feed Rate, lbs/day}}{(\text{Flow, MGD})(8.34 \text{ lbs/gal})}$$

$$= \frac{12 \text{ lbs/day}}{(0.432 \text{ MGD})(8.34 \text{ lbs/gal})}$$

$$= \frac{12}{(0.432)(8.34)}$$

$$= 3.3 \text{ mg}/L$$

4. Estimate the sludge age for an activated sludge plant with a mixed liquor suspended solids (MLSS) of 1,800 mg/L and an aeration tank volume of 0.50 MG. The flow is 2.0 MGD and primary effluent suspended solids are 110 mg/L.

Known	Unknown
MLSS, mg/L = 1,800 mg/L	Sludge Age, days
P.E. SS, mg/L = 110 mg/L	
Tank Vol, MG = 0.50 MG	
Flow, MGD = 2.0 MGD	

Calculate the sludge age for the aeration tank in days.

$$\text{Sludge Age, days} = \frac{(\text{MLSS, mg}/L)(\text{Tank Vol, MG})(8.34 \text{ lbs/gal})}{(\text{P.E. SS, mg}/L)(\text{Flow, MGD})(8.34 \text{ lbs/gal})}$$

$$= \frac{(1,800 \text{ mg}/L)(0.50 \text{ MG})(8.34 \text{ lbs/gal})}{(110 \text{ mg}/L)(2.0 \text{ MGD})(8.34 \text{ lbs/gal})}$$

$$= \frac{(1,800)(0.50)(8.34)}{(110)(2.0)(8.34)}$$

$$= 4.1 \text{ days}$$

A.16   Percentage

Expressing a number in percentage is just another, and sometimes simpler, way of writing a fraction or a decimal. It can be thought of as parts per 100 parts, since the percentage is the numerator of a fraction whose denominator is always 100. Twenty-five parts per 100 parts is more easily recognized as 25/100 or 0.25. However, it is also 25%. In this case, the symbol % takes the place of the 100 in the fraction and the decimal point in the decimal fraction.

For the above example it can be seen that changing from a fraction or a decimal to percent is not a difficult procedure.

1. To change a fraction to percent, multiply by 100%.

EXAMPLE: $\frac{2}{5} \times 100\% = \frac{200\%}{5} = 40\%$

EXAMPLE: $\frac{5}{4} \times 100\% = \frac{500\%}{4} = 125\%$

2. To change percent to a fraction, divide by 100%.

EXAMPLE: $15\% \div 100\% = 15\% \times \frac{1}{100\%} = \frac{15}{100} = \frac{3}{20}$

EXAMPLE: $0.4\% \div 100\% = 0.4\% \times \frac{1}{100\%} = \frac{0.4}{100} = \frac{4}{1,000} = \frac{1}{250}$

In these examples note that the two percent signs cancel each other.

Following is a table comparing common fractions, decimal fractions, and percent to indicate their relationship to each other:

Common Fraction	Decimal Fraction	Percent
$\frac{285}{100}$	2.85	285%
$\frac{100}{100}$	1.0	100%
$\frac{20}{100}$	0.20	20%
$\frac{1}{100}$	0.01	1%
$\frac{1}{1,000}$	0.001	0.1%
$\frac{1}{1,000,000}$	0.000001	0.0001%

A.17   Sample Problems Involving Percent

Problems involving percent are usually not complicated since their solution consists of only one or two steps. The principal error made is usually a misplaced decimal point. The most common type of percentage problem is finding:

1. *WHAT PERCENT ONE NUMBER IS OF ANOTHER*

In this case, the problem is simply one of reading carefully to determine the correct fraction and then converting to a percentage:

EXAMPLE: What percent is 20 of 25?

$$\frac{20}{25} = \frac{4}{5} = 0.8$$

$$0.8 \times 100\% = 80\%$$

EXAMPLE: Four is what percent of 14?

$$\frac{4}{14} = 0.2857$$

$$0.2857 \times 100\% = 28.57\%$$

EXAMPLE: Influent BOD to a clarifier is 200 mg/*L*. Effluent BOD is 140 mg/*L*. What is the percent removal in the clarifier?

$$\text{Removal, \%} = (\frac{\text{In} - \text{Out}}{\text{In}}) \times 100\%$$

$$= (\frac{200 \text{ mg/}L - 140 \text{ mg/}L}{200 \text{ mg/}L}) \times 100\%$$

$$= 30\%$$

2. PERCENT OF A GIVEN NUMBER

In this case the percent is expressed as a decimal, and the two numbers are multiplied together.

EXAMPLE: Find 7% of 32.

$$\frac{7\%}{100\%} \times 32 = 2.24$$

EXAMPLE: Find 90% of 5.

$$\frac{90\%}{100\%} \times 5 = 4.5$$

EXAMPLE: What is the weight of dry solids in a ton (2,000 lbs) of wastewater sludge containing 5% solids and 95% water?

> *NOTE:* 5% solids means there are 5 lbs of dry solids for every 100 lbs of wet sludge.

Therefore,

$$2,000 \text{ lbs} \times \frac{5\%}{100\%} = 100 \text{ lbs of solids}$$

A variation of the preceding problem is:

3. FINDING A NUMBER WHEN A GIVEN PERCENT OF IT IS KNOWN

Since this problem is similar to the previous problem, the solution is to convert to a decimal and divide by the decimal.

EXAMPLE: If 5% of a number is 52, what is the number?

$$(\frac{100\%}{5\%})(52) = 1,040$$

A check calculation may now be performed — what is 5% of 1,040?

$$(\frac{5\%}{100\%})(1,040) = 52 \text{ (Check)}$$

EXAMPLE: 16 is 80% of what amount?

$$(\frac{100\%}{80\%})(16) = 20$$

EXAMPLE: Percent removal of BOD in a clarifier is 35%. If 70 mg/*L* are removed, what is the influent BOD?

$$\text{Influent BOD} = (\frac{70 \text{ mg/}L}{35\%})(100\%) = 200 \text{ mg/}L$$

Check:

Original Load x % Removal = Load Removed

$$200 \text{ mg/}L \times \frac{35\%}{100\%} = 70 \text{ mg/}L$$

A.2 AREAS

A.20 Units

Areas are measured in two dimensions or in square units. In the English system of measurement the most common units are square inches, square feet, square yards, and square miles. In the metric system the units are square millimeters, square centimeters, square meters, and square kilometers.

A.21 Rectangle

The area of a rectangle is equal to its length (L) multiplied by its width (W).

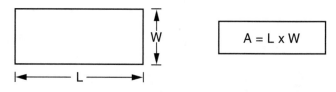

EXAMPLE: Find the area of a rectangle if the length is 5 feet and the width is 3.5 feet.

$$\text{Area, sq ft} = \text{Length, ft} \times \text{Width, ft}$$

$$= 5 \text{ ft} \times 3.5 \text{ ft}$$

$$= 17.5 \text{ ft}^2$$

$$= 17.5 \text{ sq ft}$$

EXAMPLE: The surface of a settling basin is 330 square feet. One side measures 15 feet. How long is the other side?

$$A = L \times W$$

$$330 \text{ sq ft} = L, \text{ft} \times 15 \text{ ft}$$

$$\frac{L, \text{ft} \times 15 \text{ ft}}{15 \text{ ft}} = \frac{330 \text{ sq ft}}{15 \text{ ft}} \quad \text{Divide both sides of equation by 15 ft.}$$

$$L, \text{ft} = \frac{330 \text{ sq ft}}{15 \text{ ft}}$$

$$= 22 \text{ ft}$$

A.22 Triangle

The area of a triangle is equal to one-half the base multiplied by the height. This is true for any triangle.

> *NOTE:* The area of *ANY* triangle is equal to ½ the area of the rectangle that can be drawn around it. The area of the rectangle is B x H. The area of the triangle is ½ B x H.

EXAMPLE: Find the area of triangle ABC.

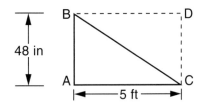

The first step in the solution is to make all the units the same. In this case, it is easier to change inches to feet.

$$48 \text{ in} = 48 \text{ in} \times \frac{1 \text{ ft}}{12 \text{ in}} = \frac{48}{12} \text{ ft} = 4 \text{ ft}$$

NOTE: All conversions should be calculated in the above manner. Since 1 ft/12 in is equal to unity, or one, multiplying by this factor changes the form of the answer but not its value.

Area, sq ft = ½(Base, ft)(Height, ft)

$$= \frac{1}{2} \times 5 \text{ ft} \times 4 \text{ ft}$$

$$= \frac{20}{2} \text{ sq ft}$$

$$= 10 \text{ sq ft}$$

NOTE: Triangle ABC is one-half the area of rectangle ABCD. The triangle is a special form called a *RIGHT TRIANGLE* since it contains a 90° angle at point A.

A.23 Circle

A square with sides of 2R can be drawn around a circle with a radius of R.

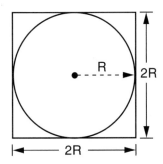

The area of the square is: A = 2R x 2R = 4R².

It has been found that the area of any circle drawn within a square is slightly more than ¾ of the area of the square. More precisely, the area of the preceding circle is:

A circle = 3 ⅐ R² = 3.14 R²

The formula for the area of a circle is usually written:

$$A = \pi R^2$$

The Greek letter π (pronounced pie) merely substitutes for the value 3.1416.

Since the diameter of any circle is equal to twice the radius, the formula for the area of a circle can be rewritten as follows:

$$A = \pi R^2 = \pi \times R \times R = \pi \times \frac{D}{2} \times \frac{D}{2} = \frac{\pi D^2}{4} = \frac{3.14}{4} D^2 = \boxed{0.785 \ D^2}$$

The type of problem and the magnitude (size) of the numbers in a problem will determine which of the two formulas will provide a simpler solution. All of these formulas will give the same results if you use the same number of digits to the right of the decimal point.

EXAMPLE: What is the area of a circle with a diameter of 20 centimeters?

In this case, the formula using a radius is more convenient since it takes advantage of multiplying by 10.

Area, sq cm = π (R, cm)²

$$= 3.14 \times 10 \text{ cm} \times 10 \text{ cm}$$

$$= 314 \text{ sq cm}$$

EXAMPLE: What is the area of a clarifier with a 50-foot radius?

In this case, the formula using diameter is more convenient.

Area, sq ft = 0.785(Diameter, ft)²

$$= 0.785 \times 100 \text{ ft} \times 100 \text{ ft}$$

$$= 7,850 \text{ sq ft}$$

Occasionally the operator may be confronted with a problem giving the area and requesting the radius or diameter. This presents the special problem of finding the square root of the number.

EXAMPLE: The surface area of a circular clarifier is approximately 5,000 square feet. What is the diameter?

A = 0.785 D², or

Area, sq ft = 0.785(Diameter, ft)²

5,000 sq ft = 0.785 D²   To solve, substitute given values in equation.

$$\frac{0.785 \ D^2}{0.785} = \frac{5,000 \ \text{sq ft}}{0.785}$$ Divide both sides by 0.785 to find D².

$$D^2 = \frac{5,000 \ \text{sq ft}}{0.785}$$

$$= 6,369 \text{ sq ft}$$

Therefore,

D = square root of 6,369 sq ft, or

Diameter, ft = $\sqrt{6,369}$ sq ft

Press the √ sign on your calculator and get

D, ft = 79.8 ft.

Sometimes a trial-and-error method can be used to find square roots. Since 80 x 80 = 6,400, we know the answer is close to 80 feet.

Try 79 x 79   = 6,241

Try 79.5 x 79.5 = 6,320.25

Try 79.8 x 79.8 = 6,368.04

The diameter is 79.8 ft, or approximately 80 feet.

A.24   Cylinder

With the formulas presented thus far, it would be a simple matter to find the number of square feet in a room that was to be painted. The length of each wall would be added together and then multiplied by the height of the wall. This would give the surface area of the walls (minus any area for doors and windows). The ceiling area would be found by multiplying length times width and the result added to the wall area gives the total area.

The surface area of a circular cylinder, however, has not been discussed. If we wanted to know how many square feet of surface area are in a tank with a diameter of 60 feet and a height of 20 feet, we could start with the top and bottom.

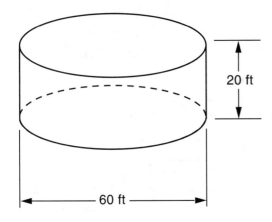

The area of the top and bottom ends are both $\pi \times R^2$.

Area, sq ft $= 2$ ends $(\pi)(\text{Radius, ft})^2$

$= 2 \times \pi \times (30 \text{ ft})^2$

$= 5,652$ sq ft

The surface area of the wall must now be calculated. If we made a vertical cut in the wall and unrolled it, the straightened wall would be the same length as the circumference of the floor and ceiling.

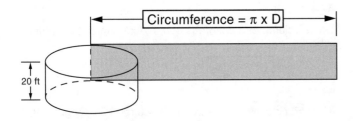

This length has been found to always be $\pi \times D$. In the case of the tank, the length of the wall would be:

Length, ft $= (\pi)(\text{Diameter, ft})$

$= 3.14 \times 60$ ft

$= 188.4$ ft

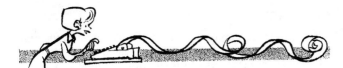

Area would be:

A_W, sq ft $= $ Length, ft $\times$ Height, ft

$= 188.4$ ft $\times 20$ ft

$= 3,768$ sq ft

Outside Surface Area to Paint, sq ft $= $ Area of Top and Bottom, sq ft $+$ Area of Wall, sq ft

$= 5,652$ sq ft $+ 3,768$ sq ft

$= 9,420$ sq ft

A container has inside and outside surfaces and you may need to paint both of them.

A.25   Cone

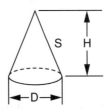

The lateral area of a cone is equal to $\frac{1}{2}$ of the slant height (S) multiplied by the circumference of the base.

$$A_L = \frac{1}{2} S \times \pi \times D = \pi \times S \times R$$

In this case the slant height is not given, but it may be calculated by:

$$S = \sqrt{R^2 + H^2}$$

EXAMPLE: Find the entire outside area of a cone with a diameter of 30 inches and a height of 20 inches.

Slant Height, in $= \sqrt{(\text{Radius, in})^2 + (\text{Height, in})^2}$

$= \sqrt{(15 \text{ in})^2 + (20 \text{ in})^2}$

$= \sqrt{225 \text{ sq in} + 400 \text{ sq in}}$

$= \sqrt{625 \text{ sq in}}$

$= 25$ in

Area of Cone, sq in $= \pi(\text{Slant Height, in})(\text{Radius, in})$

$= 3.14 \times 25 \text{ in} \times 15 \text{ in}$

$= 1,177.5$ sq in

Since the entire area was asked for, the area of the base must be added.

Area, sq in $= 0.785(\text{Diameter, in})^2$

$= 0.785 \times 30 \text{ in} \times 30 \text{ in}$

$= 706.5$ sq in

Total Area, sq in $= $ Area of Cone, sq in $+$ Area of Bottom, sq in

$= 1,177.5$ sq in $+ 706.5$ sq in

$= 1,884$ sq in

A.26 Sphere

The surface area of a sphere or ball is equal to π multiplied by the diameter squared which is four times the cross-sectional area.

$$A_s = \pi D^2$$

If the radius is used, the formula becomes:

$$A_s = \pi D^2 = \pi \times 2R \times 2R = 4\pi R^2$$

EXAMPLE: What is the surface area of a sphere-shaped water tank 20 feet in diameter?

$$\begin{aligned}
\text{Area, sq ft} &= \pi(\text{Diameter, ft})^2 \\
&= 3.14 \times 20 \text{ ft} \times 20 \text{ ft} \\
&= 1{,}256 \text{ sq ft}
\end{aligned}$$

A.3 VOLUMES

A.30 Rectangle

Volumes are measured in three dimensions or in cubic units. To calculate the volume of a rectangle, the area of the base is calculated in square units and then multiplied by the height. The formula then becomes:

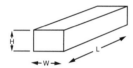

$$V = L \times W \times H$$

EXAMPLE: The length of a box is two feet, the width is 15 inches, and the height is 18 inches. Find its volume.

$$\begin{aligned}
\text{Volume, cu ft} &= \text{Length, ft} \times \text{Width, ft} \times \text{Height, ft} \\
&= 2 \text{ ft} \times \frac{15 \text{ in}}{12 \text{ in/ft}} \times \frac{18 \text{ in}}{12 \text{ in/ft}} \\
&= 2 \text{ ft} \times 1.25 \text{ ft} \times 1.5 \text{ ft} \\
&= 3.75 \text{ cu ft}
\end{aligned}$$

A.31 Prism

The same general rule that applies to the volume of a rectangle also applies to a prism.

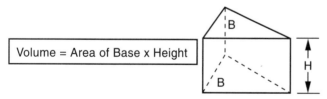

Volume = Area of Base x Height

EXAMPLE: Find the volume of a prism with a base area of 10 square feet and a height of 5 feet. (Note that the base of a prism is triangular in shape.)

$$\begin{aligned}
\text{Volume, cu ft} &= \text{Area of Base, sq ft} \times \text{Height, ft} \\
&= 10 \text{ sq ft} \times 5 \text{ ft} \\
&= 50 \text{ cu ft}
\end{aligned}$$

A.32 Cylinder

The volume of a cylinder is equal to the area of the base multiplied by the height.

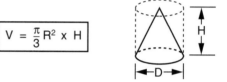

$$V = \pi R^2 \times H = 0.785 \, D^2 \times H$$

EXAMPLE: A tank has a diameter of 100 feet and a depth of 12 feet. Find the volume.

$$\begin{aligned}
\text{Volume, cu ft} &= 0.785 \times (\text{Diameter, ft})^2 \times \text{Height, ft} \\
&= 0.785 \times 100 \text{ ft} \times 100 \text{ ft} \times 12 \text{ ft} \\
&= 94{,}200 \text{ cu ft}
\end{aligned}$$

A.33 Cone

The volume of a cone is equal to ⅓ the volume of a circular cylinder of the same height and diameter.

$$V = \frac{\pi}{3} R^2 \times H$$

EXAMPLE: Calculate the volume of a cone if the height at the center is 4 feet and the diameter is 100 feet (radius is 50 feet).

$$\begin{aligned}
\text{Volume, cu ft} &= \frac{\pi}{3} \times (\text{Radius})^2 \times \text{Height, ft} \\
&= \frac{\pi}{3} \times 50 \text{ ft} \times 50 \text{ ft} \times 4 \text{ ft} \\
&= 10{,}500 \text{ cu ft}
\end{aligned}$$

A.34 Sphere

The volume of a sphere is equal to π/6 times the diameter cubed.

$$V = \frac{\pi}{6} \times D^3$$

EXAMPLE: How much gas can be stored in a sphere with a diameter of 12 feet? (Assume atmospheric pressure.)

$$\begin{aligned}
\text{Volume, cu ft} &= \frac{\pi}{6} \times (\text{Diameter, ft})^3 \\
&= \frac{\pi}{6} \times \overset{2}{\cancel{12}} \text{ ft} \times 12 \text{ ft} \times 12 \text{ ft} \\
&= 904.3 \text{ cubic feet}
\end{aligned}$$

A.4 METRIC SYSTEM

The two most common systems of weights and measures are the English system and the metric system (Le Système International d' Unités, SI). Of these two, the metric system is more popular with most of the nations of the world. The reason for this is that the metric system is based on a system of tens and is therefore easier to remember and easier to use than the English system. Even though the basic system in the United States is the English system, the scientific community uses the metric system almost exclusively. Many organizations have urged, for good reason, that the United States switch to the metric system. Today the metric system is gradually becoming the standard system of measurement in the United States.

As the United States changes from the English to the metric system, some confusion and controversy has developed. For example, which is the correct spelling of the following words:

1. Liter or litre?

2. Meter or metre?

The U.S. Government uses liter and meter and accepts no deviations. Some people argue that METRE should be used to measure LENGTH and that METER should be used to measure FLOW RATES (like a water or electric meter). Liter and Meter are used in this manual because this is most consistent with spelling in the United States.

One of the most frequent arguments heard against the U.S. switching to the metric system was that the costs of switching manufacturing processes would be excessive. Pipe manufacturers have agreed upon the use of a "soft" metric conversion system during the conversion to the metric system. Past practice in the U.S. has identified some types of pipe by external (outside) diameter while other types are classified by nominal (existing only in name, not real or actual) bore. This means that a six-inch pipe does not have a six-inch inside diameter. With the strict or "hard" metric system, a six-inch pipe would be a 152.4 mm (6 in x 25.4 mm/in) pipe. In the "soft" metric system a six-inch pipe is a 150 mm (6 in x 25 mm/in) pipe. Typical customary and "soft" metric pipe size designations are shown below:

PIPE SIZE DESIGNATIONS

Customary, in	2	4	6	8	10	12	15	18
"Soft" Metric, mm	50	100	150	200	250	300	375	450

Customary, in	24	30	36	42	48	60	72	84
"Soft" Metric, mm	600	750	900	1050	1200	1500	1800	2100

In order to study the metric system, you must know the meanings of the terminology used. Following is a list of Greek and Latin prefixes used in the metric system.

PREFIXES USED IN THE METRIC SYSTEM

Prefix	Symbol	Meaning
Micro	μ	1/1 000 000 or 0.000 001
Milli	m	1/1000 or 0.001
Centi	c	1/100 or 0.01
Deci	d	1/10 or 0.1
Unit		1
Deka	da	10
Hecto	h	100
Kilo	k	1000
Mega	M	1 000 000

A.40 Measures of Length

The basic measure of length is the meter.

1 kilometer (km) = 1,000 meters (m)
1 meter (m) = 100 centimeters (cm)
1 centimeter (cm) = 10 millimeters (mm)

Kilometers are usually used in place of miles, meters are used in place of feet and yards, centimeters are used in place of inches, and millimeters are used for inches and fractions of an inch.

LENGTH EQUIVALENTS

1 kilometer	= 0.621 mile	1 mile	= 1.61 kilometers
1 meter	= 3.28 feet	1 foot	= 0.305 meter
1 meter	= 39.37 inches	1 inch	= 0.0254 meter
1 centimeter	= 0.3937 inch	1 inch	= 2.54 centimeters
1 millimeter	= 0.0394 inch	1 inch	= 25.4 millimeters

NOTE: The above equivalents are reciprocals. If one equivalent is given, the reverse can be obtained by division. For instance, if one meter equals 3.28 feet, one foot equals 1/3.28 meter, or 0.305 meter.

A.41 Measures of Capacity or Volume

The basic measure of capacity in the metric system is the liter. For measurement of large quantities the cubic meter is sometimes used.

1 kiloliter (kL) = 1,000 liters (L) = 1 cu meter (cu m)
1 liter (L) = 1,000 milliliters (mL)

Kiloliters, or cubic meters, are used to measure capacity of large storage tanks or reservoirs in place of cubic feet or gallons. Liters are used in place of gallons or quarts. Milliliters are used in place of quarts, pints, or ounces.

CAPACITY EQUIVALENTS

1 kiloliter	= 264.2 gallons	1 gallon	= 0.003785 kiloliter
1 liter	= 1.057 quarts	1 quart	= 0.946 liter
1 liter	= 0.2642 gallon	1 gallon	= 3.785 liters
1 milliliter	= 0.0353 ounce	1 ounce	= 29.57 milliliters

A.42 Measures of Weight

The basic unit of weight in the metric system is the gram. One cubic centimeter of water at maximum density weighs one gram, and thus there is a direct, simple relation between volume of water and weight in the metric system.

1 kilogram (kg) = 1,000 grams (gm)
1 gram (gm) = 1,000 milligrams (mg)
1 milligram (mg) = 1,000 micrograms (μg)

Grams are usually used in place of ounces, and kilograms are used in place of pounds.

WEIGHT EQUIVALENTS

1 kilogram	= 2.205 pounds	1 pound	= 0.4536 kilogram
1 gram	= 0.0022 pound	1 pound	= 453.6 grams
1 gram	= 0.0353 ounce	1 ounce	= 28.35 grams
1 gram	= 15.43 grains	1 grain	= 0.0648 gram

A.43 Temperature

Just as the operator should become familiar with the metric system, you should also become familiar with the centigrade (Celsius) scale for measuring temperature. There is nothing magical about the centigrade scale—it is simply a different size than the Fahrenheit scale. The two scales compare as follows:

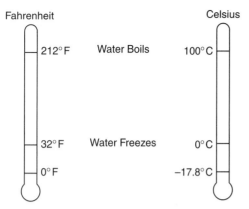

Fahrenheit   Celsius

212°F   Water Boils   100°C

32°F   Water Freezes   0°C

0°F   −17.8°C

The two scales are related in the following manner:

$$\text{Fahrenheit} = (°C \times 9/5) + 32°$$
$$\text{Celsius} = (°F − 32°) \times 5/9$$

EXAMPLE: Convert 20° Celsius to degrees Fahrenheit.

$$°F = (°C \times 9/5) + 32°$$
$$°F = (20° \times 9/5) + 32°$$
$$°F = \frac{180°}{5} + 32°$$
$$= 36° + 32°$$
$$= 68°F$$

EXAMPLE: Convert −10°C to °F.

$$°F = (−10° \times 9/5) + 32°$$
$$°F = −90°/5 + 32°$$
$$= −18° + 32°$$
$$= 14°F$$

EXAMPLE: Convert −13°F to °C.

$$°C = (°F − 32°) \times \frac{5}{9}$$
$$°C = (−13° − 32°) \times \frac{5}{9}$$
$$= −45° \times \frac{5}{9}$$
$$= −5° \times 5$$
$$= −25°C$$

A.44   Milligrams Per Liter

Milligrams per liter (mg/L) is a unit of measurement used in laboratory and scientific work to indicate very small concentrations of dilutions. Since water contains small concentrations of dissolved substances and solids, and since small amounts of chemical compounds are sometimes used in wastewater treatment processes, the term milligrams per liter is also common in treatment plants. It is a weight/volume relationship.

As previously discussed:

1,000 liters = 1 cubic meter = 1,000,000 cubic centimeters.

Therefore,

1 liter = 1,000 cubic centimeters.

Since one cubic centimeter of water weighs one gram,

1 liter of water = 1,000 grams or 1,000,000 milligrams.

$$\frac{1 \text{ milligram}}{\text{liter}} = \frac{1 \text{ milligram}}{1,000,000 \text{ milligrams}} = \frac{1 \text{ part}}{\text{million parts}} = \frac{1 \text{ part per}}{\text{million (ppm)}}$$

Milligrams per liter and parts per million (parts) may be used interchangeably as long as the liquid density is 1.0 gm/cu cm or 62.43 lbs/cu ft. A concentration of 1 milligram/liter (mg/L) or 1 ppm means that there is 1 part of substance by weight for every 1 million parts of water. A concentration of 10 mg/L would mean 10 parts of substance per million parts of water.

To get an idea of how small 1 mg/L is, divide the numerator and denominator of the fraction by 10,000. This, of course, does not change its value since 10,000 ÷ 10,000 is equal to one.

$$1 \frac{mg}{L} = \frac{1 \text{ mg}}{1,000,000 \text{ mg}} = \frac{1/10,000 \text{ mg}}{1,000,000/10,000 \text{ mg}} = \frac{0.0001 \text{ mg}}{100 \text{ mg}} = 0.0001\%$$

Therefore, 1 mg/L is equal to one ten-thousandth of a percent, or

1% is equal to 10,000 mg/L.

To convert mg/L to %, move the decimal point four places or numbers to the left.

Working problems using milligrams per liter or parts per million is a part of everyday operation in most wastewater treatment plants.

A.45   Example Problems

EXAMPLE: A plant effluent flowing at a rate of five million pounds per day contains 15 mg/L of solids. How many pounds of solids will be discharged per day?

$$15 \text{ mg/L} = \frac{15 \text{ lbs solids}}{\text{million lbs water}}$$

Solids Discharged, lbs/day = Concentration, lbs/M lbs x Flow, lbs/day

$$= \frac{15 \text{ lbs}}{\text{million lbs}} \times \frac{5 \text{ million lbs}}{\text{day}}$$
$$= 75 \text{ lbs/day}$$

There is one thing that is unusual about the above problem and that is the flow is reported in pounds per day. In most treatment plants, flow is reported in terms of gallons per minute or gallons per day. To convert these flow figures to weight, an additional conversion factor is needed. It has been found that one gallon of water weighs 8.34 pounds. Using this factor, it is possible to convert flow in gallons per day to flow in pounds per day.

EXAMPLE: A plant influent of 3.5 million gallons per day (MGD) contains 200 mg/*L* BOD. How many pounds of BOD enter the plant per day?

$$\text{Flow, lbs/day} = \text{Flow, } \frac{\text{M gal}}{\text{day}} \times \frac{8.34 \text{ lbs}}{\text{gal}}$$

$$= \frac{3.5 \text{ million gal}}{\text{day}} \times \frac{8.34 \text{ lbs}}{\text{gal}}$$

$$= 29.19 \text{ million lbs/day}$$

$$\begin{aligned}\text{BOD} \\ \text{Loading,} \\ \text{lbs/day}\end{aligned} = \text{BOD Level, mg/}L \times \text{Flow, M lbs/day}$$

$$= \frac{200 \text{ mg*}}{\text{million mg}} \times \frac{29.19 \text{ million lbs}}{\text{day}}$$

$$= 5,838 \text{ lbs/day}$$

---

* Remember that $\dfrac{1 \text{ mg}}{\text{M mg}} = \dfrac{1 \text{ lb}}{\text{M lbs}}$. They are identical ratios.

In solving the above problem, a relation was used that is most important to understand and commit to memory.

$$\boxed{\text{lbs/day} = \text{Conc, mg/}L \times \text{Flow, MGD} \times 8.34 \text{ lbs/gal}}$$

EXAMPLE: A chlorinator is set to feed 50 pounds of chlorine per day to a flow of 0.8 MGD. What is the chlorine dose in mg/*L*?

$$\begin{aligned}\text{Conc or Dose,} \\ \text{mg/}L\end{aligned} = \frac{\text{lbs/day}}{\text{MGD} \times 8.34 \text{ lbs/gal}}$$

$$= \frac{50 \text{ lbs/day}}{0.80 \text{ MG/day} \times 8.34 \text{ lbs/gal}}$$

$$= \frac{50 \text{ lbs}}{6.672 \text{ M lbs}}$$

$$= 7.5 \text{ mg/}L, \text{ or 7.5 ppm}$$

EXAMPLE: Treated effluent is pumped to a spray disposal field by a pump that delivers 500 gallons per minute. Suspended solids in the effluent average 10 mg/*L*. What is the total weight of suspended solids deposited on the spray field during a 24-hour day of continuous pumping?

$$\text{Flow, MGD} = \text{Flow, GPM} \times 60 \text{ min/hr} \times 24 \text{ hr/day}$$

$$= \frac{500 \text{ gal}}{\text{min}} \times \frac{60 \text{ min}}{\text{hr}} \times \frac{24 \text{ hr}}{\text{day}}$$

$$= 720,000 \text{ gal/day}$$

$$= 0.72 \text{ MGD}$$

$$\begin{aligned}\text{Weight of} \\ \text{Solids,} \\ \text{lbs/day}\end{aligned} = \text{Conc, mg/}L \times \text{Flow, MGD} \times 8.34 \text{ lbs/gal}$$

$$= \frac{10 \text{ mg}}{\text{M mg}} \times \frac{0.72 \text{ M gal}}{\text{day}} \times \frac{8.34 \text{ lbs}}{\text{gal}}$$

$$= 60.048 \text{ lbs/day or about 60 lbs/day}$$

A.5 WEIGHT-VOLUME RELATIONS

Another factor for the operator to remember, in addition to the weight of a gallon of water, is the weight of a cubic foot of water. One cubic foot of water weighs 62.4 lbs. If these two weights are divided, it is possible to determine the number of gallons in a cubic foot.

$$\frac{62.4 \text{ pounds/cu ft}}{8.34 \text{ pounds/gal}} = 7.48 \text{ gal/cu ft}$$

Thus we have another very important relationship to commit to memory.

$$\boxed{8.34 \text{ lbs/gal} \times 7.48 \text{ gal/cu ft} = 62.4 \text{ lbs/cu ft}}$$

It is only necessary to remember two of the above items since the third may be found by calculation. For most problems, 8⅓ lbs/gal and 7½ gal/cu ft will provide sufficient accuracy.

EXAMPLE: Change 1,000 cu ft of water to gallons.

1,000 cu ft x 7.48 gal/cu ft = 7,480 gallons

EXAMPLE: What is the weight of three cubic feet of water?

62.4 lbs/cu ft x 3 cu ft = 187.2 lbs

EXAMPLE: The net weight of a tank of water is 750 lbs. How many gallons does it contain?

$$\frac{750 \text{ lbs}}{8.34 \text{ lbs/gal}} = 90 \text{ gal}$$

A.6 FORCE, PRESSURE, AND HEAD

In order to study the forces and pressures involved in fluid flow, it is first necessary to define the terms used.

FORCE: The push exerted by water on any surface being used to confine it. Force is usually expressed in pounds, tons, grams, or kilograms.

PRESSURE: The force per unit area. Pressure can be expressed in many ways, but the most common term is pounds per square inch (psi).

HEAD: Vertical distance from the water surface to a reference point below the surface. Usually expressed in feet or meters.

An example should serve to illustrate these terms.

If water were poured into a one-foot cubical container, the *FORCE* acting on the bottom of the container would be 62.4 pounds.

The *PRESSURE* acting on the bottom would be 62.4 pounds per square foot. The area of the bottom is also 12 in x 12 in = 144 sq in. Therefore, the pressure may also be expressed as:

$$\text{Pressure, psi} = \frac{62.4 \text{ lbs}}{\text{sq ft}} = \frac{62.4 \text{ lbs/sq ft}}{144 \text{ sq in/sq ft}}$$

$$= 0.433 \text{ lb/sq in}$$

$$= 0.433 \text{ psi}$$

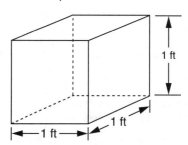

Since the height of the container is one foot, the *HEAD* would be one foot.

The pressure in any vessel at one foot of depth or one foot of head is 0.433 psi acting in any direction.

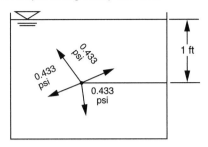

If the depth of water in the previous example were increased to two feet, the pressure would be:

$$p = \frac{2(62.4 \text{ lbs})}{144 \text{ sq in}} = \frac{124.8 \text{ lbs}}{144 \text{ sq in}} = 0.866 \text{ psi}$$

Therefore we can see that for every foot of head, the pressure increases by 0.433 psi. Thus, the general formula for pressure becomes:

$$\boxed{p, \text{psi} = 0.433(H, \text{ft})}$$

H = feet of head

p = pounds per square *INCH* of pressure

$$\boxed{P, \text{lbs/sq ft} = 62.4(H, \text{ft})}$$

H = feet of head

P = pounds per square *FOOT* of pressure

We can now draw a diagram of the pressure acting on the side of a tank. Assume a four-foot deep tank. The pressures shown on the tank are gage pressures. These pressures do not include the atmospheric pressure acting on the surface of the water.

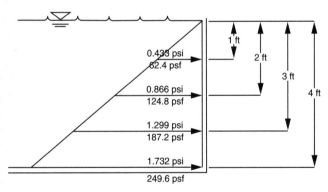

$p_0 = 0.433 \times 0 = 0.0$ psi
$p_1 = 0.433 \times 1 = 0.433$ psi
$p_2 = 0.433 \times 2 = 0.866$ psi
$p_3 = 0.433 \times 3 = 1.299$ psi
$p_4 = 0.433 \times 4 = 1.732$ psi

$P_0 = 62.4 \times 0 = 0.0$ lb/sq ft
$P_1 = 62.4 \times 1 = 62.4$ lbs/sq ft
$P_2 = 62.4 \times 2 = 124.8$ lbs/sq ft
$P_3 = 62.4 \times 3 = 187.2$ lbs/sq ft
$P_4 = 62.4 \times 4 = 249.6$ lbs/sq ft

The average *PRESSURE* acting on the tank wall is 1.732 psi/2 = 0.866 psi, or 249.6 psf/2 = 124.8 psf. We divided by two to obtain the average pressure because there is zero pressure at the top and 1.732 psi pressure on the bottom of the wall.

If the wall were 5 feet long, the pressure would be acting over the entire 20-square-foot (5 ft x 4 ft) area of the wall. The total force acting to push the wall would be:

Force, lbs = (Pressure, lbs/sq ft)(Area, sq ft)

= 124.8 lbs/sq ft x 20 sq ft

= 2,496 lbs

If the pressure in psi were used, the problem would be similar:

Force, lbs = (Pressure, lbs/sq in)(Area, sq in)

= 0.866 psi x 48 in x 60 in

= 2,494 lbs*

---

* Difference in answer due to rounding off of decimal points.

The general formula, then, for finding the total force acting on a side wall of a tank is:

$$\boxed{F = 31.2 \times H^2 \times L}$$

F = force in pounds

H = head in feet

L = length of wall in feet

31.2 = constant with units of lbs/cu ft and considers the fact that the force results from H/2 or half the depth of the water that is the average depth. The force is exerted at H/3 from the bottom.

EXAMPLE: Find the force acting on a 5-foot long wall in a 4-foot deep tank.

Force, lbs = 31.2(Head, ft)^2 (Length, ft)

= 31.2 lbs/cu ft x (4 ft)^2 x 5 ft

= 2,496 lbs

Occasionally an operator is warned: *NEVER EMPTY A TANK DURING PERIODS OF HIGH GROUNDWATER.* Why? The pressure on the bottom of the tank caused by the water surrounding the tank will tend to float the tank like a cork if the upward force of the water is greater than the weight of the tank.

$$\boxed{F = 62.4 \times H \times A}$$

F = upward force in pounds

H = head of water on tank bottom in feet

A = area of bottom of tank in square feet

62.4 = constant with units of lbs/cu ft

This formula is approximately true if the tank doesn't crack, leak, or start to float.

EXAMPLE: Find the upward force on the bottom of an empty tank caused by a groundwater depth of 8 feet above the tank bottom. The tank is 20 ft wide and 40 ft long.

Force, lbs = 62.4(Head, ft)(Area, sq ft)

= 62.4 lbs/cu ft x 8 ft x 20 ft x 40 ft

= 399,400 lbs

A.7   VELOCITY AND FLOW RATE

A.70   Velocity

The velocity of a particle or substance is the speed at which it is moving. It is expressed by indicating the length of travel and how long it takes to cover the distance. Velocity can be expressed in almost any distance and time units. For instance, a car may be traveling at a rate of 280 miles per five hours. However, it is normal to express the distance traveled per unit time. The above example would then become:

$$\text{Velocity, mi/hr} = \frac{280 \text{ miles}}{5 \text{ hours}}$$

$$= 56 \text{ miles/hour}$$

The velocity of water in a channel, pipe, or other conduit can be expressed in the same way. If the particle of water travels 600 feet in five minutes, the velocity is:

$$\text{Velocity, ft/min} = \frac{\text{Distance, ft}}{\text{Time, minutes}}$$

$$= \frac{600 \text{ ft}}{5 \text{ min}}$$

$$= 120 \text{ ft/min}$$

If you wish to express the velocity in feet per second, multiply by 1 min/60 seconds.

NOTE: Multiplying by $\dfrac{1 \text{ minute}}{60 \text{ seconds}}$ is like multiplying by $\dfrac{1}{1}$; it does not change the relative value of the answer. It only changes the form of the answer.

Velocity, ft/sec = (Velocity, ft/min)(1 min/60 sec)

$$= \frac{120 \text{ ft}}{\text{min}} \times \frac{1 \text{ min}}{60 \text{ sec}}$$

$$= \frac{120 \text{ ft}}{60 \text{ sec}}$$

$$= 2 \text{ ft/sec}$$

A.71 Flow Rate

If water in a one-foot wide channel is one foot deep, then the cross-sectional area of the channel is 1 ft x 1 ft = 1 sq ft.

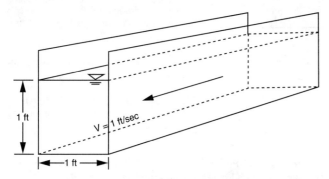

If the velocity in this channel is 1 ft per second, then each second a body of water 1 sq ft in area and 1 ft long will pass a given point. The volume of this body of water would be 1 cubic foot. Since one cubic foot of water would pass by every second, the flow rate would be equal to 1 cubic foot per second, or 1 CFS.

To obtain the flow rate in the above example the velocity was multiplied by the cross-sectional area. This is another important general formula.

$$\boxed{Q = V \times A}$$

Q = flow rate, CFS or cu ft/sec

V = velocity, ft/sec

A = area, sq ft

EXAMPLE: A rectangular channel 3 feet wide contains water 2 feet deep and flowing at a velocity of 1.5 feet per second. What is the flow rate in CFS?

Q = V x A

Flow Rate CFS = Velocity, ft/sec x Area, sq ft

$$= 1.5 \text{ ft/sec} \times 3 \text{ ft} \times 2 \text{ ft}$$

$$= 9 \text{ cu ft/sec}$$

EXAMPLE: Flow in a 2.5-foot wide channel is 1.4 ft deep and measures 11.2 CFS. What is the average velocity?

In this problem we want to find out the velocity. Therefore, we must rearrange the general formula to solve for velocity.

$$V = \frac{Q}{A}$$

$$\text{Velocity, ft/sec} = \frac{\text{Flow Rate, cu ft/sec}}{\text{Area, sq ft}}$$

$$= \frac{11.2 \text{ cu ft/sec}}{2.5 \text{ ft} \times 1.4 \text{ ft}}$$

$$= \frac{11.2 \text{ cu ft/sec}}{3.5 \text{ sq ft}}$$

$$= 3.2 \text{ ft/sec}$$

EXAMPLE: Flow in a 8-inch pipe is 500 GPM. What is the average velocity?

$$\text{Area, sq ft} = 0.785(\text{Diameter, ft})^2$$

$$= 0.785(8/12 \text{ ft})^2$$

$$= 0.785(0.67 \text{ ft})^2$$

$$= 0.785(0.67 \text{ ft})(0.67 \text{ ft})$$

$$= 0.785(0.45 \text{ sq ft})$$

$$= 0.35 \text{ sq ft}$$

$$\text{Flow, CFS} = \text{Flow, gal/min} \times \frac{\text{cu ft}}{7.48 \text{ gal}} \times \frac{1 \text{ min}}{60 \text{ sec}}$$

$$= \frac{500 \text{ gal}}{\text{min}} \times \frac{\text{cu ft}}{7.48 \text{ gal}} \times \frac{1 \text{ min}}{60 \text{ sec}}$$

$$= \frac{500 \text{ cu ft}}{448.8 \text{ sec}}$$

$$= 1.114 \text{ CFS}$$

$$\text{Velocity, ft/sec} = \frac{\text{Flow, cu ft/sec}}{\text{Area, sq ft}}$$

$$= \frac{1.114 \text{ cu ft/sec}}{0.35 \text{ sq ft}}$$

$$= 3.18 \text{ ft/sec}$$

A.8 PUMPS

A.80 Pressure

Atmospheric pressure at sea level is approximately 14.7 psi. This pressure acts in all directions and on all objects. If a tube is placed upside down in a basin of water and a 1 psi partial vacuum is drawn on the tube, the water in the tube will rise 2.31 feet.

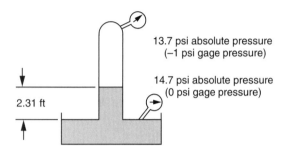

13.7 psi absolute pressure (–1 psi gage pressure)

14.7 psi absolute pressure (0 psi gage pressure)

2.31 ft

NOTE: 1 ft of water = 0.433 psi; therefore,

$$1 \text{ psi} = \frac{1}{0.433} \text{ ft} = 2.31 \text{ ft of water}$$

The action of the partial vacuum is what gets water out of a sump or well and up to a pump. It is not sucked up, but it is pushed up by atmospheric pressure on the water surface in the sump. If a complete vacuum could be drawn, the water would rise 2.31 x 14.7 = 33.9 feet; but this is impossible to achieve. The practical limit of the suction lift of a positive displacement pump is about 22 feet, and that of a centrifugal pump is 15 feet.

A.81 Work

Work can be expressed as lifting a weight a certain vertical distance. It is usually defined in terms of foot-pounds.

EXAMPLE: A 165-pound man runs up a flight of stairs 20 feet high. How much work did he do?

$$\text{Work, ft-lbs} = \text{Weight, lbs} \times \text{Height, ft}$$

$$= 165 \text{ lbs } 20 \text{ ft}$$

$$= 3,300 \text{ ft-lbs}$$

A.82 Power

Power is a rate of doing work and is usually expressed in foot-pounds per minute.

EXAMPLE: If the man is the above example runs up the stairs in three seconds, how much power has he exerted?

$$\text{Power, ft-lbs/sec} = \frac{\text{Work, ft-lbs}}{\text{Time, sec}}$$

$$= \frac{3,300 \text{ ft-lbs}}{3 \text{ sec}} \times \frac{60 \text{ sec}}{\text{minute}}$$

$$= 66,000 \text{ ft-lbs/min}$$

A.83 Horsepower

Horsepower is also a unit of power. One horsepower is defined as 33,000 ft-lbs per minute or 746 watts.

EXAMPLE: How much horsepower has the man in the previous example exerted as he climbed the stairs?

$$\text{Horsepower, HP} = (\text{Power, ft-lbs/min}) \left(\frac{\text{HP}}{33,000 \text{ ft-lbs/min}} \right)$$

$$= 66,000 \text{ ft-lbs/min} \times \frac{\text{Horsepower}}{33,000 \text{ ft-lbs/min}}$$

$$= 2 \text{ HP}$$

Work is also done by lifting water. If the flow from a pump is converted to a weight of water and multiplied by the vertical distance it is lifted, the amount of work or power can be obtained.

$$\text{Horsepower, HP} = \frac{\text{Flow, gal}}{\text{min}} \times \text{Lift, ft} \times \frac{8.34 \text{ lbs}}{\text{gal}} \times \frac{\text{Horsepower}}{33,000 \text{ ft-lbs/min}}$$

Solving the above relation, the amount of horsepower necessary to lift the water is obtained. This is called water horsepower.

$$\text{Water, HP} = \frac{(\text{Flow, GPM})(\text{H, ft})}{3,960^*}$$

$$* \frac{8.34 \text{ lbs}}{\text{gal}} \times \frac{\text{HP}}{33,000 \text{ ft-lbs/min}} = \frac{1}{3,960}$$

1 gallon weighs 8.34 pounds and 1 horsepower is the same as 33,000 ft-lbs/min.

H or Head in feet is the same as Lift in feet.

However, since pumps are not 100% efficient (they cannot transmit all the power put into them), the horsepower supplied to a pump is greater than the water horsepower. Horsepower supplied to the pump is called brake horsepower.

$$\boxed{\text{Brake, HP} = \frac{\text{Flow, GPM} \times \text{H, ft}}{3,960 \times E_p}}$$

E_p = Efficiency of Pump (Usual range 50-85%, depending on type and size of pump)

Motors are also not 100% efficient; therefore, the power supplied to the motor is greater than the motor transmits.

$$\boxed{\text{Motor, HP} = \frac{\text{Flow, GPM} \times \text{H, ft}}{3,960 \times E_p \times E_m}}$$

E_m = Efficiency of Motor (Usual range 80-95%, depending on type and size of motor)

The above formulas have been developed for the pumping of water and wastewater, which have a specific gravity of 1.0. If other liquids are to be pumped, the formulas must be multiplied by the specific gravity of the liquid.

EXAMPLE: A flow of 500 GPM of water is to be pumped against a total head of 100 feet by a pump with an efficiency of 70%. What is the pump horsepower?

$$\text{Brake, HP} = \frac{\text{Flow, GPM} \times \text{H, ft}}{3,960 \times E_p}$$

$$= \frac{500 \times 100}{3,960 \times 0.70}$$

$$= 18 \text{ HP}$$

EXAMPLE: Find the horsepower required to pump gasoline (specific gravity = 0.75) in the previous problem.

$$\text{Brake, HP} = \frac{500 \times 100 \times 0.75}{3,960 \times 0.70}$$

= 13.5 HP (gasoline is lighter and requires less horsepower)

A.84   Head

Basically, the head that a pump must work against is determined by measuring the vertical distance between the two water surfaces, or the distance the water must be lifted. This is called the static head. Two typical conditions for lifting water are shown below.

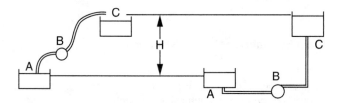

If a pump were designed in the above examples to pump only against head H, the water would never reach the intended point. The reason for this is that the water encounters friction in the pipelines. Friction depends on the roughness and length of pipe, the pipe diameter, and the flow velocity. The turbulence caused at the pipe entrance (point A); the pump (point B); the pipe exit (point C); and at each elbow, bend, or transition also adds to these friction losses. Tables and charts are available in Section A.88 for calculation of these friction losses so they may be added to the measured or static head to obtain the total head. For short runs of pipe that do not have high velocities, the friction losses are generally less than 10 percent of the static head.

EXAMPLE: A pump is to be located 8 feet above a wet well and must lift 1.8 MGD another 50 feet to a storage reservoir. If the pump has an efficiency of 75% and the motor an efficiency of 90%, what is the cost of the power consumed if one kilowatt hour cost 4 cents?

Since we are not given the length or size of pipe and the number of elbows or bends, we will assume friction to be 10% of static head.

Static Head, ft = Suction Lift, ft + Discharge Head, ft

= 8 ft + 50 ft

= 58 ft

Friction Losses, ft = 0.1(Static Head, ft)

= 0.1(58 ft)

= 5.8 ft

Total Dynamic Head, ft = Static Head, ft + Friction Losses, ft

= 58 ft + 5.8 ft

= 63.8 ft

$$\text{Flow, GPM} = \frac{1,800,000 \text{ gal}}{day} \times \frac{day}{24 \text{ hr}} \times \frac{1 \text{ hr}}{60 \text{ min}}$$

= 1,250 GPM (assuming pump runs 24 hours per day)

$$\text{Motor, HP} = \frac{\text{Flow, GPM} \times H, \text{ft}}{3,960 \times E_p \times E_m}$$

$$= \frac{1,250 \times 63.8}{3,960 \times 0.75 \times 0.9}$$

= 30 HP

Kilowatt-hr = 30 HP × 24 hr/day × 0.746 kW/HP*

= 537 kilowatt-hr/day

Cost = kWh × \$0.04/kWh

= 537 × 0.04

= \$21.48/day

---

* See Section A.10, "Basic Conversion Factors," *POWER*, page 461.

A.85   Pump Characteristics

The discharge of a centrifugal pump, unlike a positive displacement pump, can be made to vary from zero to a maximum capacity which depends on the speed, head, power, and specific impeller design. The interrelation of capacity, efficiency, head, and power is known as the characteristics of the pump.

The first relation normally looked at when selecting a pump is the head vs. capacity. The head of a centrifugal pump normally rises as the capacity is reduced. If the values are plotted on a graph they appear as follows:

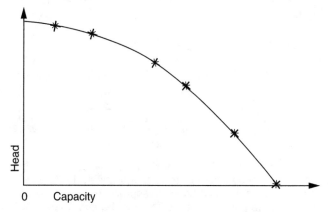

Another important characteristic is the pump efficiency. It begins from zero at no discharge, increases to a maximum, and then drops as the capacity is increased. Following is a graph of efficiency vs. capacity:

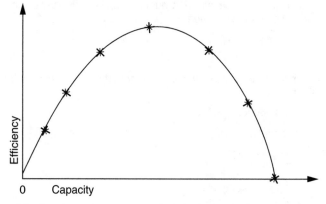

The last important characteristic is the brake horsepower or the power input to the pump. The brake horsepower usually increases with increasing capacity until it reaches a maximum, then it normally reduces slightly.

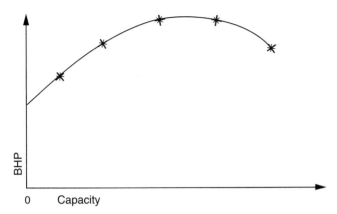

These pump characteristic curves are quite important. Pump sizes are normally picked from these curves rather than calculations. For ease of reading, the three characteristic curves are normally plotted together. A typical graph of pump characteristics is shown as follows:

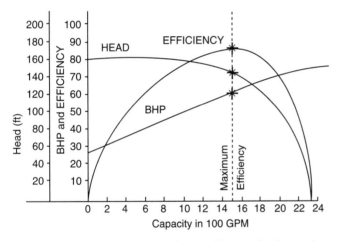

The curves show that the maximum efficiency for the particular pump in question occurs at approximately 1,475 GPM, a head of 132 feet, and a brake horsepower of 58. Operating at this point the pump has an efficiency of approximately 85%. This can be verified by calculation:

$$BHP = \frac{Flow,\ GPM \times H,\ ft}{3,960 \times E}$$

As previously explained, a number can be written over one without changing its value:

$$\frac{BHP}{1} = \frac{GPM \times H}{3,960 \times E}$$

Since the formula is now in ratio form, it can be cross multiplied.

$$BHP \times 3,960 \times E = GPM \times H \times 1$$

Solving for E,

$$E = \frac{GPM \times H}{3,960 \times BHP}$$

$$E = \frac{1,475\ GPM \times 132\ ft}{3,960 \times 58\ HP}$$

$$= 0.85\ or\ 85\%\ (Check)$$

The preceding is only a brief description of pumps to familiarize the operator with their characteristics. The operator does not normally specify the type and size of pump needed at a plant. If a pump is needed, the operator should be able to supply the information necessary for a pump supplier to provide the best possible pump for the lowest cost. Some of the information needed includes:

1. Flow range desired

2. Head conditions

   a. Suction head or lift
   b. Pipe and fitting friction head
   c. Discharge head

3. Type of fluid pumped and temperature

4. Pump location

A.86 Evaluation of Pump Performance

1. Capacity

Sometimes it is necessary to determine the capacity of a pump. This can be accomplished by determining the time it takes a pump to fill or empty a portion of a wet well or diversion box when all inflow is blocked off.

EXAMPLE:

a. Measure the size of the wet well.

   Length = 10 ft

   Width = 10 ft

   Depth = 5 ft (We will measure the time it takes
                  to lower the well a distance of 5 feet.)

   Volume, cu ft = L, ft x W, ft x D, ft

   = 10 ft x 10 ft x 5 ft

   = 500 cu ft

b. Record time for water to drop 5 feet in wet well.

   Time = 10 minutes 30 seconds
        = 10.5 minutes

c. Calculate pumping rate or capacity.

   $$Pumping\ Rate,\ GPM = \frac{Volume,\ gallons}{Time,\ minutes}$$

   $$= \frac{(500\ cu\ ft)(7.5\ gal/cu\ ft)}{10.5\ min}$$

   $$= \frac{3,750}{10.5}$$

   $$= 357\ GPM$$

If you know the total dynamic head and have the pump's performance curves, you can determine if the pump is delivering at design capacity. If not, try to determine the cause (see Chapter 15, "Maintenance," in *OPERATION OF WASTEWATER TREATMENT PLANTS*, Volume II). After a pump overhaul, the pump's actual performance (flow, head, power, and efficiency) should be compared with the pump manufacturer's performance curves. This procedure for calculating the rate of filling or emptying of a wet well or diversion box can be used to calibrate flowmeters.

2. Efficiency

To estimate the efficiency of the pump in the previous example, the total head must be known. This head may be estimated by measuring the suction and discharge pressures. Assume these were measured as follows:

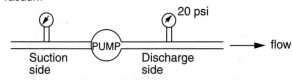

No additional information is necessary if we assume the pressure gages are at the same height and the pipe diameters are the same. Both pressure readings must be converted to feet.

Suction Lift, ft $= 2 \text{ in Mercury} \times \dfrac{1.133 \text{ ft water*}}{1 \text{ in Mercury}}$

                   $= 2.27 \text{ ft}$

Discharge Head, ft $= 20 \text{ psi} \times 2.31 \text{ ft/psi*}$

                   $= 46.20 \text{ ft}$

Total Head, ft $= \text{Suction Lift, ft} + \text{Discharge Head, ft}$

                   $= 2.27 \text{ ft} + 46.20 \text{ ft}$

                   $= 48.47 \text{ ft}$

_____

* See Section A10, "Basic Conversion Factors," *PRESSURE*, page 461.

Calculate the power output of the pump or water horsepower:

Water Horsepower, HP $= \dfrac{(\text{Flow, GPM})(\text{Head, ft})}{3{,}960}$

                      $= \dfrac{(357 \text{ GPM})(48.47 \text{ ft})}{3{,}960}$

                      $= 4.4 \text{ HP}$

To estimate the efficiency of the pump, measure the kilowatts drawn by the pump motor. Assume the meter indicates 8,000 watts or 8 kilowatts. The manufacturer claims the electric motor is 80% efficient.

Brake Horsepower, HP $= (\text{Power to elec. motor})(\text{motor eff.})$

                      $= \dfrac{(8 \text{ kW})(0.80)}{0.746 \text{ kW/HP}}$

                      $= 8.6 \text{ HP}$

Pump Efficiency, % $= \dfrac{\text{Water Horsepower, HP} \times 100\%}{\text{Brake Horsepower, HP}}$

                      $= \dfrac{4.4 \text{ HP} \times 100\%}{8.6 \text{ HP}}$

                      $= 51\%$

The following diagram may clarify the above problem:

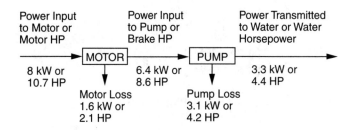

The wire-to-water efficiency is the efficiency of the power input to produce water horsepower.

Wire-to-Water Efficiency, % $= \dfrac{\text{Water Horsepower, HP}}{\text{Power Input, HP}} \times 100\%$

                             $= \dfrac{4.4 \text{ HP}}{10.7 \text{ HP}} \times 100\%$

                             $= 41\%$

The wire-to-water efficiency of a pumping system (pump and electric motor) can be calculated by using the following formula:

Efficiency, % $= \dfrac{(\text{Flow, GPM})(\text{TDH, ft})(100\%)}{(\text{Voltage, volts})(\text{Current, amps})(5.308)}$

              $= \dfrac{(357 \text{ GPM})(48.47 \text{ ft})(100\%)}{(220 \text{ volts})(36 \text{ amps})(5.308)}$

              $= 41\%$

A.87   Pump Speed—Performance Relationships

Changing the velocity of a centrifugal pump will change its operating characteristics. If the speed of a pump is changed, the flow, head developed, and power requirements will change. The operating characteristics of the pump will change with speed approximately as follows:

Flow, $Q_n = \left[\dfrac{N_n}{N_r}\right] Q_r$

Head, $H_n = \left[\dfrac{N_n}{N_r}\right]^2 H_r$ 　　　　 r = rated
　　　　　　　　　　　　　　　　　　 n = now
　　　　　　　　　　　　　　　　　　 N = pump speed

Power, $P_n = \left[\dfrac{N_n}{N_r}\right]^3 P_r$

Actually, pump efficiency does vary with speed; therefore, these formulas are not quite correct. If speeds do not vary by more than a factor of two (if the speeds are not doubled or cut in half), the results are close enough. Other factors contributing to changes in pump characteristic curves include impeller wear and roughness in pipes.

EXAMPLE: To illustrate these relationships, assume a pump has a rated capacity of 600 GPM, develops 100 ft of head, and has a power requirement of 15 HP when operating at 1,500 RPM. If the efficiency remains constant, what will be the operating characteristics if the speed drops to 1,200 RPM?

Calculate new flow rate or capacity:

$$\text{Flow, } Q_n = \left[\frac{N_n}{N_r}\right]Q_r$$

$$= \left[\frac{1,200 \text{ RPM}}{1,500 \text{ RPM}}\right](600 \text{ GPM})$$

$$= \left(\frac{4}{5}\right)(600 \text{ GPM})$$

$$= (4)(120 \text{ GPM})$$

$$= 480 \text{ GPM}$$

Calculate new head:

$$\text{Head, } H_n = \left[\frac{N_n}{N_r}\right]^2 H_r$$

$$= \left[\frac{1,200 \text{ RPM}}{1,500 \text{ RPM}}\right]^2(100 \text{ ft})$$

$$= \left(\frac{4}{5}\right)^2(100 \text{ ft})$$

$$= \left(\frac{16}{25}\right)(100 \text{ ft})$$

$$= (16)(4 \text{ ft})$$

$$= 64 \text{ ft}$$

Calculate new power requirement:

$$\text{Power, } P_n = \left[\frac{N_n}{N_r}\right]^3 P_r$$

$$= \left[\frac{1,200 \text{ RPM}}{1,500 \text{ RPM}}\right]^3(15 \text{ HP})$$

$$= \left(\frac{4}{5}\right)^3(15 \text{ HP})$$

$$= \left(\frac{64}{125}\right)(15 \text{ HP})$$

$$= \left(\frac{64}{25}\right)(3 \text{ HP})$$

$$= 7.7 \text{ HP}$$

A.88 Friction or Energy Losses

Whenever water flows through pipes, valves, and fittings, energy is lost due to pipe friction (resistance), friction in valves and fittings, and the turbulence resulting from the flowing water changing its direction. Figure A.1 can be used to convert the friction losses through valves and fittings to lengths of straight pipe that would produce the same amount of friction losses. To estimate the friction or energy losses resulting from water flowing in a pipe system, we need to know:

1. Water flow rate,

2. Pipe size or diameter and length, and

3. Number, size, and type of valve fittings.

An easy way to estimate friction or energy losses is to follow these steps:

1. Determine the flow rate;

2. Determine the diameter and length of pipe;

3. Convert all valves and fittings to equivalent lengths of straight pipe (see Figure A.1);

4. Add up total length of equivalent straight pipe; and

5. Estimate friction or energy losses by using Figure A.2. With the flow in GPM and diameter of pipe, find the friction loss per 100 feet of pipe. Multiply this value by equivalent length of straight pipe.

The procedure for using Figure A.1 is very easy. Locate the type of valve or fitting you wish to convert to an equivalent pipe length; find its diameter on the right-hand scale; and draw a straight line between these two points to locate the equivalent length of straight pipe.

EXAMPLE: Estimate the friction losses in the piping system of a pump station when the flow is 1,000 GPM. The 8-inch suction line is 10 feet long and contains a 90-degree bend (long sweep elbow), a gate valve and an 8-inch by 6-inch reducer at the inlet to the pump. The 6-inch discharge line is 30 feet long and contains a check valve, a gate valve, and three 90-degree bends (medium sweep elbows):

SUCTION LINE (8-inch diameter)

Item	Equivalent Length, ft
1. Length of pipe	10
2. 90-degree bend	14
3. Gate valve	4
4. 8-inch by 6-inch reducer	17
5. Ordinary entrance	12
Total equivalent length	57 feet

Friction loss (Figure A.2) = 1.76 ft/100 ft of pipe

DISCHARGE LINE (6-inch diameter)

Item	Equivalent Length, ft
1. Length of pipe	30
2. Check valve	38
3. Gate valve	4
4. Three 90-degree bends (3)(14)	42
Total equivalent length	114 feet

Friction loss (Figure A.2) = 7.73 ft/100 ft of pipe

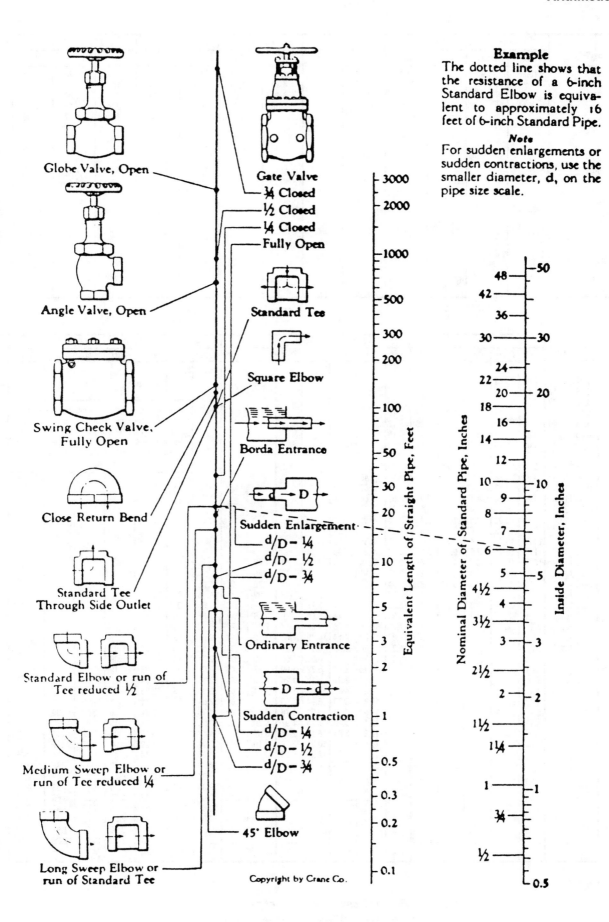

Example
The dotted line shows that the resistance of a 6-inch Standard Elbow is equivalent to approximately 16 feet of 6-inch Standard Pipe.

Note
For sudden enlargements or sudden contractions, use the smaller diameter, d, on the pipe size scale.

Fig. A.1   *Resistance of valves and fittings to flow of water*
(Reprinted by permission of Crane Co.)

U.S. GPM	0.5 in. Vel.	0.5 in. Frict.	0.75 in. Vel.	0.75 in. Frict.	1 in. Vel.	1 in. Frict.	1.25 in. Vel.	1.25 in. Frict.	1.5 in. Vel.	1.5 in. Frict.	2 in. Vel.	2 in. Frict.	2.5 in. Vel.	2.5 in. Frict.
10	10.56	95.9	6.02	23.0	3.71	6.86	2.15	1.77	1.58	.83	.96	.25	.67	.11
20	.....	.....	12.0	86.1	7.42	25.1	4.29	6.34	3.15	2.94	1.91	.87	1.34	.36
30	.....	.....	.....	.....	11.1	54.6	6.44	13.6	4.73	6.26	2.87	1.82	2.01	.75
40	.....	.....	.....	.....	14.8	95.0	8.58	23.5	6.30	10.79	3.82	3.10	2.68	1.28
50	.....	.....	.....	.....	.....	.....	10.7	36.0	7.88	16.4	4.78	4.67	3.35	1.94
60	.....	.....	.....	.....	.....	.....	12.9	51.0	9.46	23.2	5.74	6.59	4.02	2.72
70	.....	.....	.....	.....	.....	.....	15.0	68.8	11.03	31.3	6.69	8.86	4.69	3.63
80	.....	.....	.....	.....	.....	.....	17.2	89.2	12.6	40.5	7.65	11.4	5.36	4.66
90	.....	.....	.....	.....	.....	.....	.....	.....	14.2	51.0	8.60	14.2	6.03	5.82
100	.....	.....	.....	.....	.....	.....	.....	.....	15.8	62.2	9.56	17.4	6.70	7.11
120	.....	.....	.....	.....	.....	.....	.....	.....	18.9	88.3	11.5	24.7	8.04	10.0
140	.....	.....	.....	.....	.....	.....	.....	.....	.....	.....	13.4	33.2	9.38	13.5
160	.....	.....	.....	.....	.....	.....	.....	.....	.....	.....	15.3	43.0	10.7	17.4
180	.....	.....	.....	.....	.....	.....	.....	.....	.....	.....	17.2	54.1	12.1	21.9
200	.....	.....	.....	.....	.....	.....	.....	.....	.....	.....	19.1	66.3	13.4	26.7
220	.....	.....	.....	.....	.....	.....	.....	.....	.....	.....	21.0	80.0	14.7	32.2
240	.....	.....	.....	.....	.....	.....	.....	.....	.....	.....	22.9	95.0	16.1	38.1
260	.....	.....	.....	.....	.....	.....	.....	.....	.....	.....	.....	.....	17.4	44.5
280	.....	.....	.....	.....	.....	.....	.....	.....	.....	.....	.....	.....	18.8	51.3
300	.....	.....	.....	.....	.....	.....	.....	.....	.....	.....	.....	.....	20.1	58.5
350	.....	.....	.....	.....	.....	.....	.....	.....	.....	.....	.....	.....	23.5	79.2

U.S. GPM	3 in. Vel.	3 in. Frict.	4 in. Vel.	4 in. Frict.	5 in. Vel.	5 in. Frict.	6 in. Vel.	6 in. Frict.	8 in. Vel.	8 in. Frict.	10 in. Vel.	10 in. Frict.	12 in. Vel.	12 in. Frict.	14 in. Vel.	14 in. Frict.	16 in. Vel.	16 in. Frict.	18 in. Vel.	18 in. Frict.	20 in. Vel.	20 in. Frict.
20	.91	.15																				
40	1.82	.55	1.02	.13																		
50	2.72	1.17	1.53	.28	.96	.08																
80	3.63	2.02	2.04	.48	1.28	.14	.91	.06														
100	4.54	3.10	2.55	.73	1.60	.20	1.13	.10														
120	5.45	4.40	3.06	1.03	1.92	.29	1.36	.13														
140	6.35	5.93	3.57	1.38	2.25	.38	1.59	.18														
160	7.26	7.71	4.08	1.78	2.57	.49	1.82	.23														
180	8.17	9.73	4.60	2.24	2.89	.61	2.04	.28														
200	9.08	11.9	5.11	2.74	3.21	.74	2.27	.35														
220	9.98	14.3	5.62	3.28	3.53	.88	2.50	.42	1.40	.10												
240	10.9	17.0	6.13	3.88	3.85	1.04	2.72	.49	1.53	.12												
260	11.8	19.8	6.64	4.54	4.17	1.20	2.95	.57	1.66	.14												
280	12.7	22.8	7.15	5.25	4.49	1.38	3.18	.66	1.79	.16												
300	13.6	26.1	7.66	6.03	4.81	1.58	3.40	.75	1.91	.18												
350	.....	.....	8.94	8.22	5.61	2.11	3.97	1.01	2.24	.24												
400	.....	.....	10.20	10.7	6.41	2.72	4.54	1.30	2.55	.30												
450	.....	.....	11.45	13.4	7.22	3.41	5.11	1.64	2.87	.38	1.84	.12										
500	.....	.....	12.8	16.6	8.02	4.16	5.67	2.02	3.19	.46	2.04	.15	1.42	.06								
550	.....	.....	14.0	19.9	8.82	4.98	6.24	2.42	3.51	.56	2.25	.18	1.56	.07								
600	.....	.....	.....	.....	9.62	5.88	6.81	2.84	3.83	.66	2.45	.21	1.70	.08	1.25	.04						
700	.....	.....	.....	.....	11.2	7.93	7.94	3.87	4.47	.88	2.86	.29	1.99	.12	1.46	.05						
800	.....	.....	.....	.....	12.8	10.22	9.08	5.06	5.11	1.14	3.27	.37	2.27	.15	1.67	.07						
900	.....	.....	.....	.....	14.4	12.9	10.2	6.34	5.74	1.44	3.68	.46	2.55	.18	1.88	.09						
1000	.....	.....	.....	.....	.....	.....	11.3	7.73	6.68	1.76	4.09	.57	2.84	.22	2.08	.11						
1100	.....	.....	.....	.....	.....	.....	12.5	9.80	7.02	2.14	4.49	.68	3.12	.27	2.29	.13						
1200	.....	.....	.....	.....	.....	.....	13.6	11.2	7.66	2.53	4.90	.81	3.40	.32	2.50	.15	1.91	.08				
1300	.....	.....	.....	.....	.....	.....	14.7	13.0	8.30	2.94	6.31	.95	3.69	.37	2.71	.17	2.07	.09				
1400	.....	.....	.....	.....	.....	.....	.....	.....	8.93	3.40	5.72	1.09	3.97	.43	2.92	.20	2.23	.10				
1500	.....	.....	.....	.....	.....	.....	.....	.....	9.57	3.91	6.13	1.25	4.26	.49	3.13	.23	2.34	.12				
1600	.....	.....	.....	.....	.....	.....	.....	.....	10.2	4.45	6.54	1.42	4.54	.55	3.33	.25	2.55	.13	2.02	.07		
1700	.....	.....	.....	.....	.....	.....	.....	.....	10.8	5.00	6.94	1.60	4.87	.62	3.54	.29	2.71	.15	2.15	.08		
1800	.....	.....	.....	.....	.....	.....	.....	.....	11.5	5.58	7.35	1.78	5.11	.70	3.75	.32	2.87	.16	2.27	.09		
1900	.....	.....	.....	.....	.....	.....	.....	.....	12.1	6.19	7.76	1.97	5.39	.77	3.96	.35	3.03	.18	2.40	.10		
2000	.....	.....	.....	.....	.....	.....	.....	.....	12.8	6.84	8.17	2.17	5.67	.86	4.17	.39	3.19	.20	2.52	.11		
2500	.....	.....	.....	.....	.....	.....	.....	.....	.....	.....	10.2	3.38	7.10	1.33	5.21	.60	3.99	.31	3.15	.17		
3000	.....	.....	.....	.....	.....	.....	.....	.....	.....	.....	12.3	4.79	8.51	1.88	6.25	.86	4.79	.44	3.78	.24	3.06	.14
3500	.....	.....	.....	.....	.....	.....	.....	.....	.....	.....	14.3	6.55	9.93	2.56	7.29	1.16	5.58	.58	4.41	.32	3.57	.19
4000	.....	.....	.....	.....	.....	.....	.....	.....	.....	.....	.....	.....	11.3	3.31	8.34	1.50	6.38	.75	5.04	.42	4.08	.24
4500	.....	.....	.....	.....	.....	.....	.....	.....	.....	.....	.....	.....	12.8	4.18	9.38	1.88	7.18	.95	5.67	.53	4.59	.31
5000	.....	.....	.....	.....	.....	.....	.....	.....	.....	.....	.....	.....	14.7	5.13	10.4	2.30	7.98	1.17	6.30	.65	5.11	.38
6000	.....	.....	.....	.....	.....	.....	.....	.....	.....	.....	.....	.....	.....	.....	12.5	3.31	9.57	1.66	7.56	.92	6.13	.53
7000	.....	.....	.....	.....	.....	.....	.....	.....	.....	.....	.....	.....	.....	.....	14.6	4.50	11.2	2.26	8.83	1.24	7.15	.72
8000	.....	.....	.....	.....	.....	.....	.....	.....	.....	.....	.....	.....	.....	.....	.....	.....	12.8	2.96	10.09	1.61	8.17	.94
9000	.....	.....	.....	.....	.....	.....	.....	.....	.....	.....	.....	.....	.....	.....	.....	.....	14.4	3.73	11.3	2.02	9.19	1.18
10000	.....	.....	.....	.....	.....	.....	.....	.....	.....	.....	.....	.....	.....	.....	.....	.....	.....	.....	12.6	2.48	10.2	1.45

No allowance has been made for age, differences in diameter, or any other abnormal condition of interior surface. Any Factor of Safety must be estimated from the local conditions and the requirements of each particular installation. For general purposes, 15% is a responsible Factor of Safety.

Fig. A.2 Friction loss for water in feet per 100 feet of pipe

(Reprinted from the 10th Edition of the Standards of the Hydraulic
Institute, 122 East 42nd Street, New York)

Estimate the total friction losses in pumping system for a flow of 1,000 GPM.

SUCTION

Loss = (1.76 ft/100 ft)(57 ft)   = 1.0 ft

DISCHARGE

Loss = (7.73 ft/100 ft)(114 ft)   = 8.8 ft

Total Friction Losses, ft = 9.8 ft

A.9 STEPS IN SOLVING PROBLEMS

A.90 Identification of Problem

To solve any problem, you have to identify the problem, determine what kind of answer is needed, and collect the information needed to solve the problem. A good approach to this type of problem is to examine the problem and make a list of *KNOWN* and *UNKNOWN* information.

EXAMPLE: Find the theoretical detention time in a rectangular sedimentation tank 8 feet deep, 30 feet wide, and 60 feet long when the flow is 1.4 MGD.

Known	Unknown
Depth, ft = 8 ft	Detention Time, hours
Width, ft = 30 ft	
Length, ft = 60 ft	
Flow, MGD = 1.4 MGD	

Sometimes a drawing or sketch will help to illustrate a problem and indicate the knowns, unknowns, and possibly additional information needed.

A.91 Selection of Formula

Most problems involving mathematics in wastewater treatment plant operation can be solved by selecting the proper formula, inserting the known information, and calculating the unknown. In our example, we could look in Chapter 5, "Sedimentation and Flotation," or in Section A.11 of this chapter, "Basic Formulas," to find a formula for calculating detention time. From Section A.11:

$$\text{Detention Time, hr} = \frac{(\text{Tank Volume, cu ft})(7.48 \text{ gal/cu ft})(24 \text{ hr/day})}{\text{Flow, gal/day}}$$

To convert the known information to fit the terms in a formula sometimes requires extra calculations. The next step is to find the values of any terms in the formula that are not in the list of known values.

Flow, gal/day = 1.4 MGD

= 1,400,000 gal/day

From Section A.30:

$$\text{Tank Volume, cu ft} = (\text{Length, ft})(\text{Width, ft})(\text{Height, ft})$$

= 60 ft x 30 ft x 8 ft

= 14,400 cu ft

Solution of Problem:

$$\text{Detention Time, hr} = \frac{(\text{Tank Volume, cu ft})(7.48 \text{ gal/cu ft})(24 \text{ hr/day})}{\text{Flow, gal/day}}$$

$$= \frac{(14,400 \text{ cu ft})(7.48 \text{ gal/cu ft})(24 \text{ hr/day})}{1,400,000 \text{ gal/day}}$$

= 1.85 hr

The remainder of this section discusses the details that must be considered in solving this problem.

A.92 Arrangement of Formula

Once the proper formula is selected, you may have to rearrange the terms to solve for the unknown term. From Section A.71, "Flow Rate," we can develop the formula:

$$\text{Velocity, ft/sec} = \frac{\text{Flow Rate, cu ft/sec}}{\text{Cross-Sectional Area, sq ft}}$$

or $$V = \frac{Q}{A}$$

In this equation if Q and A were given, the equation could be solved for V. If V and A were known, the equation would have to be rearranged to solve for Q. To move terms from one side of an equation to another, use the following rule:

When moving a term or number from one side of an equation to the other, move the numerator (top) of one side to the denominator (bottom) of the other; or from the denominator (bottom) of one side to the numerator (top) of the other.

$$V = \frac{Q}{A} \text{ or } Q = AV \text{ or } A = \frac{Q}{V}$$

If the volume of a sedimentation tank and the desired detention time were given, the detention time formula could be rearranged to calculate the design flow.

$$\text{Detention Time, hr} = \frac{(\text{Tank Vol, cu ft})(7.48 \text{ gal/cu ft})(24 \text{ hr/day})}{\text{Flow, gal/day}}$$

By rearranging the terms

$$\text{Flow, gal/day} = \frac{(\text{Tank Vol, cu ft})(7.48 \text{ gal/cu ft})(24 \text{ hr/day})}{\text{Detention Time, hr}}$$

A.93 Unit Conversions

Each term in a formula or mathematical calculation must be of the correct units. The area of a rectangular clarifier (Area, sq ft = Length, ft x Width, ft) can't be calculated in square feet if the width is given as 246 inches or 20 feet 6 inches. The width must be converted to 20.5 feet. In the example problem, if the tank volume were given in gallons, the 7.48 gal/cu ft would not be needed. *THE UNITS IN A FORMULA MUST ALWAYS BE CHECKED BEFORE ANY CALCULATIONS ARE PERFORMED TO AVOID TIME-CONSUMING MISTAKES.*

$$\text{Detention Time, hr} = \frac{(\text{Tank Volume, cu ft})(7.48 \text{ gal/cu ft})(24 \text{ hr/day})}{\text{Flow, gal/day}}$$

$$= \frac{\cancel{\text{cu ft}}}{1} \times \frac{\text{gal}}{\cancel{\text{cu ft}}} \times \frac{\text{hr}}{\cancel{\text{day}}} \times \frac{\cancel{\text{day}}}{\cancel{\text{gal}}}$$

$$= \text{hr (all other units cancel)}$$

NOTE: We have hours = hr. One should note that the hour unit on both sides of the equation can be cancelled out and nothing would remain. This is one more check that we have the correct units. By rearranging the detention time formula, other unknowns could be determined.

If the design detention time and design flow were known, the required capacity of the tank could be calculated.

$$\text{Tank Volume, cu ft} = \frac{(\text{Detention Time, hr})(\text{Flow, gal/day})}{(7.48 \text{ gal/cu ft})(24 \text{ hr/day})}$$

If the tank volume and design detention time were known, the design flow could be calculated.

$$\text{Flow, gal/day} = \frac{(\text{Tank Volume, cu ft})(7.48 \text{ gal/cu ft})(24 \text{ hr/day})}{\text{Detention Time, hr}}$$

Rearrangement of the detention time formula to find other unknowns illustrates the need to always use the correct units.

A.94 Calculations

Sections A.12, "Multiplication," and A.13, "Division," outline the steps to follow in mathematical calculations. In general, do the calculations inside parentheses ( ) first and brackets [ ] next. Calculations should be done left to right above and below the division line before dividing.

$$\text{Detention Time, hr} = \frac{[(\text{Tank Volume, cu ft})(7.48 \text{ gal/cu ft})(24 \text{ hr/day})]}{\text{Flow, gal/day}}$$

$$= \frac{[(14,400 \text{ cu ft})(7.48 \text{ gal/cu ft})(24 \text{ hr/day})]}{1,400,000 \text{ gal/day}}$$

$$= \frac{2,585,088 \text{ gal-hr/day}}{1,400,000 \text{ gal/day}}$$

$$= 1.85, \text{ or}$$

$$= 1.9 \text{ hr}$$

A.95 Significant Figures

In calculating the detention time in the previous section, the answer is given as 1.9 hr. The answer could have been calculated:

$$\text{Detention Time, hr} = \frac{2,585,088 \text{ gal-hr/day}}{1,400,000 \text{ gal/day}}$$

$$= 1.846491429 \ldots \text{ hours}$$

How does one know when to stop dividing? Common sense and significant figures both help.

First, consider the meaning of detention time and the measurements that were taken to determine the knowns in the formula. Detention time in a tank is a theoretical value and assumes that all particles of water throughout the tank move through the tank at the same velocity. This assumption is not correct; therefore, detention time can only be a representative time for some of the water particles.

Will the flow of 1.4 MGD be constant throughout the 1.9 hours, and is the flow exactly 1.4 MGD, or could it be 1.35 MGD or 1.428 MGD? A carefully calibrated flowmeter may give a reading within 2% of the actual flow rate. Flows into a tank fluctuate and flowmeters do not measure flows extremely accurately; so the detention time again appears to be a representative or typical detention time.

Tank dimensions are probably satisfactory within 0.1 ft. A flowmeter reading of 1.4 MGD is less precise and it could be 1.3 or 1.5 MGD. A 0.1 MGD flowmeter error when the flow is 1.4 MGD is (0.1/1.4) x 100% = 7% error. A detention time of 1.9 hours, based on a flowmeter reading error of plus or minus 7%, also could have the same error or more, even if the flow was constant. Therefore, the detention time error could be 1.9 hours x 0.07 = ±0.13 hour.

In most of the calculations in the operation of wastewater treatment plants, the operator uses measurements determined in the lab or read from charts, scales, or meters. The accuracy of every measurement depends on the sample being measured, the equipment doing the measuring, and the operator reading or measuring the results. Your estimate is no better than the least precise measurement. Do not retain more than one doubtful number.

To determine how many figures or numbers mean anything in an answer, the approach called "significant figures" is used. In the example the flow was given in two significant figures (1.4 MGD), and the tank dimensions could be considered accurate to the nearest tenth of a foot (depth = 9.0 ft) or two significant figures. Since all measurements and the constants contained two significant figures, the results should be reported as two significant figures or 1.9 hours. The calculations are normally carried out to three significant figures (1.85 hours) and rounded off to two significant figures (1.9 hours).

Decimal points require special attention when determining the number of significant figures in a measurement.

Measurement	Significant Figures
0.00325	3
11.078	5
21,000.	2

EXAMPLE: The distance between two points was divided into three sections, and each section was measured by a different group. What is the distance between the two points if each group reported the distance it measured as follows?

Group	Distance, ft	Significant Figures
A	11,300.	3
B	2,438.9	5
C	87.62	4
Total Distance	13,826.52	

Group A reported the length of the section it measured to three significant figures; therefore, the distance between the two points should be reported as 13,800 feet (3 significant figures).

When adding, subtracting, multiplying, or dividing, the number of significant figures in the answer should not be more than the term in the calculations with the least number of significant figures.

A.96 Check Your Results

After completing your calculations, you should carefully examine your calculations and answer. Does the answer seem reasonable? If possible, have another operator check your calculations before making any operational changes.

A.10 BASIC CONVERSION FACTORS (ENGLISH SYSTEM)

UNITS
1,000,000 = 1 Million   1,000,000/1 Million

LENGTH
12 in = 1 ft   12 in/ft
3 ft = 1 yd   3 ft/yd
5,280 ft = 1 mi   5,280 ft/mi

AREA
144 sq in = 1 sq ft   144 sq in/sq ft
43,560 sq ft = 1 acre   43,560 sq ft/ac

VOLUME
7.48 gal = 1 cu ft   7.48 gal/cu ft
1,000 mL = 1 liter   1,000 mL/L
3.785 L = 1 gal   3.785 L/gal
231 cu in = 1 gal   231 cu in/gal

WEIGHT
1,000 mg = 1 gm   1,000 mg/gm
1,000 gm = 1 kg   1,000 gm/kg
454 gm = 1 lb   454 gm/lb
2.2 lbs = 1 kg   2.2 lbs/kg

POWER
0.746 kW = 1 HP   0.746 kW/HP

DENSITY
8.34 lbs = 1 gal   8.34 lbs/gal
62.4 lbs = 1 cu ft   62.4 lbs/cu ft

DOSAGE
17.1 mg/L = 1 grain/gal   17.1 mg/L/gpg
64.7 mg = 1 grain   64.7 mg/grain

PRESSURE
2.31 ft water = 1 psi   2.31 ft water/psi
0.433 psi = 1 ft water   0.433 psi/ft water
1.133 ft water = 1 in Mercury   1.133 ft water/in Mercury

FLOW
694 GPM = 1 MGD   694 GPM/MGD
1.55 CFS = 1 MGD   1.55 CFS/MGD

TIME
60 sec = 1 min   60 sec/min
60 min = 1 hr   60 min/hr
24 hr = 1 day   24 hr/day

NOTE: In our equations the values in the right-hand column may be written either as 24 hr/day or 1 day/24 hours depending on which units we wish to convert to obtain our desired results.

A.11 BASIC FORMULAS

FLOWS

1. $\text{Flow, MGD} = \dfrac{(\text{Flow, GPM})(60 \text{ min/hr})(24 \text{ hr/day})}{1,000,000/M}$

or $\text{Flow, GPM} = \dfrac{(\text{Flow, MGD})(1,000,000/M)}{(60 \text{ min/hr})(24 \text{ hr/day})}$

or $\text{Flow, CFS} = \dfrac{(\text{Flow, MGD})(1,000,000/M)}{(7.48 \text{ gal/cu ft})(24 \text{ hr/day})(60 \text{ min/hr})(60 \text{ sec/min})}$

$= (\text{Flow, MGD})(1.55 \text{ CFS/MGD})$

GRIT CHANNELS

2. $\text{Velocity, ft/sec} = \dfrac{\text{Distance Traveled, ft}}{\text{Time, sec}}$

or $\text{Velocity, ft/sec} = \dfrac{\text{Flow, CFS}}{\text{Area, sq ft}}$

3. $\text{Grit Removed, cu ft/MG} = \dfrac{\text{Volume of Grit, cu ft}}{\text{Volume of Flow, MG}}$

SEDIMENTATION TANKS AND CLARIFIERS

4. $\text{Detention Time, hr} = \dfrac{(\text{Tank Volume, cu ft})(7.48 \text{ gal/cu ft})(24 \text{ hr/day})}{\text{Flow, gal/day}}$

5. $\text{Surface Loading, GPD/sq ft} = \dfrac{\text{Flow, GPD}}{\text{Surface Area, sq ft}}$

6. $\text{Weir Overflow, GPD/ft} = \dfrac{\text{Flow, GPD}}{\text{Length of Weir, ft}}$

7. $\text{Solids Loading, lbs/day/sq ft} = \dfrac{\text{Solids Applied, lbs/day}}{\text{Surface Area, sq ft}}$

where

$\text{Solids Applied, lbs/day} = (\text{Flow, MGD})(\text{Solids, mg/L})(8.34 \text{ lbs/gal})$

TRICKLING FILTERS

8. $\text{Hydraulic Loading, GPD/sq ft} = \dfrac{\text{Flow, GPD}}{\text{Surface Area, sq ft}}$

9. $\text{Organic Loading, lbs BOD/day/1,000 cu ft} = \dfrac{\text{BOD Applied, lbs/day}}{\text{Volume of Media, 1,000 cu ft}}$

where

$\text{BOD Applied, lbs/day} = (\text{Flow, MGD})(\text{BOD, mg/L})(8.34 \text{ lbs/gal})$

10. $\text{Plant Efficiency, \%} = \left(\dfrac{\text{In} - \text{Out}}{\text{In}}\right)(100\%)$

ROTATING BIOLOGICAL CONTACTORS

11. $\text{Hydraulic Loading, GPD/sq ft} = \dfrac{\text{Flow, GPD}}{\text{Surface Area, sq ft}}$

12. $\text{Organic Loading, lbs BOD/day/1,000 sq ft} = \dfrac{\text{Soluble BOD Applied, lbs/day}}{\text{Surface Area of Media, 1,000 sq ft}}$

where

$\text{Soluble BOD, mg/L} = \text{Total BOD, mg/L} - (K)(\text{Suspended Solids, mg/L})$

ACTIVATED SLUDGE

13. $\text{Sludge Volume Index or SVI} = \dfrac{(\text{Settleable Solids, \%})(10,000)}{\text{MLSS, mg/L}}$

14. $\text{Aerator Solids, lbs} = (\text{Tank Vol, MG})(\text{MLSS, mg/L})(8.34 \text{ lbs/gal})$

15. $\text{Aerator Loading, lbs BOD/day} = (\text{Flow, MGD})(\text{Pri Eff BOD, mg/L})(8.34 \text{ lbs/gal})$

16. Sludge Age, days $= \dfrac{\text{Aerator or Mixed Liquor Solids, lbs}}{\text{Primary Effluent Solids, lbs/day}}$

$= \dfrac{(\text{Tank Vol, MG})(\text{MLSS, mg/}L)(8.34 \text{ lbs/gal})}{(\text{Flow, MGD})(\text{Pri Eff SS, mg/}L)(8.34 \text{ lbs/gal})}$

PONDS

17. Detention Time, days $= \dfrac{\text{Pond Volume, ac-ft}}{\text{Flow Rate, ac-ft/day}}$

where

Pond Area, acres $= \dfrac{(\text{Average Width, ft})(\text{Average Length, ft})}{43{,}560 \text{ sq ft/acre}}$

Pond Volume, ac-ft $= (\text{Avg Area, ac})(\text{Depth, ft})$

Flow Rate, ac-ft/day $= \dfrac{\text{Flow, gal/day}}{(7.48 \text{ gal/cu ft})(43{,}560 \text{ sq ft/acre})}$

18. Population Loading, persons/ac $= \dfrac{\text{Population Served, persons}}{\text{Pond Surface Area, acres}}$

19. Hydraulic Loading, inches/day $= \dfrac{\text{Depth of Pond, inches}}{\text{Detention Time, days}}$

20. Organic Loading, lbs BOD/day/ac $= \dfrac{(\text{Flow, MGD})(\text{BOD, mg/}L)(8.34 \text{ lbs/gal})}{\text{Area, acres}}$

CHLORINATION

21. Chlorine Demand, mg/L $= \text{Chlorine Dose, mg/}L - \text{Chlorine Residual, mg/}L$

22. Chlorine Feed Rate, lbs/day $= (\text{Flow, MGD})(\text{Dose, mg/}L)(8.34 \text{ lbs/gal})$

CHEMICAL DOSES

Chemical Feeder Setting

23. Chemical Dose, lbs/day $= (\text{Flow, MGD})(\text{Dose, mg/}L)(8.34 \text{ lbs/gal})$

24. Chemical Feeder Setting, mL/min $= \dfrac{(\text{Flow, MGD})(\text{Alum Dose, mg/}L)(3.785 \ L/\text{gal})(1{,}000{,}000/\text{M})}{(\text{Liquid Alum, mg/m}L)(24 \text{ hr/day})(60 \text{ min/hr})}$

25. Chemical Feeder Setting, gal/day $= \dfrac{(\text{Flow, MGD})(\text{Alum Dose, mg/}L)(8.34 \text{ lbs/gal})}{\text{Liquid Alum, lbs/gal}}$

Calibration of a Dry Chemical Feeder

26. Chemical Feed, lbs/day $= \dfrac{\text{Chemical Applied, lbs}}{\text{Length of Application, days}}$

Calibration of a Solution Chemical Feeder
(Chemical Feed Pump or a Hypochlorinator)

27. Chemical Feed, lbs/day $= \dfrac{(\text{Chem Conc, mg/}L)(\text{Vol Pumped, m}L)(60 \text{ min/hr})(24 \text{ hr/day})}{(\text{Time Pumped, min})(1{,}000 \text{ m}L/L)(1{,}000 \text{ mg/gm})(454 \text{ gm/lb})}$

28. Chemical Feed, GPM $= \dfrac{\text{Chemical Used, gal}}{(\text{Time, hr})(60 \text{ min/hr})}$

or $= \dfrac{(\text{Chemical Feed Rate, m}L/\text{sec})(60 \text{ sec/min})}{3{,}785 \text{ m}L/\text{gal}}$

29a. Chemical Solution, gal $= \dfrac{(\text{Chemical Solution, \%})(8.34 \text{ lbs/gal})}{100\%}$

29b. Feed Pump, GPD $= \dfrac{\text{Chemical Feed, lbs/day}}{\text{Chemical Solution, lbs/gal}}$

29c. Feeder Setting, % $= \dfrac{(\text{Desired Feed Pump, GPD})(100\%)}{\text{Maximum Feed Pump, GPD}}$

A.12 HOW TO USE THE BASIC FORMULAS

One clever way of using the basic formulas is to use the Davidson* Pie Method. To apply this method to the basic formula for chemical doses,

1. Chemical Feed, lbs/day $= (\text{Flow, MGD})(\text{Dose, mg/}L)(8.34 \text{ lbs/gal})$

(a) Draw a circle and draw a horizontal line through the middle of the circle;

(b) Write the Chemical Feed, lbs/day in the top half;

(c) Divide the bottom half into three parts; and

(d) Write Flow, MGD; Dose, mg/L; and 8.34 lbs/gal in the other three parts.

(e) The line across the middle of the circle represents the line in the equation. The items above the line stay above the line and those below the line stay below the line.

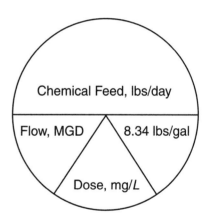

If you want to find the Chemical Feed, lbs/day, cover up the Chemical Feed, lbs/day, and what is left uncovered will give you the correct formula.

2. Chemical Feed, lbs/day $= (\text{Flow, MGD})(\text{Dose, mg/}L)(8.34 \text{ lbs/gal})$

If you know the chlorinator setting in pounds per day and the flow in MGD and would like to know the dose in mg/L, cover up the Dose, mg/L, and what is left uncovered will give you the correct formula.

3. Dose, mg/L $= \dfrac{\text{Chemical Feed, lbs/day}}{(\text{Flow, MGD})(8.34 \text{ lbs/gal})}$

Another approach to using the basic formulas is to memorize the basic formula, for example the detention time formula.

---

* Gerald Davidson, Manager, Clear Lake Oaks Water District, Clear Lake Oaks, California.

4. Detention Time, hr = $\dfrac{\text{(Tank Volume, gal)(24 hr/day)}}{\text{Flow, gal/day}}$

This formula works fine to solve for the detention time when the Tank Volume, gal, and Flow, gal/day, are given.

If you wish to determine the Flow, gal/day, when the Detention Time, hr, and Tank Volume, gal, are given, you must change the basic formula. You want the Flow, gal/day, on the left of the equal sign and everything else on the right of the equal sign. This is done by moving the terms diagonally (from top to bottom or from bottom to top) past the equal sign.

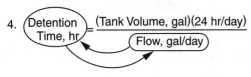

or

Flow, gal/day = $\dfrac{\text{(Tank Volume, gal)(24 hr/day)}}{\text{Detention Time, hr}}$

This same approach can be used if the Tank Volume, gal, was unknown and the Detention Time, hr, and Flow, gal/day, were given. We want Tank Volume, gal, on one side of the equation and everything else on the other side.

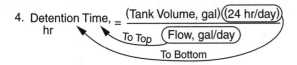

or

$\dfrac{\text{(Detention Time, hr)(Flow, gal/day)}}{\text{24 hr/day}}$ = Tank Volume, gal

or

Tank Volume, gal = $\dfrac{\text{(Detention Time, hr)(Flow, gal/day)}}{\text{24 hr/day}}$

One more check is to be sure the units in the rearranged formula cancel out correctly.

For additional information on the use of the basic formulas, refer to Sections:

A.91, "Selection of Formula,"

A.92, "Arrangement of Formula,"

A.93, "Unit Conversions," and

A.94, "Calculations."

A.13 TYPICAL WASTEWATER TREATMENT PLANT PROBLEMS (ENGLISH SYSTEM)

A.130 Flows

EXAMPLE 1

Convert a flow of 600 gallons per minute to million gallons per day.

Known	Unknown
Flow, GPM = 600 GPM	Flow, MGD

Convert flow from 600 GPM to MGD.

Flow, MGD = $\dfrac{\text{(Flow, GPM)(60 min/hr)(24 hr/day)}}{1,000,000/M}$

$= \dfrac{\text{(600 GPM)(60 min/hr)(24 hr/day)}}{1,000,000/M}$

= 0.86 MGD

NOTE: When we divide by 1,000,000/M this is just like dividing an equation by 12 inches/foot or 60 minutes/hour; all we are doing is changing units.

EXAMPLE 1a

Convert a flow from 1.2 MGD to cubic feet per second.

Known	Unknown
Flow, MGD = 1.2 MGD	Flow, CFS

Convert a flow from 1.2 MGD to CFS.

Flow, CFS = $\dfrac{\text{(Flow, MGD)(1,000,000/M)}}{\text{(7.48 gal/cu ft)(24 hr/day)(60 min/hr)(60 sec/min)}}$

$= \dfrac{\text{(1.2 MGD)(1,000,000/M)}}{\text{(7.48 gal/cu ft)(24 hr/day)(60 min/hr)(60 sec/min)}}$

= 1.86 CFS

or

Flow, CFS = (Flow, MGD)(1.55 CFS/MGD)

= (1.2 MGD)(1.55 CFS/MGD)

= 1.86 CFS

A.131 Grit Channels

EXAMPLE 2

Estimate the velocity of wastewater flowing through a grit channel if a stick travels 32 feet in 36 seconds.

Known	Unknown
Distance, ft = 32 ft	Velocity, ft/sec
Time, sec = 36 sec	

Velocity, ft/sec = $\dfrac{\text{Distance Traveled, ft}}{\text{Time, sec}}$

$= \dfrac{\text{32 ft}}{\text{36 sec}}$

= 0.89 ft/sec

EXAMPLE 3

A grit channel removed 3.2 cu ft of grit during a period when the total flow was 0.8 MG. How many cu ft of grit are removed per MG?

Known	Unknown
Vol of Grit, cu ft = 3.2 cu ft	Grit Removed, cu ft/MG
Vol of Flow, MG = 0.8 MG	

$$\text{Grit Removed, cu ft/MG} = \frac{\text{Volume of Grit, cu ft}}{\text{Volume of Flow, MG}}$$

$$= \frac{3.2 \text{ cu ft}}{0.8 \text{ MG}}$$

$$= 4.0 \text{ cu ft/MG}$$

A.132   Sedimentation Tanks and Clarifiers

EXAMPLES 4, 5, 6, and 7

A circular secondary clarifier handles a flow of 0.9 MGD and suspended solids of 3,600 mg/L. The clarifier is 50 feet in diameter and 8 feet deep. Find the detention time, surface loading rate, weir overflow rate, and solids loading.

Known	Unknown
Flow, MGD = 0.9 MGD	4. Detention Time, hours
Diameter, ft = 50 ft	5. Surface Loading, GPD/sq ft
Depth, ft = 8 ft	6. Weir Overflow, GPD/sq ft
SS, mg/L = 3,600 mg/L	7. Solids Loading, lbs/day/sq ft

4. DETENTION TIME

$$\text{Detention Time, hr} = \frac{(\text{Tank Volume, cu ft})(7.5 \text{ gal/cu ft})(24 \text{ hr/day})}{\text{Flow, gal/day}}$$

$$\text{Tank Volume, cu ft} = (\text{Area, sq ft})(\text{Depth, ft})$$

$$\text{Clarifier Area, sq ft} = 0.785(\text{Diameter, ft})^2$$

$$= 0.785(50 \text{ ft})^2$$

$$= 1,962.5 \text{ sq ft, or}$$

$$= 1,960^* \text{ sq ft}$$

$$\text{Tank Volume, cu ft} = (\text{Area, sq ft})(\text{Depth, ft})$$

$$= (1,960 \text{ sq ft})(8 \text{ ft})$$

$$= 15,680 \text{ cu ft}$$

$$\text{Detention Time, hr} = \frac{(\text{Tank Volume, cu ft})(7.5 \text{ gal/cu ft})(24 \text{ hr/day})}{\text{Flow, gal/day}}$$

$$= \frac{(15,680 \text{ cu ft})(7.5 \text{ gal/cu ft})(24 \text{ hr/day})}{900,000 \text{ gal/day}}$$

$$= \frac{2,820,000}{900,000}$$

$$= 3.1 \text{ hr}$$

---

* Rounded off to three significant figures.

5. SURFACE LOADING RATE

$$\text{Surface Loading, GPD/sq ft} = \frac{\text{Flow, GPD}}{\text{Area, sq ft}}$$

$$= \frac{900,000 \text{ GPD}}{1,960 \text{ sq ft}}$$

$$= 459 \text{ GPD/sq ft}$$

6. WEIR OVERFLOW RATE

$$\text{Weir Overflow, GPD/ft} = \frac{\text{Flow, GPD}}{\text{Length of Weir, ft}}$$

$$\text{Length of Weir, ft} = 3.14(\text{Diameter, ft})$$

$$= 3.14(50 \text{ ft})$$

$$= 157 \text{ ft}$$

$$\text{Weir Overflow, GPD/ft} = \frac{\text{Flow, GPD}}{\text{Length of Weir, ft}}$$

$$= \frac{900,000 \text{ GPD}}{157 \text{ ft}}$$

$$= 5,730 \text{ GPD/ft}$$

7. SOLIDS LOADING

$$\text{Solids Applied, lbs/day} = (\text{Flow, MGD})(\text{SS, mg/}L)(8.34 \text{ lbs/gal})$$

$$= (0.9 \text{ MGD})(3,600 \text{ mg/}L)(8.34 \text{ lbs/gal})$$

$$= 27,022 \text{ lbs/day}$$

$$\text{Solids Loading, lbs/day/sq ft} = \frac{\text{Solids Applied, lbs/day}}{\text{Surface Area, sq ft}}$$

$$= \frac{27,022 \text{ lbs/day}}{1,960 \text{ sq ft}}$$

$$= 14 \text{ lbs/day/sq ft}$$

A.133   Trickling Filters

EXAMPLES 8 and 9

A flow of 1.1 MGD is applied to a 50-ft diameter trickling filter that is 4 feet deep. The BOD of the wastewater applied to the filter is 120 mg/L. Calculate the hydraulic and organic loadings on the filter.

Known	Unknown
Flow, MGD = 1.1 MGD	8. Hydraulic Loading, GPD/sq ft
Diameter, ft = 50 ft	9. Organic Loading, $\frac{\text{lbs BOD/day}}{1,000 \text{ cu ft}}$
Depth, ft = 4 ft	
BOD, mg/L = 120 mg/L	

8. HYDRAULIC LOADING

$$\text{Hydraulic Loading, GPD/sq ft} = \frac{\text{Flow, GPD/sq ft}}{\text{Surface Area, sq ft}}$$

$$\text{Surface Area, sq ft} = 0.785(\text{Diameter, ft})^2$$

$$= 0.785(50 \text{ ft})^2$$

$$= 1,960 \text{ sq ft}$$

Hydraulic Loading, GPD/sq ft
$$= \frac{\text{Flow, GPD}}{\text{Surface Area, sq ft}}$$

$$= \frac{1,100,000 \text{ GPD}}{1,960 \text{ sq ft}}$$

$$= 561 \text{ GPD/sq ft}$$

9. ORGANIC LOADING

Organic Loading, $\frac{\text{lbs BOD/day}}{1,000 \text{ cu ft}}$
$$= \frac{\text{BOD Applied, lbs/day}}{\text{Volume of Media, 1,000 cu ft}}$$

BOD Applied, lbs/day
$$= (\text{BOD, mg/}L)(\text{Flow, MGD})(8.34 \text{ lbs/gal})$$

Volume of Media, 1,000 cu ft
$$= (\text{Surface Area, sq ft})(\text{Depth, ft})$$

BOD Applied, lbs/day
$$= (\text{BOD, mg/}L)(\text{Flow, MGD})(8.34 \text{ lbs/gal})$$

$$= (120 \text{ mg/}L)(1.1 \text{ MGD})(8.34 \text{ lbs/gal})$$

$$= 1,100 \text{ lbs BOD/day}$$

Volume of Media, 1,000 cu ft
$$= (\text{Surface Area, sq ft})(\text{Depth, ft})$$

$$= (1,960 \text{ sq ft})(4 \text{ ft})$$

$$= 7,840 \text{ cu ft}$$

$$= 7.84(1,000 \text{ cu ft})$$

Organic Loading, $\frac{\text{lbs BOD/day}}{1,000 \text{ cu ft}}$
$$= \frac{\text{BOD Applied, lbs/day}}{\text{Volume of Media, 1,000 cu ft}}$$

$$= \frac{1,100 \text{ lbs BOD/day}}{7.84(1,000 \text{ cu ft})}$$

$$= 140 \text{ lbs BOD/day/1,000 cu ft}$$

EXAMPLE 10

The influent BOD to a trickling filter plant is 200 mg/L and the effluent BOD is 20 mg/L. What is the BOD removal efficiency of the plant?

Known	Unknown
Influent BOD, mg/L = 200 mg/L	Plant Efficiency, %
Effluent BOD, mg/L = 20 mg/L	

Efficiency, %
$$= \left(\frac{\text{In} - \text{Out}}{\text{In}}\right)(100\%)$$

$$= \left(\frac{200 \text{ mg/}L - 20 \text{ mg/}L}{200 \text{ mg/}L}\right)(100\%)$$

$$= \frac{180 \text{ mg/}L}{200 \text{ mg/}L}(100\%)$$

$$= 90\%$$

A.134 Rotating Biological Contactors

EXAMPLES 11 and 12

A rotating biological contactor treats a flow of 2.4 MGD. The surface area of the media is 720,000 square feet. The influent has a total BOD of 220 mg/L and suspended solids of 240 mg/L. Assume a K value of 0.5 to calculate the soluble BOD. Calculate the hydraulic loading in gallons per day per square foot and the organic loading in pounds of soluble BOD per day per 1,000 square feet of media surface.

Known		Unknown
Flow, MGD	= 2.4 MGD	11. Hydraulic Loading, GPD/sq ft
Area, sq ft	= 720,000 sq ft	
Total BOD, mg/L	= 220 mg/L	12. Organic Loading, lbs BOD/day/1,000 sq ft
SS, mg/L	= 240 mg/L	
K	= 0.5	

11. HYDRAULIC LOADING

Hydraulic Loading, GPD/sq ft
$$= \frac{\text{Flow, GPD}}{\text{Surface Area, sq ft}}$$

$$= \frac{2,400,000 \text{ GPD}}{720,000 \text{ sq ft}}$$

$$= 3.3 \text{ GPD/sq ft}$$

12. ORGANIC LOADING

Soluble BOD, mg/L
$$= \text{Total BOD, mg/}L - (K)(\text{Suspended Solids, mg/}L)$$

$$= 220 \text{ mg/}L - (0.5)(240 \text{ mg/}L)$$

$$= 100 \text{ mg/}L$$

BOD Applied, lbs/day
$$= (\text{Flow, MGD})(\text{BOD, mg/}L)(8.34 \text{ lbs/gal})$$

$$= (2.4 \text{ MGD})(100 \text{ mg/}L)(8.34 \text{ lbs/gal})$$

$$= 2,000 \text{ lbs/day}$$

Organic Loading, lbs BOD/day/1,000 sq ft
$$= \frac{\text{Soluble BOD Applied, lbs/day}}{\text{Surface Area of Media} \cdot 1,000 \text{ sq ft}}$$

$$= \frac{2,000 \text{ lbs BOD/day}}{720 \cdot 1,000 \text{ sq ft}}$$

$$= 2.8 \text{ lbs BOD/day/1,000 sq ft}$$

A.135   Activated Sludge

EXAMPLES 13, 14, 15, and 16

Lab results and flow rate for an activated sludge plant are listed below under the known column. Information helpful to the operator in controlling the process is listed in the unknown column. The aerator or aeration tank volume is 0.50 MG.

Known

Mixed Liquor Suspended Solids (MLSS)	= 1,800 mg/L
Mixed Liquor Volatile Content	= 76%
Thirty-Minute Settleable Solids Test	= 170 mL/L, or 17%
Primary Effluent BOD	= 140 mg/L
Primary Effluent Suspended Solids	= 110 mg/L
Flow Rate	= 2.0 MGD

Unknown

13. Sludge Volume Index, SVI
14. Pounds of Solids in the Aerator
15. Pounds of BOD Applied Per Day to Aerator
16. Sludge Age, days

13. *SLUDGE VOLUME INDEX*

$$SVI = \frac{(\text{Settleable Solids, \%})(10,000)}{\text{MLSS, mg/}L} = \frac{(\text{Set Sol, m}L/L)(1,000)}{\text{MLSS, mg/}L}$$

$$= \frac{(17)(10,000)}{1,800} \qquad = \frac{(170)(1,000)}{1,800}$$

$$= 94 \qquad\qquad = 94$$

14. *POUNDS OF SOLIDS IN AERATOR*

$$\text{Aerator Solids, lbs} = (\text{MLSS, mg/}L)(\text{Tank Vol, MG})(8.34 \text{ lbs/gal})$$

$$= (1,800 \text{ mg/}L)(0.50 \text{ MG})(8.34 \text{ lbs/gal})$$

$$= 7,500 \text{ lbs}$$

15. *POUNDS OF BOD APPLIED PER DAY TO AERATOR*

$$\begin{array}{l}\text{Aerator} \\ \text{Loading,} \\ \text{lbs BOD/} \\ \text{day}\end{array} = (\text{Primary Effluent BOD, mg/}L)(\text{Flow, MGD})(8.34 \text{ lbs/gal})$$

$$= (140 \text{ mg/}L)(2.0 \text{ MGD})(8.34 \text{ lbs/gal})$$

$$= 2,335 \text{ lbs BOD/day Applied to Aerator}$$

16. *SLUDGE AGE*

Chapter 11, "Activated Sludge," and Chapter 16, "Laboratory Procedures and Chemistry," discuss the different methods of calculating sludge age and the meaning of the results.

$$\begin{array}{l}\text{Sludge} \\ \text{Age,} \\ \text{days}\end{array} = \frac{(\text{MLSS, mg/}L)(\text{Tank Volume, MG})(8.34 \text{ lbs/gal})}{(\text{SS in Primary Effl, mg/}L)(\text{Flow, MGD})(8.34 \text{ lbs/gal})}$$

$$= \frac{\text{Mixed Liquor Solids, lbs}}{\text{Primary Effluent Solids, lbs/day}} \quad \text{(or Aerator Solids, lbs)}$$

$$= \frac{7,500 \text{ lbs}}{(110 \text{ mg/}L)(2.0 \text{ MGD})(8.34 \text{ lbs/gal})}$$

$$= \frac{7,500 \text{ lbs}}{1,835 \text{ lbs/day}}$$

$$= 4.1 \text{ days}$$

A.136   Ponds

EXAMPLES 17, 18, 19, and 20

To calculate the different loadings on a pond, the information listed under known must be available.

Known

Avg Depth, ft	= 4 ft
Avg Width, ft	= 400 ft
Avg Length, ft	= 600 ft
Flow, MGD	= 0.5 MGD
BOD, mg/L	= 150 mg/L
Population, persons	= 5,000 persons

Unknown

17. Detention Time, days
18. Population Loading, persons/acre
19. Hydraulic Load, in/day
20. Organic Load, lbs BOD/day/acre

17. *DETENTION TIME*

$$\text{Detention Time, days} = \frac{\text{Pond Volume, ac-ft}}{\text{Flow Rate, ac-ft/day}}$$

$$\text{Pond Area, acres} = \frac{(\text{Avg Width, ft})(\text{Avg Length, ft})}{43,560 \text{ sq ft/acre}}$$

$$= \frac{(400 \text{ ft})(600 \text{ ft})}{43,560 \text{ sq ft/acre}}$$

$$= 5.51 \text{ acres}$$

$$\text{Pond Volume, ac-ft} = (\text{Avg Area, ac})(\text{Depth, ft})$$

$$= (5.51 \text{ acres})(4 \text{ ft})$$

$$= 22.0 \text{ ac-ft}$$

$$\text{Flow Rate, ac-ft/day} = \frac{500,000 \text{ gal/day}}{(7.48 \text{ gal/cu ft})(43,560 \text{ sq ft/ac})}$$

$$= 1.53 \text{ ac-ft/day}$$

$$\text{Detention Time, days} = \frac{\text{Pond Volume, ac-ft}}{\text{Flow Rate, ac-ft/day}}$$

$$= \frac{22.0 \text{ ac-ft}}{1.53 \text{ ac-ft/day}}$$

$$= 14.4 \text{ days}$$

18. *POPULATION LOADING*

$$\text{Population Loading, persons/ac} = \frac{\text{Population Served, persons}}{\text{Surface Pond Area, acres}}$$

$$= \frac{5,000 \text{ persons}}{5.51 \text{ acres}}$$

$$= 907 \text{ persons/acre}$$

19. *HYDRAULIC LOADING*

$$\text{Hydraulic Loading, in/day} = \frac{\text{Depth of Pond, inches}}{\text{Detention Time, days}}$$

$$= \frac{(4 \text{ ft})(12 \text{ in/ft})}{14.4 \text{ days}}$$

$$= 3.33 \text{ in/day}$$

20. *ORGANIC LOADING*

$$\text{Organic Loading,} \atop \text{lbs BOD/day/ac} = \frac{(\text{BOD, mg/}L)(\text{Flow, MGD})(8.34 \text{ lbs/gal})}{\text{Area, ac}}$$

$$= \frac{(150 \text{ mg/}L)(0.5 \text{ MGD})(8.34 \text{ lbs/gal})}{5.51 \text{ ac}}$$

$$= \frac{625.5 \text{ lbs BOD/day}}{5.51 \text{ ac}}$$

$$= 114 \text{ lbs BOD/day/ac}$$

A.137 Chlorination

EXAMPLE 21

Determine the chlorine demand* of an effluent if the chlorine dose is 10.0 mg/L and the chlorine residual is 1.1 mg/L.

Known	Unknown
Chlorine Dose, mg/L = 10.0 mg/L	Chlorine Demand, mg/L
Chlorine Residual, mg/L = 1.1 mg/L	

$$\text{Chlorine Demand,} \atop \text{mg/}L = \text{Chlor Dose, mg/}L - \text{Chlor Residual, mg/}L$$

$$= 10.0 \text{ mg/}L - 1.1 \text{ mg/}L$$

$$= 8.9 \text{ mg/}L$$

EXAMPLE 22

To maintain a satisfactory chlorine residual in a plant effluent, the chlorine dose must be 10 mg/L when the flow is 0.37 MGD. Determine the chlorinator setting (feed rate) in pounds per day.

Known	Unknown
Dose, mg/L = 10 mg/L	Chlorinator Setting, lbs/day
Flow, MGD = 0.37 MGD	

$$\text{Chlorine Feed} \atop \text{Rate, lbs/day} = (\text{Dose, mg/}L)(\text{Flow, MGD})(8.34 \text{ lbs/gal})$$

$$= (10 \text{ mg/}L)(0.37 \text{ MGD})(8.34 \text{ lbs/gal})$$

$$= 30.9, \text{ or}$$

$$= 31 \text{ lbs/day}$$

---

* *STANDARD METHODS* has used the term 'chlorine demand' when referring to stabilized water such as a domestic water supply and the term 'chlorination requirement' when referring to wastewater.

A.138 Chemical Doses

EXAMPLE 23

Determine the chlorinator setting in pounds per 24 hours to treat a flow of 2 MGD with a chlorine dose of 3.0 mg/L.

Known	Unknown
Flow, MGD = 2 MGD	Chlorinator Setting, lbs/24 hours
Chlorine Dose, mg/L = 3.0 mg/L	

Determine the chlorinator setting in pounds per 24 hours or pounds per day.

$$\text{Chemical Feed,} \atop \text{lbs/day} = (\text{Flow, MGD})(\text{Dose, mg/}L)(8.34 \text{ lbs/gal})$$

$$= (2 \text{ MGD})(3.0 \text{ mg/}L)(8.34 \text{ lbs/gal})$$

$$= 50 \text{ lbs/day}$$

NOTE: We use the "lbs/day formula" regularly in our work to calculate the setting on a chemical feeder (lbs chlorine per day) and the loading on a treatment process (lbs BOD per day). An explanation of how the units in the formula cancel helps us understand how to use and apply the formula to problems such as this one in Example 23.

Calculate the setting on a chlorinator in pounds of chlorine per day to the flow of two million gallons per day (2 MGD) at a chlorine dose of 3.0 milligrams of chlorine per liter of water (3.0 mg/L). To perform this calculation we need to realize that one liter of water weighs one million milligrams. Therefore,

$$\frac{\text{mg}}{L} = \frac{\text{mg}}{1 \text{ Million mg}} = \frac{\text{lbs}}{1 \text{ Million lbs}}$$

Calculate the chlorinator setting in pounds of chlorine per day.

$$\text{lbs Cl/day} = (\text{Flow, MGD})(\text{Dose, mg/}L)(8.34 \text{ lbs/gal})$$

$$\text{or} \qquad = (\text{Flow, MGD})(\text{Dose, lbs/M lbs})(8.34 \text{ lbs/gal})$$

$$= \frac{2 \text{ M gal H}_2\text{O}}{\text{day}} \times \frac{3.0 \text{ lbs Cl}}{1 \text{ M lbs H}_2\text{O}} \times \frac{8.34 \text{ lbs H}_2\text{O}}{\text{gal H}_2\text{O}}$$

In this formula the million (M) on the top and the bottom of the formula cancel, the gallons of water (gal H_2O) on top and bottom cancel and the pounds of water (lbs H_2O) on top and bottom cancel. This leaves us with pounds of chlorine (lbs Cl) on the top and day (day) on the bottom. The answer is the chlorinator setting in pounds of chlorine per day (lbs Cl/day).

EXAMPLE 24

The optimum liquid polymer dose from the jar tests is 12 mg/L. Determine the setting on the liquid polymer chemical feeder in milliliters per minute when the plant flow is 4.7 MGD. The liquid polymer delivered to the plant contains 642.3 milligrams of polymer per milliliter of liquid solution.

Known	Unknown
Polymer Dose, mg/L = 12 mg/L	Chemical Feeder Setting, mL/min
Flow, MGD = 4.7 MGD	
Liquid Polymer, mg/mL = 642.3 mg/mL	

Calculate the liquid polymer chemical feeder setting in milliliters per minute.

$$\text{Chemical Feeder Setting, mL/min} = \frac{\text{(Flow, MGD)(Polymer Dose, mg/}L\text{)(3.785 }L\text{/gal)(1,000,000/M)}}{\text{(Liquid Polymer, mg/m}L\text{)(24 hr/day)(60 min/hr)}}$$

$$= \frac{\text{(4.7 MGD)(12 mg/}L\text{)(3.785 }L\text{/gal)(1,000,000/M)}}{\text{(642.3 mg/m}L\text{)(24 hr/day)(60 min/hr)}}$$

$$= 231 \text{ m}L\text{/min}$$

EXAMPLE 25

The optimum liquid polymer dose from the jar tests is 12 mg/L. Determine the setting on the liquid polymer chemical feeder in gallons per day when the flow is 4.7 MGD. The liquid polymer delivered to the plant contains 5.36 pounds of polymer per gallon of liquid solution.

Known		Unknown
Polymer Dose, mg/L	= 12 mg/L	Chemical Feeder Setting, GPD
Flow, MGD	= 4.7 MGD	
Liquid Polymer, lbs/gal	= 5.36 lbs/gal	

Calculate the liquid polymer chemical feeder setting in gallons per day.

$$\text{Chemical Feeder Setting, GPD} = \frac{\text{(Flow, MGD)(Polymer Dose, mg/}L\text{)(8.34 lbs/gal)}}{\text{Liquid Polymer, lbs/gal}}$$

$$= \frac{\text{(4.7 MGD)(12 mg/}L\text{)(8.34 lbs/gal)}}{\text{5.36 lbs/gal}}$$

$$= 88 \text{ GPD}$$

EXAMPLE 26

Determine the actual chemical dose or chemical feed in pounds per day from a dry chemical feeder. A bucket placed under the chemical feeder weighed 0.3 pound empty and 2.1 pounds after 30 minutes.

Known		Unknown
Empty Bucket, lbs	= 0.3 lb	Chemical Feed, lbs/day
Full Bucket, lbs	= 2.1 lbs	
Time to Fill, min	= 30 min	

Determine the chemical feed in pounds of chemical applied per day.

$$\text{Chemical Feed, lbs/day} = \frac{\text{Chemical Applied, lbs}}{\text{Length of Application, days}}$$

$$= \frac{\text{(2.1 lbs} - \text{0.3 lb)(60 min/hr)(24 hr/day)}}{30 \text{ min}}$$

$$= 86 \text{ lbs/day}$$

EXAMPLE 27

Determine the chemical feed in pounds of polymer per day from a chemical feed pump. The polymer solution is 1.5 percent or 15,000 mg polymer per liter. Assume a specific gravity of the polymer solution of 1.0. During a test run, the chemical feed pump delivered 800 mL of polymer solution during five minutes.

Known		Unknown
Polymer Solution, %	= 1.5%	Polymer Feed, lbs/day
Polymer Conc, mg/L	= 15,000 mg/L	
Polymer Sp Gr	= 1.0	
Volume Pumped, mL	= 800 mL	
Time Pumped, min	= 5 min	

Calculate the polymer fed by the chemical feed pump in pounds of polymer per day.

$$\text{Polymer Feed, lbs/day} = \frac{\text{(Poly Conc, mg/}L\text{)(Vol Pumped, m}L\text{)(60 min/hr)(24 hr/day)}}{\text{(Time Pumped, min)(1,000 m}L/L\text{)(1,000 mg/gm)(454 gm/lb)}}$$

$$= \frac{\text{(15,000 mg/}L\text{)(800 m}L\text{)(60 min/hr)(24 hr/day)}}{\text{(5 min)(1,000 m}L/L\text{)(1,000 mg/gm)(454 gm/lb)}}$$

$$= 7.6 \text{ lbs Polymer/day}$$

EXAMPLE 28

A small chemical feed pump lowered the chemical solution in a three-foot diameter tank one foot and seven inches during an eight-hour period. Estimate the flow delivered by the pump in gallons per minute and gallons per day.

Known		Unknown
Tank Diameter, ft	= 3 ft	Flow, GPM
Chemical Drop, ft	= 1 ft 7 in	Flow, GPD
Time, hr	= 8 hr	

1. Convert the tank drop from one foot seven inches to feet.

$$\text{Tank Drop, ft} = 1 \text{ ft} + 7 \text{ inches}$$

$$= 1 \text{ ft} + \frac{7 \text{ in}}{12 \text{ in/ft}}$$

$$= 1 \text{ ft} + 0.58 \text{ ft}$$

$$= 1.58 \text{ ft}$$

2. Determine the gallons of water pumped.

$$\text{Volume Pumped, gal} = \text{(Area, sq ft)(Drop, ft)(7.48 gal/cu ft)}$$

$$= (0.785)(3 \text{ ft})^2(1.58 \text{ ft})(7.48 \text{ gal/cu ft})$$

$$= 83.5 \text{ gal}$$

3. Estimate the flow delivered by the pump in gallons per minute and gallons per day.

$$\text{Flow, GPM} = \frac{\text{Volume Pumped, gal}}{(\text{Time, hr})(60 \text{ min/hr})}$$

$$= \frac{83.5 \text{ gal}}{(8 \text{ hr})(60 \text{ min/hr})}$$

$$= 0.17 \text{ GPM}$$

or

$$\text{Flow, GPD} = \frac{(\text{Volume Pumped, gal})(24 \text{ hr/day})}{\text{Time, hr}}$$

$$= \frac{(83.5 \text{ gal})(24 \text{ hr/day})}{8 \text{ hr}}$$

$$= 250 \text{ GPD}$$

EXAMPLE 29

Determine the settings in percent stroke on a chemical feed pump for various doses of a chemical in milligrams per liter. (The chemical could be chlorine, polymer, potassium permanganate, or any other chemical solution fed by a pump.) The pump delivering the water to be treated pumps at a flow rate of 400 GPM. The solution strength of the chemical being pumped is 4.8 percent. The chemical feed pump has a maximum capacity of 92 gallons per day at a setting of 100 percent capacity.

Known		Unknown
Pump Flow, GPM	= 400 GPM	Settings, % stroke for various doses in mg/L
Solution Strength, %	= 4.8%	
Feed Pump, GPD (100% stroke)	= 92 GPD	

1. Convert the pump flow from gallons per minute to million gallons per day.

$$\text{Pump Flow, MGD} = (\text{Pump Flow, GPM})(60 \text{ min/hr})(24 \text{ hr/day})$$

$$= (400 \text{ gal/min})(60 \text{ min/hr})(24 \text{ hr/day})$$

$$= 576,000 \text{ gal/day}$$

$$= 0.576 \text{ MGD}$$

2. Change the chemical solution strength from a percent to pounds of chemical per gallon of solution. A 4.8 percent solution means we have 4.8 pounds of chemical in a solution of water and chemical weighing 100 pounds.

$$\text{Chemical Solution, lbs/gal} = \frac{4.8 \text{ lbs Chemical}}{100 \text{ lbs of Chemical and Water}}$$

$$= \frac{(4.8 \text{ lbs})(8.34 \text{ lbs/gal})}{100 \text{ lbs}}$$

$$= 0.4 \text{ lb Chemical/gallon solution}$$

3. Calculate the chemical feed in pounds per day for a chemical dose of 0.5 milligram per liter. We are going to assume various chemical doses of 0.5, 1.0, 1.5, 2.0, 2.5 mg/L and upward so that if we know the desired chemical dose, we can easily determine the setting (percent stroke) on the chemical feed pump.

$$\text{Chemical Feed, lbs/day} = (\text{Flow, MGD})(\text{Dose, mg/}L)(8.34 \text{ lbs/gal})$$

$$= (0.576 \text{ MGD})(0.5 \text{ mg/}L)(8.34 \text{ lbs/gal})$$

$$= 2.4 \text{ lbs/day}$$

4. Determine the desired flow from the chemical feed pump in gallons per day.

$$\text{Feed Pump, GPD} = \frac{\text{Chemical Feed, lbs/day}}{\text{Chemical Solution, lbs/gal}}$$

$$= \frac{2.4 \text{ lbs/day}}{0.4 \text{ lb/gal}}$$

$$= 6 \text{ GPD}$$

5. Determine the setting on the chemical feed pump as a percent. In this case we want to know the setting as a percent of the pump stroke.

$$\text{Setting, \%} = \frac{(\text{Desired Feed Pump, GPD})(100\%)}{\text{Maximum Feed Pump, GPD}}$$

$$= \frac{(6 \text{ GPD})(100\%)}{92 \text{ GPD}}$$

$$= 6.5\%$$

6. If we changed the chemical dose in Step 3 from 0.5 mg/L to 1.0 mg/L and other higher doses and repeated the remainder of the steps, we could calculate the data in Table A.1 (page 471).

7. Plot the data in Table A.1 (Chemical Dose, mg/L vs. Pump Setting, % stroke) to obtain Figure A.3. Only three points were needed since the data plotted a straight line. For any desired chemical dose in milligrams per liter, you can use Figure A.3 to determine the necessary chemical feed pump setting.

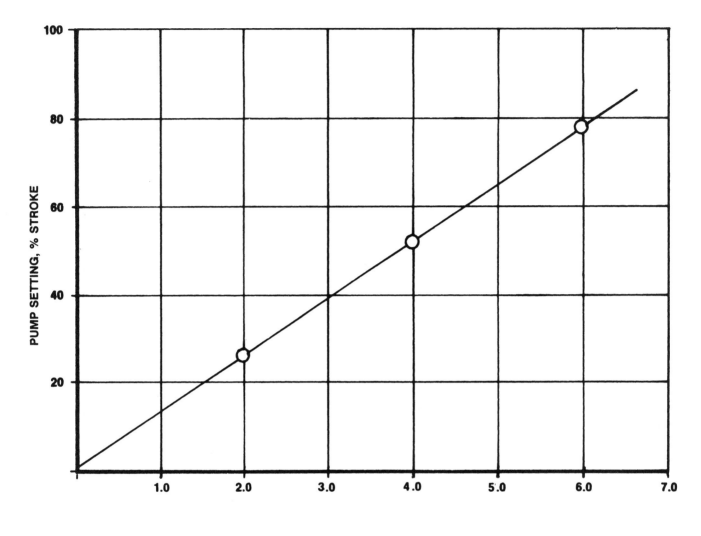

Fig. A.3   Chemical feed pump settings for various chemical doses

TABLE A.1 SETTINGS FOR CHEMICAL FEED PUMP

PUMP FLOW, GPM = 400 GPM

SOLUTION STRENGTH, % = 4.8%

Chemical Dose, mg/L	Chemical Feed, lbs/day	Feed Pump, GPD	Pump Setting, % stroke
0.5	2.4	6.0	6.5
1.0	4.8	12.0	13.0
1.5	7.2	18.0	19.5
2.0	9.6	24.0	26.1
2.5	12.0	30.0	32.6
3.0	14.4	36.0	39.1
3.5	16.8	42.0	45.6
4.0	19.2	48.0	52.2
4.5	21.6	54.0	58.7
5.0	24.0	60.0	65.2
5.5	26.4	66.0	71.7
6.0	28.8	72.0	78.2
6.5	31.2	78.0	84.8
7.0	33.6	84.0	91.3
7.5	36.0	90.0	97.8

A.139 Blueprint Reading

EXAMPLE 30

A set of blueprints for a treatment plant has a scale of $1/4$ inch = 1 foot. On the prints, the laboratory dimensions were measured and found to be 6 inches wide and 9 inches long. What is the floor area of the laboratory?

Known	Unknown
Scale: $1/4$ in = 1 ft	Area, sq ft

Length, in = 9 in

Width, in = 6 in

Area, sq ft = (Length, ft)(Width, ft)

Convert actual length and width to feet.

$$\frac{1/4 \text{ in}}{1 \text{ ft}} = \frac{9 \text{ in}}{\text{Length, ft}}$$

$$\text{Length, ft} = (9 \text{ in})\left[\frac{1 \text{ ft}}{1/4 \text{ in}}\right]$$

$$= (9)(4)$$

$$= 36 \text{ ft}$$

$$\frac{1/4 \text{ in}}{1 \text{ ft}} = \frac{6 \text{ in}}{\text{Width, ft}}$$

$$\text{Width, ft} = (6 \text{ in})\left[\frac{1 \text{ ft}}{1/4 \text{ in}}\right]$$

$$= (6)(4)$$

$$= 24 \text{ ft}$$

$$\text{Area, sq ft} = (\text{Length, ft})(\text{Width, ft})$$

$$= (36 \text{ ft})(24 \text{ ft})$$

$$= 864 \text{ sq ft}$$

A.14 BASIC CONVERSION FACTORS (METRIC SYSTEM)

LENGTH

100 cm	= 1 m	100 cm/m
3.281 ft	= 1 m	3.281 ft/m

AREA

2.4711 ac	= 1 ha*	2.4711 ac/ha
10,000 sq m	= 1 ha	10,000 sq m/ha

VOLUME

1,000 mL	= 1 liter	1,000 mL/L
1,000 L	= 1 cu m	1,000 L/cu m
3.785 L	= 1 gal	3.785 L/gal

WEIGHT

1,000 mg	= 1 gm	1,000 mg/gm
1,000 gm	= 1 kg	1,000 gm/kg

DENSITY

1 kg	= 1 liter	1 kg/L

PRESSURE

10.015 m	= 1 kg/sq cm	10.015 m/kg/sq cm
1 Pascal	= 1 N/sq m	1 Pa/N/sq m
1 psi	= 6,895 Pa	1 psi/6,895 Pa

FLOW

3,785 cu m/day	= 1 MGD	3,785 cu m/day/MGD
3.785 ML/day	= 1 MGD	3.785 ML/day/MGD

* hectare

A.15 TYPICAL WASTEWATER TREATMENT PLANT PROBLEMS (METRIC SYSTEM)

A.150 Flows

EXAMPLE 1

Convert a flow of 500 gallons per minute to liters per second and cubic meters per day.

Known	Unknown
Flow, GPM = 500 GPM	1. Flow, liters/sec
	2. Flow, cu m/day

1. Convert the flow from 500 GPM to liters per second.

$$\text{Flow, liters/sec} = \frac{(\text{Flow, gal/min})(3.785 \text{ liters/gal})}{60 \text{ sec/min}}$$

$$= \frac{(500 \text{ gal/min})(3.785 \text{ liters/gal})}{60 \text{ sec/min}}$$

$$= 31.5 \text{ liters/sec}$$

2. Convert the flow from 500 GPM to cubic meters per day.

$$\text{Flow, cu m/day} = \frac{(\text{Flow, gal/min})(3.785 \text{ L/gal})(60 \text{ min/hr})(24 \text{ hr/day})}{1,000 \text{ L/cu m}}$$

$$= \frac{(500 \text{ gal/min})(3.785 \text{ L/gal})(60 \text{ min/hr})(24 \text{ hr/day})}{1,000 \text{ L/cu m}}$$

$$= 2,725 \text{ cu m/day}$$

A.151 Grit Channels

EXAMPLE 2

Estimate the velocity of wastewater flowing through a grit channel if a stick travels 16 meters in 40 seconds.

Known	Unknown
Distance, m = 16 m	Velocity, m/sec
Time, sec = 40 sec	

$$\text{Velocity, m/sec} = \frac{\text{Distance Traveled, m}}{\text{Time, sec}}$$

$$= \frac{16 \text{ m}}{40 \text{ sec}}$$

$$= 0.4 \text{ m/sec}$$

EXAMPLE 3

A grit channel removed 100 liters of grit during a period when the total flow was 3,000 cubic meters. How many liters of grit are removed per cubic meter?

Known	Unknown
Vol of Grit, liters = 100 liters	Grit Removed, liters/cu m
Vol of Flow, cu m = 3,000 cu m	

$$\text{Grit Removed, } L\text{/cu m} = \frac{\text{Volume of Grit, liters}}{\text{Volume of Flow, cu m}}$$

$$= \frac{100 \text{ liters}}{3,000 \text{ cu m}}$$

$$= 0.033 \text{ } L\text{/cu m}$$

A.152 Sedimentation Tanks and Clarifiers

EXAMPLES 4, 5, 6, and 7

A circular secondary clarifier handles a flow of 3,400 cu m/day and suspended solids of 3,600 mg/L. The clarifier is 15 meters in diameter and 2.5 meters deep. Find the detention time, surface loading rate, weir overflow rate, and solids loading.

Known

Flow, cu m/day	= 3,400 cu m/day
Diameter, m	= 15 m
Depth, m	= 2.5 m
SS, mg/L	= 3,600 mg/L

Unknown

4. Detention Time, hours
5. Surface Loading, cu m/day/sq m
6. Weir Overflow, cu m/day/m
7. Solids Loading, kg/day/sq m

4. *DETENTION TIME*

$$\text{Detention Time, hr} = \frac{(\text{Tank Volume, cu m})(24 \text{ hr/day})}{\text{Flow, cu m/day}}$$

$$\text{Tank Volume, cu m} = (\text{Area, sq m})(\text{Depth, m})$$

$$\text{Clarifier Area, sq m} = 0.785(\text{Diameter, m})^2$$

$$= 0.785(15 \text{ m})^2$$

$$= 176.7 \text{ sq m, or}$$

$$= 180 \text{ sq m}$$

$$\text{Tank Volume, cu m} = (\text{Area, sq m})(\text{Depth, m})$$

$$= (180 \text{ sq m})(2.5 \text{ m})$$

$$= 450 \text{ cu m}$$

$$\text{Detention Time, hr} = \frac{(\text{Tank Volume, cu m})(24 \text{ hr/day})}{\text{Flow, cu m/day}}$$

$$= \frac{(450 \text{ cu m})(24 \text{ hr/day})}{3,400 \text{ cu m/day}}$$

$$= \frac{10,800}{3,400}$$

$$= 3.2 \text{ hr}$$

5. *SURFACE LOADING RATE*

$$\text{Surface Loading, } \frac{\text{cu m/day}}{\text{sq m}} = \frac{\text{Flow, cu m/day}}{\text{Area, sq m}}$$

$$= \frac{3,400 \text{ cu m/day}}{180 \text{ sq m}}$$

$$= 18.9 \text{ cu m/day/sq m}$$

6. *WEIR OVERFLOW RATE*

$$\text{Weir Overflow, } \frac{\text{cu m/day}}{\text{m}} = \frac{\text{Flow, cu m/day}}{\text{Length of Weir, m}}$$

$$\text{Length of Weir, m} = 3.14(\text{Diameter, m})$$

$$= 3.14(15 \text{ m})$$

$$= 47.1 \text{ m}$$

$$\text{Weir Overflow, } \frac{\text{cu m/day}}{\text{m}} = \frac{\text{Flow, cu m/day}}{\text{Length of Weir, m}}$$

$$= \frac{3,400 \text{ cu m/day}}{47.1 \text{ m}}$$

$$= 72 \text{ cu m/day/m}$$

7. *SOLIDS LOADING*

$$\text{Solids Applied, kg/day} = \frac{(\text{Flow, cu m/day})(\text{SS, mg/}L)(1,000 \text{ } L\text{/cu m})}{(1,000 \text{ mg/gm})(1,000 \text{ gm/kg})}$$

$$= \frac{(3,400 \text{ cu m/day})(3,600 \text{ mg/}L)(1,000 \text{ } L\text{/cu m})}{(1,000 \text{ mg/gm})(1,000 \text{ gm/kg})}$$

$$= 12,240 \text{ kg/day}$$

$$\text{Solids Loading, kg/day/sq m} = \frac{\text{Solids Applied, kg/day}}{\text{Surface Area, sq m}}$$

$$= \frac{12,240 \text{ kg/day}}{180 \text{ sq m}}$$

$$= 68 \text{ kg/day/sq m}$$

A.153   Trickling Filters

EXAMPLES 8 and 9

A flow of 4,300 cubic meters per day is applied to a 15-meter diameter trickling filter that is 1.2 meters deep. The BOD of the wastewater applied to the filter is 120 mg/L. Calculate the hydraulic and organic loadings on the filter.

Known	Unknown
Flow, cu m/day = 4,300 cu m/day	8. Hydraulic Loading, cu m/day/sq m
Diameter, m = 15 m	
Depth, m = 1.2 m	9. Organic Loading, kg BOD/day/cu m
BOD, mg/L = 120 mg/L	

8. *HYDRAULIC LOADING*

$$\text{Hydraulic Loading, cu m/day/sq m} = \frac{\text{Flow, cu m/day}}{\text{Surface Area, sq m}}$$

$$\text{Surface Area, sq m} = 0.785(\text{Diameter, m})^2$$

$$= 0.785(15 \text{ m})^2$$

$$= 180 \text{ sq m}$$

$$\text{Hydraulic Loading, cu m/day/sq m} = \frac{\text{Flow, cu m/day}}{\text{Surface Area, sq m}}$$

$$= \frac{4,300 \text{ cu m/day}}{180 \text{ sq m}}$$

$$= 24 \text{ cu m/day/sq m}$$

9. *ORGANIC LOADING*

$$\text{Organic Loading, } \frac{\text{kg BOD/day}}{\text{cu m}} = \frac{\text{BOD Applied, kg/day}}{\text{Volume of Media, cu m}}$$

$$\text{BOD Applied, kg/day} = (\text{BOD, }\frac{\text{mg}}{L})(\text{Flow, }\frac{\text{cu m}}{\text{day}})\frac{(1,000 \text{ } L)}{(1 \text{ cu m})}\frac{(1 \text{ kg})}{(1,000,000 \text{ mg})}$$

$$\text{Volume of Media, cu m} = (\text{Surface Area, sq m})(\text{Depth, m})$$

$$\text{BOD Applied, kg/day} = (\text{BOD, mg/}L)(\text{Flow, }\frac{\text{cu m}}{\text{day}})\frac{(1,000 \text{ } L)}{(1 \text{ cu m})}\frac{(1 \text{ kg})}{(1,000,000 \text{ mg})}$$

$$= (120 \text{ mg/}L)(4,300 \frac{\text{cu m}}{\text{day}})\frac{(1,000 \text{ } L)}{(1 \text{ cu m})}\frac{(1 \text{ kg})}{(1,000,000 \text{ mg})}$$

$$= 516 \text{ kg BOD/day}$$

$$\text{Volume of Media, cu m} = (\text{Surface Area, sq m})(\text{Depth, m})$$

$$= (180 \text{ sq m})(1.2 \text{ m})$$

$$= 216 \text{ cu m}$$

$$\text{Organic Loading, } \frac{\text{kg BOD/day}}{\text{cu m}} = \frac{\text{BOD Applied, kg/day}}{\text{Volume of Media, cu m}}$$

$$= \frac{516 \text{ kg BOD/day}}{216 \text{ cu m}}$$

$$= 2.4 \text{ kg BOD/day/cu m}$$

EXAMPLE 10

The influent BOD to a trickling filter plant is 200 mg/L and the effluent BOD is 20 mg/L. What is the BOD removal efficiency of the plant?

Known	Unknown
Influent BOD, mg/L = 200 mg/L	Plant Efficiency, %
Effluent BOD, mg/L = 20 mg/L	

$$\text{Efficiency, \%} = \frac{(\text{In} - \text{Out})}{\text{In}}(100\%)$$

$$= \frac{(200 \text{ mg/}L - 20 \text{ mg/}L)}{200 \text{ mg/}L}(100\%)$$

$$= \frac{180 \text{ mg/}L}{200 \text{ mg/}L}(100\%)$$

$$= 90\%$$

A.154   Rotating Biological Contactors

EXAMPLES 11 and 12

A rotating biological contactor treats a flow of 9.1 MLD (million liters per day or megaliters per day). The surface area of the media is 80,000 square meters. The influent has a total BOD of 220 mg/L and suspended solids of 240 mg/L. Assume a K value of 0.5 to calculate the soluble BOD. Calculate the hydraulic loading in liters per day per square meter and the organic loading in kilograms of soluble BOD per day per 1,000 square meters of media surface.

Known	Unknown
Flow, MLD = 9.1 MLD	11. Hydraulic Loading, L/day/sq m
Area, sq m = 80,000 sq m	
Total BOD, mg/L = 220 mg/L	12. Organic Loading, kg BOD/day/1,000 sq m
SS, mg/L = 240 mg/L	
K = 0.5	

11. *HYDRAULIC LOADING*

$$\text{Hydraulic Loading, } L\text{/day/sq m} = \frac{(\text{Flow, M}LD)(1,000,000/M)}{\text{Surface Area, sq m}}$$

$$= \frac{(9.1 \text{ M}LD)(1,000,000/M)}{80,000 \text{ sq m}}$$

$$= 114 \text{ } L\text{/day/sq m}$$

12. *ORGANIC LOADING*

$$\text{Soluble BOD, mg/}L = \text{Total BOD, mg/}L - (K)(\text{Suspended Solids, mg/}L)$$

$$= 220 \text{ mg/}L - (0.5)(240 \text{ mg/}L)$$

$$= 100 \text{ mg/}L$$

$$\text{BOD Applied, kg/day} = \frac{(\text{Flow, M}L\text{D})(\text{BOD, mg/}L)(1,000,000/\text{M})}{(1,000 \text{ mg/gm})(1,000 \text{ gm/kg})}$$

$$= \frac{(9.1 \text{ M}L\text{D})(100 \text{ mg/}L)(1,000,000/\text{M})}{(1,000 \text{ mg/gm})(1,000 \text{ gm/kg})}$$

$$= 910 \text{ kg/day}$$

$$\text{Organic Loading, kg BOD/day/1,000 sq m} = \frac{\text{Soluble BOD Applied, kg/day}}{\text{Surface Area of Media} \cdot 1,000 \text{ sq m}}$$

$$= \frac{910 \text{ kg/day}}{80 \cdot 1,000 \text{ sq m}}$$

$$= 11.4 \text{ kg BOD/day/1,000 sq m}$$

A.155   Activated Sludge

EXAMPLES 13, 14, 15, and 16

Lab results and flow rate for an activated sludge plant are listed below under the known column. Information helpful to the operator in controlling the process is listed in the unknown column. The aerator or aeration tank volume is 2,000 cu m.

Known

Mixed Liquor Suspended Solids (MLSS)	= 1,800 mg/L
Mixed Liquor Volatile Content	= 76%
Thirty-Minute Settleable Solids Test	= 170 mL/L, or 17%
Primary Effluent BOD	= 140 mg/L
Primary Effluent Suspended Solids	= 110 mg/L
Flow Rate	= 8,000 cu m/day

Unknown

13. Sludge Volume Index, SVI
14. Kilograms of Solids in the Aerator
15. Kilograms of BOD Applied Per Day to Aerator
16. Sludge Age, days

13. *SLUDGE VOLUME INDEX*

$$\text{SVI} = \frac{(\text{Settleable Solids, \%})(10,000)}{\text{MLSS, mg/}L} = \frac{(\text{Set Sol, m}L/L)(1,000)}{\text{MLSS, mg/}L}$$

$$= \frac{(17)(10,000)}{1,800} \qquad = \frac{(170)(1,000)}{1,800}$$

$$= 94 \qquad\qquad = 94$$

14. *KILOGRAMS OF SOLIDS IN AERATOR*

$$\text{Aerator Solids, kg} = (\text{MLSS, mg/}L)(\text{Tank Vol, cu m})\frac{(1,000 \text{ }L)}{(1 \text{ cu m})}\frac{(1 \text{ kg})}{(1,000,000 \text{ mg})}$$

$$= (1,800 \text{ mg/}L)(2,000 \text{ cu m})\frac{(1,000 \text{ }L)}{(1 \text{ cu m})}\frac{(1 \text{ kg})}{(1,000,000 \text{ mg})}$$

$$= 3,600 \text{ kg}$$

15. *KILOGRAMS OF BOD APPLIED PER DAY TO AERATOR*

$$\text{Aerator Loading, kg BOD/day} = (\text{P.E. BOD, mg/}L)(\text{Flow, }\frac{\text{cu m}}{\text{day}})\frac{(1,000 \text{ }L)}{(1 \text{ cu m})}\frac{(1 \text{ kg})}{(1,000,000 \text{ mg})}$$

$$= (140 \text{ mg/}L)(8,000 \frac{\text{cu m}}{\text{day}})\frac{(1,000 \text{ }L)}{(1 \text{ cu m})}\frac{(1 \text{ kg})}{(1,000,000 \text{ mg})}$$

$$= 1,120 \text{ kg BOD/day Applied to Aerator}$$

16. *SLUDGE AGE*

Chapter 11, "Activated Sludge," and Chapter 16, "Laboratory Procedures and Chemistry," discuss the different methods of calculating sludge age and the meaning of the results.

$$\text{Sludge Age, days} = \frac{(\text{MLSS, mg/}L)(\text{Tank Volume, cu m})\frac{(1,000 \text{ }L)}{(1 \text{ cu m})}\frac{(1 \text{ kg})}{(1,000,000 \text{ mg})}}{(\text{SS in P.E., mg/}L)(\text{Flow, }\frac{\text{cu m}}{\text{day}})\frac{(1,000 \text{ }L)}{(1 \text{ cu m})}\frac{(1 \text{ kg})}{(1,000,000 \text{ mg})}}$$

$$= \frac{\text{Mixed Liquor Solids, kg} \quad (\text{or Aerator Solids, kg})}{\text{Primary Effluent Solids, kg/day}}$$

$$= \frac{3,600 \text{ kg}}{(110 \text{ mg/}L)(8,000 \frac{\text{cu m}}{\text{day}})\frac{(1,000 \text{ }L)}{(1 \text{ cu m})}\frac{(1 \text{ kg})}{(1,000,000 \text{ mg})}}$$

$$= \frac{3,600 \text{ kg}}{800 \text{ kg/day}}$$

$$= 4.1 \text{ days}$$

A.156   Ponds

EXAMPLES 17, 18, 19, and 20

To calculate the different loadings on a pond, the information listed under known must be available.

Known

Average Depth, m	= 1.2 m
Average Width, m	= 120 m
Average Length, m	= 180 m
Flow, cu m/day	= 2,000 cu m/day
BOD, mg/L	= 150 mg/L
Population, persons	= 5,000 persons

Unknown

17. Detention Time, days
18. Population Loading, persons/sq m
19. Hydraulic Load, cm/day
20. Organic Load, gm BOD/day/sq m

17. *DETENTION TIME*

$$\text{Detention Time, days} = \frac{\text{Pond Volume, cu m}}{\text{Flow Rate, cu m/day}}$$

$$\text{Pond Area, sq m} = (\text{Avg Width, m})(\text{Avg Length, m})$$

$$= (120 \text{ m})(180 \text{ m})$$

$$= 21,600 \text{ sq m}$$

Pond Volume, cu m   = (Avg Area, sq m)(Depth, m)

   = (21,600 sq m)(1.2 m)

   = 25,920 cu m

Flow Rate, cu m/day   = 2,000 cu m/day

$$\text{Detention Time, days} = \frac{\text{Pond Volume, cu m}}{\text{Flow Rate, cu m/day}}$$

$$= \frac{25,920 \text{ cu m}}{2,000 \text{ cu m/day}}$$

$$= 13.0 \text{ days}$$

18. POPULATION LOADING

$$\text{Population Loading, persons/sq m} = \frac{\text{Population Served, persons}}{\text{Surface Pond Area, sq m}}$$

$$= \frac{5,000 \text{ persons}}{21,600 \text{ sq m}}$$

$$= 0.23 \text{ persons/sq m}$$

19. HYDRAULIC LOADING

$$\text{Hydraulic Loading, cm/day} = \frac{(\text{Depth of Pond, m})(100 \text{ cm/m})}{\text{Detention Time, days}}$$

$$= \frac{(1.2 \text{ m})(100 \text{ cm/m})}{13.0 \text{ days}}$$

$$= 9.2 \text{ cm/day}$$

20. ORGANIC LOADING

$$\text{Organic Loading, gm BOD/day/sq m} = \frac{(\text{BOD, mg/}L)(\text{Flow, }\frac{\text{cu m}}{\text{day}})\frac{(1,000 \, L)}{(1 \text{ cu m})}\frac{(1 \text{ gm})}{(1,000 \text{ mg})}}{\text{Area, sq m}}$$

$$= \frac{(150, \text{mg/}L)(2,000 \, \frac{\text{cu m}}{\text{day}})\frac{(1,000 \, L)}{(1 \text{ cu m})}\frac{(1 \text{ gm})}{(1,000 \text{ mg})}}{21,600 \text{ sq m}}$$

$$= \frac{300,000 \text{ gm BOD/day}}{21,600 \text{ sq m}}$$

$$= 13.9 \text{ gm BOD/day/sq m}$$

A.157 Chlorination

EXAMPLE 21

Determine the chlorine demand* of an effluent if the chlorine dose is 10.0 mg/L and the chlorine residual is 1.1 mg/L.

Known	Unknown
Chlorine Dose, mg/L = 10.0 mg/L	Chlorine Demand, mg/L
Chlorine Residual, mg/L = 1.1 mg/L	
Chlorine Demand, mg/L = Chlor Dose, mg/L – Chlor Residual, mg/L	

$$= 10.0 \text{ mg/}L - 1.1 \text{ mg/}L$$

$$= 8.9 \text{ mg/}L$$

EXAMPLE 22

To maintain a satisfactory chlorine residual in a plant effluent, the chlorine dose must be 10 mg/L when the flow is 1,400 cu m/day. Determine the chlorinator setting (feed rate) in kilograms per day.

Known	Unknown
Dose, mg/L = 10 mg/L	Chlorinator Setting, kg/day
Flow, cu m/day = 1,400 cu m/day	

$$\text{Chlorine Feed Rate, kg/day} = (\text{Dose, mg/}L)(\text{Flow, }\frac{\text{cu m}}{\text{day}})\frac{(1,000 \, L)}{(1 \text{ cu m})}\frac{(1 \text{ kg})}{(1,000,000 \text{ mg})}$$

$$= (10 \text{ mg/}L)(1,400 \, \frac{\text{cu m}}{\text{day}})\frac{(1,000 \, L)}{(1 \text{ cu m})}\frac{(1 \text{ kg})}{(1,000,000 \text{ mg})}$$

$$= 14 \text{ kg/day}$$

A.158 Chemical Doses

EXAMPLE 23

Determine the chlorinator setting in kilograms per 24 hours if 4,000 cubic meters of water per day are to be treated with a desired chlorine dose of 2.5 mg/L.

Known	Unknown
Flow, cu m/day = 4,000 cu m/day	Chlorinator Setting, kg/24 hours
Chlorine Dose, mg/L = 2.5 mg/L	

Determine the chlorinator setting in kilograms per 24 hours.

$$\text{Chlorinator Setting, kg/day} = \frac{(\text{Flow, cu m/day})(\text{Dose, mg/}L)(1,000 \, L/\text{cu m})}{(1,000 \text{ mg/gm})(1,000 \text{ gm/kg})}$$

$$= \frac{(4,000 \text{ cu m/day})(2.5 \text{ mg/}L)(1,000 \, L/\text{cu m})}{(1,000 \text{ mg/gm})(1,000 \text{ gm/kg})}$$

$$= 10 \text{ kg/day}$$

---

* *STANDARD METHODS* has used the term 'chlorine demand' when referring to stabilized water such as a domestic water supply and the term 'chlorination requirement' when referring to wastewater.

EXAMPLE 24

The optimum liquid polymer dose from the jar tests is 12 mg/L. Determine the setting on the liquid polymer chemical feeder in milliliters per minute when the plant flow is 15 megaliters per day (or million liters per day). The liquid polymer delivered to the plant contains 642.3 milligrams of polymer per milliliter of liquid solution.

Known		Unknown
Polymer Dose, mg/L	= 12 mg/L	Chemical Feeder Setting, mL/min
Flow, ML/day	= 15 ML/day	
Liquid Polymer, mg/mL	= 642.3 mg/mL	

Calculate the liquid polymer chemical feeder setting in milliliters per minute.

$$\text{Chemical Feeder Setting, mL/min} = \frac{(\text{Flow, ML/day})(\text{Polymer Dose, mg/L})(1{,}000{,}000/M)}{(\text{Liquid Polymer, mg/mL})(24\text{ hr/day})(60\text{ min/hr})}$$

$$= \frac{(15\text{ ML/day})(12\text{ mg/L})(1{,}000{,}000/M)}{(642.3\text{ mg/mL})(24\text{ hr/day})(60\text{ min/hr})}$$

$$= 195\text{ mL/min}$$

EXAMPLE 25

The optimum liquid polymer dose from the jar tests is 8 mg/L. Determine the setting on the liquid polymer chemical feeder in milliliters per minute when the flow is 12 megaliters per day. The liquid polymer delivered to the plant contains 5.36 pounds of polymer per gallon of liquid solution.

Known		Unknown
Polymer Dose, mg/L	= 8 mg/L	Chemical Feeder Setting, mL/min
Flow, ML/day	= 12 ML/day	
Liquid Polymer, lbs/gal	= 5.36 lbs/gal	

Calculate the liquid polymer chemical feeder setting in milliliters per minute.

$$\text{Chemical Feeder Setting, mL/min} = \frac{(\text{Flow, ML/day})(\text{Polymer Dose, mg/L})(3.785\text{ L/gal})(1{,}000\text{ mL/L})(1{,}000{,}000/M)}{(\text{Liquid Polymer, lbs/gal})(454\text{ gm/lb})(1{,}000\text{ mg/gm})(24\text{ hr/day})(60\text{ min/hr})}$$

$$= \frac{(12\text{ ML/day})(8\text{ mg/L})(3.785\text{ L/gal})(1{,}000\text{ mL/L})(1{,}000{,}000/M)}{(5.36\text{ lbs/gal})(454\text{ gm/lb})(1{,}000\text{ mg/gm})(24\text{ hr/day})(60\text{ min/hr})}$$

$$= 104\text{ mL/min}$$

EXAMPLE 26

Determine the actual chemical dose or chemical feed in kilograms per day from a dry chemical feeder. A bucket placed under the chemical feeder weighed 150 grams empty and 1,800 grams after 12 minutes.

Known		Unknown
Empty Bucket, gm	= 150 gm	Chemical Feed, kg/day
Full Bucket, gm	= 1,800 gm	
Time to Fill, min	= 12 min	

Determine the chemical feed in kilograms of chemical applied per day.

$$\text{Chemical Feed, kg/day} = \frac{\text{Chemical Applied, kg}}{\text{Length of Application, days}}$$

$$= \frac{(1{,}800\text{ gm} - 150\text{ gm})(60\text{ min/hr})(24\text{ hr/day})}{(1{,}000\text{ gm/kg})(12\text{ min})}$$

$$= 198\text{ kg/day}$$

EXAMPLE 27

Determine the chemical feed in kilograms of polymer per day from a chemical feed pump. The polymer solution is 1.5 percent or 15,000 mg polymer per liter. Assume a specific gravity of the polymer solution of 1.0. During a test run, the chemical feed pump delivered 800 mL of polymer solution during five minutes.

Known		Unknown
Polymer Solution, %	= 1.5%	Polymer Feed, kg/day
Polymer Conc, mg/L	= 15,000 mg/L	
Polymer Sp Gr	= 1.0	
Volume Pumped, mL	= 800 mL	
Time Pumped, min	= 5 min	

Calculate the polymer fed by the chemical feed pump in kilograms of polymer per day.

$$\text{Polymer Feed, kg/day} = \frac{(\text{Poly Conc, mg/L})(\text{Vol Pumped, mL})(60\text{ min/hr})(24\text{ hr/day})}{(\text{Time Pumped, min})(1{,}000\text{ mL/L})(1{,}000\text{ mg/gm})(1{,}000\text{ gm/kg})}$$

$$= \frac{(15{,}000\text{ mg/L})(800\text{ mL})(60\text{ min/hr})(24\text{ hr/day})}{(5\text{ min})(1{,}000\text{ mL/L})(1{,}000\text{ mg/gm})(1{,}000\text{ gm/kg})}$$

$$= 3.5\text{ kg Polymer/day}$$

EXAMPLE 28

A small chemical feed pump lowered the chemical solution in an 80-centimeter diameter tank 35 centimeters during an eight-hour period. Estimate the flow delivered by the pump in liters per minute.

Known		Unknown
Tank Diameter, cm	= 80 cm	Flow, liters/min
Chemical Drop, cm	= 35 cm	
Time, hr	= 8 hr	

Calculate the flow in liters per minute.

$$\text{Flow, liters/min} = \frac{(\text{Area, sq m})(\text{Drop, m})(1{,}000\text{ liters/cu m})}{(\text{Time, hr})(60\text{ min/hr})}$$

$$= \frac{(0.785)(0.8\text{ m})^2(0.35\text{ m})(1{,}000\text{ liters/cu m})}{(8\text{ hr})(60\text{ min/hr})}$$

$$= 0.37\text{ liters/min}$$

EXAMPLE 29

Determine the settings in percent stroke on a chemical feed pump for various doses of a chemical in milligrams per liter. (The chemical could be chlorine, polymer, potassium permanganate, or any other chemical solution fed by a pump.) The pump delivering water to be treated pumps at a flow rate of 25 liters per second. The solution strength of the chemical being pumped is five percent. The chemical feed pump has a maximum capacity of 250 milliliters per minute at a setting of 100 percent capacity.

Known	Unknown
Pump Flow, L/sec = 25 L/sec	Settings, % stroke for various doses in mg/L
Solution Strength, % = 5%	
Feed Pump, mL/min (100% stroke) = 250 mL/min	

1. Change the chemical solution strength from a percent to milligrams of chemical per liter of solution. A 5 percent solution means we have 5 milligrams of chemical in a solution of water and chemical weighing 100 milligrams.

$$\text{Chemical Solution, mg/L} = \frac{5 \text{ mg Chemical}}{100 \text{ mg of Chemical and Water}}$$

$$= \frac{(5 \text{ mg})(1,000,000 \text{ mg/L})}{100 \text{ mg}}$$

$$= 50,000 \text{ mg/L}$$

2. Calculate the chemical feed in kilograms per day for a chemical dose of 0.5 milligrams per liter. We are going to assume various chemical doses of 0.5, 1.0, 1.5, 2.0, 2.5 mg/L and upward so that if we know the desired chemical dose, we can easily determine the setting (percent stroke) on the chemical feed pump.

$$\text{Chemical Feed, kg/day} = \frac{(\text{Flow, L/sec})(\text{Dose, mg/L})(60 \text{ sec/min})(60 \text{ min/hr})(24 \text{ hr/day})}{1,000,000 \text{ mg/kg}}$$

$$= \frac{(25 \text{ L/sec})(0.5 \text{ mg/L})(60 \text{ sec/min})(60 \text{ min/hr})(24 \text{ hr/day})}{1,000,000 \text{ mg/kg}}$$

$$= 1.1 \text{ kg/day}$$

3. Determine the desired flow from the chemical feed pump in milliliters per minute.

$$\text{Feed Pump, mL/min} = \frac{\text{Chemical Feed, kg/day}}{\text{Chemical Solution, mg/L}}$$

$$= \frac{(1.1 \text{ kg/day})(1,000 \text{ mL/L})(1,000,000 \text{ mg/kg})}{(50,000 \text{ mg/L})(24 \text{ hr/day})(60 \text{ min/hr})}$$

$$= 15.3 \text{ mL/min}$$

4. Determine the setting on the chemical feed pump as a percent. In this case we want to know the setting as a percent of the pump stroke.

$$\text{Setting, \%} = \frac{(\text{Desired Feed Pump, mL/min})(100\%)}{\text{Maximum Feed Pump, mL/min}}$$

$$= \frac{(15.3 \text{ mL/min})(100\%)}{250 \text{ mL/min}}$$

$$= 6.1\%$$

5. If we changed the chemical dose in Step 2 from 0.5 mg/L to 1.0 mg/L and other higher doses and repeated the remainder of the steps, we could calculate the data in Table A.2.

TABLE A.2   SETTINGS FOR CHEMICAL FEED PUMP

PUMP FLOW, L/sec   = 25 L/sec

SOLUTION STRENGTH, % = 5.0%

Chemical Dose, mg/L	Chemical Feed, kg/day	Feed Pump, mL/min	Pump Setting, % stroke
0.5	1.1	15.3	6.1
1.0	2.2	30.6	12.2
2.0	4.3	59.7	23.9
4.0	8.6	119.4	47.8
6.0	13.0	180.6	72.2
8.0	17.3	240.3	96.0

6. The data in Table A.2 could be plotted to produce a chart similar to Figure A.3 (page 470). For any desired chemical dose in milligrams per liter, the chart could be used to determine the necessary chemical feed pump setting.

A.159   Blueprint Reading

EXAMPLE 30

A set of blueprints for a treatment plant has a scale of 1:100. On the prints, the laboratory dimensions were measured and found to be 70 mm wide and 100 mm long. What is the floor area of the laboratory?

Known	Unknown
Scale = 1:100	Area, sq m
Length, mm = 100 mm	
Width, mm = 70 mm	
Area, sq m = (Length, m)(Width, m)	

Convert length and width to meters.

$$\frac{1}{100} = \frac{100 \text{ mm}}{\text{Length, m}}$$

$$\text{Length, m} = 100 \text{ mm} \frac{(100)}{(1)} \frac{(1 \text{ m})}{(1,000 \text{ mm})}$$

$$= 10 \text{ m}$$

$$\frac{1}{100} = \frac{70 \text{ mm}}{\text{Width, m}}$$

$$\text{Width, m} = (70 \text{ mm}) \frac{(100)}{(1)} \frac{(1 \text{ m})}{(1,000 \text{ mm})}$$

$$= 7 \text{ m}$$

$$\text{Area, sq ft} = (\text{Length, m})(\text{Width, m})$$

$$= (10 \text{ m})(7 \text{ m})$$

$$= 70 \text{ sq m}$$

ABBREVIATIONS

ac	acre		km	kilometer
ac-ft	acre-feet		kN	kilonewton
af	acre feet		kW	kilowatt
amp	ampere		kWh	kilowatt-hour
°C	degrees Celsius		L	liter
CFM	cubic feet per minute		lb	pound
CFS	cubic feet per second		lbs/sq in	pounds per square inch
cm	centimeter		m	meter
cu ft	cubic feet		M	mega
cu in	cubic inch		M	million
cu m	cubic meter		mg	milligram
cu yd	cubic yard		MGD	million gallons per day
°F	degrees Fahrenheit		mg/L	milligram per liter
ft	feet or foot		min	minute
ft-lb/min	foot-pounds per minute		mL	milliliter
g	gravity		mm	millimeter
gal	gallon		N	Newton
gal/day	gallons per day		ohm	ohm
gm	gram		Pa	Pascal
GPD	gallons per day		ppb	parts per billion
gpg	grains per gallon		ppm	parts per million
GPM	gallons per minute		psf	pounds per square foot
gr	grain		psi	pounds per square inch
ha	hectare		psig	pounds per square inch gage
HP	horsepower		RPM	revolutions per minute
hr	hour		sec	second
in	inch		sq ft	square feet
k	kilo		sq in	square inches
kg	kilogram		W	watt

WASTEWATER WORDS

A Summary of the Words Defined

in

OPERATION OF WASTEWATER TREATMENT PLANTS

PROJECT PRONUNCIATION KEY

by Warren L. Prentice

The Project Pronunciation Key is designed to aid you in the pronunciation of new words. While this key is based primarily on familiar sounds, it does not attempt to follow any particular pronunciation guide. This key is designed solely to aid operators in this program.

You may find it helpful to refer to other available sources for pronunciation help. Each current standard dictionary contains a guide to its own pronunciation key. Each key will be different from each other and from this key. Examples of the difference between the key used in this program and the *WEBSTER'S NEW WORLD COLLEGE DICTIONARY*[1] "Key" are shown below.

In using this key, you should accent (say louder) the syllable that appears in capital letters. The following chart is presented to give examples of how to pronounce words using the Project Key.

WORD	SYLLABLE				
	1st	2nd	3rd	4th	5th
acid	AS	id			
coliform	COAL	i	form		
biological	BUY	o	LODGE	ik	cull

The first word, *ACID*, has its first syllable accented. The second word, *COLIFORM*, has its first syllable accented. The third word, *BIOLOGICAL*, has its first and third syllables accented.

We hope you will find the key useful in unlocking the pronunciation of any new word.

Term	Project Key	Webster Key
acid	AS-id	aś id
coliform	COAL-i-form	kō′ lə fôrm
biological	BUY-o-LODGE-ik-cull	bī ə läj′ i kəl

---

[1] *The WEBSTER'S NEW WORLD COLLEGE DICTIONARY, Fourth Edition, 1999, was chosen rather than an unabridged dictionary because of its availability to the operator. Other editions may be slightly different.*

WASTEWATER WORDS

>GREATER THAN >GREATER THAN

DO >5 mg/*L* would be read as DO GREATER THAN 5 mg/*L*.

<LESS THAN <LESS THAN

DO <5 mg/*L* would be read as DO LESS THAN 5 mg/*L*.

A

ABS ABS

Alkyl **B**enzene **S**ulfonate. A type of surfactant, or surface active agent, present in synthetic detergents in the United States before 1965. ABS was especially troublesome because it caused foaming and resisted breakdown by biological treatment processes. ABS has been replaced in detergents by **l**inear **a**lkyl **s**ulfonate (LAS) which is biodegradable.

ACEOPS ACEOPS

See **A**LLIANCE OF **C**ERTIFIED **OP**ERATOR**S**, LAB ANALYSTS, INSPECTORS, AND SPECIALISTS (ACEOPS).

ABSORPTION (ab-SORP-shun) ABSORPTION

The taking in or soaking up of one substance into the body of another by molecular or chemical action (as tree roots absorb dissolved nutrients in the soil).

ACCOUNTABILITY ACCOUNTABILITY

When a manager gives power/responsibility to an employee, the employee ensures that the manager is informed of results or events.

ACID ACID

(1) A substance that tends to lose a proton.

(2) A substance that dissolves in water with the formation of hydrogen ions.

(3) A substance containing hydrogen which may be replaced by metals to form salts.

(4) A substance that is corrosive.

ACID REGRESSION STAGE ACID REGRESSION STAGE

A time period when the production of volatile acids is reduced during anaerobic digestion. During this stage of digestion ammonia compounds form and cause the pH to increase.

ACIDITY ACIDITY

The capacity of water or wastewater to neutralize bases. Acidity is expressed in milligrams per liter of equivalent calcium carbonate. Acidity is not the same as pH because water does not have to be strongly acidic (low pH) to have a high acidity. Acidity is a measure of how much base must be added to a liquid to raise the pH to 8.2.

ACTIVATED SLUDGE (ACK-ta-VATE-ed sluj) ACTIVATED SLUDGE

Sludge particles produced in raw or settled wastewater (primary effluent) by the growth of organisms (including zoogleal bacteria) in aeration tanks in the presence of dissolved oxygen. The term "activated" comes from the fact that the particles are teeming with bacteria, fungi, and protozoa. Activated sludge is different from primary sludge in that the sludge particles contain many living organisms which can feed on the incoming wastewater.

ACTIVATED SLUDGE (ACK-ta-VATE-ed sluj) PROCESS   ACTIVATED SLUDGE PROCESS

A biological wastewater treatment process which speeds up the decomposition of wastes in the wastewater being treated. Activated sludge is added to wastewater and the mixture (mixed liquor) is aerated and agitated. After some time in the aeration tank, the activated sludge is allowed to settle out by sedimentation and is disposed of (wasted) or reused (returned to the aeration tank) as needed. The remaining wastewater then undergoes more treatment.

ACUTE HEALTH EFFECT   ACUTE HEALTH EFFECT

An adverse effect on a human or animal body, with symptoms developing rapidly.

ADSORPTION (add-SORP-shun)   ADSORPTION

The gathering of a gas, liquid, or dissolved substance on the surface or interface zone of another material.

ADVANCED WASTE TREATMENT   ADVANCED WASTE TREATMENT

Any process of water renovation that upgrades treated wastewater to meet specific reuse requirements. May include general cleanup of water or removal of specific parts of wastes insufficiently removed by conventional treatment processes. Typical processes include chemical treatment and pressure filtration. Also called TERTIARY TREATMENT.

AERATION (air-A-shun)   AERATION

The process of adding air to water. In wastewater treatment, air is added to freshen wastewater and to keep solids in suspension. With mixtures of wastewater and activated sludge, adding air provides mixing and oxygen for the microorganisms treating the wastewater.

AERATION (air-A-shun) LIQUOR   AERATION LIQUOR

Mixed liquor. The contents of the aeration tank including living organisms and material carried into the tank by either untreated wastewater or primary effluent.

AERATION (air-A-shun) TANK   AERATION TANK

The tank where raw or settled wastewater is mixed with return sludge and aerated. The same as aeration bay, aerator, or reactor.

AEROBES   AEROBES

Bacteria that must have molecular (dissolved) oxygen (DO) to survive. Aerobes are aerobic bacteria.

AEROBIC (AIR-O-bick)   AEROBIC

A condition in which atmospheric or dissolved molecular oxygen is present in the aquatic (water) environment.

AEROBIC BACTERIA (AIR-O-bick back-TEAR-e-ah)   AEROBIC BACTERIA

Bacteria which will live and reproduce only in an environment containing oxygen which is available for their respiration (breathing), namely atmospheric oxygen or oxygen dissolved in water. Oxygen combined chemically, such as in water molecules (H_2O), cannot be used for respiration by aerobic bacteria.

AEROBIC (AIR-O-bick) DECOMPOSITION   AEROBIC DECOMPOSITION

The decay or breaking down of organic material in the presence of "free" or dissolved oxygen.

AEROBIC (AIR-O-bick) DIGESTION   AEROBIC DIGESTION

The breakdown of wastes by microorganisms in the presence of dissolved oxygen. This digestion process may be used to treat only waste activated sludge, or trickling filter sludge and primary (raw) sludge, or waste sludge from activated sludge treatment plants designed without primary settling. The sludge to be treated is placed in a large aerated tank where aerobic microorganisms decompose the organic matter in the sludge. This is an extension of the activated sludge process.

AEROBIC (AIR-O-bick) PROCESS   AEROBIC PROCESS

A waste treatment process conducted under aerobic (in the presence of "free" or dissolved oxygen) conditions.

AESTHETIC (es-THET-ick)   AESTHETIC

Attractive or appealing.

AGGLOMERATION (a-GLOM-er-A-shun)   AGGLOMERATION

The growing or coming together of small scattered particles into larger flocs or particles which settle rapidly. Also see FLOC.

AIR BINDING   AIR BINDING

The clogging of a filter, pipe or pump due to the presence of air released from water. Air entering the filter media is harmful to both the filtration and backwash processes. Air can prevent the passage of water during the filtration process and can cause the loss of filter media during the backwash process.

AIR GAP   AIR GAP

An open vertical drop, or vertical empty space, between a drinking (potable) water supply and the point of use in a wastewater treatment plant. This gap prevents the contamination of drinking water by backsiphonage because there is no way wastewater can reach the drinking water supply.

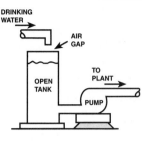

AIR LIFT PUMP   AIR LIFT PUMP

A special type of pump. This device consists of a vertical riser pipe submerged in the wastewater or sludge to be pumped. Compressed air is injected into a tail piece at the bottom of the pipe. Fine air bubbles mix with the wastewater or sludge to form a mixture lighter than the surrounding water which causes the mixture to rise in the discharge pipe to the outlet. An air-lift pump works like the center stand in a percolator coffee pot.

AIR PADDING   AIR PADDING

Pumping dry air (dew point –40°F) into a container to assist with the withdrawal of a liquid or to force a liquified gas such as chlorine or sulfur dioxide out of a container.

ALGAE (AL-gee)   ALGAE

Microscopic plants which contain chlorophyll and live floating or suspended in water. They also may be attached to structures, rocks, or other submerged surfaces. Algae produce oxygen during sunlight hours and use oxygen during the night hours. Their biological activities appreciably affect the pH, alkalinity, and dissolved oxygen of the water.

ALGAL (AL-gull) BLOOM   ALGAL BLOOM

Sudden, massive growths of microscopic and macroscopic plant life, such as green or blue-green algae, which can, under the proper conditions, develop in lakes, reservoirs, and also in ponds.

ALIQUOT (AL-li-kwot)   ALIQUOT

Representative portion of a sample. Often an equally divided portion of a sample.

ALKALI (AL-ka-lie)   ALKALI

Any of certain soluble salts, principally of sodium, potassium, magnesium, and calcium, that have the property of combining with acids to form neutral salts and may be used in chemical processes such as water or wastewater treatment.

ALKALINITY (AL-ka-LIN-it-tee)   ALKALINITY

The capacity of water or wastewater to neutralize acids. This capacity is caused by the water's content of carbonate, bicarbonate, hydroxide, and occasionally borate, silicate, and phosphate. Alkalinity is expressed in milligrams per liter of equivalent calcium carbonate. Alkalinity is not the same as pH because water does not have to be strongly basic (high pH) to have a high alkalinity. Alkalinity is a measure of how much acid must be added to a liquid to lower the pH to 4.5.

ALLIANCE OF CERTIFIED OPERATORS,   ALLIANCE OF CERTIFIED OPERATORS,
  LAB ANALYSTS, INSPECTORS,   LAB ANALYSTS, INSPECTORS,
  AND SPECIALISTS (ACEOPS)   AND SPECIALISTS (ACEOPS)

A professional organization for operators, lab analysts, inspectors, and specialists dedicated to improving professionalism; expanding training, certification, and job opportunities; increasing information exchange; and advocating the importance of certified operators, lab analysts, inspectors, and specialists. For information on membership, contact ACEOPS, 1810 Bel Air Drive, Ames, IA 50010-5125, phone (515) 663-4128 or e-mail: Info@aceops.com.

AMBIENT (AM-bee-ent) TEMPERATURE   AMBIENT TEMPERATURE
Temperature of the surroundings.

AMPEROMETRIC (am-PURR-o-MET-rick)   AMPEROMETRIC
A method of measurement that records electric current flowing or generated, rather than recording voltage. Amperometric titration is a means of measuring concentrations of certain substances in water.

ANAEROBES   ANAEROBES

Bacteria that do not need molecular (dissolved) oxygen (DO) to survive.

ANAEROBIC (AN-air-O-bick)   ANAEROBIC

A condition in which atmospheric or dissolved molecular oxygen is *NOT* present in the aquatic (water) environment.

ANAEROBIC BACTERIA (AN-air-O-bick back-TEAR-e-ah)   ANAEROBIC BACTERIA

Bacteria that live and reproduce in an environment containing no "free" or dissolved oxygen. Anaerobic bacteria obtain their oxygen supply by breaking down chemical compounds which contain oxygen, such as sulfate (SO_4^{2-}).

ANAEROBIC (AN-air-O-bick) DECOMPOSITION   ANAEROBIC DECOMPOSITION

The decay or breaking down of organic material in an environment containing no "free" or dissolved oxygen.

ANAEROBIC (AN-air-O-bick) DIGESTION   ANAEROBIC DIGESTION

Wastewater solids and water (about 5% solids, 95% water) are placed in a large tank where bacteria decompose the solids in the absence of dissolved oxygen. At least two general groups of bacteria act in balance: (1) SAPROPHYTIC bacteria break down complex solids to volatile acids, the most common of which are acetic and propionic acids; and (2) *METHANE FERMENTERS* break down the acids to methane, carbon dioxide, and water.

ANALOG READOUT   ANALOG READOUT

The readout of an instrument by a pointer (or other indicating means) against a dial or scale.

ANHYDROUS (an-HI-drous)   ANHYDROUS

Very dry. No water or dampness is present.

ANION (AN-EYE-en)   ANION

A negatively charged ion in an electrolyte solution, attracted to the anode under the influence of a difference in electrical potential. Chloride ion (Cl^-) is an anion.

ANOXIC (an-OX-ick)   ANOXIC

A condition in which atmospheric or dissolved molecular oxygen is *NOT* present in the aquatic (water) environment and nitrate is present. Oxygen deficient or lacking sufficient oxygen. The term is similar to ANAEROBIC.

ASEPTIC (a-SEP-tick)   ASEPTIC

Free from the living germs of disease, fermentation, or putrefaction. Sterile.

ASPIRATE (ASS-per-RATE)   ASPIRATE

Use of a hydraulic device (aspirator or eductor) to create a negative pressure (suction) by forcing a liquid through a restriction, such as a Venturi. An aspirator (the hydraulic device) may be used in the laboratory in place of a vacuum pump; sometimes used instead of a sump pump.

AUTHORITY   AUTHORITY

The power and resources to do a specific job or to get that job done.

AXIAL TO IMPELLER   AXIAL TO IMPELLER

The direction in which material being pumped flows around the impeller or flows parallel to the impeller shaft.

AXIS OF IMPELLER   AXIS OF IMPELLER

An imaginary line running along the center of a shaft (such as an impeller shaft).

B

BOD (pronounce as separate letters)   BOD

Biochemical **O**xygen **D**emand. The rate at which organisms use the oxygen in water or wastewater while stabilizing decomposable organic matter under aerobic conditions. In decomposition, organic matter serves as food for the bacteria and energy results from its oxidation. BOD measurements are used as a measure of the organic strength of wastes in water.

BTU (pronounce as separate letters)   BTU

British **T**hermal **U**nit. The amount of heat required to raise the temperature of one pound of water one degree Fahrenheit.

BACTERIA (back-TEAR-e-ah)   BACTERIA

Bacteria are living organisms, microscopic in size, which usually consist of a single cell. Most bacteria use organic matter for their food and produce waste products as the result of their life processes.

BACTERIAL (back-TEAR-e-al) CULTURE   BACTERIAL CULTURE

In the case of activated sludge, the bacterial culture refers to the group of bacteria classified as AEROBES, and FACULTATIVE organisms, which covers a wide range of organisms. Most treatment processes in the United States grow facultative organisms which use the carbonaceous (carbon compounds) BOD. Facultative organisms can live when oxygen resources are low. When "nitrification" is required, the nitrifying organisms are OBLIGATE AEROBES (require oxygen) and must have at least 0.5 mg/L of dissolved oxygen throughout the whole system to function properly.

BACKFLOW   BACKFLOW

A reverse flow condition, created by a difference in water pressures, which causes water to flow back into the distribution pipes of a potable water supply from any source or sources other than an intended source. Also see BACKSIPHONAGE.

BACKSIPHONAGE   BACKSIPHONAGE

A form of backflow caused by a negative or below atmospheric pressure within a water system. Also see BACKFLOW.

BAFFLE   BAFFLE

A flat board or plate, deflector, guide or similar device constructed or placed in flowing water, wastewater, or slurry systems to cause more uniform flow velocities, to absorb energy, and to divert, guide, or agitate liquids (water, chemical solutions, slurry).

BASE   BASE

(1) A substance which takes up or accepts protons.

(2) A substance which dissociates (separates) in aqueous solution to yield hydroxyl ions (OH^-).

(3) A substance containing hydroxyl ions which reacts with an acid to form a salt or which may react with metals to form precipitates.

BATCH PROCESS   BATCH PROCESS

A treatment process in which a tank or reactor is filled, the wastewater (or other solution) is treated or a chemical solution is prepared, and the tank is emptied. The tank may then be filled and the process repeated. Batch processes are also used to cleanse, stabilize or condition chemical solutions for use in industrial manufacturing and treatment processes.

BIOASSAY (BUY-o-ass-SAY)   BIOASSAY

(1) A way of showing or measuring the effect of biological treatment on a particular substance or waste, or

(2) A method of determining the relative toxicity of a test sample of industrial wastes or other wastes by using live test organisms such as fish.

BIOCHEMICAL OXYGEN DEMAND (BOD)   BIOCHEMICAL OXYGEN DEMAND (BOD)

The rate at which organisms use the oxygen in water or wastewater while stabilizing decomposable organic matter under aerobic conditions. In decomposition, organic matter serves as food for the bacteria and energy results from its oxidation. BOD measurements are used as a measure of the organic strength of wastes in water.

BIOCHEMICAL OXYGEN DEMAND (BOD) TEST   BIOCHEMICAL OXYGEN DEMAND (BOD) TEST

A procedure that measures the rate of oxygen use under controlled conditions of time and temperature. Standard test conditions include dark incubation at 20°C for a specified time (usually five days).

BIODEGRADABLE (BUY-o-dee-GRADE-able)   BIODEGRADABLE

Organic matter that can be broken down by bacteria to more stable forms which will not create a nuisance or give off foul odors is considered biodegradable.

BIODEGRADATION (BUY-o-deh-grah-DAY-shun)   BIODEGRADATION

The breakdown of organic matter by bacteria to more stable forms which will not create a nuisance or give off foul odors.

BIOFLOCCULATION (BUY-o-flock-u-LAY-shun)   BIOFLOCCULATION

The clumping together of fine, dispersed organic particles by the action of certain bacteria and algae. This results in faster and more complete settling of the organic solids in wastewater.

BIOMASS (BUY-o-MASS)   BIOMASS

A mass or clump of organic material consisting of living organisms feeding on the wastes in wastewater, dead organisms and other debris. Also see ZOOGLEAL FILM and ZOOGLEAL MASS.

BIOMONITORING   BIOMONITORING

A term used to describe methods of evaluating or measuring the effects of toxic substances in effluents on aquatic organisms in receiving waters. There are two types of biomonitoring, the BIOSURVEY and the BIOASSAY.

BIOSOLIDS   BIOSOLIDS

A primarily organic solid product, produced by wastewater treatment processes, that can be beneficially recycled. The word biosolids is replacing the word sludge.

BIOSURVEY   BIOSURVEY

A survey of the types and numbers of organisms naturally present in the receiving waters upstream and downstream from plant effluents. Comparisons are made between the aquatic organisms upstream and those organisms downstream of the discharge.

BLANK   BLANK

A bottle containing only dilution water or distilled water; the sample being tested is not added. Tests are frequently run on a *SAMPLE* and a *BLANK* and the differences are compared. The procedure helps to eliminate or reduce test result errors that could be caused when the dilution water or distilled water used is contaminated.

BLINDING   BLINDING

The clogging of the filtering medium of a microscreen or a vacuum filter when the holes or spaces in the media become clogged or sealed off due to a buildup of grease or the material being filtered.

BOND   BOND

(1)   A written promise to pay a specified sum of money (called the face value) at a fixed time in the future (called the date of maturity). A bond also carries interest at a fixed rate, payable periodically. The difference between a note and a bond is that a bond usually runs for a longer period of time and requires greater formality. Utility agencies use bonds as a means of obtaining large amounts of money for capital improvements.

(2)   A warranty by an underwriting organization, such as an insurance company, guaranteeing honesty, performance, or payment by a contractor.

BOUND WATER   BOUND WATER

Water contained within the cell mass of sludges or strongly held on the surface of colloidal particles. One of the causes of bulking sludge in the activated sludge process.

BREAKOUT OF CHLORINE   BREAKOUT OF CHLORINE

A point at which chlorine leaves solution as a gas because the chlorine feed rate is too high. The solution is saturated and cannot dissolve any more chlorine. The maximum strength a chlorine solution can attain is approximately 3,500 mg/L. Beyond this concentration molecular chlorine is present which will break out of solution and cause "off-gassing" at the point of application.

BREAKPOINT CHLORINATION   BREAKPOINT CHLORINATION

Addition of chlorine to water or wastewater until the chlorine demand has been satisfied. At this point, further additions of chlorine result in a residual that is directly proportional to the amount of chlorine added beyond the breakpoint.

BRINELLING (bruh-NEL-ing)   BRINELLING

Tiny indentations (dents) high on the shoulder of the bearing race or bearing. A type of bearing failure.

BUFFER   BUFFER

A solution or liquid whose chemical makeup neutralizes acids or bases without a great change in pH.

BUFFER ACTION   BUFFER ACTION

The action of certain ions in solution in opposing a change in hydrogen ion concentration.

BUFFER CAPACITY   BUFFER CAPACITY

A measure of the capacity of a solution or liquid to neutralize acids or bases. This is a measure of the capacity of water or wastewater for offering a resistance to changes in pH.

BUFFER SOLUTIONBUFFER SOLUTION

A solution containing two or more substances which, in combination, resist any marked change in pH following addition of moderate amounts of either strong acid or base.

BULKING (BULK-ing)BULKING

Clouds of billowing sludge that occur throughout secondary clarifiers and sludge thickeners when the sludge does not settle properly. In the activated sludge process, bulking is usually caused by filamentous bacteria or bound water.

C

CFRCFR

Code of Federal Regulations. A publication of the United States Government which contains all of the proposed and finalized federal regulations, including safety and environmental regulations.

CALL DATECALL DATE

First date a bond can be paid off.

CALORIE (KAL-o-ree)CALORIE

The amount of heat required to raise the temperature of one gram of water one degree Celsius.

CARCINOGEN (CAR-sin-o-JEN)CARCINOGEN

Any substance which tends to produce cancer in an organism.

CARBONACEOUS (car-bun-NAY-shus) STAGECARBONACEOUS STAGE

A stage of decomposition that occurs in biological treatment processes when aerobic bacteria, using dissolved oxygen, change carbon compounds to carbon dioxide. Sometimes referred to as "first-stage BOD" because the microorganisms attack organic or carbon compounds first and nitrogen compounds later. Also see NITRIFICATION STAGE.

CATHODIC (ca-THOD-ick) PROTECTIONCATHODIC PROTECTION

An electrical system for prevention of rust, corrosion, and pitting of steel and iron surfaces in contact with water, wastewater or soil. A low-voltage current is made to flow through a liquid (water) or a soil in contact with the metal in such a manner that the external electromotive force renders the metal structure cathodic. This concentrates corrosion on auxiliary anodic parts which are deliberately allowed to corrode instead of letting the structure corrode.

CATION (CAT-EYE-en)CATION

A positively charged ion in an electrolyte solution, attracted to the cathode under the influence of a difference in electrical potential. Sodium ion (Na^+) is a cation.

CATION EXCHANGE CAPACITYCATION EXCHANGE CAPACITY

The ability of a soil or other solid to exchange cations (positive ions such as calcium, Ca^{2+}) with a liquid.

CAUTIONCAUTION

This word warns against potential hazards or cautions against unsafe practices. Also see DANGER, NOTICE, and WARNING.

CAVITATION (CAV-uh-TAY-shun)CAVITATION

The formation and collapse of a gas pocket or bubble on the blade of an impeller or the gate of a valve. The collapse of this gas pocket or bubble drives water into the impeller or gate with a terrific force that can cause pitting on the impeller or gate surface. Cavitation is accompanied by loud noises that sound like someone is pounding on the impeller or gate with a hammer.

CENTRATECENTRATE

The water leaving a centrifuge after most of the solids have been removed.

CENTRIFUGECENTRIFUGE

A mechanical device that uses centrifugal or rotational forces to separate solids from liquids.

CERTIFICATION EXAMINATIONCERTIFICATION EXAMINATION

An examination administered by a state agency that operators take to indicate a level of professional competence. In most plants the Chief Operator of the plant must be "certified" (successfully pass a certification examination). In the United States, certification of operators of water treatment plants and wastewater treatment plants is mandatory.

CERTIFIED OPERATOR CERTIFIED OPERATOR

A person who has the education and experience required to operate a specific class of treatment facility as indicated by possessing a certificate of professional competence given by a state agency or professional association.

CHAIN OF CUSTODY CHAIN OF CUSTODY

A record of each person involved in the handling and possession of a sample from the person who collected the sample to the person who analyzed the sample in the laboratory and to the person who witnessed disposal of the sample.

CHEMICAL EQUIVALENT CHEMICAL EQUIVALENT

The weight in grams of a substance that combines with or displaces one gram of hydrogen. Chemical equivalents usually are found by dividing the formula weight by its valence.

CHEMICAL OXYGEN DEMAND (COD) CHEMICAL OXYGEN DEMAND (COD)

A measure of the oxygen-consuming capacity of organic matter present in wastewater. COD is expressed as the amount of oxygen consumed from a chemical oxidant in mg/L during a specific test. Results are not necessarily related to the biochemical oxygen demand (BOD) because the chemical oxidant may react with substances that bacteria do not stabilize.

CHEMICAL PRECIPITATION CHEMICAL PRECIPITATION

(1)   Precipitation induced by addition of chemicals.

(2)   The process of softening water by the addition of lime or lime and soda ash as the precipitants.

CHLORAMINES (KLOR-uh-means) CHLORAMINES

Compounds formed by the reaction of hypochlorous acid (or aqueous chlorine) with ammonia.

CHLORINATION (KLOR-uh-NAY-shun) CHLORINATION

The application of chlorine to water or wastewater, generally for the purpose of disinfection, but frequently for accomplishing other biological or chemical results.

CHLORINE CONTACT CHAMBER CHLORINE CONTACT CHAMBER

A baffled basin that provides sufficient detention time of chlorine contact with wastewater for disinfection to occur. The minimum contact time is usually 30 minutes. (Also commonly referred to as basin or tank.)

CHLORINE DEMAND CHLORINE DEMAND

Chlorine demand is the difference between the amount of chlorine added to wastewater and the amount of residual chlorine remaining after a given contact time. Chlorine demand may change with dosage, time, temperature, pH, and nature and amount of the impurities in the water.

Chlorine Demand, mg/L = Chlorine Applied, mg/L – Chlorine Residual, mg/L

CHLORINE REQUIREMENT CHLORINE REQUIREMENT

The amount of chlorine which is needed for a particular purpose. Some reasons for adding chlorine are reducing the number of coliform bacteria (Most Probable Number), obtaining a particular chlorine residual, or oxidizing some substance in the water. In each case a definite dosage of chlorine will be necessary. This dosage is the chlorine requirement.

CHLORINE RESIDUAL CHLORINE RESIDUAL

The concentration of chlorine present in water after the chlorine demand has been satisfied. The concentration is expressed in terms of the total chlorine residual, which includes both the free and combined or chemically bound chlorine residuals.

CHLORORGANIC (klor-or-GAN-ick) CHLORORGANIC

Organic compounds combined with chlorine. These compounds generally originate from, or are associated with, living or dead organic materials.

CHRONIC HEALTH EFFECT CHRONIC HEALTH EFFECT

An adverse effect on a human or animal body with symptoms that develop slowly over a long period of time or that recur frequently.

CILIATES (SILLY-ates) CILIATES

A class of protozoans distinguished by short hairs on all or part of their bodies.

CLARIFICATION (KLAIR-uh-fuh-KAY-shun) CLARIFICATION

Any process or combination of processes the main purpose of which is to reduce the concentration of suspended matter in a liquid.

CLARIFIER (KLAIR-uh-fire)   CLARIFIER

Settling Tank, Sedimentation Basin. A tank or basin in which wastewater is held for a period of time during which the heavier solids settle to the bottom and the lighter materials float to the water surface.

COAGULANT AID   COAGULANT AID

Any chemical or substance used to assist or modify coagulation.

COAGULANTS (co-AGG-you-lents)   COAGULANTS

Chemicals that cause very fine particles to clump (floc) together into larger particles. This makes it easier to separate the solids from the water by settling, skimming, draining or filtering.

COAGULATION (co-AGG-you-LAY-shun)   COAGULATION

The clumping together of very fine particles into larger particles (floc) caused by the use of chemicals (coagulants). The chemicals neutralize the electrical charges of the fine particles, allowing them to come closer and form larger clumps. This clumping together makes it easier to separate the solids from the water by settling, skimming, draining or filtering.

CODE OF FEDERAL REGULATIONS (CFR)   CODE OF FEDERAL REGULATIONS (CFR)

A publication of the United States Government which contains all of the proposed and finalized federal regulations, including environmental regulations.

COLIFORM (COAL-i-form)   COLIFORM

One type of bacteria. The presence of coliform-group bacteria is an indication of possible pathogenic bacterial contamination. The human intestinal tract is one of the main habitats of coliform bacteria. They may also be found in the intestinal tracts of warm-blooded animals, and in plants, soil, air, and the aquatic environment. Fecal coliforms are those coliforms found in the feces of various warm-blooded animals; whereas the term "coliform" also includes other environmental sources.

COLLOIDS (CALL-loids)   COLLOIDS

Very small, finely divided solids (particles that do not dissolve) that remain dispersed in a liquid for a long time due to their small size and electrical charge. When most of the particles in water have a negative electrical charge, they tend to repel each other. This repulsion prevents the particles from clumping together, becoming heavier, and settling out.

COLORIMETRIC MEASUREMENT   COLORIMETRIC MEASUREMENT

A means of measuring unknown chemical concentrations in water by *MEASURING A SAMPLE'S COLOR INTENSITY*. The specific color of the sample, developed by addition of chemical reagents, is measured with a photoelectric colorimeter or is compared with "color standards" using, or corresponding with, known concentrations of the chemical.

COMBINED AVAILABLE CHLORINE   COMBINED AVAILABLE CHLORINE

The total chlorine, present as chloramine or other derivatives, that is present in a water and is still available for disinfection and for oxidation of organic matter. The combined chlorine compounds are more stable than free chlorine forms, but they are somewhat slower in disinfection action.

COMBINED AVAILABLE CHLORINE RESIDUAL   COMBINED AVAILABLE CHLORINE RESIDUAL

The concentration of residual chlorine that is combined with ammonia, organic nitrogen, or both in water as a chloramine (or other chloro derivative) and yet is still available to oxidize organic matter and help kill bacteria.

COMBINED CHLORINE   COMBINED CHLORINE

The sum of the chlorine species composed of free chlorine and ammonia, including monochloramine, dichloramine, and trichloramine (nitrogen trichloride). Dichloramine is the strongest disinfectant of these chlorine species, but it has less oxidative capacity than free chlorine.

COMBINED RESIDUAL CHLORINATION   COMBINED RESIDUAL CHLORINATION

The application of chlorine to water or wastewater to produce a combined available chlorine residual. The residual may consist of chlorine compounds formed by the reaction of chlorine with natural or added ammonia (NH_3) or with certain organic nitrogen compounds.

COMBINED SEWER   COMBINED SEWER

A sewer designed to carry both sanitary wastewaters and storm or surface water runoff.

COMMINUTION (com-mi-NEW-shun)   COMMINUTION

Shredding. A mechanical treatment process which cuts large pieces of wastes into smaller pieces so they won't plug pipes or damage equipment. COMMINUTION and SHREDDING usually mean the same thing.

COMMINUTOR (com-mih-NEW-ter) COMMINUTOR

A device used to reduce the size of the solid chunks in wastewater by shredding (comminuting). The shredding action is like many scissors cutting or chopping to shreds all the large influent solids material in the wastewater.

COMPETENT PERSON COMPETENT PERSON

A competent person is defined by OSHA as a person capable of identifying existing and predictable hazards in the surroundings, or working conditions which are unsanitary, hazardous or dangerous to employees, and who has authorization to take prompt corrective measures to eliminate the hazards.

COMPOSITE (come-PAH-zit) (PROPORTIONAL) SAMPLE COMPOSITE (PROPORTIONAL) SAMPLE

A composite sample is a collection of individual samples obtained at regular intervals, usually every one or two hours during a 24-hour time span. Each individual sample is combined with the others in proportion to the rate of flow when the sample was collected. The resulting mixture (composite sample) forms a representative sample and is analyzed to determine the average conditions during the sampling period.

COMPOUND COMPOUND

A pure substance composed of two or more elements whose composition is constant. For example, table salt (sodium chloride, NaCl) is a compound.

CONFINED SPACE CONFINED SPACE

Confined space means a space that:

A. Is large enough and so configured that an employee can bodily enter and perform assigned work; and

B. Has limited or restricted means for entry or exit (for example, manholes, tanks, vessels, silos, storage bins, hoppers, vaults, and pits are spaces that may have limited means of entry); and

C. Is not designed for continuous employee occupancy.

(Definition from the Code of Federal Regulations (CFR) Title 29 Part 1910.146.)

CONFINED SPACE, CLASS "A" CONFINED SPACE, CLASS "A"

A confined space that presents a situation that is immediately dangerous to life or health (IDLH). These include but are not limited to oxygen deficiency, explosive or flammable atmospheres, and/or concentrations of toxic substances.

(Definition from NIOSH, "Criteria for a Recommended Standard: Working in Confined Spaces.")

CONFINED SPACE, CLASS "B" CONFINED SPACE, CLASS "B"

A confined space that has the potential for causing injury and illness, if preventive measures are not used, but not immediately dangerous to life and health.

(Definition from NIOSH, "Criteria for a Recommended Standard: Working in Confined Spaces.")

CONFINED SPACE, CLASS "C" CONFINED SPACE, CLASS "C"

A confined space in which the potential hazard would not require any special modification of the work procedure.

(Definition from NIOSH, "Criteria for a Recommended Standard: Working in Confined Spaces.")

CONFINED SPACE, NON-PERMIT CONFINED SPACE, NON-PERMIT

A non-permit confined space is a confined space that does not contain or, with respect to atmospheric hazards, have the potential to contain any hazard capable of causing death or serious physical harm.

CONFINED SPACE, PERMIT-REQUIRED CONFINED SPACE, PERMIT-REQUIRED
(PERMIT SPACE) (PERMIT SPACE)

A confined space that has one or more of the following characteristics:

• Contains or has a potential to contain a hazardous atmosphere,

• Contains a material that has the potential for engulfing an entrant,

• Has an internal configuration such that an entrant could be trapped or asphyxiated by inwardly converging walls or by a floor which slopes downward and tapers to a smaller cross section, or

• Contains any other recognized serious safety or health hazard.

(Definition from the Code of Federal Regulations (CFR) Title 29 Part 1910.146.)

CONING (CONE-ing)   CONING

Development of a cone-shaped flow of liquid, like a whirlpool, through sludge. This can occur in a sludge hopper during sludge withdrawal when the sludge becomes too thick. Part of the sludge remains in place while liquid rather than sludge flows out of the hopper. Also called coring.

CONTACT STABILIZATION   CONTACT STABILIZATION

Contact stabilization is a modification of the conventional activated sludge process. In contact stabilization, two aeration tanks are used. One tank is for separate reaeration of the return sludge for at least four hours before it is permitted to flow into the other aeration tank to be mixed with the primary effluent requiring treatment. The process may also occur in one long tank.

CONTINUOUS PROCESS   CONTINUOUS PROCESS

A treatment process in which water is treated continuously in a tank or reactor. The water being treated continuously flows into the tank at one end, is treated as it flows through the tank, and flows out the opposite end as treated water.

CONVENTIONAL TREATMENT   CONVENTIONAL TREATMENT

The preliminary treatment, sedimentation, flotation, trickling filter, rotating biological contactor, activated sludge and chlorination wastewater treatment processes.

COVERAGE RATIO   COVERAGE RATIO

The coverage ratio is a measure of the ability of the utility to pay the principal and interest on loans and bonds (this is known as "debt service") in addition to any unexpected expenses.

CROSS CONNECTION   CROSS CONNECTION

A connection between a drinking (potable) water system and an unapproved water supply. For example, if you have a pump moving nonpotable water and hook into the drinking water system to supply water for the pump seal, a cross connection or mixing between the two water systems can occur. This mixing may lead to contamination of the drinking water.

CRYOGENIC (cry-o-JEN-nick)   CRYOGENIC

Very low temperature. Associated with liquified gases (liquid oxygen).

D

DO (pronounce as separate letters)   DO

Abbreviation of **D**issolved **O**xygen. DO is the molecular (atmospheric) oxygen dissolved in water or wastewater.

DPD (pronounce as separate letters) METHOD   DPD METHOD

A method of measuring the chlorine residual in water. The residual may be determined by either titrating or comparing a developed color with color standards. DPD stands for N,N-diethyl-p-phenylene-diamine.

DANGER   DANGER

The word *DANGER* is used where an immediate hazard presents a threat of death or serious injury to employees. Also see CAUTION, NOTICE, and WARNING.

DANGEROUS AIR CONTAMINATION   DANGEROUS AIR CONTAMINATION

An atmosphere presenting a threat of causing death, injury, acute illness, or disablement due to the presence of flammable and/or explosive, toxic or otherwise injurious or incapacitating substances.

A.   Dangerous air contamination due to the flammability of a gas or vapor is defined as an atmosphere containing the gas or vapor at a concentration greater than 10 percent of its lower explosive (lower flammable) limit.

B.   Dangerous air contamination due to a combustible particulate is defined as a concentration greater than 10 percent of the minimum explosive concentration of the particulate.

C.   Dangerous air contamination due to the toxicity of a substance is defined as the atmospheric concentration immediately hazardous to life or health.

DATEOMETER (day-TOM-uh-ter)   DATEOMETER

A small calendar disc attached to motors and equipment to indicate the year in which the last maintenance service was performed.

DEBT SERVICE   DEBT SERVICE

The amount of money required annually to pay the (1) interest on outstanding debts; or (2) funds due on a maturing bonded debt or the redemption of bonds.

DECHLORINATION (dee-KLOR-uh-NAY-shun)   DECHLORINATION

The removal of chlorine from the effluent of a treatment plant. Chlorine needs to be removed because chlorine is toxic to fish and other aquatic life.

DECIBEL (DES-uh-bull)   DECIBEL

A unit for expressing the relative intensity of sounds on a scale from zero for the average least perceptible sound to about 130 for the average level at which sound causes pain to humans. Abbreviated dB.

DECOMPOSITION, DECAY   DECOMPOSITION, DECAY

Processes that convert unstable materials into more stable forms by chemical or biological action. Waste treatment encourages decay in a controlled situation so that material may be disposed of in a stable form. When organic matter decays under anaerobic conditions (putrefaction), undesirable odors are produced. The aerobic processes in common use for wastewater treatment produce much less objectionable odors.

DEGRADATION (deh-gruh-DAY-shun)   DEGRADATION

The conversion or breakdown of a substance to simpler compounds. For example, the degradation of organic matter to carbon dioxide and water.

DELEGATION   DELEGATION

The act in which power is given to another person in the organization to accomplish a specific job.

DENITRIFICATION (dee-NYE-truh-fuh-KAY-shun) (ACTIVATED SLUDGE)   DENITRIFICATION

(1)   The anoxic biological reduction of nitrate nitrogen to nitrogen gas.

(2)   The removal of some nitrogen from a system.

(3)   An anoxic process that occurs when nitrite or nitrate ions are reduced to nitrogen gas and nitrogen bubbles are formed as a result of this process. The bubbles attach to the biological floc in the activated sludge process and float the floc to the surface of the secondary clarifiers. This condition is often the cause of rising sludge observed in secondary clarifiers or gravity thickeners. Also see NITRIFICATION.

DENITRIFICATION (dee-NYE-truh-fuh-KAY-shun) (RBCs)   DENITRIFICATION

(1)   The anoxic biological reduction of nitrate nitrogen to nitrogen gas.

(2)   The removal of some nitrogen from a system.

(3)   An anoxic process that occurs when nitrite or nitrate ions are reduced to nitrogen gas and nitrogen bubbles are formed as a result of this process. The bubbles attach to the biological floc and float the floc to the surface of the secondary clarifiers. This condition is often the cause of rising sludge observed in secondary clarifiers or gravity thickeners. Also see NITRIFICATION.

DENSITY (DEN-sit-tee)   DENSITY

A measure of how heavy a substance (solid, liquid or gas) is for its size. Density is expressed in terms of weight per unit volume, that is, grams per cubic centimeter or pounds per cubic foot. The density of water (at 4°C or 39°F) is 1.0 gram per cubic centimeter or about 62.4 pounds per cubic foot.

DESICCATOR (DESS-uh-KAY-tor)   DESICCATOR

A closed container into which heated weighing or drying dishes are placed to cool in a dry environment in preparation for weighing. The dishes may be empty or they may contain a sample. Desiccators contain a substance, such as anhydrous calcium chloride, which absorbs moisture and keeps the relative humidity near zero so that the dish or sample will not gain weight from absorbed moisture.

DETENTION TIME   DETENTION TIME

The time required to fill a tank at a given flow or the theoretical time required for a given flow of wastewater to pass through a tank.

DETRITUS (dee-TRY-tus)   DETRITUS

The heavy, coarse mixture of grit and organic material carried by wastewater. Also called GRIT.

DEW POINT   DEW POINT

The temperature to which air with a given quantity of water vapor must be cooled to cause condensation of the vapor in the air.

DEWATER   DEWATER

(1)   To remove or separate a portion of the water present in a sludge or slurry. To dry sludge so it can be handled and disposed of.

(2)   To remove or drain the water from a tank or a trench.

DEWATERABLE

This is a property of sludge related to the ability to separate the liquid portion from the solid, with or without chemical conditioning. A material is considered dewaterable if water will readily drain from it.

DIAPHRAGM PUMP

A pump in which a flexible diaphragm, generally of rubber or equally flexible material, is the operating part. It is fastened at the edges in a vertical cylinder. When the diaphragm is raised suction is exerted, and when it is depressed, the liquid is forced through a discharge valve.

DIFFUSED-AIR AERATION

A diffused air activated sludge plant takes air, compresses it, and then discharges the air below the water surface of the aerator through some type of air diffusion device.

DIFFUSER

A device (porous plate, tube, bag) used to break the air stream from the blower system into fine bubbles in an aeration tank or reactor.

DIGESTER (die-JEST-er)

A tank in which sludge is placed to allow decomposition by microorganisms. Digestion may occur under anaerobic (more common) or aerobic conditions.

DIGITAL READOUT

The use of numbers to indicate the value or measurement of a variable. The readout of an instrument by a direct, numerical reading of the measured value. The signal sent to such readouts is usually an analog signal.

DISCHARGE HEAD

The pressure (in pounds per square inch or psi) measured at the centerline of a pump discharge and very close to the discharge flange, converted into feet. The pressure is measured from the centerline of the pump to the hydraulic grade line of the water in the discharge pipe.

$$\text{Discharge Head, ft} = (\text{Discharge Pressure, psi})(2.31\ \text{ft/psi})$$

DISINFECTION (dis-in-FECT-shun)

The process designed to kill or inactivate most microorganisms in wastewater, including essentially all pathogenic (disease-causing) bacteria. There are several ways to disinfect, with chlorination being the most frequently used in water and wastewater treatment plants. Compare with STERILIZATION.

DISSOLVED OXYGEN

Molecular (atmospheric) oxygen dissolved in water or wastewater, usually abbreviated DO.

DISTILLATE (DIS-tuh-late)

In the distillation of a sample, a portion is collected by evaporation and recondensation; the part that is recondensed is the distillate.

DISTRIBUTOR

The rotating mechanism that distributes the wastewater evenly over the surface of a trickling filter or other process unit. Also see FIXED SPRAY NOZZLE.

DOCTOR BLADE

A blade used to remove any excess solids that may cling to the outside of a rotating screen.

DOGS

Wedges attached to a slide gate and frame that force the gate to seal tightly.

DRIFT

The difference between the actual value and the desired value (or set point); characteristic of simple proportional controllers (without reset action). Also called OFFSET.

DUCKWEED

A small, green, cloverleaf-shaped floating plant, about one-quarter inch (6 mm) across which appears as a grainy "scum" on the surface of a pond.

DYNAMIC HEAD   DYNAMIC HEAD

When a pump is operating, the vertical distance (in feet or meters) from a point to the energy grade line. Also see TOTAL DYNAMIC HEAD and STATIC HEAD.

E

EDUCTOR (e-DUCK-ter)   EDUCTOR

A hydraulic device used to create a negative pressure (suction) by forcing a liquid through a restriction, such as a Venturi. An eductor or aspirator (the hydraulic device) may be used in the laboratory in place of a vacuum pump. As an injector, it is used to produce vacuum for chlorinators. Sometimes used instead of a suction pump.

EFFLORESCENCE (EF-low-RESS-ense)   EFFLORESCENCE

The powder or crust formed on a substance when moisture is given off upon exposure to the atmosphere.

EFFLUENT   EFFLUENT

Wastewater or other liquid—raw (untreated), partially or completely treated—flowing *FROM* a reservoir, basin, treatment process, or treatment plant.

ELECTROCHEMICAL PROCESS   ELECTROCHEMICAL PROCESS

A process that causes the deposition or formation of a seal or coating of a chemical element or compound by the use of electricity.

ELECTROLYSIS (ee-leck-TRAWL-uh-sis)   ELECTROLYSIS

The decomposition of material by an outside electric current.

ELECTROLYTE (ee-LECK-tro-LITE)   ELECTROLYTE

A substance which dissociates (separates) into two or more ions when it is dissolved in water.

ELECTROLYTIC (ee-LECK-tro-LIT-ick) PROCESS   ELECTROLYTIC PROCESS

A process that causes the decomposition of a chemical compound by the use of electricity.

ELECTROMAGNETIC FORCES   ELECTROMAGNETIC FORCES

Forces resulting from electrical charges that either attract or repel particles. Particles with opposite charges are attracted to each other. For example, a particle with positive charges is attracted to a particle with negative charges. Particles with similar charges repel each other. A particle with positive charges is repelled by a particle with positive charges and a particle with negative charges is repelled by another particle with negative charges.

ELECTRON   ELECTRON

(1)   A very small, negatively charged particle which is practically weightless. According to the electron theory, all electrical and electronic effects are caused either by the movement of electrons from place to place or because there is an excess or lack of electrons at a particular place.

(2)   The part of an atom that determines its chemical properties.

ELEMENT   ELEMENT

A substance which cannot be separated into its constituent parts and still retain its chemical identity. For example, sodium (Na) is an element.

ELUTRIATION (e-LOO-tree-A-shun)   ELUTRIATION

The washing of digested sludge with either fresh water, plant effluent or other wastewater. The objective is to remove (wash out) fine particulates and/or the alkalinity in sludge. This process reduces the demand for conditioning chemicals and improves settling or filtering characteristics of the solids.

EMULSION (e-MULL-shun)   EMULSION

A liquid mixture of two or more liquid substances not normally dissolved in one another; one liquid is held in suspension in the other.

END POINT   END POINT

Samples of water or wastewater are titrated to the end point. This means that a chemical is added, drop by drop, to a sample until a certain color change (blue to clear, for example) occurs. This is called the *END POINT* of the titration. In addition to a color change, an end point may be reached by the formation of a precipitate or the reaching of a specified pH. An end point may be detected by the use of an electronic device such as a pH meter. The completion of a desired chemical reaction.

ENDOGENOUS (en-DODGE-en-us) RESPIRATION   ENDOGENOUS RESPIRATION

A situation where living organisms oxidize some of their own cellular mass instead of new organic matter they adsorb or absorb from their environment.

ENERGY GRADE LINE (EGL)   ENERGY GRADE LINE (EGL)

A line that represents the elevation of energy head (in feet) of water flowing in a pipe, conduit or channel. The line is drawn above the hydraulic grade line (gradient) a distance equal to the velocity head ($V^2/2g$) of the water flowing at each section or point along the pipe or channel. Also see HYDRAULIC GRADE LINE.

[SEE DRAWING ON PAGE 496]

ENGULFMENT   ENGULFMENT

Engulfment means the surrounding and effective capture of a person by a liquid or finely divided (flowable) solid substance that can be aspirated to cause death by filling or plugging the respiratory system or that can exert enough force on the body to cause death by strangulation, constriction, or crushing.

ENTERIC   ENTERIC

Of intestinal origin, especially applied to wastes or bacteria.

ENTRAIN   ENTRAIN

To trap bubbles in water either mechanically through turbulence or chemically through a reaction.

ENZYMES (EN-zimes)   ENZYMES

Organic substances (produced by living organisms) which cause or speed up chemical reactions. Organic catalysts and/or biochemical catalysts.

EQUALIZING BASIN   EQUALIZING BASIN

A holding basin in which variations in flow and composition of a liquid are averaged. Such basins are used to provide a flow of reasonably uniform volume and composition to a treatment unit. Also called a balancing reservoir.

ESTUARIES (ES-chew-wear-eez)   ESTUARIES

Bodies of water which are located at the lower end of a river and are subject to tidal fluctuations.

EVAPOTRANSPIRATION (ee-VAP-o-TRANS-purr-A-shun)   EVAPOTRANSPIRATION

(1)   The process by which water vapor passes into the atmosphere from living plants. Also called TRANSPIRATION.

(2)   The total water removed from an area by transpiration (plants) and by evaporation from soil, snow and water surfaces.

EXPLOSIMETER   EXPLOSIMETER

An instrument used to detect explosive atmospheres. When the **L**ower **E**xplosive **L**imit (LEL) of an atmosphere is exceeded, an alarm signal on the instrument is activated. Also called a combustible gas detector.

F

F/M RATIO   F/M RATIO

See FOOD/MICROORGANISM RATIO.

FACULTATIVE (FACK-ul-TAY-tive)   FACULTATIVE

Facultative bacteria can use either dissolved molecular oxygen or oxygen obtained from food materials such as sulfate or nitrate ions. In other words, facultative bacteria can live under aerobic, anoxic, or anaerobic conditions.

FACULTATIVE (FACK-ul-TAY-tive) POND   FACULTATIVE POND

The most common type of pond in current use. The upper portion (supernatant) is aerobic, while the bottom layer is anaerobic. Algae supply most of the oxygen to the supernatant.

FILAMENTOUS (FILL-a-MEN-tuss) ORGANISMS   FILAMENTOUS ORGANISMS

Organisms that grow in a thread or filamentous form. Common types are *Thiothrix* and *Actinomycetes*. A common cause of sludge bulking in the activated sludge process.

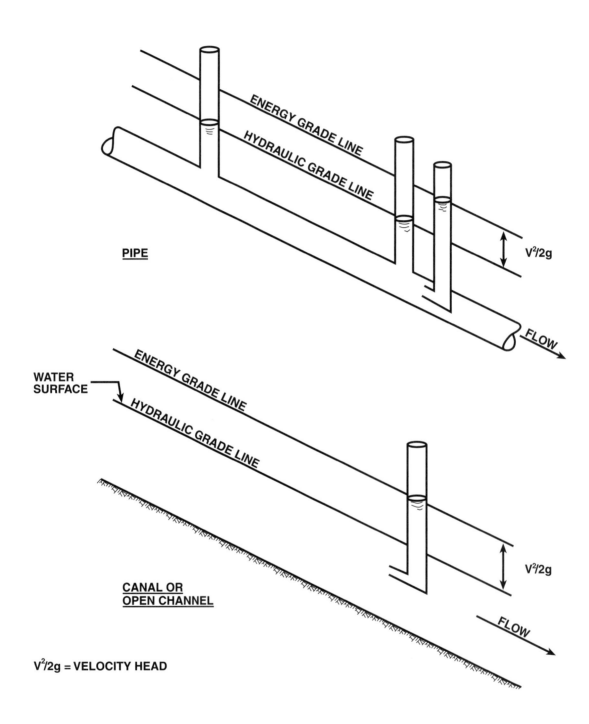

PIPE

V²/2g

FLOW

WATER SURFACE

CANAL OR OPEN CHANNEL

V²/2g

FLOW

$V^2/2g$ = VELOCITY HEAD

ENERGY GRADE LINE and HYDRAULIC GRADE LINE

FILTER AID   FILTER AID

A chemical (usually a polymer) added to water to help remove fine colloidal suspended solids.

FIXED   FIXED

A sample is "fixed" in the field by adding chemicals that prevent the water quality indicators of interest in the sample from changing before final measurements are performed later in the lab.

FIXED SPRAY NOZZLE   FIXED SPRAY NOZZLE

Cone-shaped spray nozzle used to distribute wastewater over the filter media, similar to a lawn sprinkling system. A deflector or steel ball is mounted within the cone to spread the flow of wastewater through the cone, thus causing a spraying action. Also see DISTRIBUTOR.

FLAME POLISHED   FLAME POLISHED

Melted by a flame to smooth out irregularities. Sharp or broken edges of glass (such as the end of a glass tube) are rotated in a flame until the edge melts slightly and becomes smooth.

FLIGHTS   FLIGHTS

Scraper boards, made from redwood or other rot-resistant woods or plastic, used to collect and move settled sludge or floating scum.

FLOC   FLOC

Clumps of bacteria and particles or coagulants and impurities that have come together and formed a cluster. Found in aeration tanks, secondary clarifiers and chemical precipitation processes.

FLOCCULATION (FLOCK-you-LAY-shun)   FLOCCULATION

The gathering together of fine particles after coagulation to form larger particles by a process of gentle mixing.

FLOW EQUALIZATION SYSTEM   FLOW EQUALIZATION SYSTEM

A device or tank designed to hold back or store a portion of peak flows for release during low-flow periods.

FOOD/MICROORGANISM RATIO   FOOD/MICROORGANISM RATIO

Food to microorganism ratio. A measure of food provided to bacteria in an aeration tank.

$$\frac{\text{Food}}{\text{Microorganisms}} = \frac{\text{BOD, lbs/day}}{\text{MLVSS, lbs}}$$

$$= \frac{\text{Flow, MGD} \times \text{BOD, mg/}L \times 8.34 \text{ lbs/gal}}{\text{Volume, MG} \times \text{MLVSS, mg/}L \times 8.34 \text{ lbs/gal}}$$

or by calculator math system

$$= \text{Flow, MGD} \times \text{BOD, mg/}L \div \text{Volume, MG} \div \text{MLVSS, mg/}L$$

or metric

$$= \frac{\text{BOD, kg/day}}{\text{MLVSS, kg}}$$

$$= \frac{\text{Flow, M}L\text{/day} \times \text{BOD, mg/}L \times 1 \text{ kg/M mg}}{\text{Volume, M}L \times \text{MLVSS, mg/}L \times 1 \text{ kg/M mg}}$$

Commonly abbreviated F/M Ratio.

FORCE MAIN   FORCE MAIN

A pipe that carries wastewater under pressure from the discharge side of a pump to a point of gravity flow downstream.

FREE AVAILABLE CHLORINE   FREE AVAILABLE CHLORINE

The amount of chlorine available in water. This chlorine may be in the form of dissolved gas (Cl_2), hypochlorous acid (HOCl), or hypochlorite ion (OCl^-), but does not include chlorine combined with an amine (ammonia or nitrogen) or other organic compound.

FREE AVAILABLE RESIDUAL CHLORINE   FREE AVAILABLE RESIDUAL CHLORINE

That portion of the total available residual chlorine remaining in water or wastewater at the end of a specified contact period. Residual chlorine will react chemically and biologically as hypochlorous acid (HOCl) or hypochlorite ion (OCl^-). This does not include chlorine that has combined with ammonia, nitrogen, or other compounds.

FREE CHLORINE   FREE CHLORINE

Free chlorine is chlorine (Cl_2) in a liquid or gaseous form. Free chlorine combines with water to form hypochlorous (HOCl) and hydrochloric (HCl) acids. In wastewater free chlorine usually combines with an amine (ammonia or nitrogen) or other organic compounds to form combined chlorine compounds.

FREE OXYGEN   FREE OXYGEN

Molecular oxygen available for respiration by organisms. Molecular oxygen is the oxygen molecule, O_2, that is not combined with another element to form a compound.

FREE RESIDUAL CHLORINATION   FREE RESIDUAL CHLORINATION

The application of chlorine or chlorine compounds to water or wastewater to produce a free available chlorine residual directly or through the destruction of ammonia (NH_3) or certain organic nitrogenous compounds.

FREEBOARD   FREEBOARD

The vertical distance from the normal water surface to the top of the confining wall.

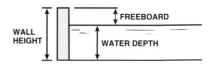

FRICTION LOSS   FRICTION LOSS

The head lost by water flowing in a stream or conduit as the result of the disturbances set up by the contact between the moving water and its containing conduit and by intermolecular friction.

G

GIS   GIS

Geographic Information System. A computer program that combines mapping with detailed information about the physical locations of structures such as pipes, valves, and manholes within geographic areas. The system is used to help operators and maintenance personnel locate utility system features or structures and to assist with the scheduling and performance of maintenance activities.

GASIFICATION (GAS-i-fi-KAY-shun)   GASIFICATION

The conversion of soluble and suspended organic materials into gas during aerobic or anaerobic decomposition. In clarifiers the resulting gas bubbles can become attached to the settled sludge and cause large clumps of sludge to rise and float on the water surface. In anaerobic sludge digesters, this gas is collected for fuel or disposed of using a waste gas burner.

GEOGRAPHIC INFORMATION SYSTEM (GIS)   GEOGRAPHIC INFORMATION SYSTEM (GIS)

A computer program that combines mapping with detailed information about the physical locations of structures such as pipes, valves, and manholes within geographic areas. The system is used to help operators and maintenance personnel locate utility system features or structures and to assist with the scheduling and performance of maintenance activities.

GRAB SAMPLE   GRAB SAMPLE

A single sample of water collected at a particular time and place which represents the composition of the water only at that time and place.

GRAVIMETRIC   GRAVIMETRIC

A means of measuring unknown concentrations of water quality indicators in a sample by *WEIGHING* a precipitate or residue of the sample.

GRIT   GRIT

The heavy material present in wastewater, such as sand, coffee grounds, eggshells, gravel and cinders.

GRIT REMOVAL   GRIT REMOVAL

Grit removal is accomplished by providing an enlarged channel or chamber which causes the flow velocity to be reduced and allows the heavier grit to settle to the bottom of the channel where it can be removed.

GROWTH RATE, Y   GROWTH RATE, Y

An experimentally determined constant to estimate the unit growth rate of bacteria while degrading organic wastes.

H

HEAD

HEAD

The vertical distance, height or energy of water above a point. A head of water may be measured in either height (feet or meters) or pressure (pounds per square inch or kilograms per square centimeter). Also see DISCHARGE HEAD, DYNAMIC HEAD, STATIC HEAD, SUCTION HEAD, SUCTION LIFT and VELOCITY HEAD.

HEAD LOSS

HEAD LOSS

An indirect measure of loss of energy or pressure. Flowing water will lose some of its energy when it passes through a pipe, bar screen, comminutor, filter or other obstruction. The amount of energy or pressure lost is called "head loss." Head loss is measured as the difference in elevation between the upstream water surface and the downstream water surface and may be expressed in feet or meters.

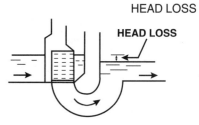

HEADER

HEADER

A large pipe to which the ends of a series of smaller pipes are connected. Also called a MANIFOLD.

HEADWORKS

HEADWORKS

The facilities where wastewater enters a wastewater treatment plant. The headworks may consist of bar screens, comminutors, a wet well and pumps.

HEPATITIS (HEP-uh-TIE-tis)

HEPATITIS

Hepatitis is an inflammation of the liver caused by an acute viral infection. Yellow jaundice is one symptom of hepatitis.

HUMUS SLUDGE

HUMUS SLUDGE

The sloughed particles of biomass from trickling filter media that are removed from the water being treated in secondary clarifiers.

HYDRAULIC GRADE LINE (HGL)

HYDRAULIC GRADE LINE (HGL)

The surface or profile of water flowing in an open channel or a pipe flowing partially full. If a pipe is under pressure, the hydraulic grade line is at the level water would rise to in a small tube connected to the pipe. To reduce the release of odors from wastewater, the water surface or hydraulic grade line should be kept as smooth as possible. Also see ENERGY GRADE LINE.

[SEE DRAWING ON PAGE 496]

HYDRAULIC JUMP

HYDRAULIC JUMP

The sudden and usually turbulent abrupt rise in water surface in an open channel when water flowing at high velocity is suddenly retarded to a slow velocity.

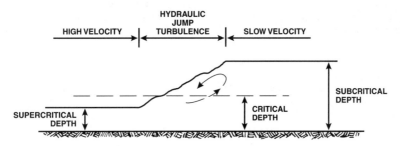

HYDRAULIC LOADING

HYDRAULIC LOADING

Hydraulic loading refers to the flows (MGD or cu m/day) to a treatment plant or treatment process. Detention times, surface loadings and weir overflow rates are directly influenced by flows.

HYDROGEN ION CONCENTRATION [H+]

HYDROGEN ION CONCENTRATION [H+]

The weight of hydrogen ion in moles per liter of solution. Commonly expressed as the pH value, which is the logarithm of the reciprocal of the hydrogen ion concentration.

$$pH = Log \frac{1}{[H^+]}$$

HYDROGEN SULFIDE GAS (H₂S)   HYDROGEN SULFIDE GAS (H₂S)

Hydrogen sulfide is a gas with a rotten egg odor. This gas is produced under anaerobic conditions. Hydrogen sulfide gas is particularly dangerous because it dulls the sense of smell so that you don't notice it after you have been around it for a while. In high concentrations, hydrogen sulfide gas is only noticeable for a very short time before it dulls the sense of smell. The gas is very poisonous to the respiratory system, explosive, flammable, colorless and heavier than air.

HYDROLOGIC (HI-dro-LOJ-ick) CYCLE   HYDROLOGIC CYCLE

The process of evaporation of water into the air and its return to earth by precipitation (rain or snow). This process also includes transpiration from plants, groundwater movement, and runoff into rivers, streams and the ocean. Also called the WATER CYCLE.

HYDROLYSIS (hi-DROLL-uh-sis)   HYDROLYSIS

(1)   A chemical reaction in which a compound is converted into another compound by taking up water.

(2)   Usually a chemical degradation of organic matter.

HYDROSTATIC SYSTEM   HYDROSTATIC SYSTEM

In a hydrostatic sludge removal system, the surface of the water in the clarifier is higher than the surface of the water in the sludge well or hopper. This difference in pressure head forces sludge from the bottom of the clarifier to flow through pipes to the sludge well or hopper.

HYGROSCOPIC (HI-grow-SKOP-ick)   HYGROSCOPIC

Absorbing or attracting moisture from the air.

HYPOCHLORINATION (HI-poe-KLOR-uh-NAY-shun)   HYPOCHLORINATION

The application of hypochlorite compounds to water or wastewater for the purpose of disinfection.

HYPOCHLORINATORS (HI-poe-KLOR-uh-NAY-tors)   HYPOCHLORINATORS

Chlorine pumps, chemical feed pumps or devices used to dispense chlorine solutions made from hypochlorites such as bleach (sodium hypochlorite) or calcium hypochlorite into the water being treated.

HYPOCHLORITE (HI-poe-KLOR-ite)   HYPOCHLORITE

Chemical compounds containing available chlorine; used for disinfection. They are available as liquids (bleach) or solids (powder, granules, and pellets) in barrels, drums, and cans. Salts of hypochlorous acid.

I

IDLH   IDLH

Immediately **D**angerous to **L**ife or **H**ealth. The atmospheric concentration of any toxic, corrosive, or asphyxiant substance that poses an immediate threat to life or would cause irreversible or delayed adverse health effects or would interfere with an individual's ability to escape from a dangerous atmosphere.

IMHOFF CONE   IMHOFF CONE

A clear, cone-shaped container marked with graduations. The cone is used to measure the volume of settleable solids in a specific volume (usually one liter) of wastewater.

IMPELLER   IMPELLER

A rotating set of vanes in a pump or compressor designed to pump or move water or air.

IMPELLER PUMP   IMPELLER PUMP

Any pump in which the water is moved by the continuous application of power to a rotating set of vanes from some rotating mechanical source.

INCINERATION   INCINERATION

The conversion of dewatered wastewater solids by combustion (burning) to ash, carbon dioxide, and water vapor.

INDICATOR (CHEMICAL)   INDICATOR (CHEMICAL)

A substance that gives a visible change, usually of color, at a desired point in a chemical reaction, generally at a specified end point.

INDOLE (IN-dole)   INDOLE

An organic compound (C_8H_7N) containing nitrogen which has an ammonia odor.

INFILTRATION (IN-fill-TRAY-shun)   INFILTRATION

The seepage of groundwater into a sewer system, including service connections. Seepage frequently occurs through defective or cracked pipes, pipe joints, connections or manhole walls.

INFLOW   INFLOW

Water discharged into a sewer system and service connections from sources other than regular connections. This includes flow from yard drains, foundation drains and around manhole covers. Inflow differs from infiltration in that it is a direct discharge into the sewer rather than a leak in the sewer itself.

INFLUENT   INFLUENT

Wastewater or other liquid—raw (untreated) or partially treated—flowing *INTO* a reservoir, basin, treatment process, or treatment plant.

INHIBITORY SUBSTANCES   INHIBITORY SUBSTANCES

Materials that kill or restrict the ability of organisms to treat wastes.

INOCULATE (in-NOCK-you-late)   INOCULATE

To introduce a seed culture into a system.

INORGANIC WASTE   INORGANIC WASTE

Waste material such as sand, salt, iron, calcium, and other mineral materials which are only slightly affected by the action of organisms. Inorganic wastes are chemical substances of mineral origin; whereas organic wastes are chemical substances usually of animal or plant origin. Also see NONVOLATILE MATTER.

INTEGRATOR   INTEGRATOR

A device or meter that continuously measures and calculates (adds) a process rate variable in cumulative fashion; for example, total flows displayed in gallons, million gallons, cubic feet, or some other unit of volume measurement. Also called a TOTALIZER.

INTERFACE   INTERFACE

The common boundary layer between two substances such as water and a solid (metal); or between two fluids such as water and a gas (air); or between a liquid (water) and another liquid (oil).

IONIC CONCENTRATION   IONIC CONCENTRATION

The concentration of any ion in solution, usually expressed in moles per liter.

IONIZATION   IONIZATION

The process of adding electrons to, or removing electrons from, atoms or molecules, thereby creating ions. High temperatures, electrical discharges, and nuclear radiation can cause ionization.

J

JAR TEST   JAR TEST

A laboratory procedure that simulates coagulation/flocculation with differing chemical doses. The purpose of the procedure is to *ESTIMATE* the minimum coagulant dose required to achieve certain water quality goals. Samples of water to be treated are placed in six jars. Various amounts of chemicals are added to each jar, stirred and the settling of solids is observed. The lowest dose of chemicals that provides satisfactory settling is the dose used to treat the water.

JOGGING   JOGGING

The frequent starting and stopping of an electric motor.

JOULE (jewel)   JOULE

A measure of energy, work or quantity of heat. One joule is the work done when the point of application of a force of one newton is displaced a distance of one meter in the direction of the force. Approximately equal to 0.7375 ft-lbs (0.1022 m-kg).

K

KJELDAHL (KELL-doll) NITROGEN   KJELDAHL NITROGEN

Nitrogen in the form of organic proteins or their decomposition product ammonia, as measured by the Kjeldahl Method.

L

LAUNDERS (LAWN-ders)   LAUNDERS

Sedimentation tank effluent troughs, consisting of overflow weir plates.

LIMIT SWITCH   LIMIT SWITCH

A device that regulates or controls the travel distance of a chain or cable.

LINEAL (LIN-e-al)   LINEAL

The length in one direction of a line. For example, a board 12 feet long has 12 lineal feet in its length.

LIQUEFACTION (LICK-we-FACK-shun)   LIQUEFACTION

The conversion of large solid particles of sludge into very fine particles which either dissolve or remain suspended in wastewater.

LOADING   LOADING

Quantity of material applied to a device at one time.

LOWER EXPLOSIVE LIMIT (LEL)   LOWER EXPLOSIVE LIMIT (LEL)

The lowest concentration of gas or vapor (percent by volume in air) that explodes if an ignition source is present at ambient temperature. At temperatures above 250°F the LEL decreases because explosibility increases with higher temperature.

LOWER FLAMMABLE LIMIT (LFL)   LOWER FLAMMABLE LIMIT (LFL)

The lowest concentration of a gas or vapor (percent by volume in air) that burns if an ignition source is present.

LYSIMETER (lie-SIM-uh-ter)   LYSIMETER

A device containing a mass of soil and designed to permit the measurement of water draining through the soil.

M

M or MOLAR   *M* or MOLAR

A molar solution consists of one gram molecular weight of a compound dissolved in enough water to make one liter of solution. A gram molecular weight is the molecular weight of a compound in grams. For example, the molecular weight of sulfuric acid (H_2SO_4) is 98. A one *M* solution of sulfuric acid would consist of 98 grams of H_2SO_4 dissolved in enough distilled water to make one liter of solution.

MBAS   MBAS

Methylene **B**lue **A**ctive **S**ubstance. Another name for surfactants, or surface active agents, is methylene blue active substances. The determination of surfactants is accomplished by measuring the color change in a standard solution of methylene blue dye.

MCRT   MCRT

Mean **C**ell **R**esidence **T**ime, days. An expression of the average time that a microorganism will spend in the activated sludge process.

$$\text{MCRT, days} = \frac{\text{Total Suspended Solids in Activated Sludge Process, lbs}}{\text{Total Suspended Solids Removed From Process, lbs/day}}$$

$$\text{or metric} = \frac{\text{Total Suspended Solids in Activated Sludge Process, kg}}{\text{Total Suspended Solids Removed From Process, kg/day}}$$

NOTE: Operators at different plants calculate the Total Suspended Solids (TSS) in the Activated Sludge Process, lbs (kg), by three different methods.

1. TSS in the Aeration Basin or Reactor Zone, lbs (kg),
2. TSS in Aeration Basin and Secondary Clarifier, lbs (kg), or
3. TSS in Aeration Basin and Secondary Clarifier Sludge Blanket, lbs (kg).

These three different methods make it difficult to compare MCRTs in days among different plants unless everyone uses the same method.

mg/*L*   mg/*L*

See MILLIGRAMS PER LITER, mg/*L*.

MPN (pronounce as separate letters)   MPN

MPN is the **M**ost **P**robable **N**umber of coliform-group organisms per unit volume of sample water. Expressed as a density or population of organisms per 100 m*L* of sample water.

MSDS   MSDS

Material **S**afety **D**ata **S**heet. A document which provides pertinent information and a profile of a particular hazardous substance or mixture. An MSDS is normally developed by the manufacturer or formulator of the hazardous substance or mixture. The MSDS is required to be made available to employees and operators whenever there is the likelihood of the hazardous substance or mixture being introduced into the workplace. Some manufacturers are preparing MSDSs for products that are not considered to be hazardous to show that the product or substance is *NOT* hazardous.

MANIFOLD   MANIFOLD

A large pipe to which the ends of a series of smaller pipes are connected. Also called a HEADER.

MANOMETER (man-NAH-mut-ter)   MANOMETER

An instrument for measuring pressure. Usually, a manometer is a glass tube filled with a liquid that is used to measure the difference in pressure across a flow measuring device such as an orifice or a Venturi meter. The instrument used to measure blood pressure is a type of manometer.

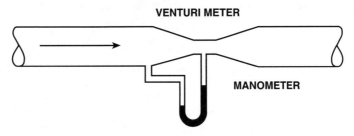

MASKING AGENTS   MASKING AGENTS

Substances used to cover up or disguise unpleasant odors. Liquid masking agents are dripped into the wastewater, sprayed into the air, or evaporated (using heat) with the unpleasant fumes or odors and then discharged into the air by blowers to make an undesirable odor less noticeable.

MATERIAL SAFETY DATA SHEET (MSDS)   MATERIAL SAFETY DATA SHEET (MSDS)

A document which provides pertinent information and a profile of a particular hazardous substance or mixture. An MSDS is normally developed by the manufacturer or formulator of the hazardous substance or mixture. The MSDS is required to be made available to employees and operators whenever there is the likelihood of the hazardous substance or mixture being introduced into the workplace. Some manufacturers are preparing MSDSs for products that are not considered to be hazardous to show that the product or substance is *NOT* hazardous.

MEAN CELL RESIDENCE TIME (MCRT)   MEAN CELL RESIDENCE TIME (MCRT)

An expression of the average time that a microorganism will spend in the activated sludge process.

$$MCRT, days = \frac{Total\ Suspended\ Solids\ in\ Activated\ Sludge\ Process,\ lbs}{Total\ Suspended\ Solids\ Removed\ From\ Process,\ lbs/day}$$

$$or\ metric = \frac{Total\ Suspended\ Solids\ in\ Activated\ Sludge\ Process,\ kg}{Total\ Suspended\ Solids\ Removed\ From\ Process,\ kg/day}$$

NOTE: Operators at different plants calculate the Total Suspended Solids (TSS) in the Activated Sludge Process, lbs (kg), by three different methods.

   1. TSS in the Aeration Basin or Reactor Zone, lbs (kg),
   2. TSS in Aeration Basin and Secondary Clarifier, lbs (kg), or
   3. TSS in Aeration Basin and Secondary Clarifier Sludge Blanket, lbs (kg).

   These three different methods make it difficult to compare MCRTs in days among different plants unless everyone uses the same method.

MECHANICAL AERATION   MECHANICAL AERATION

The use of machinery to mix air and water so that oxygen can be absorbed into the water. Some examples are: paddle wheels, mixers, or rotating brushes to agitate the surface of an aeration tank; pumps to create fountains; and pumps to discharge water down a series of steps forming falls or cascades.

MEDIA   MEDIA

The material in a trickling filter on which slime accumulates and organisms grow. As settled wastewater trickles over the media, organisms in the slime remove certain types of wastes thereby partially treating the wastewater. Also the material in a rotating biological contactor or in a gravity or pressure filter.

MEDIAN   MEDIAN

The middle measurement or value. When several measurements are ranked by magnitude (largest to smallest), half of the measurements will be larger and half will be smaller.

MEG   MEG

A procedure used for checking the insulation resistance on motors, feeders, bus bar systems, grounds, and branch circuit wiring. Also see MEGGER.

MEGGER (from megohm)   MEGGER

An instrument used for checking the insulation resistance on motors, feeders, bus bar systems, grounds, and branch circuit wiring. Also see MEG.

MEGOHM   MEGOHM

Meg means one million, so 5 megohms means 5 million ohms. A megger reads in millions of ohms.

MENISCUS (meh-NIS-cuss)   MENISCUS

The curved surface of a column of liquid (water, oil, mercury) in a small tube. When the liquid wets the sides of the container (as with water), the curve forms a valley. When the confining sides are not wetted (as with mercury), the curve forms a hill or upward bulge. When a meniscus forms in a measuring device, the top of the liquid level of the sample is determined by the bottom of the meniscus.

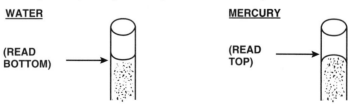

MERCAPTANS (mer-CAP-tans)   MERCAPTANS

Compounds containing sulfur which have an extremely offensive skunk-like odor; also sometimes described as smelling like garlic or onions.

MESOPHILIC (MESS-o-FILL-ick) BACTERIA   MESOPHILIC BACTERIA

Medium temperature bacteria. A group of bacteria that grow and thrive in a moderate temperature range between 68°F (20°C) and 113°F (45°C). The optimum temperature range for these bacteria in anaerobic digestion is 85°F (30°C) to 100°F (38°C).

METABOLISM   METABOLISM

All of the processes or chemical changes in an organism or a single cell by which food is built up (anabolism) into living protoplasm and by which protoplasm is broken down (catabolism) into simpler compounds with the exchange of energy.

MICRON (MY-kron)   MICRON

μm, Micrometer or Micron. A unit of length. One millionth of a meter or one thousandth of a millimeter. One micron equals 0.00004 of an inch.

MICROORGANISMS (MY-crow-OR-gan-IS-zums)   MICROORGANISMS

Very small organisms that can be seen only through a microscope. Some microorganisms use the wastes in wastewater for food and thus remove or alter much of the undesirable matter.

MICROSCREEN   MICROSCREEN

A device with a fabric straining media with openings usually between 20 and 60 microns. The fabric is wrapped around the outside of a rotating drum. Wastewater enters the open end of the drum and flows out through the rotating screen cloth. At the highest point of the drum, the collected solids are backwashed by high-pressure water jets into a trough located within the drum.

MILLIGRAMS PER LITER, mg/L (MILL-i-GRAMS per LEET-er)   MILLIGRAMS PER LITER, mg/L

A measure of the concentration by weight of a substance per unit volume. For practical purposes, one mg/L of a substance in water is equal to one part per million parts (ppm). Thus a liter of water with a specific gravity of 1.0 weighs one million milligrams. If it contains 10 milligrams of dissolved oxygen, the concentration is 10 milligrams per million milligrams, or 10 milligrams per liter (10 mg/L), or 10 parts of oxygen per million parts of water, or 10 parts per million (10 ppm).

MILLIMICRON (MILL-uh-MY-kron)   MILLIMICRON

A unit of length equal to $10^{-3}\mu$ (one thousandth of a micron), 10^{-6} millimeters, or 10^{-9} meters; correctly called a nanometer, nm.

MIXED LIQUOR   MIXED LIQUOR

When the activated sludge in an aeration tank is mixed with primary effluent or the raw wastewater and return sludge, this mixture is then referred to as mixed liquor as long as it is in the aeration tank. Mixed liquor also may refer to the contents of mixed aerobic or anaerobic digesters.

MIXED LIQUOR SUSPENDED SOLIDS (MLSS)   MIXED LIQUOR SUSPENDED SOLIDS (MLSS)

Suspended solids in the mixed liquor of an aeration tank.

MIXED LIQUOR VOLATILE SUSPENDED SOLIDS   MIXED LIQUOR VOLATILE SUSPENDED SOLIDS
    (MLVSS)   (MLVSS)

The organic or volatile suspended solids in the mixed liquor of an aeration tank. This volatile portion is used as a measure or indication of the microorganisms present.

MOLECULAR OXYGEN   MOLECULAR OXYGEN

The oxygen molecule, O_2, that is not combined with another element to form a compound.

MOLECULAR WEIGHT   MOLECULAR WEIGHT

The molecular weight of a compound in grams is the sum of the atomic weights of the elements in the compound. The molecular weight of sulfuric acid (H_2SO_4) in grams is 98.

Element	Atomic Weight	Number of Atoms	Molecular Weight
H	1	2	2
S	32	1	32
O	16	4	64
			98

MOLECULE (MOLL-uh-KULE)   MOLECULE

The smallest division of a compound that still retains or exhibits all the properties of the substance.

MOTILE (MO-till)   MOTILE

Capable of self-propelled movement. A term that is sometimes used to distinguish between certain types of organisms found in water.

MOVING AVERAGE   MOVING AVERAGE

To calculate the moving average for the last 7 days, add up the values for the last 7 days and divide by 7. Each day add the most recent day's value to the sum of values and subtract the oldest value. By using the 7-day moving average, each day of the week is always represented in the calculations.

MUFFLE FURNACE   MUFFLE FURNACE

A small oven capable of reaching temperatures up to 600°C. Muffle furnaces are used in laboratories for burning or incinerating samples to determine the amounts of volatile solids and/or fixed solids in samples of wastewater.

MULTI-STAGE PUMP   MULTI-STAGE PUMP

A pump that has more than one impeller. A single-stage pump has one impeller.

N

N or NORMAL   *N* or NORMAL

A normal solution contains one gram equivalent weight of reactant (compound) per liter of solution. The equivalent weight of an acid is that weight which contains one gram atom of ionizable hydrogen or its chemical equivalent. For example, the equivalent weight of sulfuric acid (H_2SO_4) is 49 (98 divided by 2 because there are two replaceable hydrogen ions). A one *N* solution of sulfuric acid would consist of 49 grams of H_2SO_4 dissolved in enough water to make one liter of solution.

NIOSH (NYE-osh)   NIOSH

The **N**ational **I**nstitute of **O**ccupational **S**afety and **H**ealth is an organization that tests and approves safety equipment for particular applications. NIOSH is the primary federal agency engaged in research in the national effort to eliminate on-the-job hazards to the health and safety of working people. The NIOSH Publications Catalog, Seventh Edition, NIOSH Pub. No. 87-115, lists the NIOSH publications concerning industrial hygiene and occupational health. To obtain a copy of the catalog, write to National Technical Information Service (NTIS), 5285 Port Royal Road, Springfield, VA 22161. NTIS Stock No. PB88-175013, price, $141.00, plus $5.00 shipping and handling per order.

NPDES PERMIT   NPDES PERMIT

National **P**ollutant **D**ischarge **E**limination **S**ystem permit is the regulatory agency document issued by either a federal or state agency which is designed to control all discharges of potential pollutants from point sources and storm water runoff into U.S. waterways. NPDES permits regulate discharges into navigable waters from all point sources of pollution, including industries, municipal wastewater treatment plants, sanitary landfills, large agricultural feedlots and return irrigation flows.

NAMEPLATE   NAMEPLATE

A durable metal plate found on equipment which lists critical operating conditions for the equipment.

NEUTRALIZATION (new-trall-i-ZAY-shun)   NEUTRALIZATION

Addition of an acid or alkali (base) to a liquid to cause the pH of the liquid to move toward a neutral pH of 7.0.

NITRIFICATION (NYE-truh-fuh-KAY-shun)   NITRIFICATION

An aerobic process in which bacteria change the ammonia and organic nitrogen in wastewater into oxidized nitrogen (usually nitrate). The second-stage BOD is sometimes referred to as the "nitrogenous BOD" (first-stage BOD is called the "carbonaceous BOD"). Also see DENITRIFICATION.

NITRIFICATION STAGE   NITRIFICATION STAGE

A stage of decomposition that occurs in biological treatment processes when aerobic bacteria, using dissolved oxygen, change nitrogen compounds (ammonia and organic nitrogen) into oxidized nitrogen (usually nitrate). The second-stage BOD is sometimes referred to as the "nitrification stage" (first-stage BOD is called the "carbonaceous stage").

NITRIFYING BACTERIA   NITRIFYING BACTERIA

Bacteria that change the ammonia and organic nitrogen in wastewater into oxidized nitrogen (usually nitrate).

NITROGENOUS (nye-TRAH-jen-us)   NITROGENOUS

A term used to describe chemical compounds (usually organic) containing nitrogen in combined forms. Proteins and nitrate are nitrogenous compounds.

NOMOGRAM (NOME-o-gram)   NOMOGRAM

A chart or diagram containing three or more scales used to solve problems with three or more variables instead of using mathematical formulas.

NONCORRODIBLE   NONCORRODIBLE

A material that resists corrosion and will not be eaten away by wastewater or chemicals in wastewater.

NONSPARKING TOOLS   NONSPARKING TOOLS

These tools will not produce a spark during use. They are made of a nonferrous material, usually a copper-beryllium alloy.

NONVOLATILE MATTER   NONVOLATILE MATTER

Material such as sand, salt, iron, calcium, and other mineral materials which are only slightly affected by the actions of organisms and are not lost on ignition of the dry solids at 550°C. Volatile materials are chemical substances usually of animal or plant origin. Also see INORGANIC WASTE and VOLATILE MATTER or VOLATILE SOLIDS.

NOTICE   NOTICE

This word calls attention to information that is especially significant in understanding and operating equipment or processes safely. Also see CAUTION, DANGER, and WARNING.

NUTRIENT CYCLE   NUTRIENT CYCLE

The transformation or change of a nutrient from one form to another until the nutrient has returned to the original form, thus completing the cycle. The cycle may take place under either aerobic or anaerobic conditions.

NUTRIENTS   NUTRIENTS

Substances which are required to support living plants and organisms. Major nutrients are carbon, hydrogen, oxygen, sulfur, nitrogen, and phosphorus. Nitrogen and phosphorus are difficult to remove from wastewater by conventional treatment processes because they are water soluble and tend to recycle. Also see NUTRIENT CYCLE.

O

O & M MANUAL   O & M MANUAL

Operation and **M**aintenance Manual. A manual that describes detailed procedures for operators to follow to operate and maintain a specific wastewater treatment or pretreatment plant and the equipment of that plant.

OSHA (O-shuh)   OSHA

The Williams-Steiger **O**ccupational **S**afety and **H**ealth **A**ct of 1970 (OSHA) is a federal law designed to protect the health and safety of industrial workers and treatment plant operators. It regulates the design, construction, operation and maintenance of industrial plants and wastewater treatment plants. The Act does not apply directly to municipalities, *EXCEPT* in those states that have approved plans and have asserted jurisdiction under Section 18 of the OSHA Act. *HOWEVER, CONTRACT OPERATORS AND PRIVATE FACILITIES DO HAVE TO COMPLY WITH OSHA REQUIREMENTS.* Wastewater treatment plants have come under stricter regulation in all phases of activity as a result of OSHA standards. OSHA also refers to the federal and state agencies which administer the OSHA regulations.

OBLIGATE AEROBES   OBLIGATE AEROBES

Bacteria that must have atmospheric or dissolved molecular oxygen to live and reproduce.

ODOR PANEL   ODOR PANEL

A group of people used to measure odors.

OFFSET   OFFSET

The difference between the actual value and the desired value (or set point); characteristic of proportional controllers that do not incorporate reset action. Also called DRIFT.

OLFACTOMETER (ol-FACT-toe-meter)   OLFACTOMETER

A device used to measure odors in the field by diluting odors with odor-free air.

OLFACTORY (ol-FAK-tore-ee) FATIGUE

OLFACTORY FATIGUE

A condition in which a person's nose, after exposure to certain odors, is no longer able to detect the odor.

OPERATING RATIO

OPERATING RATIO

The operating ratio is a measure of the total revenues divided by the total operating expenses.

ORGANIC WASTE

ORGANIC WASTE

Waste material which comes mainly from animal or plant sources. Organic wastes generally can be consumed by bacteria and other small organisms. Inorganic wastes are chemical substances of mineral origin.

ORGANISM

ORGANISM

Any form of animal or plant life. Also see BACTERIA.

ORGANIZING

ORGANIZING

Deciding who does what work and delegating authority to the appropriate persons.

ORIFICE (OR-uh-fiss)

ORIFICE

An opening (hole) in a plate, wall, or partition. In a trickling filter distributor, the wastewater passes through an orifice to the surface of the filter media. An orifice flange or plate placed in a pipe consists of a slot or a calibrated circular hole smaller than the pipe diameter. The difference in pressure in the pipe above and at the orifice may be used to determine the flow in the pipe.

ORTHOTOLIDINE (or-tho-TOL-uh-dine)

ORTHOTOLIDINE

Orthotolidine is a colorimetric indicator of chlorine residual. If chlorine is present, a yellow-colored compound is produced. This reagent is no longer approved for tests of effluent chlorine residual.

OUCH PRINCIPLE

OUCH PRINCIPLE

This principle says that as a manager when you delegate job tasks you must be **O**bjective, **U**niform in your treatment of employees, **C**onsistent with utility policies, and **H**ave job relatedness.

OUTFALL

OUTFALL

(1) The point, location or structure where wastewater or drainage discharges from a sewer, drain, or other conduit.

(2) The conduit leading to the final disposal point or area.

OVERFLOW RATE

OVERFLOW RATE

One of the guidelines for the design of settling tanks and clarifiers in treatment plants. Used by operators to determine if tanks and clarifiers are hydraulically (flow) over- or underloaded. Also called SURFACE LOADING.

$$\text{Overflow Rate, GPD/sq ft} = \frac{\text{Flow, gallons/day}}{\text{Surface Area, sq ft}}$$

OVERTURN

OVERTURN

The almost spontaneous mixing of all layers of water in a reservoir or lake when the water temperature becomes similar from top to bottom. This may occur in the fall/winter when the surface waters cool to the same temperature as the bottom waters and also in the spring when the surface waters warm after the ice melts. This is also called "turnover."

OXIDATION (ox-uh-DAY-shun)

OXIDATION

Oxidation is the addition of oxygen, removal of hydrogen, or the removal of electrons from an element or compound. In wastewater treatment, organic matter is oxidized to more stable substances. The opposite of REDUCTION.

OXIDATION STATE/OXIDATION NUMBER

OXIDATION STATE/OXIDATION NUMBER

In a chemical formula, a number accompanied by a polarity indication (+ or −) that together indicate the charge of an ion as well as the extent to which the ion has been oxidized or reduced in a REDOX REACTION.

Due to the loss of electrons, the charge of an ion that has been oxidized would go from negative toward or to neutral, from neutral to positive, or from positive to more positive. As an example, an oxidation number of 2+ would indicate that an ion has lost two electrons and that its charge has become positive (that it now has an excess of two protons).

Due to the gain of electrons, the charge of the ion that has been reduced would go from positive toward or to neutral, from neutral to negative, or from negative to more negative. As an example, an oxidation number of 2− would indicate that an ion has gained two electrons and that its charge has become negative (that it now has an excess of two electrons). As an ion gains electrons, its oxidation state (or the extent to which it is oxidized) lowers; that is, its oxidation state is reduced.

OXIDATION-REDUCTION POTENTIAL (ORP)

OXIDATION-REDUCTION POTENTIAL (ORP)

The electrical potential required to transfer electrons from one compound or element (the oxidant) to another compound or element (the reductant); used as a qualitative measure of the state of oxidation in wastewater treatment systems. ORP is measured in milli-volts, with negative values indicating a tendency to reduce compounds or elements and positive values indicating a tendency to oxi-dize compounds or elements.

OXIDATION-REDUCTION (REDOX) REACTION

OXIDATION-REDUCTION (REDOX) REACTION

A two-part reaction between two substances involving a transfer of electrons from one substance to the other. Oxidation is the loss of electrons by one substance, and reduction is the acceptance of electrons by the other substance. Reduction refers to the lowering of the OXIDATION STATE or OXIDATION NUMBER of the substance accepting the electrons.

In a redox reaction, the substance that gives up the electrons (that is oxidized) is called the reductant because it causes a reduction in the oxidation state or number of the substance that accepts the transferred electrons. The substance that receives the electrons (that is reduced) is called the oxidant because it causes oxidation of the other substance. Oxidation and reduction always occur simultaneously.

OXIDIZED ORGANICS

OXIDIZED ORGANICS

Organic materials that have been broken down in a biological process. Examples of these materials are carbohydrates and proteins that are broken down to simple sugars.

OXIDIZING AGENT

OXIDIZING AGENT

Any substance, such as oxygen (O_2) or chlorine (Cl_2), that will readily add (take on) electrons. When oxygen or chlorine is added to wastewater, organic substances are oxidized. These oxidized organic substances are more stable and less likely to give off odors or to contain disease-causing bacteria. The opposite is a REDUCING AGENT.

OXYGEN DEFICIENCY

OXYGEN DEFICIENCY

An atmosphere containing oxygen at a concentration of less than 19.5 percent by volume.

OXYGEN ENRICHMENT

OXYGEN ENRICHMENT

An atmosphere containing oxygen at a concentration of more than 23.5 percent by volume.

OZONATION (O-zoe-NAY-shun)

OZONATION

The application of ozone to water, wastewater, or air, generally for the purposes of disinfection or odor control.

P

POTW

POTW

Publicly **O**wned **T**reatment **W**orks. A treatment works which is owned by a state, municipality, city, town, special sewer district or other publicly owned and financed entity as opposed to a privately (industrial) owned treatment facility. This definition includes any devices and systems used in the storage, treatment, recycling and reclamation of municipal sewage (wastewater) or industrial wastes of a liquid nature. It also includes sewers, pipes and other conveyances only if they carry wastewater to a POTW treatment plant. The term also means the municipality (public entity) which has jurisdiction over the indirect discharges to and the discharges from such a treatment works.

PACKAGE TREATMENT PLANT

PACKAGE TREATMENT PLANT

A small wastewater treatment plant often fabricated at the manufacturer's factory, hauled to the site, and installed as one facility. The package may be either a small primary or a secondary wastewater treatment plant.

PARALLEL OPERATION

PARALLEL OPERATION

Wastewater being treated is split and a portion flows to one treatment unit while the remainder flows to another similar treatment unit. Also see SERIES OPERATION.

PARASITIC (PAIR-a-SIT-tick) BACTERIA

PARASITIC BACTERIA

Parasitic bacteria are those bacteria which normally live off another living organism, known as the "host."

PATHOGENIC (PATH-o-JEN-ick) ORGANISMS

PATHOGENIC ORGANISMS

Bacteria, viruses, cysts, or protozoa which can cause disease (giardiasis, cryptosporidiosis, typhoid, cholera, dysentery) in a host (such as a person). There are many types of organisms which do *NOT* cause disease and which are *NOT* called pathogenic. Many beneficial bacteria are found in wastewater treatment processes actively cleaning up organic wastes.

PERCENT SATURATION   PERCENT SATURATION

The amount of a substance that is dissolved in a solution compared with the amount dissolved in the solution at saturation, expressed as a percent.

$$\text{Percent Saturation, \%} = \frac{\text{Amount of Substance That Is Dissolved x 100\%}}{\text{Amount Dissolved in Solution at Saturation}}$$

PERCOLATION (PURR-co-LAY-shun)   PERCOLATION

The movement or flow of water through soil or rocks.

PERISTALTIC (PAIR-uh-STALL-tick) PUMP   PERISTALTIC PUMP

A type of positive displacement pump.

pH (pronounce as separate letters)   pH

pH is an expression of the intensity of the basic or acidic condition of a liquid. Mathematically, pH is the logarithm (base 10) of the reciprocal of the hydrogen ion activity.

$$pH = \text{Log} \frac{1}{[H^+]}$$

The pH may range from 0 to 14, where 0 is most acidic, 14 most basic, and 7 neutral. Natural waters usually have a pH between 6.5 and 8.5.

PHENOL (FEE-noll)   PHENOL

An organic compound that is a derivative of benzene.

PHENOLPHTHALEIN (FEE-nol-THAY-leen) ALKALINITY   PHENOLPHTHALEIN ALKALINITY

A measure of the hydroxide ions plus one-half of the normal carbonate ions in aqueous suspension. Measured by the amount of sulfuric acid required to lower the water to a pH value of 8.3, as indicated by a change in color of phenolphthalein. It is expressed as milligrams per liter of equivalent calcium carbonate.

PHOTOSYNTHESIS (foe-toe-SIN-thuh-sis)   PHOTOSYNTHESIS

A process in which organisms, with the aid of chlorophyll (green plant enzyme), convert carbon dioxide and inorganic substances into oxygen and additional plant material, using sunlight for energy. All green plants grow by this process.

PHYSICAL WASTE TREATMENT PROCESS   PHYSICAL WASTE TREATMENT PROCESS

Physical waste treatment processes include use of racks, screens, comminutors, clarifiers (sedimentation and flotation) and filtration. Chemical or biological reactions are important treatment processes, but NOT part of a physical treatment process.

PILLOWS   PILLOWS

Plastic tubes shaped like pillows that contain exact amounts of chemicals or reagents. Cut open the pillow, pour the reagents into the sample being tested, mix thoroughly and follow test procedures.

PLANNING   PLANNING

Management of utilities to build the resources and financial capability to provide for future needs.

PLUG FLOW   PLUG FLOW

A type of flow that occurs in tanks, basins or reactors when a slug of wastewater moves through a tank without ever dispersing or mixing with the rest of the wastewater flowing through the tank.

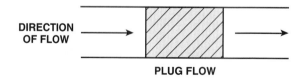

PLUG FLOW

POLE SHADER   POLE SHADER

A copper bar circling the laminated iron core inside the coil of a magnetic starter.

POLLUTION   POLLUTION

The impairment (reduction) of water quality by agricultural, domestic or industrial wastes (including thermal and radioactive wastes) to a degree that the natural water quality is changed to hinder any beneficial use of the water or render it offensive to the senses of sight, taste, or smell or when sufficient amounts of wastes create or pose a potential threat to human health or the environment.

POLYELECTROLYTE (POLY-ee-LECK-tro-lite)   POLYELECTROLYTE

A high-molecular-weight substance that is formed by either a natural or synthetic process. Natural polyelectrolytes may be of biological origin or derived from starch products, cellulose derivatives, and alignates. Synthetic polyelectrolytes consist of simple substances that have been made into complex, high-molecular-weight substances. Often called a POLYMER.

POLYMER (POLY-mer)   POLYMER

A long chain molecule formed by the union of many monomers (molecules of lower molecular weight). Polymers are used with other chemical coagulants to aid in binding small suspended particles to larger chemical flocs for their removal from water.

POLYSACCHARIDE (polly-SAC-a-ride)   POLYSACCHARIDE

A carbohydrate, such as starch, insulin or cellulose, composed of chains of simple sugars.

PONDING   PONDING

A condition occurring on trickling filters when the hollow spaces (voids) become plugged to the extent that water passage through the filter is inadequate. Ponding may be the result of excessive slime growths, trash, or media breakdown.

POPULATION EQUIVALENT   POPULATION EQUIVALENT

A means of expressing the strength of organic material in wastewater. In a domestic wastewater system, microorganisms use up about 0.2 pound of oxygen per day for each person using the system (as measured by the standard BOD test). May also be expressed as flow (100 gallons per day per person) or suspended solids (0.2 lb SS/day/person).

$$\text{Population Equivalent, persons} = \frac{\text{Flow, MGD x BOD, mg/}L\text{ x 8.34 lbs/gal}}{0.2 \text{ lb BOD/day/person}}$$

POSITIVE PRESSURE   POSITIVE PRESSURE

A positive pressure is a pressure greater than atmospheric. It is measured as pounds per square inch (psi) or as inches of water column. A negative pressure (vacuum) is less than atmospheric and is sometimes measured in inches of mercury. In the metric system pressures are measured in kg/sq m, kg/sq cm or pascals (1 psi = 6,895 Pa = 6.895 kN/sq m).

POSTCHLORINATION   POSTCHLORINATION

The addition of chlorine to the plant discharge or effluent, *FOLLOWING* plant treatment, for disinfection purposes.

POTABLE (POE-tuh-bull) WATER   POTABLE WATER

Water that does not contain objectionable pollution, contamination, minerals, or infective agents and is considered satisfactory for drinking.

PRE-AERATION   PRE-AERATION

The addition of air at the initial stages of treatment to freshen the wastewater, remove gases, add oxygen, promote flotation of grease, and aid coagulation.

PRECHLORINATION   PRECHLORINATION

The addition of chlorine in the collection system serving the plant or at the headworks of the plant *PRIOR TO* other treatment processes mainly for odor and corrosion control. Also applied to aid disinfection, to reduce plant BOD load, to aid in settling, to control foaming in Imhoff units and to help remove oil.

PRECIPITATE (pre-SIP-uh-TATE)   PRECIPITATE

(1)   An insoluble, finely divided substance which is a product of a chemical reaction within a liquid.

(2)   The separation from solution of an insoluble substance.

PRECOAT   PRECOAT

Application of a free-draining, noncohesive material such as diatomaceous earth to a filtering media. Precoating reduces the frequency of media washing and facilitates cake discharge.

PRELIMINARY TREATMENT   PRELIMINARY TREATMENT

The removal of metal, rocks, rags, sand, eggshells, and similar materials which may hinder the operation of a treatment plant. Preliminary treatment is accomplished by using equipment such as racks, bar screens, comminutors, and grit removal systems.

PRESENT WORTH   PRESENT WORTH

The value of a long-term project expressed in today's dollars. Present worth is calculated by converting (discounting) all future benefits and costs over the life of the project to a single economic value at the start of the project. Calculating the present worth of alternative projects makes it possible to compare them and select the one with the largest positive (beneficial) present worth or minimum present cost.

PRETREATMENT FACILITY   PRETREATMENT FACILITY

Industrial wastewater treatment plant consisting of one or more treatment devices designed to remove sufficient pollutants from wastewaters to allow an industry to comply with effluent limits established by the U.S. EPA General and Categorical Pretreatment Regulations or locally derived prohibited discharge requirements and local effluent limits. Compliance with effluent limits allows for a legal discharge to a POTW.

PRIMARY TREATMENT   PRIMARY TREATMENT

A wastewater treatment process that takes place in a rectangular or circular tank and allows those substances in wastewater that readily settle or float to be separated from the water being treated.

PROCESS VARIABLE   PROCESS VARIABLE

A physical or chemical quantity which is usually measured and controlled in the operation of a wastewater treatment plant or an industrial plant.

PROGRAMMABLE LOGIC CONTROLLER (PLC)   PROGRAMMABLE LOGIC CONTROLLER (PLC)

A small computer that controls process equipment (variables) and can control the sequence of valve operations.

PROTEINACEOUS (PRO-ten-NAY-shus)   PROTEINACEOUS

Materials containing proteins which are organic compounds containing nitrogen.

PROTOZOA (pro-toe-ZOE-ah)   PROTOZOA

A group of motile microscopic organisms (usually single-celled and aerobic) that sometimes cluster into colonies and generally consume bacteria as an energy source.

PRUSSIAN BLUE   PRUSSIAN BLUE

A blue paste or liquid (often on a paper like carbon paper) used to show a contact area. Used to determine if gate valve seats fit properly.

PSYCHROPHILIC (sy-kro-FILL-ick) BACTERIA   PSYCHROPHILIC BACTERIA

Cold temperature bacteria. A group of bacteria that grow and thrive in temperatures below 68°F (20°C).

PURGE   PURGE

To remove a gas or vapor from a vessel, reactor or confined space, usually by displacement or dilution.

PUTREFACTION (PEW-truh-FACK-shun)   PUTREFACTION

Biological decomposition of organic matter with the production of foul-smelling products associated with anaerobic conditions.

PUTRESCIBLE (pew-TRES-uh-bull)   PUTRESCIBLE

Material that will decompose under anaerobic conditions and produce nuisance odors.

PYROMETER (pie-ROM-uh-ter)   PYROMETER

An apparatus used to measure high temperatures.

Q

(NO LISTINGS)

R

RACK   RACK

Evenly spaced parallel metal bars or rods located in the influent channel to remove rags, rocks, and cans from wastewater.

RADIAL TO IMPELLER   RADIAL TO IMPELLER

Perpendicular to the impeller shaft. Material being pumped flows at a right angle to the impeller.

RAW WASTEWATER   RAW WASTEWATER

Plant influent or wastewater *BEFORE* any treatment.

REAGENT (re-A-gent)   REAGENT

A pure chemical substance that is used to make new products or is used in chemical tests to measure, detect, or examine other substances.

RECALCINATION (re-CAL-sin-NAY-shun)   RECALCINATION

A lime recovery process in which the calcium carbonate in sludge is converted to lime by heating at 1,800°F (980°C).

RECARBONATION (re-CAR-bun-NAY-shun)   RECARBONATION

A process in which carbon dioxide is bubbled into the water being treated to lower the pH.

RECEIVING WATER   RECEIVING WATER

A stream, river, lake, ocean, or other surface or groundwaters into which treated or untreated wastewater is discharged.

RECHARGE RATE   RECHARGE RATE

Rate at which water is added beneath the ground surface to replenish or recharge groundwater.

RECIRCULATION   RECIRCULATION

The return of part of the effluent from a treatment process to the incoming flow.

REDOX (ree-DOCKS)   REDOX

Oxidation-reduction reactions in which the oxidation state of at least one reactant is raised while that of another is lowered.

REDOX REACTION   REDOX REACTION

(See OXIDATION-REDUCTION (REDOX) REACTION)

REDUCING AGENT   REDUCING AGENT

Any substance, such as base metal (iron) or the sulfide ion (S^{2-}), that will readily donate (give up) electrons. The opposite is an OXIDIZING AGENT.

REDUCTION (re-DUCK-shun)   REDUCTION

Reduction is the addition of hydrogen, removal of oxygen, or the addition of electrons to an element or compound. Under anaerobic conditions (no dissolved oxygen present), sulfur compounds are reduced to odor-producing hydrogen sulfide (H_2S) and other compounds. The opposite of OXIDATION.

REFLUX   REFLUX

Flow back. A sample is heated, evaporates, cools, condenses, and flows back to the flask.

REFRACTORY (re-FRACK-toe-ree) MATERIALS   REFRACTORY MATERIALS

Materials difficult to remove entirely from wastewater such as nutrients, color, taste- and odor-producing substances and some toxic materials.

RELIQUEFACTION (re-LICK-we-FACK-shun)   RELIQUEFACTION

The return of a gas to the liquid state; for example, a condensation of chlorine gas to return it to its liquid form by cooling.

REPRESENTATIVE SAMPLE   REPRESENTATIVE SAMPLE

A sample portion of material or wastestream that is as nearly identical in content and consistency as possible to that in the larger body of material or wastestream being sampled.

RESIDUAL CHLORINE   RESIDUAL CHLORINE

The concentration of chlorine present in water after the chlorine demand has been satisfied. The concentration is expressed in terms of the total chlorine residual, which includes both the free and combined or chemically bound chlorine residuals.

RESPIRATION   RESPIRATION

The process in which an organism uses oxygen for its life processes and gives off carbon dioxide.

RESPONSIBILITY   RESPONSIBILITY

Answering to those above in the chain of command to explain how and why you have used your authority.

RETENTION TIME   RETENTION TIME

The time water, sludge or solids are retained or held in a clarifier or sedimentation tank. See DETENTION TIME.

RIPRAP   RIPRAP

Broken stones, boulders, or other materials placed compactly or irregularly on levees or dikes for the protection of earth surfaces against the erosive action of waves.

RISING SLUDGE   RISING SLUDGE

Rising sludge occurs in the secondary clarifiers of activated sludge plants when the sludge settles to the bottom of the clarifier, is compacted, and then starts to rise to the surface, usually as a result of denitrification, or anaerobic biological activity that produces carbon dioxide and/or methane.

ROTAMETER (RODE-uh-ME-ter)   ROTAMETER

A device used to measure the flow rate of gases and liquids. The gas or liquid being measured flows vertically up a tapered, calibrated tube. Inside the tube is a small ball or bullet-shaped float (it may rotate) that rises or falls depending on the flow rate. The flow rate may be read on a scale behind or on the tube by looking at the middle of the ball or at the widest part or top of the float.

ROTARY PUMP   ROTARY PUMP

A type of displacement pump consisting essentially of elements rotating in a pump case which they closely fit. The rotation of these elements alternately draws in and discharges the water being pumped. Such pumps act with neither suction nor discharge valves, operate at almost any speed, and do not depend on centrifugal forces to lift the water.

ROTATING BIOLOGICAL CONTACTOR (RBC)   ROTATING BIOLOGICAL CONTACTOR (RBC)

A secondary biological treatment process for domestic and biodegradable industrial wastes. Biological contactors have a rotating "shaft" surrounded by plastic discs called the "media." The shaft and media are called the "drum." A biological slime grows on the media when conditions are suitable and the microorganisms that make up the slime (biomass) stabilize the waste products by using the organic material for growth and reproduction.

ROTIFERS (ROE-ti-fers)   ROTIFERS

Microscopic animals characterized by short hairs on their front end.

S

SAR (Sodium Adsorption Ratio)   SAR

This ratio expresses the relative activity of sodium ions in the exchange reactions with soil. The ratio is defined as follows:

$$SAR = \frac{Na}{[\frac{1}{2}(Ca + Mg)]^{1/2}}$$

where Na, Ca, and Mg are concentrations of the respective ions in milliequivalents per liter of water.

$$Na, meq/L = \frac{Na, mg/L}{23.0 \ mg/meq} \qquad Ca, meq/L = \frac{Ca, mg/L}{20.0 \ mg/meq} \qquad Mg, meq/L = \frac{Mg, mg/L}{12.15 \ mg/meq}$$

SCADA (ss-KAY-dah) SYSTEM   SCADA SYSTEM

Supervisory Control And Data Acquisition system. A computer-monitored alarm, response, control and data acquisition system used by operators to monitor and adjust their wastewater treatment processes and facilities.

SCFM   SCFM

Cubic Feet of air per Minute at Standard conditions of temperature, pressure, and humidity (0°C, 14.7 psia, and 50% relative humidity).

SVI (**S**ludge **V**olume **I**ndex)   SVI

This is a calculation which indicates the tendency of activated sludge solids (aerated solids) to thicken or to become concentrated during the sedimentation/thickening process. SVI is calculated in the following manner: (1) allow a mixed liquor sample from the aeration basin to settle for 30 minutes; (2) determine the suspended solids concentration for a sample of the same mixed liquor; (3) calculate SVI by dividing the measured (or observed) wet volume (mL/L) of the settled sludge by the dry weight concentration of MLSS in grams/L.

$$\text{SVI, m}L/\text{gm} = \frac{\text{Settled Sludge Volume/Sample Volume, m}L/L}{\text{Suspended Solids Concentration, mg}/L} \times \frac{1{,}000 \text{ mg}}{\text{gram}}$$

SACRIFICIAL ANODE   SACRIFICIAL ANODE

An easily corroded material deliberately installed in a pipe or tank. The intent of such an installation is to give up (sacrifice) this anode to corrosion while the water supply facilities remain relatively corrosion free.

SANITARY SEWER   SANITARY SEWER

A pipe or conduit (sewer) intended to carry wastewater or waterborne wastes from homes, businesses, and industries to the POTW (**P**ublicly **O**wned **T**reatment **W**orks). Storm water runoff or unpolluted water should be collected and transported in a separate system of pipes or conduits (storm sewers) to natural watercourses.

SAPROPHYTIC (SAP-row-FIT-ick) ORGANISMS   SAPROPHYTIC ORGANISMS

Organisms living on dead or decaying organic matter. They help natural decomposition of the organic solids in wastewater.

SCREEN   SCREEN

A device used to retain or remove suspended or floating objects in wastewater. The screen has openings that are generally uniform in size. It retains or removes objects larger than the openings. A screen may consist of bars, rods, wires, gratings, wire mesh, or perforated plates.

SEALING WATER   SEALING WATER

Water used to prevent wastewater or dirt from reaching moving parts. Sealing water is at a higher pressure than the wastewater it is keeping out of a mechanical device.

SECCHI (SECK-key) DISC   SECCHI DISC

A flat, white disc lowered into the water by a rope until it is just barely visible. At this point, the depth of the disc from the water surface is the recorded Secchi disc transparency.

SECONDARY TREATMENT   SECONDARY TREATMENT

A wastewater treatment process used to convert dissolved or suspended materials into a form more readily separated from the water being treated. Usually the process follows primary treatment by sedimentation. The process commonly is a type of biological treatment process followed by secondary clarifiers that allow the solids to settle out from the water being treated.

SEED SLUDGE   SEED SLUDGE

In wastewater treatment, seed, seed culture or seed sludge refers to a mass of sludge which contains populations of microorganisms. When a seed sludge is mixed with wastewater or sludge being treated, the process of biological decomposition takes place more rapidly.

SEIZING   SEIZING

Seizing occurs when an engine overheats and a component expands to the point where the engine will not run. Also called freezing.

SEPTIC (SEP-tick)   SEPTIC

A condition produced by anaerobic bacteria. If severe, the sludge produces hydrogen sulfide, turns black, gives off foul odors, contains little or no dissolved oxygen, and the wastewater has a high oxygen demand.

SEPTICITY (sep-TIS-it-tee)   SEPTICITY

Septicity is the condition in which organic matter decomposes to form foul-smelling products associated with the absence of free oxygen. If severe, the wastewater produces hydrogen sulfide, turns black, gives off foul odors, contains little or no dissolved oxygen, and the wastewater has a high oxygen demand.

SERIES OPERATION   SERIES OPERATION

Wastewater being treated flows through one treatment unit and then flows through another similar treatment unit. Also see PARALLEL OPERATION.

SET POINT   SET POINT

The position at which the control or controller is set. This is the same as the desired value of the process variable. For example, a thermostat is set to maintain a desired temperature.

SEWAGE   SEWAGE

The used household water and water-carried solids that flow in sewers to a wastewater treatment plant. The preferred term is WASTEWATER.

SEWER GAS   SEWER GAS

(1) Gas in collection lines (sewers) that results from the decomposition of organic matter in the wastewater. When testing for gases found in sewers, test for lack of oxygen and also for explosive and toxic gases.

(2) Any gas present in the wastewater collection system, even though it is from such sources as gas mains, gasoline, and cleaning fluid.

SHEAR PIN   SHEAR PIN

A straight pin that will fail when a certain load or stress is exceeded. The purpose of the pin is to protect equipment from damage due to excessive loads or stresses.

SHOCK LOAD (ACTIVATED SLUDGE)   SHOCK LOAD

The arrival at a plant of a waste which is toxic to organisms in sufficient quantity or strength to cause operating problems. Possible problems include odors and bulking sludge which will result in a high loss of solids from the secondary clarifiers into the plant effluent and a biological process upset that may require several days to a week to recover. Organic or hydraulic overloads also can cause a shock load.

SHOCK LOAD (TRICKLING FILTERS)   SHOCK LOAD

The arrival at a plant of a waste which is toxic to organisms in sufficient quantity or strength to cause operating problems. Possible problems include odors and sloughing off of the growth or slime on the trickling filter media. Organic or hydraulic overloads also can cause a shock load.

SHORT-CIRCUITING   SHORT-CIRCUITING

A condition that occurs in tanks or basins when some of the flowing water entering a tank or basin flows along a nearly direct pathway from the inlet to the outlet. This is usually undesirable since it may result in shorter contact, reaction, or settling times in comparison with the theoretical (calculated) or presumed detention times.

SHREDDING   SHREDDING

Comminution. A mechanical treatment process which cuts large pieces of wastes into smaller pieces so they won't plug pipes or damage equipment. SHREDDING and COMMINUTION usually mean the same thing.

SIDESTREAM   SIDESTREAM

Wastewater flows that develop from other storage or treatment facilities. This wastewater may or may not need additional treatment.

SIGNIFICANT FIGURE   SIGNIFICANT FIGURE

The number of accurate numbers in a measurement. If the distance between two points is measured to the nearest hundredth and recorded as 238.41 feet, the measurement has five significant figures.

SINGLE-STAGE PUMP   SINGLE-STAGE PUMP

A pump that has only one impeller. A multi-stage pump has more than one impeller.

SKATOLE (SKATE-tole)   SKATOLE

An organic compound (C_9H_9N) that contains nitrogen and has a fecal odor.

SLAKE   SLAKE

To mix with water so that a true chemical combination (hydration) takes place, such as in the slaking of lime.

SLOUGHINGS (SLUFF-ings)   SLOUGHINGS

Trickling filter slimes that have been washed off the filter media. They are generally quite high in BOD and will lower effluent quality unless removed.

SLUDGE (sluj) SLUDGE

(1) The settleable solids separated from liquids during processing.

(2) The deposits of foreign materials on the bottoms of streams or other bodies of water.

SLUDGE AGE SLUDGE AGE

A measure of the length of time a particle of suspended solids has been retained in the activated sludge process.

$$\text{Sludge Age, days} = \frac{\text{Suspended Solids Under Aeration, lbs or kg}}{\text{Suspended Solids Added, lbs/day or kg/day}}$$

SLUDGE DENSITY INDEX (SDI) SLUDGE DENSITY INDEX (SDI)

This calculation is used in a way similar to the Sludge Volume Index (SVI) to indicate the settleability of a sludge in a secondary clarifier or effluent. The weight in grams of one milliliter of sludge after settling for 30 minutes. SDI = 100/SVI. Also see SLUDGE VOLUME INDEX (SVI).

SLUDGE DIGESTION SLUDGE DIGESTION

The process of changing organic matter in sludge into a gas or a liquid or a more stable solid form. These changes take place as microorganisms feed on sludge in anaerobic (more common) or aerobic digesters.

SLUDGE GASIFICATION SLUDGE GASIFICATION

A process in which soluble and suspended organic matter are converted into gas by anaerobic decomposition. The resulting gas bubbles can become attached to the settled sludge and cause large clumps of sludge to rise and float on the water surface.

SLUDGE VOLUME INDEX (SVI) SLUDGE VOLUME INDEX (SVI)

This is a calculation which indicates the tendency of activated sludge solids (aerated solids) to thicken or to become concentrated during the sedimentation/thickening process. SVI is calculated in the following manner: (1) allow a mixed liquor sample from the aeration basin to settle for 30 minutes; (2) determine the suspended solids concentration for a sample of the same mixed liquor; (3) calculate SVI by dividing the measured (or observed) wet volume (mL/L) of the settled sludge by the dry weight concentration of MLSS in grams/L.

$$\text{SVI, mL/gm} = \frac{\text{Settled Sludge Volume/Sample Volume, mL/L}}{\text{Suspended Solids Concentration, mg/L}} \times \frac{1{,}000 \text{ mg}}{\text{gram}}$$

SLUDGE-VOLUME RATIO (SVR) SLUDGE-VOLUME RATIO (SVR)

The volume of sludge blanket divided by the daily volume of sludge pumped from the thickener.

SLUGS SLUGS

Intermittent releases or discharges of industrial wastes.

SLURRY (SLUR-e) SLURRY

A thin, watery mud or any substance resembling it (such as a grit slurry or a lime slurry).

SODIUM ADSORPTION RATIO (SAR) SODIUM ADSORPTION RATIO (SAR)

This ratio expresses the relative activity of sodium ions in the exchange reactions with soil. The ratio is defined as follows:

$$\text{SAR} = \frac{\text{Na}}{[\frac{1}{2}(\text{Ca} + \text{Mg})]^{1/2}}$$

where Na, Ca, and Mg are concentrations of the respective ions in milliequivalents per liter of water.

$$\text{Na, meq/L} = \frac{\text{Na, mg/L}}{23.0 \text{ mg/meq}} \qquad \text{Ca, meq/L} = \frac{\text{Ca, mg/L}}{20.0 \text{ mg/meq}} \qquad \text{Mg, meq/L} = \frac{\text{Mg, mg/L}}{12.15 \text{ mg/meq}}$$

SOFTWARE PROGRAMS SOFTWARE PROGRAMS

Computer programs; the list of instructions that tell a computer how to perform a given task or tasks. Some software programs are designed and written to monitor and control municipal wastewater treatment processes.

SOLIDS CONCENTRATION SOLIDS CONCENTRATION

The solids in the aeration tank which carry microorganisms that feed on wastewater. Expressed as milligrams per liter of Mixed Liquor Volatile Suspended Solids (MLVSS, mg/L).

SOLUBLE BOD SOLUBLE BOD

Soluble BOD is the BOD of water that has been filtered in the standard suspended solids test. The soluble BOD is a measure of food for microorganisms that is dissolved in the water being treated.

SOLUTE SOLUTE

The substance dissolved in a solution. A solution is made up of the solvent and the solute.

SOLUTION SOLUTION

A liquid mixture of dissolved substances. In a solution it is impossible to see all the separate parts.

SPECIFIC GRAVITY SPECIFIC GRAVITY

(1) Weight of a particle, substance or chemical solution in relation to the weight of an equal volume of water. Water has a specific gravity of 1.000 at 4°C (39°F). Wastewater particles or substances usually have a specific gravity of 0.5 to 2.5.

(2) Weight of a particular gas in relation to the weight of an equal volume of air at the same temperature and pressure (air has a specific gravity of 1.0). Chlorine has a specific gravity of 2.5 as a gas.

SPLASH PAD SPLASH PAD

A structure made of concrete or other durable material to protect bare soil from erosion by splashing or falling water.

SPOIL SPOIL

Excavated material such as soil from the trench of a sewer.

STABILIZE STABILIZE

To convert to a form that resists change. Organic material is stabilized by bacteria which convert the material to gases and other relatively inert substances. Stabilized organic material generally will not give off obnoxious odors.

STABILIZED WASTE STABILIZED WASTE

A waste that has been treated or decomposed to the extent that, if discharged or released, its rate and state of decomposition would be such that the waste would not cause a nuisance or odors in the receiving water.

STANDARD METHODS STANDARD METHODS

STANDARD METHODS FOR THE EXAMINATION OF WATER AND WASTEWATER, 20th Edition. A joint publication of the American Public Health Association (APHA), American Water Works Association (AWWA), and the Water Environment Federation (WEF) which outlines the accepted laboratory procedures used to analyze the impurities in water and wastewater. Available from Water Environment Federation, Publications Order Department, 601 Wythe Street, Alexandria, VA 22314-1994. Order No. S82010. Price to members, $164.75; nonmembers, $209.75; price includes cost of shipping and handling.

STANDARD SOLUTION STANDARD SOLUTION

A solution in which the exact concentration of a chemical or compound is known.

STANDARDIZE STANDARDIZE

To compare with a standard.

(1) In wet chemistry, to find out the exact strength of a solution by comparing it with a standard of known strength. This information is used to adjust the strength by adding more water or more of the substance dissolved.

(2) To set up an instrument or device to read a standard. This allows you to adjust the instrument so that it reads accurately, or enables you to apply a correction factor to the readings.

STASIS (STAY-sis) STASIS

Stagnation or inactivity of the life processes within organisms.

STATIC HEAD STATIC HEAD

When water is not moving, the vertical distance (in feet or meters) from a specific point to the water surface is the static head.

STATOR STATOR

That portion of a machine which contains the stationary (non-moving) parts that surround the moving parts (rotor).

STEP-FEED AERATION

STEP-FEED AERATION

Step-feed aeration is a modification of the conventional activated sludge process. In step aeration, primary effluent enters the aeration tank at several points along the length of the tank, rather than all of the primary effluent entering at the beginning or head of the tank and flowing through the entire tank in a plug flow mode.

STERILIZATION (STARE-uh-luh-ZAY-shun)

STERILIZATION

The removal or destruction of all microorganisms, including pathogenic and other bacteria, vegetative forms and spores. Compare with DISINFECTION.

STETHOSCOPE

STETHOSCOPE

An instrument used to magnify sounds and convey them to the ear.

STOP LOG

STOP LOG

A log or board in an outlet box or device used to control the water level in ponds and also the flow from one pond to another pond or system.

STORM SEWER

STORM SEWER

A separate pipe, conduit or open channel (sewer) that carries runoff from storms, surface drainage, and street wash, but does not include domestic and industrial wastes. Storm sewers are often the recipients of hazardous or toxic substances due to the illegal dumping of hazardous wastes or spills created by accidents involving vehicles and trains transporting these substances. Also see SANITARY SEWER.

STRIPPED GASES

STRIPPED GASES

Gases that are released from a liquid by bubbling air through the liquid or by allowing the liquid to be sprayed or tumbled over media.

STRIPPED ODORS

STRIPPED ODORS

Odors that are released from a liquid by bubbling air through the liquid or by allowing the liquid to be sprayed or tumbled over media.

STRUVITE (STREW-vite)

STRUVITE

A deposit or precipitate of magnesium ammonium phosphate hexahydrate found on the rotating components of centrifuges and centrate discharge lines. Struvite can be formed when anaerobic sludge comes in contact with spinning centrifuge components rich in oxygen in the presence of microbial activity. Struvite can also be formed in digested sludge lines and valves in the presence of oxygen and microbial activity. Struvite can form when the pH level is between 5 and 9.

STUCK

STUCK

Not working. A stuck digester does not decompose organic matter properly. The digester is characterized by low gas production, high volatile acid to alkalinity relationship, and poor liquid-solids separation. A digester in a stuck condition is sometimes called a "sour" or "upset" digester.

SUBSTRATE

SUBSTRATE

(1) The base on which an organism lives. The soil is the substrate of most seed plants; rocks, soil, water, or other plants or animals are substrates for other organisms.

(2) Chemical used by an organism to support growth. The organic matter in wastewater is a substrate for the organisms in activated sludge.

SUCTION HEAD

SUCTION HEAD

The *POSITIVE* pressure [in feet (meters) or pounds per square inch] on the suction side of a pump. The pressure can be measured from the centerline of the pump *UP TO* the elevation of the hydraulic grade line on the suction side of the pump.

SUCTION LIFT

SUCTION LIFT

The *NEGATIVE* pressure [in feet (meters) or inches (centimeters) of mercury vacuum] on the suction side of the pump. The pressure can be measured from the centerline of the pump *DOWN TO* (lift) the elevation of the hydraulic grade line on the suction side of the pump.

SUPERNATANT (sue-per-NAY-tent)

SUPERNATANT

Liquid removed from settled sludge. Supernatant commonly refers to the liquid between the sludge on the bottom and the scum on the surface of an anaerobic digester. This liquid is usually returned to the influent wet well or to the primary clarifier.

SURFACE-ACTIVE AGENT

SURFACE-ACTIVE AGENT

The active agent in detergents that possesses a high cleaning ability. Also called SURFACTANTS.

SURFACE LOADING

SURFACE LOADING

One of the guidelines for the design of settling tanks and clarifiers in treatment plants. Used by operators to determine if tanks and clarifiers are hydraulically (flow) over- or underloaded. Also called OVERFLOW RATE.

$$\text{Surface Loading, GPD/sq ft} = \frac{\text{Flow, gallons/day}}{\text{Surface Area, sq ft}}$$

SURFACTANT (sir-FAC-tent)

SURFACTANT

Abbreviation for surface-active agent. The active agent in detergents that possesses a high cleaning ability.

SUSPENDED SOLIDS

SUSPENDED SOLIDS

(1) Solids that either float on the surface or are suspended in water, wastewater, or other liquids, and which are largely removable by laboratory filtering.

(2) The quantity of material removed from wastewater in a laboratory test, as prescribed in *STANDARD METHODS FOR THE EXAMINATION OF WATER AND WASTEWATER*, and referred to as Total Suspended Solids Dried at 103–105°C.

T

TOC (pronounce as separate letters)

TOC

Total **O**rganic **C**arbon. TOC measures the amount of organic carbon in water.

TAILGATE SAFETY MEETING

TAILGATE SAFETY MEETING

Brief (10 to 20 minutes) safety meetings held every 7 to 10 working days. The term *TAILGATE* comes from the safety meetings regularly held by the construction industry around the tailgate of a truck.

TERTIARY (TER-she-AIR-ee) TREATMENT

TERTIARY TREATMENT

Any process of water renovation that upgrades treated wastewater to meet specific reuse requirements. May include general cleanup of water or removal of specific parts of wastes insufficiently removed by conventional treatment processes. Typical processes include chemical treatment and pressure filtration. Also called ADVANCED WASTE TREATMENT.

THERMOPHILIC (thur-moe-FILL-ick) BACTERIA

THERMOPHILIC BACTERIA

Hot temperature bacteria. A group of bacteria that grow and thrive in temperatures above 113°F (45°C). The optimum temperature range for these bacteria in anaerobic decomposition is 120°F (49°C) to 135°F (57°C). Aerobic thermophilic bacteria thrive between 120°F (49°C) and 158°F (70°C).

THIEF HOLE

THIEF HOLE

A digester sampling well which allows sampling of the digester contents without venting digester gas.

THRESHOLD ODOR

THRESHOLD ODOR

The minimum odor of a sample (gas or water) that can just be detected after successive odorless (gas or water) dilutions.

THRUST BLOCK

THRUST BLOCK

A mass of concrete or similar material appropriately placed around a pipe to prevent movement when the pipe is carrying water. Usually placed at bends and valve structures.

TIME LAG

TIME LAG

The time required for processes and control systems to respond to a signal or to reach a desired level.

TIME WEIGHTED AVERAGE (TWA)

TIME WEIGHTED AVERAGE (TWA)

The average concentration of a pollutant based on the times and levels of concentrations of the pollutant. The time weighted average is equal to the sum of the portion of each time period (as a decimal, such as 0.25 hour) multiplied by the pollutant concentration during the time period divided by the hours in the workday (usually 8 hours). 8TWA PEL is the Time Weighted Average permissible exposure limit, in parts per million, for a normal 8-hour workday and a 40-hour workweek to which nearly all workers may be repeatedly exposed, day after day, without adverse effect.

TITRATE (TIE-trate)

TITRATE

To *TITRATE* a sample, a chemical solution of known strength is added drop by drop until a certain color change, precipitate, or pH change in the sample is observed (end point). Titration is the process of adding the chemical reagent in small increments (0.1 – 1.0 milliliter) until completion of the reaction, as signaled by the end point.

TOTAL CHLORINE   TOTAL CHLORINE

The total concentration of chlorine in water, including the combined chlorine (such as inorganic and organic chloramines) and the free available chlorine.

TOTAL CHLORINE RESIDUAL   TOTAL CHLORINE RESIDUAL

The total amount of chlorine residual (value for residual chlorine, including both free chlorine and chemically bound chlorine) present in a water sample after a given contact time.

TOTAL DYNAMIC HEAD (TDH)   TOTAL DYNAMIC HEAD (TDH)

When a pump is lifting or pumping water, the vertical distance (in feet or meters) from the elevation of the energy grade line on the suction side of the pump to the elevation of the energy grade line on the discharge side of the pump.

TOTALIZER   TOTALIZER

A device or meter that continuously measures and calculates (adds) a process rate variable in cumulative fashion; for example, total flows displayed in gallons, million gallons, cubic feet, or some other unit of volume measurement. Also called an INTEGRATOR.

TOXIC (TOX-ick)   TOXIC

A substance which is poisonous to a living organism.

TOXICITY (tox-IS-it-tee)   TOXICITY

The relative degree of being poisonous or toxic. A condition which may exist in wastes and will inhibit or destroy the growth or function of certain organisms.

TRANSPIRATION (TRAN-spur-RAY-shun)   TRANSPIRATION

The process by which water vapor is released to the atmosphere by living plants. This process is similar to people sweating. Also see EVAPOTRANSPIRATION.

TRICKLING FILTER   TRICKLING FILTER

A treatment process in which the wastewater trickles over media that provide the opportunity for the formation of slimes or biomass which contain organisms that feed upon and remove wastes from the water being treated.

TRICKLING FILTER MEDIA   TRICKLING FILTER MEDIA

Rocks or other durable materials that make up the body of the filter. Synthetic (manufactured) media have been used successfully.

TRUNK SEWER   TRUNK SEWER

A sewer that receives wastewater from many tributary branches or sewers and serves a large territory and contributing population.

TURBID   TURBID

Having a cloudy or muddy appearance.

TURBIDITY METER   TURBIDITY METER

An instrument for measuring and comparing the turbidity of liquids by passing light through them and determining how much light is reflected by the particles in the liquid. The normal measuring range is 0 to 100 and is expressed as Nephelometric Turbidity Units (NTUs).

TURBIDITY UNITS (TU)   TURBIDITY UNITS (TU)

Turbidity units are a measure of the cloudiness of water. If measured by a nephelometric (deflected light) instrumental procedure, turbidity units are expressed in nephelometric turbidity units (NTU) or simply TU. Those turbidity units obtained by visual methods are expressed in Jackson Turbidity Units (JTU) which are a measure of the cloudiness of water; they are used to indicate the clarity of water. There is no real connection between NTUs and JTUs. The Jackson turbidimeter is a visual method and the nephelometer is an instrumental method based on deflected light.

TWO-STAGE FILTERS   TWO-STAGE FILTERS

Two filters are used. Effluent from the first filter goes to the second filter, either directly or after passing through a clarifier.

U

ULTRAFILTRATION   ULTRAFILTRATION

A membrane filter process used for the removal of some organic compounds in an aqueous (watery) solution.

UPPER EXPLOSIVE LIMIT (UEL)   UPPER EXPLOSIVE LIMIT (UEL)

The point at which the concentration of a gas in air becomes too great to allow an explosion upon ignition due to insufficient oxygen present.

UPSET   UPSET

An upset digester does not decompose organic matter properly. The digester is characterized by low gas production, high volatile acid/alkalinity relationship, and poor liquid-solids separation. A digester in an upset condition is sometimes called a "sour" or "stuck" digester.

V

VELOCITY HEAD   VELOCITY HEAD

The energy in flowing water as determined by a vertical height (in feet or meters) equal to the square of the velocity of flowing water divided by twice the acceleration due to gravity ($V^2/2g$).

VISCOSITY (vis-KOSS-uh-tee)   VISCOSITY

A property of water, or any other fluid, which resists efforts to change its shape or flow. Syrup is more viscous (has a higher viscosity) than water. The viscosity of water increases significantly as temperatures decrease. Motor oil is rated by how thick (viscous) it is; 20 weight oil is considered relatively thin while 50 weight oil is relatively thick or viscous.

VOLATILE (VOL-uh-tull)   VOLATILE

(1) A volatile substance is one that is capable of being evaporated or changed to a vapor at relatively low temperatures. Volatile substances also can be partially removed by air stripping.

(2) In terms of solids analysis, volatile refers to materials lost (including most organic matter) upon ignition in a muffle furnace for 60 minutes at 550°C. Natural volatile materials are chemical substances usually of animal or plant origin. Manufactured or synthetic volatile materials such as ether, acetone, and carbon tetrachloride are highly volatile and not of plant or animal origin. Also see NONVOLATILE MATTER.

VOLATILE ACIDS   VOLATILE ACIDS

Fatty acids produced during digestion which are soluble in water and can be steam-distilled at atmospheric pressure. Also called organic acids. Volatile acids are commonly reported as equivalent to acetic acid.

VOLATILE LIQUIDS   VOLATILE LIQUIDS

Liquids which easily vaporize or evaporate at room temperature.

VOLATILE MATTER   VOLATILE MATTER

Matter in water, wastewater, or other liquids that is lost on ignition of the dry solids at 550°C.

VOLATILE SOLIDS   VOLATILE SOLIDS

Those solids in water, wastewater, or other liquids that are lost on ignition of the dry solids at 550°C.

VOLUMETRIC   VOLUMETRIC

A measurement based on the volume of some factor. Volumetric titration is a means of measuring unknown concentrations of water quality indicators in a sample *BY DETERMINING THE VOLUME* of titrant or liquid reagent needed to complete particular reactions.

VOLUTE (vol-LOOT)   VOLUTE

The spiral-shaped casing which surrounds a pump, blower, or turbine impeller and collects the liquid or gas discharged by the impeller.

W

WARNING   WARNING

The word *WARNING* is used to indicate a hazard level between *CAUTION* and *DANGER*. Also see CAUTION, DANGER, and NOTICE.

WASTEWATER   WASTEWATER

A community's used water and water-carried solids that flow to a treatment plant. Storm water, surface water, and groundwater infiltration also may be included in the wastewater that enters a wastewater treatment plant. The term "sewage" usually refers to household wastes, but this word is being replaced by the term "wastewater."

WATER CYCLE

WATER CYCLE

The process of evaporation of water into the air and its return to earth by precipitation (rain or snow). This process also includes transpiration from plants, groundwater movement, and runoff into rivers, streams and the ocean. Also called the HYDROLOGIC CYCLE.

WATER HAMMER

WATER HAMMER

The sound like someone hammering on a pipe that occurs when a valve is opened or closed very rapidly. When a valve position is changed quickly, the water pressure in a pipe will increase and decrease back and forth very quickly. This rise and fall in pressures can cause serious damage to the system.

WEIR (weer)

WEIR

(1) A wall or plate placed in an open channel and used to measure the flow of water. The depth of the flow over the weir can be used to calculate the flow rate, or a chart or conversion table may be used to convert depth to flow.

(2) A wall or obstruction used to control flow (from settling tanks and clarifiers) to ensure a uniform flow rate and avoid short-circuiting.

WEIR (weer) DIAMETER

WEIR DIAMETER

Many circular clarifiers have a circular weir within the outside edge of the clarifier. All the water leaving the clarifier flows over this weir. The diameter of the weir is the length of a line from one edge of a weir to the opposite edge and passing through the center of the circle formed by the weir.

WEIR (weer), PROPORTIONAL

WEIR, PROPORTIONAL

A specially shaped weir in which the flow through the weir is directly proportional to the head.

WET OXIDATION

WET OXIDATION

A method of treating or conditioning sludge before the water is removed. Compressed air is blown into the liquid sludge. The air and sludge mixture is fed into a pressure vessel where the organic material is stabilized. The stabilized organic material and inert (inorganic) solids are then separated from the pressure vessel effluent by dewatering in lagoons or by mechanical means.

WET WELL

WET WELL

A compartment or tank in which wastewater is collected. The suction pipe of a pump may be connected to the wet well or a submersible pump may be located in the wet well.

X

(NO LISTINGS)

Y

Y, GROWTH RATE

Y, GROWTH RATE

An experimentally determined constant to estimate the unit growth rate of bacteria while degrading organic wastes.

Z

ZOOGLEAL (ZOE-glee-al) FILM

ZOOGLEAL FILM

A complex population of organisms that form a "slime growth" on the trickling filter media and break down the organic matter in wastewater. These slimes consist of living organisms feeding on the wastes in wastewater, dead organisms, silt, and other debris. "Slime growth" is a more common term.

ZOOGLEAL (ZOE-glee-al) MASS

ZOOGLEAL MASS

Jelly-like masses of bacteria found in both the trickling filter and activated sludge processes. These masses may be formed for or function as the protection against predators and for storage of food supplies. Also see BIOMASS.

SUBJECT INDEX

A

AIDS, 20, 64, 346
Abbreviations, 478
Abnormal operation
  activated sludge, 265
  aerated grit chambers, 88
  chlorination, 365
  chlorinators, 365, 368
  grit channels, 87
  oxidation ditches, 275
  ponds, 307
  primary treatment, 119
  rotating biological contactors, 229
  sedimentation, 119
  sulfonator, 415
  trickling filters, 182, 186
Activated carbon, dechlorination, 403
Activated sludge
  Also see Package aeration plants, Chapter 11 in Volume II
    and Chapter 2, *ADVANCED WASTE TREATMENT*
  abnormal operation, 265
  aeration methods, 261
  aeration tank, 48, 254
  aerobic digestion, 257
  bulking, 265
  centrifuge tests, 254
  chlorination, 265, 401
  clarifiers, 137, 142, 249, 254, 265, 266
  cold weather, 265, 283
  complete mix, 257, 260, 261
  contact stabilization, 257, 260
  contact time, 254, 257
  control of process, 254, 255, 265
  conventional activated sludge, 260
    Also see Volume II, Chapter 11
  description, 47, 249
  diffusers, 261, 263, 264, 282
  dissolved oxygen, 254
  efficiency of process, 254, 256
  effluent, 254, 255, 265, 266
  energy use, 261
  extended aeration, 257, 260
  floc mass, 254, 255
  flow diagram, 38, 252, 253, 258, 259, 260
  foam, 264, 265, 266
  food/microorganism ratio, 249, 257, 280, 281
  housekeeping, 266, 267
  hydraulic loading, 254, 255, 265
  industrial waste treatment, 256
    Also see Chapter 3, *INDUSTRIAL WASTE*
      TREATMENT, Volume II
  influent, 249, 256
  inspecting new facilities, 261
  laboratory testing, 266
  layout, 38, 253, 258, 259, 260
  maintenance, 266, 267

  mechanical aeration, 257, 261, 262
  microorganisms, 249, 250, 254, 265
  mixed liquor suspended solids, 257, 261, 264, 267
  mixing, 261
  odors, 254, 264, 265
  operation, 255, 264, 265
  operational strategy, 255, 266
  organisms, 249, 250, 254, 265
  oxidation, 249
  oxidation ditch, 269
  oxygen requirements, 249, 254, 255, 264
  package plants, 47, 257-267
  plans and specifications, 282
  preliminary treatment requirements, 249
  pure oxygen, 47, 239
    Also see Chapter 2, *ADVANCED WASTE TREATMENT*
  purpose, 249, 257
  recordkeeping, 264, 266, 267
  removal efficiencies, 47
  removal of sludge, 47
  return sludge, 142, 264, 265
  safety, 267
  sampling, 266
  secondary clarifiers, 47, 249, 254, 265, 267
  settleability tests, 254, 266
  shutdown, 266
  sludge age, 257, 280, 281
  sludge blanket, 254
  sludge pumps, 138
  sludge removal, 138
  sludge volume index (SVI), 142
  sludge wasting, 47, 142, 254, 261, 264, 266
  solids, 136, 142, 254
  stabilize, 249
  start-up, 261, 264
  storm flows, 265
  temperature, 265
  toxic wastes, 255, 265
  trend chart, 266
  troubleshooting, 265
  types of processes, 249
  visual inspection, 265, 266
  wasting sludge, 47, 142, 254, 261, 264, 266
  zoogleal mass, 249, 250
Advanced waste treatment, 41, 54
Aerated grit chamber, 88
Aerated pond, 297
Aeration systems, activated sludge, 47, 261
Aeration tank appearance, activated sludge, 264, 265
Aeration tanks, activated sludge, 48, 254
Aerobic bacteria, 19
Aerobic digestion
  Also see Volume II, Chapter 12
  description, 49, 257
  package plant, 257
  sludge disposal, 49, 55
  sludge handling, 49

NOTES

NOTES

NOTES

NOTES

NOTES

NOTES

NOTES